Handbook of
Cluster Analysis

T0225622

Chapman & Hall/CRC
Handbooks of Modern Statistical Methods

Series Editor

Garrett Fitzmaurice

Department of Biostatistics
Harvard School of Public Health
Boston, MA, U.S.A.

Aims and Scope

The objective of the series is to provide high-quality volumes covering the state-of-the-art in the theory and applications of statistical methodology. The books in the series are thoroughly edited and present comprehensive, coherent, and unified summaries of specific methodological topics from statistics. The chapters are written by the leading researchers in the field, and present a good balance of theory and application through a synthesis of the key methodological developments and examples and case studies using real data.

The scope of the series is wide, covering topics of statistical methodology that are well developed and find application in a range of scientific disciplines. The volumes are primarily of interest to researchers and graduate students from statistics and biostatistics, but also appeal to scientists from fields where the methodology is applied to real problems, including medical research, epidemiology and public health, engineering, biological science, environmental science, and the social sciences.

Published Titles

Handbook of Mixed Membership Models and Their Applications
Edited by Edoardo M. Airoldi, David M. Blei,
Elena A. Erosheva, and Stephen E. Fienberg

Handbook of Markov Chain Monte Carlo
Edited by Steve Brooks, Andrew Gelman,
Galin L. Jones, and Xiao-Li Meng

Handbook of Discrete-Valued Time Series
Edited by Richard A. Davis, Scott H. Holan,
Robert Lund, and Nalini Ravishanker

Handbook of Design and Analysis of Experiments
Edited by Angela Dean, Max Morris,
John Stufken, and Derek Bingham

Longitudinal Data Analysis
Edited by Garrett Fitzmaurice, Marie Davidian,
Geert Verbeke, and Geert Molenberghs

Handbook of Spatial Statistics
Edited by Alan E. Gelfand, Peter J. Diggle,
Montserrat Fuentes, and Peter Guttorp

Handbook of Cluster Analysis
Edited by Christian Hennig, Marina Meila,
Fionn Murtagh, and Roberto Rocci

Handbook of Survival Analysis
Edited by John P. Klein, Hans C. van Houwelingen,
Joseph G. Ibrahim, and Thomas H. Scheike

Handbook of Missing Data Methodology
Edited by Geert Molenberghs, Garrett Fitzmaurice,
Michael G. Kenward, Anastasios Tsiatis, and Geert Verbeke

Chapman & Hall/CRC
Handbooks of Modern Statistical Methods

Handbook of Cluster Analysis

Edited by

Christian Hennig
University College London, UK

Marina Meila
University of Washington, Seattle, USA

Fionn Murtagh
University of Derby, UK
Goldsmiths, University of London, UK

Roberto Rocci
University of Rome Tor Vergata, Italy

CRC Press
Taylor & Francis Group
Boca Raton London New York

CRC Press is an imprint of the
Taylor & Francis Group, an **informa** business
A CHAPMAN & HALL BOOK

First published 2016 by Chapman & Hall

Published 2019 by CRC Press
Taylor & Francis Group
6000 Broken Sound Parkway NW, Suite 300
Boca Raton, FL 33487-2742

First issued in paperback 2020

ISBN 13: 978-0-367-57040-8 (pbk)
ISBN 13: 978-1-4665-5188-6 (hbk)

Library of Congress Cataloging-in-Publication Data

Names: Hennig, Christian M., editor. | Meilæa, Marina, 1962- editor |
Murtagh, Fionn, editor. | Rocci, Roberto, editor.
Title: Handbook of cluster analysis / Christian M Hennig, Marina Meila, Fionn
Murtagh, Roberto Rocci, editors.
Description: Boca Raton : Taylor & Francis, 2016. | Series: Chapman &
Hall/CRC handbooks of modern statistical methods ; 9 | "A CRC title." |
Includes bibliographical references and index.
Identifiers: LCCN 2015021473 | ISBN 9781466551886 (alk. paper)
Subjects: LCSH: Cluster analysis. | Spatial analysis (Statistics)
Classification: LCC QA278 .H3456 2016 | DDC 519.5/3--dc23
LC record available at http://lccn.loc.gov/2015021473

Visit the Taylor & Francis Web site at
http://www.taylorandfrancis.com

and the CRC Press Web site at
http://www.crcpress.com

Contents

Section I Optimization Methods

Section II Dissimilarity-Based Methods

Section III Methods Based on Probability Models

Section VI   Cluster Validation and Further General Issues

Preface

This Handbook intends to give a comprehensive, structured, and unified account of the central developments in current research on cluster analysis.

The book is aimed at researchers and practitioners in statistics, and all the scientists and engineers who are involved in some way in data clustering and have a sufficient background in statistics and mathematics. Recognizing the interdisciplinary nature of cluster analysis, most parts of the book were written in a way that is accessible to readers from various disciplines. How much background is required depends to some extent on the chapter. Familiarity with mathematical and statistical reasoning is very helpful, but an academic degree in mathematics or statistics is not required for most parts. Occasionally some knowledge of algorithms and computation will help to make most of the material.

Since we wanted this book to be immediately useful in practice, the clustering methods we present are usually described in enough detail to be directly implementable. In addition, each chapter comprises the general ideas, motivation, advantages and potential limits of the methods described, signposts to software, theory and applications, and a discussion of recent research issues.

For those already experienced with cluster analysis, the book offers a broad and structured overview. For those starting to work in this field, it offers an orientation and an introduction to the key issues. For the many researchers who are only temporarily or marginally involved with cluster analysis problems, the book chapters contain enough algorithmic and practical detail to give them a working knowledge of specific areas of clustering. Furthermore, the book should help scientists, engineers, and other users of clustering methods to make informed choices of the most suitable clustering approach for their problem, and to make better use of the existing cluster analysis tools.

Cluster analysis, also sometimes known as unsupervised classification, is about finding groups in a set of objects characterized by certain measurements. This task has a very wide range of applications such as delimitation of species in biology, data compression, classification of diseases or mental illnesses, market segmentation, detection of patterns of unusual Internet use, delimitation of communities, or classification of regions or countries for administrative use. Unsupervised classification can be seen as a basic human learning activity, connected to issues as basic as the development of stable concepts in language.

Formal cluster analysis methodology has been developed, among others, by mathematicians, statisticians, computer scientists, psychologists, social scientists, econometrists, biologists, and geoscientists. Some of these branches of development existed independently for quite some time. As a consequence, cluster analysis as a research area is very heterogeneous. This makes sense, because there are also various different relevant concepts of what constitutes a cluster. Elements of a cluster can be connected by being very similar to each other and distant from nonmembers of the cluster, by having a particular characterization in terms of few (potentially out of many) variables, by being appropriately represented by the same centroid object, by constituting some kind of distinctive shape or pattern, or by being generated from a common homogeneous probabilistic process.

Cluster analysis is currently a very popular research area and its popularity can be expected to grow more connected to the growing availability and relevance of data collected in all areas of life, which often come in unstructured ways and require some processing in order to become useful. Unsupervised classification is a central technique to structure such data.

Research on cluster analysis faces many challenges. Cluster analysis is applied to ever new data formats; many approaches to cluster analysis are computer intensive and their application to large databases is difficult; there is little unification and standardization in the field of cluster analysis, which makes it difficult to compare different approaches in a systematic manner. Even the investigation of properties such as statistical consistency and stability of traditional elementary cluster analysis techniques is often surprisingly hard.

Cluster analysis as a research area has grown so much in recent years that it is all but impossible to cover everything that could be considered relevant in a handbook like this. We have chosen to organize this book according to the traditional core approaches to cluster analysis, tracing them from the origins to recent developments. The book starts with an overview of approaches (Chapter 1), followed by a quick journey through the history of cluster analysis (Chapter 2). The next four sections of the book are devoted to four major approaches toward cluster analysis, all of which go back to the beginnings of cluster analysis in the 1950s and 1960s or even further. (Probably Pearson's paper on fitting Gaussian mixtures in 1894, see Chapter 2, was the first publication of a method covered in this Handbook, although Pearson's use of it is not appropriately described as "cluster analysis.")

Section I is about methods that aim at optimizing an objective function that describes how well data is grouped around centroids. The most popular of these methods and probably the most popular clustering method in general is K-means. The efficient optimization of the K-means and other objective functions of this kind is still a hard problem and a topic of much recent research.

Section II is concerned with dissimilarity-based methods, formalizing the idea that objects within clusters should be similar and objects in different clusters should be dissimilar. Chapters treat the traditional hierarchical methods such as single linkage, and more recent approaches to analyze dissimilarity data such as spectral clustering and graph-based approaches.

Section III covers the broad field of clustering methods based on probability models for clusters, that is, mixture models and partitioning models. Such models have been analyzed for many different kinds of data, including standard real-valued vector data, categorical, ordinal and mixed data, regression-type data, functional and time-series data, spatial data, and network data. A related issue, treated in Chapter 15, is to test for the existence of clusters.

Section IV deals with clustering methods inspired by nonparametric density estimation. Instead of setting up specific models for the clusters in the data, these approaches identify clusters with the "islands" of high density in the data, no matter what shape these have, or they aim at finding the modes of the data density, which are interpreted as "attractors" or representatives for the rest of the points. Most of these methods also have a probabilistic background, but their nature is nonparametric; they formalize a cluster by characterizing in terms of the density or distribution of points instead of setting it up.

Section V collects a number of further approaches to cluster analysis, partly analyzing specific data types such as symbolic data and ensembles of clusterings, partly presenting specific problems such as constrained and semi-supervised clustering and two-mode and multipartitioning, fuzzy and rough set clustering.

By and large, Sections I through V are about methods for clustering. But having a clustering method is not all that is needed in cluster analysis. Section VI treats further relevant issues, many of which can be grouped under the headline "cluster validation," evaluating the quality of a clustering. Aspects include indexes to measure cluster validity (which are often also used for choosing the number of clusters), comparing different clusterings, measuring cluster stability and robustness of clustering methods, cluster visualization, and the general strategy in carrying out a cluster analysis and the choice of an appropriate method.

Given the limited length of the book, there are a number of topics that some readers may expect in the *Handbook of Cluster Analysis*, but that are not covered. We see most of the presented material as essential; some decisions were motivated by individual preferences and the chosen focus, some by the difficulty of finding good authors for certain topics. Much of what is missing are methods for further types of data (such as text clustering), some more recent approaches that are currently used by rather limited groups of users, some of the recent progress in computational issues for large data sets including some of the clustering methods, of which the main motivation is to be able to deal with large amounts of data, and some hybrid approaches that piece together various elementary ideas from clustering and classification. We have weighted an introduction to the elementary approaches (on which there is still much research and that still confronts us with open problems) higher than the coverage of as many branches as possible of current specialized cutting-edge research, although some of this was included by chapter authors, all of whom are active and well-distinguished researchers in the area.

It has been a long process to write this book and we are very grateful for the continuous support by Chapman & Hall/CRC and particularly by Robert Calver, who gave us a lot of encouragement and pushed us when necessary.

Editors

Christian Hennig is senior lecturer at the Department of Statistical Science, University College London. Previous affiliations were the Seminar für Statistik, ETH Zürich and the Faculty of Mathematics, University of Hamburg. He is currently secretary of the International Federation of Classification Societies. He is associate editor of *Statistics and Computing, Computational Statistics and Data Analysis, Advances in Data Analysis and Classification*, and *Statistical Methods and Applications*. His main research interests are cluster analysis, philosophy of statistics, robust statistics, multivariate analysis, data visualization, and model selection.

Marina Meila is professor of statistics at the University of Washington. She earned an MS in electrical engineering from the Polytechnic University of Bucharest, and a PhD in computer science and electrical engineering from the Massachusetts Institute of Technology. She held appointments at the Bucharest Research Institute for Computer Technology, the Polytechnic University of Bucharest, and the Robotics Institute of Carnegie Mellon University. Her long-term interests are in machine learning and reasoning in uncertainty, and how these can be performed efficiently on large complex data sets.

Fionn Murtagh earned degrees in engineering science, mathematics, computer science, a PhD in mathematical statistics, and habilitation in computational astronomy. He works in the field of data science and big data analytics. He served the Space Science Department of the European Space Agency for 12 years. He also held professorial chairs in computer science in a number of universities in the United Kingdom. He currently is a professor of data science. He is a fellow of the International Association for Pattern Recognition, a fellow of the British Computer Society, and an elected member of the Royal Irish Academy and of Academia Europaea. He is a member of the editorial boards of many journals, and has been editor-in-chief of the *Computer Journal* for more than 10 years.

Roberto Rocci is full professor of statistics at the Department of Economics and Finance, University of Rome Tor Vergata. He earned his PhD in statistics in 1994 at the Department of Statistical Science, Probability and Applied Probability, University of Rome La Sapienza. The topic of his dissertation was on multilinear models for multiway data. His field of interests are cluster analysis, mixture models, and latent variable models. He is the author of many papers published in international journals. Recently, he was the secretary of the Italian Statistical Society (SIS). Currently, he is associate editor of the *Statistical Methods and Applications Journal* and board member of SIS-CLADAG (SIS-CLassification and Data Analysis Group).

Contributors

Ayan Acharya
Department of Electrical and Computer
   Engineering
University of Texas at Austin
Austin, Texas

Marco Alfó
Department of Statistical Sciences
Sapienza University of Rome
Rome, Italy

Pranjal Awasthi
Department of Computer Science
Rutgers University
New Brunswick, New Jersey

Adelchi Azzalini
Senior Scholar
University of Padua
Padua, Italy

Maria Florina Balcan
School of Computer Science
Carnegie Mellon University
Pittsburgh, Pennsylvania

Paula Brito
Faculdade de Economia
   and LIAAD-INESC TEC
Universidade do Porto
Porto, Portugal

Jorge Caiado
CEMAPRE/ISEG
University of Lisbon
Lisbon, Portugal

Miguel Á. Carreira-Perpiñán
Electrical Engineering and
   Computer Science
University of California, Merced
Merced, California

G. Celeux
Orsay
Île-de-France, France

Radha Chitta
Department of Computer Science
   and Engineering
Michigan State University
East Lansing, Michigan

Pedro Contreras
Thinking Safe Limited
Egham, United Kingdom

Sébastien Déjean
Institut de Mathématiques
   UMR CNRS et Université
   de Toulouse
Université Paul Sabatier
Toulouse, France

Ivo Düntsch
Department of Computer Science
Brock University
St. Catharines, Ontario, Canada

Pierpaolo D'Urso
Department of Social Sciences and
   Economics
Sapienza University of Rome
Rome, Italy

L.A. García-Escudero
Departamento de Estadística e
   Investigación Operativa and IMUVA
Universidad de Valladolid
Valladolid, Spain

Günther Gediga
Fachbereich Psychologie
Universität Münster
Münster, Germany

Joydeep Ghosh
Department of ECE
University of Texas at Austin
Austin, Texas

A. Gordaliza
Departamento de Estadística e
  Investigación Operativa
  and IMUVA
Universidad de Valladolid
Valladolid, Spain

Gérard Govaert
Saclay
Île-de-France, France

and

CNRS and Université Technologique de
  Compiègne
Compiègne, France

Mark C. Greenwood
Department of Mathematical Sciences
Montana State University
Bozeman, Montana

Maria Halkidi
Department of Digital Systems
University of Piraeus
Piraeus, Greece

Julia Handl
Manchester Business School
University of Manchester
Manchester, United Kingdom

Lisa Handl
Institute of Stochastics
Ulm University
Ulm, Germany

David Neil Hayes
Lineberger Comprehensive
  Cancer Center
Department of Internal Medicine
University of North Carolina
Chapel Hill, North Carolina

Christian Hennig
Department of Statistical Science
University College London
London, United Kingdom

Christian Hirsch
Institute of Stochastics
Ulm University
Ulm, Germany

David B. Hitchcock
Department of Mathematical
  Sciences
Montana State University
Bozeman, Montana

Hanwen Huang
Department of Epidemiology and
  Biostatistics
University of Georgia
Athens, Georgia

Anil Jain
Department of Computer Science and
  Engineering
Michigan State University
East Lansing, Michigan

Rong Jin
Department of Computer Science and
  Engineering
Michigan State University
East Lansing, Michigan

Joshua Knowles
School of Computer Science
University of Manchester
Manchester, United Kingdom

Friedrich Leisch
Institute of Applied Statistics and
  Computing
University of Natural Resources and Life
  Sciences
Vienna, Austria

Yufeng Liu
Department of Statistics and Operations
    Research
Carolina Center for Genome Sciences
Lineberger Comprehensive Cancer Center
University of North Carolina
Chapel Hill, North Carolina

Elizabeth Ann Maharaj
Department of Econometrics and
    Business Statistics
Monash University
Melbourne, Australia

J.S. Marron
Department of Statistics and Operations
    Research
Lineberger Comprehensive Cancer Center
University of North Carolina
Chapel Hill, North Carolina

C. Matrán
Departamento de Estadística e
    Investigación Operativa and
    IMUVA
Universidad de Valladolid
Valladolid, Spain

A. Mayo-Iscar
Departamento de Estadística e
    Investigación Operativa
    and IMUVA
Universidad de Valladolid
Valladolid, Spain

Geoffrey J. McLachlan
Department of Mathematics
University of Queensland
St. Lucia, Australia

Marina Meila
Department of Statistics
University of Washington
Seattle, Washington

Boris Mirkin
Department of Computer Science
Birkbeck University of London
London, United Kingdom

and

Department of Data Analysis and
    Machine Intelligence
National Research University Higher
    School of Economics
Moscow, Russia

Josiane Mothe
Ecole Supérieure du Professorat et
    de L'éducation Académie de Toulouse
Institut de Recherche en Informatique
    de Toulouse
Université de Toulouse
Toulouse, France

Thomas Brendan Murphy
School of Mathematical Sciences
    Complex and Adaptive Systems
    Laboratory and Insight Research Centre
University College Dublin
Dublin, Ireland

Fionn Murtagh
Department of Computing
Goldsmiths University of London
London, United Kingdom

and

Department of Computing and
    Mathematics
University of Derby
Derby, United Kingdom

Andrew Nobel
Department of Statistics and
    Operations Research
University of North Carolina
Chapel Hill, North Carolina

Vinayak Rao
Department of Statistics
Purdue University
West Lafayette, Indiana

Suren I. Rathnayake
Department of Mathematics
University of Queensland
St. Lucia, Australia

Roberto Rocci
Department of Economics and Finance
University of Tor Vergata
Rome, Italy

Volker Schmidt
Institute of Stochastics
Ulm University
Ulm, Germany

Douglas Steinley
Department of Psychological Sciences
University of Missouri
Columbia, Missouri

Michalis Vazirgiannis
Department of Informatics
Athens University of
 Economics and Business
Athens, Greece

Maurizio Vichi
Department of Statistical Sciences
Sapienza University of Rome
Rome, Italy

Sara Viviani
Department of Statistical Sciences
Sapienza University of Rome
Rome, Italy

1

Cluster Analysis: An Overview

Christian Hennig and Marina Meila

CONTENTS

Abstract

This chapter gives an overview of the basic concepts of cluster analysis, including some references to aspects not covered in this Handbook. It introduces general definitions of a clustering, for example, partitions, hierarchies, and fuzzy clusterings. It distinguishes objects × variables data from dissimilarity data and the parametric and nonparametric clustering regimes. A general overview of principles for clustering data is given, comprising centroid-based clustering, hierarchical methods, spectral clustering, mixture model and other probabilistic methods, density-based clustering, and further methods. The chapter then reviews methods for cluster validation, that is, assessing the quality of a clustering, which includes the decision about the number of clusters. It then briefly discusses variable selection, dimension reduction, and the general strategy of cluster analysis.

1.1 Introduction

Informally speaking, clustering means finding groups in data. Aristotle's classification of living things was one of the first-known clusterings, a hierarchical clustering. The knowledge of biology has grown, yet the schematic organization of all known species remains in the form of a (hierarchical) clustering. Doctors defining categories of tumors by their properties, astronomers grouping galaxies by their shapes, companies observing that users of their products group according to behavior, programs that label the pixels of an image by the object they belong to, other programs that segment a video stream into scenes, recommender systems that group products into categories, all are performing clustering.

Cluster analysis has been developed in several different fields with very diverse applications in mind. Therefore, there is a wide range of approaches to cluster analysis and an even wider range of styles of presentation and notation in the literature that may seem rather bewildering. In this chapter, we try to provide the reader with a systematic if somewhat simplified view of cluster analysis. References are given to the Handbook chapters in which the presented ideas are treated in detail.

More formally, if we are given a set of objects, also called a *dataset*, $\mathcal{D} = \{x_1, x_2, \ldots, x_n\}$ containing n *data points*, the task of clustering is to group them into K disjoint subsets of $\mathcal{D}$, denoted by $C_1, C_2, \ldots, C_K$. A clustering is, in first approximation, the partition obtained, that is, $\mathcal{C} = \{C_1, C_2, \ldots, C_K\}$. If a data point x_i belongs to *cluster C_k*, then we say that the *label* of x_i is k. Hence, labels range from 1 to K, the number of clusters. Obviously, not every such $\mathcal{C}$ qualifies as a "good" or "useful" clustering, and what is demanded of a "good" $\mathcal{C}$ can depend on the particular approach one takes to cluster analysis.

As we shall soon see, this first definition can be extended in various ways. In fact, the reader should be aware that clustering, or grouping of data, can mean different things in different contexts, as well as in different areas of data analysis. There is no unique definition of what a cluster is, or what the "best" clustering of an abstract $\mathcal{D}$ should be. Hence, the cornerstone of any rigorous cluster analysis is an appropriate and clear definition of what a "good" clustering is in the specific context. For a practitioner, cluster analysis should start with the question: which of the existing clustering paradigms best fits my intuitive knowledge and my needs?

Research in clustering has produced over the years a vast number of different paradigms, approaches, and methods for clustering to choose from. This book is a guide to the field, which, rather than being exhaustive, aims at being systematic, explaining and describing the main topics and tools. Limits on the size of the book combined with the vast amount of existing work on cluster analysis (Murtagh and Kurtz (2015) count more than 404,000 documents related to cluster analysis in the literature) meant that some results or issues could not make their way into the specialized Handbook chapters. In such cases, we give references to the literature.

This chapter provides a guide through the rest of the book, simplifying and summarizing even more. In what follows we define clustering, describe its various dimensions, and list the main paradigms developed and covered by the book. We close by highlighting issues like validation and method selection, as well as underscoring the distinction between clustering and *classification,** a task that may often seem related to clustering.

1.2 Dimensions (Dichotomies) of Clustering

Here, we give a first overview of the main dimensions along which one can describe clustering paradigms and methods. For the sake of clarity and simplicity, we present as dichotomies what in reality is almost a continuum. A more detailed picture will be filled in by the sections following this one.

1.2.1 By Type of Clustering: Hard vs. Soft

A clustering that is a partition of the data $\mathcal{D}$, so that each object x_i belongs to one and only one cluster $C \in \mathcal{C}$, is called a *hard* or *categorical* clustering. The K-means algorithm is a widely known algorithm that outputs hard clusterings.

By contrast, in a *soft* clustering, a data point x_i is allotted a *degree of membership* γ_{ik} to each of K clusters. Interpreting $(\gamma_{i1}, \ldots, \gamma_{iK})$ as a probability distribution over the label values $1, \ldots, K$ is particularly useful, and has led to the *mixture models* paradigm for clustering. A soft clustering $\mathcal{C}$ is a collection of soft assignments $\{\gamma_{ik}, x_i \in \mathcal{D}, k = 1, \ldots, K\}$.

1.2.2 By Type of Clustering: Flat vs. Hierarchical

Both types of clusterings described above are *flat* clusterings. To describe hierarchical clusterings, consider for simplicity the hard clustering case, $\mathcal{C} = \{C_1, C_2, \ldots, C_K\}$. If each cluster is further partitioned into subclusters, we have a *hierarchical clustering*. More generally, each subcluster could be further partitioned, adding new levels to the hierarchy. Or the clusters themselves can be grouped into "superclusters," and so on. Taxonomies (of genes, species,...) represent hierarchical clusterings. More on hierarchical clustering can be found in Section 1.3.

1.2.3 By Data Type or Format

Sometimes the type of data that is observed can profoundly influence how we think about clustering. Suppose, we want to cluster the students in a class. We could collect data about each student's grades, study habits, and so on, and use these variables for the purpose of clustering. Alternatively, we could collect data about how students interact with each other, like the network of friendships and collaboration in the class. In the former case, we describe the data points by a vector of features; in the latter, we describe the (pairwise)

---

* The use of the term "classification" in the literature is not unified; here we use it for what is called "supervised classification" in some literature, in which "classification" is a more general term, as opposed to "unsupervised classification," that is, clustering.

relations between data points. Hence, the latter is distinguished by terms such as *relational* clustering or *similarity-based* clustering. Other data formats exist, see Section 1.4.

1.2.4 By Clustering Criterion: (Probabilistic) Model-Based vs. Cost-Based

Some of the most successful formulations of clustering treat the data labels as latent variables in a probabilistic model of the data. Among these are the finite mixture approaches and the nonparametric Bayesian approaches (see Section 1.5.4). There are also approaches that define clusters in relation to the data distribution (e.g., as prominent peaks of the data density), for example, the level sets and mean-shift methods, see Section 1.5.5. Other successful paradigms view clustering as minimizing a cost, with no probabilistic interpretation. For instance, centroid-based algorithms like K-means (Section 1.5.1) are concerned with the dispersion within a cluster, whereas spectral clustering methods (Section 1.5.3) are concerned with the similarity between data points in different clusters.

1.2.5 By Regime: Parametric (K Is Input) vs. Nonparametric
(Smoothness Parameter Is Input)

This distinction can be seen superficially as a choice of algorithm.* But it is really a distinction at the level of (our prior beliefs about) the data generating process. If the data contains few ($K \approx 10$) clusters, of roughly equal sizes, then we are in the regime of *parametric clustering*. For instance, a *finite mixture* and an algorithm like K-means embody the concept of parametric clustering. In this regime, the appropriate algorithms to use typically require K as input, although K is typically not known, and found by methods akin to model selection, see Section 1.6.2.

But there are cases when the clusters' sizes vary greatly, over orders of magnitude. In such cases, there necessarily will be many clusters (perhaps hundreds) and the boundary between a small cluster and outlier data will become blurred. Very often, there is structure in the data labeled as "outliers"—with collecting more data, new clusters emerge where before were just "outliers." Hence, it is more natural to view "outliers" as very small clusters, possibly containing a single data point x. Consequently, the "number of clusters" K stops being meaningful. A more natural way to describe "number of clusters" is by the "level of granularity" in the clustering, quantified by a *smoothness parameter*. This is the *nonparametric clustering* regime. The archetypal nonparametric clustering paradigm is represented, not surprisingly, by nonparametric mixtures (see, e.g., Chapter 10).

Thus, in both regimes, the clustering obtained will depend on some user input governing the "number of clusters." While the terms "parametric/nonparametric" clustering are not universally accepted (actually nonparametric clustering requires a parameter, too), the distinction between the two regimes they designate is recognized as fundamental, affecting the choice of algorithm, the validation method, and the interpretation of the results, across various data formats and clustering types.

We close by reminding the reader that the dichotomies presented here can be called so only in first approximation. The long history of clustering has produced methods that cover more or less continuously the space spanned by the above dimensions.

---

* We stress that the terms "parametric (nonparametric) clustering" are in some of the literature and in the following used in a way specific to cluster analysis, and, while closely related to the broader terms of "parametric (nonparametric) model/statistics/estimation," should not be confused with the latter.

1.3 Types of Clusterings

First, different approaches of cluster analysis differ in how the different clusters relate to each other and to the object set $\mathcal{D}$.

- Many cluster analysis approaches aim at finding *partitions*, in which $C_j \cap C_k = \emptyset$ for $j \neq k$ (on the other hand, *overlapping clusterings* are those that violate this condition), which are in most cases (but not always) required to be *exhaustive*, that is, $\bigcup_{j=1}^{K} C_j = \mathcal{D}$. Some methods, particularly those meant to deal with outliers in the object set (see Chapter 29), allow that some objects are not assigned to any cluster.

- A special case of overlapping clusterings are *hierarchies*. A hierarchy is a sequence of partitions $\mathcal{C} = \bigcup_{j=1}^{m} \mathcal{C}_j$, where $\mathcal{C}_j$, $i = j, \ldots, m$ are partitions with $K_1 = |\mathcal{C}_1| > \cdots > K_m = |\mathcal{C}_m|$ ($|C|$ denoting the number of elements of C) so that for $C_j \in \mathcal{C}_j$ and $C_k \in \mathcal{C}_k$ with $j < k$ either $C_j \cap C_k = C_j$ or $C_j \cap C_k = \emptyset$, so that the sets on lower levels are partitions of the sets of the higher levels. Hierarchies can be visualized as trees ("dendrograms"), which show how the clusters in the "finer" lower level partitions are merged in order to arrive at the higher level partitions (whether the finer partitions are called of "lower" or "higher" level is somewhat arbitrary; here we use these terms in agreement with the most widespread convention to draw dendrograms). Figure 1.1 shows a hierarchy of sets and the corresponding dendrogram for a toy dataset.

  Originally, the use of hierarchies in cluster analysis was motivated by phylogenetic trees, that is, biological ideas of the evolutionary relationship between species. Imposing hierarchical structures can be useful in many fields; for example, they can be useful for efficient searches in databases or texts.

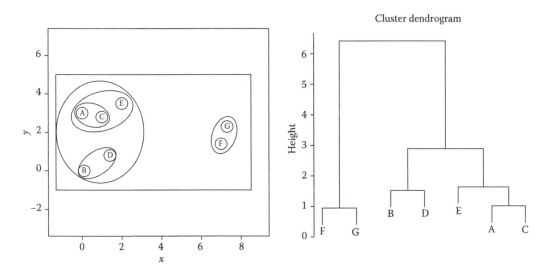

FIGURE 1.1
Hierarchy of sets and corresponding dendrogram (Average Linkage based on Euclidean distance) on toy dataset.

Some methods that construct hierarchies construct *full hierarchies*, in which $m = n$ and $K_i = n + 1 - i$, $i = 1, \ldots, n$, so that C_1 consists of all the one-object sets $\{x_i\}$, $i = 1, \ldots, n$, and C_n only consists of the whole of $\mathcal{D}$ as a single cluster.

- Some other set structures allowing for overlap have been proposed for clustering, such as pyramids and weak hierarchies (Bertrand and Janowitz 2002); other work for nonhierarchical overlapping clustering (e.g., Shepard and Arabie (1979); sometimes referred to as "clumping" in the literature) does not add additional restrictions to the set structure of the clustering.

- Clusters in *fuzzy clustering* are fuzzy sets, that is, there is a membership function that assigns a value in $[0, 1]$ to every object in $\mathcal{D}$ indicating the degree of membership to a cluster.

- In *probabilistic clusterings*, the objects have probabilities to belong to the clusters that add up to one for each object. Probabilistic clusterings can be interpreted as fuzzy clusterings (with the probabilities interpreted as membership values). But the probability approach formalizes that for every object there is a single true (but unknown) cluster, so that there is an underlying "crisp" partition. The data can be partitioned by assigning them to the cluster that maximizes the objects' membership probability.

- Clusters in *rough set clustering* are pairs of "upper approximations" and "lower approximations" of sets, that is, a set of objects which are possibly but not certainly members and a core set of sure members the cluster.

The main focus of this Handbook is on structures following the "partitioning paradigm" in some sense, that is, partitions, hierarchies, and probabilistic clusterings, with chapters on fuzzy and rough set clustering (Chapters 24 and 25) and some side remarks hinting at other nonpartitioning techniques.

A further important distinction in cluster analysis is whether data are interpreted as sample from an underlying probability distribution, in which case cluster analysis is based on estimating features of the probability distribution (e.g., parameters of mixture distributions or high density level sets). Researchers would usually make such an assumption if they aim at generalizing the results of the cluster analysis in some sense to a wider population.

In other clustering problems (such as data compression or the analysis of specific social networks), generalization may not be required, and only clustering the available objects may be of interest. Correspondingly, many clustering methods do not refer to underlying probability models that are assumed to be true but unknown. The delimitation between these two classes of methods is not always clear cut; for example, the K-means clustering method is often applied without reference to probability models, but can also be derived from a model with spherical Gaussian distributions for the clusters, see Chapter 3 (although Pollard (1981) showed that what is consistently estimated by the K-means are not exactly the means of the Gaussian distributions). On the other hand, some probability model-based clustering models may bring forth good clusterings even in situations in which their model assumptions are violated, in which case the model assumptions rather serve to clarify what kind of clusters the methods are looking for. It is therefore misleading to suggest that it is generally better to use methods that are not based on probability models for the only reason that they are not subject to restrictive model assumptions. Every cluster analysis method is implicitly based on some kind of "cluster concept," whether or not this is formalized by a probability model, see Chapter 31.

1.4 Data Formats

The objects to be clustered can be characterized in various ways. Two characterizations are used most widely:

1. *Objects × variables data* is probably most common, in which the objects are characterized by their values on a number of variables (features), which may be nominal, ordinal (this type and sometimes the previous are also called *vector data*), interval scaled, or of mixed type. Many clustering methods rely on (continuous) interval scaled data requiring some kind of score or dummy variables for the categories of nominal and ordinal data, but there are exceptions (i.e., Chapter 9).

   In some situations, the values of the variables are more complex such as in *Symbolic Data Analysis* or *functional clustering* (Chapters 13 and 21), or there is some additional structure in the variables such as a distinction between explanatory and response variable as in regression or time series clustering (Chapters 11 and 12).

2. The aim of cluster analysis is often described as collecting similar objects in the same cluster and having large dissimilarity between objects in different clusters. Therefore, some cluster analysis methods are based on a matrix of pairwise *similarities or dissimilarities* between objects. A dissimilarity is a function $d : \mathcal{X}^2 \mapsto \mathbb{R}$, $\mathcal{X}$ being the object space, so that $d(\mathbf{x}, \mathbf{y}) = d(\mathbf{y}, \mathbf{x}) \geq 0$ and $d(\mathbf{x}, \mathbf{x}) = 0$ for $\mathbf{x}, \mathbf{y} \in \mathcal{X}$. Similarities have complementary characteristics, and there are various ways to translate dissimilarities into similarities, the simplest one being

$$s(\mathbf{x}, \mathbf{x}') = \max_{x_1, x_2 \in \mathcal{X}} (d(\mathbf{x}_1, \mathbf{x}_2)) - d(\mathbf{x}, \mathbf{x}')$$

   so that approaches for dissimilarities and similarities are mostly equivalent.

   Another equivalent way of interpreting similarity data is as weights of edges in a network or graph, in which the objects are the nodes or vertices. This means that methods for *graph and network clustering* (Chapters 7 and 16) are also implicitly based on similarity data. Such similarities are sometimes obtained in a special way, e.g., for unweighted graphs they can take only the values 1 (two nodes are connected) or 0 (not connected).

Dissimilarities can be obtained from objects × variables data in various ways (the most widely used way is the Euclidean distance, but there are many alternatives, see Chapter 31). By suitable definition of a dissimilarity measure, dissimilarity-based clustering methods can handle any type of data. On the other hand, multidimensional scaling techniques (Borg and Groenen 2005) approximate dissimilarities by points in Euclidean space, making dissimilarity data accessible to clustering techniques for Euclidean data.

Often, cluster analysis depends heavily on details of the data format. The specific definition of the dissimilarity measure can have a large impact on the clustering solution, as can transformation and standardization of variables and variable selection. An important consideration in this respect is that the chosen characterization of the data should correspond to the subject matter meaning of "similarity" and in which sense objects in the same cluster are meant to "belong together." These choices are discussed in Chapter 31. On the other hand, the task of learning a distance measure from the data is discussed in Chapter 20. Another important issue regarding what data to use is variable selection and dimension reduction, see Section 1.6.3.

1.5 Clustering Approaches

This section gives an overview of the basic approaches to clustering, that is, the principles behind the clustering methods.

1.5.1 Centroid-Based Clustering

In case the number of clusters K is fixed and known (where to get K from is discussed in Section 1.6.2 for all approaches for which this needs to be specified), the idea behind the K-means method to find K "centroid" objects in order to represent all objects in an optimal manner is one of the earliest approaches to cluster analysis, see Chapters 2 through 5. Usually this is formalized as minimizing

$$S(\mathcal{D}, \mathbf{m}_1, \ldots, \mathbf{m}_K) = \sum_{i=1}^{n} d(\mathbf{x}_i, \mathbf{m}_{c(i)}), \text{ where} \tag{1.1}$$

$$c(i) = \operatorname*{argmin}_{j \in \{1, \ldots, K\}} d(\mathbf{x}_i, \mathbf{m}_j), \ i = 1, \ldots, n$$

by choice of the centroids $\mathbf{m}_1, \ldots, \mathbf{m}_K$, where d is a dissimilarity measure. The centroids $\mathbf{m}_1, \ldots, \mathbf{m}_K$ may be required to be objects in $\mathcal{D}$ (in which case they are sometimes called "exemplars,"), or they may stem from the data space $\mathcal{X}$. d may be the given dissimilarity measure that characterizes the objects or a transformation. For the K-means method, $\mathbf{x}_1, \ldots, \mathbf{x}_n \in \mathbb{R}^p$, $\mathbf{m}_1, \ldots, \mathbf{m}_K$ are not required to be exemplars, and d is the squared Euclidean distance, which implies that $\mathbf{m}_1, \ldots, \mathbf{m}_K$ have to be mean vectors of their respective clusters in order to minimize Equation 1.1, thus the name "K-means." There are fuzzy and rough set versions of K-means, see Chapters 24 and 25.

As mentioned before (and in more detail in Chapters 3), the use of the squared Euclidean distance is connected to the likelihood of the Gaussian distribution. For more general dissimilarity measures d, there is often not such a simple characterization of the centroid objects, and no obvious connection to parameters of a distribution. If the analysis starts from a given dissimilarity matrix that is not derived from an available objects × variables matrix locating the objects on some kind of data space, centroids have to be exemplars because there is no data space available from which to find centroids such as means, although the K-means objective function can be expressed in terms of Euclidean distances alone without involving centroids, see Chapter 26.

Clusterings of this type require every object in a cluster to be close to the centroid, which tends to restrict the cluster shapes (depending on the data space and dissimilarity, clusters tend to be convex). This is often not appropriate for elongated and/or nonlinear clusters (see Figure 1.2), although it is often possible to represent such data in a different space (for example applying spectral clustering to transformed dissimilarities, see Chapter 7) so that centroid-based methods such as K-means can find clusters that are irregularly shaped but well separated in the original representation among the representations of the objects in the resulting space.

An important problem with Equation 1.1 is that finding a global optimum is computationally hard, and therefore there is much work on efficient computation of good locally optimal solutions, see particularly Chapters 5.

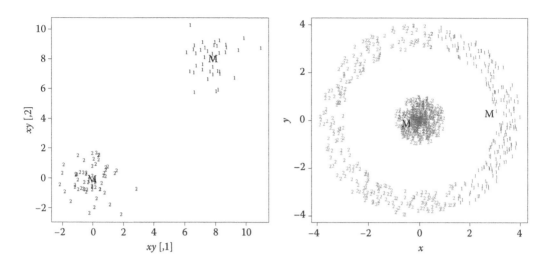

FIGURE 1.2
A dataset in which 2-means works well ("M" indicates the 2 cluster mean vectors), and a dataset in which it gives a solution that runs counter to many people's intuition.

1.5.2 Agglomerative Hierarchical Methods

As K-means, agglomerative hierarchical methods such as Single Linkage can be traced back to the 1950s and are therefore among the oldest cluster analysis methods (see Chapters 2 and 6). There are two main ways to build a hierarchy step by step. The first one is agglomerative, the second one is divisive. Agglomerative hierarchical clustering starts from a clustering in which every object forms its own cluster, so that there are n clusters. At each step, the two most similar clusters are merged, so that there is a new clustering with the number of clusters reduced by one, until at the top level all objects are joined into a single cluster. Different agglomerative hierarchical methods are characterized by different ways to compute the dissimilarity D between two clusters from the dissimilarities d between the member objects. The most famous methods are

Single Linkage (or "nearest neighbor"), where

$$D(C_1, C_2) = \min_{\mathbf{x}_1 \in C_1, \mathbf{x}_2 \in C_2} d(\mathbf{x}_1, \mathbf{x}_2)$$

Complete Linkage (or "furthest neighbor"), where

$$D(C_1, C_2) = \max_{\mathbf{x}_1 \in C_1, \mathbf{x}_2 \in C_2} d(\mathbf{x}_1, \mathbf{x}_2)$$

Average Linkage (or "UPGMA" for "Unweighted Pair Group Method with Arithmetic Mean"), where

$$D(C_1, C_2) = \frac{1}{|C_1| \, |C_2|} \sum_{\mathbf{x}_1 \in C_1, \mathbf{x}_2 \in C_2} d(\mathbf{x}_1, \mathbf{x}_2)$$

These methods are based on dissimilarity data and are not connected to probability models. An exception is Ward's method, which merges clusters at each step in such a way that the

K-means criterion as introduced in Section 1.5.1 is optimally improved, thus producing some kind of locally optimal solution of Equation 1.1.

Partitions into K clusters can be obtained from hierarchies by cutting the hierarchy at the appropriate level. In some respects the different agglomerative hierarchical methods are quite different from each other, emphasizing, for example, separation of clusters over homogeneity in the most extreme manner (Single Linkage), or the opposite (Complete Linkage), but on the other hand the agglomeration process makes partitions obtained by them rather more similar on some datasets than to what can be obtained from other clustering approaches. Chapter 6 gives more details and introduces further hierarchical methods.

Divisive hierarchical methods start with all objects in a single cluster and proceed downwards by splitting up one of the existing clusters at each step. Computationally, it is much more difficult to find optimal splits as required for divisive clustering than to find optimal merges as required for agglomerative clustering (there are $2^n - 1$ possible splits for the top-level cluster with all n objects, whereas agglomerative clustering in the first step only needs to consider $\binom{n}{2}$ possible merges). Therefore, agglomerative hierarchical clustering is much more widely used. Divisive clusters are mainly used for "monothetic" clusterings in which only a single variable at a time is used to split clusters, for example, the method "MONA" in Kaufman and Rousseeuw (1990).

1.5.3 Spectral Clustering

Spectral clustering originally deals with (dis-)similarity data, given in the form of an $n \times n$ similarity matrix $\mathbf{S}$. However, this method is frequently used with other data formats, after having obtained similarities from the original data. For example, a matrix of objects $\times$ variables can be transformed into a square matrix S of similarities between pairs of objects by one of the methods described in Chapter 7 or in Section 1.4. The advantage of using this method for data lying in Euclidean space is that spectral clustering allows clusters to have arbitrary shapes, like those in Figure 1.2.

The term "spectral clustering" refers to the family of techniques based on the spectral decomposition of the so-called *Laplacian* matrix, a matrix derived from the similarity matrix $\mathbf{S}$. The K principal eigenvectors of the Laplacian matrix provide a mapping of the objects into K dimensions. To obtain clusters, the resulting K-dimensional vectors are clustered by standard methods, usually K-means.

There are various interpretations of this. In the framework of graph or network clustering, a spectral clustering can be viewed as finding "cuts" of low similarity in the weighted graph represented by $\mathbf{S}$. Viewing $\mathbf{S}$ as coming from a (reversible) Markov chain, objects associate if they lie along paths of high probability in the state space of the Markov chain. This is the primary interpretation for clusterings obtained from data which originally is Euclidean.* For these data, spectral clustering acts as a remarkably robust linkage method. Spectral clustering is treated in detail in Chapter 7.

1.5.4 Mixture Probability Models

Clustering via mixtures of parametric probability models is sometimes in the literature referred to as "model-based clustering." Most statisticians trained in probability tend to

---

* Note that the original dimension of the objects $\times$ variable data can be either smaller or larger than K.

find this the most "natural" way of thinking about cluster analysis. Usually, this approach assumes objects × variables data, but it has been applied also to infer such models on an assumed latent data space for objects that come as similarity or network data, see Chapter 16.

The general form of a mixture model is that data are assumed independently identically distributed (i.i.d.) according to a distribution with density

$$f(\mathbf{x}) = \sum_{j=1}^{K} \pi_j f_{\theta_j}(\mathbf{x}) \tag{1.2}$$

where f_θ, $\theta \in \Theta$ defines a parametric family of distributions (Θ may be more than one-dimensional, for example it could contain pairs of mean vectors and covariance matrices of Gaussian distributions), and $\pi_j \geq 0$, $j = 1, \ldots, K$, $\sum_{j=1}^{K} \pi_j = 1$ for the proportions of the mixture components. The parameters $(\pi_1, \theta_1), \ldots, (\pi_K, \theta_K)$ are often estimated by Maximum Likelihood. Given estimators $(\hat{\pi}_1, \hat{\theta}_1), \ldots, (\hat{\pi}_K, \hat{\theta}_K)$, and assuming a two-step mechanism for generating points from f in which first a component membership $c(i)$ for $\mathbf{x}_i$, $i = 1, \ldots, n$ is generated according to a multinomial distribution with component probabilities $\pi_1, \ldots, \pi_K$, probabilities can be estimated for a given object $\mathbf{x}_i$ to have been generated by the mixture component with parameter θ_j:

$$p(c(i) = j \mid \mathbf{x}_i) = \frac{\hat{\pi}_j f_{\hat{\theta}_j}(\mathbf{x}_i)}{\sum_{k=1}^{K} \hat{\pi}_k f_{\hat{\theta}_k}(\mathbf{x}_i)} \tag{1.3}$$

This defines a probabilistic clustering, and by maximizing $p(c(i) = j \mid \mathbf{x}_i)$, a crisp partition can be obtained. Usually, every mixture component f_{θ_j} is interpreted as defining a cluster, which means that the family of parametric densities f_θ, $\theta \in \Theta$ defines the shapes of what is interpreted as a "cluster." This is not always appropriate; there may be a mixture component for "outliers," or several mixture components taken together may define a cluster, see Chapter 29.

Maximum Likelihood estimation in mixture models enables statistical theory such as consistency and asymptotic normality and allows therefore for statistical inference.

The approach defined by Equation 1.2 can be used for various data formats. There are mixture models for interval scaled or continuous data assuming various distributional shapes for clusters (Chapter 8), categorical or mixed type data (Chapter 9; the mixture approach is often called "latent class clustering" in this case), and data from plain or generalized linear models, time series models or functional data (Chapters 11 through 13).

Most often, Maximum Likelihood estimators for the parameters are computed for fixed K, and K is estimated by fitting the model for several different values of K and then optimizing information-based criteria such as the Bayesian information criterion (BIC), see Chapter 8. There is also a possibility to implicitly estimate K by the so-called "non-parametric Maximum Likelihood" (Lindsay 1995), although this usually produces more mixture components than what can be reasonably interpreted as clusters.

An alternative approach to estimating mixtures is to assume a Bayesian prior for the mixture parameters and the number of components (Richardson and Green 1997). A recently popular Bayesian approach is to assume a Dirichlet process prior, which does not assume the number of mixture components K to be fixed, but can in principle generate an unlimited number of new components for new observations. For a given dataset, this makes

Bayesian inference possible about how many mixture components were already observed, see Chapter 10.

1.5.5 Density-Based Clustering

An intuitive definition of a cluster based on probability models is to assume a nonparametric model of density for the whole dataset, and to identify clusters either with regions of high density, or with density modes, assigning an observation to the density mode that "attracts" it. Such methods are much more flexible regarding the cluster shapes than mixtures of parametric models (although such flexibility is not always desirable, particularly in applications in which large within-cluster distances should be avoided).

Chapter 18 is about estimating the data density first by kernel density estimators, and then about methods to find modes. Chapter 17 presents an approach to identify high density level-sets in data. In some datasets, density levels may be quite different in different regions of the data space, and density levels at which regions are seen to qualify as clusters may differ, in which case a mode-based approach may be preferable. On the other hand, weak density modes in the data may often be spurious and unstable.

The Single Linkage technique, although not directly based on probabilistic modeling, has been interpreted as a technique to find clusters between which there are density valleys, and there are a number of further nonprobabilistic techniques that also aim at finding clusters that can be interpreted as dense point clouds, such as DBSCAN and DENCLUE, see Chapter 6. For larger and higher dimensional data, a popular approach is to first define a grid, that is, to partition the data space uniformly into a finite number of cells, and then to connect neighboring cells to clusters according to the number of points within the cells, that is, the within-cell density, see Cheng et al. (2014). Usually, the idea of density-based clustering refers to continuous data, but there are density-based methods for categorical and mixed type data as well, based on organizing the data into a network and assessing connectedness, see Andreopoulos (2014).

Some "self-organizing" techniques that implement clustering as a dynamic process in which objects group themselves according to certain rules are also driven by object neighborhoods and attraction by density, see Chapter 19.

1.5.6 Other Probability Models

There are alternative modeling approaches for clustering, one being a model in which instead of having mixture proportions, the cluster memberships of the observations are treated as fixed parameters. This is sometimes called "fixed partition model." Maximum Likelihood estimation of such models is often similar to mixture models; the K-means technique (Chapter 3) can be interpreted in this way.

Chapter 14 treats the clustering of data from spatial point processes; the main method proposed there is a generalization of Single Linkage hierarchical clustering, which is not in itself a probabilistic method.

Chapter 15 uses probability models not for modeling a clustering structure in the first place, but for modeling homogeneous nonclustered data, in order to test homogeneity against clustering.

PD-clustering of Ben-Israel and Iyigun (2008) models probabilities for objects to belong to fixed clusters (to be estimated) as inversely proportional to the distance from the cluster's centroid, without making parametric model assumptions for the distributions of points within clusters.

1.5.7 Further Clustering Approaches

There are a number of further approaches to clustering. Some of them are presented in Chapter 19, namely methods mimicking self-organizing processes in nature, optimization of certain noncentroid-based objective functions such as cluster validation indexes (see below), and multiobjective clustering, attempting to account for the fact that the usual criteria for clustering such as Equation 1.1 focus on a specific aspect of what is expected of a clustering (e.g., cluster homogeneity) at the expense of others (e.g., separation).

In some applications, there is information available of the type that, for example, certain pairs of objects have to be in the same cluster or should not be in the same cluster. This leads to a constrained ("semi-supervised") clustering problem and is treated in Chapter 20.

Given that different clustering algorithms often yield strikingly different clusterings for the same dataset, research has been devoted to define "consensus clustering," constructing a clustering from a set of clusterings obtained by different approaches, or from different resampled versions of the dataset, see Chapter 22.

Chapter 23 is about simultaneous clustering of objects and variables. The "double K-means" method applies K-means-type criteria simultaneously to the objects and the variables.

Outliers are an endemic problem in cluster analysis as in the whole of statistics. Chapter 29 reviews clustering methods that allow to classify some observations as "not belonging to any cluster," along with a discussion of the impact of outliers and violations of model assumptions on many clustering methods.

Finally, we mention one more principle that is not covered in this book. The *information bottleneck* principle (Tishby and Slonim 2001) applies to a broader class of problems, which includes, in addition to clustering, regression and classification. This principle states that to optimize a predictor between an input variable X and an output variable Y (which for clustering would be the label $c(x)$) one estimates an intermediate variable X', whose role is to embody the information from X that is relevant to Y. "Fitting the model" corresponds to minimizing the *mutual information* (defined in Chapter 27) between X and X' while maximizing the mutual information between X' and Y. The Multivariate Information Bottleneck method (Slonim and Friedman 2006) has been applied successfully to two-way clustering.

1.6 Cluster Validation and Further Issues

1.6.1 Approaches for Cluster Validation

The term "cluster validation" usually refers to the assessment of the quality of a clustering. Given the many approaches outlined above, often leading to different results on the same dataset, there is a need for techniques that can help the researcher to decide between the different approaches and to assess to what extent the clustering results are informative and reliable.

There are a number of different approaches to this problem:

- In many applications it is of interest to *test* whether the dataset is actually clustered or rather homogeneous (in which case most clustering approaches still will return a clustering). Such tests are treated in Chapter 15.

- There is a number of *cluster validation indexes*, that is, statistics that measure the quality of a clustering. In principle, objective functions defining clustering methods such as Equation 1.1 measure the quality of a clustering, too, but cluster validation measures are constructed in order to deliver a fair comparison of clusterings with, for example, different numbers of clusters (Equation 1.1 can always be improved adding clusters), or from different methods. Such measures are introduced in Chapter 26. Chapter 19 has some thoughts on optimizing such indexes, or using them for multicriterion approaches.

- Chapter 27 is on *indexes to compare different clusterings* that can be used for exploring the diversity or stability of solutions for a given dataset, or for comparing a clustering with an external grouping that is expected to have some connection to the clustering. Comparing a clustering with external information, which may come as another grouping of the data, is a general cluster validation technique.

- The evaluation of the *stability* of a clustering is often relevant, here meaning that a similar dataset should yield a similar clustering, if the clustering is deemed reliable. Stability can be evaluated by *resampling* the dataset and by studying the similarity of the resulting clusterings. This is done in Chapter 28.

- A major technique for cluster validation and actually also for exploring the dataset a priori and sometimes for helping with the clustering task is *data visualization*. Some ways to use data visualization for clustering and cluster validation are presented in Chapter 30.

1.6.2 Number of Clusters

In *parametric clustering*, most methods assume the number of clusters as fixed in their most basic form, for example, Section 1.5.1; this also includes the use of hierarchical methods for obtaining a partition by cutting the hierarchy at some level. Choosing between different numbers of clusters can be seen as a problem of cluster validation, and some literature uses the term "cluster validation" for methods to decide about the number of clusters.

Popular "general purpose" techniques for determining the number of clusters are the optimization of cluster validation indexes (Chapter 26; some of these indexes are adapted to specific clustering methods such as K-means) or stability optimization (Chapter 28). Some clustering approaches come with their own theory and specific methods for estimating the number of clusters, such as probability mixture models, for which the number of clusters can be interpreted as a model parameter to be estimated. Unfortunately, the number of clusters as a model parameter behaves in a very nonstandard way, and therefore theoretical results are hard to obtain (Keribin 2000), underscoring the general difficulty of the number of clusters problem.

Occasionally, literature appears in which certain clustering methods are advertised as having the data making all the decisions without any input of the researcher. Such advertisement should not be trusted because making these decisions in a fully automatic way basically means that the method makes the required decisions internally hidden from the researcher, with no guarantee that the interpretation of the resulting clusters matches what the researcher is interested in. In some cases (e.g., Rodriguez and Laio 2014), a tuning decision is indeed required, although not advertised. In other cases, the rationale for automatic decision and its limits need to be well understood by the researcher. For example, using the BIC with Gaussian mixture models (Chapter 8) will pick more than one cluster for datasets or data subsets that cannot be well fitted by a single Gaussian distribution, although it

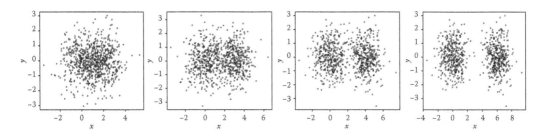

FIGURE 1.3
Data from two Gaussian populations with distance between mean vectors equal to 1.9, 3, 4, 6. Any borderline distance value from which upward this is interpreted as "two clusters" has to depend on the context and aim of clustering and the interpretation of the clusters.

may well be fitted by another distribution that is unimodal and homogeneous enough (in Figure 1.3 for example the BIC will pick two clusters for all four example datasets). We acknowledge, though, that in some applications such as signal segmentation or network traffic monitoring, in which clustering is routinely carried out many times, automatic decisions are required.

Nonparametric clustering methods do not require the number of clusters to be fixed, and the determination of the number of clusters is integrated in the determination of the clustering (e.g., Chapters 17 and 18 and the DBSCAN-method in Chapter 6). These methods require other tuning constants such as a kernel bandwidth or a neighborhood size in which to look for objects to merge into the same cluster.

There is a good reason for this necessity. Figure 1.3 illustrates that some kind of decision is needed in cluster analysis about how separated clusters are supposed to be. This decision can be supplied in various ways, one of them being fixing the number of clusters, another being cluster granularity.

1.6.3 Variable Selection, Dimension Reduction, Big Data Issues

In many applications of cluster analysis, the collection of variables on which the clusters are to be found is not fixed. In some situations there are so many variables that some kind of dimension reduction or variable selection is required for performing a reasonable cluster analysis. Many clustering methods become computationally prohibitive in high dimensions and those that still give results are often very unstable. Dissimilarity-based techniques do not have specific computational issues with many variables because the size of the dissimilarity matrix does not depend on the dimension of the data space from which it is computed; however, using standard dissimilarity measures such as the Euclidean distance, in high dimensions typically all objects are very dissimilar from all (or most) other objects, making the task of finding reliable clusters very difficult.

Approaches for dimension reduction in cluster analysis can be classified in at least three ways. First, dimension reduction can be carried out before clustering, or integrated with clustering. Second, dimension reduction can be done by selecting a subset of the available variables, or by defining new variables from the existing ones, for example, by using linear combinations. Third, the whole clustering can be defined on the same lower dimensional space, or there can be different lower dimensional subspaces for different clusters.

For carrying out dimension reduction before clustering, standard techniques such as principal component analysis can be used, although it cannot be guaranteed that the data subspace picked by such methods without using clustering information is indeed the most suitable for clustering, see Chapters 30 and 31, mentioning also some alternatives. An alternative strategy, as mentioned in Chapter 31, is to use subject matter knowledge to either pick the most relevant variables or to define indexes from the existing variables that capture best the aspects of the data that are relevant for the clustering aim.

Integrating dimension reduction with clustering is mentioned in the part on mixtures of factor analyzers in Chapter 8. Spectral clustering (Chapter 7) involves mapping the objects to be clustered into a suitable low dimensional space for clustering. If the original objects are high dimensional vectors, this can be thought of as a dimension reduction of the data. In this case, one needs to compute all pairwise distances from the objects × variables data. In two-mode clustering (Chapter 23), clusters are defined by not only a subset of objects, but also a subset of variables. Methods for reweighting variables for objective functions such as Equation 1.1 as treated in Chapter 3 are also related.

As variable selection and dimension reduction are not in the main focus of this Handbook, we add some references to further literature. Witten and Tibshirani (2010) use a Lasso-type penalty on the L_1-norm of variable weights to propose sparse versions of K-means and hierarchical clustering. Friedman and Meulman (2004) weight variables involving a similar penalty for the definition of a dissimilarity and clustering based on a dissimilarity-based objective function so that different clusters have different variable weights. Rocci et al. (2011) and Vichi et al. (2007) combine linear factor analysis type dimension reduction with simultaneous K-means clustering in a simultaneous technique for two and three way data, respectively. Mixture model-based techniques for variable selection have also been proposed, for example, Raftery and Dean (2006), Maugis et al. (2009), using modified versions of the BIC, or, in the framework of Bayesian Dirichlet process mixtures, Kim et al. (2006). Ritter (2015) treats variable selection combined with robust cluster analysis.

Apart from dimension reduction, there are further issues dealing with large datasets. Particularly, many clustering methods can become hard to compute with many objects as well as with many dimensions. Suitable approaches to computation are treated, for example, in Chapters 5 and 19.

To those with a particular interest in variable selection, dimension reduction, algorithms, and streaming data, we recommend the book Aggarwal and Reddy (2014), which focuses more than the present volume on big data issues in cluster analysis.

1.6.4 General Clustering Strategy and Choice of Method

Chapter 31 gives a general overview on clustering strategy, which encompasses data preprocessing, the construction of an appropriate dissimilarity measure, the choice and construction of suitable variables, the choice of the clustering method, the number of clusters, and cluster validation. Regarding the comparison of different clustering approaches, the chapter hints at work on the theoretical axiomatization of the clustering problem and at benchmarking studies comparing different methods. It is argued that the choice of the clustering method and clustering strategy needs to depend strongly on the aim of clustering and on the context. No method is optimal in a general sense, and knowledge of the different characteristics of the methods is important for making the required decisions.

1.6.5 Clustering Is Different from Classification

Both in clustering and in classification, one assigns labels, say from 1 to K, to observed objects. Both deal with putting observations, characterized by features, or measurements, or attributes, in a finite set of categories.

Yet from a statistical perspective, and often from a user's perspective as well, these problems are different. For statistically meaningful results, clustering and classification should be approached with different tools. Table 1.1 summarizes their main differences.

Clustering is often an exploratory task. This means that we perform it on data to find out about it, to "understand" the data better, without a precisely formulated ultimate objective. We postulate that the data may not be homogeneous, that it could contain distinct groups, but we do not know yet how many groups there are or what is their meaning. In contrast, classification is a *predictive* task. The set of categories is known, and the objective is to assign labels to data points that do not have labels yet.

In many a data analysis process, clustering precedes classification. The former discovers the classes, the latter automatically assigns new observations to the appropriate class. In other words, one first uses cluster analysis of a representative sample to discover if categories exist in the data; once these categories acquire meaning, one finds it useful to train a classifier on the collected sample, that will be used *out of sample* to automatically classify/label new observations from the same data source. See Figure 1.4 for an example. Further uses of clustering in data analysis are exemplified in Chapter 31.

TABLE 1.1

Comparison between Classification and Clustering

	Classification	Clustering
Goal	Prediction of output y given $\mathbf{x}$	Exploration, information organization, and many more
Supervised/unsupervised	Supervised	Unsupervised
Generalization	Performance on new data is what matters	Performance on current data is what matters
Cost (or loss) function	Expected classification error or expected loss	Many loss functions, not all probabilistic
Number classes/clusters K	Known	Unknown

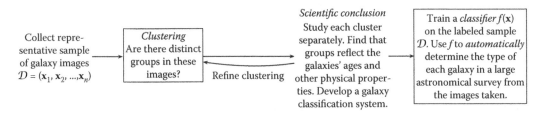

FIGURE 1.4

Clustering and classification within the flow of scientific research; a made-up example.

In the parlance of machine learning, classification is *supervised* and clustering is *unsupervised*.* The former means that when we fit the model we are given a sample of data which already have labels, presumably assigned by a "teacher," or "supervisor," so the classifier is built using information about what its target output should be. In the latter case, the data is processed without regard to a target output and without benefiting from a "teacher."

It follows that in classification, as in any supervised learning task, the predictive accuracy is the natural criterion by which to measure success. There is no such clear cut criterion for clustering. Instead, a variety of criteria, definitions and validation approaches are available.

Because the ultimate goal of classification is prediction, the concern is with future data, and with probabilistic assumptions about the data distribution. State-of-the-art classification is solidly founded on probability and statistics. The concern of clustering is primarily with the actual data at hand. Hence, a variety of probabilistic formulations for the clustering problem coexist with successful nonprobabilistic ones.

A major difference between clustering and classification is that in classification operations such as variable transformations, variable selection, the definition of dissimilarity measures (where required), and the choice of the classification method are ultimately dominated by the prediction goal; whatever serves prediction quality best is to be preferred. In clustering, however, all such operations change the implicit meaning of the clusters, which may be appropriate for serving different aims. For example, in a dataset giving both socioeconomic and environmental measurements for a number of locations, it is equally legitimate to cluster the locations using all variables, or only the socioeconomic ones, or only the environmental ones, but the clusters obtained by these three approaches have to be interpreted in different ways, making reference to the information involved. Which one is preferable depends on the aim of clustering. Which variables are the most suitable for clustering cannot be decided by a purely data-based quality measure alone.

Finally, classification as a statistical decision task is concerned ultimately not with the distinctiveness of the classes, but with the utility of the decisions made. Clustering on the other hand implicitly assumes that the objects inside a cluster are in some sense similar to each other, and differ to some extent from object in other clusters. For example, even though a population of students may represent a continuum of skill levels, a teacher will have to decide which students pass her class and which do not. Some passing students may be more similar to the failing students than to the top students in the class. It would be wrong to conclude that the class contains two different populations ("clusters"), "passing" and "failing," even though the teacher's decision assigns students these labels. However, groups may exist in the class, and they may be weakly or not at all related to the pass/fail decision. For example, students could indeed fall into distinct groups by gender, by their favorite band, or by whether they learn best in groups or alone.

---

* As some readers will know, this is a simplified presentation. In reality, there are various hybrid paradigms that bridge supervised and unsupervised learning. In particular, *semi-supervised learning* is a paradigm where in a sample $\mathcal{D}$ of x observations, some labels y are given, and the others are to be inferred. Conversely, supervised "learning to cluster" problems have been defined, where from a given clustering $\mathcal{C}$ of a sample, one learns for instance a similarity function (see Chapter 7) that will be applied to cluster other samples similar to this one.

References

Aggarwal, C. C. and C. K. Reddy (Eds.) 2014. *Data Clustering: Algorithms and Applications*. Boca Raton, FL: CRC Press.

Andreopoulos, B. 2014. Clustering categorical data. In C. C. Aggarwal and C. K. Reddy (Eds.), *Data Clustering: Algorithms and Applications*, 277–304. Boca Raton, FL: CRC Press.

Ben-Israel, A. and C. Iyigun 2008. Probabilistic d-clustering. *Journal of Classification 25*, 5–26.

Bertrand, P. and M. F. Janowitz 2002. Pyramids and weak hierarchies in the ordinal model for clustering. *Discrete Applied Mathematics 122*, 55–81.

Borg, I. and P. Groenen 2005. *Modern Multidimensional Scaling: Theory and Applications* (2nd ed.). New York, NY: Springer.

Cheng, W., W. Wang, and S. Batista 2014. Grid-based clustering. In C. C. Aggarwal and C. K. Reddy (Eds.), *Data Clustering: Algorithms and Applications*, 127–148. Boca Raton, FL: CRC Press.

Friedman, J. H. and J. J. Meulman 2004. Clustering objects on subsets of attributes (with discussion). *Journal of the Royal Statistical Society B 66*, 815–849.

Kaufman, L. and P. Rousseeuw 1990. *Finding Groups in Data*. New York: Wiley.

Keribin, C. 2000. Consistent estimation of the order of a mixture model. *Sankhya A 62*, 49–66.

Kim, S., M. G. Tadesse, and M. Vannucci 2006. Variable selection in clustering via Dirichlet process mixture models. *Biometrika 93*, 877–893.

Lindsay, B. G. 1995. *Mixture Models: Theory, Geometry and Applications*. Hayward, CA: Institute of Mathematical Statistics and the American Statistical Association.

Maugis, C., G. Celeux, and M.-L. Martin-Magniette 2009. Variable selection in model-based clustering: A general variable role modeling. *Computational Statistics and Data Analysis 53*, 3872–3882.

Murtagh, F. and M. J. Kurtz in press. The classification society's bibliography over four decades: History and content analysis. *Journal of Classification*.

Pollard, D. 1981. Strong consistency of k-means clustering. *Annals of Statistics 9*, 135–140.

Raftery, A. and N. Dean 2006. Variable selection for model-based clustering. *Journal of the American Statistical Assocation 101*, 168–178.

Richardson, S. and P. J. Green 1997. On Bayesian analysis of mixtures with an unknown number of components (with discussion). *Journal of the Royal Statistical Society B 59*, 731–792.

Ritter, G. 2015. *Robust Cluster Analysis and Variable Selection*. Boca Raton, FL: CRC Press.

Rocci, R., S. A. Gattone, and M. Vichi 2011. A new dimension reduction method: Factor discriminant k-means. *Journal of Classification 28*, 210–226.

Rodriguez, A. and A. Laio 2014. Clustering by fast search and find of density peaks. *Science 344*, 1492–1496.

Shepard, R. N. and P. Arabie 1979. Additive clustering representation of similarities as combinations of discrete overlapping properties. *Psychological Review 86*, 87–123.

Slonim, N. and N. Friedman 2006. Multivariate information bottleneck. *Neural Computation 18*, 1739–1789.

Tishby, N. and N. Slonim 2001. Data clustering by Markovian relaxation via the information bottleneck method. In T. K. Leen, T. G. Dietterich, and V. Tresp (Eds.), *Advances in Neural Information Processing Systems*, Volume 13. Cambridge, MA: MIT Press, pp. 929–936.

Vichi, M., R. Rocci, and H. A. L. Kiers 2007. Simultaneous component and clustering models for three-way data: Within and between approaches. *Journal of Classification 24*, 71–98.

Witten, D. M. and R. Tibshirani 2010. A framework for feature selection in clustering. *Journal of the American Statistical Association 105*, 713–726.

2

A Brief History of Cluster Analysis

Fionn Murtagh

CONTENTS

Abstract

Beginning with some statistics on the remarkable growth of cluster analysis research and applications over many decades, we proceed to view cluster analysis in terms of its major methodological and algorithmic themes. We then review the early, influential domains of application. We conclude with a short list of surveys of the area, and an online resource with scanned copies of early pioneering books.

2.1 Introduction

Clustering as a problem and as a practice in many different domains has proven to be quite perennial. Testifying to this is the presence of "clustering" or "cluster analysis" as a term in an important classification system. The premier professional organization in computing research, the Association for Computing Machinery (ACM), has a standard classification labeling system for publications. Released in 1998, a major update was released in September 2012. ACM Computing Classification System (2012), part of the category tree is as follows: "Mathematics of Computing," "Probability and Statistics," "Statistical Paradigms," "Cluster Analysis." Computing Classification System (1998) had clustering included in category H.3.3, and I.5.3 was another category "Clustering."

Figures 2.1 and 2.2, using the Google Scholar content-searchable holdings, present a view of this perennial and mostly ever-growing use of clustering. The term "cluster analysis" was used. Documents retrieved, that use that term in the title or body, increased to 404,000 in the decade 2000–2009. Of course lots of other closely related terms, or more specific terms, could additionally be availed of. These figures present no more than an expression of the growth of the field of cluster analysis. The tremendous growth in activity post-2000 is looked at in more detail in Figure 2.2. Time will tell if there is a decrease in use of the

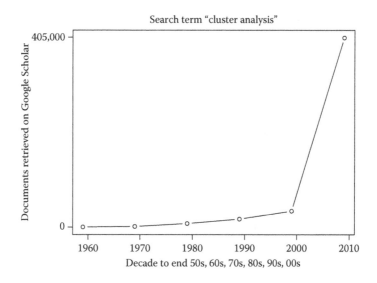

FIGURE 2.1
Google Scholar retrievals using search term "cluster analysis," for the years 1950–1959, 1960–1960, etc., up to 2000–2009. (Data collected in September 2012.)

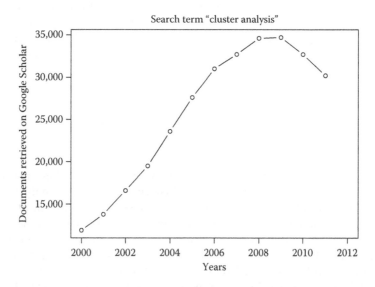

FIGURE 2.2
Google Scholar retrievals using search term "cluster analysis," for the years 2000, 2001, 2002, …, 2011. (Data collected in September 2012.)

term "cluster analysis." Again we note that this is just one term and many other related terms are relevant too.

A sampling of historical overviews of clustering follow. Kurtz (1983) presents an overview for the astronomer and space scientist. The orientation toward computer science is strong (Murtagh 2008), which takes in linkages to the Benzécri school of data analysis, and also current developments that have led Google's Peter Norvig to claim (with some justification, albeit very debatable) that similarity-based clustering has led to a correlation basis coming to the fore in science, potentially replacing entirely the causation principle. A general history of cluster analysis is also presented in (Murtagh 2014).

Using four decades of bibliographic data (Murtagh and Kurtz in press) an analysis is presented of the changes over time, and differences in orientation and focus of published work, that have come about over time. This study of bibliographical data from 1994 to date used 135,088 core cluster analysis publications that cited one or more of about (approximately, only due to some changes over time) 82 standard journal or book publications in the field of cluster analysis.

From the point of view of algorithms, and mathematical underpinnings, the encyclopaedic treatizes of Benzécri (1979) and Bock (1974) remain very topical.

2.2 Methods

2.2.1 Hierarchical Methods

Much published work in cluster analysis involves the use of either of two classes of clustering algorithm: hierarchical or nonhierarchical (often partitioning) algorithms. Hierarchical algorithms in particular have been dominant in the literature. Each of the many clustering methods—and of the many hierarchical methods—which have been proposed over the last few decades have possibly advantageous properties.

The single linkage hierarchical clustering approach outputs a set of clusters (to use graph theoretic terminology, a set of maximal connected subgraphs) at each level—or for each threshold value which produces a new partition. The single linkage method is one of the oldest and most widely used methods, its usage being traced to the early 1950s. "Wroclaw taxonomy" was the term applied to this method when used in those early years (Graham and Hell 1985). The mathematics of the single link method is very rich (including, e.g., yielding the so-called subdominant or maximal inferior ultrametric of an input dissimilarity measure). Jardine and Sibson (1971) favored it, in part because of its continuity properties in respect of the ultrametric or dendrogram distance that was created in the agglomerative hierarchical construction. This created some controversy at the time, because of its less than ideal properties from a practical point of view of general, synoptic clustering (see further on this point below, when we discuss the minimum variance agglomerative hierarchical criterion). However, the work of Jardine and Sibson was enormously influential. A recent work that shows its continuing influence is Janowitz (2010) that covers not only hierarchical clustering and graph-based clustering, but also Formal Concept Analysis, a major lattice-based (rather than tree-based, or more general graph-based) clustering and data analysis methodology.

About 75% of all published work on clustering has employed hierarchical algorithms (according to Blashfield and Aldenderfer (1978)).

Much early work on hierarchic clustering was in the field of biological taxonomy, from the 1950s and more so from the 1960s onwards. The central reference in this area, the first edition of which dates from the early 1960s, is Sneath and Sokal (1973). One major interpretation of hierarchies has been the evolution relationships between the organisms under study.

Hierarchic agglomerative algorithms may be conveniently broken down into two groups of methods. The first group is that of linkage methods—the single, complete, weighted, and unweighted average linkage methods. These are methods for which a graph representation can be used. There is a particularly close relationship between the single linkage hierarchy and the minimal spanning tree (MST), whereby the latter is easily transformed into the former. Sneath and Sokal (1973) may be consulted for many other graph representations of the stages in the construction of hierarchic clusterings.

The second group of hierarchic clustering methods are methods which allow the cluster centers to be specified (as an average or a weighted average of the member vectors of the cluster). These methods include the centroid, median, and minimum variance methods.

The latter group of agglomerative criteria may be specified either in terms of dissimilarities, alone, or alternatively in terms of cluster centre coordinates and dissimilarities. A very convenient formulation, in dissimilarity terms, which embraces all the hierarchical methods mentioned so far, is the *Lance–Williams dissimilarity update formula* (Lance and Williams 1967).

For cluster centre methods, and with suitable alterations for graph methods, a "stored data approach" is an alternative algorithm to the general dissimilarity-based algorithm the latter may be described as a "stored dissimilarities approach." For large datasets, this leads a great economy of storage.

For such agglomerative hierarchical clustering algorithms, an efficiency improvement came through the nearest neighbor chain, and reciprocal nearest neighbors, algorithm family. A survey is presented in Murtagh (1985).

The variance or spread of a set of points (i.e., the sum of squared distances from the center) has been the point of departure for specifying many clustering algorithms. Many of these algorithms—iterative, optimization algorithms as well as the hierarchical, agglomerative algorithms—are briefly described and appraised in Wishart (1969).

The search for clusters of maximum homogeneity leads to the minimum variance criterion. Since no coordinate axis is privileged by the Euclidean distance, the resulting clusters will be approximately hyperspherical. Such ball-shaped clusters will therefore be very unsuitable for examining straggly patterns of points. However, in the absence of information about such patterns in the data, homogeneous clusters will provide the most useful condensation of the data. This makes the minimum variance agglomerative strategy particularly suitable for synoptic clustering.

The use of variance in a clustering criterion links the resulting clustering to other data-analytic techniques which involve a decomposition of variance. Principal components analysis is one example. Another example is correspondence analysis that involves a decomposition of clouds of observations and of attributes into principal moments of inertia. In that case, the minimum variance is generalized easily to an agglomerative hierarchical clustering algorithm that minimizes the change in inertia from one agglomeration level to the next. Weights can be taken into account for the observations.

2.2.2 Minimal Spanning Tree

Graph representation, and graph models, has been both a natural and widely used analysis framework. We have already considered hierarchies expressing rooted, possibly binary (i.e., each tree node contains at most two child nodes), possible labeled trees.

Aspects of the MST are covered in most texts on graph theory, and on many other areas besides (see Graham and Hell 1985).

For MST algorithms, see Tucker (1980). Breaking up the MST, and thereby automatically obtaining components, is a problem addressed by Zahn (1971). Zahn applied these approaches to point pattern recognition—obtaining what he termed "Gestalt patterns" among sets of planar points; picking out bubble chamber particle tracks, indicated by curved sequences of points; and detecting density gradients, where differing clusters of points have different densities associated with them and hence are distinguishable to the human eye. The MST provides a useful starting point for undertaking such pattern recognition problems.

The MST is also often suitable for outlier detection. Since, outlying data items will be of greater than average distance from their neighbors in the MST, they may be detected by drawing a histogram of edge lengths. Unusually, large lengths will indicate the abnormal points (or data items) sought. Rohlf (1975) gave a statistical gap test, under the assumption that the edge lengths in the MST were normally distributed.

2.2.3 Partitioning Methods

We will next have a short look at other nonhierarchical clustering methods. A large number of assignment algorithms have been proposed. The single-pass approach usually achieves computational efficiency at the expense of precision, and there are many iterative approaches for improving on crudely derived partitions.

As an example of a single-pass algorithm, the following one is given in Salton and McGill (1983). The general principle followed is: make one pass through the data, assigning each object to the first cluster which is close enough, and making a new cluster for objects that are not close enough to any existing cluster.

As a nonhierarchic strategy, it is hardly surprising that the variance criterion has always been popular (for some of the same reasons as were seen above for the hierarchical approach based on this criterion). We may, for instance, minimize the within-class variance. Iterative refinement is used to furnish a suboptimal solution. Such an algorithm may be understood as an EM, expectation–maximization algorithm, where the redefinition of the cluster centers is the E-step, and the reassignment to the closer center comprises the M-step. For a history of k-means clustering, obtained through optimization in this way, see Bock (2008). Bock (2008) refers to the indirectly related work of Thorndike (1953) in regard to k-means. Steinley's survey may also be referred to for history of the k-means iterative optimization algorithm (furnishing suboptimal partitions) (Steinley 2006).

Another approach to optimizing the same minimum variance criterion is the exchange method (Späth 1985), which guarantees that no cluster becomes empty.

A difficulty with iterative algorithms, in general, is the requirement for parameters to be set in advance. Anderberg (1973) describes a version of the Iterative Self-Organizing Data Analysis Technique (ISODATA) iterative clustering method which requires seven preset parameters. As a broad generalization, it may thus be asserted that iterative algorithms ought to be considered when the problem is clearly defined in terms of numbers and other characteristics of clusters; but hierarchical routines often offer a more general-purpose and user-friendly option.

The influential Ball and Hall ISODATA procedure (see (Ball and Hall 1965)) was an early k-means algorithm. So too was Forgy's algorithm (Forgy 1965), and MacQueen's (MacQueen 1967).

Other variants of k-means have been developed in competitive learning, with learning variants ranging over "online" and "batch" algorithms for their implementation. A comprehensive albeit short review can be found in Darken and Moody (1990).

In signal processing, partitioning obtained through iterative optimization is termed vector quantization (Lloyd 1982).

With a well-constrained output representation, the Kohonen self-organizing feature map is in the k-means family of algorithms (Kohonen 1984). This is also referred to as SOM or the self-organizing map method. The k output representational nodes have a separate representational set of relationships, such as a regular planar grid, see Murtagh and Hernández-Pajares (1995). Vectors associated with the output representational grid assignment operation is referred to as a "winner takes all" assignment, as opposed to a fuzzy assignment.

2.2.4 Mixture Modeling

Mixture modeling approaches, as well as the addressing of some major themes in cluster analysis, are described very comprehensively in Fraley and Raftery (1998). One strand in this work was the flexible modeling (parametrizing the covariance structure of the Gaussian, thereby expanding greatly the patterns that could be fitted with Gaussian mixtures) that originated in character recognition work Murtagh and Raftery (1984).

The work of Pearson in the late nineteenth century involved fitting data with Gaussian components (Pearson 1894).

Mixture modeling as a well-deployed software tool for clustering originated in the work of John Wolfe in the 1960s. See Wolfe (1970) and his web site at http://alumnus. caltech.edu/~wolfe containing NORMIX software and bibliographies.

2.2.5 Spectral Clustering

Spectral reduction, that is, the determining of eigenvectors and eigenvalues of a positive semidefinite matrix, has become in recent years an important basis for clustering. This is so because of a natural ability to handle very large dimensionalities (cf. Murtagh et al. 2000), to address clustering by graph operations such as graph cuts (von Luxburg 2007) and also to trace out linkages with many other vantage points on clustering including k-means partitioning (von Luxburg 2007).

An even more integral linkage between spectral reduction and cluster analysis has long been known, implying that the levels of a hierarchical clustering are given by the eigenvalues, and the corresponding partition's cluster members are read off the corresponding eigenvectors. This is the case for spectral reduction on a semiring (semi due to use of positive values used, viz. dissimilarities) with "addition" and "multiplication" operations given by min and max, see Gondran (1976).

Metric multidimensional scaling, also known as Gower's or Torgerson's scaling (Torgerson 1958), is based on a direct spectral reduction of a matrix defined from dissimilarities. If the latter is positive semidefinite, then a Euclidean embedding and nonnegative eigenvalues are guaranteed.

2.3 Applications

The first use of the term "cluster analysis" seems to have been in Tryon (1939). Tryon was a behavioral psychologist, who worked at the University of California, Berkeley.

According to Jain (2010), the term "data clustering" first appeared in 1954, and also points to synonyms such as "Q-analysis," clumping, taxonomy, and typology. "Polish taxonomy" that flourished in Wrocław in the 1950s (Florek 1951) was based on the MST and closely associated single link hierarchical clustering.

From the earliest times, clustering has been a broad church. Gavin Ross, working in the Statistics Department at Rothamsted Experimental Station from 1961, recalled his work in this way (Murtagh 2014): "... we had several requests for classification jobs, mainly agricultural and biological at first, such as classification of nematode worms, bacterial strains, and soil profiles. On this machine and its faster successor, the Ferranti Orion, we performed numerous jobs, for archaeologists, linguists, medical research laboratories, the Natural History Museum, ecologists, and even the Civil Service Department."

The Classification Society was established on April 17, 1964 as an interdisciplinary sharing concerns in relation to methodology and to approaches to data analysis. Its justification was that "it became clear that there are many aspects of classification common to such widely separated disciplines as biology, librarianship, soil science, and anthropology, and that opportunities for joint discussion of these aspects would be of value to all the disciplines concerned" (Classification Society 1964).

In the UK, the major motivation for development of clustering algorithms and data analysis through cluster analysis and unsupervised classification lay in language engineering and linguistics, in computing. The work of Roger Needham (1935–2003) and Karen Spärck Jones (1935–2007) in Cambridge typifies this (e.g., http://www.cl.cam.ac.uk/archive/ksj21.) This too was the case in France with Jean-Paul Benzécri (born in 1932), where linguistics was an early driver in his work in Rennes in the 1960s and later in Paris. In Benzécri (1979), there is as frontpiece a line drawing of Linnaeus, shown in Figure 2.3.

The Computer Journal, established in 1958, is the second oldest computer science journal (the oldest is JACM, *Journal of the Association for Computing Machinery*). Early articles published in it include the following:

- Vol. 1, No. 3, 1958, J.C. Gower, "A note on an iterative method for root extraction," 142–143.
- Vol. 4, No. 4, 1962, J.C. Gower, "The handling of multiway tables on computers," 280–286.
- Vol. 7, No. 3, 1964, M.J. Rose, "Classification of a set of elements," 208–211. Abstract: "The paper describes the use of a computer in some statistical experiments on weakly connected graphs. The work forms part of a statistical approach to some classification problems."

A past editor-in-chief of this journal was C.J. (Keith) van Rijsbergen who contributed a great deal to cluster analysis algorithms and to the use of cluster analysis in information retrieval. The current editor-in-chief is the author, having been editor-in-chief also in the years 2000–2007. Further historical perspectives are traced out in Murtagh (2014).

2.4 Surveys of Clustering and an Online Digital Resource

We note a number of broad-ranging surveys of clustering. Among many surveys and books, there are the following ranging over three decades:

FIGURE 2.3
From Benzécri's *Taxinomie*: a line drawing of Carl Linnaeus, 1707–1778, botanist and zoologist.

- In statistics: Allan Gordon (1981 and 1987).
- In pattern recognition, computer science and engineering: Jain and Dubes (1988), Jain et al. (1999).
- In economics, computer science, and other fields: Mirkin (1996).
- In electrical and computer engineering Xu and Wunsch (2008).

The following books have been scanned (or in the cases of Jain and Dubes, and van Rijsbergen, were made available digitally) and are online in their entirety with access to members of the Classification Society and the British Classification Society. This digital resource was formerly part of the *Classification Literature Automated Search Service*, an annual bibliography of cluster analysis, that was on CD and was distributed formerly with the *Journal of Classification* and is now available online. See http://www.classicationsociety.org/clsoc (the Classification Society) for details, or also http://brclasssoc.org.uk (the British Classification Society).

1. *Algorithms for Clustering Data* (1988), A.K. Jain and R.C. Dubes.
2. *Automatische Klassifikation* (1974), H.-H. Bock.

3. *Classification et Analyse Ordinale des Données* (1981), I.C. Lerman.

4. *Clustering Algorithms* (1975), J.A. Hartigan.

5. *Information Retrieval* (1979, 2nd ed.), C.J. van Rijsbergen.

6. *Multidimensional Clustering Algorithms* (1985), F. Murtagh.

7. *Principles of Numerical Taxonomy* (1963), R.R. Sokal and P.H.A. Sneath.

8. *Numerical Taxonomy: the Principles and Practice of Numerical Classification* (1973), P.H.A. Sneath and R.R. Sokal.

References

M.R. Anderberg, *Cluster Analysis for Applications*, Academic Press, New York, 1973.

G.H. Ball and D.J. Hall, *Isodata: A Method of Data Analysis and Pattern Classification*, Stanford Research Institute, Menelo Park, 1965.

J.P. Benzécri, *L'Analyse des Données. I. La Taxinomie*, Dunod, Paris, France, 1979 (2nd ed.).

R.K. Blashfield and M.S. Aldenderfer, The literature on cluster analysis, *Multivariate Behavioral Research*, 13, 271–295, 1978.

H.-H. Bock, *Automatische Klassifikation*, Vandenhoek und Rupprecht, Göttingen, Germany, 1974.

H.-H. Bock, Origins and extensions of the k-means algorithm in cluster analysis, *Electronic Journal for History of Probability and Statistics*, 4 (2), 26, 2008.

Classification Society, minutes of Inaugural Meeting, 17 April 1964, www.classification-society.org/clsoc/ClassSoc1964.pdf

Computing Classification System (CCS), Association for Computing Machinery, ACM, 1998. http://www.acm.org/about/class/ccs98-html

Computing Classification System (CCS), Association for Computing Machinery, ACM, 2012. http://dl.acm.org/ccs.cfm

C. Darken and J. Moody, Fast, adaptive K-means clustering: Some empirical results, *Proceedings of the IEEE IJCNN (International Joint Conference on Neural Networks) Conference*, San Diego, IEEE Press, Piscataway, NJ, 1990.

K. Florek, J. Łukaszewicz, J. Perkal, H. Steinhaus and S. Zarycki, Sur la liaison et la division des points d'un ensemble fini, *Colloquium Mathematicum* (Wrocław), 2, 282–285, 1951.

E. Forgy, Cluster analysis of multivariate data: Efficiency vs. interpretability of classifications, *Biometrics*, 21, 768–769, 1965.

C. Fraley and A.E. Raftery, How many clusters? Which clustering methods? Answers via model-based cluster analysis, *Computer Journal*, 41, 578–588, 1998.

M. Gondran, Valeurs propres et vecteurs propres en classification hiérarchique, *RAIRO, Revue Française d'Automation, Informatique, Recherche Opérationnelle, Informatique Théorique*, 10, 39–46, 1976.

A.D. Gordon, *Classification*, Chapman & Hall, London, 1981.

A.D. Gordon, A review of hierarchical classification, *Journal of the Royal Statistical Society A*, 150, 119–137, 1987.

R.L. Graham and P. Hell, On the history of the minimum spanning tree problem, *Annals of the History of Computing*, 7, 43–57, 1985.

A.K. Jain, Data clustering: 50 years beyond k-means, *Pattern Recognition Letters*, 31, 651–666, 2010.

A.K. Jain and R.C. Dubes, *Algorithms For Clustering Data*, Prentice-Hall, Englwood Cliffs, 1988.

A.K. Jain, M.N. Murty and P.J. Flynn, Data clustering: A review, *ACM Computing Surveys*, 31, 264–323, 1999.

M.F. Janowitz, *Ordinal and Relational Clustering*, World Scientific Publishing Company, Hackensack, NJ, 2010.

N. Jardine and R. Sibson, *Mathematical Taxonomy*, Wiley, New York, 1971.

T. Kohonen, *Self-Organization and Associative Memory*, Springer-Verlag, Berlin, Germany, 1984.

M.J. Kurtz, Classification methods: An introductory survey, in *Statistical Methods in Astronomy*, European Space Agency Special Publication, ESA, Noordwijk, 201, pp. 47–58, 1983.

G.N. Lance and W.T. Williams, A general theory of classificatory sorting strategies. 1. Hierarchical systems, *Computer Journal*, 9 (4), 373–380, 1967.

S.P. Lloyd, Least-squares quantization in PCM, *IEEE Transactions on Information Theory*, IT-28, 129–137, 1982.

J. MacQueen, Some methods for classification and analysis of multivariate observations, Vol. 1, *Proceedings of the Fifth Berkeley Symposium on Mathematical Statistics and Probability*, pp. 281–297, University of California Press, 1967.

B. Mirkin, *Mathematical Classification and Clustering*, Kluwer, Dordrecht, 1996.

F. Murtagh, *Multidimensional Clustering Algorithms*, COMPSTAT Lectures Volume 4, Physica-Verlag, Vienna, Austria, 1985.

F. Murtagh, Origins of modern data analysis linked to the beginnings and early development of computer science and information engineering, *Electronic Journal for History of Probability and Statistics*, 4 (2), 26, 2008.

F. Murtagh, History of cluster analysis, in J. Blasius and M. Greenacre, Eds., *The Visualization and Verbalization of Data*, Chapman & Hall, Boca Raton, FL, pp. 117–133, 2014.

F. Murtagh and M. Hernández-Pajares, The Kohonen self-organizing map method: An assessment, *Journal of Classification*, 12, 165–190, 1995.

F. Murtagh and M.J. Kurtz, A history of cluster analysis using the Classification Society's bibliography over four decades, *Journal of Classification*, in press.

F. Murtagh and A.E. Raftery, Fitting straight lines to point patterns, *Pattern Recognition*, 17, 479–483, 1984.

F. Murtagh, J.-L. Starck and M. Berry, Overcoming the curse of dimensionality in clustering by means of the wavelet transform, *Computer Journal*, 43, 107–120, 2000.

K. Pearson, Contributions to the mathematical theory of evolution, *Philosphical Transactions of the Royal Society A*, 185, 71–110, 1894.

F.J. Rohlf, Generalization of the gap test for the detection of multivariate outliers, *Biometrics*, 31, 93–101, 1975.

G. Salton and M.J. McGill, *Introduction to Modern Information Retrieval*, McGraw-Hill, New York, 1983.

P.H.A. Sneath and R.R. Sokal, *Numerical Taxonomy*, Freeman, San Francisco, CA, 1973.

H. Späth, *Cluster Dissection and Analysis: Theory, Fortran Programs, Examples*, Ellis Horwood, Chichester, 1985.

D. Steinley, K-means clustering: A half-century synthesis, *British Journal of Mathematical and Statistical Psychology*, 59, 1–34, 2006.

R.L. Thorndike, Who belongs to the family?, *Psychometrika*, 18, 267–276, 1953.

W.S. Torgerson, *Theory and Methods of Scaling*, Wiley, New York, 1958.

R.C. Tryon, *Cluster Analysis: Correlation Profile and Orthometric (Factor) Analysis for the Isolation of Unities in Mind and Personality*, Edwards Brother, Ann Arbor, MI, 122 p., 1939.

A. Tucker, *Applied Combinatorics*, Wiley, New York, 1980.

U. von Luxburg, A tutorial on spectral clustering, *Statistics and Computing*, 17, 395–416, 2007.

D. Wishart, Mode analysis: A generalization of nearest neighbour which reduces chaining effects, in A.J. Cole, Ed., *Numerical Taxonomy*, Academic Press, New York, pp. 282–311, 1969.

J.H. Wolfe, Pattern clustering by multivariate mixture analysis, *Multivariate Behavioral Research*, 5, 329–350, 1970.

R. Xu and D.C. Wunsch, *Clustering*, IEEE Computer Society Press, 2008.

C.T. Zahn, Graph–theoretical methods for detecting and describing Gestalt clusters, *IEEE Transactions on Computers*, C–20, 68–86, 1971.

Section I

Optimization Methods

3

Quadratic Error and k-Means

Boris Mirkin

CONTENTS

Abstract

This chapter presents an updated review of k-means clustering, arguably the most popular clustering method. First, the square-error k-means criterion and method are introduced in three frameworks: a naive one, data recovery, and mixture of distributions. Then several equivalent reformulations are given as leading to different local optimization strategies. A number of challenges and ways for addressing them are discussed as related to both the properties of solutions and ways for optimum finding. A few extensions are mentioned including fuzzy clustering and feature weighting.

3.1 Conventional k-Means Clustering

3.1.1 General

Currently, k-means is the most popular clustering technique. Amazingly, when looking through Google web browser on July 2, 2012, query "k-means" returned 234 million pages, whereas query "clustering" has led to less than 26 million web pages, which has been somewhat alleviated by 208 million pages returned for query "cluster." Such a popularity probably is because thousands of data analysis practitioners in banks, marketing research, pharmaceuticals, etc. are using "k-means" software without much regard for more general aspects of clustering.

The method became known in late sixties after a first theoretical result was proved in MacQueen (1967). Later predecessors have been found; two histories of the method, with different emphases, are now available (see Bock (2008) and Steinley (2006)). The method is present, in various forms, in major statistics packages such as SPSS (Green and Salkind 2003) and SAS (Der and Everitt 2001) and data mining packages such as Clementine (Clementine 2003) and DBMiner (Han et al. 2011), as well as freeware like Weka (Witten 2011). It is described in all materials on clustering including most classical texts such as in Hartigan (1975) and Jain and Dubes (1988). The algorithm is appealing in many aspects. It is computationally easy, fast, and memory-efficient. Conceptually, this method may be considered a model for the cognitive process of making a data-conditioned typology. Also, it has nice mathematical properties. However, there are some issues too, most urgently with respect to the initial setting and stability of results. Some questions arise with respect to the meaning and relevance of the k-means criterion.

3.1.2 Three Frameworks for the Square-Error Criterion

The square-error clustering criterion can be applied in various settings, for various dataset and cluster structure formats. The generic k-means applies to a most common data format, a set of entities $i \in I$ presented as m-dimensional space points $x_i = (x_{i1}, x_{i2}, \ldots, x_{im})$ (see rows of Table 3.1).

TABLE 3.1

Illustrative Data of Seven
Two-Dimensional Points

Points	v_1	v_2
$p1$	1.0	2.0
$p2$	1.5	2.2
$p3$	3.0	1.3
$p4$	2.5	1.1
$p5$	0.9	2.1
$p6$	2.0	1.5
$p7$	2.5	1.4

A two cluster clustering at k-means ($K = 2$) can be variously defined by specifying (i) two cluster centers, c_1 and c_2, or (ii) a two cluster partition on I, $S = \{S_1, S_2\}$ so that $S_1 \cup S_2 = I$ and $S_1 \cap S_2 = \emptyset$, or (iii) by both. In general, centers can belong to a different space, as for example, when they characterize a parametric regression function over a cluster. Here, they are assumed to belong to the space of points, as usual, for example, $c_1 = (1, 2.5)$ and $c_2 = (3, 2)$. To score the goodness of a clustering (S, c) with K clusters S_k and centers c_k ($k = 1, 2, \ldots, K$), different frameworks can be used of which the following three prevail: (i) a naive one, (ii) data recovery (approximation), and (iii) probabilistic modeling.

3.1.2.1 Naive Framework

It assumes that there is a distance or similarity function defined between entity points x_i and centers c_k, the most popular choice being the squared Euclidean distance

$$d(x_i, c_k) = \sum_{h=1}^{m} (x_{ih} - c_{kh})^2 \tag{3.1}$$

The k-means square-error criterion is defined as the summary distance between entities and their cluster centers:

$$W(S, c) = \sum_{k=1}^{K} \sum_{i \in S_k} d(x_i, c_k) = \sum_{k=1}^{K} \sum_{i \in S_k} \sum_{h=1}^{m} (x_{ih} - c_{kh})^2 \tag{3.2}$$

This criterion is illustrated in Figure 3.1: each of the points x_i relates to the center of its cluster so that the total number of distances in the sum is $n = |I|$ and does not depend on the number of clusters K.

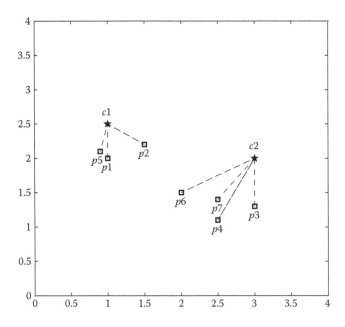

FIGURE 3.1
The set of distances taken into account in the square-error criterion for partition $\{\{p1, p2, p5\}, \{p3, p4, p6, p7\}\}$ with corresponding centers $c1$ and $c2$.

The smaller the value $W(S, c)$, the better the clustering (S, c). Although, finding optimal pair (S, c) is quite a computationally intensive task, finding either optimal S or optimal c is fairly easy, as the following statements say.

3.1.2.1.1 Centers Set

Given partition $S = \{S_k\}$, the optimal centers c_k, with respect to Equation 3.2, are computed as the mean points in S_k:

$$c_k = \sum_{i \in S_k} x_i / n_k \tag{3.3}$$

where n_k is the number of elements in S_k, $k = 1, \ldots, K$.

The means are aggregate representations of clusters and as such they are referred to as standard points or centroids or prototypes or seeds—these are considered synonymous in clustering.

With the optimal centers in Equatoin 3.3, a controversial issue of simultaneously processing quantitative and categorical features gets a positive resolution with the k-means criterion. Indeed, given a category v such that a subset $S_v \subset I$ falls in it, let us introduce a dummy variable vv such that $vv(i) = 1$ if $i \in S_v$ and $vv(i) = 0$, otherwise. Obviously, the sum of vv one-zero values within S_k is equal to the size of the intersection of S_k and S_v, $n_{kv} = |S_k \cap S_v|$, and the average of vv within S_k is

$$c_{vv} = n_{kv} / n_k = p(v \mid k) \tag{3.4}$$

the conditional probability of v at S_k. Therefore, given a nominal feature with categories v, it is to be enveloped into a set of corresponding dummy variables, so that the part of the center of cluster S_k corresponding to the feature recoded as the set of category-related dummy variables is but a vector of conditional probabilities of the v categories. This is exactly the characteristic of a nominal feature which is used by those who treat the numerical and categorical parts of data separately. Equation 3.4 leads us therefore to recommend, at clustering with k-means, to recode the categorical part of the data into the dummy variable format and process the mixed data in such a way as this is just a numeric data.

3.1.2.1.2 Clusters Set

Given a set of centers c_k, $k = 1, \ldots, K$, the optimal with respect to Equation 3.2 clusters S_k are determined according to the so-called *minimum distance rule*: each entity x_i is assigned to its nearest center c_k so that $S_k = \{i : d(x_i, c_k) = \min_l d(x_i, c_l)\}$. When two or more of the distances $d(x_i, c_k)$, $k = 1, \ldots, K$, coincide, the assignment is done arbitrarily among the nearest candidates, for example, to that with the minimum number of elements. The cluster set process for an $x_i = p2$ is illustrated in Figure 3.2.

The minimum distance rule is popular in data analysis and can be found in many approaches such as Voronoi diagrams and vector learning quantization.

Consider, for example, pair (S, c) where $S_1 = \{p1, p2, p3, p4\}$ and $S_2 = \{p5, p6, p7\}$ and $c_1 = (1, 1), c_2 = (2, 2)$.

Table 3.2, top, presents the squared Euclidean distances to c_1 and c_2 from all the seven entities. The criterion $W(S, c)$ value is therefore the summary distance of $p1 - p4$ to $c1$, $1.00 + 1.69 + 4.09 + 2.26 = 9.04$, and $p5 - p7$ to $c2$, $1.22 + 0.25 + 0.61 = 2.08$, so that $W(S, c) = 9.04 + 2.08 = 11.12$. Let us find the optimal centers at the given S according to the centers set procedure, see the means c'_1 and c'_2 in the bottom part of Table 3.2.

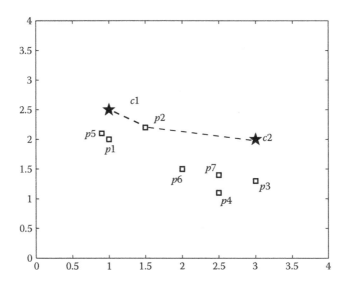

FIGURE 3.2
Point $p2$ is assigned to $c1$ because it is nearer than $c2$.

TABLE 3.2

Squared Euclidean Distances of the Seven Two-Dimensional
Points from Centers

Center	$p1$	$p2$	$p3$	$p4$	$p5$	$p6$	$p7$
$c_1 = (1, 1)$	1.00	1.69	4.09	2.26	1.22	1.25	2.41
$c_2 = (2, 2)$	1.00	0.29	1.49	1.06	1.22	0.25	0.61
$c'_1 = (2.00, 1.65)$	1.12	0.55	1.12	0.55	1.41	0.02	0.31
$c'_2 = (1.80, 1.67)$	0.75	0.37	1.57	0.81	1.00	0.07	0.56

With these centers, the summary distance is $W(S, c') = 1.12 + 0.55 + 1.12 + 0.55 + 1.00 + 0.07 + 0.56 = 4.97$, more than a 50% decrease. Similarly, given c_1 and c_2, let us set clusters according to the minimum distance rule applied to the distances in the upper part of Table 3.2. Rather unexpectedly, the distances from c_2 to each of the seven entities are less than or equal to those from c_1 – which happens because c_1 is rather far away from the entities. Such an effect can generate a "degenerate clustering case" when one or more clusters get empty because of "wrong" centroids. Yet, in this case, we can define S'_2 to consist of all the entities that are nearer to $c2$ than to $c1$, $S'_2 = \{p2, p3, p4, p6, p7\}$ and $S'_1 = \{p1, p5\}$. Therefore, the summary distance in this case will be $W(S', c) = 1.00 + 0.29 + 1.49 + 1.06 + 1.22 + 0.25 + 0.61 = 5.92$ which is not as small as the previous value, yet is much smaller than the original $W(S, c) = 11.12$. The clusterings (S, c') and (S', c) can be further improved with moves to a better partition or centers set. The best possible criterion value will be achieved at the partition with clusters $\{p1, p2, p5\}$ and $\{p3, p4, p6, p7\}$. It is not difficult to see that, at this partition and the optimal centers $g_1 = (1.13, 2.10)$ and $g_2 = (2.50, 1.32)$, $W(S, c) = 0.82$. No better 2-cluster partition of I ever exists.

The optimal settings can be used for iterative improvement of a given partition, or set of centers, or both. By specifying a starting configuration, the optimal settings constitute what is called the batch version of the k-means algorithm. Most formulations of k-means start by initializing with specifying both K and initial centers $c_1, \ldots, c_K$:

Batch k-means

    Input: K and $c_1, \ldots, c_K$.

1. *Clusters update:* Given K centers, find clusters using *Clusters set* above.
2. *Centers update:* Given clusters, find cluster means using *Centers set* above.
3. *Stopping test:* If the set of found means do not coincide with the set of centers, set the means as new centers and go back to 1.

    Output Both centroids and clusters.

Because of the optimality of S_k at each step 1 and c_k, at each step 2, the criterion may only decrease from step to step, so that no loop in the process may occur. Since the number of partitions on I is finite, this warrants the process to converge starting from any initial centers (seeds). The number of steps usually is rather small, in single or double digits. To avoid the degenerate cluster case of empty clusters, initial centers are set to coincide with K observed data points.

3.1.2.2 Approximation Framework

This focuses on the transformation of the dataset that is undergoing as a result of the clustering process. Indeed, the k-means clustering goal is to simplify data by representing them with K centroids rather than the original n entities. Therefore, k-means clustering can be considered as a device to change the original $n \times m$ data matrix X with a simplified matrix at which each row x_i is substituted by row c_k where c_k is the center of cluster S_k containing i.

Returning to our illustrative dataset of seven entities in Table 3.1 and clustering (S, c) where $S_1 = \{p1, p2, p3, p4\}$, $S_2 = \{p5, p6, p7\}$ and $c_1 = (1, 1)$, $c_2 = (2, 2)$, we can counterpose the original and simplified matrices in the following matrix equation:

$$
\begin{bmatrix}
1.0 & 2.0 \\
1.5 & 2.2 \\
3.0 & 1.3 \\
2.5 & 1.1 \\
0.9 & 2.1 \\
2.0 & 1.5 \\
2.5 & 1.4
\end{bmatrix}
=
\begin{bmatrix}
1 & 1 \\
1 & 1 \\
1 & 1 \\
1 & 1 \\
2 & 2 \\
2 & 2 \\
2 & 2
\end{bmatrix}
+
\begin{bmatrix}
0 & 1.0 \\
0.5 & 1.2 \\
2.0 & 0.3 \\
1.5 & 0.1 \\
-1.1 & 0.1 \\
0 & -0.5 \\
0.5 & -0.6
\end{bmatrix}
$$

The matrix on the right is the difference between the original matrix X and the cluster-expressing matrix Y, which shows how well the clustering fits the data: the smaller are the differences, the better the fit. The goodness of fit can be measured by the sum of squares of the differences. There should be no wonder that the sum of square differences here is the same as the sum of Euclidean squared distances in Equation 3.2, 11.12.

This can be put formally, for the general case, by introducing $K \times m$ matrix $C = (c_{kh})$ whose rows are cluster centers c_k, $k = 1, \ldots, K$, and binary $n \times K$ cluster membership matrix $Z = (z_{ik})$ where $z_{ik} = 1$ if $i \in S_k$ and $z_{ik} = 0$, otherwise. Then the general formula for the matrix equation above is:

$$X = ZC + E \tag{3.5}$$

where E is the matrix of differences $e_{ih} = x_{ih} - \sum_{k=1}^{K} z_{ik} c_{kh}$.

The k-means square-error criterion, in this framework, is the sum of squared differences, $\|E\|^2 = \sum_{i \in I} \sum_{h=1}^{m} e_{ih}^2$ which is, obviously, equal to $W(S, c)$. Why then put old wine into new bottles?

Well, the approximation perspective allows to see a wider picture of clustering, that of fitting a mathematical model of the data. For example, model in Equation 3.5 much resembles the celebrated singular-value decomposition (SVD) of matrix X, except that Z here must be a binary partition membership matrix, which is related to both principal component analysis and spectral properties of related similarity matrices XX^T and $X^T X$. This leads to a number of ideas for tackling the problem of minimization of the k-means criterion in the spectral analysis and similar frameworks. Moreover, for any partition membership matrix Z, if C is filled in with the within-cluster means, the Equation 3.5 leads to a Pythagorean decomposition of the quadratic data scatter,

$$\|X\|^2 = B(S, c) + \|E\|^2 \tag{3.6}$$

where

$$B(S, c) = \sum_{k=1}^{K} \sum_{h=1}^{m} n_k c_{kh}^2 \tag{3.7}$$

and $\|X\|^2 = \sum_{i \in I} \sum_{h=1}^{m} x_{ih}^2$, the so-called data scatter.

Criterion $B(S, c)$ is the part of the data scatter taken into account by the clustering (S, c). Therefore, data scatter $\|X\|^2$ is decomposed in two parts: the part $B(S, c)$ explained by the cluster structure (S, c), and the unexplained part $\|E\|^2 = W(S, c)$. The larger the explained part, the better the match between clustering (S, c) and data. The decomposition Equation 3.6 allows to score the proportion of the data scatter taken into account by the clustering (S, c), $R^2 = B(S, c)/\|X\|^2$, that is an index akin to the determinacy coefficient in regression analysis. It can be used for deriving different algorithms for fitting the approximation criterion of k-means.

3.1.2.3 Probabilistic Modeling Framework

According to this approach, each of the yet unknown clusters k is modeled by a density function $f(x; a_k)$ which represents a family of density functions over x defined up to a parameter vector a_k. A one-dimensional density function $f(x)$, for any small $dx > 0$, assigns probability $f(x)dx$ to the interval between x and $x + dx$; multidimensional density functions have similar interpretation.

Usually, the density $f(x; a_k)$ is considered unimodal (the mode corresponding to a cluster center), such as the Gaussian density function defined by its mean vector μ_k and covariance matrix Σ_k:

$$f(x; a_k) = (2^m \pi^m |\Sigma_k|)^{-1/2} \exp\{-(x - \mu_k)^T \Sigma_k^{-1} (x - \mu_k)/2\} \tag{3.8}$$

The shape of Gaussian clusters is ellipsoidal because any surface at which $f(x; a_k)$ Equation 3.8 is constant satisfies equation $(x - \mu_k)^T \Sigma_k^{-1} (x - \mu_k) = const$ defining an ellipsoid. The mean vector μ_k specifies the k-th cluster's location; the covariance matrix, its spread.

The mixture of distribution approach makes a flexible model-based framework for clustering described at length in Chapter 3.1. Here we consider only an aspect related to the k-means criterion.

According to the mixture of distributions clustering model, the row points $x_1, \ldots, x_n$ are considered an independent random sample of m-dimensional observations from a population with density function $f(x)$ which is a mixture of individual cluster density functions $f(x; \mu_k, \Sigma_k)$ $(k = 1, \ldots, K)$ so that $f(x) = \sum_{k=1}^{K} p_k f(x; \mu_k, \Sigma_k)$ where $p_k > 0$ are the cluster probabilities, $\sum_k p_k = 1$.

To estimate the individual cluster parameters, the maximum likelihood principle applies: those events that have really occurred are those that are most likely. Therefore, the estimates must maximize the logarithm of the likelihood of the observed data under the assumption that the data come from a mixture of distributions. There can be several different formulations of the principle. Let us assume what is referred to as a fixed partition model at which the observations come indeed from individual clusters so that among unknowns is a cluster partition $S = \{S_1, S_2, \ldots, S_K\}$ for $i \in S_k$, x_i comes from the individual distribution $f(x; a_k)$. Then the likelihood of the sample, up to a power of the constant 2π in Equation 3.8, is the product of individual probabilities (Celeux and Govaert 1992)

$$L = \prod_{k=1}^{K} |\Sigma_k|^{-n_k/2} \exp\left\{ -\frac{1}{2} \sum_{i \in S_k}^{K} (x_i - \mu_k)^T \Sigma_k^{-1} (x_i - \mu_k) \right\}$$

Its logarithm is

$$\log L = \sum_{k=1}^{K} -\frac{n_k}{2} \log |\Sigma_k| - \frac{1}{2} \sum_{k=1}^{K} \sum_{i \in S_k} (x_i - \mu_k)^T \Sigma_k^{-1} (x_i - \mu_k)$$

Assume that all covariance matrices Σ_k are diagonal and have the same variance value σ^2 on the diagonal, that is, all clusters have the same spherical distribution so that observations in each cluster S_k form approximately spherical "balls" of the same size. This is the situation of our interest because the maximum likelihood criterion here leads to $W(S, c)$ criterion of k-means and, moreover, there is a certain homology between the EM and Batch k-means algorithms, although there are differences as well. For example, the variance σ^2 is to be estimated in EM, but this has nothing to do with k-means.

Indeed, under this assumption the log-likelihood function is

$$l(\{X \mid \mu_k, \sigma^2, S_k\}) = -nm \log(\sigma) - \sum_{k=1}^{K} \sum_{i \in S_k} (x_i - \mu_k)^T (x_i - \mu_k)/\sigma^2 \qquad (3.9)$$

This function, to be maximized, is a theoretic counterpart to k-means criterion $W(S, \mu)$ applied to vectors x_i normalized by the standard deviation. Some may conclude from this that the k-means criterion and algorithm can be applied only in situations at which the clusters to be found are expected to be of a spherical shape and of an equal spread. Yes, indeed—but only in the case when the user believes that the data come from a random sample of a mixture of Gaussians, under the fixed partition model, and is interested in estimating parameters of the function. Then k-means is applicable only for spherical equal-sized Gaussians. Under a different hypothesis of the probabilistic distribution, they should use a different method. However, if the user is not interested in the distribution, and wants just to find a clustering (S, c) minimizing the criterion $W(S, c)$ in Equation 3.2, neither the principle of maximum likelihood nor the constraints on the shape and size of clusters are relevant anymore. Moreover, some of the conclusions can be wrong because they are derived from a model that might be at odds with data. Take for example the advice of application of "z-scoring" transformation of features. This comes from the assumption of equal variances of the variables, which can be satisfied only if the features are normalized by their standard deviations. However, this normalization gives an unjustified advantage to single-modal features against those multimodal ones, which is counterintuitive at clustering. The experimental findings in Chiang and Mirkin (2010) show that applying k-means is adequate at different cluster spreads and elongated, not necessarily spherical, shapes.

The three different frameworks for k-means may be viewed as corresponding to different levels of domain knowledge by the user and data collectors. When the knowledge is very poor, the features are superficial, and measurement methods are poorly defined, it is the naive approach, which is convenient in such a situation. As domain knowledge becomes stronger, more consistent, measurements are more precise and reproducible, this is a stage for the approximation approach so that characteristics such as contribution of clustering or feature to clustering become useful. The probabilistic modeling for k-means should be used only when there is a perspective of equally sized spherical clusters to be recovered as a mixture of distributions. Otherwise, more subtle criteria involved in the EM-method for maximization the likelihood criterion are utilized (see details in Chapter 3.1).

3.2 Equivalent Reformulations of k-Means Criterion

The k-means criterion admits a number of reformulations, which give different perspectives leading to different approaches to optimization of the criterion. It is ironic that, although known for some time already, these have not been picked up by the research community as a system. Even those that have been explored, like spectral clustering or cosine-based k-means, are developed as purely heuristic, unrelated to k-means, approaches.

The explained part of the data scatter $B(S,c)$ in Equation 3.7 complements the k-means criterion to a constant, the data scatter. Therefore, a maximum of $B(S,c)$ corresponds to a minimum of $W(S,c)$. The following sections present four different reformulations of $B(S,c)$; when c_k is computed as the mean points in S_k, then these criteria are equivalent to the square-error clustering criterion $W(S,c)$ with the only difference that they are to be maximized rather than minimized. The fact that all four are different expressions for the same $B(S,c)$ in Equation 3.7 can be easily proven with little algebraic transformations.

3.2.1 Anomalous Clustering Criterion

Consider

$$B(S,c) = \sum_{k=1}^{K} n_k d(0, c_k) \tag{3.10}$$

where $n_k = |S_k|$ is the number of entities in S_k; 0, all-zero vector, the space origin; and d, the squared Euclidean distance.

Indeed, to maximize Equation 3.10, the clusters should be as far away from 0 as possible. Therefore, when the origin is shifted into a reference point such as the grand mean, the criterion means that clusters should be anomalous indeed—as far away from the "norm," expressed in 0, as possible. An algorithm for finding anomalous clusters one-by-one according to Equation 3.10 was developed in Mirkin (1996), see also Chiang and Mirkin (2010), Mirkin (2012); no algorithm for simultaneously finding anomalous clusters has been developed so far.

3.2.2 Inner-Product k-Means Criterion

Another reformulation:

$$B(S,c) = \sum_{k=1}^{K} \sum_{i \in S_k} \langle x_i, c_k \rangle \tag{3.11}$$

Equation 3.11 is similar to that of $W(S,c)$ itself, except that the relation between x_i and c_k is scored with the inner product here, not the distance. Therefore, Equation 3.11 is to be maximized rather than minimized. Note, however, that the distance-based criterion does not depend on the choice of the space origin, whereas the inner product-based criterion seemingly depends on that. Yet it is easy to derive that the value of Equation 3.11 changes just by a constant under any shift of the origin. There is a claim in the literature that the inner product is beneficial at higher dimensions of the feature set if the data is prenormalized in such a way that the rows, corresponding to entities, are normed so that each has its Euclidean norm equal to unity (note the change of the emphasis in normalization here, from features to entities), for example, France et al. (2012); the distances are more or less useless in the higher dimensions.

3.2.3 Kernel-Wise Criterion of Maximization of the Semi-Averaged Internal Similarities

Centers vanish from the following equation:

$$B(S,c) = \sum_{k=1}^{K} \sum_{i,j \in S_k} \langle x_i, x_j \rangle / n_k \tag{3.12}$$

Equation 3.12 expresses the criterion in terms of similarities, the inner products $\langle x_i, x_j \rangle$. As now is well recognized, a criterion in which the data are present only through the inner products can be subject to the so-called "kernel trick." The trick is to change the inner products $\langle x_i, x_j \rangle$ for values of a so-called positive semidefinite kernel function $K(x_i, x_j)$ under the assumption that the function expresses the inner product in a higher dimension space where the original features have been mapped, possibly, in a nonlinear way. Among the popular kernel functions is the so-called Gaussian kernel $K(x_i, x_j) = \exp(-d(x_i, x_j)/s)$, where d is the squared Euclidean distance and s a parameter. The criterion Equation 3.12 is frequently used for clustering similarity data, with no relation to k-means. It should be mentioned that the kernel trick can be applied to a modified form of the criterion $W(S,c)$ itself (Girolami 2002).

3.2.4 Spectral Rayleigh Quotient Formulation

The formulation here is:

$$B(S,c) = \sum_{k=1}^{K} \frac{z_k^T A z_k}{z_k^T z_k} \tag{3.13}$$

where $A = XX^T$ and $z_k = (z_{ik})$ is the binary membership vector so that $z_{ik} = 1$ if $i \in S_k$ and $z_{ik} = 0$, otherwise.

Equation 3.13 is but a matrix form of Equation 3.12, yet it has a special meaning because of its relation to the theory of eigenvalues and eigenvectors of a square matrix. Indeed, the ratio $r(z) = z^T A z / z^T z$ is well known as the so-called Rayleigh quotient. Given a symmetric matrix A, the Rayleigh quotient's maximum with respect to arbitrary vectors z is equal to the maximum eigenvalue of A and it is reached at the corresponding eigenvector. Moreover, the maximum of criterion Equation 3.13 with respect to arbitrary $z_1, z_2, \ldots, z_K$ is reached at the eigenvectors of A corresponding to its maximum K eigenvalues. Recall that here z_k must be binary cluster membership vectors. Therefore, the spectral approach is applicable. According to this approach, the binarity constraints are relaxed, so that the eigenvectors form a solution of the relaxed problem. The clusters are to be found as fragments of the natural orderings of the entity set according to the descending ordering of eigenvectors. The spectral form of the clustering criterion Equation 3.13 did not attract much interest before recently. We can indicate only paper (Zha et al. 2001) giving a node to this. An effective heuristic formulation involving eigenvectors of a Laplacian transformation of a square kernel similarity matrix proposed in Ng et al. (2002) should be counted as a step in the same direction. Yet recently, Kumar and Kannan (2010) proposed an interesting condition of separation between the centers which can lead to useful theoretic results on the relation between k-means and SVD.

3.3 Challenges for k-Means Criterion and Algorithm

3.3.1 Properties of the Criterion

There are a couple of geometric properties of the k-means criterion, which are quite easy to prove. The other properties mentioned below concern its ability to recover the clusters rather than the data.

3.3.1.1 Convexity of Cluster Domains

Each k-means cluster is located in a convex polytope linearly separated from polytopes of the other clusters.

This follows from the minimum distance rule. The k-th cluster lies in a halfspace Ω_l separated from any l-th cluster by the hyperplane orthogonally crossing the interval between c_k and c_l in its middle point, and therefore, in the intersection of these subspaces Ω_l over all $l \neq k$.

3.3.1.2 Monotonicity of the Minimum with Respect to the Number of Clusters K

Given a dataset, denote by w_K the minimum value of $W(S, c)$ with respect to all possible K-cluster partitions; then $w_{K+1} \leq w_K$ for all $K = 1, 2, \ldots, n - 1$.

3.3.1.3 Dependence on the Feature Measurement Scales

Since k-means heavily relies on the squared Euclidean distance, it is highly dependent on the measurement scales of the variables. If, say, the scale of a feature changes 10 times, the

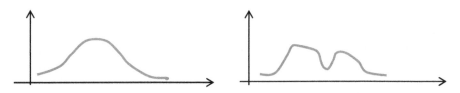

FIGURE 3.3
Two-modal variable on the right leads to two natural clusters, yet after normalization by the standard deviations it contributes less than the one-modal variable on the left.

contribution of the feature to the squared distance changes 100 times. This leads to the idea of balancing the feature contributions either preliminarily or during the process of iterations in k-means. An approach to latter will be described in Section 3.3.6. The preliminary balancing usually is done with the so-called normalization of features so that afterwards their contributions become comparable. The most straightforward and frequently advised normalization is performed by dividing features by their standard deviations, so that all the normalized features have the same standard deviation.

Although quite sound in many cases, this normalization should not be applied when clustering data (Mirkin 2012). Indeed, the contribution of a feature with a multimodal distribution would relatively decrease under this normalization because the standard deviation of such a feature is greater than that of a unimodal feature (see Figure 3.3). This would be counterintuitive, because multimodal features are more useful at clustering than those unimodal. The cause is that the value of standard deviation depends both on the scale and shape of the distribution. Normalization over feature ranges in some cases would better suit the balancing purposes. Indeed, experimental evidence supports the use of feature ranges rather than standard deviations for normalization (Milligan and Cooper 1988, Steinley and Brusco 2007).

3.3.1.4 Propensity to Balance the Sizes of Clusters

Consider one-dimensional points A, B, C with coordinates 1, 2, 6, respectively. Assume that 50 entities are put at A, the other 50 are put at B, and only one entity is in C. Take $K = 2$, and see which split minimizes the k-means criterion: A and B versus C or A versus B and C. In the former case, A and B are in one cluster so that the center of the 100-point merged cluster will be at point 1.5. Therefore, the squared Euclidean distance from any point in A or B to the center is 0.25, the square of the difference 0.5. The summary distance from all 100 entities in A merged with B to the center is $100 \times 0.25 = 25$, which is exactly the value of k-means criterion for the partition $\{A \cup B, C\}$. The other split puts together 51 entities in B and C, with their center being at $CB = (50 \times 2 + 6)/51 = 2.0784$. The summary squared Euclidean distance from this point to the 51 entities in the merged cluster is equal to $d = 50 \times 0.784^2 + 3.9216^2 = 15.6863$. This clearly shows that k-means favors the latter partition, $\{A, B \cup C\}$ over the former one, which contradicts the intuition that the nearest points A and B should be put together to make a good 2-clustering.

3.3.1.5 Square Error Versus Consensus

The fact above that the square-error criterion can be counterintuitive made the current author to look at consensus clustering (see Kuncheva and Vetrov (2005)) as a potential supplement to the square-error criterion. In Shestakov and Mirkin (2012), Gaussian cluster sets of different degrees of intermix were generated; given such a dataset, k-means starting

from random K entities as centers was run and then postprocessed with two options:
(i) BSE: choosing the best of obtained clusterings according to the k-means square-error
criterion $W(S, c)$, (ii) ECC: finding an ensemble consensus clustering according to the algorithm described in Mirkin (2012). The ensemble consensus clustering criterion operates
with incidence matrices of partitions: given a partition S on I, its incidence matrix $Z = (z_{ik})$
has its entries $z_{ik} = 1$ if i belongs kth class of S and $z_{ik} = 0$, otherwise. Given a number of
partitions on I, an ECC partition is defined as such a partition S on I that projections of
the incidence matrices of all given partitions on the linear subspace $L(Z)$ spanning its incidence matrix Z are as close to themselves as possible. When the ECC goodness of fit is
the summary squared error, this criterion is equivalent to the kernel-wise criterion Equation 3.12 in which the inner product $\langle x_i, x_j \rangle$ is substituted by the consensus matrix value
a_{ij} (Mirkin 2012). The consensus matrix value a_{ij} is the number of those among the given
partitions in which i and j belong to the same cluster. The ECC algorithm outperformed its
competition in a series of experiments (Shestakov and Mirkin 2012). It appears, on average,
that the ECC clusterings are much closer to those generated than the corresponding BSE
clusterings (see Shestakov and Mirkin (2012) and Section 5.4.2 in Mirkin (2012)).

3.3.2 Different Clustering Structures

Method k-means partitions the dataset into K nonoverlapping clusters. The clusters should
be nonempty, although, if an initial center is far away from both the other centers and the
entity points, it might remain empty since no entity is nearer to it than to other centers.
To warrant that this may not happen, it is usually sufficient to put initial centers in data
points. In some problems, partition can be considered an overly rigid structure: it requires
each entity point to belong to one and only one cluster. There can be situations when entities could belong to more than one cluster or to none of them. There are at least four types
of structures that can deal with the multiple memberships:

1. Mixture of distributions mentioned in Section 3.1.2—this will be extensively
   covered in Chapter 3.1.

2. Fuzzy cluster partition: the full membership value of an entity can be divided
   among several clusters so that each cluster gets a corresponding share of the membership and the shares total to unity. An extension of k-means to this will be
   covered in Chapter 5.5.

3. Overlapping clusters: clusters may be crisp and still may have nonempty intersection. An extension of the model Equation 3.5 to that case is considered in Depril
   et al. (2008).

4. Hierarchical clusters: clusters may be crisp and form a hierarchical tree, so that each
   cluster, which is a proper subset of all the entity set, is part of a larger cluster in the
   tree. Then k-means can be a viable tool for building such a hierarchy divisively
   with what is called Bisecting k-means (Mirkin 1996, 2012, Steinbach et al. 2000).
   This subject is out of scope of this chapter.

The case of incomplete clustering can be covered with the model in Equation 3.5 by
using an incomplete matrix Z, similarly to the case of Anomalous Pattern (AP) clustering
in which Z consists of just one column (see Section 3.3.3).

Another type of different clustering structures emerges when a cluster center is not just
a point in the feature space but forms a structure of its own (regression-wise clustering,
PCA-wise clustering, etc.). These are out of scope of this chapter too.

3.3.3 Initialization of k-Means

To initialize k-means, one needs to specify:

1. The number of clusters, K

2. Initial centers (sometimes referred to as seeds), $c_1, c_2, \ldots, c_K$

This generated scores of different proposals that essentially boil down to two generic ideas: (1) investigate the data structure (Steinley and Brusco 2007), (2) use multiple data of the process under consideration, including domain knowledge, as well as a combination of these.

The data structure can be effectively investigated at all the stages of clustering process: [A] preclustering, [B] while-clustering, and [C] postclustering.

Most efforts by researchers in finding a "right" number of clusters K have been devoted to option [C], postprocessing, in the format of multiple runs of k-means starting from random initial centers with a follow-up analysis of relative changes in the criterion values when moving from K to $K = K + 1$ or $K - 1$; see more on this in Chapter 6.1.

With option [A], preprocessing, a viable idea is to locate "anomalous" patterns in the dataset and put initial centers in those. This concurs with the equivalent k-means clustering criterion in Equation 3.10. A relatively simple method starting from the two most distant entities and adding one by one further farthest entities does not always work, probably because it can dwell on outliers which are not characteristic for the structure (Mirkin 2012). A good method should warrant not only an "anomalous" point but a pattern, that is, a "dense" set of points, around it as well.

Such is the AP method that finds APs one by one by locally maximizing criterion $r(z) = z^T A z / z^T z = n(S_1) a(S_1)$, an item in Equation 3.13, where S_1 is the sought cluster, $a(S_1)$, the average within-S_1 similarity a_{ij}, $n(S_1)$, the number of entities in S_1, and $A = (a_{ij}) = (< x_i, x_j >)$, the matrix of similarities (Chiang and Mirkin 2010, Mirkin 2012). It puts the space origin into the grand mean of the dataset or any other "reference" points, and the initial center of AP into the point that is farthest away from the origin. Then it applies iterations of 2-Means to the origin and the center, at which the origin is never moved. The stabilized cluster around the center is the AP sought (see Figure 3.4). After it is removed from the dataset, the next AP is found at the remainder, etc. When no entities remain in the remainder, the singleton patterns are removed, and those larger than that are used to both set K and initialize k-means (ik-means). In a series of experiments with a wide set of cluster structures, this method has appeared superior over many other methods in terms of the cluster recovery, although the number of clusters it generates is overly high sometimes (Chiang and Mirkin 2010).

A popular procedure Build (Kaufman and Rousseeuw 1990) for selecting initial seeds proceeds in a manner resembling that of the iterated AP. The seeds here must belong to the dataset, the number K is prespecified, and the center of gravity is taken as the first seed.

Option [B], while-clustering, has received a boost recently in (Tasoulis et al. 2010) and Kovaleva and Mirkin (2013). It appears, a process of divisive clustering can be stopped at the "right" clusters by applying an additional criterion of stopping the division process if a (proportion of the) unidimensional projection(s) of the within-cluster density function has no visible minima.

Consider the second generic idea, the usage of multiple data of the process under consideration, including domain knowledge. This aspect was neglected previously, probably, because of lacking multiple data in most cases. Yet the availability of multiple data suggests a formalized view of a most important clustering criterion, quite well known

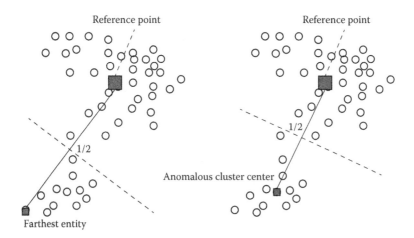

FIGURE 3.4
Extracting an "Anomalous Pattern" cluster with the origin in the gravity center: the initial iteration is illustrated on the left side, and the final on the right.

to all practitioners in the field—consistency between clusters and other aspects of the phenomenon in question. Long considered as a purely intuitive and thus thoroughly unscientific matter, this emerges currently as a powerful device, first of all in bioinformatics studies. Dotan-Cohen et al. (2009) and Freudenberg et al. (2009) show how the knowledge of biomolecular functions embedded in the so-called Gene Ontology can be used to cut functionally meaningful clusters. Mirkin et al. (2010) derives the number and location of clusters by combining three different sources of information on genes: protein sequence, gene arrangement, and reconstructed evolutionary histories of genes. The more knowledge of different aspects of real-world phenomena emerges, the greater importance of the consistency criterion in deciding of the right number of clusters.

3.3.4 Getting a Deeper Minimum

Although some recent mathematics studies suggest that the number of iterations in a run of k-means can be very large (Vattani 2009), this is not what happens in practical computations. Typically, the number of iterations is small and the final centers are not far away from the initial ones. That is, the stationary "minima" of $W(S, c)$ achieved with k-means are not deep.

There are two different views on this. One relates to the perspective in which clustering is just a tool for building classifications. In this perspective, k-means expresses stages of typology making, at which the criterion is considered not as something that must be minimized at any cost but rather a search direction. The centers should express prototypes that are to be domain knowledge driven. k-means is but a tool to adjust them to real data so that the closer the final prototypes to the initial ones, the better the knowledge reflects the reality. What is important in this perspective, though, is defining an appropriate, rather than random, initial setting.

The second perspective is not so much concerned with typology making but rather with minimization of the criterion. In this perspective, the goal is to globally minimize the criterion. This is a hard problem (Drineas et al. 2004). This is why various heuristics are used to get a better algorithm than the batch k-means. First of all, attempts should be mentioned that change the structure of neighborhoods:

1. Incremental version of k-means by moving one entity at a time (MacQueen 1967).
2. Variable neighborhood versions such as J-Means (Hansen and Mladenovic 2001).
3. Nature inspired approaches:
   a. Genetic algorithms in which partitions are taken as the "chromosomes" (Chang et al. 2009, Krishna and Murty 1999, Lu 2004);
   b. Evolution and Differential Evolution algorithms (Naldi et al. 2011, Paterlini and Krink 2006) in which the population is represented by cluster centers rather than partitions, which allows for smoother changes in it through additions of small random changes
   c. Particle Swarm Optimization algorithms (Lam et al. 2013, Van der Merwe and Engelbrecht 2003) in which the cluster centers are subject to moving with taking into account the best places the population had encountered (as bees supposedly do)
   d. Other population behavior modeling like ant-colony optimization (Saatchi and Hung 2005, Tsai et al. 2011).

Xavier and Xavier (2011) proposes considering a smoothed version of criterion $W(S, c)$ which allows to lessen the number of local minima in the function. It also makes a useful criterion-based distinction between "boundary" and "core" regions; the latter being rather stable in the iterative process.

3.3.5 Interpretation Aids

Two conventional tools for interpreting k-means clustering results, partition S and cluster centroids $c = \{c_1, \ldots, c_K\}$, are

1. Analysis of cluster centroids c_k, or "representative" entities
2. Analysis of bivariate distributions between cluster partition $S = \{S_k\}$ and categorical features related to the dataset

In fact, (2) can be considered part of (1) since, with the zero-one dummy coding of categories, cross-classification frequencies are but cluster centroids.

The decomposition Equation 3.6 leads to a number of less conventional aids based on contributions of cluster-feature pairs $B_{kh} = n_k c_{kh}^2$ to the data scatter (Mirkin 2012). Such a value has a simple intuitive meaning if the features have been preliminarily centered. Indeed, in this case, the value $n_k c_{kh}^2$ is proportional to the squared difference between the grand mean and within-cluster mean of variable h: the further away the within cluster mean from the grand mean, the greater the contribution. Curiously, for centered dummy variables representing categories, these sum to conventional cross-classification association measures such as Pearson chi-squared, depending on the dummies' normalization (Mirkin 2012).

Somewhat less intuitive is the choice of a cluster representative, a "prototypic" entity, based on the decomposition Equation 3.6. In empirical domains, such as mineralogy or oenology, indication of a "prototypic" entity makes much sense. The contribution of an entity i to the explained part of the data scatter, $B(S, c)$ Equation 3.7, appears to be equal to $\langle x_i, c_k \rangle$ where c_k is the center of cluster containing i. Therefore, the prototype should maximize the inner product, rather than minimize the distance, to its cluster's center. In the case when the data have been centered, this choice better follows tendencies represented in c_k versus the grand mean than the conventional choice according to the distance.

3.3.6 Feature Weighting, Three-Stage k-Means, and Minkowski Metric

Drawbacks of k-means method have been mentioned above such as the lack of advice on the choice of the number of clusters and location of the initial centers. The latter is aggravated by the fact that k-means usually reaches very superficial minima of the criterion. One more drawback of the method is that k-means method cannot distinguish noise, or irrelevant, features from those useful ones and, therefore, is defenseless against them. To address this, a three-stage version of k-means involving feature weighting was proposed in Makarenkov and Legendre (2001) and further developed in (Huang et al. 2008). These have been integrated in a three-stage Minkowski k-means approach (Amorim and Mirkin 2012). The approach adjusts the feature scales in such a way that those supporting the found partition get greater weights over those less relevant ones. Here is an outline of the approach.

In data analysis, p-th power of Minkowski distance between vectors $x = (x_v)$ and $y = (y_v)$ is defined by formula

$$d^p(x, y) = \sum_{v \in V} |x_v - y_v|^p$$

Minkowski metric k-means clustering criterion is specified as

$$W_p(S, c) = \sum_{k=1}^{K} \sum_{i \in S_k} d^p(x_i, c_k) = \sum_{k=1}^{K} \sum_{i \in S_k} \sum_{v \in V} |x_{iv} - c_{kv}|^p \qquad (3.14)$$

Obviously, at $p = 2$ this is the conventional k-means criterion Equation 3.2.

To formulate the Batch k-means algorithm in this case, one needs an algorithm for finding Minkowski's center, that is, c minimizing

$$D_p(c) = \sum_{i=1}^{n} |x_i - c|^p \qquad (3.15)$$

for any series of reals $x_1, x_2, \ldots, x_n$ (see Figure 3.5).

To include the issue of feature weighting in k-means, let us reformulate the criterion as follows. Given a standardized data matrix $X = (x_{ih})$, where $i \in I$ are entities and $h \in H$

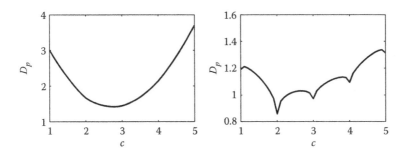

FIGURE 3.5
Graphs of Minkowski $D_p(c)$ function for the series $I = \{2, 2, 3, 4, 5, 1\}$ at $p = 1.5$, on the left, and $p = 0.5$, on the right.

features, the nonnegative weights w_h, satisfying condition $\sum_{h \in H} w_h = 1$, serve as rescaling factors so that the criterion Equation 3.14 is reformulated over the rescaled data $x'_{ih} = w_h x_{ih}$ and rescaled centroids $c'_{kh} = w_h c_{kh}$:

$$W_p(S, c, w) = \sum_{k=1}^{K} \sum_{i \in S_k} d^p(x'_i, c'_k) = \sum_{k=1}^{K} \sum_{i \in S_k} \sum_{h \in H} w_h^p |x_{ih} - c_{kh}|^p \tag{3.16}$$

Criterion Equation 3.16 holds to the generic form of k-means criterion, which is in stark contrast to other approaches (for a review, see (Amorim and Mirkin 2012)). The alternating minimization of the criterion Equation 3.16 over three groups of variables, corresponding to S, c, and w, respectively, leads to a three-step version of k-means. The computation starts similarly to batch k-means, from K tentative centroids $c_1, c_2, \ldots, c_K$ and initial weights all equal to each other, $w_h = 1/|H|$.

Here is a formulation of the algorithm, referred to as MWK-Means in (Amorim and Mirkin 2012), for alternating minimization of criterion Equation 3.16 at a given p:

1. Given centroids c_k and weights w_h, update the cluster assignment of entities by using the minimum distance rule with distance defined as p-power of Minkowski metric $d^p(x_i, c_k) = \sum_{h \in H} w_h^p |x_{ih} - c_{kh}|^p$;

2. Given clusters S_k and weights w_h, update centroid $c_k = (c_{kh})$ of each cluster S_k as its Minkowski center so that, at each h, c_{kh} is defined as $w_h c$ where c minimizes an item in Minkowski's distance power, $\sum_{i=1}^{n} w_h |x_{ih} - c|^p$ according to Equation 3.15;

3. Given clusters S_k and centroids c_k, update weights according to formula

$$w_h = \frac{1}{\sum_{h' \in H} [D_{hp}/D_{h'p}]^{\frac{1}{p-1}}} \tag{3.17}$$

where $D_{hp} = \sum_{k=1}^{K} \sum_{i \in S_k} |x_{ih} - c_{kh}|^p$.

The formulation of the MWK-Means algorithm suggests that the feature weights can be made cluster-specific both in the criterion Equation 3.16 and the algorithm (Amorim and Mirkin 2012).

It appears that cluster recovery can improve indeed, at some datasets drastically, at an appropriate p depending on the dataset. Moreover, an appropriate version of anomalous pattern method can be used to define the number of clusters K. For example, at the celebrated 150×4 Iris dataset, the Minkowski three-stage clustering method makes only five errors at $p = 1.2$—an absolute minimum of all the supervised and unsupervised methods reported (Amorim and Mirkin 2012). The issue of choosing an appropriate value of Minkowski exponent p can be addressed by learning in a semi-supervised manner if cluster labels are supplied at about 5% of the entities (Amorim and Mirkin 2012). In fact, (Amorim and Mirkin 2012) also reports of a successful experiment in determining the right p in a nonsupervised manner.

3.4 Notes on Software for k-Means

Although the k-means method deserves to be coded as a stand-alone application supplied with powerful tools for getting a meaningful cluster structure by using a wide spectrum of computation and interpretation devices, the method currently appears in a generic version in most popular computational platforms and packages. In this section, a brief description of implementations of the method will be given in computational platforms:

- Matlab (see http://www.mathworks.com/products/matlab/),
- Weka (see http://www.cs.waikato.ac.nz/ml/weka/),
- R (see http://www.r-project.org/),
- SPSS (see http://www-01.ibm.com/software/analytics/spss/).

In the end, an outline of currently available platforms for applying k-means for the analysis of "big data" will be given.

Matlab k-means clustering takes data matrix X and the number of clusters K as its inputs and performs a run of batch k-means clustering with the squared Euclidean distances by default. It allows using some different distances, first of all the city block and cosine, as well. Other user selected parameters include:

- An initialization method: selecting K random rows from X, or a matrix of user-specified K points in the feature space, or results of preliminary clustering of a 10% random subsample of the dataset.
- Number of runs with different initializations.
- The way of reacting to the fact that some cluster happens to get empty: this can be an "error" message or removal of empty clusters or making a singleton instead.
- The maximum number of iterations so that the computation stops upon reaching that number, 100, by default.
- A possibility of further minimization of the criterion by moving individual entities between clusters.

In Weka, k-means is implemented as SimpleKMeans command. This command automatically handles a mixture of categorical and numerical attributes. To this end, the program converts all nominal attributes into binary numeric attributes and normalizes scales of all the numeric features in the dataset to lie within the interval [0, 1]. When prompted, the program replaces all missing values for nominal and numeric attributes with the modes and means of features in the training data. Initialization is made randomly from among the observations. Instance weighting can be included if needed. The Weka SimpleKMeans algorithm uses Euclidean distance measure to compute distances between instances and centers. It can visualize the results with figures drawn for each cluster so that features are represented by lines radiating from the center.

In R, Project for statistical computing, k-means clustering is available in either batch version or incremental one. It works either with random K observations taken as initial centers or with a user-specified set of centers. The maximum number of iterations can be specified, as well as the number of runs from random initializations. In the Weka version, $K = 1$ is allowed. If an empty cluster emerges, an "error" message appears.

The statistical package SPSS allows to use its rich system for both preprocessing of data and interpretation of results. Its implementation of k-means is similar to that in Matlab—all the parameters available in Matlab are available in SPSS, except for the way for handling missing values. There is a different system for handling missing data in SPSS. Additional features include the possibility for storing the clusters as a nominal variable, ANOVA table for clusters to allow to see what variables contribute most to a cluster, and the matrix of distances between cluster centers.

Currently, some efforts are devoted to parallelization of k-means computations to apply it to big data, that is, data with millions or more entities. The method is well suited for distributed computations. Say, a method in Zhao et al. (2009) is based on the idea that the dataset is partitioned into portions processed at different processors. An iteration is performed as follows: a central processor supplies a set of centers so that each processor computes distances from its portion entities to the centers and within cluster sums, after which it supplies the central processor with the sums and number of entities. Then the central processor sums all the received sums within clusters and computes the cluster centers. Further development of cloud computations as a commodity will show what versions of the algorithm will be required in the future.

Acknowledgments

This work has been supported by the Laboratory of Decision Choice and Analysis NRU HSE Moscow and, also, the Laboratory of Algorithms and Technologies for Networks Analysis NRU HSE Nizhny Novgorod (by means of RF government grant, ag. 11.G34.31.0057), as well as by grants from the Academic Fund of the NRU HSE in 2010–2013.

References

R.C. Amorim, B. Mirkin 2012 Minkowski metric, feature weighting and anomalous cluster initializing in K-Means clustering, *Pattern Recognition*, 45, 1061–1074.

R.C. Amorim, B. Mirkin 2014 Selecting the Minkowski exponent for intelligent K-Means with feature weighting, In F. Aleskerov, B. Goldengorin, P. Pardalos (Eds.) *Clusters, Orders, Trees: Methods and Applications, Springer Optimization and Its Applications*, Springer Science+Business Media, New York, 92, 103–117.

H.-H. Bock 2008 Origins and extensions of the k-means algorithm in cluster analysis, *E-Journal for History of Probability and Statistics*, 4, 2, http://www.jehps.net/Decembre2008/Bock.pdf

G. Celeux, G. Govaert 1992 A classification EM algorithm and two stochastic versions, *Computational Statistics and Data Analysis*, 14, 315–332.

D.-X. Chang, X.-D. Zhang, C.-W. Zheng 2009 A genetic algorithm with gene rearrangement for k-means clustering, *Pattern Recognition*, 42(7), 1210–1222.

M. Chiang, B. Mirkin 2010 Intelligent choice of the number of clusters in K-Means clustering: An experimental study with different cluster spreads, *Journal of Classification*, 27(1), 3–41.

Clementine 7.0 User's Guide Package 2003 Chicago, SPSS Inc.

D. Depril, I. Van Mechelen, B. Mirkin 2008 Algorithms for additive clustering of rectangular data tables, *Computational Statistics and Data Analysis*, 52(11), 4923–4938.

G. Der, B.S. Everitt 2001 *Handbook of Statistical Analyses Using SAS*, 2nd Edition, CRC Press, Boca Raton, FL.

D. Dotan-Cohen, S. Kasif, A.A. Melkman 2009 Seeing the forest for the trees: Using the Gene Ontology to restructure hierarchical clustering, *Bioinformatics*, 25(14), 1789–1795.

P. Drineas, R. Kannan, A. Frieze, S. Vempala, V. Vinay 2004 Clustering large graphs via the singular value decomposition, *Machine Learning*, 56, 9–33.

S.L. France, C.D. Carroll, H. Xiong 2012 Distance metrics for high dimensional nearest neighborhood recovery: Compression and normalization, *Information Sciences*, 184, 92–110.

J. Freudenberg, V. Joshi, Z. Hu, M. Medvedovic 2009 CLEAN: Clustering enrichment analysis, *BMC Bioinformatics*, 10, 234.

M. Girolami 2002 Mercer kernel based clustering in feature space, *IEEE Transactions on Neural Networks*, 13, 780–784.

S.B. Green, N.J. Salkind 2003 *Using SPSS for the Windows and Macintosh: Analyzing and Understanding Data*, 6th Edition, Prentice-Hall, Upper Saddle River, NJ.

J. Han, M. Kamber, J. Pei 2011 *Data Mining: Concepts and Techniques*, 3rd Edition, Morgan Kaufmann Publishers, Waltham, MA.

P. Hansen, N. Mladenovic 2001 J-means: A new local search heuristic for minimum sum-of-squares clustering, *Pattern Recognition*, 34, 405–413.

J.A. Hartigan 1975 *Clustering Algorithms*, New York: J.Wiley & Sons.

J.Z. Huang, J. Xu, M. Ng, Y. Ye 2008 Weighting method for feature selection in K-Means, In H. Liu, H. Motoda (Eds.) *Computational Methods of Feature Selection*, Chapman & Hall/CRC, Boca Raton, FL, 193–209.

A.K. Jain, R.C. Dubes 1988 *Algorithms for Clustering Data*, Prentice-Hall, Englewood Cliffs, NJ.

L. Kaufman, P. Rousseeuw 1990 *Finding Groups in Data: An Introduction to Cluster Analysis*, Wiley, New York.

E.V. Kovaleva, B. Mirkin 2015 Bisecting K-means and 1D projection divisive clustering: A unified framework and experimental comparison, *Journal of Classification*, 32.

K. Krishna, M. Murty 1999 Genetic K-means algorithm, *IEEE Transactions on Systems, Man and Cybernetics, Part B: Cybernetics*, 29(3), 433–439.

A. Kumar, R. Kannan 2010 Clustering with spectral norm and the k-means algorithm, *51th Annual IEEE Symposium on Foundations of Computer Science*, FOCS 2010, October 23–26, 2010, Las Vegas, Nevada, 299–308.

L.I. Kuncheva, D.P. Vetrov 2005 Evaluation of stability of k-means cluster ensembles with respect to random initialization, *IEEE Transactions on Pattern Analysis and Machine Intelligence*, 28, 1798–1808.

Y.-K. Lam, P.W.M. Tsang, C.-S. Leung 2013 PSO-based K-means clustering with enhanced cluster matching for gene expression data, *Neural Computing and Applications*, 22, 1349–1355.

Y. Lu, S. Lu, F. Fotouhi, Y. Deng, S.J. Brown 2004 Incremental genetic K-means algorithm and its application in gene expression data analysis, *BMC Bioinformatics* 5, 172, http://www.biomedcentral.com/1471-2105/5/172.

J.B. MacQueen 1967 Some methods for classification and analysis of multivariate observations, In L. Lecam, J. Neymen (Eds.) *Proceedings of 5th Berkeley Symposium*, 2, 281–297, University of California Press, Berkeley.

V. Makarenkov, P. Legendre 2001 Optimal variable weighting for ultrametric and additive trees and K-Means partitioning, *Journal of Classification*, 18, 245–271.

G.W. Milligan, M.C. Cooper 1988 A study of standardization of the variables in cluster analysis, *Journal of Classification*, 5, 181–204.

B. Mirkin 1996 *Mathematical Classification and Clustering*, Kluwer Academic Press, Dordrecht.

B. Mirkin 2012 *Clustering: A Data Recovery Approach*, 2nd Edition, Chapman & Hall/CRC Press, Boca Raton, FL.

B. Mirkin, R. Camargo, T. Fenner, G. Loizou, R. Kellam 2010 Similarity clustering of proteins using substantive knowledge and reconstruction of evolutionary gene histories in herpesvirus, *Theoretical Chemistry Accounts*, 125(3–6), 569–581.

M.C. Naldi, R.J.G.B. Campello, E.R. Hruschka, A.C.P.L.F. Carvalho 2011 Efficiency issues of evolutionary k-means, *Applied Soft Computing*, 11(2), 1938–1952.

A.Y. Ng, M.I. Jordan, Y. Weiss 2002. On spectral clustering: Analysis and an algorithm. *Advances in Neural Information Processing Systems*, 2, 849–856.

S. Paterlini, T. Krink 2006 Differential evolution and particle swarm optimisation in partitional clustering, *Computational Statistics & Data Analysis*, 50(5), 1220–1247.

S. Saatchi, C.-C. Hung 2005 Using ant colony optimization and self-organizing map for image segmentation, *Image Analysis Lecture Notes in Computer Science*, 3540, 283–293, Springer.

A. Shestakov, B. Mirkin 2013 Least square consensus clustering: Criteria, methods, experiments, advances in information retrieval, *Lecture Notes in Computer Science*, Volume 7814, Springer, Berlin-Heidelberg, 764–767.

M. Steinbach, G. Karypis, V. Kumar 2000 A comparison of document clustering techniques, KDD Workshop on Text Mining.

D. Steinley 2006. K-means clustering: A half-century synthesis, *British Journal of Mathematical and Statistical Psychology*, 59, 1–34.

D. Steinley, M. Brusco 2007 Initializing K-Means batch clustering: A critical evaluation of several techniques, *Journal of Classification*, 24, 99–121.

S.K. Tasoulis, D.K. Tasoulis, V.P. Plagianakos 2010 Enhancing principal direction divisive clustering, *Pattern Recognition*, 43(10), 3391–3411.

C.-W. Tsai, K.-C. Hu, M.-C. Chiang, C.-S. Yang 2011 Ant colony optimization with dual pheromone tables for clustering, Fuzzy Systems (FUZZ), *2011 IEEE International Conference*, 2916–2921, DOI: 978-1-4244-7315-1.

D.W. Van der Merwe, A.P. Engelbrecht 2003 Data clustering using Particle Swarm Optimization, *The 2003 Congress on Evolutionary Computation (CEC '03)*, 1, 215–220.

A. Vattani 2009 K-Means requires exponentially many iterations even in the plane, *Proceedings of the 25th Annual Symposium on Computational Geometry*, ACM New-York, 324–332.

I.H. Witten, E. Frank, M.A. Hall 2011 *Data Mining: Practical Machine Learning Tools and Techniques*, Morgan Kaufman, Burlington, MA, 680 p.

A.E. Xavier, V.L. Xavier 2011 Solving the minimum sum-of-squares clustering problem by hyperbolic smoothing and partition into boundary and gravitational regions, *Pattern Recognition*, 44(1), 70–77.

H. Zha, X. He, C.H.Q. Ding, M. Gu, H.D. Simon 2001 Spectral relaxation for K-means clustering, In *Proceedings of NIPS'2001*, 1057–1064.

W. Zhao, H. Ma, Q. He 2009 Parallel K-Means clustering based on mapReduce, *Cloud Computing*, Lecture Notes in Computer Science, V. 5931, Springer, 674–679.

4

K-Medoids and Other Criteria for Crisp Clustering

Douglas Steinley

CONTENTS

Abstract

This chapter addresses crisp clustering techniques other than *K*-means, which was covered in the previous chapter. Primarily, focus is given to the *K*-medoids (e.g., p-median) approach to assigning observations to clusters. In addition to *K*-medoids clustering, a brief discussion of *K*-midranges and *K*-modes clustering is provided.

4.1 Introduction

The most common approaches for partitioning a dataset are related to minimizing the sum-of-squares error for each cluster and generally appear under the guise of *K*-means clustering (Hartigan and Wong, 1979; MacQueen, 1967; Steinhaus, 1956; Thorndike, 1953, see Steinley (2006a) for an extensive review of *K*-means clustering and Brusco and Steinley

(2007) for a comparison of heuristic methods for this problem). However, there are several reasons that alternatives to K-means clustering may be desired, and, in terms of clustering algorithms (versus a statistical-based clustering approach, such as mixture modeling), the most common choice is the p-median (e.g., K-medoids) clustering algorithm. The K-medoids approach has at least three distinct differences/advantages over the standard K-means approach. First, both the K-means and K-medoids clustering procedures provide "examplars" as cluster centers, in the former's case, the exemplar of a cluster is the mean of the cluster; however, in the case of the latter, the exemplar is an actual object from the cluster. Second, as seen below when the K-medoids algorithm is formally introduced, it is based on the standard Euclidean distance, rather than like the squared Euclidean distance used in K-means clustering; consequently, K-medoids is generally more robust to outliers than K-means (Kaufman and Rousseeuw, 1990). Third, whereas K-means is generally applied directly to a standard data matrix structured as objects by variables, the K-medoids approach can be used for any distance measure derived from the matrix or assessed directly, not to mention situations where there are violations of the triangle inequality and/or symmetry.

4.2 K-Medoids

4.2.1 The Problem

Historically, K-medoids clustering has its roots in operations research for attempts to optimize the planning of facility locations (Hanjoul and Peeters, 1985; Kuehn and Hamburger, 1963). Exemplar-based clustering via the K-medoids model is derived from this optimization problem, which includes two primary tasks: (i) identify a set of K objects (e.g., medoids) to serve as cluster centers, and (ii) assign the remaining $N - K$ objects (e.g., satellites) to the chosen medoids such that the objective function of the total sum of medoid-to-satellite Euclidean distances is minimized. The K-medoids procedure has been proposed several times in the literature (Kaufman and Rousseeuw, 1990; Massart et al., 1983; Späth, 1985; Vinod, 1969), with Kaufman and Rousseeuw's book and subsequent discussion of the partitioning around medoids (PAM) for conducting the clustering being the primary reference. The PAM algorithm proceeds as:

1. Initialize: randomly select (without replacement) K of the N data points as the initial medoids (as Steinley, notes in the context of K-means clustering, this type of initialization strategy can be problematic (Steinley, 2003)).

2. Assign each observation to the medoid with which it is closest, where closest is based on a specific distance measure (the most common distance to minimize is the Euclidean distance) and compute the total cost across all observations, where the cost is the sum of the distance of each observation to its associated medoid.

3. For each medoid k, for $k = 1, \ldots, K$ consider all $N - K$ nonmedoid, o. Swap k and o and recompute the total cost.

4. Select the solution with the lowest cost.

5. Repeat steps 2–4 until the set of medoids does not change.

In terms of formal representation, Köhn et al. (2010) represent the K-medoid as the following linear integer programming problem

$$IP1 : min \left\{ f(\mathbf{X}) = \sum_{i=1}^{N} \sum_{j=1}^{N} d_{ij} b_{ij} \right\} \tag{4.1}$$

subject to to the constraints:

$$C1 : \sum_{j=1}^{N} b_{jj} = K,$$

$$C2 : b_{ij} \leq b_{jj} \quad \forall i, j,$$

$$C3 : \sum_{j=1}^{N} b_{ij} = 1 \quad \forall i,$$

$$C4 : b_{ij} \in \{0, 1\} \quad \forall i, j$$

where d_{ij} represents the given input dissimilarities and b_{ij} denotes binary decision variables that only take values of zero or unity (e.g., the constraint in C4). As seen in the cost function, each of the binary b_{ij}'s are multiplied by d_{ij}'s, which represent the distance between the ith and jth observation and are collected in the $N \times N$ matrix, $\mathbf{D}_{N \times N}$. Obviously, the b_{ij}'s are collected in a comparative matrix, $\mathbf{B}$. The first constraint (C1) operates on the diagonal of $\mathbf{B}$, indicating that each observation is a candidate for being a medoid; however, the number of medoids is constrained to be exactly equal to the number of clusters (hence, the summation to K). The second constraint (C2) indicates that the ith observation can only be assigned to medoid j if and only if the jth observation has been selected as a medoid. Constraint three (C3) indicates that each observation can be assigned to only one cluster—this is the constraint that enforces mutually exclusive and exhaustive clusters.

Rewriting K-medoids as a combinatorial optimization (e.g., discrete optimization) problem provides several insights. First, we know that a globally optimal solution always exists, which follows from the fact that, in technical terms, the set of all possible solutions is finite. However, finding such an optimal solution can be very difficult as the number of possible candidate solutions for the K-medoids problem is $\binom{N}{K}$, creating a prohibitive approach to the strategy of conducting a complete search when the sample size becomes too larger. The search for an optimal solution is made more challenging by the fact that combinatorial optimization problems have nonsmooth objective functions. As such, several heuristic algorithms have been developed for finding solutions to the K-medoids problem, including a multistart fast interchange methods (Teitz and Bart, 1968), simulated annealing (Chiyoshi and Galvao, 2000), tabu search (Rolland et al., 1996), genetic algorithms (Alba and Dominguez, 2006), and Lagrangian relaxation (Mulvey and Crowder, 1979).

4.2.2 Algorithms

4.2.2.1 Heuristics

Brusco and Köhn (2009) developed a simulated annealing heuristic that compared favorably to existing simulated annealing algorithms (Chiyoshi and Galvao, 2000), the affinity propagation algorithm (Frey and Dueck, 2007), and a vertex substitution heuristic (Mladenovic et al., 2007), outperforming each across a wide range of test problems. As such, Brusco and Köhn's algorithm is presented and reviewed here. The simulated annealing algorithm proceeds as

1. Step 0. Initialization: Randomly generate an initial set of K exemplars. Set $U = N/J$, $J^* = J$, compute $z_1 = IP1$, and set $z^* = z_1$. Choose $0 < c < 1$, G, and an initial value for T. Set $g = 0$ and $b = 0$. Note that N is the set of observations and J is the set of K exemplars.

2. Step 1. Generate a trial solution: Set $g = g + 1$. Modify J to create a trial set of exemplars, J', and compute the objective value for the trial solution z_1'.

3. Step 2. Evaluate the trial solution: Compute $\Delta = z_1' - z_1$. If $\Delta > 0$, then go to Step 4; otherwise, go to Step 3.

4. Step 3. Accept an improved trial solution: Set $J = J'$, $U = N/J$, $z_1 = z_1'$, $b = 1$, and, if $z_1' < z^*$, then also set $J^* = J$ and $z^* = z_1'$. Go to Step 5.

5. Step 4. Accept an inferior trial solution?: Generate a uniform random number, r, on the interval $[0, 1]$. If $r < exp(\Delta/T)$, then set $b = 1$, $J = J'$, $U = I/J$, and $z_1 = z_1'$. Go to Step 5.

6. Step 5. Update Temperature?: If $g < G$, then set $g = g + 1$ and go to Step 1; otherwise, go to Step 6.

7. Step 6. Termination?: If $b = 1$, then set $b = 0$, $T = cT$, $g = 0$ and return to Step 1; otherwise, STOP.

Step 0 is the initial definition of all elements evaluated in the simulated annealing program. As stated, J is the initial set of exemplars, while J^* represents the best-found set of exemplars. Similarly, z^* represents the best-found objective value that corresponds to J^*, with $U = N/J$ being set theory notation for the set of objects that are not selected as exemplars. Common to simulated annealing algorithms is a temperature and a cooling schedule. The temperature, T, controls the probability of accepting a solution that worsens the objective function (e.g., choosing an inferior objective value), with the notion behind accepting a worse solution being to avoid potentially locally optimal solutions. Related to the temperature is the cooling factor, c, which periodically reduces the temperature (e.g., the probability of accepting a worse solution decreases the longer the algorithm runs). The final component of the cooling schedule is the temperature length, G, which represents the number of trial solutions to evaluate at each temperature. Brusco and Köhn indicated that the two most important aspects for developing an effective simulated annealing algorithm were the selection of the parameters that govern the cooling schedule, and the generation of trial solutions (and readers are referred to their original article for a more in depth treatment). The values for the cooling schedule chosen by Brusco and Köhn were $G = 10N$, T was set to the maximum difference in the objective function when 200 randomly generated exchanges of exemplars in J with nonexemplars in U, while c is commonly chosen to be $0.85 \leq c \leq 0.90$. The choice of G being 10 times the sample size allows larger problems to have more trial solutions, and the 200 randomly generated exchanges allows for enough inferior solutions for the solution space to be explored.

Step 1 generates a trial solution by changing the exemplar set in J, with the new set noted as J'. Brusco and Köhn's primary innovation over Chiyoshi and Galvao (2000) is the manner in which J' is created. The new procedure randomly chooses an object, $u \in U$, that is not an exemplar and replaces the exemplar in J, $j \in J$, such that the objective function is improved by the maximal amount.

Step 2 compares the new objective function, z', with the original objective function, z, via a difference score, Δ. If Δ is positive it indicates that $z' > z$ and the perturbed set of exemplars *is not* an improvement over the original set; on the other hand, if $z' < z$, Δ is negative and the objective function has been improved.

Step 3 governs the acceptance of an improved trial solution. The exemplars and objective function are updated.

Step 4 governs the decision of whether to accept an inferior solution. The acceptance is based on comparing a number generated from a random uniform distribution to the quantity $exp(\Delta/T)$. If r is less than this quantity, then the inferior solution is accepted and the exemplars and objective function are updated.

Step 5 updates the temperature function. If $g < G$, then the algorithm iterates back to Step 1 and proceeds with the evaluation of another trial solution. Once the temperature length, $10N$, has been reached, then the algorithm goes to the final termination step.

Step 6 terminates the algorithm if a set of exemplars is able to make it through an entire pass of the temperature schedule without any replacements, whether they are resultant from an improved solution or from accepting an inferior solution. Each pass through this step decreases T, which in turn makes accepting inferior solutions less likely.

4.2.2.2 Exact Algorithms

Commonly, heuristics are used as exact solutions are computationally infeasible for many datasets—in fact, it should be clear from combinatorics that the number of possible ways to choose K medoids from N observations is $N!/(K!(N-K)!)$. However, there are some instances when a globally optimal solution can be obtained. Obviously, from the mathematical programming formulation given above, this problem could be solved via integer programming (Mulvey and Crowder, 1979); however, Brusco and Köhn (2008a) indicated that an integer programming solution is quite inefficient. Rather, Brusco and Köhn propose a three-stage solution:

1. Stage 1. The first stage relies on the classic vertex substitution heuristic (Teitz and Bart, 1968) to establish an upper bound, $IP1_{UB1}$ for $IP1_{opt}$ (e.g., the optimal value of the objective function). In this step, 20 replications of the vertex substitution are conducted.

2. Stage 2. The second step passes $IP1_{UB1}$ to a Lagrangian relaxation algorithm (Hanjoul and Peeters, 1985) that attempts to find a globally optimal solution.

3. Stage 3. If a globally optimal solution cannot be verified in Stage 2, then a branch-and-bound procedure is used to obtain the globally optimal solution.

4.2.3 Choosing the Number of Clusters

As with most clustering algorithms, choosing the number of clusters is a daunting task (see Steinley and Brusco (2011c), for a discussion with the difficulties in K-means clustering).

The generally accepted procedure for choosing the number of clusters in K-medoids clustering is the silhouette index (Rousseeuw, 1987), given as

$$SI_K = \frac{1}{N} \sum_{i=1}^{N} \frac{b(i) - a(i)}{\max\{a(i), b(i)\}}, \tag{4.2}$$

where $a(i)$ denotes the average dissimilarity between object i and all objects in its cluster and $b(i)$ represents the minimum average dissimilarity of object i assigned to a different cluster, the minimum taken over all other clusters. SI_K is bounded between zero and unity, with the basic strategy for choosing the number of clusters to be fitting the K-medoids model for several different values of K and then choosing K to be the value which corresponds the maximum value of SI_K. See Chapter 26 for alternatives.

4.2.4 Advantages of *K*-Medoids

The three properties that are often given in support of K-medoids over competing algorithms, commonly K-means, is that: (i) K-medoids is generally more robust, (ii) K-medoids can be applied to general proximities and is not reliant on Euclidean (or squared Euclidean in the case of K-means), and (iii) K-medoids can, under the right conditions, be more predisposed to finding a provably globally optimal solution. Each of these is addressed in turn.

First, K-medoids has been proposed as a more robust version of K-means; the argument being that the objective function will be less influenced by outliers or influential observations. However, this view is not universally shared, and some contend that K-medoids is only insensitive to outliers in special cases and, in general, the objective function can be as affected by outliers as K-means (Garcia-Escudero and Gordaliza, 1999). While the median may be a robust centrality measure for one random variable, it is unlikely that the "joint" selection of two medians will be robust for two random variables (Cuesta-Albertos et al., 1997), and, by extension, it is likely that the desirable properties exhibited at the univariate level will continue to degrade as the dimensionality of the median increases. Thus, somewhat counterintuitively, K-medoids does not exhibit the same robustness properties as the median possesses in location theory (Huber, 1981). That being said, from a practical point of view, there still seems to be many gains when using K-medoids for outlier mitigation.

While it is noted that both K-means and K-medoids are prone to extreme outliers, K-medoids will be less influenced by outliers *within* the clusters themselves. This is readily seen by replacing d_{ij} in IP1 by d_{ij}^2 (which would be, in the most common instance, replacing Euclidean distance with squared Euclidean distance), highlighting the fact that largest set of d_{ij}'s will dominate the K-means objective function. An additional implication of using d_{ij}^2 in K-means instead of the d_{ij} commonly used in K-medoids relates back to the shape of the clusters that will tend to be discovered by the respective procedure. For K-means clustering, the d_{ij}^2 can be rewritten in the traditional form of the K-means algorithm, where it is the squared distance between each observation and its cluster mean. Aggregated across all the observations within the cluster creates an objective function which is basically the within-cluster sum of squares, or the numerator of the variance term of the cluster. Consequently, K-means clustering can also be conceptualized, like Ward's hierarchical method (Ward, 1963), as a minimum variance clustering approach. As has been pointed out repeatedly, see Steinley and Brusco (2011a) for instance, this results in the tendency of K-means clustering to favor cluster structures that are spherical in nature. Unfortunately, this is not only a tendency to favor finding spherical clusters when they exist, but also the propensity to impose a spherical cluster structure on non-spherical clusters. This inappropriate imposition of sphericity obscures, and occasionally, completely misrepresents the true structure of the clusters. As indicated by Kaufman and Rousseeuw, an additional advantage of using K-medoids is the added flexibility of being able to discover nonspherical cluster structures.*

Second, there are often claims that K-medoids has an advantage over K-means in that it can handle general proximity data and is not reliant on squared Euclidean distance; however, Steinley and Hubert (2008) showed that K-means clustering can be applied to square proximity data as well by reformulating the structure of the within-cluster sum-of-squares and cross products matrix. While not a one-to-one mapping of the situations that K-medoids can be applied to, it does demonstrate a greater flexibility of K-means than is usually indicated. Third, recent development in K-means clustering has shown that globally optimal solutions are obtainable. For example, Steinley and Hubert (2008) use dynamic programming to find optimal clustering under the K-means criteria in the presence of order constraints. Furthermore, Aloise et al. (2012) have developed a column generation algorithm for the K-means criteria that is able to optimally partition more than 2300 observations in six dimensions.

4.3 Example and Software

4.3.1 Example: Classifying Animals

Steinley and Hubert (2008) provided an example, using a constrained K-means algorithm, of classifying a set of 100 animals based on the number of legs the animal has and the following binary (1 = possesses, 0 = does not possess) characteristics: hair, feathers, eggs, milk, airborne, aquatic, predator, toothed, backbone, breathes, venomous, fins, tail, hoof, and horns. Here, the animal data is reanalyzed using the simulated annealing algorithm for K-medoids proposed by Brusco and Köhn (2009). For comparative purposes with Steinley and Hubert (2008), a 13-cluster solution was chosen.

The final clustering solution (with the "exemplar" bolded in group) is:

1. {aardvark, bear, boar, cheetah, human girl, **leopard**, lion, lynx, mink, mole, mongoose, opossum, platypus, polecat, puma, pussycat, raccoon, wolf}
2. {cow, cavy, goat, **hamster**, pony, reindeer}
3. {crow, **hawk**, kiwi, rhea, vulture}
4. {bass, catfish, chub, dogfish, herring, **pike**, piranha, seasnake, stingray, tuna}
5. {**flea**, gnat, ladybird, slug, termite, tortoise, worm}
6. {carp, **haddock**, seahorse, sole, toad}
7. {**gull**, penguin, skimmer, skua}

---

* However, we do note that this is not as much of a concern for K-means clustering if an appropriate variable weighting scheme is used, such as the variance-to-range ratio suggested by Steinley and Brusco (2008a).

8. {antelope, buffalo, **deer**, elephant, fruitbat, giraffe, gorilla, hare, oryx, squirrel, vampire bat, vole, wallaby}

9. {honeybee, **housefly**, moth, wasp}

10. {clam, crab, **crayfish**, lobster, octopus, seawasp, starfish}

11. {frog, newt, pitviper, scorpion, slowworm, **tuatara**}

12. {chicken, dove, duck, flamingo, lark, ostrich, parakeet, pheasant, **sparrow**, swan, wren}

13. {dolphin, **porpoise**, seal, sealion}

Obviously, one advantage is that choosing a medoid provides an immediate understanding of the cluster. In this case we could interpret (loosely) the clusters as wild mammals (leopard), domesticated mammals (hamster), predator/scavenger bird (hawk), fish (pike), insect (flea), fish (haddock), sea birds (gull), nonpredator mammals (deer), flying insect (housefly), shellfish (crayfish), reptile (tuatara), bird (sparrow), and aquatic mammal (porpoise).

In comparison to the K-means clustering conducted by Steinley and Hubert (2008), the solutions are quite different, having a modest adjusted Rand index of 0.66—a value Steinley (2004) indicates is only a "moderate" level of agreement, where good agreement and excellent agreement have values of 0.80 and 0.90, respectively. The primary source of disagreement is the "deer" cluster, where Steinley and Hubert (2008)'s solution makes a distinction of mammals with and without hoofs.

From a practical point of view, this indicates that the nature of the solution and interpretation of the resultant clusters will be *highly* dependent on the clustering algorithm used to obtain the data. Prima facia, this type of claim seems obvious; however, it is often the case that researchers attempt to "validate" their final solution by observing consistency across methods. While observing a set of core consistency may point to a level of stability that lends itself to increased confidence in the results, inspecting the discontinuities across solutions is more likely to illuminate interesting structure within the data.

4.3.2 Software Implementations

Standard implementation of K-medoids is not supported by SAS or SPSS. However, routines are readily available in both R and Matlab, with Kaufman and Rousseeuw (1990)'s implementation of the vertex substitution heuristic available in R and Köhn et al. (2010) simulated annealing-based program available in Matlab. Naturally, programs with faster computational are available in scientific programming languages such as Fortran (Hansen and Mladenovic, 1997) and C++ (Levanova and Loresh, 2004; Resende and Werneck, 2004).

4.4 Other Crisp Clustering Methods

In addition to K-medoids and K-means, there are two other popular (although less so) criteria for hard partitioning problems: K-midranges and K-modes. While K-medoids is based on the L_1-norm, K-midranges and K-modes are based on the L_∞- and L_0-norms, respectively. Given their limited popularity, only brief description of these two criteria is provided.

4.4.1 *K*-Midranges

The *K*-midranges algorithm (Carroll and Chaturvedi, 1998; Späth, 1985) follows the same algorithmic process as the *K*-means method, the only difference being that *K*-midranges minimizes a loss function based on the L_∞-norm (also referred to as the 'max metric' and 'dominance metric', Borg and Groenen (1997)), replacing the *K*-medoids objective function with

$$\sum_{j=1}^{p}\sum_{k=1}^{K}\sum_{i \in \mathcal{C}_k} |x_{ij} - mid_j^{(k)}|, \tag{4.3}$$

where $mid_j^{(k)}$ is the midrange of the *j*th variable on the *k*th cluster and is calculated by the average of the two most extreme values for the *j*th variable in the *k*th cluster

$$mid_j^{(k)} = \frac{1}{2}(max(x_j^{(k)}) + min(x_j^{(k)})). \tag{4.4}$$

As with *K*-means and *K*-medoids, initial midranges are chosen and objects are assigned to the cluster with the closest midrange. Midranges are then recalculated and the observations are reassigned to clusters, with the process repeating (e.g., the midranges are fixed with observations reassigned, and then the midranges are recomputed) until the objective function can no longer be reduced. Like other crisp clustering criteria and algorithms, *K*-midranges is plagued by the local optima problem. Späth (1985) indicated that *K*-midranges is extremely sensitive to outliers (more so than *K*-means and *K*-medoids) and use in practical applications should be approached cautiously; however, Carroll and Chaturvedi (1998) posited that *K*-midranges might find use in detecting outlying clusters.

Further, from the objective function of *K*-midranges, it is clear that the most focus will be given to the maximum distance within a cluster. Where *K*-means clustering is attempting to minimize the distance of each observation within a cluster to its mean (and consequently having spherical clusters without within cluster outliers), *K*-midranges is attempting to place each observation close to the midrange on each variable within the cluster. This goal then translates into attempting to minimize, on average, the observed range of each cluster across a set of variables. This is closely related to the classic problem of minimum diameter partitioning (MDP), where the diameter of the cluster is the largest distance between any pair of points within the cluster.* However, this type of distance measure is fraught with its own set of problems. It is well known that locally optimal solutions plague *K*-means clustering (Steinley, 2003, 2006b); however, there is usually one unique globally optimal solution. In the context of MDP, Brusco and Steinely (2014) showed that there are numerous equivalent globally optimal solutions. Given the similarities between MDP and *K*-midranges, it is strongly suspected that *K*-midranges may suffer from the same indeterminacy.

4.4.2 *K*-Modes

K-modes clustering (Chaturvedi et al., 2001; Huang, 1998; Huang and Ng, 2003) is a *K*-means-like algorithm designed to derive clusters from original, unprocessed categor-

---

* The MDP problem has been traditionally approximated by complete linkage clustering; however, Brusco and Steinely (2014) provide a branch-and-bound procedure that improves on complete linkage clustering.

ical data. The K-modes algorithm uses the L_0-norm-based loss function, which is the limiting case of the L_p metric as p approaches zero, minimizing the distance between each observation and the mode of its parent cluster. If $\mathbf{M}_{N \times K}$ is a binary indicator matrix where $m_{ik} = 1$ if observation i is in cluster k and $\mathbf{Z}_{K \times P}$ matrix representing the K-modes for the K-clusters, the objective function can be written as

$$\sum_{i=1}^{N} \sum_{j=1}^{P} (x_{ij} - \sum_{k=1}^{K} w_{ik} z_{kj})^0, \tag{4.5}$$

where $(\cdot)^0$ is the "counting metric."

It has been suggested that K-modes has fewer problems with locally optimal solutions than the usual maximum likelihood approach to finding groups in binary data: latent class analysis (Chaturvedi et al., 2001). Furthermore, a small simulation study in the same article found that K-modes outperformed latent class analysis under a variety of conditions. This result mirrors numerous studies of K-means clustering outperforming finite mixture models (Steinley and Brusco, 2008b, 2011a).

4.5 Conclusion and Open Problems

Brusco and Köhn (2008a,b, 2009) have addressed many of the recent outstanding concerns by developing an enhanced optimal method for p-median clustering, an augmented simulated annealing heuristic that outperforms existing heuristics, and showing that recent developments in the biological sciences actually perform quite poorly in comparison to other known heuristics.

Given the the p-median clustering problem seems to have made great strides in "internal" advancement, the primary problem seems to be one of promoting wider adoption of the procedure among users. As Köhn et al. (2010) indicate, p-medians is little known in the behavioral and social sciences, with some of this lack of exposure likely due to the dearth of available software, although, this is expected to become less of a problem as R becomes more widely adopted in nonstatistics disciplines. Nonetheless, given recent advancements in efficiency in K-means clustering, broader comparisons between the two optimization procedures should be made—both in terms of the types of clusters they are designed to uncover and their computational efficiency.

Much focus has been given to K-means and p-median clustering; however, there has been little investigation into the performance or properties of K-midranges and K-modes clustering. As with the foundational simulation studies by Milligan (1980) for hierarchical clustering and Steinley (2006b) for K-means clustering, such broad simulation studies would be interesting for these other types of crisp clustering algorithms. In fact, one could expect numerous studies could be conducted, with a natural study being a comparison of latent class analysis and K-modes clustering that echoes Steinley and Brusco (2011a,b)'s recent comparison of K-means clustering and finite mixture modeling.

References

Alba, E. and E. Dominguez 2006. Comparative analysis of modern optimization tools for the p-median problem. *Statistics and Computing 16*, 251–260.

Aloise, D., P. Hansen, and L. Liberti 2012. An improved column generation algorithm for minimum sum-of-squares clustering. *Mathematical Programming 131*, 195–220.

Borg, I. and P. Groenen 1997. *Modern Multidimensional Scaling: Theory and Applications*. New York, NY: Springer.

Brusco, M. J. and H.-F. Köhn 2008a. Optimal partitioning of a data set based on the p-median model. *Psychometrika 73*, 89–105.

Brusco, M. J. and H.-F. Köhn 2008b. Comment on 'clustering by passing messages between data points'. *Science 319*, 726.

Brusco, M. J. and H.-F. Köhn 2009. Exemplar-based clustering via simulated annealing. *Psychometrika 74*, 457–475.

Brusco, M. J. and D. Steinley 2007. A comparison of heuristic procedures for minimum within-cluster sums of squares partitioning. *Psychometrika 72*, 583–600.

Brusco, M. J. and D. Steinely 2014. Model selection for minimum-diameter partitioning. *British Journal of Mathematical and Statistical Psychology 67*, 417–495.

Carroll, J. D. and A. Chaturvedi 1998. K-midranges clustering. In *Advances in Data Science and Classification*, A. Rizzi, M. Vichi, H.-H. Bock (eds.). pp. 3–14. Berlin: Springer.

Chaturvedi, A. D., P. E. Green, and J. D. Carroll 2001. K-modes clustering. *Journal of Classification 18*, 35–55.

Chiyoshi, F. and R. Galvao 2000. A statistical analysis of simulated annealing applied to the p-median problem. *Annals of Operations Research 96*, 61–74.

Cuesta-Albertos, J. A., A. Gordaliza, and C. Matran 1997. Trimmed k-means: An attempt to robustify quantizers. *Annals of Statistics 25*, 553–576.

Frey, B. and D. Dueck 2007. Clustering by passing messages between data points. *Science 315*, 972–976.

Garcia-Escudero, L. A. and A. Gordaliza 1999. Robustness of properties of k-means and trimmed k-means. *Journal of the American Statistical Association 94*, 956–969.

Hanjoul, P. and D. Peeters 1985. A comparison of two dual-based procedures for solving the p-median problem. *European Journal of Operational Research 20*, 387–396.

Hansen, P. and N. Mladenovic 1997. Variable neighborhood search for the p-median. *Location Science 5*, 207–226.

Hartigan, J. and M. Wong 1979. Algorithm as136: A k-means clustering program. *Applied Statistics 5*, 100–128.

Huang, Z. 1998. Extensions to the k-means algorithm for clustering large data sets with categorical values. *Data Mining and Knowledge Discovery 2*, 283–304.

Huang, Z. and M. K. Ng 2003. A note on k-modes clustering. *Journal of Classification 20*, 257–261.

Huber, P. J. 1981. *Robust statistics*. New York, NY: Wiley.

Kaufman, L. and P. Rousseeuw 1990. *Finding Groups in Data: An Introduction to Cluster Analysis*. New York, NY: Wiley.

Köhn, H.-F., D. Steinley, and M. J. Brusco 2010. The p-median model as a tool for clustering psychological data. *Psychological Methods 15*, 87–95.

Kuehn, A. and M. J. Hamburger 1963. A heuristic program for locating warehouses. *Management Science 9*, 643–666.

Levanova, T. and M. A. Loresh 2004. Algorithms of ant system and simulated annealing for the p-median problem. *Automation and Remote Control 65*, 431–438.

MacQueen, J. 1967. Some methods of classification and analysis of multivariate observations. In *Proceedings of the Fifth Berkeley Symposium on Mathematical Statistics and Probability*, Volume 1, pp. 281–297. Berkeley, CA: University of California Press.

Massart, D., F. Plastria, and L. Kaufman 1983. Non-hierarchical clustering with masloc. *Pattern Recognition 16*, 507–516.

Milligan, G. W. 1980. An examination of the effect of six types of error perturbation on fifteen clustering algorithms. *Psychometrika 45*, 325–342.

Mladenovic, N., J. Brimberg, P. Hansen, and J. Moreno-Pérez 2007. The p-median problem: A survey of metaheuristic approaches. *European Journal of Operational Research 179*, 927–939.

Mulvey, J. M. and H. P. Crowder 1979. Cluster analysis: An application of lagrangian relaxation. *Management Science 25*, 329–340.

Resende, M. G. C. and R. F. Werneck 2004. A hybrid heuristic for the p-median problem. *Journal of Heuristics 10*, 59–88.

Rolland, E., D. A. Schilling, and J. R. Current 1996. An efficient tabu search procedure for the p-median problem. *Journal of Operational Research 96*, 329–342.

Rousseeuw, P. J. 1987. Silhouettes: A graphical aid to the interpretation and validation of cluster analysis. *Computational and Applied Mathematics 20*, 53–65.

Späth, H. 1985. *Cluster Dissection and Analysis: Theory, FORTRAN Programs, Examples*. New York, NY: Wiley.

Steinhaus 1956. Sur la division des corps materials en parties. *Bulletin de l'Academie Polonaise des Sciences 12*, 801–804.

Steinley, D. 2003. Local optima in *k*-means clustering: What you don't know may hurt you. *Psychological Methods 8*, 294–304.

Steinley, D. 2004. Properties of the hubert-arabie adjusted rand index. *Psychological Methods 9*, 386–396.

Steinley, D. 2006a. K-means clustering: A half-century synthesis. *British Journal of Mathematical and Statistical Psychology 59*, 1–34.

Steinley, D. 2006b. Profiling local optima in k-means clustering: Developing a diagnostic technique. *Psychological Methods 11*, 178–192.

Steinley, D. and M. J. Brusco 2008a. A new variable weighting and selection procedure for k-means cluster analysis. *Multivariate Behavioral Research 43*, 77–108.

Steinley, D. and M. J. Brusco 2008b. Selection of variables in cluster analysis: An empirical comparison of eight procedures. *Psychometrika 73*, 125–144.

Steinley, D. and M. J. Brusco 2011a. Evaluating the performance of model-based clustering: Recommendations and cautions. *Psychological Methods 16*, 63–79.

Steinley, D. and M. J. Brusco 2011b. K-means clustering and model-based clustering: Reply to mclachlan and vermunt. *Psychological Methods 16*, 89–92.

Steinley, D. and M. J. Brusco 2011c. Testing for validity and choosing the number of clusters in k-means clustering. *Psychological Methods 16*, 285–297.

Steinley, D. and L. Hubert 2008. Order constrained solutions in *k*-means clustering: Even better than being globally optimal. *Psychometrika 73*, 647–664.

Teitz, M. B. and P. Bart 1968. Heuristic methods for estimating the generalized vertex median of a weighted graph. *Operations Research 16*, 955–961.

Thorndike, R. 1953. Who belongs in the family? *Psychometrika 18*, 267–276.

Vinod, H. D. 1969. Integer programming and the theory of grouping. *Journal of the American Statistical Association 64*, 506–519.

Ward, J. H. 1963. Hierarchical grouping to optimize an objective function. *Journal of the American Statistical Association 58*, 236–244.

5

Foundations for Center-Based Clustering: Worst-Case Approximations and Modern Developments

Pranjal Awasthi and Maria Florina Balcan

CONTENTS

Abstract

In the first part of this chapter, we detail center-based clustering methods, namely methods based on finding a "best" set of center points and then assigning data points to their nearest center. In particular, we focus on k-means and k-median clustering which are two of the most widely used clustering objectives. We describe popular heuristics for these methods and theoretical guarantees associated with them. We also describe how to design worst case approximately optimal algorithms for these problems. In the second part of the chapter, we describe recent work on how to improve on these worst case algorithms even further by using insights from the nature of real world clustering problems and datasets. Finally, we also summarize theoretical work on clustering data generated from mixture models such as a mixture of Gaussians.

5.1 Approximation Algorithms for k-Means and k-Median

One of the most popular approaches to clustering is to define an objective function over the data points and find a partitioning which achieves the optimal solution, or an approximately optimal solution to the given objective function. Common objective functions include center-based objective functions such as k-median and k-means where one selects k center points and the clustering is obtained by assigning each data point to its closest center point. Here closeness is measured in terms of a pairwise distance function $d()$, which the clustering algorithm has access to, encoding how dissimilar two data points are. For instance, the data could be points in Euclidean space with $d()$ measuring Euclidean distance, or it could be strings with $d()$ representing an edit distance, or some other dissimilarity score. For mathematical convenience, it is also assumed that the distance function $d()$ is a metric. In k-median clustering, the objective is to find center points $c_1, c_2, \ldots, c_k$, and a partitioning of the data so as to minimize $\mathcal{L}_{k-median} = \sum_x \min_i d(x, c_i)$. This objective is historically very useful and well studied for facility location problems (Jain et al. 2002; Arya et al. 2004). Similarly the objective in k-means is to minimize $\mathcal{L}_{k-means} = \sum_x \min_i d(x, c_i)^2$. Optimizing this objective is closely related to fitting the maximum likelihood mixture model for a given dataset. For a given set of centers, the optimal clustering for that set is obtained by assigning each data point to its closest center point. This is known as the Voronoi partitioning of the data. Unfortunately, exactly optimizing the k-median and the k-means objectives is a notoriously hard problem. Intuitively this is expected since the objective function is a non-convex function of the variables involved. This apparent hardness can also be formally justified by appealing to the notion of NP completeness (Jain et al. 2002; Dasgupta 2008; Aloise et al. 2009). At a high level the notion of NP completeness identifies a wide class of problems which are in principle equivalent to each other. In other words, an efficient algorithm for exactly optimizing one of the problems in the class on all instances would also lead to algorithms for all the problems in the class. This class contains many optimization problems that are believed to be hard* to exactly optimize in the worst case and not surprisingly, k-median and k-means also fall into the class. Hence it is unlikely that one would be able to optimize these objectives exactly using efficient

---

* This is the famous P vs NP problem, and there is a whole area called Computational Complexity Theory that studies this and related problems (Arora and Barak 2009).

algorithms. Naturally, this leads to the question of recovering approximate solutions and a lot of the work in the theoretical community has focused on this direction (Arora et al. 1999; Charikar et al. 1999; Jain et al. 2002; Kanungo et al. 2002; Fernandez de la Vega et al. 2003; Arya et al. 2004; Kumar et al. 2004; Ostrovsky et al. 2006; Balcan et al. 2009). Such works typically fall into two categories, (i) providing formal worst case guarantees on all instances of the problem, and (ii) providing better guarantees suited for nicer, stable instances. In this chapter we discuss several stepping stone results in these directions, focusing our attention on the k-means objective. A lot of the the ideas and techniques mentioned apply in a straightforward manner to the k-median objective as well. We will point out crucial differences between the two objectives as and when they appear. We will additionally discuss several practical implications of these results.

We will begin by describing a very popular heuristic for the k-means problem known as Lloyd's method. Lloyd's method (Lloyd 1982) is an iterative procedure which starts out with a set of k seed centers and at each step computes a new set of centers with a lower k-means cost. This is achieved by computing the Voronoi partitioning of the current set of centers and replacing each center with the center of the corresponding partition. We will describe the theoretical properties and limitations of Lloyd's method, which will also motivate the need for good worst case approximation algorithms for k-means and k-median. We will see that the method is very sensitive to the choice of the seed centers. Next, we will describe a general method based on local search which achieves constant factor approximations for both the k-means and the k-median objectives. Similar to Lloyd's method, the local search heuristic starts out with a set of k seed centers and at each step swaps one of the centers for a new one resulting in a decrease in the k-means cost. Using a clever analysis, it can be shown that this procedure outputs a good approximation to the optimal solution (Kanungo et al. 2002). This is interesting, since, as mentioned above, optimizing the k-means is NP-complete, in fact it is NP-complete even for $k = 2$, for points in the Euclidean space (Dasgupta 1999).*

In the second part of the chapter, we will describe some of the recent developments in the study of clustering objectives. These works take a non-worst case analysis approach to the problem. The basic theme is to design algorithms which give good solutions to clustering problems only when the underlying optimal solution has a meaningful structure. We will call such clustering instances as *stable* instances. We would describe in detail two recently studied notions of stability. The first one called *separability* was proposed by Ostrovsky et al. (2006). According to this notion a k-clustering instance is stable if it is much more expensive to cluster the data using $(k - 1)$ or fewer clusters. For such instances Ostrovsky et al. show that one can design a simple Lloyd's type algorithm which achieves a constant factor approximation. A different notion called *approximation stability* was proposed by Balcan et al. (2009). The motivation comes from the fact that often in practice optimizing an objective function acts as a proxy for the real problem of getting close to the correct unknown ground truth clustering. Hence it is only natural to assume that any good approximation to the proxy function such as k-means or k-median will also be close to the ground truth clustering in terms of structure. Balcan et al. show that under this assumption one can design algorithms that solve the end goal of getting close to the ground truth clustering. More surprisingly this is true even in cases where it is NP-hard to achieve a good approximation to the proxy objective.

---

* If one restricts centers to be data points, then k-means can be solved optimally in time $O(n^{k+1})$ by trying all possible k-tuples of centers and choosing the best. The difficulty of k-means for $k = 2$ in Euclidean space comes from the fact that the optimal centers need not be data points.

In the last part of the chapter, we briefly review existing theoretical work on clustering data generated from mixture models. We mainly focus on Gaussian mixture models (GMMs) which are the most widely studied distributional model for clustering. We will study algorithms for clustering data from a GMM under the assumption that the mean vectors of the component Gaussians are well separated. We will also see the effectiveness of spectral techniques for GMMs. Finally, we will look at recent work on estimating the parameters of a GMM under minimal assumptions.

5.2 Lloyd's Method for k-Means

Consider a set A of n points in the d-dimensional Euclidean space. We start by formally defining Voronoi partitions.

Definition 5.1 (Voronoi Partition)

Given a clustering instance $C \subset \mathbb{R}^d$ and k points $c_1, c_2, \ldots c_k$, a Voronoi partitioning using these centers consists of k disjoint clusters. Cluster i consists of all the points $x \in C$ satisfying $d(x, c_i) \leq d(x, c_j)$ for all $j \neq i$.*

Lloyd's method, also known as the k-means algorithm, is the most popular heuristic for k-means clustering in the Euclidean space which has been shown to be one of the top 10 algorithms in data mining (Wu et al. 2008). The method is an iterative procedure which is described below.

Algorithm LLOYD'S K-MEANS

1. **Seeding:** Choose k seed points $c_1, c_2, \ldots c_k$. Set $\mathcal{L}_{old} = \infty$. Compute the current k-means cost $\mathcal{L}$ using seed points as centers, i.e.

$$\mathcal{L}_{curr} = \sum_{i=1}^{n} \min_{j} d^2(x_i, c_j)$$

2. **While $\mathcal{L}_{curr} < \mathcal{L}_{old}$,**

   a. **Voronoi partitioning:** Compute the Voronoi partitioning of the data based on the centers $c_1, c_2, \ldots, c_k$. In other words, create k clusters $C_1, C_2, \ldots, C_k$ such that $C_i = \{x : d(x, c_i) \leq \min_{j \neq i} d(x, c_j)\}$. Break ties arbitrarily.

   b. **Reseeding:** Compute new centers $\hat{c}_1, \hat{c}_2, \ldots, \hat{c}_k$, where

   $$\hat{c}_j = \text{mean}(C_j) = 1/|C_j| \sum_{x \in C_j} x.$$

   Set $\mathcal{L}_{old} = \mathcal{L}_{curr}$. Update the current k-means cost $\mathcal{L}_{curr}$ using the new centers.

   c. **Output:** The current set of centers $c_1, c_2, \ldots, c_k$.

---

* Ties can be broken arbitrarily.

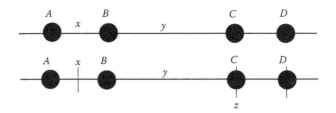

FIGURE 5.1
Consider four points $\{A, B, C, D\}$ on a line separated by distances x, y, and z such that $z < x < y$. Let $k = 3$. The optimal solution has centers at A, B and the centroid of C, D with a total cost of $z^2/2$. When choosing random seeds, there is a constant probability that we choose $\{A, C, D\}$. In this case, the final centers will be C, D and the centroid of A, B with a total cost of $x^2/2$. This ratio can be made arbitrarily bad.

We would like to stress that although Lloyd's method is popularly known as the k-means algorithm, there is a difference between the underlying k-means objective (which is usually hard to optimize) and the k-means algorithm which is a heuristic to solve the problem. An attractive feature of Lloyd's method is that the k-means cost of the clustering obtained never increases. This follows from the fact that for any set of points, the 1-means cost is minimized by choosing the mean of the set as the center. Hence for any cluster C_i in the partitioning, choosing $\text{mean}(C_i)$ will never lead to a solution of higher cost. Hence, if we repeat this method until there is no change in the k-means cost, we will reach a local optimum of the k-means cost function in finite time. In particular the number of iterations will be at most $n^{O(kd)}$ which is the maximum number of Voronoi partitions of a set of n points in $\mathbb{R}^d$ (Inaba et al. 1994). The basic method mentioned above leads to a class of algorithms depending upon the choice of the seeding method. A simple way is to start with k randomly chosen data points. This choice, however, can lead to arbitrarily bad solution quality as shown in Figure 5.1. In addition it is also known that the Lloyd's method can take upto 2^n iterations to converge even in two-dimensions (Arthur and Vassilvitskii 2006; Vattani 2009).

In sum, from a theoretical standpoint, k-means with random/arbitrary seeds is not a good clustering algorithm in terms of efficiency or quality. Nevertheless, the speed and simplicity of k-means are quite appealing in practical applications. Therefore, recent work has focused on improving the initialization procedure: deciding on a better way to initialize the clustering dramatically changes the performance of the Lloyd's iteration, both in terms of quality and convergence properties. For example, Arthur and Vassilvitskii (2007) showed that choosing a good set of seed points is crucial and if done carefully can itself be a good candidate solution without the need for further iterations. Their algorithm called KMEANS++ uses the following seeding procedure: it selects only the first center uniformly at random from the data and each subsequent center is selected with a probability proportional to its contribution to the overall error given the previous selections. See Algorithm KMEANS++ for a formal description:

Arthur and Vassilvitskii (2007) showed that Algorithm KMEANS++ is a $\log k$ approximation algorithm for the k-means objective. We say that an algorithm is an α-approximation for a given objective function $\mathcal{L}$ if for every clustering instance the algorithm outputs a solution of expected cost at most α times the cost of the best solution. The design of approximation algorithms for NP-hard problems has been a fruitful research direction and has led to a wide array of tools and techniques. Formally, in Theorem 5.1, Arthur and Vassilvitskii (2007) show that:

Algorithm KMEANS++

1. **Initialize:** A set S by choosing a data point at random.
2. **While** $|S| < k$,

   a. Choose a data point x with probability proportional to $min_{z \in S} d(x, z)^2$, and add it to S.

3. **Output:** The clustering obtained by the Voronoi partitioning of the data using the centers in S.

Theorem 5.1 (Arthur and Vassilvitskii 2007)

Let S be the set of centers output by the above algorithm and $\mathcal{L}(S)$ be the k-means cost of the clustering obtained using S as the centers. Then $E[\mathcal{L}(S)] \leq O(\log k)\mathcal{L}^*$, where $\mathcal{L}^*$ is the cost of the optimal k-means solution. ∎

We would like to point out that in general the output of k-means++ is not a local optimum. Hence it might be desirable in practice to run a few steps of the Lloyd's method starting from this solution. This could only lead to a better solution.

Subsequent work of Ailon et al. (2009) introduced a streaming algorithm inspired by the k-means++ algorithm that makes a single pass over the data. They show that if one is allowed to cluster using a little more than k centers, specifically $O(k \log k)$ centers, then one can achieve a constant-factor approximation in expectation to the k-means objective. The approximation guarantee was improved in Aggarwal et al. (2009). Such approximation algorithms which use more than k centers are also known as bi-criteria approximations.

As mentioned earlier, Lloyd's method can take up to exponential iterations in order to converge to a local optimum. However Arthur et al. (2011) showed that the method converges quickly on an "average" instance. In order to formalize this, they study the problem under the smoothed analysis framework of (Spielman and Teng 2004). In the smoothed analysis framework, the input is generated by applying a small Gaussian perturbation to an adversarial input. Spielman and Teng (2004) showed that the simplex method takes polynomial number of iterations on such smoothed instances. In a similar spirit, Arthur et al. (2011) showed that for smoothed instances Lloyd's method runs in time polynomial in n, the number of points and $\frac{1}{\sigma}$, the standard deviation of the Gaussian perturbation. However, these works do not provide any guarantee on the quality of the final solution produced.

We would like to point out that in principle the Lloyd's method can be extended to the k-median objective. A natural extension would be to replace the mean computation in the Reseeding step with computing the median of a set of points X in the Euclidean space, that is, a point $c \in \mathbb{R}^d$ such that $\sum_{x \in X} d(x, c)$ is minimized. However, this problem turns out to be NP-complete (Megiddo and Supowit 1984). For this reason, the Lloyd's method is typically used only for the k-means objective.

5.3 Properties of the k-Means Objective

In this section we provide some useful facts about the k-means clustering objective. We will use $\mathcal{C}$ to denote the set of n points which represent a clustering instance. The first fact can be used to show that given a Voronoi partitioning of the data, replacing a given center with the mean of the corresponding partition can never increase the k-means cost Fact 5.1.

Fact 5.1

Consider a finite set $X \subset \mathbb{R}^d$ and $c = \text{mean}(X)$. For any $y \in \mathbb{R}^d$, we have that, $\sum_{x \in X} d(x, y)^2 = \sum_{x \in X} d(x, c)^2 + |X| d(c, y)^2$.

Proof. Representing each point in the coordinate notation as $x = (x_1, x_2, \ldots, x_d)$, we have that

$$\sum_{x \in X} d(x, y)^2 = \sum_{x \in X} \sum_{i=1}^{d} |x_i - y_i|^2$$

$$= \sum_{x \in X} \sum_{i=1}^{d} (|x_i - c_i|^2 + |c_i - y_i|^2 + 2(x_i - c_i)(c_i - y_i))$$

$$= \sum_{x \in X} d(x, c)^2 + |X| d(c, y)^2 + \sum_{i=1}^{d} 2(c_i - y_i) \sum_{x \in X} (x_i - c_i)$$

$$= \sum_{x \in X} d(x, c)^2 + |X| d(c, y)^2$$

Here the last equality follows from the fact that for any i, $c_i = \sum_{x \in X} x_i / n$. ■

An easy corollary of the above fact is the following:

Corollary 5.1

Consider a finite set $X \subset \mathbb{R}^d$ and let $c = \text{mean}(X)$. We have $\sum_{x, y \in X} d(x, y)^2 = 2|X| \sum_{x \in X} d(x, c)^2$.

Below we prove another fact which will be useful later.

Fact 5.2

Let $X \subset \mathbb{R}^d$ be finite set of points. Let $\mathcal{L}_1(X)$ denote the 1-means cost of X. Given a partition of X into X_1 and X_2 such that $c = \text{mean}(X)$, $c_1 = \text{mean}(X_1)$ and $c_2 = \text{mean}(X_2)$, we have that a) $\mathcal{L}_1(X) = \mathcal{L}_1(X_1) + \mathcal{L}_1(X_2) + \frac{|X_1||X_2|}{|X|} d(c_1, c_2)^2$ and b) $d(c, c_1)^2 \leq \frac{\mathcal{L}_1(X)|X_2|}{|X||X_1|}$.

Proof. We can write $\mathcal{L}_1(X) = \sum_{x \in X_1} d(x, c)^2 + \sum_{x \in X_2} d(x, c)^2$. Using Fact 5.1, we can write

$$\sum_{x \in X_1} d(x, c)^2 = \mathcal{L}_1(X_1) + |X_1| d(c, c_1)^2$$

Similarly, $\sum_{x \in X_2} d(x, c)^2 = \mathcal{L}_1(X_2) + |X_2| d(c, c_2)^2$. Hence we have

$$\mathcal{L}_1(X) = \mathcal{L}_1(X_1) + \mathcal{L}_1(X_2) + |X_1| d(c, c_1)^2 + |X_2| d(c, c_2)^2$$

Part (a) follows by substituting $c = \frac{|X_1|c_1 + |X_2|c_2}{|X_1| + |X_2|}$ in the above equation.

From Part (a) we have that

$$\mathcal{L}_1(X) \geq \frac{|X_1||X_2|}{|X|} d(c_1, c_2)^2$$

Part (b) follows by substituting $c_2 = \frac{(|X_1| + |X_2|)}{X_2} c - \frac{|X_1|}{|X_2|} c_1$ above.   ■

5.4 Local Search-Based Algorithms

In the previous section, we saw that a carefully chosen seeding can lead to a good approximation for the k-means objective. In this section, we will see how to design much better (constant factor) approximation algorithms for k-means (as well as k-median). We will describe a very generic approach based on local search. These algorithms work by making local changes to a candidate solution and improving it at each step. They have been successfully used for a variety of optimization problems (Hansen and Jaumard 1990; Papadimitriou 1991; Chandra et al. 1994; Alimonti 1995; Drake and Hougardy 2003; Schuurman and Vredeveld 2007). Kanungo et al. (2002) analyzed a simple local search-based algorithm for k-means as described below.

Algorithm K-MEANS LOCAL SEARCH

1. **Initialization:** Choose k data points $\{c_1, c_2, \ldots, c_k\}$ arbitrarily from the dataset $\mathcal{D}$. Let this set be T. Let $\mathcal{L}(T)$ denote the cost of the k-means solution using T as centers, i.e., $\mathcal{L}(T) = \sum_{i=1}^{n} \min_j d^2(x_i, c_j)$. Set $T_{old} = \phi$, $T_{curr} = T$.
2. **While** $\mathcal{L}(T_{curr}) < \mathcal{L}(T_{old})$,
   a. For $x \in T_{curr}$ and $y \in \mathcal{D} \setminus T_{curr}$:
      if $\mathcal{L}((T_{curr} \setminus \{x\}) \cup \{y\}) < \mathcal{L}(T_{curr})$, update $T_{old} = T_{curr}$ and $T_{curr} \leftarrow (T_{curr} \setminus \{x\}) \cup \{y\}$.
3. **Output:** $S = T_{curr}$ as the set of final centers.

We would like to point out that in order to make the above algorithm run in polynomial time, one needs to change the criteria in the while loop to be $\mathcal{L}(T_{curr}) < (1 - \epsilon)\mathcal{L}(T_{old})$. The running time will then depend polynomially in n and $1/\epsilon$. For simplicity of analysis, we will prove the following theorem for the idealized version of the algorithm with no ϵ.

Theorem 5.2 (Kanungo et al. 2002)

Let S be the final set of centers returned by the above procedure. Then, $\mathcal{L}(S) \leq 50\mathcal{L}^*$.   ■

In order to prove the above theorem, we start by building up some notation. Let T be the set of k data points returned by the local search algorithm as candidate centers. Let O be the set of k data points which achieve the minimum value of the k-means cost function among all sets of k data points. Note that the centers in O do not necessarily represent the optimal solution as the optimal centers might not be data points. However, using the next lemma one can show that using data points as centers is only twice as bad as the optimal solution.

Lemma 5.1

Given $C \subseteq \mathbb{R}^d$, and the optimal k-means clustering of C, $\{C_1, C_2, \ldots, C_k\}$, there exists a set S of k data points such that $\mathcal{L}(S) \leq 2\mathcal{L}^*$.

Proof. For a given set $C \subseteq \mathbb{R}^d$, let $\mathcal{L}_1$ represent the 1-means cost of C. From Fact 5.1, it is easy to see that this cost is achieved by choosing the mean of C as the center. In order to prove the above lemma it is enough to show that, for each optimal cluster C_i with mean c_i, there exists a data point $x_i \in C_i$ such that $\sum_{x \in C_i} d(x, x_i)^2 \leq 2\mathcal{L}_1(C_i)$. Let x_i be the data point in C_i which is closest to c_i. Again using Fact 5.1, we have $\sum_{x \in C_i} d(x, x_i)^2 = \mathcal{L}_1(C_i) + |C_i| d(x, c_i)^2 \leq 2\mathcal{L}_1(C_i)$. ∎

Hence it is enough to compare the cost of the centers returned by the algorithm to the cost of the optimal centers using data points. In particular, we will show that $\mathcal{L}(T) \leq 25\mathcal{L}(O)$. We start with the simple observation that by the property of the local search algorithm, for any $t \in T$, and $o \in O$, swapping t for o results in an increase in cost. In other words

$$\mathcal{L}(T - t + o) - \mathcal{L}(T) \geq 0 \tag{5.1}$$

The main idea is to add up Equation 5.1 over a carefully chosen set of swaps $\{o, t\}$ to get the desired result. In order to describe the set of swaps chosen, we start by defining a cover graph.

Definition 5.2

A cover graph is a bipartite graph with the centers in T on one side and the centers in O on the other side. For each $o \in O$, let t_o be the point in T which is closest to o. The cover graph contains edges of the form o, t_o for all $o \in O$ as shown in Figure 5.2.

Next we use the cover graph to generate the set of useful swaps. For each $t \in T$ which has degree 1 in the cover graph, we output the swap pair $\{t, o\}$ where o is the point connected to t. Let T' be the degree 0 vertices in the cover graph. We pair the remaining vertices $o \in O$ with the vertices in T' such that each vertex in O has degree 1 and each vertex in T' has degree at most 2. To see that such a pairing will exist notice that for any $t \in T$ of degree $k > 1$, there will exist $k - 1$ distinct zero vertices in T; these vertices can be paired to vertices in O connected to t maintaining the above property. We then output all the edges in this pairing as the set of useful swaps.

FIGURE 5.2
An example cover graph.

5.4.1 Bounding the Cost of a Swap

Consider a swap $\{o, t\}$ output by using the cover graph. We will apply Equation 5.1 to this pair. We will explicitly define a clustering using centers in $T - t + o$ and upper bound its cost. We will then use the lower bound of $\mathcal{L}(T)$ from Equation 5.1 to get the kind of equations we want to sum up over. Let the clustering given by centers in T be $C_1, C_2, \ldots, C_k$. Let C_o^* be the cluster corresponding to center o in the optimal clustering given by O. Let o_x be the closest point in O to x. Similarly let t_x be the closest point in T to x. The key property satisfied by any output pair $\{o, t\}$ is the following.

Fact 5.3

Let $\{o, t\}$ be a swap pair output using the cover graph. Then we have that for any $x \in C_t$ either $o_x = o$ or $t_{o_x} \neq t$.

Proof. Assume that for some $x \in C_t$, $o_x = o' \neq o$. By the procedure used to output swap pairs, we have that t has degree 1 or 0 in the cover graph. In addition, if t has degree 1 then $t_o = t$. In both the cases, we have that $t_{o'} \neq t$. ∎

Next we create a new clustering by swapping o for t and assigning all the points in C_o^* to o. Next we reassign points in $C_t \setminus C_o^*$. Consider a point $x \in C_t \setminus C_o^*$. Clearly $o_x \neq o$. Let t_{o_x} be the point in T which is connected to o_x in the cover graph. We assign x to t_{o_x}. One needs to ensure here that $t_{o_x} \neq t$ which follows from Fact 5.3. From Equation 5.1, the increase in cost due to this reassignment must be non-negative. In other words, we have

$$\sum_{x \in C_o^*} (d(x, o)^2 - d(x, t_x)^2) + \sum_{x \in C_t \setminus C_o^*} (d(x, t_{o_x})^2 - d(x, t)^2) \geq 0 \tag{5.2}$$

We will add up Equation 5.2 over the set of all good swaps.

5.4.2 Adding It All Up

In order to sum up over all swaps notice that in the first term in Equation 5.2 every point $x \in C$ appears exactly once by being in C_o^* for some $o \in O$. Hence, the sum over all swaps of the first term can be written as $\sum_{x \in C}(d(x, o_x)^2 - d(x, t_x)^2)$. Consider the second term in Equation 5.2. We have that $(d(x, t_{o_x})^2 - d(x, t)^2) \geq 0$ since x is in C_t. Hence, we can replace the second summation over all $x \in C_t$ without affecting the inequality. Also every point $x \in C$ appears at most twice in the second term by being in C_t for some $t \in T$. Hence, the sum over all swaps of the second term is at most $\sum_{x \in C}(d(x, t_{o_x})^2 - d(x, t_x)^2)$. Adding these up and rearranging, we get that

$$\mathcal{L}(O) - 3\mathcal{L}(T) + 2R \geq 0 \tag{5.3}$$

Here $R = \sum_{x \in C} d(x, t_{o_x})^2$.

In the last part we will upper bound the quantity R. R represents the cost of assigning every point x to a center in T but not necessarily the closest one. Hence, $R \geq \mathcal{L}(T) \geq \mathcal{L}(O)$. However, we next show that this reassignment cost is not too large.

Notice that R can also be written as $\sum_{o \in O} \sum_{x \in C_o^*} d(x, t_o)^2$. Also $\sum_{x \in C_o^*} d(x, t_o)^2 = \sum_{x \in C_o^*} d(x, o)^2 + |C_o^*| d(o, t_o)^2$. Hence, we have that $R = \sum_{o \in O} \sum_{x \in C_o^*} (d(x, o)^2 + d(o, t_o)^2)$. Also note that $d(o, t_o) \leq d(o, t_x)$ for any x. Hence

$$R \leq \sum_{o \in O} \sum_{x \in C_o^*} (d(x, o)^2 + d(o, t_x)^2)$$

$$= \sum_{x \in C} (d(x, o_x)^2 + d(o_x, t_x)^2)$$

Using triangle inequality we know that $d(o_x, t_x) \leq d(o_x, x) + d(x, t_x)$. Substituting above and expanding, we get that

$$R \leq 2\mathcal{L}(O) + \mathcal{L}(T) + 2 \sum_{x \in C} d(x, o_x) d(x, t_x) \tag{5.4}$$

The last term in the above equation can be bounded using Cauchy–Schwarz inequality as $\sum_{x \in C} d(x, o_x) d(x, t_x) \leq \sqrt{\mathcal{L}(O)} \sqrt{\mathcal{L}(S)}$. So we have that $R \leq 2\mathcal{L}(O) + \mathcal{L}(T) + 2\sqrt{\mathcal{L}(O)}\sqrt{\mathcal{L}(S)}$. Substituting this in Equation 5.3 and solving we get the desired result that $\mathcal{L}(T) \leq 25\mathcal{L}(O)$. Combining this with Lemma 5.1 proves Theorem 5.2.

A natural generalization of Algorithm K-MEANS LOCAL SEARCH is to swap more than one centers at each step. This could potentially lead to a much better local optimum. This multiswap scheme was analyzed by Kanungo et al. (2002) and using a similar analysis as above one can show the following.

Theorem 5.3

Let S be the final set of centers by the local search algorithm which swaps upto p centers at a time. Then we have that $\mathcal{L}(S) \leq 2(3 + \frac{2}{p})^2 \mathcal{L}^*$, where $\mathcal{L}^*$ is the cost of the optimal k-means solution. ∎

For the case of k-median, the same algorithm and analysis gives Arya et al. (2004).

Theorem 5.4

Let S be the final set of centers by the local search algorithm which swaps upto p centers at a time. Then we have that $\mathcal{L}(S) \leq (3 + \frac{2}{p})\mathcal{L}^*$, where $\mathcal{L}^*$ is the cost of the optimal k-median solution. ∎

This approximation factor for k-median has recently been improved to $(1 + \sqrt{3} + \epsilon)$ (Li and Svensson 2013). For the case of k-means in Euclidean space (Kumar et al. 2004) give an algorithm which achieves a $(1 + \epsilon)$ approximation to the k-means objective for any constant $\epsilon > 0$. However, the runtime of the algorithm depends exponentially in k and hence it is only suitable for small instances.

5.5 Clustering of *Stable* Instances

In this part of the chapter, we delve into some of the more modern research in the theory of clustering. In recent past, there has been an increasing interest in designing clustering algorithms that enjoy strong theoretical guarantees on non-worst case instance. This is of significant interest for two reasons: (i) From a theoretical point of view, this helps us understand and characterize the class of problems for which one can get optimal or close to optimal guarantees, (ii) From a practical point of view, real world instances often have additional structure that could be exploited to get better performance.

Compared to worst case analysis, the main challenge here is to formalize well motivated and interesting additional structures of clustering instances under which good algorithms exist. In this section, we present two popular interesting notions.

5.5.1 ϵ-Separability

This notion of stability was proposed by Ostrovsky et al. (2006). Given an instance of k-means clustering, let $\mathcal{L}^*(k)$ denote the cost of the optimal k-means solution. We can also decompose $\mathcal{L}^*(k)$ as $\mathcal{L}^* = \sum_{i=1}^{k} \mathcal{L}_i^*$, where $\mathcal{L}_i^*$ denotes the 1-means cost of cluster C_i, i.e., $\sum_{x \in C_i} d(x, c_i)^2$. Such an instance is called ϵ-*separable* if it satisfies $\mathcal{L}^*(k-1) > \frac{1}{\epsilon^2} \mathcal{L}^*(k)$.

The definition is motivated by the following issue: when approaching a clustering problem, one typically has to decide how many clusters one wants to partition the data in, that is, the value of k. If the k-means objective is the underlying criteria being used to judge the quality of a clustering, and the optimal $(k-1)$-means clustering is comparable to the optimal k-means clustering, then one can in principle also use $(k-1)$ clusters to describe the dataset. In fact this particular method is a very popular heuristic to find out the number of hidden clusters in the dataset. In other words choose the value of k at which there is a significant increase in the k-means cost when going from k to $k-1$. As an illustrative example, consider the case of a mixture of k spherical unit variance Gaussians in d dimensions whose pairwise means are separated by a distance $D \gg 1$. Given n points from each Gaussian, the optimal k-means cost with high probability is nkd. On the other hand, if we try to cluster this data using $(k-1)$ clusters, the optimal cost will now become $n(k-1)d + n(D^2 + d)$. Hence, taking the ratio of the two costs, this instance will be ϵ-separable for $\frac{1}{\epsilon^2} = \frac{(k-1)d + D^2 + d}{kd} = 1 + (\frac{D^2}{kd})$, so $\epsilon = (1 + \frac{D^2}{kd})^{-1/2}$. Hence, if $D \gg \sqrt{kd}$, then the instance will be highly separable (the separability parameter ϵ will be $o(1)$).

It was shown by Ostrovsky et al. (2006) that one can design much better approximation algorithms for ϵ-separable instances.

Theorem 5.5 (Ostrovsky et al. 2006)

An algorithm that runs in polynomial time which given any ϵ-separable 2-means instance returns a clustering of cost at most $\frac{\mathcal{L}^*}{1-\rho}$ with probability at least $1 - O(\rho)$ where $c_2 \epsilon^2 \leq \rho \leq c_1 \epsilon^2$ for some constants $c_1, c_2 > 0$.   ∎

Theorem 5.6 (Ostrovsky et al. 2006)

There is a polynomial time algorithm which given any ϵ-separable k-means instance a clustering of cost at most $\frac{\mathcal{L}^*}{1-\rho}$ with probability $1 - O((\rho)^{1/4})$ where $c_2 \epsilon^2 \leq \rho \leq c_1 \epsilon^2$ for some constants $c_1, c_2 > 0$.   ∎

5.5.2 Proof Sketch and Intuition for Theorem 5.5

Notice that the above algorithm does not need to know the value of ϵ from the separability of the instance. Define r_i to be the radius of cluster C_i in the optimal k-means clustering, i.e., $r_i^2 = \frac{\mathcal{L}_i^*}{|C_i|}$. The main observation is that under the ϵ-separability condition, the optimal k-means clustering is "spread out." In other words, the radius of any cluster is much smaller than the inter cluster distances. This can be formulated in the following lemma.

Lemma 5.2

$\forall i, j, d(c_i, c_j)^2 \geq \frac{1-\epsilon^2}{\epsilon^2} max(r_i^2, r_j^2)$.

Proof. Given an ϵ-separable instance of k-means, consider any two clusters C_i and C_j in the optimal clustering with centers c_i and c_j, respectively. Consider the $(k-1)$ clustering obtained by deleting c_j and assigning all the points in C_j to C_i. By ϵ-separability, the cost of this new clustering must be at least $\frac{\mathcal{L}^*}{\epsilon^2}$. However, the increase in the cost will be exactly $|C_j| d(c_i, c_j)^2$. This follows from the simple observation stated in Fact 5.1. Hence, we have that $|C_j| d(c_i, c_j)^2 > (\frac{1}{\epsilon^2} - 1)\mathcal{L}^*$. This gives us that $r_j^2 = \frac{\mathcal{L}^*}{|C_j|} \leq \frac{\epsilon^2}{1-\epsilon^2} d(c_i, c_j)^2$. Similarly, if we delete c_i and assign all the points in $C_i - C_j$, we get that $r_i^2 \leq \frac{\epsilon^2}{1-\epsilon^2} d(c_i, c_j)^2$. ∎

When dealing with the two means problem, if one could find two initial candidate center points which are close to the corresponding optimal centers, then we could hope to run a Lloyd's type step and improve the solution quality. In particular if we could find $\bar{c}_1$ and $\bar{c}_2$ such that $d(c_1, \bar{c}_1)^2 \leq \alpha r_1^2$ and $d(c_2, \bar{c}_2)^2 \leq \alpha r_2^2$, then we know from Fact 5.1 that using these center points will give us a $(1+\alpha)$ approximation to $\mathcal{L}^*$. Lemma 5.2 suggests the following approach: pick data points x, y with probability proportional to $d(x,y)^2$. We will show that this will lead to seed points $\hat{c}_1$ and $\hat{c}_2$ not too far from the optimal centers. Applying a Lloyd type reseeding step will then lead us to the final centers which will be much closer to the optimal centers. We start by defining the core of a cluster.

Definition 5.3 (Core of a cluster)

Let $\rho < 1$ be a constant. We define $X_i = \{x \in C_i : d(x, c_i)^2 \leq \frac{r_i^2}{\rho}\}$. We call X_i as the core of the cluster C_i.

We next show that if we pick initial seeds $\{\hat{c}_1, \hat{c}_2\} = \{x, y\}$ with probability proportional to $d(x,y)^2$ then with high probability the points lie within the core of different clusters.

Lemma 5.3

For sufficiently small ϵ and $\rho = \frac{100\epsilon^2}{1-\epsilon^2}$, we have $Pr[\{\hat{c}_1, \hat{c}_2\} \cap X_1 \neq \emptyset$ and $\{x, y\} \cap X_2 \neq \emptyset] = 1 - O(\rho)$.

Proof Sketch. For simplicity assume that the sizes of the two clusters are the same, that is, $|C_i| = |C_j| = n/2$. In this case, we have $r_1^2 = r_2^2 = \frac{2\mathcal{L}^*}{n} = r^2$. Also, let $d^2(c_1, c_2) = d^2$. From

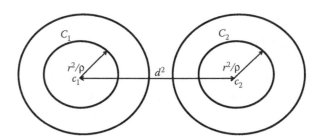

FIGURE 5.3
An ϵ-separable 2-means instance.

ϵ-separability, we know that $d^2 > \frac{1-\epsilon^2}{\epsilon^2}r^2$. Also, from the definition of the core, we know that at least $(1 - \rho)$ fraction of the mass of each cluster lies within the core. Hence, the clustering instance looks like the one showed in Figure 5.3. Let $A = \sum_{x \in X_1, y \in X_2} d(x, y)^2$ and $B = \sum_{x, y \in \mathcal{C}} d(x, y)^2$. Then the probability of the event is exactly $\frac{A}{B}$. Let's analyze quantity B first. The proof goes by arguing that the pairwise distances between X_1 and X_2 will dominate B. This is because of Lemma 5.2 which says that d^2 is much greater than r^2, the average radius of a cluster. More formally, from Corollary 5.1 and from Fact 5.2, we can get that $B = n\mathcal{L}_1(\mathcal{C}) = n\mathcal{L}_2(\mathcal{C}) + n^2/4d^2$. In addition ϵ-separability tells us that $\mathcal{L}_1(\mathcal{C}) > 1/\epsilon^2\mathcal{L}_2(\mathcal{C})$. Hence we get that $B \leq \frac{n^2}{4(1-\epsilon^2)}d^2$.

Let's analyze $A = \sum_{x \in X_1, y \in X_2} d(x, y)^2$. From triangle inequality, we have that for any $x \in X_1$, $y \in X_2$, $d^2(x, y) \geq (d - 2r/\sqrt{\rho})^2$. Hence $A \geq \frac{1}{4}(1 - \rho)^2 n^2 (d - 2r/\sqrt{\rho})^2$. Substituting these bounds and using the fact that $\rho O(\epsilon^2)$ gives us that $A/B \geq (1 - O(\rho))$.   ∎

Using these initial seeds we now show that a single step of a Lloyd's type method can yield good a solution. Define $r = d(\hat{c}_1, \hat{c}_2)/3$. Define $\bar{c}_1$ as the mean of the points in $B(\hat{c}_1, r)$ and $\bar{c}_2$ as the mean of the points in $B(\hat{c}_2, r)$. Notice that instead of taking the mean of the Voronoi partition corresponding to $\hat{c}_1$ and $\hat{c}_2$, we take the mean of the points within a small radius of the given seeds.

Lemma 5.4

Given $\hat{c}_1 \in X_1$ and $\hat{c}_2 \in X_2$, the clustering obtained using $\bar{c}_1$ and $\bar{c}_2$ as centers has 2-means cost at most $\frac{\mathcal{L}^*}{1-\rho}$.

Proof. We will first show that $X_1 \subseteq B(\hat{c}_1, r) \subseteq C_1$. Using Lemma 5.2 we know that $d(\hat{c}_1, c_1) \leq \frac{\epsilon}{\rho(1-\epsilon^2)}d(c_1, c_2) \leq d(c_1, c_2)/10$ for sufficiently small ϵ. Similarly $d(\hat{c}_2, c_2) \leq d(c_1, c_2)/10$. Hence, we get that $4/5 \leq r \leq 6/5$. So for any $z \in B(\hat{c}_1, r)$, $d(z, c_1) \leq d(c_1, c_2)/2$. Hence $z \in C_1$. Also for any $z \in X_1$, $d(z, \hat{c}_1) \leq 2\frac{r_1^2}{\rho} \leq r$. Similarly one can show that $X_2 \subseteq B(\hat{c}_2, r) \subseteq C_2$. Now applying Fact 5.2, we can claim that $d(\bar{c}_1, c_1) \leq \frac{\rho}{1-\rho}r_1^2$ and $d(\bar{c}_2, c_2) \leq \frac{\rho}{1-\rho}r_2^2$. So using $\bar{c}_1$ and $\bar{c}_2$ as centers, we get a clustering of cost at most $\mathcal{L}^* + \frac{\rho}{1-\rho}\mathcal{L}^* = \frac{\mathcal{L}^*}{1-\rho}$.   ∎

Summarizing the discussion above, we have the following simple algorithm for the 2-means problem.

Algorithm 2-MEANS

1. **Seeding:** Choose initial seeds x, y with probability proportional to $d(x, y)^2$.
2. Given seeds $\hat{c}_1, \hat{c}_2$, let $r = d(\hat{c}_1, \hat{c}_2)/3$. Define $\bar{c}_1 = \text{mean}(B(\hat{c}_1, r))$ and $\bar{c}_2 = \text{mean}(B(\hat{c}_2, r))$.
3. **Output:** $\bar{c}_1$ and $\bar{c}_2$ as the cluster centers.

5.5.3 Proof Sketch and Intuition for Theorem 5.6

In order to generalize the above argument to the case of k clusters, one could follow a similar approach and start with k initial seed centers. Again, we start by choosing x, y with probability proportional to $d(x, y)^2$. After choosing a set of U of points, we choose the next point z with probability proportional to $min_{\hat{c}_i \in U} d(z, \hat{c}_i)^2$. Using a similar analysis as in Lemma 5.3, one can show that if we pick k seeds then with probability $(1 - O(\rho))^k$ they will lie with the cores of different clusters. However, this probability of success is exponentially small in k and is not good for our purpose. The approach taken in Ostrovsky et al. (2006) is to sample a larger set of points and argue that with high probability it is going to contain k seed points from the "outer" cores of different clusters. Here, we define outer core of a cluster as $X_i{}^{out} = \{x \in C_i : d(x, c_i)^2 \leq \frac{r_i^2}{\rho^3}\}$—so this notion is similar to the core notion for $k = 2$ except that the radius of the core is bigger by a factor of $1/(\rho)$ than before. We would like to again point out a similar seeding procedure as the one described above is used in the k-means++ algorithm (Arthur and Vassilvitskii 2007) (see Section 5.2). One can show that using k seed centers in this way gives an $O(\log(k))$-approximation to the k-means objective in the worst case.

Lemma 5.5 (Ostrovsky et al. 2006)

Let $N = \frac{2k}{1-5\rho} + \frac{2\ln(2/\delta)}{(1-5\rho)^2}$, where $\rho = \sqrt{\epsilon}$. If we sample N points using the sampling procedure, then $Pr[\forall j = 1 \cdots k$, there exists some $\hat{x}_i \in X_j{}^{out}] \geq 1 - \delta$.

Since we sample more than k points in the first step, one needs to extract k good seed points out of this set before running the Lloyd step. This is achieved by the following greedy procedure:

Algorithm GREEDY DELETION K-MEANS INITIALIZATION

1. Let S denote the current set of candidate centers. Let $\mathcal{L}(S)$ denote the k-means cost of the Voronoi partition using S. Similarly, for $x \in S$ denote $\mathcal{L}(S_x)$ be the k-means cost of the Voronoi partition using $S \setminus \{x\}$.
2. **While** $|S| > k$,
   a. Remove a point x from S, such that $\mathcal{L}(S_x) - \mathcal{L}(S)$ is minimum.
   b. fOR every remaining point $x \in S$, let $R(x)$ denote the Voronoi set corresponding to x. Replace x by $mean(R(x))$.
   c. **Output:** S.

At the end of the greedy procedure, we have the following guarantee.

Lemma 5.6

For every optimal center c_i, there is a point $\hat{c}_i \in S$, such that $d(c_i, \hat{c}_i) \leq \frac{D_i}{10}$. Here $D_i = min_{j \neq i} d(c_i, c_j)$.

Using the above lemma and applying the same Lloyd step as in the 2-means problem, we get a set of k good final centers. These centers have the property that for each i, $d(c_i, \bar{c}_i) \leq \frac{\rho}{1-\rho} r_i^2$. Putting the above argument formally, we get the desired result.

5.5.4 Approximation Stability

In Balcan et al. (2009) introduce and analyze a class of approximation stable instances for which they provide polynomial time algorithms for finding accurate clustering. The starting point of this work is that for many problems of interest to machine learning, such as as clustering proteins by function, images by subject, or documents by topic, there is some unknown correct "target" clustering. In such cases the implicit hope when pursuing an objective-based clustering approach (k-means or k-median) is that approximately optimizing the objective function will in fact produce a clustering of low clustering error, that is, a clustering that is point wise close to the target clustering. Balcan et al. have shown that by making this implicit assumption explicit, one can efficiently compute a low-error clustering even in cases when the approximation problem of the objective function is NP-complete! This is quite interesting since it shows that by exploiting the properties of the problem at hand one can solve the desired problem and bypass worst case hardness results. A similar stability assumption, regarding additive approximations, was presented in Meilă (2006). The work of (Meilă 2006) studied sufficient conditions under which the stability assumption holds true. Formally, the *approximation stability* notion is defined as follows:

Definition 5.4 $((1 + \alpha, \epsilon)$-approximation-stability))

Let X be a set of n points residing in a metric space $\mathcal{M}$. Given an objective function $\mathcal{L}$ (such as k-median, k-means, or min-sum), we say that instance $(\mathcal{M}, X)$ *satisfies $(1 + \alpha, \epsilon)$-approximation-stability for $\mathcal{L}$* if all clusterings $\mathcal{C}$ with $\mathcal{L}(\mathcal{C}) \leq (1 + \alpha) \cdot \mathcal{L}_{\mathcal{L}}^*$ are ϵ-close to the target clustering $\mathcal{C}_T$ for $(\mathcal{M}, S)$.

Here the term "target" clustering refers to the ground truth clustering of X which one is trying to approximate. It is also important to clarify what we mean by an ϵ-close clustering. Given two k clusterings $\mathcal{C}$ and $\mathcal{C}^*$ of n points, the distance between them is measured as $dist(\mathcal{C}, \mathcal{C}^*) = \min_{\sigma \in S_k} \frac{1}{n} \sum_{i=1}^{k} |C_i \setminus C^*_{\sigma(i)}|$. We say that $\mathcal{C}$ is ϵ-close to $\mathcal{C}^*$ if the distance between them is at most ϵ. Interestingly, this approximation stability condition implies a lot of structure about the problem instance which could be exploited algorithmically. For example, we can show the following.

Theorem 5.7 (Balcan et al. 2009)

If the given instance $(\mathcal{M}, S)$ satisfies $(1 + \alpha, \epsilon)$-approximation-stability for the k-median or the k-means objective, then we can efficiently produce a clustering that is $O(\epsilon + \epsilon/\alpha)$-close to the target clustering $\mathcal{C}_T$.   ∎

Notice that the above theorem is valid even for values of α for which getting a $(1 + \alpha)$-approximation to k-median and k-means is NP-hard! In a recent paper, (Agarwal et al. 2013), it is shown that running the Algorithm KMEANS++ algorithm for approximation stable instances of k-means gives a constant factor approximation with probability $\Omega(\frac{1}{k})$. In the following, we will provide a sketch of the proof of Theorem 5.7 for k-means clustering.

5.5.5 Proof Sketch and Intuition for Theorem 5.7

Let $C_1, C_2, \ldots, C_k$ be an optimal k-means clustering of C. Let $c_1, c_2, \ldots, c_k$ be the corresponding cluster centers. For any point $x \in C$, let $w(x)$ be the distance of x to its cluster center. Similarly let $w_2(x)$ be the distance of x to the second closest center. The value of the optimal solution can then be written as $\mathcal{L}^* = \sum_x w(x)^2$. The main implication of approximation stability is that most of the points are much closer to their own center than to the centers of other clusters. Specifically:

Lemma 5.7

If the instance $(\mathcal{M}, X)$ satisfies $(1 + \alpha, \epsilon)$-approximation-stability, then less than $6\epsilon n$ points satisfy $w_2(x)^2 - w(x)^2 \leq \frac{\alpha \mathcal{L}^*}{2\epsilon n}$.

Proof. Let C^* be the optimal k-means clustering. First notice that by approximation-stability $dist(C^*, C_T) = \epsilon^* \leq \epsilon$. Let B be the set of points that satisfy $w_2(x)^2 - w(x)^2 \leq \frac{\alpha \mathcal{L}^*}{2\epsilon n}$. Let us assume that $|B| > 6\epsilon n$. We will create a new clustering C' by transferring some of the points in B to their second closest center. In particular it can be shown that there exists a subset of size $|B|/3$ such that for each point reassigned in this set, the distance of the clustering to C^* increases by $1/n$. Hence, we will have a clustering C' which is 2ϵ away from C^* and at least ϵ away from C_T. However, the increase in cost in going from C^* to C' is at most $\alpha \mathcal{L}^*$. This contradicts the approximation stability assumption. ∎

Let us define $d_{\text{crit}} = \sqrt{\frac{\alpha \mathcal{L}^*}{50 \epsilon n}}$ as the critical distance. We call a point x *good* if it satisfies $w(x)^2 < d_{\text{crit}}^2$ and $w_2(x)^2 - w(x)^2 > 25 d_{\text{crit}}^2$. Otherwise, we call x as a *bad* point. Let B be the set of all bad points and let G_i be the good points in target cluster i. By Lemma 5.7 at most $6\epsilon n$ points satisfy $w_2(x)^2 - w(x)^2 > 25 d_{\text{crit}}^2$. Also from Markov's inequality at most $\frac{50 \epsilon n}{\alpha}$ points can have $w(x)^2 > d_{\text{crit}}^2$. Hence $|B| = O(\epsilon/\alpha)$.

Given Lemma 5.7, if we then define the τ-*threshold graph* $G_\tau = (S, E_\tau)$ to be the graph produced by connecting all pairs $\{x, y\} \in \binom{C}{2}$ with $d(x, y) < \tau$, and consider $\tau = 2d_{\text{crit}}$ we get the following two properties:

1. For $x, y \in C_i^*$ such that x and y are good points, we have $\{x, y\} \in E(G_\tau)$.
2. For $x \in C_i^*$ and $y \in C_j^*$ such that x and y are good points, $\{x, y\} \notin E(G_\tau)$.
3. For $x \in C_i^*$ and $y \in C_j^*$, x and y do not have any good point as a common neighbor.

Hence, the threshold graph has the structure as shown in Figure 5.4, where each G_i is a clique representing the set of good points in cluster i. This suggests the following algorithm for k-means clustering. Notice that unlike the algorithm for ϵ-separability, the

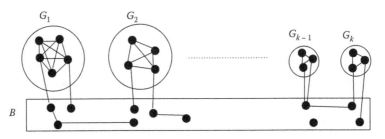

FIGURE 5.4
The structure of the threshold graph.

algorithm for approximation stability mentioned below needs to know the values of the stability parameters α and ϵ.[*]

Algorithm STABLE K-MEANS

Input: $\epsilon \leq 1, \alpha > 0, k$.

1. **Initialization:** Define $d_{\text{crit}} = \sqrt{\frac{\alpha \mathcal{L}^*}{50 \epsilon n}}$ [a]

2. Construct the τ-threshold graph G_τ with $\tau = 2d_{\text{crit}}$.

3. **For** $j = 1$ to k do:
   Pick the vertex v_j of highest degree in G_τ.
   Remove v_j and its neighborhood from G_τ and call this cluster $C(v_j)$.

4. **Output:** the k clusters $C(v_1), \ldots, C(v_{k-1}), S - \cup_{i=1}^{k-1} C(v_i)$.

---

[a]For simplicity we assume here that one knows the value of $\mathcal{L}^*$. If not, one can run a constant-factor approximation algorithm to produce a sufficiently good estimate.

The authors in Balcan et al. (2009) use the properties of the threshold graph to show that the greedy method of Step 3 of the algorithm produces an accurate clustering. In particular, if the vertex v_j we pick is a good point in some cluster C_i, then we are guaranteed to extract the whole set G_i of good points in that cluster and potentially some bad points as well (see Figure 5.5a). If on the other hand the vertex v_j we pick is a bad point, then we might extract only a part of a good set G_i and miss some good points in G_i, which might lead to some errors. (Note that by property (Agarwal et al. 2013) we never extract parts of two different good sets G_i and G_j.) However, since v_j was picked to be the vertex of the highest degree in G_τ, we are guaranteed to extract at least as many bad points as the number of missed good points in G_i, see Figure 5.5b. These then imply that overall we can charge the errors to the bad points, so the distance between the target clustering and the resulting clustering is $O(\epsilon/\alpha)n$, as desired.

---

[*] This is specifically for the goal of finding a clustering that nearly matches an unknown target clustering, because one may not in general have a way to identify which of two-proposed solutions is preferable. On the other hand, if the goal is to find a solution of low cost, then one does not need to know α or ϵ: one can just try all possible values for d_{crit} in the algorithm and take the solution of least total cost.

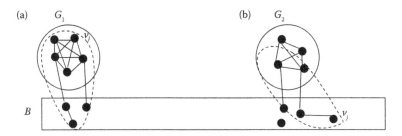

FIGURE 5.5
If the greedy algorithm chooses a good vertex v_j as in (a), we get the entire good set of points from that cluster. If v_j is a bad point as in (b), the missed good points can be charged to bad points.

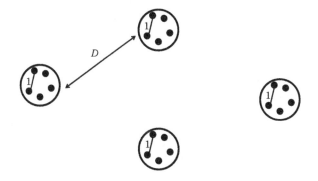

FIGURE 5.6
Suppose ϵ is a small constant, and consider a clustering instance in which the target consists of $k = \sqrt{n}$ clusters with $\sqrt{n}$ points each, such that all points in the same cluster have distance 1 and all points in different clusters have distance $D + 1$ where D is a large constant. Then, merging two clusters increases the cost additively by $\Theta(\sqrt{n})$, since D is a constant. Consequently, the optimal $(k - 1)$-means/median solution is just a factor $1 + O(1/\sqrt{n})$ more expensive than the optimal k-means/median clustering. However, for D sufficiently large compared to $1/\epsilon$, this example satisfies $(2, \epsilon)$-approximation-stability or even $(1/\epsilon, \epsilon)$-approximation-stability—see Balcan et al. (2013) for formal details.

5.5.6 Other Notions of Stability and Relations between Them

This notion of ϵ-separability is in fact related to (c, ϵ)-approximation-stability. Indeed, in Theorem 5.1 of their paper, (Ostrovsky et al. 2006), they show that their ϵ-separatedness assumption implies that any near-optimal solution to k-means is $O(\epsilon^2)$-close to the k-means optimal clustering. However, the converse is not necessarily the case: an instance could satisfy approximation-stability without being ϵ-separated.* Balcan et al. (2013) presents a specific example of points in Euclidean space with $c = 2$. In fact, for the case that k is much larger than $1/\epsilon$, the difference between the two properties can be more substantial. See Figure 5.6 for an example. In addition, algorithms for approximation stability have been successfully applied in clustering problems arising in computational biology (Voevodski et al. 2010) (see Section 5.5.8 for details).

---

* Ostrovsky et al. (2006) shows an implication in this direction (Theorem 5.2); however, this implication requires a substantially stronger condition, namely that data satisfy (c, ϵ)-approximation-stability for $c = 1/\epsilon^2$ (and that target clusters be large). In contrast, the primary interest of Balcan et al. (2013) is the case, where c is below the threshold for existence of worst-case approximation algorithms.

Awasthi et al. (2010) studies center-based clustering objectives and defines a notion of stability called α-*weak deletion* stability. A clustering instance is stable under this notion if in the optimal clustering merging any two clusters into one increases the cost by a multiplicative factor of $(1 + \alpha)$. This a broad notion of stability that generalizes both the ϵ-separability notion studied in Section 5.5.1 and the approximation stability in the case of large cluster sizes. Remarkably, Awasthi et al. (2010) shows that for such instances of k-median and k-means one can design a $(1 + \epsilon)$ approximation algorithm for any $\epsilon > 0$. This leads to immediate improvements over the works of Balcan et al. (2009) (for the case of large clusters) and of Ostrovsky et al. (2006). However, the runtime of the resulting algorithm depends polynomially in n and k and exponentially in the parameters $1/\alpha$ and $1/\epsilon$, so the simpler algorithms of Awasthi et al. (2010) and Balcan et al. (2009) are more suitable for scenarios where one expects the stronger properties to hold. See Section 5.5.8 for further discussion. Ackerman and Ben-David (2009) also studies various notions of clusterability of a dataset and presents algorithms for such stable instances.

Kumar and Kannan (2010) consider the problem of recovering a target clustering under deterministic separation conditions that are motivated by the k-means objective and by Gaussian and related mixture models. They consider the setting of points in Euclidean space, and show that if the projection of any data point onto the line joining the mean of its cluster in the target clustering to the mean of any other cluster of the target is $\Omega(k)$ standard deviations closer to its own mean than the other mean, then they can recover the target clusters in polynomial time. This condition was further analyzed and reduced by work of Awasthi et al. (2012). This separation condition is formally incomparable to approximation-stability (even restricting to the case of k-means with points in Euclidean space). In particular, if the dimension is low and k is large compared to $1/\epsilon$, then this condition can require more separation than approximation-stability (e.g., with k well-spaced clusters of unit radius approximation-stability would require separation only $O(1/\epsilon)$ and independent of k—see Balcan et al. (2013) for an example). On the other hand if the clusters are high-dimensional, then this condition can require less separation than approximation-stability since the ratio of projected distances will be more pronounced than the ratios of distances in the original space.

Bilu and Linial (2010) consider inputs satisfying the condition that the optimal solution to the objective remains optimal even after bounded perturbations to the input weight matrix. This condition is known as *perturbation resilience*. Bilu and Linial (2010) give an algorithm for a different clustering objective known as maxcut. The maxcut objective asks for a two partitioning of a graph such that the total number of edges going between the two pieces is maximized. The authors show that the maxcut objective is easy under the assumption that the optimal solution is stable to $O(n^{2/3})$-factor multiplicative perturbations to the edge weights. The work of Makarychev et al. (2014) subsequently reduced the required resilience factor to $O(\sqrt{\log n})$. In Awasthi et al. (2012), the authors study perturbation resilience for center-based clustering objectives such as k-median and k-means, and give an algorithm that finds the optimal solution when the input is stable to only factor-3 perturbations. This factor is improved to $1 + \sqrt{2}$ by Balcan and Liang (2012), who also design algorithms under a relaxed (c, ϵ)-stability to perturbations condition in which the optimal solution need not be identical on the c-perturbed instance, but may change on an ϵ fraction of the points (in this case, the algorithms require $c = 4$). Note that for the k-median objective, (c, ϵ)-approximation-stability with respect to C^* implies (c, ϵ)-stability to perturbations because an optimal solution in a c-perturbed instance is guaranteed to be a c-approximation on the original instance;* so, (c, ϵ)-stability to perturbations is a weaker condition. Similarly, for k-means, (c, ϵ)-stability to perturbations is implied by (c^2, ϵ)-approximation-stability. However, as noted above, the values of c known to lead to efficient clustering in the case of stability to perturbations are larger than for approximation-stability, where any constant $c > 1$ suffices.

5.5.7 Runtime Analysis

Below we provide the run time guarantees of the various algorithms discussed so far. While these may be improved with appropriate data structures, we assume here a straight-forward implementation in which computing the distance between two data points takes time $O(d)$, as does adding or averaging two data points. For example, computing a step of Lloyd's algorithm requires assigning each of the n data points to its nearest center, which in turn requires taking the minimum of k distances per data point (so $O(nkd)$ time total), and then resetting each center to the average of all data points assigned to it (so $O(nd)$ time total). This gives Lloyd's algorithm a running time of $O(nkd)$ per iteration. The k-means++ algorithm has only a seed-selection step, which can be run in time $O(nd)$ per seed by remembering the minimum distances of each point to the previous seeds, so it has a total time of $O(nkd)$.

For the ϵ-separability algorithm, to obtain the sampling probabilities for the first two seeds one can compute all pairwise distances at cost of $O(n^2d)$. Obtaining the rest of the seeds is faster since one only needs to compute distances to previous seeds, so this takes time $O(ndk)$. Finally, there is a greedy deletion initialization at time $O(ndk)$ per step for $O(k)$ steps. So the overall time is $O(n^2d + ndk^2)$.

For the approximation-stability algorithm, creating a graph of distances takes time $O(n^2d)$, after which creating the threshold graph takes time $O(n^2)$ if one knows the value of d_{crit}. For the rest of the algorithm, each step takes time $O(n)$ to find the highest-degree vertex, and then time proportional to the number of edges examined to remove the vertex and its neighbors. Over the entire remainder of the algorithm, this takes time $O(n^2)$ total. If the value of d_{crit} is not known, one can try $O(n)$ values, taking the best solution. This gives an overall time of $O(n^3 + n^2d)$.

Finally, for local search, one can first create a graph of distances in time $O(n^2d)$. Each local swap step has $O(nk)$ pairs (x, y) to try, and for each pair one can compute its cost in time $O(nk)$ by computing the minimum distance of each data point to the proposed k centers. So, the algorithm can be run in time $O(n^2k^2)$ per iteration. The total number of iterations is at most $poly(n)$† so the overall running time is at most $O(n^2d + n^2k^2 poly(n))$. As can be seen from Table 5.1 below, the algorithms become more and more computationally expensive if one needs formal guarantees on a larger instance space. For example, the local search algorithm provides worst case approximation guarantees on all instances but is very slow. On the other hand, Lloyd's method and k-means++ are very fast but provide bad worst case guarantees, especially when the number of clusters k is large. Algorithms based on stability notions aim to provide the best of both worlds by being fast and provably good on well-behaved instances. In the conclusion Section 5.7, we outline a guideline for practitioners when working with the various clustering assumptions.

---

* In particular, a c-perturbed instance $\tilde{d}$ satisfies $d(x, y) \leq \tilde{d}(x, y) \leq cd(x, y)$ for all points x, y. So, using $\mathcal{L}$ to denote cost in the original instance, $\tilde{\mathcal{L}}$ to denote cost in the perturbed instance and using $\tilde{\mathcal{C}}$ to denote the optimal clustering under $\tilde{\mathcal{L}}$, we have $\mathcal{L}(\tilde{\mathcal{C}}) \leq \tilde{\mathcal{L}}(\tilde{\mathcal{C}}) \leq \tilde{\mathcal{L}}(\mathcal{C}^*) \leq c\mathcal{L}(\mathcal{C}^*)$.

† The actual number of iterations depends upon the cost of the initial solution and the stopping condition.

TABLE 5.1

A Run Time Analysis of Various Algorithms Discussed in
the Chapter. The Running Time Degrades as One Requires
Formal Guarantees on Larger Instance Spaces

Method	Runtime
LLOYD'S K-MEANS	$O(nkd) \times (\#\,iterations)$
KMEANS++	$O(nkd)$
ϵ-Separability	$O(n^2 d + ndk^2)$
Approximation stability	$O(n^3 + n^2 d)$
K-MEANS LOCAL SEARCH	$O(n^2 d + n^2 k^2 \, poly(n))$

5.5.8 Extensions

5.5.8.1 Variants of the k-Means Objective

k-means clustering is the most popular methods for vector quantization, which is used
in encoding speech signals and data compression (Gersho and Gray 1991). There have
been variants of the k-means algorithm called fuzzy k-means, which allow each point to
have a degree of membership into various clusters (Bezdek 1981). This modified k-means
objective is popular for image segmentation (Nguyen and Cohen 1993, Solberg et al. 1996).
There have also been experiments on speeding up the Lloyd's method by updating centers at each step by only choosing a random sample of the entire dataset (Faber 1994).
Bottou and Bengio (1995) presents an empirical study on the convergence properties of
Lloyd's method. Rasmussen (1992) contains a discussion of k-means clustering for information retrieval. Dhillon (2001) presents an empirical comparison of k-means and spectral
clustering methods. Pelleg and Moore (2000) studies a modified k-means objective with an
additional penalty for the number of clusters chosen. They motivate the new objective as a
way to solve the cluster selection problem. This approach is inspired by the Bayesian model
selection procedures (Schwarz 1978). For further details on the applications of k-means,
refer to Chapters 1.2 and 2.3.

5.5.8.2 k-Means++: Streaming and Parallel Versions of k-Means

As we saw in Section 5.2, careful seeding is crucial in order for the Lloyd's method to
succeed. One such method is proposed in the k-means++ algorithm. Using the seed centers output by k-means++, one can immediately guarantee an $O(\log k)$ approximation to
the k-means objective. However k-means++ is an iterative method, which needs to be
repeated k times in order to get a good set of seed points. This makes it undesirable for
use in applications involving massive datasets with thousands of clusters. This problem is
overcome in Bahmani et al. (2012), where the authors propose a scalable and parallel version of KMEANS++. The new algorithm runs in much fewer iterations and chooses more
than one seed point at each step. The authors experimentally demonstrate that this leads
to much better computational performance in practice without losing out on the solution quality. In Ailon et al. (2009), the authors design an algorithm for k-means, which
makes a single pass over the data. This makes it much more suitable for applications
where one needs to process data in the streaming model. The authors show that if one is
allowed to store a little more than k centers ($O(k \log k)$) then one can also achieve good
approximation guarantees and at the same time have an extremely efficient algorithm.

They experimentally demonstrate that the proposed method is much faster than known implementations of the Lloyd's method. There has been subsequent work on improving the approximation factors and making the algorithms more practical Shindler et al. (2011).

5.5.8.3 Approximation Stability in Practice

Motivated by clustering applications in computational biology, Voevodski et al. (2010) analyzes (c, ϵ)-approximation-stability in a model with unknown distance information where one can only make a limited number of *one versus all* queries. Voevodski et al. (2010) designs an algorithm that given (c, ϵ)-approximation-stability for the k-median objective finds a clustering that is very close to the target by using only $O(k)$ one-versus-all queries in the large cluster case, and in addition is faster than the algorithm we present here. In particular, the algorithm for the large clusters case described in Balcan et al. (2009) (similar to the one we described in Section 5.5.4 for the k-means objective) can be implemented in $O(|S|^3)$ time, while the one proposed in Voevodski et al. (2010) runs in time $O(|S|k(k + \log |S|))$. Voevodski et al. (2010) uses their algorithm to cluster biological datasets in the Pfam (Finn et al. 2010) and SCOP (Murzin et al. 1995) databases, where the points are proteins and distances are inversely proportional to their sequence similarity. This setting nicely fits the one-versus-all queries model because one can use a fast sequence database search program to query a sequence against an entire dataset. The Pfam (Finn et al. 2010) and SCOP (Murzin et al. 1995) databases are used in biology to observe evolutionary relationships between proteins and to find close relatives of particular proteins. Voevodski et al. (2010) finds that for one of these sources they can obtain clusterings that almost exactly match the given classification, and for the other the performance of their algorithm is comparable to that of the best-known algorithms using the full distance matrix.

5.6 Mixture Models

In the previous sections, we saw worst case approximation algorithms for various clustering objectives. We also saw examples of how assumptions on the nature of the optimal solution can lead to much better approximation algorithms. In this section, we will study a different assumption on how the data is generated in the first place. In the machine learning literature, such assumptions take the form of a probabilistic model for generating a clustering instance. The goal is to cluster correctly (with high probability) an instance generated from the particular model. The most famous and well-studied example of this is the GMM (Kannan and Vempala 2009). This will be the main focus of this section. We will illustrate conditions under which datasets arising from such a mixture model can be provably clustered.

Gaussian Mixture Model A univariate Gaussian random variable X, with mean μ and variance σ^2 has the density function $f(x) = \frac{1}{\sigma\sqrt{2\pi}}e^{\frac{-(x-\mu)^2}{\sigma^2}}$. Similarly, a multivariate Gaussian random variable, $\mathbf{X} \in \mathbb{R}^n$, has the density function

$$f(\mathbf{x}) = \frac{1}{|\Sigma|^{1/2}(2\pi)^{n/2}}e^{\left(\frac{-1}{2}(\mathbf{x}-\mathbf{\mu})^T \Sigma^{-1}(\mathbf{x}-\mathbf{\mu})\right)}$$

Here $\mu \in \mathbb{R}^n$ is called the mean vector and Σ is the $n \times n$ covariance matrix. A special case is the spherical Gaussian for which $\Sigma = \sigma^2 I_n$. Here σ^2 refers to the variance of the Gaussian in any given direction. Consider k n-dimensional Gaussian distributions, $\mathcal{N}(\mu_1, \Sigma_1), \mathcal{N}(\mu_2, \Sigma_2), \ldots, \mathcal{N}(\mu_k, \Sigma_k)$. A GMM $\mathcal{M}$ refers to the distribution obtained from a convex combination of such Gaussian. More specifically

$$\mathcal{M} = w_1 \mathcal{N}(\mu_1, \Sigma_1) + w_2 \mathcal{N}(\mu_2, \Sigma_2) + \cdots + w_k \mathcal{N}(\mu_k, \Sigma_k)$$

Here $w_i \geq 0$ are called the mixing weights and satisfy $\sum_i w_i = 1$. One can think of a point being generated from $\mathcal{M}$ by first choosing a component Gaussian i, with probability w_i, and then generating a point from the corresponding Gaussian distribution $\mathcal{N}(\mu_i, \Sigma_i)$. Given a dataset of m points coming from such a mixture model, a fairly natural question is to recover the individual components of the mixture model. This is a clustering problem where one wants to cluster the points into k clusters such that the points drawn from the same Gaussian are in a single partition. Notice that unlike in the previous sections, the algorithms designed for mixture models will have probabilistic guarantees. In other words, we would like the clustering algorithm to recover, with high probability, the individual components. Here the probability is over the draw of the m sample points. Another problem one could ask is to approximate the parameters (mean, variance) of each individual component Gaussian. This is known as the parameter estimation problem. It is easy to see that if one could solve the clustering problem approximately optimally, then estimating the parameters of each individual component is also easy. Conversely, after doing parameter estimation, one can easily compute the Bayes optimal clustering. To study the clustering problem, one typically assumes separation conditions among the component Gaussians which limit the amount of overlap between them. The most common among them is to assume that the mean vectors of the component Gaussians are far apart. However, there are also scenarios when such separation conditions do not hold (consider two Gaussian which are aligned in an 'X' shape), yet the data can be clustered well. In order to do this, one first does parameter estimation which needs much weaker assumptions. After estimating the parameters, the optimal clustering can be recovered. This is an important reason to study parameter estimation. In the next section we will see examples of some separation conditions and the corresponding clustering algorithms that one can use. Later, we will also look at recent work on parameter estimation under minimal separation conditions.

5.6.1 Clustering Methods

In this section, we will look at distance-based clustering algorithms for learning a mixture of Gaussians. For simplicity, we will start with the case of k spherical Gaussians in $\mathbb{R}^n$ with means $\{\mu_1, \mu_2, \ldots, \mu_k\}$ and variance $\Sigma = \sigma^2 I_n$. The algorithms we describe will work under the assumption that the means are far apart. We will call this as the center separation property:

Definition 5.5 (Center Separation)

A mixture of k identical spherical Gaussians satisfies center separation if $\forall i \neq j$,

$$\Delta_{i,j} = \|\mu_i - \mu_j\| > \beta_{i,j} \sigma$$

The quantity $\beta_{i,j}$ typically depends on k the number of clusters, n the dimensionality of the dataset and w_{min}, the minimum mixing weight. If the spherical Gaussians have different variances σ_i's, the R.H.S. is replaced by $\beta_{i,j}(\sigma_i + \sigma_j)$. For the case of general Gaussians, σ_i will denote the maximum variance of Gaussian i in any particular direction. One of the earliest results using center separation for clustering is by Dasgupta (1999). We will start with a simple condition that $\beta_{i,j} = C\sqrt{n}$, for some constant $C > 4$ and will also assume that $w_{min} = \Omega(1/k)$. Let's consider a typical point x from a particular Gaussian $\mathcal{N}(\mu_i, \sigma^2 I_n)$. We have $E[||X - \mu_i||^2] = E[\sum_{d=1}^{n} |x_d - \mu_{id}|^2] = n\sigma^2$. Now consider two typical points x and y from two different Gaussians $\mathcal{N}(\mu_i, \sigma^2 I_n)$ and $\mathcal{N}(\mu_j, \sigma^2 I_n)$. We have

$$E[||X - Y||^2] = E[||X - \mu_i + \mu_i - \mu_j - (Y - \mu_j)||^2]$$
$$= E[||X - \mu_i||^2] + E[||Y - \mu_j||^2] + ||\mu_i - \mu_j||^2$$
$$\geq 2n\sigma^2 + C^2\sigma^2 n$$

For C large enough (say $C > 4$), we will have that for any two typical points x, y in the same cluster, $||x - y||^2 \leq 2\sigma^2 n$. And for any two points in different clusters $||x - y||^2 > 18\sigma^2 n$. Using standard concentration bounds, we can say that for a sample of size $poly(n)$, with high probability, all points from a single Gaussian will be closer to each other, than to points from other Gaussians. In this case one could simply create a graph by connecting any two points x, y such that $||x - y||^2 \leq 2\sigma^2 n$. It is easy to see that the connected components in this graph will correspond precisely to the individual components of the mixture model. If C is smaller, say 2, one needs a stronger concentration result (Arora and Kannan 2009) mentioned below:

Lemma 5.8

If x, y are picked independently from $N(\mu_i, \sigma^2 I_n)$, then with probability $1 - 1/n^3$, $||x - y||^2 \in [2\sigma^2 n(1 - 4\log(n)/\sqrt{n}), 2\sigma^2 n(1 + 5\log(n)/\sqrt{n})]$.

Also, as before, one can show that with high probability, for x and y from two different Gaussians, we have $||x - y||^2 > 2\sigma^2 n(1 + 4\log(n)/\sqrt{n})$. From this it follows that if r is the minimum distance between any two points in the sample, then for any x in Gaussian i and any y in the same Gaussian, we have $||x - y||^2 \leq (1 + 4.5\log(n)/\sqrt{n})r$. And for a point z in any other Gaussian, we have $||x - z||^2 > (1 + 4.5\log(n)/\sqrt{n})r$. This suggests the following algorithm:

Algorithm CLUSTER SPHERICAL GAUSSIANS

1. Let $\mathcal{D}$ be the set of all sample points.

2. **For:** $i = 1$ to k,

   a. Let x_0 and y_0 be such that $||x_0 - y_0||^2 = r = \min_{x,y \in \mathcal{D}} ||x - y||^2$.

   b. Let $T = \{y \in \mathcal{D} : ||x_0 - y||^2 \leq r(1 + \frac{4.5\log n}{\sqrt{n}})\}$.

   c. **Output:** T as one of the clusters.

5.6.1.1 Handling Smaller C

For smaller values of C, for example $C < 1$, one cannot in general say that the above strong concentration will hold true. In fact, in order to correctly classify the points, we might need to see points which are much closer to the center of a Gaussain (say at distance less than $\frac{1}{2}\sigma\sqrt{n}$). However, most of the mass of a Gaussian lies in a thin shell around radius of $\sigma\sqrt{n}$. Hence, one might have to see exponentially many samples in order to get a good classification. Dasgupta (1999) solves this problem by first projecting the data onto a random $d = O(\log(k)/\epsilon^2)$ dimensional subspace. This has the effect that the center separation property is still preserved up to a factor of $(1 - \epsilon)$. One can now do distance-based clustering in this subspace as the number of samples needed will be proportional to 2^d instead of 2^n.

5.6.1.2 General Gaussians

The results of Dasgupta were extended by Arora and Kannan (2009) to the case of general Gaussians. They also managed to reduce the required separation between means. They assumed that $\beta_{i,j} = \Omega(\log(n))(R_i + R_j)(\sigma_i + \sigma_j)$. As mentioned before, σ_i denotes the maximum variance of Gaussian i in any direction. R_i denotes the median radius of Gaussian i.[*] For the case of spherical Gaussians, this separation becomes $\Omega(n^{1/4}\log(n)(\sigma_i + \sigma_j))$. Arora and Kannan use isoperimetric inequalities to get strong concentration results for such Gaussians. In particular they show that:

Theorem 5.8

Given $\beta_{i,j} = \Omega(\log(n)(R_i + R_j))$, there exists a polynomial time algorithm which given at least $m = \frac{n^2 k^2}{\delta^2 w^6_{min}}$ samples from a mixture of k general Gaussians solves the clustering problem exactly with probability $(1 - \delta)$.   ∎

Proof Intuition: The first step is to generalize Lemma 5.8 for the case of general Gaussians. In particular one can show that for x, y are picked at random from a general Gaussian i, with median radius R_i and maximum variance σ_i, we have with high probability

$$2R_i^2 - 18\log(n)\sigma_i R_i \leq ||x - y||^2 \leq 2(R_i + 20\log(n)\sigma_i)^2$$

Similarly, for x, y from different Gaussians i and j, we have with high probability

$$||x - y||^2 > 2\min(R_i^2, R_j^2) + 120\log(n)(\sigma_i + \sigma_j)(R_i + R_j) + \Omega(\log(n))^2(\sigma_i^2 + \sigma_j^2).$$

The above concentration results imply (w.h.p.) that pairwise distances within points from a Gaussian i lie in an interval I_i and distances between Gaussians $I_{i,j}$ lie in the interval $I_{i,j}$. Furthermore, $I_{i,j}$ will be disjoint from the interval corresponding to the Gaussian with smaller value of R_i. In particular, if one looks at balls of increasing radius around a point from the Gaussian with minimum radius, σ_i, there will be a stage when there exists a gap: i.e., increasing the radius slightly does not include any more points. From the above lemmas, this gap will be roughly $\Omega(\sigma_i)$. Hence, at this stage, we can remove this Gaussian from the data and recurse. This property suggests the following algorithm outline.

---

[*] The radius such that the probability mass within R_i equals $1/2$.

Algorithm CLUSTER GENERAL GAUSSIANS

1. Let r be the smallest radius such that $|B(x,r)| > \frac{3}{4}w_{min}|S|$, for some $x \in \mathcal{D}$.
   Here $|S|$ is the size of dataset and $B(x,r)$ denotes the ball of radius r around x.

2. Let σ denote the maximum variance of the Gaussian with the least radius.
   Let $\gamma = O(\sqrt{w_{min}}\sigma)$.

3. **While** $\mathcal{D}$ is non-empty,

   a. Let s be such that $|B(x,r+s\gamma)| \cap \mathcal{D} = |B(x,r+(s-1)\gamma)|$.

   b. Remove a set T containing all the points from S which are in $B(x,r+s\gamma\log(n))$.

   c. **Output:** T as one of the cluster.

One point to mention is that one does not really know beforehand the value of sigma at each iteration. Arora and Kannan (2009) get around this by estimating the variance from the data in the ball $B(x,r)$. They then show that this estimate is good enough for the algorithm to work.

5.6.2 Spectral Algorithms

The algorithms mentioned in the above section need the center separation to grow polynomially with n. This is prohibitively large especially in cases when $k \ll n$. In this section, we look at how spectral techniques can be used to only require the separation to grow with k instead of n.

5.6.2.1 Algorithmic Intuition

In order to remove the dependence on n, we would like to project the data such that points from the same Gaussian become much closer while still maintaining the large separation between means. One idea is to do a random projection. However, random projections from n to d dimensions scale each squared distance equally (by factor d/n) and will not give us any advantage. However, consider the case of two spherical Gaussians with means μ_1 and μ_2 and variance $\sigma^2 I_n$. Consider projecting all the points to the line joining μ_1 and μ_2. Now consider any random point x from the first Gaussian. For any unit vector along the line joining μ_1 and μ_2, we have that $(x - \mu_1).v$ behaves like a one-dimensional Gaussian with mean 0 and variance σ^2. Hence the expected distance of a point x from its mean becomes σ^2. This means that for any two points in the same Gaussian, the expected squared distance becomes $4\sigma^2$ (as opposed to $2n\sigma^2$). However, the distance between the means remains the same. In fact, the above claim is true if we project onto any subspace containing the means. This subspace is exactly characterized by the singular value decomposition (SVD) of the data matrix. This suggests the following algorithm:

Algorithm SPECTRAL CLUSTERING FOR GAUSSIAN MIXTURES

1. Compute the SVD decomposition of the data.

2. Project the data onto the space of top-k right singular vectors.

3. Run a distance-based clustering method in this projected space.

Such spectral algorithms were proposed by Vempala and Wang (2004) who reduced the separation for spherical Gaussians to $\beta_{i,j} = \Omega(k^{1/4}(\log(n/w_{min}))^{1/4})$. The case of general Gaussians was studied in Achlioptas and McSherry (2005) who give efficient clustering algorithms for $\beta_{i,j} = (\frac{1}{\sqrt{min(w_i, w_j)}} + \sqrt{k\log(k_{min}(2^k, n))})$. Kannan et al. (2005) gives algorithms for general Gaussians for $\beta_{i,j} = \frac{k^{3/2}}{w_{min}^2}$.

5.6.3 Parameter Estimation

In the previous sections, we looked at the problem of clustering points from a GMM. Another important problem is that of estimating the parameters of the component Gaussians. These parameters refer to the mixture weights w_i's, mean vectors μ_i's and the covariance matrices Σ_i's. As mentioned before, if one could do efficiently get a good clustering, then the parameter estimation problem is solved by simply producing empirical estimates from the corresponding clusters. However, there could be scenarios when it is not possible to produce a good clustering. For example consider two one-dimensional Gaussians with mean 0 and variance σ^2 and $2\sigma^2$. These Gaussians have a large overlap and any clustering method will inherently have a large error. On the other hand, let's look at the statistical distance between the two Gaussians, that is, $\int_x |f_1(x) - f_2(x)| dx$. This measures how much one distribution dominates the other one. It is easy to see that in this case the Gaussian with the higher variance will dominate the other Gaussian almost everywhere. Hence the statistical distance is close to 1. This suggests that information theoretically, one should be able to estimate the parameters of these two mixtures. In this section, we will look at some recent work of Kalai et al. (2010), and Moitra and Valiant (2010) in efficient algorithms for estimating the parameters of a GMM. These works make minimal assumption on the nature of the data, namely, that the component Gaussians have noticeable statistical distance. Similar results were proven in Belkin and Sinha (2010) who also gave algorithms for more general distributions.

5.6.3.1 The Case of Two Gaussians

We will first look at the case of two Gaussians in $\mathbb{R}^n$. We will assume that the statistical distance between the Gaussians, $D(\mathcal{N}_1, \mathcal{N}_2)$, is noticeable, that is, $\int_x |f_1(x) - f_2(x)| dx > \alpha$. Kalai et al. (2010) show the following theorem:

Theorem 5.9

Let $\mathcal{M} = w_1 \mathcal{N}_1(\mu_1, \Sigma_1) + w_2 \mathcal{N}_2(\mu_2, \Sigma_2)$ be an isotropic GMM where $D(\mathcal{N}_1, \mathcal{N}_2) > \alpha$. Then, there is an algorithm which outputs $\mathcal{M}' = w_1' \mathcal{N}'_1(\mu_1', \Sigma_1') + w_2' \mathcal{N}'_2(\mu_2', \Sigma_2')$ such that for some permutation $\pi : \{0, 1\} \mapsto \{0, 1\}$ we have,

$$|w_i - w'_{\pi(i)}| \leq \epsilon$$

$$||\mu_i - \mu'_{\pi(i)}|| \leq \epsilon$$

$$||\Sigma_i - \Sigma'_{\pi(i)}|| \leq \epsilon$$

The algorithm runs in time $poly(n, 1/\epsilon, 1/\alpha, 1/w_1, 1/w_2)$.   ∎

The condition on the mixture being isotropic is necessary to recover a good additive approximation for the means and the variances since otherwise, one could just scale the data and the estimates will scale proportionately.

5.6.3.2 Reduction to a One-Dimensional Problem

In order to estimate the mixture parameters, Kalai et al. reduce the problem to a series of one-dimensional learning problems. Consider an arbitrary unit vector v. Suppose, we project the data onto the direction of v and let the means of the Gaussians in this projected space be μ'_1 and μ'_2. Then, we have that $\mu_1 = E[x \cdot v] = E[(x - \mu_1) \cdot v] = \mu_1 \cdot v$. Hence, the parameters of the original mean vector are linearly related to the mean in the projected space. Similarly, let's perturb v to get $v' = v + \epsilon(e_i + e_j)$. Here, e_i and e_j denote the basis vectors corresponding to coordinates i and j. Let σ'^2_1 be the variance of the Gaussian in the projected space v'. Then writing $\sigma'^2_1 = E[(x \cdot v')^2]$ and expanding, we get that $E[x_i x_j]$ will be linearly related to σ'^2_1, σ^2_1, and the μ_i's. Hence, by estimating the parameters correctly over a series of n^2, one-dimensional vectors, one can efficiently recover the original parameters (by solving a system of linear equations).

5.6.3.3 Solving the One-Dimensional Problem

The one-dimensional problem is solved by the method of moments. In particular, define $L_i[\mathcal{M}]$ to be the ith moment for the mixture model $\mathcal{M}$, that is, $L_i[\mathcal{M}] = E_{x \sim \mathcal{M}}[x^i \mathcal{M}(x)]$. Also, define $\hat{L}_i$ to be the empirical ith moment of the data. The algorithm in Kalai et al. (2010) does a brute force search over the parameter space for the two Gaussians and for a given candidate model $\mathcal{M}'$ computes the first six moments. If all the moments are within ϵ of the empirical moments, then the analysis in Kalai et al. (2010) shows that the parameters will be $\epsilon^{1/67}$ close to the parameters of the two Gaussians. The same claim is also true for learning a mixture of k one-dimensional Gaussians if one goes upto $(4k - 2)$ moments (Moitra and Valiant 2010). The search space, however, will be exponential in k. It is shown in Moitra and Valiant (2010) that for learning k one-dimensional Gaussians, this exponential dependence is unavoidable.

5.6.3.4 Solving the Labeling Problem

As noted above, the learning algorithm will solve n^2, one-dimensional problems and get parameter estimates for the two Gaussians for each one-dimensional problem. In order to solve for the parameters of the original Gaussians, we need to identify for each Gaussian, the corresponding n^2 parameters for each of the subproblems. Kalai et al. do this by arguing that if one projects the two Gaussians onto a random direction v, with high enough probability, the corresponding parameters for the two projected Gaussians will differ by $poly(\alpha)$. Hence, if one takes small random perturbations of this vector v, the corresponding parameter estimates will be easily distinguishable.

The overall algorithm has the following structure:

Algorithm LEARN TWO GAUSSIANS

1. Choose a random vector v and choose n^2 random perturbations $v_{i,j}$.
2. For each i, j, project the data onto $v_{i,j}$ and solve the one dimensional problem using the method of moments.
3. Solve the labeling problem to identify the n^2 parameter sets corresponding to a single Gaussian.
4. Solve a system of linear equations on this parameter set to obtain the original parameters.

For the case of more than two Gaussians, Moitra and Valiant (2010) extend the ideas mentioned above to provide an algorithm for estimating the parameters of a mixture of k Gaussians. For the case of k Gaussians, additional complications arise as it is not true anymore that projecting the k Gaussians to a random one-dimensional subspace maintains the statistical distance. For example, consider Figure 5.7. Here, projecting the data onto a random direction will almost surely collapse components 2 and 3. Moitra and Valiant (2010) solves this problem by first running a clustering algorithm to separate components 2 and 3 from component 1 and recursively solving the two subinstances. Once, 2 and 3 have been separated, one can scale the space to ensure that they remain separated over a random projection. The algorithm from Moitra and Valiant (2010) has the sample complexity, which depends exponentially on k. They also show that this dependence is necessary. One could use the algorithm from Moitra and Valiant (2010) to also cluster the points into component Gaussians under minimal assumptions. The sample complexity, however, will depend exponentially in k. In contrast, one could use algorithms from previous sections to cluster in polynomial time under stronger separation assumptions. The work of Anandkumar et al. (2012), Hsu and Kakade (2013) removes the exponential dependence on k and designs polynomial time algorithms for clustering data from a GMM under minimal separation assuming only that the mean vectors span a k dimensional subspace. However, their algorithm which is based on Tensor decompositions only works in the case when all the component Gaussians are spherical. It is an open question to get similar result for general Gaussians. There has also been work on clustering points from a mixture of other distributions. Chaudhuri and Rao (2008a,b) gave algorithms for clustering a mixture of heavy tailed distributions. Brubaker and Vempala (2008) gave algorithms

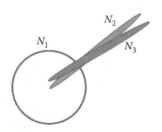

FIGURE 5.7
The case of three Gaussians.

for clustering a mixture of two Gaussians assuming only that the two distributions are separated by a hyperplane. The recent work of Kumar and Kannan (2010) studies a deterministic separation condition on a set of points and shows that any set of points satisfying this condition can be clustered accurately. Using this they easily derive many previously known results for clustering mixture of Gaussians as a corollary.

5.7 Conclusion

In this chapter, we presented a selection of recent work on clustering problems in the computer science community. As is evident, the focus of all these works is on providing efficient algorithms with rigorous guarantees for various clustering problems. In many cases, these guarantees depend on the specific structure and properties of the instance at hand which are captured by stability assumptions and/or distributional assumptions. The study of different stability assumptions also provides insights into the structural properties of real world data and in some cases also leads to practically useful algorithms (Voevodski et al. 2010). As discussed in Section 5.5.6, different assumptions are suited for different kinds of data and they relate to each other in interesting ways. For instance, perturbation resilience is a much weaker assumption than both ϵ-separability and approximation stability. However, we have algorithms with much stronger guarantees for the latter two. As a practitioner one is often torn between using algorithms with formal guarantees (which are typically slower) versus fast heuristics like the Lloyd's method. When dealing with data which may satisfy any of the stability notions proposed in this chapter, a general rule of thumb we suggest is to run the algorithms proposed in this chapter on a smaller random subset of the data and use the solution obtained to initialize fast heuristics like the Lloyd's method. Current research on clustering algorithms continues to explore more realistic notions of data stability and their implications for practical clustering scenarios.

References

D. Achlioptas and F. McSherry. On spectral learning of mixtures of distributions. In *Proceedings of the Eighteenth Annual Conference on Learning Theory*, Bertinoro, Italy, 2005.

M. Ackerman and S. Ben-David. Clusterability: A theoretical study. *Journal of Machine Learning Research–Proceedings Track*, 5:1–8, 2009.

M. Agarwal, R. Jaiswal, and A. Pal. k-means++ under approximation stability. *The 10th Annual Conference on Theory and Applications of Models of Computation*, Hong Kong, China, 2013.

A. Aggarwal, A. Deshpande, and R. Kannan. Adaptive sampling for k-means clustering. In *Proceedings of the 12th International Workshop and 13th International Workshop on Approximation, Randomization, and Combinatorial Optimization. Algorithms and Techniques*, APPROX '09 / RANDOM '09, 2009.

N. Ailon, R. Jaiswal, and C. Monteleoni. Streaming k-means approximation. In *Advances in Neural Information Processing Systems*, Y. Bengio (ed.), NIPS, Vancouver, Canada, 2009.

P. Alimonti. Non-oblivious local search for graph and hypergraph coloring problems. In *Graph-Theoretic Concepts in Computer Science*, Lecture Notes in Computer Science, 1995.

D. Aloise, A. Deshpande, P. Hansen, and P. Popat. Np-hardness of euclidean sum-of-squares clustering. *Machine Learning*, 75:245–248, 2009.

A. Anandkumar, R. Ge, D. Hsu, S. M. Kakade, and M. Telgarsky. Tensor decompositions for learning latent variable models. Technical report, http://arxiv.org/abs/1210.7559, 2012.

S. Arora and B. Barak. *Computational Complexity: A Modern Approach*. Cambridge University Press, New York, 2009.

S. Arora and R. Kannan. Learning mixtures of arbitrary Gaussians. In *Proceedings of the 37th ACM Symposium on Theory of Computing*, Baltimore, USA, 2005.

S. Arora, P. Raghavan, and S. Rao. Approximation schemes for Euclidean k-medians and related problems. In *Proceedings of the Thirty-First Annual ACM Symposium on Theory of Computing*, Atlanta, USA, 1999.

D. Arthur, B. Manthey, and H. Röglin. Smoothed analysis of the k-means method. *Journal of the ACM*, 58(5):1–31, 2011.

D. Arthur and S. Vassilvitskii. How slow is the k-means method? In *Proceedings of the Twenty-Second Annual Symposium on Computational Geometry*, Sedona, USA, 2006.

D. Arthur and S. Vassilvitskii. k-means++: The advantages of careful seeding. In *Proceedings of the Eighteenth Annual ACM-SIAM Symposium on Discrete Algorithms*, New Orleans, USA, 2007.

V. Arya, N. Garg, R. Khandekar, A. Meyerson, K. Munagala, and V. Pandit. Local search heuristics for k-median and facility location problems. *SIAM Journal on Computing*, 33(3):544–562, 2004.

P. Awasthi, A. Blum, and O. Sheffet. Stability yields a PTAS for k-median and k-means clustering. In *Proceedings of the 2010 IEEE 51st Annual Symposium on Foundations of Computer Science*, Las Vegas, USA, 2010.

P. Awasthi, A. Blum, and O. Sheffet. Center-based clustering under perturbation stability. *Information Processing Letters*, 112(1–2):49–54, 2012.

B. Bahmani, B. Moseley, A. Vattani, R. Kumar, and S. Vassilvitskii. Scalable k-means++. In *Proceedings of the 38th International Conference on Very Large Databases*, Istanbul, Turkey, 2012.

M.-F. Balcan, A. Blum, and A. Gupta. Approximate clustering without the approximation. In *Proceedings of the ACM-SIAM Symposium on Discrete Algorithms*, New York, USA, 2009.

M.-F. Balcan, A. Blum, and A. Gupta. Clustering under approximation stability. *Journal of the ACM*, 60:8, 2013.

M.-F. Balcan and Y. Liang. Clustering under perturbation resilience. *Proceedings of the 39th International Colloquium on Automata, Languages and Programming*, Warwick, UK, 2012.

M. Belkin and K. Sinha. Polynomial learning of distribution families. In *Proceedings of the 51st Annual IEEE Symposium on Foundations of Computer Science*, Las Vegas, USA, 2010.

J. C. Bezdek. *Pattern Recognition with Fuzzy Objective Function Algorithms*. Kluwer Academic Publishers, Norwell, MA 1981.

Y. Bilu and N. Linial. Are stable instances easy? In *Proceedings of the First Symposium on Innovations in Computer Science*, Beijing, China, 2010.

L. Bottou and Y. Bengio. Convergence properties of the k-means algorithms. In *Advances in Neural Information Processing Systems 7*, D.S. Touretzky (ed.), pp. 585–592. MIT Press, Cambridge, 1995.

S. C. Brubaker and S. Vempala. Isotropic PCA and affine-invariant clustering. In *Proceedings of the 2008 49th Annual IEEE Symposium on Foundations of Computer Science*, Philadelphia, USA, 2008.

B. Chandra, H. Karloff, and C. Tovey. New results on the old k-opt algorithm for the tsp. In *Proceedings of the Fifth Annual ACM-SIAM Symposium on Discrete Algorithms*, Arlington, USA, 1994.

M. Charikar, S. Guha, E. Tardos, and D. B. Shmoy. A constant-factor approximation algorithm for the k-median problem. In *Proceedings of the Thirty-First Annual ACM Symposium on Theory of Computing*, Atlanta, USA, 1999.

K. Chaudhuri and S. Rao. Beyond gaussians: Spectral methods for learning mixtures of heavy-tailed distributions. In *Proceedings of the 21st Annual Conference on Learning Theory*, Helsinki, Finland, 2008a.

K. Chaudhuri and S. Rao. Learning mixtures of product distributions using correlations and independence. In *Proceedings of the 21st Annual Conference on Learning Theory*, Helsinki, Finland, 2008b.

S. Dasgupta. Learning mixtures of gaussians. In *Proceedings of The 40th Annual Symposium on Foundations of Computer Science*, New York, USA, 1999.

S. Dasgupta. *The Hardness of k-means Clustering*. Technical report, University of California, San Diego, 2008.

I. S. Dhillon. Co-clustering documents and words using bipartite spectral graph partitioning. In *Proceedings of the Seventh ACM SIGKDD International Conference on Knowledge Discovery and Data Mining*, San Francisco, USA, 2001.

D. E. Drake and S. Hougardy. Linear time local improvements for weighted matchings in graphs. In *Proceedings of the 2nd International Conference on Experimental and Efficient Algorithms*, Ascona, Switzerland, 2003.

V. Faber. *Clustering and the Continuous k-Means Algorithm*, 1994.

W. Fernandez de la Vega, M. Karpinski, C. Kenyon, and Y. Rabani. Approximation schemes for clustering problems. In *Proceedings of the Thirty-Fifth Annual ACM Symposium on Theory of Computing*, San Diego, USA, 2003.

R. D. Finn, J. Mistry, J. Tate, P. Coggill, A. Heger, J. E. Pollington, O. L. Gavin, P. Gunesekaran, G. Ceric, K. Forslund, L. Holm, E. L. Sonnhammer, S. R. Eddy, and A. Bateman. The pfam protein families database. *Nucleic Acids Research*, 38:D211–222, 2010.

A. Gersho and R. M. Gray. *Vector Quantization and Signal Compression*. Kluwer Academic Publishers, Norwell, MA 1991.

P. Hansen and B. Jaumard. Algorithms for the maximum satisfiability problem. *Computing*, 44:279–303, 1990.

D. Hsu and S. M. Kakade. Learning mixtures of spherical gaussians: Moment methods and spectral decompositions. In *Proceedings of the 4th Innovations in Theoretical Computer Science Conference*, Berkeley, USA, 2013.

M. Inaba, N. Katoh, and H. Imai. Applications of weighted voronoi diagrams and randomization to variance-based k-clustering: (Extended abstract). In *Proceedings of the Tenth Annual Symposium on Computational Geometry*, Kyoto, Japan, 1994.

K. Jain, M. Mahdian, and A. Saberi. A new greedy approach for facility location problems. In *Proceedings of the 34th Annual ACM Symposium on Theory of Computing*, Montreal, Canada, 2002.

A. T. Kalai, A. Moitra, and G. Valiant. Efficiently learning mixtures of two gaussians. In *Proceedings of the 42th ACM Symposium on Theory of Computing*, Cambridge, USA, 2010.

R. Kannan, H. Salmasian, and S. Vempala. The spectral method for general mixture models. In *Proceedings of The Eighteenth Annual Conference on Learning Theory*, Bertinoro, Italy, 2005.

R. Kannan and S. Vempala. Spectral algorithms. *Foundations and Trends in Theoretical Computer Science*, 4(3–4):157–228, 2009.

T. Kanungo, D. M. Mount, N. S. Netanyahu, C. D. Piatko, R. Silverman, and A. Y. Wu. A local search approximation algorithm for k-means clustering. In *Proceedings of the Eighteenth Annual Symposium on Computational Geometry*, New York, NY, 2002.

A. Kumar and R. Kannan. Clustering with spectral norm and the k-means algorithm. In *Proceedings of the 51st Annual IEEE Symposium on Foundations of Computer Science*, Las Vegas, USA, 2010.

A. Kumar, Y. Sabharwal, and S. Sen. A simple linear time $(1 + \epsilon)$-approximation algorithm for k-means clustering in any dimensions. In *Proceedings of the 45th Annual IEEE Symposium on Foundations of Computer Science*, Washington, DC, 2004.

S. Li and O. Svensson. Approximating k-median via pseudo-approximation. In *Proceedings of the 45th ACM Symposium on Theory of Computing*, Palo Alto, USA, 2013.

S. P. Lloyd. Least squares quantization in PCM. *IEEE Transactions on Information Theory*, 28(2): 129–137, 1982.

K. Makarychev, Y. Makarychev, and A. Vijayaraghavan. Bilu-linial stable instances of max cut and minimum multiway cut. In *SODA*, Chandra Chekuri (ed.), pp. 890–906. SIAM, Portland, USA, 2014.

N. Megiddo and K. Supowit. On the complexity of some common geometric location problems. *SIAM Journal on Computing*, 13(1):182–196, 1984.

M. Meilă. The uniqueness of a good optimum for K-means. In *Proceedings of the International Machine Learning Conference*, Pittsburgh, USA, pp. 625–632, 2006.

A. Moitra and G. Valiant. Settling the polynomial learnability of mixtures of gaussians. In *Proceedings of the 51st Annual IEEE Symposium on Foundations of Computer Science*, Las Vegas, USA, 2010.

A. G. Murzin, S. E. Brenner, T. Hubbard, and C. Chothia. Scop: A structural classification of proteins database for the investigation of sequences and structures. *Journal of Molecular Biology*, 247: 536–540, 1995.

H. H. Nguyen and P. Cohen. Gibbs random fields, fuzzy clustering, and the unsupervised segmentation of textured images. *CVGIP: Graphical Models and Image Processing*, 55(1): 1–19, 1993.

R. Ostrovsky, Y. Rabani, L. Schulman, and C. Swamy. The effectiveness of lloyd-type methods for the k-means problem. In *Proceedings of the 47th Annual IEEE Symposium on Foundations of Computer Science*, Berkeley, USA, 2006.

C. H. Papadimitriou. On selecting a satisfying truth assignment (extended abstract). In *Proceedings of the 32nd Annual Symposium on Foundations of Computer Science*, San Juan, Puerto Rico, 1991.

D. Pelleg and A. W. Moore. X-means: Extending k-means with efficient estimation of the number of clusters. In *Proceedings of the Seventeenth International Conference on Machine Learning*, Stanford, USA, 2000.

E. M. Rasmussen. Clustering algorithms. *Information Retrieval: Data Structures & Algorithms*, pp. 419–442. 1992.

P. Schuurman and T. Vredeveld. Performance guarantees of local search for multiprocessor scheduling. *INFORMS Journal on Computing*, 19:52–63, 2007.

G. Schwarz. Estimating the dimension of a model. *The Annals of Statistics*, 6(2):461–464, 1978.

M. Shindler, A. Wong, and A. Meyerson. Fast and accurate k-means for large datasets. In *Proceedings of the 25th Annual Conference on Neural Information Processing Systems*, Granada, Spain, 2011.

A. H. S. Solberg, T. Taxt, and A. K. Jain. A markov random field model for classification of multisource satellite imagery. *IEEE Transactions on Geoscience and Remote Sensing*, 34:100–113, 1996.

D. A. Spielman and S.-H. Teng. Smoothed analysis of algorithms: Why the simplex algorithm usually takes polynomial time. *Journal of the ACM*, 51(3):385–463, 2004.

A. Vattani. k-means requires exponentially many iterations even in the plane. In *Proceedings of the 25th Annual Symposium on Computational Geometry*, Aarhus, Denmark, 2009.

S. Vempala and G. Wang. A spectral algorithm for learning mixture models. *Journal of Computer and System Sciences*, 68(2):841–860, 2004.

K. Voevodski, M. F. Balcan, H. Roeglin, S. Teng, and Y. Xia. Efficient clustering with limited distance information. In *Proceedings of the 26th Conference on Uncertainty in Artificial Intelligence*, Catalina Island, USA, 2010.

X. Wu, V. Kumar, J. Ross Quinlan, J. Ghosh, Q. Yang, H. Motoda, G. J. McLachlan, A. Ng, B. Liu, P. S. Yu, Z.-H. Zhou, M. Steinbach, D. J. Hand, and D. Steinberg. The top ten algorithms in data mining. *Knowledge and Information Systems*, 14:1–37, 2008.

Section II

Dissimilarity-Based Methods

6

Hierarchical Clustering

Pedro Contreras and Fionn Murtagh

CONTENTS

Abstract

To begin with, we review dissimilarity, metric, and ultrametric. Next, we introduce hierarchical clustering using the single link agglomerative criterion. Then we present agglomerative hierarchical clustering in full generality. Storage and computational properties are reviewed. This includes the state-of-the art agglomerative hierarchical clustering algorithm that uses a nearest-neighbor chain and reciprocal nearest neighbors. We then review various, recently developed, hierarchical clustering algorithms that use density or grid-based approaches. That includes a linear time algorithm. A number of examples, and R implementation, completes this chapter.

6.1 Introduction

Agglomerative hierarchical clustering has been the dominant approach to constructing embedded classification schemes. We also survey relatively recently developed divisive hierarchical clustering algorithms. We consider both efficiency properties (the latter, from the computational and storage points of view) and effectiveness properties (from the application point of view). It is often helpful to distinguish between *method*, involving a compactness criterion and the target structure of a two-way tree representing the partial order on subsets of the power set; as opposed to an *implementation*, which relates to the detail of the algorithm used.

As with many other multivariate techniques, the objects to be classified have numerical measurements on a set of variables or attributes. Hence, the analysis is carried out on the rows of an array or matrix. If we do not have a matrix of numerical values to begin with, then it may be necessary to appropriately construct such a matrix. The objects, or rows of the matrix, can be viewed as vectors in a multidimensional space (the dimensionality of this space being the number of variables or columns). A geometric framework of this type is not the only one which can be used to formulate clustering algorithms. Suitable alternative forms of storage of a rectangular array of values are not inconsistent with viewing the problem in geometric terms (and in matrix terms—e.g., expressing the adjacency relations in a graph).

Motivation for clustering in general including hierarchical clustering and applications encompass: analysis of data and pattern recognition, storage, search, and retrieval.

Surveys of clustering with coverage also of hierarchical clustering include Gordon (1981), March (1983), Jain and Dubes (1988), Gordon (1987), Mirkin (1996), Jain et al. (1999), and Xu and Wunsch (2005). Lerman (1981) and Janowitz (2010) present overarching reviews of clustering including through use of lattices that generalize trees. The case for the central role of hierarchical clustering in information retrieval was made by van Rijsbergen (1979) and continued in the work of Willett and others. Various mathematical views of hierarchy, all expressing symmetry in one way or another, are explored in Murtagh (2009). This chapter is organized as follows. We begin, in Section 6.2, with input data, prior to inducing a hierarchy on the data.

In Section 6.3 there is a pedagogical introduction using the single link agglomerative criterion. In Section 6.4, we discuss the Lance–Williams formulation of a wide range of algorithms, and how these algorithms can be expressed in graph theoretic terms and in geometric terms. In Section 6.5, we describe the principles of the reciprocal nearest neighbor and nearest neighbor chain algorithm, to support building a hierarchical clustering in a more efficient way compared to the Lance–Williams or general geometric approaches.

Section 6.6 surveys developments in grid- and density-based clustering. We also present a recent algorithm of this type, which is particularly suitable for the hierarchical clustering of massive datasets. Finally Section 6.7 looks at some examples, using R.

6.2 Input Data, Dissimilarity, Metric, and Ultrametric

To group data, we need a way to measure the elements and their distances relative to each other in order to decide which elements belong to a group. This is also called similarity, although on many occasions a dissimilarity measurement is used. Note that not any arbitrary measurement is of use to us here, and in practice usually this measurement will be a metric distance.

When working in a vector space a traditional way to measure distances is a Minkowski distance, which is a family of metrics defined as follows:

$$L_p(\mathbf{x}_a, \mathbf{x}_b) = \left(\sum_{i=1}^{n} |\mathbf{x}_{i,a} - \mathbf{x}_{i,b}|^p \right)^{1/p} \quad ; \quad \forall \, p \geq 1, \; p \in \mathbb{Z} \tag{6.1}$$

where $\mathbb{Z}$ is the set of integers.

The Manhattan, Euclidean, and Chebyshev distances (the latter is also called maximum distance) are special cases of the Minkowski distance when $p = 1$, $p = 2$, and $p \to \infty$.

Additionally, we can mention the *cosine* similarity, which gives the angle between two vectors. This is widely used in text retrieval to match vector queries to the dataset. The smaller the angle between a query vector to the document set vectors, the closer is a query to a document. The normalized cosine similarity is defined as follows:

$$s(\mathbf{x}_a, \mathbf{x}_b) = \cos(\theta) = \frac{\mathbf{x}_a \cdot \mathbf{x}_b}{\|\mathbf{x}_a\| \, \|\mathbf{x}_b\|} \tag{6.2}$$

where $\mathbf{x}_a \cdot \mathbf{x}_b$ is the dot product and $\| \cdot \|$ the norm.

Other relevant distances are the Hellinger, Variational, Mahalanobis, and Hamming distances. Anderberg (1973) gives a good review of measurement and metrics, where their interrelationships are also discussed. Also Deza and Deza (2009) have produced a comprehensive list of distances in their *Encyclopedia of Distances*.

By mapping our input data into a Euclidean space, where each object is equiweighted, we can use a Euclidean distance for the clustering that follows. Correspondence analysis is very versatile in determining a Euclidean, factor space from a wide range of input data types, including frequency counts, mixed qualitative and quantitative data values, ranks or scores, and others. Further reading on this is to be found in Benzécri (1979), Le Roux and Rouanet (2004), and Murtagh (2005).

A hierarchical clustering can be viewed as mapping a metric space (e.g.) into an ultra-metric space. Figure 6.1, left, illustrates the triangular inequality, which defines a metric, together with the properties of positive semidefiniteness (for all observations i and j, their distance, $d(i, j)$, satisfies $d(i, j) \geq 0$ and $d(i, j) = 0$ only if $i = j$); and symmetry ($d(i, j) = d(j, i)$).

Figure 6.1, right, illustrates the strong triangular inequality, or ultrametric inequality, that is a defining property of a metric defined on a tree.

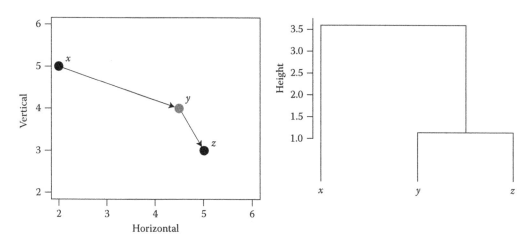

FIGURE 6.1

Left: The triangular inequality defines a metric: every triplet of points satisfies the relationship: $d(x, z) \leq d(x, y) + d(y, z)$ for distance d. Right: The strong triangular inequality defines an ultrametric: every triplet of points satisfies the relationship: $d(x, z) \leq \max\{d(x, y), d(y, z)\}$ for distance d. Cf. by reading off the hierarchy, how this is verified for all x, y, z: $d(x, z) = 3.5; d(x, y) = 3.5; d(y, z) = 1.0$. In addition, the symmetry and positive definiteness conditions hold for any pair of points.

6.3 Introduction to Hierarchical Clustering Using the Single Linkage Agglomerative Criterion

Agglomerative hierarchical clustering algorithms can be characterized as *greedy*. A sequence of irreversible algorithm steps is used to construct the desired data structure. Assume that a pair of clusters, including possibly singletons, is merged or agglomerated at each step of the algorithm. Then the following are equivalent views of the same output structure constructed on n objects: a set of $n-1$ partitions, starting with the fine partition consisting of n classes and ending with the trivial partition consisting of just one class, the entire object set; a binary tree (one or two child nodes at each nonterminal node) commonly referred to as a dendrogram; a partially ordered set (poset) which is a subset of the power set of the n objects; and an ultrametric topology on the n objects. We proceed now with the short pedagogical example.

The single linkage hierarchical clustering approach outputs a set of clusters (to use graph theoretic terminology, a set of maximal connected subgraphs) at each level—or for each threshold value which produces a new partition. The following algorithm, in its general structure, is relevant for a wide range of hierarchical clustering methods which vary only in the update formula used in step 2. These methods may, for example, define a criterion of compactness in step 2 to be used instead of the connectivity criterion used here. The single linkage method with which we begin is one of the oldest methods, its origins being traced to Polish researchers in the 1950s (Graham and Hell, 1985). An example is shown in Figure 6.2. Note that the dissimilarity coefficient is assumed to be symmetric, and so the clustering algorithm is implemented on half the dissimilarity matrix.

Single linkage hierarchical clustering

> *Input* An $n(n-1)/2$ set of dissimilarities.
> *Step 1* Determine the smallest dissimilarity, d_{ik}.
> *Step 2* Agglomerate objects i and k: that is, replace them with a new object, $i \cup k$; update dissimilarities such that, for all objects $j \neq i, k$:
>
> $$d_{i \cup k, j} = \min\{d_{ij}, d_{kj}\}$$
>
> Delete dissimilarities d_{ij} and d_{kj}, for all j, as these are no longer used.
> *Step 3* While at least two objects remain, return to step 1.

Equal dissimilarities may be treated in an arbitrary order. There are precisely $n-1$ agglomerations in step 2 (allowing for arbitrary choices in step 1 if there are identical dissimilarities). It may be convenient to index the clusters found in step 2 by $n+1$, $n+2, \ldots, 2n-1$, or an alternative practice is to index cluster $i \cup k$ by the lower of the indices of i and k.

The title *single linkage* arises since, in step 2, the interconnecting dissimilarity between two clusters ($i \cup k$ and j) or components is defined as the least interconnecting dissimilarity between a member of one and a member of the other. Other hierarchical clustering methods are characterized by other functions of the interconnecting linkage dissimilarities.

Since there are $n-1$ agglomerations, and hence iterations, and since step 2 requires $< n$ operations, the algorithm for the single linkage hierarchical clustering described here is of time complexity $O(n^2)$.

	1	2	3	4	5
1	0	4	9	5	8
2	4	0	6	3	6
3	9	6	0	6	3
4	5	3	6	0	5
5	8	6	3	5	0

Agglomerate 2 and 4
at dissimilarity 3

	1	2∪4	3	5
1	0	4	9	8
2∪4	4	0	6	5
3	9	6	0	3
5	8	5	3	0

Agglomerate 3 and 5
at dissimilarity 3

	1	2∪4	3∪5
1	0	4	8
2∪4	4	0	5
3∪5	8	5	0

Agglomerate 1 and 2∪4
at dissimilarity 4

	1∪2∪4	3∪5
1∪2∪4	0	5
3∪5	5	0

Agglomerate 1∪2∪4 and 3∪5 at
dissimilarity 5

Resulting dendrogram

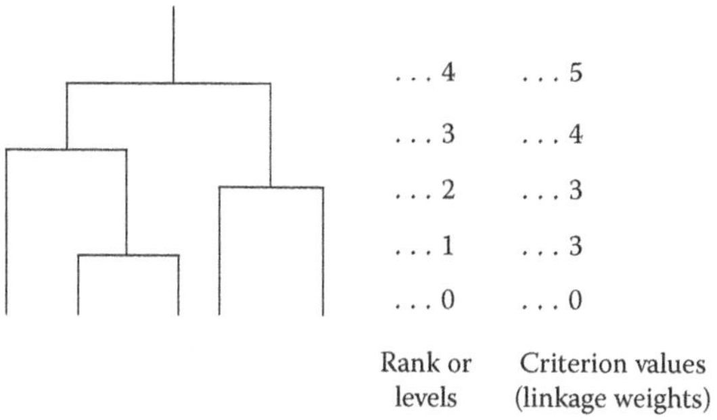

Rank or levels	Criterion values (linkage weights)
. . . 4	. . . 5
. . . 3	. . . 4
. . . 2	. . . 3
. . . 1	. . . 3
. . . 0	. . . 0

FIGURE 6.2
Construction of a dendrogram by the single linkage method.

Compared to other hierarchical clustering techniques, the single linkage method can give rise to a notable disadvantage for summarizing interrelationships. This is known as *chaining*. An example is to consider four subject-areas, which it will be supposed are characterized by certain attributes: computer science, statistics, probability, and measure theory. It is conceivable that "computer science" is connected to "statistics" at some threshold value, "statistics" to "probability," and "probability" to "measure theory," thereby giving rise to the fact that "computer science" and "measure theory" find themselves, undesirably, in the same cluster. This is due to the intermediaries "statistics" and "probability."

As early as the 1970s, it was held that about 75% of all published work on clustering employed hierarchical algorithms (Blashfield and Aldenderfer, 1978). Interpretation of the information contained in a dendrogram is often of one or more of the following kinds:

- Set inclusion relationships
- Partition of the object-sets
- Significant clusters

Much early work on hierarchical clustering was in the field of biological taxonomy, from the 1950s and more so from the 1960s onwards. The central reference in this area, the first edition of which dates from the early 1960s, is Sneath and Sokal (1973). One major interpretation of hierarchies has been the evolution relationships between the organisms under study. It is hoped, in this context, that a dendrogram provides a sufficiently accurate model of underlying evolutionary progression.

The most common interpretation made of hierarchical clustering is to derive a partition: a line is drawn horizontally through the hierarchy, to yield a set of classes. These clusters are precisely the connected components in the case of the single linkage method. A line drawn just above rank 3 (or criterion value 4) on the dendrogram in Figure 6.2 yields classes $\{1, 2, 4\}$ and $\{3, 5\}$. Generally, the choice of where "to draw the line" is arrived at on the basis of large changes in the criterion value. However, the changes in criterion value increase (usually) toward the final set of agglomerations, which renders the choice of best partition on this basis difficult. Since every line drawn through the dendrogram defines a partition, it may be expedient to choose a partition with convenient features (number of classes, number of objects per class).

A further type of interpretation is to dispense with the requirement that the classes chosen constitute a partition, and instead detect maximal (i.e., disjoint) clusters of interest at varying levels of the hierarchy. Such an approach is used by Rapoport and Fillenbaum (1972) in a clustering of colors based on semantic attributes. Lerman (1981) developed an approach for finding significant clusters at varying levels of a hierarchy, which has been widely applied. See also Murtagh (2007) which, based on a wavelet transform on a dendrogram, is used to find the important, that is best approximating, clusters.

In summary, a dendrogram expresses many of the proximity and classificatory relationships in a body of data. It is a convenient representation which answers such questions as: "How many groups are in this data?" "What are the salient interrelationships present?" But it should be stressed that differing answers can feasibly be provided by a dendrogram for most of these questions, depending on the application.

6.4 Agglomerative Hierarchical Clustering Algorithms

In the last section, a general agglomerative algorithm was discussed. A wide range of these algorithms have been proposed at one time or another. Hierarchical agglomerative algorithms may be conveniently broken down into two groups of methods. The first group is that of linkage methods—the single, complete, weighted, and unweighted average linkage methods. These are methods for which a graph representation can be used. Sneath and Sokal (1973) may be consulted for many other graph representations of the stages in the construction of hierarchical clusterings.

The second group of hierarchical clustering methods are methods which allow the cluster centers to be specified (as an average or a weighted average of the member vectors of the cluster). These methods include the centroid, median, and minimum variance methods.

The latter may be specified either in terms of dissimilarities, alone, or alternatively in terms of cluster center coordinates and dissimilarities. A very convenient formulation, in dissimilarity terms, which embraces all the hierarchical methods mentioned so far, is the *Lance–Williams dissimilarity update formula*. If points (objects) i and j are agglomerated into cluster $i \cup j$, then we must simply specify the new dissimilarity between the cluster and all other points (objects or clusters). The formula is:

$$d(i \cup j, k) = \alpha_i d(i,k) + \alpha_j d(j,k) + \beta d(i,j) + \gamma |d(i,k) - d(j,k)|$$

where α_i, α_j, β, and γ define the agglomerative criterion. Values of these are listed in the second column of Table 6.1. In the case of the single link method, using $\alpha_i = \alpha_j = \frac{1}{2}$, $\beta = 0$,

TABLE 6.1

Specifications of Seven Hierarchical Clustering Methods

Hierarchical Clustering Methods (and Aliases)	Lance and Williams Dissimilarity Update Formula	Coordinates of Center of Cluster, Which Agglomerates Clusters i and j	Dissimilarity between Cluster Centers g_i and g_j
Single link (nearest neighbor)	$\alpha_i = 0.5$ $\beta = 0$ $\gamma = -0.5$ (More simply: $\min\{d_{ik}, d_{jk}\}$)		
Complete link (diameter)	$\alpha_i = 0.5$ $\beta = 0$ $\gamma = 0.5$ (More simply: $\max\{d_{ik}, d_{jk}\}$)		
Group average (average link, UPGMA)	$\alpha_i = \dfrac{|i|}{|i| + |j|}$ $\beta = 0$ $\gamma = 0$		
McQuitty's method (WPGMA)	$\alpha_i = 0.5$ $\beta = 0$ $\gamma = 0$		
Median method (Gower's, WPGMC)	$\alpha_i = 0.5$ $\beta = -0.25$ $\gamma = 0$	$g = \dfrac{g_i + g_j}{2}$	$\|g_i - g_j\|^2$
Centroid (UPGMC)	$\alpha_i = \dfrac{|i|}{|i| + |j|}$ $\beta = -\dfrac{|i||j|}{(|i| + |j|)^2}$ $\gamma = 0$	$g = \dfrac{|i|g_i + |j|g_j}{|i| + |j|}$	$\|g_i - g_j\|^2$
Ward's method (minimum variance, error sum of squares)	$\alpha_i = \dfrac{|i| + |k|}{|i| + |j| + |k|}$ $\beta = -\dfrac{|k|}{|i| + |j| + |k|}$ $\gamma = 0$	$g = \dfrac{|i|g_i + |j|g_j}{|i| + |j|}$	$\dfrac{|i||j|}{|i| + |j|}\|g_i - g_j\|^2$

Note: $|i|$ is the number of objects in cluster i; g_i is a vector in m-space (m is the set of attributes)—either an initial point or a cluster center; $\| \cdot \|$ is the norm in the Euclidean metric; the names UPGMA, etc. are due to Sneath and Sokal (1973). The Lance and Williams recurrence formula is:

$$d_{i \cup j, k} = \alpha_i d_{ik} + \alpha_j d_{jk} + \beta d_{ij} + \gamma |d_{ik} - d_{jk}|$$

and $\gamma = -\frac{1}{2}$ gives us

$$d(i \cup j, k) = \frac{1}{2}d(i,k) + \frac{1}{2}d(j,k) - \frac{1}{2}|d(i,k) - d(j,k)|$$

which, it may be verified by taking a few simple examples of three points, i, j, and k, can be rewritten as

$$d(i \cup j, k) = \min\{d(i,k), d(j,k)\}$$

This was exactly the update formula used in the agglomerative algorithm given in Section 6.3. Using other update formulas, as given in column 2 of Table 6.1, allows the other agglomerative methods to be implemented in a very similar way to the implementation of the single link method.

In the case of the methods which use cluster centers, we have the center coordinates (in column 3 of Table 6.1) and dissimilarities as defined between cluster centers (column 4 of Table 6.1). The Euclidean distance must be used for equivalence between the two approaches. In the case of the *median method*, for instance, we have the following (cf. Table 6.1).

Let **a** and **b** be two points (i.e., m-dimensional vectors: these are objects or cluster centers) which have been agglomerated, and let **c** be another point. From the Lance–Williams dissimilarity update formula, using squared Euclidean distances, we have:

$$\begin{aligned}
d^2(a \cup b, c) &= \frac{d^2(a,c)}{2} + \frac{d^2(b,c)}{2} - \frac{d^2(a,b)}{4} \\
&= \frac{\|\mathbf{a} - \mathbf{c}\|^2}{2} + \frac{\|\mathbf{b} - \mathbf{c}\|^2}{2} - \frac{\|\mathbf{a} - \mathbf{b}\|^2}{4}
\end{aligned} \tag{6.3}$$

The new cluster center is $(\mathbf{a} + \mathbf{b})/2$, so that its distance to point **c** is

$$\left\| \mathbf{c} - \frac{\mathbf{a} + \mathbf{b}}{2} \right\|^2 \tag{6.4}$$

That these two expressions are identical is readily verified. The correspondence between these two perspectives on the one agglomerative criterion is similarly proved for the centroid and minimum variance methods.

The single linkage algorithm discussed in Section 6.3, duly modified for the use of the Lance-Williams dissimilarity update formula, is applicable for all agglomerative strategies. The update formula listed in Table 6.1 is used in step 2 of the algorithm.

For cluster center methods, and with suitable alterations for graph methods, the following algorithm is an alternative to the general dissimilarity-based algorithm. The latter may be described as a "stored dissimilarities approach" (Anderberg, 1973).

Stored data approach

Step 1 Examine all interpoint dissimilarities, and form cluster from two closest points.

Step 2 Replace two points clustered by representative point (centre of gravity) or by cluster fragment.

Step 3 Return to step 1, treating clusters as well as remaining objects, until all objects are in one cluster.

In steps 1 and 2, "point" refers either to objects or clusters, both of which are defined as vectors in the case of cluster center methods. This algorithm is justified by storage considerations, since we have $O(n)$ storage required for n initial objects and $O(n)$ storage for the $n-1$ (at most) clusters. In the case of linkage methods, the term "fragment" in step 2 refers (in the terminology of graph theory) to a connected component in the case of the single link method and to a clique or complete subgraph in the case of the complete link method. The overall complexity of the above algorithm is $O(n^3)$: the repeated calculation of dissimilarities in step 1, coupled with $O(n)$ iterations through steps 1, 2, and 3. Note, however, that this does not take into consideration the extra processing required in a linkage method, where "closest" in step 1 is defined with respect to graph fragments.

While the stored data algorithm is instructive, it does not lend itself to efficient implementations. In the section to follow, we look at the reciprocal nearest neighbor and mutual nearest neighbor algorithms which have to be used in practice for implementing agglomerative hierarchical clustering algorithms.

Before concluding this overview of the agglomerative hierarchical clustering, we will describe briefly the minimum variance method.

The variance or spread of a set of points (i.e., the sum of squared distances from the center) has been the point of departure for specifying many clustering algorithms. Many of these algorithms—iterative, optimization algorithms as well as the hierarchical, agglomerative algorithms—are briefly described and appraised in Wishart (1969). The use of variance in a clustering criterion links the resulting clustering to other data-analytic techniques which involve a decomposition of variance, and make the minimum variance agglomerative strategy particularly suitable for synoptic clustering. Hierarchies are also more balanced with this agglomerative criterion, which is often of practical advantage.

The minimum variance method produces clusters which satisfy compactness and isolation criteria. These criteria are incorporated into the dissimilarity. We seek to agglomerate two clusters, c_1 and c_2, into cluster c such that the within-class variance of the partition thereby obtained is minimum. Alternatively, the between-class variance of the partition obtained is to be maximized. Let P and Q be the partitions prior to, and subsequent to, the agglomeration; let $p_1, p_2, \ldots$ be classes of the partitions:

$$P = \{p_1, p_2, \ldots, p_k, c_1, c_2\}$$
$$Q = \{p_1, p_2, \ldots, p_k, c\}$$

Letting V denote *variance*, then in agglomerating two classes of P, the variance of the resulting partition (i.e., $V(Q)$) will necessarily decrease: therefore in seeking to minimize this decrease, we simultaneously achieve a partition with maximum between-class variance. The criterion to be optimized can then be shown to be:

$$V(P) - V(Q) = V(c) - V(c_1) - V(c_2)$$
$$= \frac{|c_1||c_2|}{|c_1| + |c_2|} \|\mathbf{c_1} - \mathbf{c_2}\|^2$$

which is the dissimilarity given in Table 6.1. This is a dissimilarity which may be determined for any pair of classes of partition P; and the agglomerands are those classes, c_1 and c_2, for which it is minimum.

It may be noted that if c_1 and c_2 are singleton classes, then $V(\{c_1, c_2\}) = \frac{1}{2}\|\mathbf{c_1} - \mathbf{c_2}\|^2$ (i.e., the variance of a pair of objects is equal to half their Euclidean distance).

6.5 Efficient Hierarchical Clustering Algorithms Using Nearest Neighbor Chains

Early, efficient algorithms for hierarchical clustering are due to Sibson (1973), Rohlf (1973), and Defays (1977). Their $O(n^2)$ implementations of the single link method and of a (nonunique) complete link method, respectively, have been widely cited.

In the early 1980s a range of significant improvements (de Rham, 1980; Juan, 1982) were made to the Lance–Williams, or related, dissimilarity update schema, which had been in wide use since the mid-1960s. Murtagh (1983, 1985, 1992) presents a survey of these algorithmic improvements. We will briefly describe them here. The new algorithms, which have the potential for *exactly* replicating results found in the classical but more computationally expensive way, are based on the construction of *nearest neighbor chains* and *reciprocal* or *mutual NNs* (NN-chains and RNNs).

An NN-chain consists of an arbitrary point (a in Figure 6.3); followed by its NN (b in Figure 6.3); followed by the NN from among the remaining points (c, d, and e in Figure 6.3) of this second point; and so on until we necessarily have some pair of points which can be termed reciprocal or mutual NNs. (Such a pair of RNNs may be the first two points in the chain; and we have assumed that no two dissimilarities are equal.)

In constructing an NN-chain, irrespective of the starting point, we may agglomerate a pair of RNNs as soon as they are found. What guarantees that we can arrive at the same hierarchy as if we used traditional "stored dissimilarities" or "stored data" algorithms? Essentially, this is the same condition as that under which no inversions or reversals are produced by the clustering method. Figure 6.4 gives an example of this, where s is agglomerated at a lower criterion value (i.e., dissimilarity) than was the case at the previous agglomeration between q and r. Our ambient space has thus contracted because of the agglomeration. This is due to the algorithm used—in particular the agglomeration criterion—and it is something we would normally wish to avoid.

This is formulated as:

$$\text{Inversion impossible if: } d(i,j) < d(i,k) \quad \text{or} \quad d(j,k) \quad \Rightarrow \quad d(i,j) < d(i \cup j,k)$$

This is one form of Bruynooghe's *reducibility property* (Bruynooghe, 1977; see also Murtagh, 1984). Using the Lance–Williams dissimilarity update formula, it can be shown that the minimum variance method does not give rise to inversions; neither do the linkage methods; but the median and centroid methods cannot be guaranteed not to have inversions.

To return to Figure 6.3, if we are dealing with a clustering criterion which precludes inversions, then c and d can justifiably be agglomerated, since no other point (e.g., b or e) could have been agglomerated to either of these.

The processing required, following an agglomeration, is to update the NNs of points such as b in Figure 6.3 (and on account of such points, this algorithm was dubbed *algorithme des*

a   b   c d   e

FIGURE 6.3
Five points, showing NNs and RNNs.

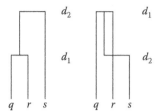

FIGURE 6.4
Alternative representations of a hierarchy with an inversion. Assuming dissimilarities, as we go vertically up, criterion values (d_1, d_2) decrease. But here, undesirably, $d_2 > d_1$.

célibataires, or batchelors' algorithm, in de Rham, 1980). The following is a summary of the algorithm:

NN-chain algorithm

Step 1 Select a point arbitrarily.

Step 2 Grow the NN-chain from this point until a pair of RNNs are obtained.

Step 3 Agglomerate these points (replacing with a cluster point, or updating the dissimilarity matrix).

Step 4 From the point which preceded the RNNs (or from any other arbitrary point if the first two points chosen in steps 1 and 2 constituted a pair of RNNs), return to step 2 until only one point remains.

In Murtagh (1983, 1984, 1985) and Day and Edelsbrunner (1984), one finds discussions of $O(n^2)$ time and $O(n)$ space implementations of Ward's minimum variance (or error sum of squares) method and of the centroid and median methods. The latter two methods are termed the UPGMC and WPGMC criteria by Sneath and Sokal (1973). Now, a problem with the cluster criteria used by these latter two methods is that the reducibility property is not satisfied by them. This means that the hierarchy constructed may not be unique as a result of inversions or reversals (nonmonotonic variation) in the clustering criterion value determined in the sequence of agglomerations.

Murtagh (1983, 1985) describes $O(n^2)$ time and $O(n^2)$ space implementations for the single link method, the complete link method and for the weighted and unweighted group average methods (WPGMA and UPGMA). This approach is quite general vis à vis the dissimilarity used and can also be used for hierarchical clustering methods other than those mentioned.

Day and Edelsbrunner (1984) prove the exact $O(n^2)$ time complexity of the centroid and median methods using an argument related to the combinatorial problem of optimally packing hyperspheres into an m-dimensional volume. They also address the question of metrics: results are valid in a wide class of distances including those associated with the Minkowski metrics.

The construction and maintenance of the nearest neighbor chain as well as the carrying out of agglomerations whenever reciprocal nearest neighbors meet, both offer possibilities for distributed implementation. Implementations on a parallel machine architecture were described by Willett (1989).

Evidently both coordinate data and graph (e.g., dissimilarity); data can be input to these agglomerative methods. Gillet et al. (1998) in the context of clustering chemical structure databases refer to the common use of the Ward method, based on the reciprocal nearest neighbors algorithm, on datasets of a few hundred thousand molecules.

Applications of hierarchical clustering to bibliographic information retrieval are assessed in Griffiths et al. (1984). Ward's minimum variance criterion is favored.

From details in White and McCain (1997), the Institute of Scientific Information clusters citations (science and social science) by first clustering highly cited documents based on a single linkage criterion, and then four more passes are made through the data to create a subset of a single linkage hierarchical clustering.

In the CLUSTAN and R statistical data analysis packages (other than `hclust` in R, see `flashClust` due to P. Langfelder and available on CRAN, "Comprehensive R Archive Network," cran.r-project.org), there are implementations of the NN-chain algorithm for the minimum variance agglomerative criterion. A property of the minimum variance agglomerative hierarchical clustering method is that we can use weights on the objects on which we will induce a hierarchy. By default, these weights are identical and equal to 1. Such weighting of observations to be clustered is not widely available.

6.6 Density and Grid-Based Clustering Techniques

Many modern clustering techniques focus on large scale data, in Xu and Wunsch (2008, p. 215) these are classified as follows:

- Random sampling
- Data condensation
- Density-based approaches
- Grid-based approaches
- Divide and conquer
- Incremental learning

From the point of view of this chapter, density and grid-based approaches are of interest because they either look for data densities or split the data space into cells when looking for groups. Thus in this section, we now take a look at these two families of methods.

The main idea is to use a grid like structure to split the information space, separating the dense grid regions from the less dense ones to form groups.

In general, a typical approach within this category will consist of the following steps as presented by Grabusts and Borisov (2002):

1. Creating a grid structure, that is, partitioning the data space into a finite number of nonoverlapping cells.
2. Calculating the cell density for each cell.
3. Sorting of the cells according to their densities.
4. Identifying cluster centers.
5. Traversal of neighbor cells.

6.6.1 Grid-Based Hierarchical Clustering Algorithms: A Survey

Some of the most important algorithms within this category are the following:

- *STING*: STatistical INformation Grid-based clustering was proposed by Wang et al. (1997), who divide the spatial area into rectangular cells represented by a hierarchical structure. The root is at hierarchical level 1, its children at level 2, and so on. This algorithm has a computational complexity of $O(K)$, where K is the number of cells in the bottom layer. This implies that scaling this method to higher dimensional spaces is difficult (Hinneburg and Kelim, 1999). For example, if in high dimensional data space each cell has four children, then the number of cells in the second level will be 2^m, where m is the dimensionality of the database.

- *OptiGrid*: Optimal Grid-Clustering was introduced by Hinneburg and Keim (1999) as an efficient algorithm to cluster high-dimensional databases with noise. It uses data partitioning based on divisive recursion by multidimensional grids, focusing on separation of clusters by hyperplanes. A cutting plane is chosen which goes through the point of minimal density, therefore splitting two dense half-spaces. This process is applied recursively with each subset of data. This algorithm is hierarchical, with time complexity of $O(n \cdot m)$ (Gan et al., 2007, pp. 210–212).

- *GRIDCLUS*: proposed by Schikuta (1996) is a hierarchical algorithm for clustering very large datasets. It uses a multidimensional data grid to organize the space surrounding the data values rather than organize the data themselves. Thereafter patterns are organized into blocks, which in turn are clustered by a topological neighbor search algorithm. Five main steps are involved in the GRIDCLUS method: (i) insertion of points into the grid structure, (ii) calculation of density indices, (iii) sorting the blocks with respect to their density indices, (iv) identification of cluster centres, and (v) traversal of neighbor blocks.

- *WaveCluster*: This clustering technique proposed by Sheikholeslami (2000) defines a uniform two-dimensional grid on the data and represents the data points in each cell by the number of points. Thus, the data points become a set of grey-scale points, which is treated as an image. Then the problem of looking for clusters is transformed into an image segmentation problem, where wavelets are used to take advantage of their multiscaling and noise reduction properties. The basic algorithm is as follows: (i) create a data grid and assign each data object to a cell in the grid, (ii) apply the wavelet transform to the data, (iii) use the average sub-image to find connected clusters (i.e., connected pixels), and (iv) map the resulting clusters back to the points in the original space.

Additional information about grid-based clustering can be found in the following work.

- *DBSCAN*: Density-Based Spatial Clustering of Applications with Noise was proposed by Ester et al. (1996) to discover arbitrarily shaped clusters. Since it finds clusters based on density, it does not need to know the number of clusters at initialization time. This algorithm has been widely used and counts with many variations (e.g., see GDBSCAN by Sander et al., 1998; PDBSCAN by Xu et al., 1999; and DBCluC by Zaïane and Lee, 2002).

- *BRIDGE*: Proposed by Dash et al. (2001) uses a hybrid approach integrating k-means to partition the dataset into k clusters, and then density-based algorithm DBSCAN is applied to each partition to find dense clusters.

- *DBCLASD*: Distribution-based Clustering of LArge Spatial Databases (see Xu et al., 1998) assumes that data points within a cluster are uniformly distributed. The produced cluster is defined in terms of the nearest neighbor distance.
- *DENCLUE*: DENsity-based CLUstering aims to cluster large multimedia data. It can find arbitrarily shaped clusters and at the same time deals with noise in the data. This algorithm has two steps, first a precluster map is generated, the data is divided in hypercubes where only the populated are considered. The second step takes the highly populated cubes and cubes that are connected to a highly populated cube to produce the clusters. For a detailed presentation of these steps, see Hinneburg and Keim (1998).
- *CUBN*: It has three steps. First an erosion operation is carried out to find border points. Second, the nearest neighbor method is used to cluster the border points. Finally, the nearest neighbor is used to cluster the inner points. This algorithm is capable of finding nonspherical shapes and wide variations in size. Its computational complexity is $O(n)$ with n being the size of the dataset. For a detailed presentation of this algorithm, see Wang and Wang (2003).

Further grid-based clustering algorithms can be found in the following: Chang and Jin (2002), Park and Lee (2004), Gan et al. (2007), and Xu and Wunsch (2008).

6.6.2 Linear Time Grid-Based Hierarchical Clustering

In the last section we have seen a number of clustering methods that split the data space into cells, cubes, or dense regions to locate high density areas that can be further studied to find clusters.

For large datasets, clustering via an m-adic (m integer, which if a prime is usually denoted as p) expansion is possible, with the advantage of doing so in linear algorithmic time. The usual base 10 system for numbers is none other than the case of $m = 10$ and the base 2 or binary system can be referred to as 2-adic where $p = 2$. Let us consider the following distance relating to the case of vectors x and y with 1 attribute, hence unidimensional:

$$B_m(x_K, y_K) = \begin{cases} 1 & \text{if } x_1 \neq y_1 \\ \inf m^{-k} & x_k = y_k \quad 1 \leq k \leq |K| \end{cases} \tag{6.5}$$

This distance defines the longest common prefix of strings, taken as located in a Baire space. Thus we call this the Baire distance: here the longer the common prefix, the closer a pair of sequences. What is of interest to us here is this longest common prefix metric, which is an ultrametric (Murtagh et al., 2008).

For example, let us consider two such values, x and y. We take x and y to be bounded by 0 and 1. Each is of some precision, and we take the integer $|K|$ to be the maximum precision.

Thus we consider ordered sets x_k and y_k for $k \in K$. In line with our notation, we can write x_k and y_k for these numbers, with the set K now ordered. So, $k = 1$ is the index of the first decimal place of precision; $k = 2$ is the index of the second decimal place; and on on, until $k = |K|$ is the index of the $|K|th$ decimal place. The cardinality of the set K is the precision with which a number, x_k, is (or can be: we can pad with 0s if necessary) measured.

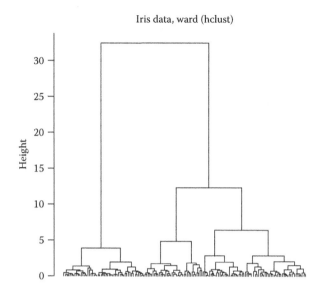

FIGURE 6.5
Hierarchical clustering of the 150 iris flowers based on their four measured features. R command `hclust` used.

Consider as examples $x_K = 0.478$; and $y_K = 0.472$. In these cases, $|K| = 3$. Start from the first decimal position. For $k = 1$, we find $x_k = y_k = 4$. For $k = 2$, $x_k = y_k$. But for $k = 3$, $x_k \neq y_k$. Baire distance, $B_m(x_K, y_K) = 10^{-2}$.

It is seen that this distance splits a unidimensional string of values into a 10-way hierarchy, in which each leaf can be seen as a grid cell. From Equation 6.5, we can read off

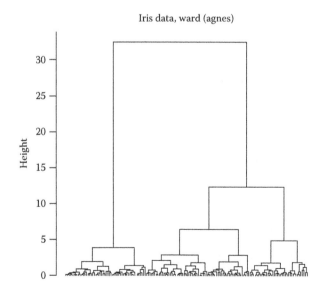

FIGURE 6.6
Hierarchical clustering of the 150 iris flowers based on their four measured features. R command `agnes` from package `cluster` used.

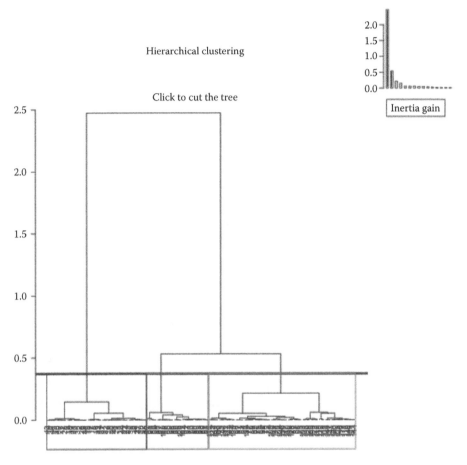

FIGURE 6.7
Hierarchical clustering of the 150 iris flowers based on their four measured features. R command HCPC from package FactoMineR used.

the distance that points assigned to the same grid cell. All pairwise distances of points assigned to the same cell are the same. From Equation 6.5, the distance can be shown to be an ultrametric. This last fact is intuitively clear in that the grid structure built up ensues from (in our example) a 10-way tree.

This Baire distance has been successfully applied to areas such a chemoinformatics (Murtagh et al., 2008), astronomy, and text retrieval (Contreras and Murtagh, 2012).

6.7 Examples and Software

We will use the R open source statistical analysis platform. The hclust program performs hierarchical clustering with a range of agglomerative criteria supported. Murtagh and Legendre (2014) may be referred to for discussion of implementations of the Ward minimum variance criterion. We also use the agnes method, see Kaufman and Rousseeuw (1990).

We use Fisher's well known iris data.

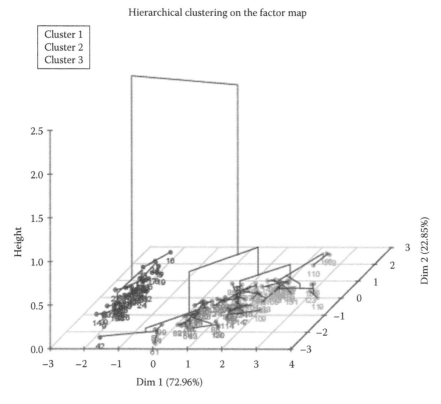

FIGURE 6.8

Hierarchical clustering of the 150 iris flowers based on their four measured features. Displayed in the principal component plane.

Figure 6.5 is produced with the following code.

```
h1 <- hclust(dist(iris[,1:4])^2, method="ward") # Sqd. Eucl. dist. input
h1$height <- sqrt(h1$height)                    # Sqrtof agglom. levels
plclust(h1, labels=FALSE, sub="", xlab="", hang=-1, main="Iris data,
Ward (hclust)")
```

Figure 6.6, which is morphologically equivalent to Figure 6.5, is produced with the following code.

```
install.packages("cluster")
library(cluster)                                # For "agnes"
h2 <- agnes(iris[,1:4], method="ward")
# Convert dendrogram structure using the "as.hclust" method.
plclust(as.hclust(h2), labels=FALSE, sub="", xlab="", hang=-1,
  main="Iris data, Ward (agnes)")
```

Figures 6.7 through 6.9 are all produced with the following code.

```
install.packages("FactoMineR")
library(FactoMineR)
```

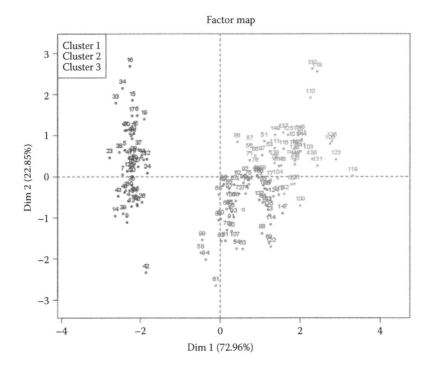

FIGURE 6.9
Hierarchical clustering of the 150 iris flowers based on their four measured features, in the principal component plane.

```
iris.res <- PCA(iris[,1:4])    #PCA on correlations
iris.hc <- HCPC(iris.res)      #Ward on principal components
```

6.8 Conclusions

Hierarchical clustering methods, with roots going back to the 1960s and 1970s, are continually replenished with new challenges. As a family of algorithms, they are central to the addressing of many important problems. Their deployment in many application domains testifies to how hierarchical clustering methods will remain crucial for a long time to come.

We have looked at both traditional agglomerative hierarchical clustering, and more recent developments in grid- or cell-based approaches. We have discussed various algorithmic aspects, including well-definedness (e.g., inversions) and computational properties. Finally, we looked at some standard commands in R, and how to obtain useful displays.

References

Anderberg MR. *Cluster Analysis for Applications*. Academic Press, New York, NY, 1973.
Benzécri JP. *L'Analyse des Données. I. La Taxinomie*. Dunod, Paris, 1979 (3rd ed.).

Blashfield RK and Aldenderfer MS. The literature on cluster analysis. *Multivariate Behavioral Research* 1978, 13: 271–295.

Bruynooghe M. Méthodes nouvelles en classification automatique des données taxinomiques nombreuses. *Statistique et Analyse des Données* 1977, 3: 24–42.

Chang J-W and Jin D-S. A new cell-based clustering method for large, high-dimensional data in data mining applications, in *SAC '02: Proceedings of the 2002 ACM Symposium on Applied Computing*. ACM, New York, NY, 2002, pp. 503–507.

Contreras P and Murtagh F. Fast, linear time hierarchical clustering using the Baire metric. *Journal of Classification* 2012, 29: 118–143.

Dash M, Liu H, and Xu X. 1 + 1 > 2: Merging distance and density based clustering, in *DASFAA '01: Proceedings of the 7th International Conference on Database Systems for Advanced Applications*. IEEE Computer Society, Washington, DC, 2001, pp. 32–39.

Day WHE and Edelsbrunner H. Efficient algorithms for agglomerative hierarchical clustering methods. *Journal of Classification* 1984, 1: 7–24.

Defays D. An efficient algorithm for a complete link method. *Computer Journal* 1977, 20: 364–366.

de Rham C. La classification hiérarchique ascendante selon la méthode des voisins réciproques. *Les Cahiers de l'Analyse des Données* 1980, V: 135–144.

Deza MM and Deza E. *Encyclopedia of Distances*. Springer, Berlin, 2009.

Ester M, Kriegel H-P, Sander J, and Xu X. A density-based algorithm for discovering clusters in large spatial databases with noise, in *2nd International Conference on Knowledge Discovery and Data Mining*. AAAI Press, Palo Alta, CA, 1996, pp. 226–231.

Gan G, Ma C, and Wu J. *Data Clustering Theory, Algorithms, and Applications*. Society for Industrial and Applied Mathematics, SIAM, Philadelphia, PA, 2007.

Gillet VJ, Wild DJ, Willett P, and Bradshaw J. Similarity and dissimilarity methods for processing chemical structure databases. *Computer Journal* 1998, 41: 547–558.

Gordon AD. *Classification*. Chapman and Hall, London, 1981.

Gordon AD. A review of hierarchical classification. *Journal of the Royal Statistical Society A* 1987, 150: 119–137.

Grabusts P and Borisov A. Using grid-clustering methods in data classification, in *PARELEC '02: Proceedings of the International Conference on Parallel Computing in Electrical Engineering*. IEEE Computer Society, Washington, DC, 2002.

Graham RH and Hell P. On the history of the minimum spanning tree problem. *Annals of the History of Computing* 1985, 7: 43–57.

Griffiths A, Robinson LA, and Willett P. Hierarchic agglomerative clustering methods for automatic document classification. *Journal of Documentation* 1984, 40: 175–205.

Hinneburg A and Keim DA. A density-based algorithm for discovering clusters in large spatial databases with noise, in *Proceedings of the 4th International Conference on Knowledge Discovery and Data Mining*. AAAI Press, New York, NY, 1998, pp. 58–68.

Hinneburg A and Keim D. Optimal grid-clustering: Towards breaking the curse of dimensionality in high-dimensional clustering, in *VLDB '99: Proceedings of the 25th International Conference on Very Large Data Bases*. Morgan Kaufmann Publishers Inc., San Francisco, CA, 1999, pp. 506–517.

Jain AK and Dubes RC. *Algorithms For Clustering Data*. Prentice-Hall, Englwood Cliffs, NJ, 1988.

Jain AK, Murty, MN, and Flynn PJ. Data clustering: A review. *ACM Computing Surveys* 1999, 31: 264–323.

Janowitz, MF. *Ordinal and Relational Clustering*. World Scientific, Singapore, 2010.

Juan J. Programme de classification hiérarchique par l'algorithme de la recherche en chaîne des voisins réciproques. *Les Cahiers de l'Analyse des Données* 1982, VII: 219–225.

Kaufman L and Rousseeuw PJ. *Finding Groups in Data: An Introduction to Cluster Analysis*. Wiley, New York, NY, 1990.

Lerman IC. *Classification Automatique et Analyse Ordinale des Données*. Dunod, Paris, 1981.

Le Roux B and Rouanet H. *Geometric Data Analysis: From Correspondence Analysis to Structured Data Analysis*. Kluwer, Dordrecht, 2004.

March ST. Techniques for structuring database records. *ACM Computing Surveys* 1983, 15: 45–79.

Mirkin B. *Mathematical Classification and Clustering*. Kluwer, Dordrecht, 1996.

Murtagh F. A survey of recent advances in hierarchical clustering algorithms. *Computer Journal* 1983, 26: 354–359.

Murtagh F. Complexities of hierarchic clustering algorithms: State of the art. *Computational Statistics Quarterly* 1984, 1: 101–113.

Murtagh F. *Multidimensional Clustering Algorithms*. Physica-Verlag, Würzburg, 1985.

Murtagh F. *Correspondence Analysis and Data Coding with Java and R*. Chapman and Hall, Boca Raton, FL, 2005.

Murtagh F. The Haar wavelet transform of a dendrogram. *Journal of Classification* 2007, 24: 3–32.

Murtagh F. Symmetry in data mining and analysis: A unifying view based on hierarchy. *Proceedings of Steklov Institute of Mathematics* 2009, 265: 177–198.

Murtagh F, Downs G, and Contreras P. Hierarchical clustering of massive, high dimensional data sets by exploiting ultrametric embedding. *SIAM Journal on Scientific Computing* 2008, 30(2): 707–730.

Murtagh F and Legendre P. Ward's hierarchical agglomerative clustering method: Which algorithms implement Ward's criterion? *Journal of Classification* 2014, 31(3): 274–295.

Murtagh F, Taskaya T, Contreras P, Mothe J, and Englmeier K. Interactive visual user interfaces: A survey. *Artificial Intelligence Review* 2003, 19: 263–283.

Park NH and Lee WS. Statistical grid-based clustering over data streams. *SIGMOD Record* 2004, 33(1): 32–37.

Rapoport A and Fillenbaum S. An experimental study of semantic structures, in eds. AK Romney, RN Shepard, and SB Nerlove, *Multidimensional Scaling; Theory and Applications in the Behavioral Sciences. Vol. 2, Applications*, Seminar Press, New York, 1972, pp. 93–131.

Rohlf FJ. Algorithm 76: Hierarchical clustering using the minimum spanning tree. *Computer Journal* 1973, 16: 93–95.

Rui Xu and Wunsch D. Survey of clustering algorithms. *IEEE Transactions on Neural Networks* 2005, 16: 645–678.

Rui Xu and Wunsch DC. *Clustering*. IEEE Computer Society Press, Washington, DC, 2008.

Sander J, Ester M, Kriegel H-P, and Xu X. Density-based clustering in spatial databases: The algorithm GDBSCAN and its applications. *Data Mining Knowledge Discovery* 1998, 2(2): 169–194.

Schikuta E. Grid-clustering: An efficient hierarchical clustering method for very large data sets, in *ICPR '96: Proceedings of the 13th International Conference on Pattern Recognition*. IEEE Computer Society, Washington, DC, 1996, pp. 101–105.

Sheikholeslami G, Chatterjee S, and Zhang A, Wavecluster: A wavelet based clustering approach for spatial data in very large databases. *The VLDB Journal* 2000, (3–4): 289–304.

Sibson R. SLINK: An optimally efficient algorithm for the single link cluster method. *Computer Journal* 1973, 16: 30–34.

Sneath PHA and Sokal RR. *Numerical Taxonomy*, Freeman, San Francisco, 1973.

van Rijsbergen CJ. *Information Retrieval*. Butterworths, London, 1979 (2nd ed.).

Wang L and Wang Z-O. CUBN: A clustering algorithm based on density and distance, in *Proceedings of the 2003 International Conference on Machine Learning and Cybernetics*. IEEE Press, Washington, DC, 2003, pp. 108–112.

Wang W, Yang J, and Muntz R. STING: A statistical information grid approach to spatial data mining, in *VLDB '97: Proceedings of the 23rd International Conference on Very Large Data Bases*. Morgan Kaufmann Publishers Inc., San Francisco, CA, 1997, pp. 18–195.

White HD and McCain KW. Visualization of literatures, in ed. M.E. Williams, *Annual Review of Information Science and Technology (ARIST)*, Information Today Inc., Medford, NJ, Vol. 32, pp. 99–168, 1997.

Willett P. Efficiency of hierarchic agglomerative clustering using the ICL distributed array processor. *Journal of Documentation* 1989, 45: 1–45.

Wishart D. Mode analysis: A generalization of nearest neighbor which reduces chaining effects, in ed. A.J. Cole, *Numerical Taxonomy*, Academic Press, New York, NY, pp. 282–311, 1969.

Xu X, Ester M, Kriegel H-P, and Sander J. A distribution-based clustering algorithm for mining in large spatial databases, in *ICDE '98: Proceedings of the Fourteenth International Conference on Data Engineering*. IEEE Computer Society, Washington, DC, 1998, pp. 324–331.

Xu X, Jäger J, and Kriegel H-P. A fast parallel clustering algorithm for large spatial databases. *Data Mining Knowledge Discovery* 1999, 3(3), 263–290.

Zaïane OR and Lee C-H. Clustering spatial data in the presence of obstacles: A density-based approach, in *IDEAS '02: Proceedings of the 2002 International Symposium on Database Engineering and Applications*. IEEE Computer Society, Washington, DC, 2002, pp. 214–223.

7

Spectral Clustering

Marina Meila

CONTENTS

Abstract

Spectral clustering is a family of methods to find K clusters using the eigenvectors of a matrix. Typically, this matrix is derived from a set of pairwise similarities S_{ij} between the points to be clustered. This task is called similarity-based clustering, graph clustering, or clustering of diadic data. One remarkable advantage of spectral clustering is its ability to cluster "points" which are not necessarily vectors, and to use for this a "similarity," which is less restrictive than a distance. A second advantage of spectral clustering is its flexibility; it can find clusters of arbitrary shapes, under realistic separations. This chapter introduces the similarity-based clustering paradigm, describes the algorithms used, and sets the foundations for understanding these algorithms. Practical aspects, such as obtaining the similarities, are also discussed.

7.1 Similarity-Based Clustering: Definitions and Criteria

7.1.1 What Is Similarity-Based Clustering?

Clustering, when the data are similar between pairs of points is called *similarity-based clustering*. A typical example of similarity-based clustering is community detection in social

networks (White and Smyth 2005) (see also Chapter 16), where the observations are individual links between people, which may be due to friendship, shared interests, and work
relationships. The "strength" of a link can be the frequency of interactions, for example,
communications by e-mail, phone or other social media, co-authorships, or citations.

In this clustering paradigm, the points to be clustered are not assumed to be part of a
vector space. Their attributes (or features) are incorporated into a single dimension, the link
strength, or *similarity*, which takes a numerical value S_{ij} for each pair of points i, j. Hence,
the natural representation for this problem is by means of the *similarity matrix* $\mathbf{S} = [S_{ij}]_{i,j=1}^{n}$.
The similarities are symmetric ($S_{ij} = S_{ji}$), and nonnegative ($S_{ij} \geq 0$).

Less obvious domains where similarity-based clustering is used include image segmentation, where the points to be clustered are pixels in an image, and text analysis, where
words appearing in the same context are considered similar.

The goal of similarity-based clustering is to find the global clustering of the data set that
emerges from the pairwise interactions of its points. Namely, we want to put points that
are similar to each other in the same cluster, dissimilar points in different clusters.

7.1.2 Similarity-Based Clustering and Cuts in Graphs

It is useful to cast similarity-based clustering in the language of graph theory. Let the points
to be clustered $V = \{1, \ldots, n\}$ be the nodes of a graph $\mathcal{G}$, and the graph edges be represented
by the pairs i, j with $S_{ij} > 0$. The similarity itself is the weight of edge ij.

$$\mathcal{G} = (V, E), \quad E = \{(i, j), S_{ij} > 0\} \subseteq V \times V \tag{7.1}$$

Thus, $\mathcal{G}$ is an *undirected* and *weighted* graph. A partition of the nodes of a graph into K clusters is known as a (K-way) *graph cut*, therefore similarity-based clustering can be viewed
as finding a cut in the graph $\mathcal{G}$. The following definitions will be helpful. We denote

$$d_i = \sum_{j \in V} S_{ij} \tag{7.2}$$

the *degree* of node $i \in V$. The volume of V is $\mathrm{Vol}\, V = \sum_{i \in V} d_i$. Similarly, we define the
volume of cluster $C \subseteq V$ by

$$d_C = \sum_{i \in C} d_i$$

Note that the volume of a single node is d_i.

The value of the cut between subsets $C, C' \subseteq V, C \cap C' = \emptyset$, briefly called the *cut* of C, C'
is the sum of the edge weigths that cross between C and C'.

$$Cut(C, C') = \sum_{i \in C} \sum_{j \in C'} S_{ij}$$

Now we define the K-way *Cut* and respectively *Normalized Cut* associated to a partition
$\mathcal{C} = (C_1, \ldots, C_K)$ of V as

$$Cut(\mathcal{C}) = \frac{1}{2} \sum_{k=1}^{K} Cut(C_k, V \setminus C_k) \tag{7.3}$$

$$NCut(\mathcal{C}) = \sum_{k=1}^{K} \frac{Cut(C_k, V \setminus C_k)}{d_{C_k}} \tag{7.4}$$

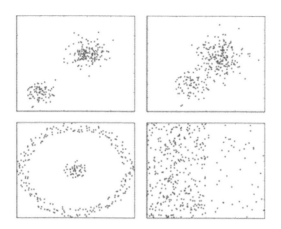

FIGURE 7.1
Four cases in which the minimum *NCut* partition agrees with human intuition.

In particular, for $K = 2$,

$$NCut(C,C') = Cut(C,C') \left(\frac{1}{d_C} + \frac{1}{d_{C'}} \right)$$

Intuitively, a small $Cut(C)$ is indicative of a "good" clustering, as most of the removed edges must have zero or low similarity S_{ij}. For $K = 2$, $\text{argmin}_{|C|=2} Cut(C)$ can be found tractably by the MINCUT/MAXFLOW algorithm (Papadimitriou and Steiglitz 1998). For $K \geq 3$, minimizing the cut is NP-hard; in practice one applies the MINCUT/MAXFLOW recursively to obtain K-cuts of low value. Unfortunately, like the better known Single Linkage criterion, the Cut criterion is very sensitive to outliers; on most realistic dataset, the smallest cut will be between an outlier and the rest of the data. Consequently, clustering by minimizing Cut is found empirically to produce very *imbalanced* partitions.[*]

This prompted (Shi and Malik 2000) to introduce the *NCut* (which they called *balanced cut*). A partition can have small *NCut* only if it has both a small cut value and if all its clusters have sufficiently large volumes d_C. As Figure 7.1 shows, *NCut* is a very flexible criterion, capturing our intuitive notion of clusters in a variety of situations.

7.1.3 The Laplacian and Other Matrices of Spectral Clustering

In addition to the similarity matrix **S**, a number of other matrices derived from it matrices play a central role in spectral clustering. These are listed in Table 7.1.

One such matrix is **P**, the *random walk* matrix of $\mathcal{G}$, sometimes called the *random walk Laplacian* of $\mathcal{G}$. **P** is obtained by normalizing the rows of **S** to sum to 1.

$$\mathbf{P} = \mathbf{D}^{-1}\mathbf{S} \tag{7.5}$$

with **D** being the diagonal matrix of the node degrees

$$\mathbf{D} = \text{diag}(d_1, \ldots, d_n) \tag{7.6}$$

---

[*] An interesting *randomizing and averaging* algorithm using MINCUT/MAXFLOW was proposed by Gdalyahu et al. (1999).

TABLE 7.1

The Relevant Matrices in Spectral Clustering

Matrix	Name	Dim	Definition	Properties
S	Similarity matrix	$n \times n$		$S_{ij} = S_{ji} \geq 0$
D	Degree matrix	$n \times n$	$D = \mathrm{diag}(d_1, \ldots, d_n)$	$D_{ii} = d_i > 0$, $D_{ij} = 0$, $j \neq i$
P	Random walk matrix	$n \times n$	$P = D^{-1}S$	$P_{ij} \geq 0$, $\sum_{j=1}^n P_{ij} = 1$
L	Laplacian matrix	$n \times n$	$L = I - D^{-1/2}SD^{-1/2}$	$L_{ij} = L_{ji}$, $L \geq 0$
$\hat{P}$	Transition matrix btw. clusters	$K \times K$	$\hat{P}_{kl} = \sum_{i \in C_k} \sum_{j \in C_l} S_{ij}/d_{C_k}$	

Thus, **P** is a stochastic matrix, satisfying $P_{ij} \geq 0$, $\sum_{j=1}^n P_{ij} = 1$. Another matrix of interest is **L**, the *Normalized Laplacian* (Chung 1997) of $\mathcal{G}$, which we will call for brevity the *Laplacian*.

$$L = I - D^{-1/2}SD^{-1/2} \tag{7.7}$$

where **I** is the unit matrix.

Proposition 7.1 (Relationship between L and P)

Denote by $1 = \lambda_1 \geq \lambda_2 \geq \cdots \lambda_n \geq -1$ the eigenvalues of **P** and by $\mathbf{v}^1, \ldots, \mathbf{v}^n$ the corresponding eigenvectors. Denote by $\mu_1 \leq \mu_2 \leq \cdots \mu_n$ the eigenvalues of **L** and by $\mathbf{u}^1, \ldots, \mathbf{u}^n$ the corresponding eigenvectors. Then,

1.

$$\mu_i = 1 - \lambda_i, \quad \mathbf{u}^i = D^{1/2}\mathbf{v}^i, \quad \text{for all } i = 1, \ldots, n \tag{7.8}$$

2. $\lambda_1 = 1$ and $\mu_1 = 0$.

3. The multiplicity of $\lambda_1 = 1$ (or, equivalently, of $\mu_1 = 0$) is $K > 1$ iff **P** (**L**) is block diagonal with K blocks.

This proposition has two consequences. Because $\lambda_j \leq 1$, it follows that $\mu_j \geq 0$; in other words, that **L** is positive semidefinite, with $\mu_1 = 0$. Moreover, Proposition 7.1 ensures that the eigenvalues of **P** are always real and its eigenvectors linearly independent.

7.1.4 Four Bird's Eye Views of Spectral Clustering

We can approach the problem of similarity-based clustering from multiple perspectives.

1. We can view each data point i as the row vector $S_{i:}$ in $\mathbb{R}^n$, and find a low dimensional embedding of these vectors. Once this embedding is found, one could proceed to cluster the data by e.g K-means algorithm, in the low-dimensional space. This view is captured by Algorithm 7.2 in Equation 7.2.

2. We can view the data points as states of a Markov chain defined by **P**. We group states by their *pattern of high-level connections*. This view is described in Section 7.3.1.

3. We can view the data points as nodes of graph $\mathcal{G} = (V, E, S)$ as in Section 7.1.2. We can remove a set of edges with small total weight, so that none of the connected

components of the remaining graph is too small, in other words we can cluster by minimizing the *NCut*. This view is further explored in Section 7.3.2.

4. We can view a cluster C as its $\{0,1\}$-valued *indicator function* $\mathbf{x}_C$. We can find the partition whose K indicator functions are "smoothest" with respect to the graph $\mathcal{G}$, that is, stay constant between nodes with high similarity. This view is described in Section 7.3.3.

As we shall see, the four paradigms above are equivalent, when the data is "well clustered," are all implemented by the same algorithm, which we describe in the next section.

7.2 Spectral Clustering Algorithms

The workflow of a typical spectral clustering algorithm is shown in the top row of Figure 7.2.

The algorithm we recommend is based on Meila and Shi (2001a,b) and Ng et al. (2002).

Algorithm SPECTRAL CLUSTERING

Input Similarity matrix $\mathbf{S}$, number of clusters K

1. *Transform* $\mathbf{S}$
   Calculate $d_i \leftarrow \sum_{j=1}^{n} S_{ij}$ the *node degrees*, for $i = 1 : n$.
   Form the *transition matrix* $\mathbf{P}$ with $P_{ij} \leftarrow S_{ij}/d_i$, for $i, j = 1 : n$

2. *Eigendecomposition*
   Compute the largest K eigenvalues $\lambda_1 \geq \cdots \geq \lambda_K$ and eigenvectors $\mathbf{v}^1, \ldots, \mathbf{v}^K$ of $\mathbf{P}$.

3. *Embed the data in K-th principal subspace*
   Let $\mathbf{x}_i = [\mathbf{v}_i^2 \ \mathbf{v}_i^3 \ \ldots \ \mathbf{v}_i^K] \in \mathbb{R}^{K-1}$, for $i = 1, \ldots, n$.

4. Run the K-MEANS algorithm on the "data" $\mathbf{x}_{1:n}$

Output the clustering $\mathcal{C}$ obtained in step 4.

Note that in step 3 we discard the first eigenvector, as this is usually constant and is not informative of the clustering.

Some useful variations and improvements of SPECTRAL CLUSTERING are:

- *Orthogonal initialization* (Ng et al. 2002) Find the K initial centroids $\bar{\mathbf{x}}_{1:K}$ of K-MEANS in step 4 by

---

Algorithm ORTHOGONAL INITIALIZATION

1. Choose $\bar{\mathbf{x}}_1$ randomly from $\mathbf{x}_1, \ldots, \mathbf{x}_n$
2. For $k = 2, \ldots, K$ set $\bar{\mathbf{x}}_k = \text{argmin}_{\mathbf{x}_i} \max_{k' < k} |\cos(\bar{\mathbf{x}}_{k'}, \mathbf{x}_i)|$.

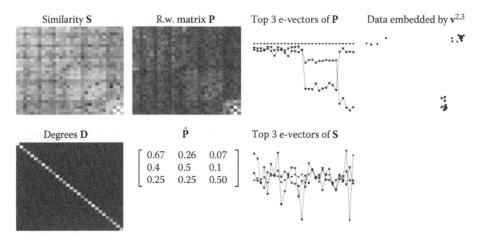

FIGURE 7.2
Spectral clustering of a synthetic data set with $n = 30$ points and $K = 3$ clusters of sizes 15, 10 and 5; the data are sorted so that points in the same cluster are consecutive. The *top row*, from left to right, displays the similarity matrix **S**, the random walk matrix **P**, the entries in the top 3 eigenvectors of **P**, plotted versus the index $i = 1, \ldots, 30$, and finally, the embedding $\mathbf{x}_{1:n}$ of the data obtained from the eigenvectors. The similarity **S** is a perfect similarity matrix to which noise was added; hence in the second and third eigenvectors of **P** corresponding to a cluster have approximately but not exactly the same value; the first eigenvector of **P** is proportional to **1** and hence has exactly equal entries for all i. Since $\mathbf{v}^{2,3}$ are almost piecewise-constant, in the embedding the points $\mathbf{x}_{1:n}$ are well clustered. The *bottom row* displays the node degrees on the diagonal of **D**, the $\hat{P}_{kl}$ values of the transition probabilities between blocks, and the top 3 eigenvectors of **S**. Note that this is not a case of nearly block diagonal **S**: the probabilities of transitioning between clusters are significantly away from 0, and the minimum *NCut* is not small (its value is $1.33 = 3 - \text{trace}\,\hat{\mathbf{P}}$). Yet the data is very "well clustered," if one uses the eigenvectors of **P** for clustering. In contrast, the top 3 eigenvectors of the untransformed **S** are not informative (nor are the other eigenvectors of **S**). The *Cut* corresponding to the clustering found by SPECTRAL CLUSTERING is 140.3 (which represents 0.23 of the total Vol$V = 614.5$); in contrast removing the point of smallest degree has *Cut* equal to 11.7.

This initialization is a variant of the FASTEST FIRST TRAVERSAL algorithm (Hochbaum and Shmoys 1985); FASTEST FIRST TRAVERSAL is part of one the best EM and K-MEANS initialization algorithms known to date (Dasgupta and Schulman 2007; Bubeck et al. 2012).

- *Rescaling* $\mathbf{x}_i$ *to have unit length* in step 3 was recommended by Ng et al. (2002) and was found empirically to have good noise reduction effects.

- *Rescaling* $\mathbf{v}^{2:K}$ *by the eigenvalues (diffusion distance rescaling)* in step 2. When **P** is almost block diagonal, or close to perfect, this rescaling will have almost no effect. But in the noisier situations, it can put more weight on the first eigenvectors which are more robust to noise (see also Section 7.5). Moreover, Nadler et al. (2006) showed that setting $\mathbf{v}^k \leftarrow \lambda_k^{2t} \mathbf{v}^k$, with some $t > 1$, is related to the *diffusion distance*, a true metric on the nodes of a graph. The parameter t is a *smoothing parameter*, with larger t causing more smoothing.

- *Using* **S** *instead of* **P** in step 2 (and skipping the transformation in step 1). This algorithm variant can be shown to (approximately) minimize a criterion call *Ratio Cut (RCut)*.

$$RCut(\mathcal{C}) = \sum_{k=1}^{K} \frac{Cut(C_k, V \setminus C_k)}{|C_k|} \tag{7.9}$$

The *RCut* differs from the *NCut* only in the denominators, which are the cluster cardinalities, instead of the cluster volumes. The discussion in Sections 7.3.1–7.3.3 applies with only small changes to this variant of SPECTRAL CLUSTERING, w.r.t. the *RCut* criterion. However, it can be shown that whenever **S** has piecewise constant eigenvectors (see Section 7.3.1) then **P** will have piecewise constant eigenvectors as well, but the converse is not true (Verma and Meila 2003). Hence, whenever this algorithm variant can find a good clustering, the original 7.2 can find it too. Moreover, the eigenvectors and values of **P** converge to well-defined limits when $n \to \infty$, whereas those of **S** may not.

The most significant variant of Algorithm 7.2 is its original recursive form (Shi and Malik 2000) given below.

Algorithm TWO-WAY SPECTRAL CLUSTERING

Input Similarity matrix **S**
   1. *Transform* **S**
      Calculate $d_i = \sum_{j=1}^{n} S_{ij}$, $j = 1 : n$ the *node degrees*.
      Form the *transition matrix* **P** with $P_{ij} \leftarrow S_{ij}/d_i$ for $i, j = 1, \ldots, n$.
   2. Compute the eigenvector **v** corresponding to the second largest eigenvalue λ_2 of **P**.
   3. *Sort*
      Let $\mathbf{v}^{sort} = [v_{i_1}\, v_{i_2}\, \ldots\, v_{i_n}]$ be the entries of **v** sorted in increasing order and denote $C_j = \{i_1, i_2, \ldots, i_j\}$ for $j = 1, \ldots, n-1$.
   4. *Cut*
      For $j = 1, \ldots, n-1$ compute $NCut(C_j, V \setminus C_j)$ and find $j_0 = \operatorname{argmin}_j NCut(C_j, V \setminus C_j)$.
Output clustering $\mathcal{C} = \{C_{j_0}, V \setminus C_{j_0}\}$

TWO-WAY SPECTRAL CLUSTERING is called recursively on each of the two resulting clusters, if one wishes to obtain a clustering with $K > 2$ clusters.

Finally, an observation related to numerical implementation that is too important to omit. From Proposition 7.1, it follows that steps 1 and 2 of SPECTRAL CLUSTERING can be implemented equivalently as

---

Algorithm STABLE SPECTRAL EMBEDDING

1. $\tilde{L}_{ij} \leftarrow S_{ij}/\sqrt{d_i d_j}$ for $i, j = 1 : n$ (note that $\tilde{\mathbf{L}} = \mathbf{I} - \mathbf{L}$)
2. Compute the largest K eigenvalues $\lambda_1 = 1 \geq \lambda_2 \geq \cdots \geq \lambda_K$ and eigenvectors $\mathbf{u}^1, \ldots, \mathbf{u}^k$ of $\tilde{\mathbf{L}}$ (these are the eigenvalues of **P** and the eigenvectors of **L**).
   Rescale $\mathbf{v}^k \leftarrow \mathbf{D}^{-1/2} \mathbf{u}^k$ (obtain the eigenvectors of **P**).

Eigenvector computations for symmetric matrices like $\tilde{\mathbf{L}}$ are much more stable numerically than for general matrices like **P**. This modification guarantees that the eigenvalues will be real and the eigenvectors orthogonal.

7.3 Understanding the Spectral Clustering Algorithms

7.3.1 Random Walk/Markov Chain View

Recall the stochastic matrix $\mathbf{P}$ defines a Markov chain (or random walk) on the nodes V. Remarkably, the stationary distribution $\boldsymbol{\pi}$ of this chain has the explicit and simple form*

$$\pi_i = \frac{d_i}{\mathrm{Vol}V} \quad \text{for } i \in V \tag{7.10}$$

Indeed, it is easy to verify that

$$[\pi_1 \ldots \pi_n]\mathbf{P} = \frac{1}{\mathrm{Vol}V}\left[\sum_i d_i P_{i1} \ldots \sum_i d_i P_{in}\right] = \frac{1}{\mathrm{Vol}V}\left[\sum_{i=1}^n S_{i1} \ldots \sum_{i=1}^n S_{in}\right] = [\pi_1 \ldots \pi_n] \tag{7.11}$$

If the Markov chain is ergodic, then $\boldsymbol{\pi}$ is the unique stationary distribution of $\mathbf{P}$, otherwise, uniqueness is not guaranteed, yet property 7.11 still holds.

Now let's consider the Algorithm 7.2 and ask when are the points $\mathbf{x}_i \in \mathbb{R}^K$ well clustered? Is there a case when the $\mathbf{x}_i$'s are identical for all the nodes i that belong to the same cluster k? If this happens we say that $\mathbf{S}$ (and $\mathbf{P}$) are *perfect*. In the perfect case, the K-MEANS algorithm (or, by that matter, any clustering algorithm) will be guaranteed to find the same clustering.

Thus, to understand what is a "good" clustering from the point of view of spectral clustering, it is necessary to understand what the perfect case represents.

Definition 7.1

If $\mathcal{C} = (C_1, \ldots, C_K)$ is a partition of V, we say that a vector $\mathbf{x}$ is *piecewise constant* w.r.t. $\mathcal{C}$ if for all pairs i, j in the same cluster C_k we have $x_i = x_j$.

Proposition 7.2 (Lumpability Lemma (Meila and Shi 2001))

Let $\mathbf{P}$ be a matrix with rows and columns indexed by V that has independent eigenvectors. Let $\mathcal{C}^* = (C_1, C_2, \ldots, C_k)$ be a partition of V. Then, $\mathbf{P}$ has K eigenvectors that are piecewise constant w.r.t. $\mathcal{C}$ and correspond to non-zero eigenvalues if and only if the sums $P_{ik} = \sum_{j \in C_k} P_{ij}$ are constant for all $i \in C_l$ and all $k, l = 1, \ldots, K$ and the matrix $\hat{\mathbf{P}} = [\hat{P}_{kl}]_{k,l=1,\ldots,K}$ (with $\hat{P}_{kl} = \sum_{j \in C_l} P_{ij}$, $i \in C_k$) is nonsingular. We say that (the Markov chain represented by) $\mathbf{P}$ is *lumpable* w.r.t. $\mathcal{C}^*$.

Corrolary 7.3

If stochastic matrix $\mathbf{P}$ obtained in step 1 is lumpable w.r.t. $\mathcal{C}^*$ with piecewise constant eigenvectors $\mathbf{v}^1, \ldots, \mathbf{v}^K$ corresponding to the K largest eigenvalues of $\mathbf{P}$, then Algorithm 7.2 will output $\mathcal{C}^*$.

---

* This is true for any *reversible* Markov chain.

Corrolary 7.3 shows that spectral clustering will find clusterings for which points i, i' are in the same cluster k if they have the same probability $\hat{P}_{kl}$ of transitioning to cluster l, for all $l = 1, \ldots, K$.

A well-known special case of lumpability is the case when the clusters are *completely separated*, that is when $S_{ij} = 0$ whenever i, j are in different clusters. Then, **S** and **P** are block diagonal with K blocks, each block representing a cluster. From Proposition 7.2, it follows that **P** has K eigenvalues equal to 1, and that $\hat{\mathbf{P}} = \mathbf{I}$. What can be guaranteed in the vicinity of this case has been intensely studied in the literature. In particular, Ng et al. (2002) and later Balakrishnan et al. (2011) give theoretical results showing that if **S** is nearly block diagonal, the clusters representing the blocks of **S** can be recovered by spectral clustering.

The Lumpability Lemma shows however that having an approximately block diagonal **S** is not necessary, and that spectral clustering algorithms will work in a much broader range of cases, namely, as long as "the points in the same cluster behave approximately in the same way" in the sense of Proposition 7.2.

This interpretation relates spectral clustering to a remarkable fact about Markov chains. It is well known that if one groups the states of a Markov chain in clusters $C_1, \ldots, C_K$, a sequence of states $i_1, i_2, \ldots, i_t$ implies a sequence of cluster labels $k_1, k_2, \ldots, k_t \in \{1, \ldots, K\}$. From the transition matrix **P** and the clustering $C_1, \ldots, C_K$, one can calculate the transition matrix at the cluster level $Pr[C_k \to C_l | C_k] = \hat{P}_{kl}$, as well as the stationary distribution w.r.t. the clusters by

$$\hat{P}_{kl} = \sum_{i \in C_k} \sum_{j \in C_l} S_{ij} / d_{C_k}, \quad \hat{\pi}_k = \frac{d_{C_k}}{\text{Vol} V}, \quad k, l = 1, \ldots, K \tag{7.12}$$

However, it can be easily shown that the chain $k_1, k_2, \ldots, k_t, \ldots$ is in general not Markov; that is, $Pr[k_{t+1} | k_t, k_{t-1}] \neq Pr[k_{t+1} | k_t]$, or knowing past states *can* give information about future states even when the present state k_t is known. *Lumpability* in Markov chain terminology means that there exists a clustering $\mathcal{C}^*$ of the nodes in V so that the chain defined by $\hat{\mathbf{P}}$ is Markov. Proposition 7.2 shows that lumpability holds essentially iff **P** has piecewise-constant eigenvectors. Hence, spectral clustering Algorithm 7.2 finds *equivalence classes* of nodes (when they exist) so that all nodes in an equivalence class C_k contain the same information about the future.

The following proposition underscores the discussion about lumpability, showing that the eigenvectors of **P**, when they are piecewise constant, are "stretched versions" of the eigenvectors of $\hat{\mathbf{P}}$.

Proposition 7.4 (Relationship between P and $\hat{\mathbf{P}}$ (Telescope Lemma))

Assume that the conditions of Proposition 7.2 hold. Let $\mathbf{v}^1, \ldots, \mathbf{v}^K \in \mathbb{R}^n$ and $1 = \lambda_1 \geq \lambda_2 \geq \cdots \lambda_K$ be the piecewise constant eigenvectors of **P** and their eigenvalues and $1 = \hat{\lambda}_1 \geq \hat{\lambda}_2 \geq \cdots \hat{\lambda}_K$ and $\hat{\mathbf{v}}^1, \ldots, \hat{\mathbf{v}}^K \in \mathbb{R}^K$ the eigenvalues and eigenvectors of $\hat{\mathbf{P}}$. Then, for any $k \in 1, \ldots, K$

$$\hat{\lambda}_k = \lambda_k \quad \text{and} \tag{7.13}$$

$$\hat{v}_l^k = v_i^k \quad \text{for } l = 1, \ldots, K \quad \text{and} \quad i \in C_l \tag{7.14}$$

7.3.2 Spectral Clustering as Finding a Small Balanced Cut in $\mathcal{G}$

We now explain the relationship between spectral clustering algorithms like 7.2 and minimizing the K-way normalized cut.

First, we show that the *NCut* defined in (7.4) can be rewritten in terms of probabilities $\hat{P}_{kl}$ of transitioning between clusters in the random walk defined by **P**.

Proposition 7.5 (*NCut* as conditional probability of leaving a cluster)

The K-way normalized cut associated to a partition $\mathcal{C} = (C_1, \ldots, C_K)$ of V is equal to

$$NCut(\mathcal{C}) = \sum_{k=1}^{K} \left[1 - \frac{\sum_{i \in C_k} \pi_i \sum_{j \in C_k} P_{ij}}{\sum_{i \in C_k} \pi_i} \right] = \sum_{k=1}^{K} \left[1 - \hat{P}_{kk} \right] = K - \text{trace } \hat{\mathbf{P}} \qquad (7.15)$$

The denominators $\sum_{i \in C_k} \pi_i$ above represent $d_{C_k}/\text{Vol}C_k = \pi_{C_k}$, the probability of being in cluster C_k under the stationary distribution π. Consequently each term of the sum represents the probability of leaving cluster C_k given that the Markov chain is in C_k, under the stationary distribution.

In the perfect case, from Proposition 7.4, $\hat{\lambda}_{1:K}$ are also the top K eigenvalues of **P**, hence

$$NCut(\mathcal{C}^*) = K - \sum_{k=1}^{K} \lambda_k \quad \text{for all } k, l = 1, \ldots, K \qquad (7.16)$$

Next, we show that the value $K - \sum_{k=1}^{K} \lambda_k$ is the lowest possible *NCut* value for any K-clustering $\mathcal{C}$ in any graph.

Proposition 7.6 (Multicut Lemma)

Let **S**, **L**, **P**, $\mathbf{v}^1, \ldots, \mathbf{v}^K$ and $\lambda_1, \ldots, \lambda_K$ be defined as before, and let $\mathcal{C}$ be a partition of V into K disjoint clusters. Then,

$$NCut(\mathcal{C}) \geq \min\{\text{trace } \mathbf{Y}^T \mathbf{L} \mathbf{Y} \mid \mathbf{Y} \in \mathbb{R}^{n \times K}, \mathbf{Y} \text{ has orthonormal columns}\} \qquad (7.17)$$

$$= K - (\lambda_1 + \lambda_2 + \cdots + \lambda_K) \qquad (7.18)$$

The proof is both simple and informative so we will present it here. Consider, an arbitrary partition $\mathcal{C} = (C_1, \ldots, C_K)$. Denote by $\mathbf{x}^k \in \{0, 1\}^n$ the indicator vector of cluster C_k for $k = 1, \ldots, K$.

We start with rewriting, again, the expression of *NCut* . From Proposition 7.5, noting that $\sum_{i \in C_k} d_i = \sum_{i \in V} (x_i^k)^2 d_i$ and

$$\sum_{i,j \in C_k} S_{ij} = \sum_{i,j \in V} S_{ij} x_i^k x_j^k = \sum_{i \in V} (x_i^k)^2 d_i - \sum_{ij \in E} S_{ij} (x_i^k - x_j^k)^2 \qquad (7.19)$$

we obtain that

$$NCut(\mathcal{C}) = K - \sum_{k=1}^{K} \frac{\sum_{i,j \in C_k} S_{ij}}{\sum_{i \in C_k} d_i} = \sum_{k=1}^{K} \frac{\sum_{ij \in E} S_{ij} (x_i^k - x_j^k)^2}{\sum_{i \in V} (x_i^k)^2 d_i} = \sum_{k=1}^{K} R(x^k) \qquad (7.20)$$

In the sums above, $i, j \in C_k$ means summation over the ordered pairs (i, j) while $ij \in E$ means summation over all "edges," that is all unordered pairs (i, j) with $i \neq j$. Next, we substitute

$$\mathbf{y}^k = \mathbf{D}^{1/2}\mathbf{x}^k \tag{7.21}$$

obtaining

$$R(\mathbf{x}^k) = \frac{(\mathbf{y}^k)^T \mathbf{L} \mathbf{y}^k}{(\mathbf{y}^k)^T \mathbf{y}^k} = \tilde{R}(\mathbf{y}^k) \tag{7.22}$$

and

$$NCut(\mathcal{C}) = \sum_{k=1}^{K} \tilde{R}(\mathbf{y}^k) \tag{7.23}$$

The expression $\tilde{R}(\mathbf{y})$ represents the *Rayleigh quotient* (Chung 1997) for the symmetric matrix **L** of Equation 7.13. Recall a classic Rayleigh–Ritz theorem in linear algebra (Strang 1988), stating that the sum of K Rayleigh quotients depending on orthogonal vectors $\mathbf{y}^1 \cdots \mathbf{y}^K$ is minimized by the eigenvectors of **L** corresponding to its smallest K eigenvalues $\mu_1 \le \mu_2 \le \cdots \mu_K$. As $\mathbf{y}^k, \mathbf{y}^l$ defined by 7.21 are orthogonal, the expression 7.23 cannot be smaller than $\sum_{k=1}^{K} \tilde{R}(\mathbf{u}^k) = \sum_{k=1}^{K} \mu_k = K - \sum_{k=1}^{K} \lambda_k$, which completes the proof.

Hence, if **S** is perfect with respect to some K-clustering $\mathcal{C}^*$, then $\mathcal{C}^*$ is the minimum *NCut* clustering, and Algorithm 7.2 returns $\mathcal{C}^*$.

Recall that finding the clustering $\mathcal{C}^\dagger$ that minimizes *NCut* is NP-hard. Formulated in terms of $\mathbf{y}^{1:K}$, this problem is

$$\min_{\mathbf{y}^1,\dots,\mathbf{y}^K \in \mathbb{R}^n} \sum_{k=1}^{K} (\mathbf{y}^k)^T \mathbf{L} \mathbf{y}^k \quad \text{s.t.} \quad (\mathbf{y}^l)^T \mathbf{y}^k = \delta_{kl} \quad \text{for all } k,l = 1,\dots,K \tag{7.24}$$

$$\text{there exist } \mathbf{x}^{1:K} \in \{0,1\}^n \text{ so that 7.21 holds} \tag{7.25}$$

By dropping constraint 7.25, we obtain

$$\min_{\mathbf{y}^1,\dots,\mathbf{y}^K \in \mathbb{R}^n} \sum_{k=1}^{K} (\mathbf{y}^k)^T L \mathbf{y}^k \quad \text{s.t.} \quad (\mathbf{y}^l)^T \mathbf{y}^k = \delta_{kl} \quad \text{for all } k,l = 1,\dots,K \tag{7.26}$$

whose solution is given by the eigenvectors $\mathbf{u}^1,\dots,\mathbf{u}^K$ and smallest eigenvalues $\mu^1,\dots,\mu^K$ of **L**. Applying 7.21 and Proposition 7.1 to $\mathbf{u}^{1:K}$, we see that the $\mathbf{x}^{1:K}$ correspondig to the solution of 7.26 are no other than the eigenvectors $\mathbf{v}^{1:K}$ of **P**. Equation 7.26 can be formulated directly in the **x** variables as

$$\min_{\mathbf{x}^1,\dots,\mathbf{x}^K} \sum_{k=1}^{K} R(\mathbf{x}^k) \quad \text{s.t.} \quad \mathbf{x}^k \perp \mathbf{D}\mathbf{x}^l \quad \text{for } k \ne l \quad \text{and} \quad ||\mathbf{x}^k|| = 1 \quad \text{for all } k \tag{7.27}$$

Equation 7.27 is called a *relaxation* of the original minimization Equation 7.24. Intuitively, the solution of the relaxed problem is an approximation to the original Problem 7.25 when the latter has a clustering with cost near the lower bound. This intuition was proved formally by Bach and Jordan (2006) and Meila (2014). Hence, spectral clustering algorithms are an approximate way to find the minimum *NCut*.

We have shown here that (i) when **P** is perfect, Algorithm 7.2 minimizes the *NCut* exactly and that (ii) otherwise, the algorithm solves the relaxed Equation 7.27 and *rounds* the results by K-means to obtain an approximately optimal *NCut* clustering.

7.3.3 Spectral Clustering as Finding Smooth Embeddings

Here, we explore further the connection between the normalized cut of a clustering C and the Laplacian matrix $\mathbf{L}$ seen as an operator applied to functions on the set V, and the functional $||\mathbf{f}||_{\Delta}^2$ defined below as a *smoothness functional*.

Proposition 7.7

Let $\mathbf{L}$ be the normalized Laplacian defined by Equation 7.7 and $\mathbf{f} \in \mathbb{R}^n$ be any vector indexed by the set of nodes V. Then

$$\Delta f \stackrel{def}{=} \mathbf{f}^T \mathbf{L} \mathbf{f} = \sum_{ij \in E} S_{ij} \left(\frac{f_i}{\sqrt{d_i}} - \frac{f_j}{\sqrt{d_j}} \right)^2 \tag{7.28}$$

The proof follows closely the steps 7.19–7.20.

Now, consider the *NCut* expression 7.24 and replace $\mathbf{y}^k$ by $\mathbf{D}^{-1/2}\mathbf{x}^k$ according to 7.21. We obtain*

$$\tilde{R}(\mathbf{y}^k) = R(\mathbf{x}^k) = \sum_{ij \in E} S_{ij}(x_i^k - x_j^k)^2 \tag{7.29}$$

This shows that a clustering that has low *NCut* is one whose indicator functions $\mathbf{x}^{1:K}$ are *smooth w.r.t. the graph* $\mathcal{G}$. In other words, the functions $\mathbf{x}^k$ must be almost constant on groups of nodes that are very similar, and are allowed to make abrupt changes only along edges with $S_{ij} \approx 0$.

The symbol Δ and the name "Laplacian" indicate that $\mathbf{L}$ and $\mathbf{f}^T \mathbf{L} \mathbf{f}$ are the graph analogues of the well-known Laplace operator on $\mathbb{R}^d$, while Proposition 7.28 corresponds to the relationship $< f, \Delta f >= \int_{\text{dom } f} |\nabla f|^2 dx$ in real analysis. The relationship between the continuous Δ and the graph Laplacian has been studied by Belkin and Niyogi (2002), Coifman and Lafon (2006), and Hein et al. (2007).

7.4 Where Do the Similarities Come From?

If the original data are vectors in $\mathbf{x}_i \in \mathbb{R}^d$ (note the abusive notation $\mathbf{x}$ in this section only), then the similarity is typically the *Gaussian kernel* (also called *heat kernel*)

$$S_{ij} = \exp \left(-\frac{||\mathbf{x}_i - \mathbf{x}_j||^2}{\sigma^2} \right) \tag{7.30}$$

This similarity gives raise to a complete graph $\mathcal{G}$, as $S_{ij} > 0$ always. Alternatively, one can define graphs that are dense only over *local neighborhoods*. For example, one can set S_{ij} by 7.30 if $||\mathbf{x}_i - \mathbf{x}_j|| \leq c\sigma$ and 0 otherwise, with the constant $c \approx 3$. This construction leads to a sparse graph, which is, however, a good approximation of the complete graph obtained

---

* This expression is almost identical to 7.20; the only difference is that in 7.20 the indicator vectors $\mathbf{x}^k$ take values in $\{0, 1\}$ while here they are normalized by $(\mathbf{x}^k)^T \mathbf{D} \mathbf{x}^k = 1$.

by the heat kernel (Ting 2010). A variant of the above is zero out all S_{ij} except for the m nearest neighbors of data point i. This method used without checks can produce matrices that are not symmetric.

Even though the two graph construction methods appear to be very similar, it has been shown theoretically and empiricaly (Hein 2007; Maier et al. 2008) that the spectral clustering results they produce can be very different, both in high and in low dimensions. With the fixed m-nearest neighbor graphs, the clustering results strongly favor balanced cuts, even if the cut occurs in regions of higher density; the radius-neighbor graph construction favors finding cuts of low density. This is explained by the observation below, that the graph density in the latter graphs reflects the data density stronger than in the former type of graph.

It was pointed out that when the data density varies much, there is no unique radius that correctly reflects "locality," while the K-nearest neighbor graphs adapt to the varying density. A simple and widely used way to "tune" the similarity function to the local density (Zelnik-Manor and Perona 2004) is to set

$$S_{ij} = \exp\left(-\frac{||\mathbf{x}_i - \mathbf{x}_j||^2}{\sigma_i \sigma_j}\right) \tag{7.31}$$

where σ_i is the distance from $\mathbf{x}_i$ to its m-th nearest neighbor. Another simple heuristic to choose σ is to try various σ values and to pick the one that produces the smallest K-means cost in step 4 (Ng et al. 2002).

If the features in the data $\mathbf{x}$ have different units, or come from different modalities of measuring similarity, then it is useful to give each feature x_f a different kernel width σ_f. Hence, the similarity becomes

$$S_{ij} = \exp\left(\sum_{f=1}^{d} \frac{(x_{if} - x_{jf})^2}{\sigma_f^2}\right) \tag{7.32}$$

Clustering by similarities is not restricted to points in vector spaces. This represents one of the strengths of spectral clustering. If a distance $dist(i, j)$ can be defined on the data, then $dist(i, j)^2$ can substitute $||\mathbf{x}_i - \mathbf{x}_j||^2$ in 7.30; $dist$ can be obtained from the kernel trick (Schölkopf and Smola 2002). Hence, spectral clustering can be applied to a variety of classes of nonvector data for which Mercer kernels have been designed, like trees, sequences or phylogenies (Schölkopf and Smola 2002; Clark et al. 2011; Shin et al. 2011).

Several methods for *learning* the similarities as a function of data features in a supervised setting exist (Meila and Shi 2001; Meila et al. 2005; Bach and Jordan 2006); the method of Meila et al. (2005) has been extended to the unsupervised setting by Shortreed and Meila (2005).

7.5 Practical Considerations

The main advantage of spectral clustering is that it does not make any assumptions about the cluster shapes, and even allows clusters to "touch," as long as the clusters have sufficient overall separation and internal coherence (e.g., Figure 7.2 right panels).

The method is computationally expensive compared to, for example, center-based clustering, as it needs to store and manipulate similarities/distances between all pairs of points instead of only distances to centers. The eigendecomposition step can also be computationally intensive. However, with a careful implementation, for example using sparse neighborhood graphs as in Section 7.4 instead of all pairwise similarities, and sparse matrix representations, the memory and computational requirements can be made tractable for sample sizes in the tens of thousands or larger. Several fast and approximate methods for spectral clustering have been proposed (Fowlkes et al. 2004; Chen et al. 2006; Liu et al. 2007; Wauthier et al. 2012).

It is known from matrix perturbation theory (Stewart and Sun 1990) that eigenvectors with smaller λ_k are more affected by numerical errors and noise in the similarities. This can be a problem when the number of clusters K is not small. In such a case, one can either (i) use only the first $K_0 < K$, eigenvectors of $\mathbf{P}$ or, (ii) use the diffusion distance type rescaling $\mathbf{v}^k$ by λ_k^α, with $\alpha > 1$ which will smoothly decrease the effect of the noisier eigenvectors or (iii) use TWO-WAY SPECTRAL CLUSTERING recursively.

One drawback of spectral clustering is the dependence of the eigenvectors $\mathbf{v}^k$ on the similarity $\mathbf{S}$ in ways that are not intuitive. For example, monotonic transformations of S_{ij}, even shift by a constant, can change a perfect $\mathbf{S}$ into one that is not perfect.

Outliers in spectral clustering need special treatment. An outlier is a point which has very low similarity with all other points (e.g., because it is far away from them). An outlier will produce a spurious eigenvalue very close to 1 with an eigenvector which approximates an indicator vector for the outlier. So, l outliers in a data set will cause the l principal eigenvectors to be outliers, not clusters. Thus, it is *strongly recommended* that outliers be detected and removed *before* the eigendecomposition is performed. This is done easiest by removing all points for which $\sum_{j \neq i} S_{ij} \leq \epsilon$ for some ϵ which is small w.r.t. the average d_i. Also before the eigendecomposition, one should detect if $\mathcal{G}$ is disconnected by a *connected components* algorithm (see Chapter 18).

7.6 Conclusions

The tight relationship between K-MEANS and SPECTRAL CLUSTERING hints at the situations when SPECTRAL CLUSTERING is recommended. Namely, SPECTRAL CLUSTERING returns hard, nonoverlaping clusterings, requires the number of clusters K as input, and works best when this number is not too large (up to $K = 10$). For larger K, recursive partitioning based on TWO-WAY SPECTRAL CLUSTERING is more robust. The relationship with K-MEANS is even deeper than we have presented it here (Ding and He 2004). As mentioned above, the algorithm is sensitive to outliers and transformations of $\mathbf{S}$, but it is very robust to the shapes of clusters, to small amounts of data "spilling" from one cluster to the another, and can balance well cluster sizes and their internal coherence.

For chosing the number of clusters K, there are two important indicators: the *eigengap* $\lambda_K - \lambda_{K+1}$, and the *gap* $NCut(\mathcal{C}_K) - (K - \sum_{k=1}^K \lambda_K)$, where we have denoted by $\mathcal{C}_K$ the clustering returned by a spectral clustering algorithm with input $\mathbf{S}$ and K. Ideally, the former should be large, indicating a stable principal subspace, and the latter should be near zero, indicating almost perfect $\mathbf{P}$ for that K and $\mathcal{C}_K$. A heuristic proposed by Meila and Xu (2003) is to find the knee in the graph of gap vs. K, or in the graph of gap divided by the eigengap, as suggested by the theory in Meila (2014); Azran and Ghahramani (2006) proposes heuristic based on the eigengaps $\lambda_k^t - \lambda_{k+1}^t$ for $t > 1$ that can find clusterings at different granularity levels and works well for matrices that are almost block diagonal.

Other formulations of clustering that aim to minimize the same Normalized Cut criterion are based on Semidefinite Programming (Xing and Jordan 2003), and on submodular function optimization (Boykov et al. 2001; Kolmogorov and Zabih 2004; Narasimhan and Bilmes 2007).

Spectral clustering has been extended to directed graphs (Pentney and Meila 2005; Andersen et al. 2007; Meila and Pentney 2007) as well as finding *the local cluster* of a data point in a large graph (Spielman and Teng 2008).

Clusterability for spectral clustering, i.e., the problem of defining what is a "good" clustering, has been studied by Ackerman and Ben-David (2009), Balcan and Braverman (2009), Kannan et al. (2000), Meila (2014), and Meila (2006); some of these references also introduced new algorithms with guarantees that depend on how clusterable is the data.

Finally, the ideas and algorithms presented here have deep connections with the fast growing areas of *nonlinear dimension reduction*, also known as *manifold learning* (Belkin and Niyogi 2002) and of solving very large linear systems (Batson et al. 2013).

References

M. Ackerman and S. Ben-David. Clusterability: A theoretical study. In D. A. Van Dyk and M. Welling, editors, *Proceedings of the Twelfth International Conference on Artificial Intelligence and Statistics, AISTATS 2009, Clearwater Beach, FL, April 16–18, 2009*, volume 5 of *JMLR Proceedings*, pp. 1–8. JMLR.org, 2009.

R. Andersen, F. R. K. Chung, and K. J. Lang. Local partitioning for directed graphs using pagerank. In *WAW*, pp. 166–178, 2007.

A. Azran and Z. Ghahramani. Spectral methods for automatic multiscale data clustering. In *Computer Vision and Pattern Recognition*, pp. 190–197. IEEE Computer Society, 2006.

F. Bach and M. I. Jordan. Learning spectral clustering with applications to speech separation. *Journal of Machine Learning Research*, 7:1963–2001, 2006.

S. Balakrishnan, M. Xu, A. Krishnamurthy, and A. Singh. Noise thresholds for spectral clustering. In *Advances in Neural Information Processing Systems 24: 25th Annual Conference on Neural Information Processing Systems 2011. Proceedings of a Meeting Held December 12–14, 2011, Granada*, pp. 954–962, 2011.

M.-F. Balcan and M. Braverman. Finding low error clusterings. In *COLT 2009—The 22nd Conference on Learning Theory, Montreal, Quebec, June 18–21, 2009*, 2009.

J. D. Batson, D. A. Spielman, N. Srivastava, and S.-H. Teng. Spectral sparsification of graphs: Theory and algorithms. *Communications of the ACM*, 56(8):87–94, 2013.

M. Belkin and P. Niyogi. Laplacian eigenmaps and spectral techniques for embedding and clustering. In T. G. Dietterich, S. Becker, and Z. Ghahramani, editors, *Advances in Neural Information Processing Systems 14*, Cambridge, MA, 2002. MIT Press.

Y. Boykov, O. Veksler, and R. Zabih. Fast approximate energy minimization via graph cuts. *IEEE Transactions on Pattern Analysis and Machine Intelligence*, 23(11):1222–1239, 2001.

S. Bubeck, M. Meila, and U. von Luxburg. How the initialization affects the stability of the k-means algorithm. *ESAIM: Probability and Statistics*, 16:436–452, 2012.

B. Chen, B. Gao, T.-Y. Liu, Y.-F. Chen, and W.-Y. Ma. Fast spectral clustering of data using sequential matrix compression. In *Proceedings of the 17th European Conference on Machine Learning, ECML*, pp. 590–597, 2006.

F. R. K. Chung. *Spectral Graph Theory*. Number Regional Conference Series in Mathematics in 92. American Mathematical Society, Providence, RI, 1997.

A. Clark, C. C. Florêncio, and C. Watkins. Languages as hyperplanes: Grammatical inference with string kernels. *Machine Learning*, 82(3):351–373, 2011.

R. R. Coifman and S. Lafon. Diffusion maps. *Applied and Computational Harmonic Analysis*, 30(1): 5–30, 2006.

S. Dasgupta and L. Schulman. A probabilistic analysis of EM for mixtures of separated, spherical gaussians. *Journal of Machine Learnig Research*, 8:203–226, 2007.

C. Ding and X. He. K-means clustering via principal component analysis. In C. E. Brodley, editor, *Proceedings of the International Machine Learning Conference (ICML)*. Morgan Kauffman, 2004.

C. Fowlkes, S. Belongie, F. Chung, and J. Malik. Spectral grouping using the Nyström method. *IEEE Transactions on Pattern Analysis and Machine Intelligence*, 26(2):214–225, 2004.

Y. Gdalyahu, D. Weinshall, and M. Werman. Stochastic image segmentation by typical cuts. In *Computer Vision and Pattern Recognition*, volume 2, pp. 2596. IEEE Computer Society, IEEE, 1999.

M. Hein, J.-Y. Audibert, and U. von Luxburg. Graph laplacians and their convergence on random neighborhood graphs. *Journal of Machine Learning Research*, 8:1325–1368, 2007.

D. S. Hochbaum and D. B. Shmoys. A best possible heuristic for the k-center problem. *Mathematics of Operations Research*, 10(2):180–184, 1985.

R. Kannan, S. Vempala, and A. Vetta. On clusterings: Good, bad and spectral. In *Proceedings of 41st Symposium on the Foundations of Computer Science, FOCS 2000*, 2000.

V. Kolmogorov and R. Zabih. What energy functions can be minimized via graph cuts? *IEEE Transactions on Pattern Analysis and Machine Intelligence*, 26(2):147–159, 2004.

T.-Y. Liu, H.-Y. Yang, X. Zheng, T. Qin, and W.-Y. Ma. Fast large-scale spectral clustering by sequential shrinkage optimization. In *Proceedings of the 29th European Conference on IR Research*, pp. 319–330, 2007.

M. Maier, U. von Luxburg, and M. Hein. Influence of graph construction on graph-based clustering measures. In *Advances in Neural Information Processing Systems 21, Proceedings of the Twenty-second Annual Conference on Neural Information Processing Systems, Vancouver, British Columbia, Canada, December 8–11, 2008*, pp. 1025–1032, 2008.

M. Meila. The uniqueness of a good optimum for K-means. In A. Moore and W. Cohen, editors, *Proceedings of the International Machine Learning Conference (ICML)*, pp. 625–632. International Machine Learning Society, 2006.

M. Meila. The stability of a good clustering. Technical Report 624, University of Washington, 2014.

M. Meila and W. Pentney. Clustering by weighted cuts in directed graphs. In *Proceedings of the Seventh SIAM International Conference on Data Mining, April 26–28, 2007, Minneapolis, MN*, pp. 135–144, 2007.

M. Meila and J. Shi. Learning segmentation by random walks. In T. K. Leen, T. G. Dietterich, and V. Tresp, editors, *Advances in Neural Information Processing Systems*, volume 13, pp. 873–879, Cambridge, MA, 2001. MIT Press.

M. Meila and J. Shi. A random walks view of spectral segmentation. In T. Jaakkola and T. Richardson, editors, *Proceedings of the Eighth International Workshop on Artificial Intelligence and Statistics AISTATS*, 2001. Available at http://www.gatsby.ucl.ac.uk/aistats/aistats2001/papers.html.

M. Meila, S. Shortreed, and L. Xu. Regularized spectral learning. In R. Cowell and Z. Ghahramani, editors, *Proceedings of the Tenth International Workshop on Artificial Intelligence and Statistics (AISTATS 05)*, pp. 230–237, 2005.

M. Meila and L. Xu. Multiway cuts and spectral clustering. Technical Report 442, University of Washington, Department of Statistics, 2003.

B. Nadler, S. Lafon, R. Coifman, and I. Kevrekidis. Diffusion maps, spectral clustering and eigenfunctions of fokker-planck operators. In Y. Weiss, B. Schölkopf, and J. Platt, editors, *Advances in Neural Information Processing Systems 18*, pp. 955–962, Cambridge, MA, 2006. MIT Press.

M. Narasimhan and J. Bilmes. Local search for balanced submodular clusterings. In M. M. Veloso, editor, *IJCAI 2007, Proceedings of the 20th International Joint Conference on Artificial Intelligence, Hyderabad, January 6–12, 2007*, pp. 981–986, 2007.

A. Y. Ng, M. I. Jordan, and Y. Weiss. On spectral clustering: Analysis and an algorithm. In T. G. Dietterich, S. Becker, and Z. Ghahramani, editors, *Advances in Neural Information Processing Systems 14*, Cambridge, MA, 2002. MIT Press.

C. Papadimitriou and K. Steiglitz. *Combinatorial optimization. Algorithms and complexity.* Dover Publication, Inc., Minneola, NY, 1998.

W. Pentney and M. Meila. Spectral clustering of biological sequence data. In M. Veloso and S. Kambhampati, editors, *Proceedings of Twentieth National Conference on Artificial Intelligence (AAAI-05)*, pp. 845–850, Menlo Park, CA, 2005. The AAAI Press.

B. Schölkopf and A. J. Smola. *Learning with Kernels*, Cambridge, MA, 2002. MIT Press.

J. Shi and J. Malik. Normalized cuts and image segmentation. *PAMI*, 2000.

K. Shin, M. Cuturi, and T. Kuboyama. Mapping kernels for trees. In L. Getoor and T. Scheffer, editors, *Proceedings of the 28th International Conference on Machine Learning, ICML 2011, Bellevue, WA, June 28–July 2, 2011*, pp. 961–968. Omnipress, 2011.

S. Shortreed and M. Meila. Unsupervised spectral learning. In T. Jaakkola and F. Bachhus, editors, *Proceedings of the 21st Conference on Uncertainty in AI*, pp. 534–544, Arlington, VA, 2005. AUAI Press.

D. Spielman and S.-H. Teng. A local clustering algorithm for massive graphs and its application to nearly-linear time graph partitioning. Technical Report:0809.3232v1 [cs.DS], arXiv, 2008.

G. W. Stewart and J.-G. Sun. *Matrix Perturbation Theory*. Academic Press, San Diego, CA, 1990.

G. Strang. *Linear Algebra and Its Applications, 3rd Edition*. Saunders College Publishing, San Diego, CA, 1988.

D. Ting, L. Huang, and M. I. Jordan. An analysis of the convergence of graph laplacians. In *Proceedings of the 27th International Conference on Machine Learning (ICML-10), June 21–24, 2010, Haifa*, pp. 1079–1086, 2010.

D. Verma and M. Meila. A comparison of spectral clustering algorithms. TR 03-05-01, University of Washington, 2003.

F. Wauthier, N. Jojic, and M. Jordan. Active spectral clustering via iterative uncertainty reduction. In *18th ACM SIGKDD International Conference on Knowledge Discovery and Data Mining*, pp. 1339–1347, 2012.

S. White and P. Smyth. A spectral clustering approach to finding communities in graphs. In *Proceedings of SIAM International Conference on Data Mining*, 2005.

E. P. Xing and M. I. Jordan. On semidefinite relaxation for normalized k-cut and connections to spectral clustering. Technical Report UCB/CSD-03-1265, EECS Department, University of California, Berkeley, 2003.

L. Zelnik-Manor and P. Perona. Self-tuning spectral clustering. In L. K. Saul, Y. Weiss, and L. Bottou, editors, *Advances in Neural Information Processing Systems*, volume 17, pp. 1601–1608, MIT Press, Cambridge, MA, 2004.

Section III

Methods Based
on Probability Models

8

Mixture Models for Standard p-Dimensional
Euclidean Data

Geoffrey J. McLachlan and Suren I. Rathnayake

CONTENTS

Abstract

In this chapter, we consider the use of finite mixture models for the clustering of multivariate data observed from a random sample. Such models can be fitted by maximum likelihood via the expectation–maximization (EM) algorithm. The focus is on the use of mixtures of normal component distributions with attention also given to component t-distributions for clusters with tails longer than the normal. There is also coverage of recent developments on the use of mixtures of skew normal and skew t-distributions for nonelliptical-shaped clusters that may possibly contain outliers.

8.1 Introduction

Clustering procedures based on finite mixture models are being increasingly used due to their sound mathematical basis and to the interpretability of their results. Mixture model-based procedures provide a probabilistic clustering that allows for overlapping clusters corresponding to the components of the mixture model. The uncertainties that the observations belong to the clusters are provided in terms of the fitted values for their posterior probabilities of component membership of the mixture. Typically, each component in a finite mixture model corresponds to a cluster. In this case, the problem of choosing an appropriate clustering method can be recast as the choice of a suitable statistical model. Also, in this case, it allows the important question of how many clusters are there in the data to be approached through an assessment of how many components are needed in the mixture model. These questions of model choice and number of clusters can be considered in terms of the likelihood function.

Scott and Symons (1971) were one of the first to adopt a model-based approach to clustering, which is equivalent to the so-called classification maximum likelihood approach (Ganesalingam and McLachlan 1979). Assuming that the data were normally distributed within a cluster, Scott and Symons (1971) considered the maximization of the likelihood function over all values of the parameters in the cluster distributions as well as over all possible values of the cluster indicator variables. They subsequently showed their approach is equivalent to some commonly used clustering criteria with various constraints on the cluster covariance matrices. However, from an estimation point of view, this approach yields inconsistent estimators of the parameters (Bryant and Williamson 1978; McLachlan 1982).

This inconsistency can be avoided by working with the mixture likelihood formed under the assumption that the observed data are from a mixture of classes corresponding to the clusters to be imposed on the data, as proposed by Day (1969), Wolfe (1965), Wolfe (1970). Finite mixture models have since been increasingly used to model the distributions of a wide variety of random phenomena and to cluster data sets (Böhning 1999; McLachlan and Peel 2000; Frühwirth-Schnatter 2006; Mengersen et al. 2011), and the references therein. Earlier references on mixture models may be found in the previous books by Banfield and Raftery (1993), Everitt and Hand (1981), Fraley and Raftery (2002), Lindsay (1995), McLachlan and Basford (1988), Titterington et al. (1985).

8.2 Definition of Mixture Models

We let Y denote a random vector consisting of p feature variables associated with the random phenomenon of interest. We let $y_1, \ldots, y_n$ denote n independent observations on Y. With the finite mixture model-based approach to density estimation and clustering, the density of Y is modeled as a mixture of a number (g) of component densities $f_i(y)$ in some unknown proportions $\pi_1, \ldots, \pi_g$. That is, each data point is taken to be a realization of the mixture probability density function (p.d.f.),

$$f(y; \Psi) = \sum_{i=1}^{g} \pi_i f_i(y), \qquad (8.1)$$

where the mixing proportions π_i are non-negative and sum to one. In density estimation, the number of components g can be taken sufficiently large for (8.1) to provide an arbitrarily accurate estimate of the underlying density function (Li and Barron 2000). For clustering purposes, each component in the mixture model (8.1) is usually taken to correspond to a cluster. The posterior probability that an observation with feature vector y_j belongs to the ith component of the mixture is given by

$$\tau_i(y_j) = \pi_i f_i(y_j) / f(y_j) \qquad (8.2)$$

for $i = 1, \ldots, g$. A probabilistic clustering of the data into g clusters can be obtained in terms of the fitted posterior probabilities of component membership for the data.

An outright partitioning of the observations into g nonoverlapping clusters $C_1, \ldots, C_g$ is effected by assigning each observation to the component to which it has the highest estimated posterior probability of belonging. Thus, the ith cluster C_i contains those observations assigned to group G_i. That is, C_i contains those observations j with $\hat{z}_{ij} = (\hat{z}_j)_i = 1$, where

$$\hat{z}_{ij} = 1, \qquad \text{if } i = \arg \max_h \hat{\tau}_h(y_j),$$

$$= 0, \qquad \text{otherwise}, \qquad (8.3)$$

and $\hat{\tau}_i(y_j)$ is an estimate of $\tau_i(y_j)$ ($i = 1, \ldots, g; \ j = 1, \ldots, n$). As the notation implies, $\hat{z}_{ij}$ can be viewed as an estimate of z_{ij} which, under the assumption that the observations come from a mixture of g groups $G_1, \ldots, G_g$, is defined to be one or zero according as the jth observation does or does not come from G_i ($i = 1, \ldots, g; \ j = 1, \ldots, n$).

8.3 Maximum-Likelihood Estimation

On specifying a parametric form $f_i(y_j; \theta_i)$ for each component density, we can fit this parametric mixture model

$$f(y_j; \Psi) = \sum_{i=1}^{g} \pi_i f_i(y_j; \theta_i) \qquad (8.4)$$

by maximum likelihood (ML). Here, $\Psi = (\xi^T, \pi_1, \ldots, \pi_{g-1})^T$ is the vector of unknown parameters, where ξ consists of the elements of the θ_i known *a priori* to be distinct. We let Ω denote the parameter space for Ψ. In order to estimate Ψ from the observed data, it must be identifiable. This will be so if the representation (8.4) is unique up to a permutation of the component labels. The maximum-likelihood estimate (MLE) of Ψ, $\hat{\Psi}$, is given by an appropriate root of the likelihood equation,

$$\partial \log L(\Psi)/\partial \Psi = 0, \tag{8.5}$$

where $L(\Psi)$ denotes the likelihood function for Ψ,

$$L(\Psi) = \prod_{j=1}^{n} f(y_j; \Psi).$$

Solutions of (8.5) corresponding to local maximizers of $\log L(\Psi)$ can be obtained via the expectation–maximization (EM) algorithm of Dempster et al. (1977) and McLachlan and Krishnan (1997). Let $\hat{\Psi}$ denote the estimate of Ψ so obtained.

8.4 Fitting Mixture Models via the EM Algorithm

Now, we consider the ML fitting of the mixture model (8.4) via the EM algorithm. It is straightforward, at least in principle, to find solutions of (8.5) using the EM algorithm. It is easy to program for this problem and proceeds iteratively in two steps, E (for expectation) and M (for maximization).

For the purpose of the application of the EM algorithm, the observed data are regarded as being incomplete. The complete data are taken to be the observed feature vectors $y_1, \ldots, y_n$, along with their component-indicator vectors $z_1, \ldots, z_n$, which are unobservable in the framework of the mixture model being fitted. Consistent with the notation introduced in the last section, the ith element z_{ij} of z_j is defined to be one or zero, according as the jth with feature vector y_j does or does not come from the ith component of the mixture, that is, from group G_i ($i = 1, \ldots, g$; $j = 1, \ldots, n$). Thus, the data are conceptualized to have come from g groups $G_1, \ldots, G_g$, irrespective of whether these groups do externally exist.

For this specification, the complete data log likelihood is

$$\log L_c(\Psi) = \sum_{i=1}^{g} \sum_{j=1}^{n} z_{ij} \log \pi_i + \sum_{i=1}^{g} \sum_{j=1}^{n} z_{ij} \log f_i(y_j; \theta_i). \tag{8.6}$$

8.4.1 E-Step

The addition of the unobservable data to the problem (here the z_j) is handled by the E-step, which takes the conditional expectation of the complete-data log likelihood, $\log L_c(\Psi)$, given the observed data

$$y_{\text{obs}} = (y_1^T, \ldots, y_n^T)^T,$$

using the current fit for Ψ. Let $\Psi^{(0)}$ be the value specified initially for Ψ. Then, on the first iteration of the EM algorithm, the E-step requires the computation of the conditional expectation of $\log L_c(\Psi)$ given y_{obs}, using $\Psi^{(0)}$ for Ψ, which can be written as

$$Q(\Psi; \Psi^{(0)}) = E_{\Psi^{(0)}}\{\log L_c(\Psi) \mid y_{obs}\}. \tag{8.7}$$

The expectation operator E has the subscript $\Psi^{(0)}$ to explicitly convey that this expectation is being effected using $\Psi^{(0)}$ for Ψ.

It follows that on the $(k+1)$th iteration, the E-step requires the calculation of $Q(\Psi; \Psi^{(k)})$, where $\Psi^{(k)}$ is the value of Ψ after the kth EM iteration. As the complete data log likelihood, $\log L_c(\Psi)$, is linear in the unobservable data z_{ij}, the E-step (on the $(k+1)$th iteration) simply requires the calculation of the current conditional expectation of Z_{ij} given the observation y_j, where Z_{ij} is the random variable corresponding to z_{ij}. Now

$$E_{\Psi^{(k)}}(Z_{ij} \mid y_j) = \text{pr}_{\Psi^{(k)}}\{Z_{ij} = 1 \mid y_j\}$$

$$= \tau_i(y_j; \Psi^{(k)}), \tag{8.8}$$

where, corresponding to (8.2),

$$\tau_i(y_j; \Psi^{(k)}) = \pi_i^{(k)} f_i(y_j; \theta_i^{(k)})/f(y_j; \Psi^{(k)})$$

$$= \pi_i^{(k)} f_i(y_j; \theta_i^{(k)})/\sum_{h=1}^{g} \pi_h^{(k)} f_h(y_j; \theta_h^{(k)}) \tag{8.9}$$

for $i = 1, \ldots, g$; $j = 1, \ldots, n$. The quantity $\tau_i(y_j; \Psi^{(k)})$ is the posterior probability that the jth member of the sample with observed value y_j belongs to the ith component of the mixture. Using (8.8), we have on taking the conditional expectation of (8.6) given y_{obs} that

$$Q(\Psi; \Psi^{(k)}) = \sum_{i=1}^{g}\sum_{j=1}^{n} \tau_i(y_j; \Psi^{(k)})\{\log \pi_i + \log f_i(y_j; \theta_i)\}. \tag{8.10}$$

8.4.2 M-Step

The M-step on the $(k+1)$th iteration requires the global maximization of $Q(\Psi; \Psi^{(k)})$ with respect to Ψ over the parameter space Ω to give the updated estimate $\Psi^{(k+1)}$. For the finite mixture model, the updated estimates $\pi_i^{(k+1)}$ of the mixing proportions π_i are calculated independently of the updated estimate $\xi^{(k+1)}$ of the parameter vector ξ containing the unknown parameters in the component densities.

If the z_{ij} were observable, then the complete-data MLE of π_i would be given simply by

$$\hat{\pi}_i = \sum_{j=1}^{n} z_{ij}/n \quad (i = 1, \ldots, g). \tag{8.11}$$

As the E-step simply involves replacing each z_{ij} with its current conditional expectation $\tau_i(y_j; \Psi^{(k)})$ in the complete-data log likelihood, the updated estimate of π_i is given by replacing each z_{ij} in (8.11) by $\tau_i(y_j; \Psi^{(k)})$ to give

$$\pi_i^{(k+1)} = \sum_{j=1}^{n} \tau_i(y_j; \Psi^{(k)})/n \qquad (i = 1, \ldots, g). \tag{8.12}$$

Thus in forming the estimate of π_i on the $(k + 1)$th iteration, there is a contribution from each observation y_j equal to its (currently assessed) posterior probability of membership of the ith component of the mixture model.

Concerning the updating of ξ on the M-step of the $(k + 1)$th iteration, it can be seen from (8.10) that $\xi^{(k+1)}$ is obtained as an appropriate root of

$$\sum_{i=1}^{g} \sum_{j=1}^{n} \tau_i(y_j; \Psi^{(k)}) \partial \log f_i(y_j; \theta_i)/\partial \xi = 0. \tag{8.13}$$

One nice feature of the EM algorithm is that the solution of (8.13) often exists in closed form, as is to be demonstrated for the normal mixture model in Section 8.8.

The E- and M-steps are alternated repeatedly until the difference

$$\log L(\Psi^{(k+1)}) - \log L(\Psi^{(k)})$$

changes by an arbitrarily small amount in the case of convergence of the sequence of likelihood values $\{L(\Psi^{(k)})\}$. Dempster et al. (1977) showed that the (incomplete-data) likelihood function $L(\Psi)$ is not decreased after an EM iteration; that is,

$$L(\Psi^{(k+1)}) \geq L(\Psi^{(k)}) \tag{8.14}$$

for $k = 0, 1, 2, \ldots$. Hence, convergence must be obtained with a sequence of likelihood values $\{L(\Psi^{(k)})\}$ that are bounded above. In almost all cases, the limiting value L^* is a local maximum. In any event, if an EM sequence $\{\Psi^{(k)}\}$ is trapped at some stationary point Ψ^* that is not a local or global maximizer of $L(\Psi)$ (e.g., a saddle point), a small random perturbation of Ψ away from the saddle point Ψ^* will cause the EM algorithm to diverge from the saddle point. Further details may be found in McLachlan and Krishnan (1997) (Chapter 3). We let $\hat{\Psi}$ be the chosen solution of the likelihood equation.

8.5 Choice of Starting Values for the EM Algorithm

McLachlan and Peel (2000) provide an in-depth account of the fitting of finite mixture models. Briefly, with mixture models the likelihood typically will have multiple maxima; that is, the likelihood equation will have multiple roots. For an observed sample, $\hat{\Psi}$ is usually taken to be the root of (8.5) corresponding to the largest of the local maxima located. That is, in those cases, where $L(\Psi)$ has a global maximum in the interior of the parameter space, $\hat{\Psi}$ is the global maximizer, assuming that the global maximum has been located.

Thus, the EM algorithm needs to be started from a variety of initial values for the parameter vector Ψ or for a variety of initial partitions of the data into g groups. The latter can be obtained by randomly dividing the data into g groups corresponding to the g components of the mixture model. With random starts, the effect of the central limit theorem tends to have the component parameters initially being similar at least in large samples. Nonrandom partitions of the data can be obtained via some clustering procedure such as k-means. Also, Coleman et al. (1999) have proposed some procedures for obtaining nonrandom starting partitions.

8.6 Advantages of Mixture Model-Based Clustering

It can be seen that this mixture likelihood-based approach to clustering is model based in that the form of each component density of an observation has to be specified in advance. Hawkins et al. (1982) commented that most writers on cluster analysis "lay more stress on algorithms and criteria in the belief that intuitively reasonable criteria should produce good results over a wide range of possible (and generally unstated) models." For example, the trace W criterion, where W is the pooled within-cluster sums of squares and products matrix, is predicated on normal groups with (equal) spherical covariance matrices; but as they pointed out, many users apply this criterion even in the face of evidence of non-spherical clusters or, equivalently, would use Euclidean distance as a metric. They strongly supported the increasing emphasis on a model-based approach to clustering. Indeed, as remarked by Aitkin et al. (1981) in the reply to the discussion of their paper, "when clustering samples from a population, no cluster method is, *a priori* believable without a statistical model." Concerning the use of mixture models to represent nonhomogeneous populations, they noted in their paper that "Clustering methods based on such mixture models allow estimation and hypothesis testing within the framework of standard statistical theory." Previously, Marriott (1974) had noted that the mixture likelihood-based approach "is about the only clustering technique that is entirely satisfactory from the mathematical point of view. It assumes a well-defined mathematical model investigates it by well-established statistical techniques, and provides a test of significance for the results." More recently, in the context of the analysis of gene expression data, Yeung et al. (2001) commented that "in the absence of a well-grounded statistical model, it seems difficult to define what is meant by a 'good' clustering algorithm or the 'right' number of clusters."

8.7 Choice of the Number of Components in a Mixture Model

With a mixture model-based approach to clustering, the question of how many clusters there are can be considered in terms of the smallest number of components needed for the mixture model to be compatible with the data. The estimation of the order of a mixture model has been considered mainly by consideration of the likelihood, using two main ways. One way is based on a penalized form of the log likelihood. The other main way is based on a resampling approach. In a Bayesian framework, Richardson and Green (1997) have used reversible jump Markov chain Monte Carlo methods to handle the case where the dimension of the parameter space is of varying dimension.

8.7.1 Bayesian Information Criterion and Related Methods

The main Bayesian-based information criteria use an approximation to the integrated likelihood, as in the original proposal by Schwarz (1978) leading to his Bayesian information criterion (BIC). Available general theoretical justifications of this approximation rely on the same regularity conditions that break down for inference on the number of components in a frequentist framework; see McLachlan and Peel (2000).

In the literature, the information criteria so formed are generally expressed in terms of twice the negative difference between the log likelihood and the penalty term. This negative difference for the Bayesian information criterion (BIC) is given by

$$-2 \log L(\hat{\boldsymbol{\Psi}}) + d \log n \qquad (8.15)$$

where d is the number of parameters in the model. The intent is to minimize the criterion (8.15) in model selection, including the present situation for the number of components g in a mixture model. Under certain conditions, Keribin (2000) has shown that BIC performs consistently in choosing the true number of components in a mixture model (Leroux 1992). Much research remains to be done in this area.

So far as assessing the number of clusters, it has been observed that BIC tends to favour models with enough components in order to provide a good estimate of the mixture density. Hence it tends to overestimate the number of clusters (Biernacki et al. 1998). This led Biernacki et al. (1998) to develop the integrated classification criterion (ICL). An approximation to this criterion is given by

$$-2 \log L(\hat{\boldsymbol{\Psi}}) + d \log n + EN(\hat{\boldsymbol{\tau}}), \qquad (8.16)$$

where

$$EN(\hat{\boldsymbol{\tau}}) = -\sum_{i=1}^{g} \sum_{j=1}^{n} \hat{\tau}_i(\boldsymbol{y}_j) \log \hat{\tau}_i(\boldsymbol{y}_j) \qquad (8.17)$$

is the entropy of the fuzzy classification matrix $((\hat{\tau}_i(\boldsymbol{y}_j)))$. Here, $\hat{\tau}_i(\boldsymbol{y}_j) = \tau_i(\boldsymbol{y}_j; \hat{\boldsymbol{\Psi}})$ and

$$\hat{\boldsymbol{\tau}} = (\hat{\boldsymbol{\tau}}_1^T, \ldots, \hat{\boldsymbol{\tau}}_n^T)^T, \qquad (8.18)$$

where

$$\hat{\boldsymbol{\tau}}_j = (\hat{\tau}_1(\boldsymbol{y}_j), \ldots, \hat{\tau}_g(\boldsymbol{y}_j))^T$$

is the vector of the estimated posterior probabilities of component membership of $\boldsymbol{y}_j$ $(j = 1, \ldots, n)$. That is, the ICL criterion uses the entropy term $EN(\hat{\boldsymbol{\tau}})$ to penalize the model for its complexity (too many components and hence clusters).

Another approach to refining the number of clusters has been given recently by Baudry et al. (2009), who have suggested a way in which the components can be recombined. Another interesting paper on how to recombine components is Hennig (2010).

8.7.2 Resampling Approach

A formal test of the null hypothesis $H_0 : g = g_0$ versus the alternative $H_1 : g = g_1 (g_1 > g_0)$ can be undertaken using a resampling method, as described in McLachlan (1987). With this approach, bootstrap samples are generated from the mixture model fitted under the null hypothesis of g_0 components. That is, the bootstrap samples are generated from the g_0-component mixture model with the vector Ψ of unknown parameters replaced by its ML estimate $\hat{\Psi}_{g_0}$ computed by consideration of the log likelihood formed from the original data under H_0. The value of $-2\log\lambda$, where λ is the likelihood ratio statistic, is computed for each bootstrap sample after fitting mixture models for $g = g_0$ and g_1 to it in turn. The process is repeated independently B times and the replicated values of $-2\log\lambda$ formed from the successive bootstrap samples provide an assessment of the bootstrap, and hence of the true, null distribution of $-2\log\lambda$. An account of other resampling approaches including the Gap statistic of Tibshirani et al. (2001) may be found in McLachlan and Khan (2004).

8.8 Clustering via Normal Mixtures

Frequently, in practice, the clusters in the case of Euclidean data are essentially elliptical, so that it is reasonable to consider fitting mixtures of elliptically symmetric component densities. Within this class of component densities, the multivariate normal density is a convenient choice given its computational tractability.

8.8.1 Heteroscedastic Components

Under the assumption of multivariate normal components, the ith component-conditional density $f_i(y; \theta_i)$ is given by

$$f_i(y; \theta_i) = \phi(y; \mu_i, \Sigma_i), \tag{8.19}$$

where θ_i consists of the elements of μ_i and the $\frac{1}{2}p(p+1)$ distinct elements of Σ_i ($i = 1, \ldots, g$). Here

$$\phi(y; \mu_i, \Sigma_i) = (2\pi)^{-\frac{p}{2}} |\Sigma_i|^{-1/2} \exp\left\{-\frac{1}{2}(y-\mu_i)^T \Sigma_i^{-1}(y-\mu_i)\right\}. \tag{8.20}$$

It follows that on the M-step of the $(k+1)$th iteration, the updates of the component means μ_i and component-covariance matrices Σ_i are given explicitly by

$$\mu_i^{(k+1)} = \sum_{j=1}^{n} \tau_{ij}^{(k)} y_j \Big/ \sum_{j=1}^{n} \tau_{ij}^{(k)} \tag{8.21}$$

and

$$\Sigma_i^{(k+1)} = \sum_{j=1}^{n} \tau_{ij}^{(k)}(y_j - \mu_i^{(k+1)})(y_j - \mu_i^{(k+1)})^T \Big/ \sum_{j=1}^{n} \tau_{ij}^{(k)} \tag{8.22}$$

for $i = 1, \ldots, g$, where

$$\tau_{ij}^{(k)} = \tau_i(y_j; \Psi^{(k)}) \qquad (i = 1, \ldots, g; \; j = 1, \ldots, n).$$

The updated estimate of the ith mixing proportion π_i is given by (8.12).

One attractive feature of adopting mixture models with elliptically symmetric components such as the normal or t-densities is that the implied clustering is invariant under affine transformations of the data; that is, invariant under transformations of the feature vector y of the form,

$$y \rightarrow Cy + a, \tag{8.23}$$

where C is a nonsingular matrix. If the clustering of a procedure is invariant under (8.23) for only diagonal C, then it is invariant under change of measuring units but not rotations. But as commented upon by Hartigan (1975), this form of invariance is more compelling than affine invariance.

8.8.2 Homoscedastic Components

Often in practice, the component-covariance matrices Σ_i are restricted to being the same,

$$\Sigma_i = \Sigma \qquad (i = 1, \ldots, g), \tag{8.24}$$

where Σ is unspecified. In this case of homoscedastic normal components, the updated estimate of the common component-covariance matrix Σ is given by

$$\Sigma^{(k+1)} = \sum_{i=1}^{g} \pi_i^{(k)} \Sigma_i^{(k+1)} / n, \tag{8.25}$$

where $\Sigma_i^{(k+1)}$ is given by (8.22), and the updates of π_i and μ_i are as above in the heteroscedastic case.

8.8.3 Spherical Components

A further simplification is to take the component-covariance matrices to have a common spherical form, where the covariance matrix of each component is taken to be a (common) multiple of the $p \times p$ identity matrix I_p, namely

$$\Sigma_i = \sigma^2 I_p \qquad (i = 1, \ldots, g). \tag{8.26}$$

However, under the constraint (8.26), the normal mixture model loses its invariance property. This constraint means that the clusters produced are spherical. If we also take the mixing proportions to be equal, then it is equivalent to a "soft" version of k-means clustering. It is a soft version in contrast to k-means where the observations are assigned outright at each of the iterations.

8.8.4 Spectral Representation of Component-Covariance Matrices

It can be seen from (8.20) that the mixture model with unrestricted group-covariance matrices in its normal component distributions is a highly parameterized one with $(1/2)p(p+1)$ parameters for each component-covariance matrix Σ_i $(i = 1, \ldots, g)$. As an alternative to taking the component-covariance matrices to be the same or diagonal, we can adopt some model for the component-covariance matrices that is intermediate between homoscedasticity and the unrestricted model, as in the approach of Banfield and Raftery (1993) and (2002).

An alternative approach is to adopt mixtures of factor analyzers as to be presented in Section 8.10. With this approach, the number of free parameters is controlled through the dimension of the latent factor space. By working in this reduced space, it allows a model for each component-covariance matrix with complexity lying between that of the isotropic and full covariance structure models without any restrictions on the covariance matrices.

8.8.5 Choice of Root

The choice of root of the likelihood equation in the case of homoscedastic normal components is straightforward in the sense that the ML estimate exists as the global maximizer of the likelihood function. The situation is less straightforward in the case of heteroscedastic normal components as the likelihood function is unbounded. It is known that as the sample size goes to infinity, there exists a sequence of roots of the likelihood equation that is consistent and asymptotically efficient. With probability tending to one, these roots correspond to local maxima in the interior of the parameter space; see McLachlan and Peel (2000). Usually, the intent is to choose as the ML estimate of the parameter vector $\boldsymbol{\Psi}$ the local maximizer corresponding to the largest of the local maxima located. But in practice, consideration has to be given to the problem of relatively large local maxima that occur as a consequence of a fitted component having a very small (but nonzero) variance for univariate data or generalized variance (the determinant of the covariance matrix) for multivariate data. Such a component corresponds to a cluster containing a few data points either relatively close together or some points almost lying in a lower-dimensional subspace in the case of multivariate data. There is thus a need to monitor the relative size of the fitted mixing proportions and of the component variances for univariate observations, or of the generalized component variances for multivariate data, in an attempt to identify these spurious local maximizers; see also Ingrassia and Rocci (2007) and Seo and Kim (2012).

8.9 Multivariate t-Distribution

8.9.1 Definition of t-Distribution

The mixture model with normal components (8.20) is sensitive to outliers since it adopts the multivariate normal family for the distributions of the errors. An obvious way to improve the robustness of this model for data which have longer tails than the normal or atypical observations is to consider using the multivariate t-family of elliptically symmetric distributions. It has an additional parameter called the degrees of freedom that controls the length of the tails of the distribution. Although the number of outliers needed for breakdown is almost the same as with the normal distribution, the outliers have to be much larger. This point is made more precise in Hennig (2004) who has provided an excellent account of breakdown points for ML estimation of location-scale mixtures with a fixed number of components g.

The t-distribution for the ith component-conditional distribution of Y_j is obtained by embedding the normal $N_p(\boldsymbol{\mu}_i, \boldsymbol{\Sigma}_i)$ distribution in a wider class of elliptically symmetric distributions with an additional parameter ν_i called the degrees of freedom. This

t-distribution can be characterized by letting W_j denote a random variable distributed as

$$W_j \sim \text{gamma}\left(\frac{1}{2}\nu_i, \frac{1}{2}\nu_i\right), \tag{8.27}$$

where the gamma (α, β) density function is equal to

$$f_G(w; \alpha, \beta) = \{\beta^\alpha w^{\alpha-1} / \Gamma(\alpha)\} \exp(-\beta w) I_{[0,\infty)}(w) \qquad (\alpha, \beta > 0), \tag{8.28}$$

and $I_A(w)$ denotes the indicator function that is 1 if w belongs to A and is zero otherwise. Then, if the conditional distribution of Y_j given $W_j = w_j$ is specified to be

$$Y_j \mid w_j \sim N_p(\mu_i, \Sigma_i/w_j), \tag{8.29}$$

the unconditional distribution of Y_j has a (multivariate) t-distribution with mean μ_i, scale matrix Σ_i, and degrees of freedom ν_i. The mean of this t-distribution is μ_i and its covariance matrix is $\{\nu_i/(\nu_i - 2)\}\Sigma_i$. We write

$$Y_j \sim t_p(\mu_i, \Sigma_i, \nu_i), \tag{8.30}$$

and we let $f_t(y_j; \mu_i, \Sigma_i, \nu_i)$ denote the corresponding density. As ν_i tends to infinity, the t-distribution approaches the normal distribution. Hence, this parameter ν_i may be viewed as a robustness tuning parameter. It can be fixed in advance or it can be inferred from the data for each component. A detailed account of the t-distribution is given in the monograph by Kotz and Nadarajah (2004), which is devoted to this topic.

Recently, Forbes and Wraith (2013) have considered an extension of the t-distribution to allow each variable to have its own degrees of freedom controlling its long-tailedness. Also, Finegold and Drton (2011) have considered a similar extension in the context of graphical modeling.

8.9.2 ML estimation of Mixtures of t-Distributions

McLachlan and Peel (1998) first suggested the use of mixtures of t-distributions to provide a robust extension to mixtures of normals; see also Peel and McLachlan (2000). They implemented the E- and M-steps of the EM algorithm and its variant, the ECM (expectation–conditional maximization) algorithm for the ML estimation of multivariate t-components. The ECM algorithm proposed by Meng and Rubin (1993) replaces the M-step of the EM algorithm by a number of computationally simpler conditional maximization (CM) steps.

In the EM framework for this problem, the unobservable variable w_j in the characterization (8.29) of the t-distribution for the ith component of the t-mixture model and the component-indicator labels z_{ij} are treated as being the "missing" data.

It can be shown that the conditional expectation of W_j given y_j and $z_{ij} = 1$ can be expressed as

$$E\{W_j \mid y_j, z_{ij} = 1\} = w_i(y_j; \Psi),$$

where

$$w_i(y_j; \Psi) = \frac{\nu_i + p}{\nu_i + d(y_j, \mu_i; \Sigma_i)} \tag{8.31}$$

and where

$$d(\mathbf{y}_j, \boldsymbol{\mu}_i; \boldsymbol{\Sigma}_i) = (\mathbf{y}_j - \boldsymbol{\mu}_i)^T \boldsymbol{\Sigma}_i^{-1}(\mathbf{y}_j - \boldsymbol{\mu}_i) \tag{8.32}$$

denotes the squared Mahalanobis distance between $\mathbf{y}_j$ and $\boldsymbol{\mu}_i$ ($i = 1, \ldots, g$; $j = 1, \ldots, n$).

On the $(k+1)$th iteration of the EM algorithm, the updated estimates of the mixing proportion, the mean vector $\boldsymbol{\mu}_i$, and the scale matrix $\boldsymbol{\Sigma}_i$ are given by

$$\pi_i^{(k+1)} = \sum_{j=1}^{n} \tau_{ij}^{(k)}/n, \tag{8.33}$$

$$\boldsymbol{\mu}_i^{(k+1)} = \sum_{j=1}^{n} \tau_{ij}^{(k)} w_{ij}^{(k)} \, \mathbf{y}_j \, / \sum_{j=1}^{n} \tau_{ij}^{(k)} w_{ij}^{(k)} \tag{8.34}$$

and

$$\boldsymbol{\Sigma}_i^{(k+1)} = \frac{\sum_{j=1}^{n} \tau_{ij}^{(k)} w_{ij}^{(k)} (\mathbf{y}_j - \boldsymbol{\mu}_i^{(k+1)})(\mathbf{y}_j - \boldsymbol{\mu}_i^{(k+1)})^T}{\sum_{j=1}^{n} \tau_{ij}^{(k)}}. \tag{8.35}$$

In the above,

$$\tau_{ij}^{(k)} = \frac{\pi_i^{(k)} f(\mathbf{y}_j; \boldsymbol{\mu}_i^{(k)}, \boldsymbol{\Sigma}_i^{(k)}, v_i^{(k)})}{f(\mathbf{y}_j; \boldsymbol{\Psi}^{(k)})} \tag{8.36}$$

is the posterior probability that $\mathbf{y}_j$ belongs to the ith component of the mixture, using the current fit $\boldsymbol{\Psi}^{(k)}$ for $\boldsymbol{\Psi}$ ($i = 1, \ldots, g$; $j = 1, \ldots, n$). Also,

$$w_{ij}^{(k)} = \frac{v_i^{(k)} + p}{v_i^{(k)} + d(\mathbf{y}_j, \boldsymbol{\mu}_i^{(k)}; \boldsymbol{\Sigma}_i^{(k)})}, \tag{8.37}$$

which is the current estimate of the conditional expectation of W_j given $\mathbf{y}_j$ and $z_{ij} = 1$.

The updated estimate $v_i^{(k+1)}$ of v_i does not exist in closed form, but is given as a solution of the equation

$$\left\{ -\psi(\tfrac{1}{2}v_i) + \log(\tfrac{1}{2}v_i) + 1 + \frac{1}{n_i^{(k)}} \sum_{j=1}^{n} \tau_{ij}^{(k)} (\log w_{ij}^{(k)} - w_{ij}^{(k)}) \right.$$

$$\left. + \psi\left(\frac{v_i^{(k)} + p}{2}\right) - \log\left(\frac{v_i^{(k)} + p}{2}\right) \right\} = 0, \tag{8.38}$$

where $n_i^{(k)} = \sum_{j=1}^{n} \tau_{ij}^{(k)}$ ($i = 1, \ldots, g$) and $\psi(\cdot)$ is the Digamma function.

Following the proposal of Kent et al. (1994) in the case of a single-component t-distribution, we can replace the divisor $\sum_{j=1}^{n} \tau_{ij}^{(k)}$ in (8.35) by

$$\sum_{j=1}^{n} \tau_{ij}^{(k)} w_{ij}^{(k)},$$

which should improve the speed of convergence. It corresponds to an application of the parameter-expanded EM (PX-EM) algorithm (Liu et al. 1998).

These E- and M-steps are alternated until the changes in the estimated parameters or the log likelihood are less than some specified threshold. It can be seen that if the degrees of freedom ν_i is fixed in advance for each component, then the M-step exists in closed form. In this case where ν_i is fixed beforehand, the estimation of the component parameters is a form of M-estimation. However, an attractive feature of the use of the t-distribution to model the component distributions is that the degrees of robustness as controlled by ν_i can be inferred from the data by computing its MLE.

Work has continued on various aspects of the fitting of mixtures of t-distributions; see, for example, Shoham (2002), Greselin and Ingrassia (2010), Andrews et al. (2011). The readers are referred to Cuesta-Albertos et al. (2008), García-Escudero et al. (2010), and Neykov et al. (2007) and the references therein for other approaches to providing a robust approach to the fitting of normal mixture models.

8.10 Factor Analysis Model for Dimension Reduction

As remarked earlier, the g-component normal mixture model with unrestricted component-covariance matrices is a highly parameterized model with $(1/2)p(p+1)$ parameters for each component-covariance matrix Σ_i $(i = 1, \ldots, g)$. As discussed in Section 8.8.4, Banfield and Raftery (1993) introduced a parameterization of the component-covariance matrix Σ_i based on a variant of the standard spectral decomposition of Σ_i $(i = 1, \ldots, g)$. However, if p is large relative to the sample size n, it may not be possible to use this decomposition to infer an appropriate model for the component-covariance matrices. Even if it is possible, the results may not be reliable due to potential problems with near-singular estimates of the component-covariance matrices when p is large relative to n.

A common approach to reducing the number of dimensions is to perform a principal component analysis (PCA). But as is well known, projections of the feature data y_j onto the first few principal axes are not always useful in portraying the group structure; see McLachlan and Peel (2000) (Chapter 8). This point was also stressed by Chang (1993), who showed in the case of two groups that the principal component of the feature vector that provides the best separation between groups in terms of Mahalanobis distance is not necessarily the first component.

Another approach for reducing the number of unknown parameters in the forms for the component-covariance matrices is to adopt the mixture of factor analyzers model, as considered in McLachlan and Peel (2000a,b). This model was originally proposed by Ghahramani and Hinton (1997) and Hinton et al. (1997) for the purposes of visualizing high dimensional data in a lower dimensional space to explore for group structure; see also Tipping and Bishop (1997) who considered the related model of mixtures of principal component analyzers for the same purpose. Further references may be found in Baek and McLachlan (2011), McLachlan et al. (2007), McLachlan et al. (2003).

8.10.1 Factor Analysis Model for a Single Component

Factor analysis is commonly used for explaining data, in particular, correlations between variables in multivariate observations. It can be used also for dimensionality reduction.

In a typical factor analysis model, each observation Y_j is modeled as

$$Y_j = \mu + BU_j + e_j \quad (j = 1, \ldots, n), \tag{8.39}$$

where U_j is a q-dimensional $(q < p)$ vector of latent or unobservable variables called factors and B is a $p \times q$ matrix of factor loadings (parameters). The U_j are assumed to be i.i.d. as $N(0, I_q)$, independently of the errors e_j, which are assumed to be i.i.d. as $N(0, D)$, where D is a diagonal matrix,

$$D = \text{diag}(\sigma_1^2, \ldots, \sigma_p^2),$$

and where I_q denotes the $q \times q$ identity matrix. Thus, conditional on $U_j = u_j$, the Y_j are independently distributed as $N(\mu + B u_j, D)$. Unconditionally, the Y_j are distributed according to a normal distribution with mean μ and covariance matrix

$$\Sigma = BB^T + D. \tag{8.40}$$

If q is chosen sufficiently smaller than p, the representation (8.40) imposes some constraints on the component-covariance matrix Σ and thus reduces the number of free parameters to be estimated. Note that in the case of $q > 1$, there is an infinity of choices for B, since (8.40) is still satisfied if B is replaced by BC, where C is any orthogonal matrix of order q. One (arbitrary) way of uniquely specifying B is to choose the orthogonal matrix C so that $B^T D^{-1} B$ is diagonal (with its diagonal elements arranged in decreasing order) Lawley and Maxwell (1971) (Chapter 1). Assuming that the eigenvalues of BB^T are positive and distinct, the condition that $B^T D^{-1} B$ is diagonal as above imposes $(1/2)q(q-1)$ constraints on the parameters. Hence then the number of free parameters is $pq + p - (1/2)q(q-1)$.

Unlike the PCA model, the factor analysis model (8.39) enjoys a powerful invariance property: changes in the scales of the feature variables in y_j appear only as scale changes in the appropriate rows of the matrix B of factor loadings.

8.10.2 Mixtures of Factor Analyzers

A global nonlinear approach can be obtained by postulating a finite mixture of linear submodels for the distribution of the full observation vector Y_j given the (unobservable) factors u_j. That is, we can provide a local dimensionality reduction method by assuming that, conditional on its membership of the ith component of a mixture, the distribution of the observation Y_j can be modelled as

$$Y_j = \mu_i + B_i U_{ij} + e_{ij} \quad \text{with prob. } \pi_i \quad (i = 1, \ldots, g) \tag{8.41}$$

for $j = 1, \ldots, n$, where the factors $U_{i1}, \ldots, U_{in}$ are distributed independently $N(0, I_q)$, independently of the e_{ij}, which are distributed independently $N(0, D_i)$, where D_i is a diagonal matrix $(i = 1, \ldots, g)$.

Thus in the normal mixture model given by (8.19), the ith component-covariance matrix Σ_i has the form

$$\Sigma_i = B_i B_i^T + D_i \quad (i = 1, \ldots, g), \tag{8.42}$$

where B_i is a $p \times q$ matrix of factor loadings and D_i is a diagonal matrix $(i = 1, \ldots, g)$. The parameter vector Ψ now consists of the elements of the μ_i, the B_i, and the D_i, along with the mixing proportions π_i $(i = 1, \ldots, g-1)$, on putting $\pi_g = 1 - \sum_{i=1}^{g-1} \pi_i$.

As $(1/2)q(q-1)$ constraints are needed for $\boldsymbol{B}_i$ to be uniquely defined, the number of free parameters in (8.42) is

$$pq + p - \tfrac{1}{2}q(q-1). \tag{8.43}$$

Thus with this representation (8.42), the reduction in the number of parameters for $\boldsymbol{\Sigma}_i$ is

$$
\begin{aligned}
r &= \tfrac{1}{2}p(p+1) - pq - p + \tfrac{1}{2}q(q-1) \\
&= \tfrac{1}{2}\{(p-q)^2 - (p+q)\},
\end{aligned} \tag{8.44}
$$

assuming that q is chosen sufficiently smaller than p so that this difference is positive. The total number of parameters is

$$d_1 = (g-1) + 2gp + g\left\{pq - \frac{1}{2}q(q-1)\right\}. \tag{8.45}$$

McLachlan and Peel (2000) referred to this approach as MFA (mixtures of factor analyzers).

We can think of the use of this mixture of factor analyzers model as being purely a method of regularization, but in several applications, it is possible to make a case for it being a reasonable model for the correlation structure between the variables within a cluster.

The mixture of factor analyzers model can be fitted by using the alternating expectation–conditional maximization (AECM) algorithm (Meng and van Dyk 1997). The AECM algorithm is an extension of the ECM algorithm, where the specification of the complete data is allowed to be different on each CM-step. Meng and van Dyk (1997) established that monotone convergence of the sequence of likelihood values is retained with the AECM algorithm.

8.10.3 Choice of the Number of Factors q

In practice, consideration has to be given to the number of components g and the number of factors q in the mixture of factor analyzers model. One obvious approach is to use the Bayesian Information Criterion (BIC) of (Schwarz 1978). An alternative approach is to use the likelihood ratio statistic for tests on g and q. For tests on g, it is well known that regularity conditions do not hold for the usual chi-squared approximation to the asymptotic null distribution of the likelihood ratio test statistic to be valid. This is also the case for tests on q at a given level of g (Drton 2009; Ninomiya et al. 2008; Bai and Li 2012; Tu and Xu 2012). Previously, in the case of a single component ($g = 1$), Geweke and Singleton (1980) identified situations where regularity conditions do hold for the asymptotic distribution of the likelihood ratio statistic to apply in testing the goodness of fit of a factor model.

8.10.4 Mixtures of t-Factor Analyzers

The mixture of factor analyzers model is sensitive to outliers since it uses normal errors and factors. McLachlan et al. (2007) have considered the use of mixtures of t-analyzers in an attempt to make the model less sensitive to outliers. Zhao and Jiang (2006) have independently considered this problem in the special case of spherical $\boldsymbol{D}_i$.

Even with the MFA approach, the number of parameters still might not be manageable, particularly if the number of dimensions p is large and/or the number of components (clusters) g is not small. Baek and McLachlan (2008) and Baek et al. (2010) considered how this factor-analytic approach can be modified to provide a greater reduction in the number of parameters. They termed their approach mixtures of common factor analyzers (MCFA). This is because the matrix of factor loadings is common to the components before the component-specific rotation of the component factors to make them white noise. Note that the component-factor loadings are not common after this rotation, as in McNicholas and Murphy (2008). Recently, Baek and McLachlan (2011) have extended the approach to mixtures of common t-factor analyzers.

In another approach with common factor loadings adopted recently by Galimberti et al. (2008) for the data after mean centring, the factors in the component distributions are taken to have a mixture distribution with the constraints that its mean is the null vector and its covariance matrix is the identity matrix. As their program applies only to mean-centred data, it cannot be used to visualize the original data.

In other related work, Sanguinetti (2008) has considered a method of dimensionality reduction in a cluster analysis context. However, its underlying model assumes sphericity in the specification of the variances/covariances of the factors in each cluster. The MCFA approach allows for oblique factors, which provides the extra flexibility needed to cluster more effectively high-dimensional data sets in practice. Xie et al. (2010) have considered a penalized version of mixture of factor analyzers. The MCFA approach can be formed also by placing a mixture distribution on the factors in the model (8.39) (Galimberti 2008; Montanari and Viroli 2010; Viroli 2010). A fuller discussion of related work may be found in McLachlan et al. (2011).

8.11 Some Recent Extensions for High-Dimensional Data

In situations where the sample size n is very large relative to the dimension p, it might not be practical to fit mixtures of factor analyzers, as it would involve a considerable amount of computation time. Thus, initially, some of the variables may have to be removed. Indeed, the simultaneous use of too many variables in the cluster analysis may serve only to create noise that masks the effect of a smaller number of variables. Also, the intent of the cluster analysis may not be to produce a clustering of the observations on the basis of all the available genes, but rather to discover and study different clusterings of the observations corresponding to different subsets of the variables.

Therefore, McLachlan et al. (2002) developed the so-called EMMIX-GENE procedure that has two optional steps before the final step of clustering the observations. The first step considers the selection of a subset of relevant variables from the available set of variables by screening the variables on an individual basis to eliminate those which are of little use in clustering the observations. The usefulness of a given variable to the clustering process can be assessed formally by a test of the null hypothesis that it has a single-component normal distribution over the observations. A faster but *ad hoc* way is to make this decision on the basis of the interquartile range. Even after this step has been completed, there may still remain too many variables. Thus, there is a second step in EMMIX-GENE in which the retained variables are clustered (after standardization) into a number of groups on the basis of Euclidean distance so that variables with similar profiles are put into the same group. In general, care has to be taken with the scaling of variables before clustering of the observations, as the nature of the variables can be intrinsically different. Also, as noted above, the clustering of the observations via normal mixture models is invariant under changes in scale and location. The clustering of the observations can be carried out on the basis of the groups considered individually using some or all of the variables within a group or collectively. For the latter, we can replace each group by a representative (a metavariable) such as the sample mean as in the EMMIX-GENE procedure. Another way to proceed with the fitting of mixture models to high-dimensional data is to use a penalized approach as adopted by Pan and Shen (2007) and Zhou and Pan (2009). Recently, Witten and Tibshirani (2010) have provided a framework for feature selection in a clustering context; see also the references therein, including Raftery and Dean (2006) and Maugis et al. (2009) who considered the feature selection problem in terms of model selection.

In the above account of model-based clustering, we have made the usual assumptions that hold in a typical cluster analysis, namely that:

1. There are no replications on any particular entity specifically identified as such;
2. All the observations on the entities are independent of one another.

These assumptions should hold, for example, with the clustering of, say, tissue samples from some microarray experiments, although the tissue samples have been known to be correlated for different tissues due to flawed experimental conditions. However, condition (2) will not hold for the clustering of gene profiles, since not all the genes are independently distributed, and condition (1) will generally not hold either as the gene profiles may be measured over time or on technical replicates. While this correlated structure can be incorporated into the normal mixture model (8.19) by appropriate specification of the component-covariance matrices Σ_i, it is difficult to fit the model under such specifications. For example, the M-step may not exist in closed form.

Accordingly, Ng et al. (2006) have developed the procedure called EMMIX-WIRE (**EM**-based **MIX**ture analysis **W**ith **R**andom **E**ffects) to handle the clustering of correlated data that may be replicated. They adopted conditionally a mixture of linear mixed models to specify the correlation structure between the variables and to allow for correlations among the observations. It also enables covariate information to be incorporated into the clustering process.

8.12 Mixtures of Skew Distributions

The past few years have seen an increasing use of skew mixture distributions to provide improved modelling and clustering of data that consists of asymmetric clusters with outliers as, for example, in Pyne et al. (2009). In particular, the skew normal and skew t-mixture models are emerging as promising extensions to the traditional normal and t-mixture models. Most of these parametric families of skew symmetric distributions are closely related. Recently, Lee and McLachlan (2013) classified them into four classes, namely, the restricted, unrestricted, extended, and generalized classes.

8.12.1 Classification of Skew Normal Distributions

Azzalini (1985) introduced the so-called skew normal distribution for modeling asymmetry in univariate data sets, while a multivariate version was proposed by Azzalini and Dalla Valle (1996). Since then numerous "extensions" of the so-called skew normal distribution have appeared in rapid succession. Most of these developments can be considered as special cases of the fundamental skew normal (FUSN) distribution Arellano-Valle and Genton (2005).

The FUSN distribution can be generated by conditioning a multivariate normal variable on another (univariate or multivariate) random variable. Suppose $Y_1 \sim N_p(0, \Sigma)$ and Y_0 is a q-dimensional random vector. Adopting the notation as used in Azzalini and Dalla Valle (1996), we let $Y_1 \mid Y_0 > 0$ be the vector Y_1 if all elements of Y_0 are positive and $-Y_1$ otherwise. Then $Y = \mu + (Y_1 \mid Y_0 + \alpha > 0)$ has a FUSN distribution. The parameter $\alpha \in \mathbb{R}^q$, known as the extension parameter, can be viewed as a location shift for the latent variable Y_0. When the joint distribution of Y_1 and Y_0 is multivariate normal, the FUSN distribution reduces to a LOCATION-SCALE VARIANT OF THE CANONICAL FUSN (CFUSN) DISTRIBUTION, given by

$$Y = \mu + (Y_1 \mid Y_0 > 0), \tag{8.46}$$

where

$$\begin{bmatrix} Y_0 \\ Y_1 \end{bmatrix} \sim N_{q+p}\left(\begin{bmatrix} \alpha \\ 0 \end{bmatrix}, \begin{bmatrix} \Gamma & \Delta^T \\ \Delta & \Sigma \end{bmatrix} \right), \tag{8.47}$$

where α is a q-dimensional vector, μ is p-dimensional vector, Γ is a $q \times q$ scale matrix, Δ is an arbitrary $p \times q$ matrix of skewness parameters, and Σ is a $p \times p$ scale matrix.

In the classification scheme of Lee and McLachlan (2013), the restricted case corresponds to a highly specialized form of (8.47), where Y_0 is restricted to be univariate (that is, $q = 1$), $\alpha = 0$, and $\Gamma = 1$. In the unrestricted case, both Y_0 and Y_1 have a p-dimensional normal distribution (that is, $q = p$) and $\alpha = 0$. The extended form has no restriction on the dimension q of Y_0, and α is not specified to be the zero vector, while in the generalized case, the assumption of normality for the distribution of Y_0 is also relaxed.

8.12.2 Restricted Skew Normal Distribution

The restricted skew normal (rMSN) distribution as termed by Lee and McLachlan (2013) has density given by

$$f(y; \mu, \Sigma, \delta) = 2\phi_p(y; \mu, \Sigma)\, \Phi_1\left(\delta^T \Sigma^{-1}(y - \mu); 0, 1 - \delta^T \Sigma^{-1}\delta \right), \tag{8.48}$$

where μ is a location vector, Σ is a scale matrix, and δ is a skewness vector. Here, we let $\Phi_p(.; \mu, \Sigma)$ be the distribution function corresponding to the p-variate normal density $\phi_p(.; \mu, \Sigma)$ with mean vector μ and covariance matrix Σ. The form (8.48) is a reparameterization of the form of the skew normal distribution proposed by Azzalini and Dalla Valle (1996).

It follows from (8.46) and (8.47) that the conditioning-type representation of the rMSN distribution is given by

$$Y = \mu + (Y_1 \mid Y_0 > 0), \tag{8.49}$$

where

$$\begin{bmatrix} Y_0 \\ Y_1 \end{bmatrix} \sim N_{1+p}\left(\begin{bmatrix} 0 \\ 0 \end{bmatrix}, \begin{bmatrix} 1 & \delta^T \\ \delta & \Sigma \end{bmatrix} \right). \tag{8.50}$$

The corresponding convolution-type representation is

$$Y = \mu + \delta \left| \tilde{Y}_0 \right| + \tilde{Y}_1, \tag{8.51}$$

where $\tilde{Y}_0$ and $\tilde{Y}_1$ are independent variables distributed as $\tilde{Y}_0 \sim N_1(0, 1)$ and $\tilde{Y}_1 \sim N_p(\mathbf{0}, \tilde{\Sigma})$, respectively, and where $\tilde{\Sigma} = \Sigma - \delta\delta^T$.

8.12.3 Unrestricted Skew Normal Distribution

The unrestricted skew normal (uMSN) distribution so termed by Lee and McLachlan (2013) belongs to the unrestricted class in their classification scheme with the additional restriction that Δ is a diagonal matrix, $\Delta = \text{diag}(\delta)$, where δ is the p-dimensional vector of skewness parameters. This additional restriction implies that the rMSN distribution is not nested within the family of uMSN distributions.

The uMSN density function is given by

$$f(y; \mu, \Sigma, \delta) = 2^p \phi_p(y; \mu, \Sigma) \Phi_p\left(\Delta\Sigma^{-1}(y - \mu); \mathbf{0}, \Lambda\right), \tag{8.52}$$

where $\Lambda = I_p - \Delta\Sigma^{-1}\Delta$. The uMSN distribution belongs to the class of skew normal distributions introduced by Sahu et al. (2003).

The conditioning-type stochastic representation of (8.52) is given by $Y = \mu + (Y_1 \mid Y_0 > 0)$, where

$$\begin{bmatrix} Y_0 \\ Y_1 \end{bmatrix} \sim N_{2p}\left(\begin{bmatrix} \mathbf{0} \\ \mathbf{0} \end{bmatrix}, \begin{bmatrix} I_p & \Delta \\ \Delta & \Sigma \end{bmatrix}\right), \tag{8.53}$$

and the convolution-type representation is given by

$$Y = \mu + \Delta|\tilde{Y}_0| + \tilde{Y}_1, \tag{8.54}$$

where $\tilde{Y}_0$ and $\tilde{Y}_1$ are independent variables distributed as $\tilde{Y}_0 \sim N_p(\mathbf{0}, I_p)$ and $\tilde{Y}_1 \sim N_p(\mathbf{0}, \tilde{\Sigma})$, respectively, and where $\tilde{\Sigma} = \Sigma - \Delta^2$.

8.12.4 Skew *t*-Distributions

The multivariate skew *t*-distribution is an important member of the family of skew-elliptical distributions. Like the skew normal distributions, there exists various different versions of the multivariate skew *t*-(MST) distribution, which can be classified into four broad forms (Lee and McLachlan 2013).

Following (Pyne et al. 2009), the p-variate density of the restricted MST (rMST) distribution is given by

$$f(y; \mu, \Sigma, \delta, \nu) = 2t_p(y; \mu, \Sigma, \nu)T_1\left(\delta^T\Sigma^{-1}(y - \mu)\sqrt{\frac{\nu + p}{\nu + d(y)}}; 0, \lambda, \nu + p\right), \tag{8.55}$$

where $\lambda = 1 - \delta^T\Sigma^{-1}\delta$, $d(y) = (y - \mu)^T\Sigma^{-1}(y - \mu)$ is the squared Mahalanobis distance between y and μ with respect to Σ. Here, we let $t_p(.; \mu, \Sigma, \nu)$ denote the p-dimensional t-distribution with location vector μ, scale matrix Σ, and degrees of freedom ν. Also, we let $T_p(.; \mu, \Sigma, \nu)$ be the corresponding distribution function.

Corresponding to the specification of the rMSN distribution by (8.49) and (8.50), the rMST distribution can be specified via each of the two equivalent stochastic mechanisms, the conditioning of t-variables, and the convolution of t- and truncated t-variables.

The restricted MST distribution has a conditioning-type stochastic representation given by

$$Y = \mu + (Y_1 \mid Y_0 > 0),$$ (8.56)

where

$$\begin{bmatrix} Y_0 \\ Y_1 \end{bmatrix} \sim t_{1+p} \left(\begin{bmatrix} 0 \\ 0 \end{bmatrix}, \begin{bmatrix} 1 & \delta^T \\ \delta & \Sigma \end{bmatrix}, \nu \right).$$ (8.57)

The equivalent convolution-type representation is given by

$$Y = \mu + \delta |\tilde{Y}_0| + \tilde{Y}_1,$$ (8.58)

where the two random variables $\tilde{Y}_0$ and $\tilde{Y}_1$ have a joint multivariate central t-distribution with scale matrix $\begin{bmatrix} 1 & 0 \\ 0 & \tilde{\Sigma} \end{bmatrix}$ and ν degrees of freedom, where $\tilde{\Sigma} = \Sigma - \delta\delta^T$.

The density of the unrestricted skew t-(uMST) distribution is given by

$$f(y; \mu, \Sigma, \delta, \nu) = 2^p t_p(y; \mu, \Sigma, \nu) \, T_p \left(\Delta \Sigma^{-1}(y - \mu) \sqrt{\frac{\nu + p}{\nu + d(y)}}; 0, \Lambda, \nu + p \right),$$ (8.59)

where $\Lambda = I_p - \Delta \Sigma^{-1} \Delta$. This form of the MST distribution is studied in detail in Sahu and Branco (2003).

The conditioning- and convolution-type stochastic representations for the unrestricted case extend directly from (8.53) and (8.54). The unrestricted MST distribution can be generated by

$$Y = \mu + (Y_1 \mid Y_0 > 0),$$ (8.60)

where

$$\begin{bmatrix} Y_0 \\ Y_1 \end{bmatrix} \sim t_{2p} \left(\begin{bmatrix} 0 \\ 0 \end{bmatrix}, \begin{bmatrix} I_p & \Delta \\ \Delta & \Sigma \end{bmatrix}, \nu \right).$$ (8.61)

The analogous convolution-type representation is given by

$$Y = \mu + \Delta |\tilde{Y}_0| + \tilde{Y}_1,$$ (8.62)

where the two random vectors $\tilde{Y}_0$ and $\tilde{Y}_1$ are jointly distributed as

$$\begin{bmatrix} \tilde{Y}_0 \\ \tilde{Y}_1 \end{bmatrix} \sim t_{2p} \left(\begin{bmatrix} 0 \\ 0 \end{bmatrix}, \begin{bmatrix} I_p & 0 \\ 0 & \tilde{\Sigma} \end{bmatrix}, \nu \right),$$ (8.63)

where $\tilde{\Sigma} = \Sigma - \Delta^2$.

8.12.5 Skew Symmetric Distributions

An asymmetric density can be generated by perturbing a symmetric density, yielding a so-called multivariate skew symmetric (MSS) density (Azzalini and Capitanio 2003). Typically, an MSS density can be expressed as a product of a (multivariate) symmetric function $f_p(\cdot)$ and a perturbation (or skewing) function $h_q(\cdot)$, where $h_q(\cdot)$ is a q-variate function that maps its argument into the unit interval; that is,

$$f_p(\boldsymbol{y}; \boldsymbol{\mu}) h_q(\cdot), \tag{8.64}$$

where $f_p(\cdot)$ is symmetric around $\boldsymbol{\mu}$.

It can be seen from the forms (8.48) and (8.55) for the restricted skew normal and t-densities that they use a univariate skewing function. In contrast, the unrestricted versions (8.52) and (8.59) use a p-variate skewing function that is proportional to the distribution function corresponding to the density function.

8.12.6 Mixtures of Skew Normal and t-Distributions

The EM algorithm can be used to fit mixtures of skew normal and skew t-distributions as covered in Lee and McLachlan (2014). The implementation is rather straightforward for mixtures of restricted skew normal and t-distributions since the E-step can be undertaken in closed form Pyne et al. (2009). The reader is referred to Lin (2010) and Lee and McLachlan (2013, 2014). Lee and McLachlan (2013) have provided several examples on the modelling and clustering for a variety data sets via mixtures of skew normal and skew t-distributions.

8.12.7 Other Non-Normal Distributions

While skew symmetric distributions, in particular, the skew normal and skew t-distributions, have played a central role in the development of non-normal models and mixtures of them, there are other models that have been receiving attention. For example, Karlis and Santourian (2009) have considered mixtures of the multivariate normal-inverse-Gaussian (MNIG) distribution, while Franczak et al. (2012) have considered mixtures of the mean shifted asymmetric Laplace (SAL) distribution. More recently, Browne and McNicholas (2015) have studied mixtures of the multivariate generalized hyperbolic distribution.

8.13 Available Software

The reader is referred to the appendix in McLachlan and Peel (2000) for the availability of software for the fitting of mixture models, including the EMMIX program of McLachlan and Peel (1999). The current version of EMMIX, including a package for the R open source software environment, is available from the World Wide Web address

```
http://www.maths.uq.edu.au/~gjm/mix_soft/index.html
```

On the availability of other mixture modelling packages, there is the well-known `mclust` R package of Fraley and Raftery (2002). For the fast prototyping of new types of mixture models, the R package `flexmix` (Leisch 2004) provides a general framework for fitting finite mixture models via EM algorithm or one of its variants. The R software package `mixtools` (Benaglia et al. 2009) provides various tools for analyzing a variety of finite mixture models, including finite mixture models with univariate and multivariate normal distributions. The following web page maintained by Friedrich Leisch and Bettina Gruen

```
http://cran.r-project.org/web/views/Cluster.html
```

lists a wide range of R packages for the analysis of finite mixture models, including for model-based clustering.

Lee and McLachlan (2013) have provided an R package EMMIX-uskew for the fitting of mixtures of unrestricted skew t-distributions. Previously, Wang et al. (2009) have provided an R package EMMIX-skew for the fitting of mixtures of restricted skew normal and t-distributions.

8.14 Conclusion

We have considered how the clustering of an Euclidean data set into g clusters can be effected by fitting a g-component mixture distribution in which the component distributions are selected to model the distribution of the observations in the corresponding clusters. Commonly, the component distributions are specified to belong to the multivariate normal family or the t-family in the case of observations with longer tails than the normal. In some cases as with the clustering of flow cytometry data, the clusters are not elliptically symmetric and may contain outliers. Such data can still be clustered by the fitting of normal or t-mixture models with a greater number of components than clusters and then combining some of the components to form the desired number of clusters. However, one has the problem of trying to identify those clusters that correspond to multiple components. Thus, if the component distributions can be chosen to be a suitable model for the observations in the clusters, then the one-to-one correspondence between the components and the clusters can be maintained without the consequent need to identify asymmetric clusters with more than one component in the mixture model. For example, Pyne et al. (2009) showed that mixtures of skew multivariate normal and skew t-distributions provided a very good model for the distributions of markers on cells in flow cytometry. There are various versions of the skew normal and t-distributions, and there has since been much interest in the use of mixtures of these skew distributions to partition data into clusters that are not elliptically symmetric with tails that may be longer than the normal (Lin 2010; Lee and McLachlan 2011, 2014; Ho et al. 2012).

References

M. Aitkin, D. Anderson, and J. Hinde. 1981. Statistical modelling of data on teaching styles. *Journal of the Royal Statistical Society A*, 144, 419–461.

J.L. Andrews, P.D. McNicholas, and S. Subedi. 2011. Model-based classification via mixtures of multivariate t-distributions. *Computational Statistics & Data Analysis*, 55, 520–529.

R.B. Arellano-Valle and M.G. Genton. 2005. On fundamental skew distributions. *Journal of Multivariate Analysis*, 96, 93–116.

A. Azzalini. 1985. A class of distributions which includes the normal ones. *Scandinavian Journal of Statistics*, 12, 171–178.

A. Azzalini and A. Capitanio. 2003. Distribution generated by perturbation of symmetry with emphasis on a multivariate skew *t* distribution. *Journal of the Royal Statistical Society B*, 65, 367–389.

A. Azzalini and A. Dalla Valle. 1996. The multivariate skew-normal distribution. *Biometrika*, 83, 715–726.

J. Baek and G.J. McLachlan. 2008. *Mixtures of Factor Analyzers with Common Factor Loadings for the Clustering and Visualisation of High-Dimensional Data.* Technical Report NI08018-SCH, Preprint Series of the Isaac Newton Institute for Mathematical Sciences, Cambridge.

J. Baek and G.J. McLachlan. 2011. Mixtures of common *t*-factor analyzers for clustering high-dimensional microarray data. *Bioinformatics*, 27(9), 1269–1276.

J. Baek, G.J. McLachlan, and L.K. Flack. 2010. Mixtures of factor analyzers with common factor loadings: Applications to the clustering and visualization of high-dimensional data. *IEEE Transactions on Pattern Analysis and Machine Intelligence*, 32, 1298–1309.

J. Bai and K. Li. 2012. Statistical analysis of factor models of high dimension. *The Annals of Statistics*, 40, 436–465.

J.D. Banfield and A.E. Raftery. 1993. Model-based Gaussian and non-Gaussian clustering. *Biometrics*, 49, 803–821.

J.P. Baudry, A.E. Raftery, G. Celeux, K. Lo, and R. Gottardo. 2009. Combining mixture components for clustering. *Journal of Computational and Graphical Statistics*, 9, 323–353.

T. Benaglia, D. Chauveau, D.R. Hunter, and D.S. Young. 2009. Mixtools: an R package for analyzing mixture models. *Journal of Statistical Software*, 32, 1–29.

C. Biernacki, G. Celeux, and G. Govaert. 1998. Assessing a mixture model for clustering with the integrated classification likelihood. *IEEE Transactions on Pattern Analysis and Machine Intelligence*, 22, 719–725.

D. Böhning. 1999. *Computer-Assisted Analysis of Mixtures and Applications: Meta-Analysis, Disease Mapping and Others.* Chapman & Hall/CRC, New York.

R.P. Browne and P.D. McNicholas. 2015. A mixture of generalized hyperbolic distributions. *Canadian Journal of Statistics*, 43, 176–198.

P. Bryant and J.A. Williamson. 1978. Asymptotic behaviour of classification maximum likelihood estimates. *Biometrika*, 65, 273–281.

W.C. Chang. 1983. On using principal components before separating a mixture of two multivariate normal distributions. *Applied Statistics*, 32, 267–275.

D. Coleman, X. Dong, J. Hardin, D.M. Rocke, and D.L. Woodruff. 1999. Some computational issues in cluster analysis with no a priori metric. *Computational Statistics & Data Analysis*, 31, 1–11.

J.A. Cuesta-Albertos, C. Matrán, and A. Mayo-Iscar. 2008. Robust estimation in the normal mixture model based on robust clustering. *Journal of the Royal Statistical Society B*, 70, 779–802.

N.E. Day. 1969. Estimating the components of a mixture of normal distributions. *Biometrika*, 56, 463–474.

A.P. Dempster, N.M. Laird, and D.B. Rubin. 1977. Maximum likelihood from incomplete data via the EM algorithm (with discussion). *Journal of the Royal Statistical Society B*, 39, 1–38.

M. Drton. 2009. Likelihood ratio tests and singularities. *The Annals of Statistics*, 37, 979–1012.

B.S. Everitt and D.J. Hand. 1981. *Finite Mixture Distributions.* Chapman & Hall, London.

M. Finegold and M. Drton. 2011. Robust graphical modeling of gene networks using classical and alternative *t*-distributions. *The Annals of Applied Statistics*, 5, 1057–1080.

F. Forbes and D. Wraith. 2013. A new family of multivariate heavy-tailed distributions with variable marginal amounts of tailweight: Application to robust clustering. *Statistics and Computing*, 24, 971–984.

C. Fraley and A.E. Raftery. 2002. Model-based clustering, discriminant analysis, and density estimation. *Journal of the American Statistical Association*, 97, 611–631.

B.C. Franczak, R.P. Browne, and P.D. McNicholas. 2014. Mixtures of shifted asymmetric Laplace distributions. *IEEE Transactions on Pattern Analysis & Machine Intelligence*, 36, 1149–1157.

S. Frühwirth-Schnatter. 2006. *Finite Mixture and Markov Switching Models*. Springer, New York.

G. Galimberti, A. Montanari, and C. Viroli. 2008. Latent classes of objects and variable selection. In P. Brito, editor, *COMPSTAT 2008*, Physica-Verlag HD. pp. 373–383.

S. Ganesalingam and G.J. McLachlan. 1979. A case study of two clustering methods based on maximum likelihood. *Statistica Neerlandica*, 33, 81–90.

L.A. García-Escudero, A. Gordaliza, C. Matrán, and A. Mayo-Iscar. 2010. A review of robust clustering methods. *Advances in Data Analysis and Classification*, 4, 89–109.

J.F. Geweke and K.J. Singleton. 1980. Interpreting the likelihood ratio statistic in factor models when sample size is small. *Journal of the American Statistical Association*, 75, 133–137.

Z. Ghahramani and G.E. Hinton. 1997. *The EM Algorithm for Factor Analyzers*. Technical Report, The University of Toronto, Toronto.

F. Greselin and S. Ingrassia. 2010. Constrained monotone EM algorithms for mixtures of multivariate t distributions. *Statistics and Computing*, 20, 9–22.

J.A. Hartigan. 1975. *Clustering Algorithms*. Wiley, New York.

D.M. Hawkins, M.W. Muller, and J.A. ten Krooden. 1982. *Topics in Applied Multivariate Analysis*. Cambridge University Press, Cambridge.

C. Hennig. 2004. Breakdown points for maximum likelihood estimators of location-scale mixtures. *The Annals of Statistics*, 32, 1313–1340.

C. Hennig. 2010. Methods for merging Gaussian mixture components. *Advances in Data Analysis and Classification*, 4, 3–34.

G.E. Hinton, P. Dayan, and M. Revow. 1997. Modeling the manifolds of images of handwritten digits. *IEEE Transactions on Neural Networks*, 8, 65–74.

H.J. Ho, T.I. Lin, H.Y. Chen, and W.L. Wang. 2012. Some results on the truncated multivariate t distribution. *Journal of Statistical Planning and Inference*, 142, 25–40.

S. Ingrassia and R. Rocci. 2007. Constrained monotone EM algorithms for finite mixture of multivariate Gaussians. *Computational Statistics & Data Analysis*, 51, 5339–5351.

D. Karlis and A. Santourian. 2009. Model-based clustering with non-elliptically contoured distributions. *Statistics and Computing*, 19, 73–83.

J.T. Kent, D.E. Tyler, and Y. Vard. 1994. A curious likelihood identity for the multivariate t-distribution. *Communications in Statistics—Simulation and Computation*, 23, 441–453.

C. Keribin. 2000. Consistent estimation of the order of mixture models. *Sankhyā: The Indian Journal of Statistics A*, 62, 49–66.

S. Kotz and S. Nadarajah. 2004. *Multivariate t-distributions and their applications*. Cambridge, Cambridge University Press.

D.N. Lawley and A.E. Maxwell. 1971. *Factor analysis as a statistical method*. Butterworths, London, 2nd edition.

S.X. Lee and G.J. McLachlan. 2011. On the fitting of mixtures of multivariate skew t-distributions via the EM algorithm. *arXiv e-Prints*.

S.X. Lee and G.J. McLachlan. 2013. EMMIXuskew: An R package for fitting mixtures of multivariate skew t-distributions via the EM algorithm. *Journal of Statistical Software*, 55(12), 1–22.

S.X. Lee and G.J. McLachlan. 2013. Model-based clustering with non-normal mixture distributions (with discussion). *Statistical Methods & Applications*, 22, 427–479.

S.X. Lee and G.J. McLachlan. 2013. On mixtures of skew normal and skew t-distributions. *Advances in Data Analysis and Classification*, 10, 241–266.

S.X. Lee and G.J. McLachlan. 2014. Finite mixtures of multivariate skew t-distributions: Some recent and new results. *Statistics and Computing*, 24, 181–202.

F. Leisch. 2004. FlexMix: A general framework for finite mixture models and latent class regression in R. *Journal of Statistical Software*, 11, 1–18.

B.G. Leroux. 1992. Consistent estimation of a mixing distribution. *Annals of Statistics*, 20, 1350–1360.

J.Q. Li and A.R. Barron. 2000. *Mixture Density Estimation*. Technical Report, Department of Statistics, Yale University, New Haven, Connecticut.

T.-I. Lin. 2010. Robust mixture modeling using multivariate skew t distributions. *Statistics and Computing*, 20, 343–356.

B.G. Lindsay. 1995. Mixture Models: Theory, Geometry and Applications. In *NSF-CBMS Regional Conference Series in Probability and Statistics*, Institute of Mathematical Statistics and the American Statistical Association. Volume 5, Alexandria, VA.

C. Liu, D.B. Rubin, and Y.N. Wu. 1998. Parameter expansion to accelerate EM: The PX-EM algorithm. *Biometrika*, 85, 755–770.

F.H.C. Marriott. 1974. *The Interpretation of Multiple Observations*. Academic Press, London.

C. Maugis, G. Celeux, and M.L. Martin-Magniette. 2009. Variable selection for clustering with Gaussian mixture models. *Biometrics*, 65, 701–709.

G.J. McLachlan. 1982. The classification and mixture maximum likelihood approaches to cluster analysis. In P.R. Krishnaiah and L. Kanal, editors, *Handbook of Statistics*, volume 2, North-Holland, Amsterdam. pp. 199–208.

G.J. McLachlan. 1987. On bootstrapping the likelihood ratio test statistic for the number of components in a normal mixture. *Applied Statistics*, 36, 318–324.

G.J. McLachlan, J. Baek, and S.I. Rathnayake. 2011. *Mixtures of Factor Analyzers for the Analysis of High-Dimensional Data*. In K.L. Mengersen, C.P. Robert, and D.M. Titterington, editors, *Mixtures: Estimation and Applications*, Wiley, Hoboken, New Jersey. pp. 171–191.

G.J. McLachlan and K.E. Basford. 1988. *Mixture Models: Inference and Applications to Clustering*. Marcel Dekker, New York.

G.J. McLachlan, R.W. Bean, and L. Ben-Tovim Jones. 2007. Extension of the mixture of factor analyzers model to incorporate the multivariate t-distribution. *Computational Statistics & Data Analysis*, 51, 5327–5338.

G.J. McLachlan, R.W. Bean, and D. Peel. 2002. A mixture model-based approach to the clustering of microarray expression data. *Bioinformatics*, 18, 413–422.

G.J. McLachlan and N. Khan. 2004. On a resampling approach for tests on the number of clusters with mixture model-based clustering of tissue samples. *Journal of Multivariate Analysis*, 90, 90–105.

G.J. McLachlan and T. Krishnan. 1997. *The EM Algorithm and Extensions*. Wiley, New York.

G.J. McLachlan and D. Peel. 1998. Robust cluster analysis via mixtures of multivariate t-distributions. In *Lecture Notes in Computer Science*, volume 1451, pp. 658–666.

G.J. McLachlan and D. Peel. 1999. The EMMIX algorithm for the fitting of normal and t-components. *Journal of Statistical Software*, 4, 1–14.

G.J. McLachlan and D. Peel. 2000. *Finite Mixture Models*. Wiley, New York.

G.J. McLachlan and D. Peel. 2000. Mixtures of factor analyzers. In *Proceedings of the Seventeenth International Conference on Machine Learning*, Morgan Kaufmann. pp. 599–606.

G.J. McLachlan, D. Peel, and Bean R.W. 2003. Modelling high-dimensional data by mixtures of factor analyzers. *Computational Statistics & Data Analysis*, 41, 379–388.

P. McNicholas and T. Murphy. 2008. Parsimonious Gaussian mixture models. *Statistics and Computing*, 18, 285–296.

X.L. Meng and D.B. Rubin. 1993. Maximum likelihood estimation via the ECM algorithm: A general framework. *Biometrika*, 80, 267–278.

X.L. Meng and D. van Dyk. 1997. The EM algorithm—An old folk-song sung to a fast new tune. *Journal of the Royal Statistical Society B*, 59, 511–567.

K.L. Mengersen, C.P. Robert, and D.M. Titterington, Editors. 2011. *Mixtures: Estimation and Applications*. Wiley, Hoboken, New Jersey.

A. Montanari and C. Viroli. 2010. Heteroscedastic factor mixture analysis. *Statistical Modelling*, 10, 441–460.

N. Neykov, P. Filzmoser, R. Dimova, and P. Neytchev. 2007. Robust fitting of mixtures using the trimmed likelihood estimator. *Computational Statistics & Data Analysis*, 52, 299–308.

S.K. Ng, G.J. Mclachlan, K. Wang, L. Ben-Tovim Jones, and S.W. Ng. 2006. A mixture model with random-effects components for clustering correlated gene-expression profiles. *Bioinformatics*, 22, 1745–1752.

Y. Ninomiya, H. Yanagihara, and K.-H. Yuan. 2008. *Selecting the Number of Factors in Exploratory Factor Analysis via Locally Conic Parameterization.* Technical Report No. 1078, ISM Research Memorandum.

W. Pan and X. Shen. 2007. Penalized model-based clustering with application to variable selection. *Journal of Machine Learning Research*, 8, 1145–1164.

D. Peel and G.J. McLachlan. 2000. Robust mixture modelling using the t distribution. *Statistics and Computing*, 10, 339–348.

S. Pyne, X. Hu, K. Wang, E. Rossin, T.-I. Lin, L.M. Maier, C. Baecher-Allan, G.J. McLachlan, P. Tamayo, D.A. Hafler, P.L. De Jager, and J.P. Mesirow. 2009. Automated high-dimensional flow cytometric data analysis. *Proceedings of the National Academy of Sciences*, 106, 8519–8524.

A. Raftery and N. Dean. 2006. Variable selection for model-based clustering. *Journal of the American Statistical Association*, 101, 168–178.

S. Richardson and P.J. Green. 1997. On Bayesian analysis of mixtures with an unknown number of components (with discussion). *Journal of the Royal Statistical Society B*, 59, 731–792.

S.K. Sahu, D.K. Dey, and M.D. Branco. 2003. A new class of multivariate skew distributions with applications to Bayesian regression models. *The Canadian Journal of Statistics*, 31, 129–150.

G. Sanguinetti. 2008. Dimensionality reduction of clustered data sets. *IEEE Transactions Pattern Analysis and Machine Intelligence*, 30, 535–540.

G. Schwarz. 1978. Estimating the dimension of a model. *The Annals of Statistics*, 6, 461–464.

A.J. Scott and M.J. Symons. 1971. Clustering methods based on likelihood ratio criteria. *Biometrics*, 27, 387–397.

B. Seo and D. Kim. 2012. Root selection in normal mixture models. *Computational Statistics & Data Analysis*, 56, 2454–2470.

S. Shoham. 2002. Robust clustering by deterministic agglomeration EM of mixtures of multivariate t-distributions. *Pattern Recognition*, 35, 1127–1142.

R. Tibshirani, G. Walther, and T. Hastie. 2001. Estimating the number of clusters in a data set via the gap statistic. *Journal of the Royal Statistical Society B*, 63, 411–423.

M.E. Tipping and C.M. Bishop. 1997. *Mixtures of Probabilistic Principal Component Analysers.* Technical Report NCRG/97/003, Neural Computing Research Group, Aston University, Birmingham.

D.M. Titterington, A.F.M. Smith, and U.E. Makov. 1985. *Statistical Analysis of Finite Mixture Distributions.* Wiley, New York.

S. Tu and L. Xu. 2012. A theoretical investigation of several model selection criteria for dimensionality reduction. *Pattern Recognition Letters*, 33, 1117–1126.

C. Viroli. 2010. Dimensionally reduced model-based clustering through mixtures of factor mixture analyzers. *Journal of Classification*, 27, 363–388.

K. Wang, G.J. McLachlan, S.K. Ng, and D. Peel. 2009. EMMIX-skew: EM algorithm for mixture of multivariate skew normal/t distributions. http://www.maths.uq.edu.au/~gjm/mix_soft/EMMIX-skew.

D.M. Witten and R. Tibshirani. 2010. A framework for feature selection in clustering. *Journal of the American Statistical Association*, 105, 713–726.

J.H. Wolfe. 1965. *A Computer Program for the Computation of Maximum Likelihood Analysis of Types.* Technical report, Research Memo. SRM 65-12, San Diego: U.S. Naval Personnel Research Activity.

J.H. Wolfe. 1970. Pattern clustering by multivariate mixture analysis. *Multivariate Behavioral Research*, 5, 329–350.

B. Xie, W. Pan, and X. Shen. 2010. Penalized mixtures of factor analyzers with application to clustering high-dimensional microarray data. *Bioinformatics*, 26, 501–508.

K.Y. Yeung, C. Fraley, A. Murua, A.E. Raftery, and W.L. Ruzzo. 2001. Model-based clustering and data transformations for gene expression data. *Bioinformatics*, 17, 977–987.

J. Zhao and Q. Jiang. 2006. Probabilistic PCA for t distributions. *Neurocomputing*, 69, 2217–2226.

H. Zhou and W. Pan. 2009. Penalized model-based clustering with unconstrained covariance matrices. *Electronic Journal of Statistics*, 3, 1473–1496.

9

Latent Class Models for Categorical Data

G. Celeux and Gérard Govaert

CONTENTS

Abstract

This chapter deals with mixture models for clustering categorical and mixed-type data, which are in the literature often referred to as latent class models. The chapter introduces the maximum-likelihood approach and the EM algorithm. It introduces Bayesian approaches and has a comprehensive discussion of parsimonious models and methods for model selection and estimating the number of clusters such as information criteria. This chapter concludes with techniques for ordinal- and mixed-type data.

9.1 Introduction

Model-based clustering (MBC) offers a powerful framework to tackle properly the questions arising in cluster analysis. MBC consists of assuming that the data come from a finite mixture of parametric probability distributions and that each cluster is arising from one of the mixture components.

This chapter is essentially concerned with the situation where the observations to be clustered are categorical. In such a case, the central role in MBC is played by the multivariate multinomial distribution, which involves a local independence assumption within the clusters. Before entering the mathematical details of MBC for categorical data, it is worthwhile to remark that the terms latent class analysis (LCA) and latent class model (LCM) are often used to refer to a mixture model in which all the observed variables are categorical and assume local independence.

Thus, the observations to be classified are described by d categorical variables. Each variable j has m_j response levels. Data are represented in the following way: $(x_1, \ldots, x_n)$ where x_i is coded by the vector $(x_i^j; j = 1, \ldots, d)$ of response level h for each variable j or, equivalently, by the binary vector $(x_i^{jh}; j = 1, \ldots, d; h = 1, \ldots, m_j)$ with

$$\begin{cases} x_i^{jh} = 1 & \text{if } x_i^j = h \\ x_i^{jh} = 0 & \text{otherwise.} \end{cases}$$

Latent class (LC) analysis was originally introduced by Lazarsfeld (1950) as a way of explaining respondent heterogeneity in survey response patterns involving dichotomous items. The standard LCM was first proposed by Goodman (1974) and was as follows. Data are supposed to arise from a mixture of g multivariate multinomial distributions with pdf

$$f(\mathbf{x}_i; \theta) = \sum_{k=1}^{g} p_k M_k(\mathbf{x}_i; \alpha_k) = \sum_k p_k \prod_{j,h} (\alpha_k^{jh})^{x_i^{jh}},$$

where α_k^{jh} is denoting the probability that variable j has level h if object i is in cluster k, and $\alpha_k = (\alpha_k^{jh}; j = 1, \ldots, d; h = 1, \ldots, m_j)$, $p = (p_1, \ldots, p_g)$ is denoting the vector of mixing proportions of the g latent clusters, $\theta = (p_k, \alpha_k, k = 1, \ldots, g)$ being the vector parameter of the latent class model to be estimated. It is important to notice that the LCM assumes that the variables are conditionally independent knowing the latent clusters. This assumption is also known as the local independence assumption. This local independence assumption basically decomposes the dependence between the variables into mixture components. It is remarkable that such an assumption is rather universally made for latent class models, but would count as rather restrictive for mixtures on continuous data. The main reason is probably that there is no simple coefficient as the linear correlation coefficient to measure dependence between discrete variables. Some effort has been made in order to avoid the local independence assumption. For instance, Bock (1986) proposed a mixture of loglinear models. But such more flexible models require more parameters to be estimated and did not find widespread use. In the same spirit, mixtures of latent trait models, which are particular cases of the Rasch model (Agresti 2002), have been proposed (Uebersax and Grove 1993, Gollini and Murphy 2014). For instance, the latent trait model used in Gollini and Murphy (2014) assumes that there is a D-dimensional continuous latent variable $\mathbf{y}$

underlying the behavior of the binary categorical variables x. It assumes that

$$f(x) = \int f(x \mid y) d\mathbf{y},$$

where the conditional distribution of x given y is

$$f(x \mid y) = \prod_{j=1}^{d} \pi_j(y)^{x_j} (1 - \pi_j(y))^{1-x_j},$$

and π_j is a logistic function

$$\pi_j(y) = f(x_j = 1 \mid y) = \frac{1}{1 + \exp[(-b_j + w_j^T y)]}, \quad 0 \le \pi(y) \le 1,$$

b_j and w_j being the intercept and slope parameters in the logistic function. Estimating the parameters of the mixture of this latent trait models involves difficulties that Gollini and Murphy (2014) circumvent using a variational EM approximation. These authors show illustrations where this model, which avoids the local independence assumption, provides good and interpretable clusterings.

Obviously *ad hoc* practices, such as pooling variables for which the local independence assumption should be avoided, could be used. But, many applications, for instance Anderlucci and Hennig (2014), imply to some extent that local independence is an assumption that can be quite successful at bringing similar observations together, which one typically aims at in cluster analysis.

9.1.1 Identifiability of the LCM

The identifiability of the LCM is an important issue which required attention. First, a necessary condition for the LCM to be identifiable is that the number of possible values of the categorical variable table is greater than the number of parameters in the LCM at hand (Goodman 1974). For instance, for a standard LCM with g clusters on a binary table with d variables, this necessary condition is $2^d - 1 \ge (g - 1) + dg$. Strictly speaking, this necessary condition is not sufficient to ensure the identifiability of the LCM (Gyllenberg et al. 1994). Examples of simple binary mixtures leading to the same distribution are in Carreira-Perpiñán and Renals (2000). Those authors argue that this lack of identifiability of the LCM is not an issue from the practical point of view. This claim has been confirmed by the work of Allman et al. (2009) who gave sufficient conditions under which the LCM is generically identifiable, namely, that the set of non identifiable parameters has measure zero. It means that, as guessed by Carreira-Perpiñán and Renals (2000), the statistical inference remains meaningful even if the LCM is not strictly identifiable. For binary mixtures with d variables and g clusters, the condition of Allman et al. (2009) is $d \ge 2\lceil \log_2 g \rceil + 1$, $\lceil a \rceil$ being the smallest integer at least as large as a. For a mixture of d variables with m levels each, their condition is $d \ge 2\lceil \log_m g \rceil + 1$.

9.2 Maximum-Likelihood Estimation

The loglikelihood of the LCM is

$$L(\theta) = L(\theta; x) = \sum_{i=1}^{n} \log \left(\sum_{k=1}^{g} p_k \prod_{j=1}^{d} \prod_{h=1}^{m_j} (\alpha_k^{jh})^{x_i^{jh}} \right).$$

Denoting $z = (z_1, \ldots, z_g)$ with $z_k = (z_{1k}, \ldots, z_{nk})$ and $z_{ik} = 1$ if x_i arose from cluster k, $z_{ik} = 0$ otherwise, the unknown indicator vectors of the clusters, the completed loglikelihood is

$$L_C(\theta) = L(\theta; x, z) = \sum_{i=1}^{n} \sum_{k=1}^{g} z_{ik} \log \left(p_k \prod_{j=1}^{d} \prod_{h=1}^{m_j} (\alpha_k^{jh})^{x_i^{jh}} \right).$$

The updating formulas of the EM algorithm to derive the maximum-likelihood (ML) estimates of this standard model by maximizing the conditional expectation of the complete likelihood are straightforward.

Starting from an initial position $\theta^{(0)} = (p^{(0)}, \alpha^{(0)})$, the two steps of the EM algorithm are as follows:

- E-step: calculation of $t^{(r)} = (t_{ik}^{(r)}, i = 1, \ldots, n, k = 1, \ldots, g)$, where $t_{ik}^{(r)}$ is the conditional probability that x_i arose from cluster k

$$t_{ik}^{(r)} = \frac{p_k^{(r)} M_k(x_i; \alpha_k^{(r)})}{\sum_{\ell=1}^{g} p_\ell^{(r)} M_\ell(x_i; \alpha_\ell^{(r)})}.$$

- M-step: Updating of the mixture parameter estimates,

$$p_k^{(r+1)} = \frac{n_k^{(r)}}{n}, \quad k = 1, \ldots, g$$

$$(\alpha_k^{jh})^{(r+1)} = \frac{(u_k^{jh})^{(r)}}{n_k^{(r)}}, \quad j = 1, \ldots, d; \quad h = 1, \ldots, m_j; \quad k = 1, \ldots, g,$$

where $n_k = \sum_{i=1}^{n} t_{ik}$ and $u_k^{jh} = \sum_{i=1}^{n} t_{ik} x_i^{jh}$.

Analyzing multivariate categorical data is made difficult because of the curse of dimensionality. The standard latent class model requires $(g - 1) + g \sum_j (m_j - 1)$ parameters to be estimated. It is more parsimonious than the saturated loglinear model which requires $\prod_j m_j$ parameters. For example, with $g = 5$, $d = 10$, $m_j = 4$ for all variables, the latent class model is defined with 154 parameters, whereas the saturated loglinear model requires about 10^6 parameters. However, the standard LCM can appear to be too complex in regard of the sample size n and even involves numerical difficulties ("divide by zero" occurrences or parameters α_k^{jh} estimated to zero). Such drawbacks can be avoided using the regularized formula:

$$(\alpha_k^{jh})^{(r+1)} = \frac{(u_k^{jh})^{(r)} + c - 1}{n_k^{(r)} + m_j(c - 1)},$$

c being some fixed positive number. The formula can be recommended for updating the estimation of the α_k^{jh}s. The formula can be sensitive to choosing c, and considering this regularization issue in a Bayesian setting could be beneficial (Section 9.5). In this perspective, more parsimonious LCMs could be thought of as desirable. Such models have been proposed in Celeux and Govaert (1991) and are now presented.

9.3 Parsimonious Latent Class Models

These parsimonious LCMs arise from a model parametrization, which is now presented.

9.3.1 A Parametrization

Denoting, for any variable j, a_k^j the most frequent response level in cluster k, the parameter α_k can be replaced by $(\mathbf{a}_k, \varepsilon_k)$ with $\mathbf{a}_k = (a_k^1, \ldots, a_k^d)$ and $\varepsilon_k = (\varepsilon_k^{11}, \ldots, \varepsilon_k^{dm_d})$ where

$$a_k^j = \arg\max_h \alpha_k^{jh} \quad \text{and} \quad \varepsilon_k^{jh} = \begin{cases} 1 - \alpha_k^{jh} & \text{if } h = a_k^j \\ \alpha_k^{jh} & \text{otherwise.} \end{cases}$$

For instance, for two variables with $m_1 = 3$ and $m_2 = 2$, if $\alpha_k = (0.7, 0.2, 0.1; 0.2, 0.8)$, the new parameters are $\mathbf{a}_k = (1, 2)$ and $\varepsilon_k = (0.3, 0.2, 0.1; 0.2, 0.2)$. Vector $\mathbf{a}_k$ provides the modal levels in cluster k for the variables and the elements of vector ε_k can be regarded as scatter values.

9.3.2 Five Latent Class Models

Using this form, it is possible to impose various constraints to the scatter parameters ε_k^{jh}. The models we consider are the following:

- The standard LCM $[\varepsilon_k^{jh}]$: The scatter depends upon clusters, variables, and levels.
- $[\varepsilon_k^j]$: The scatter depends upon clusters and variables but not upon levels.
- $[\varepsilon_k]$: The scatter depends upon clusters, but not upon variables.
- $[\varepsilon^j]$: The scatter depends upon variables, but not upon clusters and levels.
- $[\varepsilon]$: The scatter is constant over variables and clusters.

Obviously, these constrained models can be thought of as unrealistic. In particular, the model $[\varepsilon]$ can appear to be too simplistic. However, it could be useful when the number of observations is dramatically small with respect to the number of variables. Moreover, this simple model provides a probabilistic interpretation of the K-medoids clustering with the Manhattan distance (Hennig and Liao 2013). From our experience, the model $[\varepsilon_k^j]$, which is equivalent to the general ε_k^{jh} model when all the variables are binary, is of particular interest. It often offers a good compromise between parsimony and different within-cluster variances for different variables and clusters.

9.3.3 A Short Focus on Model $[\varepsilon_k^j]$

With this model, it can be remarked that the scatter vector parameter is completely characterized by the scatter value ε_k^{jh} of the modal level $h = a_k^j$. Denoting this value by ε_k^j, we have

$$
\varepsilon_k^{jh} = \begin{cases} \varepsilon_k^j & \text{if } h = a_k^j \\[2ex] \dfrac{\varepsilon_k^j}{m_j - 1} & \text{otherwise.} \end{cases}
$$

This model has $(g - 1) + gd$ parameters to be compared to the $(g - 1) + g * \sum_j (m_j - 1)$ parameters of the standard latent class model $[\varepsilon_k^{jh}]$. Notice that the model $[\varepsilon_k^j]$ has been proposed by Aitchison and Aitken (1976) in a supervised classification context.

The model can be written

$$
f(\mathbf{x}_i; \theta) = \sum_{k=1}^{g} p_k \prod_{j=1}^{m_j} (1 - \varepsilon_k^j) \left(\frac{\varepsilon_k^j}{(m_j - 1)(1 - \varepsilon_k^j)} \right)^{1 - \delta(x_i^j, a_k^j)},
$$

where δ is the Kronecker function and the complete(d) loglikelihood can be written as

$$
L_C(\mathbf{t}, \theta) = \sum_{k=1}^{g} n_k \log p_k + \sum_{k=1}^{g} \sum_{j=1}^{d} \log \left(\frac{\varepsilon_k^j}{(m_j - 1)(1 - \varepsilon_k^j)} \right) \left(n_k - \sum_{i=1}^{n} t_{ik} \delta(x_i^j, a_k^j) \right)
$$
$$
+ \sum_{k=1}^{g} n_k \sum_{j=1}^{d} \log(1 - \varepsilon_k^j).
$$

The ML estimation of these parsimonious model parameters with the EM algorithm does not involve difficulties. The E-step remains unchanged and the M-step is as follows.

9.3.4 M-step

Updating the mixing proportions is unchanged and updating the a_k^j does not differ over the four parsimonious models.

$$
(a_k^j)^{(r+1)} = \arg \max_h (u_k^{jh})^{(r)}.
$$

Changes occur when updating the scatter parameters ε.

We will use the notation $u^{jh} = \sum_{k=1}^{g} u_k^{jh}$, and we will distinguish the modal level a_k^j by denoting $v_k^j = u_k^{jh}$, $v_k = \sum_{j=1}^{d} v_k^j$, $v^j = \sum_{k=1}^{g} v_k^j$ and $v = \sum_{j=1}^{d} \sum_{k=1}^{g} v_k^j$ when h is the modal level a_k^j.

Updating the scatter estimate for model $[\varepsilon_k^j]$. We get for the model $[\varepsilon_k^j]$

$$
(\varepsilon_k^j)^{(r+1)} = \frac{n_k^{(r)} - (v_k^j)^{(r)}}{n_k^{(r)}}.
$$

This means that $(\varepsilon_k^j)^{(r)}$ is the percentage of times the level of an object in cluster k is different from its modal level for variable j.

Scatter estimate for models $[\varepsilon_k]$, $[\varepsilon^j]$ *and* $[\varepsilon]$

- $(\varepsilon_k)^{(r+1)} = (n_k^{(r)}d - v_{k.}^{(r)})/(n_k^{(r)}d)$: This is the percentage of times the level of an object in cluster k level is different from its modal level.
- $\varepsilon^j = (n - (v^j)^{(r)})/n$: This is the percentage of times the level of an object is different from its modal level for variable j.
- $\varepsilon = (nd - v^{(r)})/nd$: This is the percentage of times the level of an object is different from its modal level.

9.3.5 Limitations

As noted by Vandewalle (2009) in his thesis (Chapter 5 page 124), when the categorical variables do not have the same number of levels, models $[\varepsilon_k]$ and $[\varepsilon]$ suffer from a possible inconsistency: for those models where the scatter is not depending on the levels, the estimated probability of the modal level for a variable with few levels can be smaller than the estimated probability of the minority levels. Vandewalle (2009) proposed an alternative parametrization avoiding such an inconsistency. Unfortunately, the M-step of the resulting EM-algorithm can no longer be written down in closed form. Our opinion is that when the categorical variables have different number of levels, models with scatter parameter not depending on the variables (as models $[\varepsilon_k]$ and $[\varepsilon]$) are not relevant. In the following, we will suppose m_j to be constant for these two models.

9.4 The Latent Class Model as a Cluster Analysis Tool

The LCM is a parsimonious hidden structure model using a conditional independence assumption to analyze relations between categorical variables. This model does not necessarily involve a clustering task. Obviously, in the context of the present handbook, a focus on using the LCM for cluster analysis is desirable and this section is devoted to this point of view.

First, it is to be stated that an LCM, estimated for instance with the EM algorithm, leads in a straightforward manner to a clustering of the observed data by using a Classification step which consists of assigning each object x_i to the cluster maximizing the conditional probability t_{ik} that x_i arose from cluster k, $1 \leq k \leq g$ from a Maximum A Posteriori (MAP) principle. Moreover, an alternative Classification EM (CEM) can be considered in order to estimate the LCM parameters and the clustering $\mathbf{z}$ in the same exercise, see for instance Celeux and Govaert (1995). In the following description of the CEM algorithm, the quantities n_k, u_k^{jh}, and u^{jh} are as previously defined except that the conditional probabilities t_{ik} are replaced with the labels z_{ik}.

9.4.1 CEM Algorithm

- E-step: As in EM, calculation of the conditional probabilities $t_{ik}^{(r)}$ that x_i arose from cluster k ($i = 1, \ldots, n, k = 1, \ldots, g$).

- C-step: Assign each observation to one of the g clusters using the MAP (Maximum A Posterior) operator: $\mathbf{z}^{(r+1)} = \text{MAP}(\mathbf{t}^{(r)})$:

$$z_{ik}^{(r+1)} = 1 \quad \text{if } k = \arg\max_k t_{ik}^{(r)} \quad \text{and} \quad 0 \text{ otherwise.}$$

- M-step: Update the parameter estimate by maximizing the completed likelihood $L(\theta; \mathbf{x}, \mathbf{z}^{r+1})$. For the standard LCM, this leads to

$$p_k^{(r+1)} = \frac{n_k^{(r+1)}}{n} \quad \text{and} \quad (\alpha_k^{jh})^{(r+1)} = \frac{(u_k^{jh})^{(r+1)}}{n_k^{(r+1)}}.$$

Note that α_k^{jh} is simply the current relative frequency of level h for variable j in cluster k.

The CEM algorithm maximizes the completed likelihood $L_C(\theta)$ rather than the actual likelihood $L(\theta)$. Therefore, it produces biased estimates of the LCM parameters. On the other hand, it converges in a finite (small) number of iterations and when the mixture components are well separated with similar proportions, the CEM algorithm is expected to provide a relevant clustering.

A benefit of the classification approach to the LCM is to allow to highlight underlying relations between the LCM and distance-based methods. Thus, Celeux and Govaert (1991) proved that maximizing the completed likelihood of the standard LCM is equivalent to maximizing the information criterion

$$H(\mathbf{z}) = \sum_{k=1}^g \sum_{j=1}^d \sum_{h=1}^{m_j} u_k^{jh} \ln u_k^{jh} - d \sum_{k=1}^g n_k \ln n_k.$$

Moreover, several authors, including Benzecri (1973) or Govaert (1983), have shown that maximizing this information criterion $H(\mathbf{z})$ is practically equivalent to minimizing the popular χ^2 criterion

$$W(\mathbf{z}) = \sum_{k=1}^g \sum_{j=1}^d \sum_{h=1}^{m_j} \frac{(nu_k^{jh} - n_k u^{jh})^2}{nn_k u^{jh}},$$

and minimizing this χ^2 criterion can be performed with a k-means like algorithm where the distance between observations is the χ^2 distance.

Comparing the LCM and distance-based methods in clustering is a subject of interest. Anderlucci and Hennig (2014) conducted an extensive simulation study for comparing the standard LCM estimated with the EM algorithm and a k-means type algorithm where the distance between the objects was the Manhattan distance. The conclusions of this study are in accordance with the difference between the maximum-likelihood approach and the classification maximum-likelihood approach for the LCM. When the clusters are well separated with similar proportions, the two approaches give similar performances and, in such cases, the distance-based approach could be preferred since it is more rapid. Otherwise (overlapping clusters or unequal proportions), the LCM gives more satisfying results, in particular the adjusted rand index (ARI) of Hubert and Arabie (1985) between the true clustering and the clustering derived with the EM algorithm applied to the LCM is better. See Anderlucci (2012) for details.

9.5 Bayesian Inference

Contrary to the continuous distribution mixture model setting, fully noninformative Bayesian analysis is possible for LCM. As a matter of fact, noninformative prior distributions for a multinomial distribution $\mathcal{M}_r(q_1, \ldots, q_r)$ are conjugate Dirichlet distribution $\mathcal{D}(a, \ldots, a)$ with density

$$f(q_1, \ldots, q_r) = \frac{\Gamma(ra)}{\Gamma(a) \cdots \Gamma(a)} q_1^{a-1} \cdots q_r^{a-1},$$

Γ denoting the gamma function. The mean value and variance of each q_ℓ are $1/r$ and $(r-1)[/r^2(ra+1)]$, respectively. Thus, a Gibbs sampling implementation of noninformative Bayesian inference for the five latent class models can be derived in a straightforward way from the full conditional distributions which are now presented.

9.5.1 Gibbs Sampling

In this paragraph, as in the Section 9.4, conditional probabilities t_{ik} are replaced by the labels z_{ik} in the definition of $n_k, u_k^{jh}, v_k^j, v_k, v^j$ and v. For instance, $u_k^{jh} = \sum_{i=1}^n z_{ik} x_i^{jh}$ is now the number of occurrences of level h of variable j in cluster k.

For all the models, the prior distribution of the mixing weights is a $\mathcal{D}(b, \ldots, b)$ distribution. Then the full conditional distribution of $(p_k, k = 1, \ldots, g)$ is a Dirichlet distribution $\mathcal{D}(b + n_1, \ldots, b + n_g)$.

For the standard LCM, the prior distribution of $(\alpha_k^{j1}, \ldots, \alpha_k^{jm_j})$ is a $\mathcal{D}(c, \ldots, c)$ for $k = 1, \ldots, g$ and $j = 1, \ldots, d$. Thus, the full conditional distribution for $(\alpha_k^{j1}, \ldots, \alpha_k^{jm_j})$, $j = 1, \ldots, d; k = 1, \ldots, g;$ is

$$(\alpha_k^{j1}, \ldots, \alpha_k^{jm_j}) \sim \mathcal{D}(c + u_k^{j1}, \ldots, c + u_k^{jm_j}).$$

The choice of the prior hyperparameters b and c can have an impact. Noting that when b or c equals one the Dirichlet prior distribution is the uniform distribution, this could be thought of as a natural choice. But, as discussed later and shown in Frühwirth-Schnatter (2011), choosing b and c greater than one could be helpful to avoid the numerical difficulties of the LCM.

For the parsimonious models presented in Section 9.3, the prior distribution will be defined by $p(\alpha_1, \ldots, \alpha_g) = p(\mathbf{a})p(\varepsilon)$ where $\mathbf{a} = (a_1^1, \ldots, a_g^d)$ and $\varepsilon = (\varepsilon_1^1, \ldots, \varepsilon_g^d)$.

A quite natural choice for the prior distribution of the modal levels $\mathbf{a}$ is $p(\mathbf{a}) = \prod_{k=1}^g \prod_{j=1}^d p(a_k^j)$ where the $p(a_k^j)$'s are discrete uniform distributions on $\{1, \ldots, m_j\}$. Thus, the full conditional posterior probabilities $p(a_k^j = h \mid \ldots)$ are $\rho_k^{jh} = (\gamma_k^{jh})/(\sum_{h=1}^{m_j} \gamma_k^{jh})$ where $\gamma_k^{jh} = \left(((m_j - 1)(1 - \varepsilon_k^j))/(\varepsilon_k^j) \right)^{u_{kj}^h}$, for $k = 1, \ldots, g, j = 1, \ldots, d$ and $h = 1, \ldots, m_j$.

Conditionally to the a_k^js and z, it can be shown that the distributions of the weighted means are binomial distributions:

- Model $[\varepsilon_k^j]$: $v_k^j \sim \mathcal{B}_i(n_k, 1 - \varepsilon_k^j)$ for $k = 1, \ldots, g$ and $j = 1, \ldots, d$;
- Model $[\varepsilon^j]$: $v^j \sim \mathcal{B}_i(n, 1 - \varepsilon^j)$ for $j = 1, \ldots, d$;

- Model $[\varepsilon_k]$: $v_k \sim \mathcal{B}_i(n_k d, 1 - \varepsilon_k)$ for $k = 1, \ldots, g$;
- Model $[\varepsilon]$: $v \sim \mathcal{B}_i(nd, 1 - \varepsilon)$.

Choosing a noninformative prior distribution is less simple for the scatter parameters and different views are possible. However, it seems that the most reasonable choice is to ensure the exchangeability of the levels for each considered models. This invites us not to favor the modal level and leads to the following developments.

Denoting $\mathcal{B}t_{[a,b]}(p, q)$ the truncated Beta distribution of parameters p and q defined on the interval $[a, b] \subseteq [0, 1]$, the prior and full conditional posterior distributions of the scatter parameters for the different parsimonious models are

- Model $[\varepsilon_k^j]$. For $k = 1, \ldots, g$ and $j = 1, \ldots, d$
    - Prior: $\varepsilon_k^j \sim \mathcal{B}t_{[0,(m_j-1)/m_j]}((m_j - 1)(c - 1) + 1, c)$.
    - Full conditional posterior: $\varepsilon_k^j \sim \mathcal{B}t_{[0,(m_j-1)/m_j]}(n_k - v_k^j + (m_j - 1)(c - 1) + 1, v_k^j + c)$.
- Model $[\varepsilon^j]$. For $j = 1, \ldots, d$
    - Prior: $\varepsilon^j \sim \mathcal{B}t_{[0,(m_j-1)/m_j]}(g(m_j - 1)(c - 1) + 1, g(c - 1) + 1)$.
    - Full conditional posterior: $\varepsilon^j \sim \mathcal{B}t_{[0,(m_j-1)/m_j]}(n - v^j + g(m_j - 1)(c - 1) + 1, v^j + g(c - 1) + 1)$.
- Model $[\varepsilon_k]$. For $k = 1, \ldots, g$
    - Prior: $\varepsilon_k \sim \mathcal{B}t_{[0,(m-1)/m]}(d(m - 1)(c - 1) + 1, d(c - 1) + 1)$.
    - Full conditional posterior: $\varepsilon_k \sim \mathcal{B}t_{[0,(m-1)/m]}(n_k d - v_k + d(m - 1)(c - 1) + 1, v_k + d(c - 1) + 1)$.
- Model $[\varepsilon]$:
    - Prior: $\varepsilon \sim \mathcal{B}t_{[0,(m-1)/m]}(gd(m - 1)(c - 1) + 1, d(c - 1) + 1)$.
    - Full conditional posterior: $\varepsilon \sim \mathcal{B}t_{[0,(m-1)/m]}(nd - v + gd(m - 1)(c - 1) + 1, v + dg(c - 1) + 1)$.

Recall that for the two last models the number of levels is supposed to be the same for each variable, $m = m_j$ for $j = 1, \ldots, d$.

The choice of the hyperparameter c is important. Note that, for each model, the comparison of the ML and Bayesian formulas shows that c can be regarded as the number of virtual observations that has been added to the observed sample in each variable level and each cluster by the Bayesian "expert." Thus, the choice $c = 1$ could be regarded as a (minimal) default choice. And, as for the standard LCM with Dirichlet prior distributions, taking $c > 1$ could be beneficial to avoid numerical problems.

9.5.2 The Label Switching Problem

Because the prior distribution is symmetric in the components of the mixture, the posterior distribution is invariant under a permutation of the component labels (see for instance McLachlan and Peel (2000), Chapter 4). This lack of identifiability of a mixture is known as the so-called label switching problem. In order to deal with this problem, many authors from Stephens (2000) or Celeux et al. (2000) to Papastamoulis and Iliopoulos (2010) have proposed relatively efficient clustering-like procedures to possibly change the component labels of the simulated values for θ during a Gibbs sampling. However, these procedures have to choose among $g!$ permutations of the labels and become difficult to perform as soon as $g > 6$. Alternative procedures using a point-process representation avoid this difficulty but require data analytic skills (Frühwirth-Schnatter 2011). In any case, label switching remains a difficulty when dealing with full Bayesian inference for mixture analysis, and thus, ML can be preferred to Bayesian inference through an MCMC-algorithm for estimating the parameters of an LCM (Uebersax 2010).

9.5.3 Regularized Maximum Likelihood through Bayesian Inference

However, Bayesian inference can be quite useful as an alternative to ML in order to avoid the numerical traps related to the curse of dimensionality. Actually, Bayesian inference is introducing additional information which can be useful in order to solve ill-posed problems. In this perspective, maximizing the posterior probability $p(\theta \mid x)$, using the non informative prior distributions defined previously, can be performed with the EM algorithm. In this setting, the E-step remains unchanged, and updating the mixture parameter estimates in the M-step leads to

$$p_k^{(r+1)} = \frac{n_k^{(r)} + b - 1}{n + g(b-1)}, \quad k = 1,\ldots,g,$$

for the mixing proportions, and to

$$(\alpha_k^{jh})^{(r+1)} = \frac{(u_k^{jh})^{(r)} + c - 1}{n_k^{(r)} + m_j(c-1)},$$

with $k = 1,\ldots,g$, $j = 1,\ldots,d$ and $h = 1,\ldots,m_j$ for the standard LCM model ($[\varepsilon_k^{jh}]$).
For the parsimonious models, the update of the modal level is unchanged

$$(a_k^j)^{(r+1)} = \arg\max_h (u_k^{jh})^{(r)},$$

as for the EM algorithm. The updates of the scatter parameters are as follows

- Model $[\varepsilon_k^j]$. For $k = 1,\ldots,g$ and $j = 1,\ldots,d$

$$\varepsilon_k^j = \frac{n_k - v_k^j + (m_j - 1)(c-1)}{n_k + m_j(c-1)},$$

- Model $[\varepsilon^j]$. For $j = 1,\ldots,d$

$$\varepsilon^j = \frac{n - v^j + g(m_j - 1)(c-1)}{n + gm_j(c-1)},$$

- Model $[\varepsilon_k]$. For $k = 1,\ldots,g$

$$\varepsilon_k = \frac{n_k d - v_k + d(m-1)(c-1)}{n_k d + dm(c-1)},$$

- Model $[\varepsilon]$.

$$\varepsilon = \frac{nd - v + gd(m-1)(c-1)}{nd + gdm(c-1)}.$$

Note the strong similarity between this M-step and the M-step of the standard EM algorithm. This algorithm does not yield the ML estimators, but provides a regularized alternative to the EM algorithm. Moreover, it is not jeopardized by the label switching problem since it is restricted to find the maximum a posteriori estimates of the model parameters. The role of the hyperparameters b and c is important: the greater they are, the more the estimates are shrunk to a common value. We refer to Frühwirth-Schnatter (2011) for a thorough analysis of the role of these hyperparameters, and we note that she advocates to take them greater than four. In any case, looking at the formulas of the EM-Bayes algorithm, it is clear that it is necessary to take b and c greater than one to get regularized parameter estimates. As a matter of fact, taking b and c equals to one produces no regularization, and taking b or c smaller than one is worse than useless. Typically, this EM-Bayes algorithm, which avoids the label switching problem, can be preferred to EM when the sample size is not large with respect to the number of parameters of the LCM to be estimated.

9.6 Model Selection

9.6.1 Information Criteria

In the mixture context, the criteria BIC (Schwarz 1978) and ICLbic (Biernacki et al. 2000) are popular for choosing a mixture model and especially the number of mixture components. BIC is recommended to select a model in a density estimation purpose, and ICLbic could be preferred when the focus of the mixture analysis is clustering. Both criteria are asymptotic approximations of integrated likelihoods: BIC is approximating the integrated observed-data likelihood, while ICLbic is approximating the completed integrated likelihood.

The integrated observed-data likelihood of an LCM is

$$p(\mathbf{x}) = \int_{\Theta} f(x; \theta) p(\theta) d\theta, \tag{9.1}$$

Θ being the whole unconstrained parameter space and $p(\theta)$ being the prior distribution of the parameter θ. The asymptotic approximation of $\ln p(x)$ is given by

$$\ln p(x) = \ln f(x; \hat{\theta}) - \frac{v}{2} \ln n + O_p(1), \tag{9.2}$$

where $\hat{\theta}$ is the ML estimator of the model parameter and v is the number of free parameters of the LCM at hand. This leads to the BIC criterion to be maximized

$$\text{BIC} = \ln f(x; \hat{\theta}) - \frac{v}{2} \ln n. \tag{9.3}$$

It should be noted that although BIC is motivated from a Bayesian perspective, it does not depend on the prior distribution and its use is not relevant for Bayesian inference.

The integrated complete-data likelihood of an LCM is defined by

$$p(x, z) = \int_\Theta p(x, z; \theta) p(\theta) d\theta, \tag{9.4}$$

and the asymptotic BIC-like approximation of $\ln p(x, z)$ is

$$\ln p(x, z) = \ln p(x, z; \hat{\theta}) - \frac{\nu}{2} \ln n + O_p(1). \tag{9.5}$$

Replacing the missing labels z by their maximum A Posteriori (MAP) values $\hat{z}$ for $\hat{\theta}$ defined as

$$\hat{z}_{ik} = \begin{cases} 1 & \text{if } \arg\max_\ell t_{i\ell}(\hat{\theta}) = k \\ 0 & \text{otherwise,} \end{cases} \tag{9.6}$$

leads to the following ICLbic criterion (Biernacki et al. 2000):

$$\text{ICLbic} = \ln p(x, \hat{z}; \hat{\theta}) - \frac{\nu}{2} \ln n. \tag{9.7}$$

This criterion aims at favoring mixture situations giving rise to a clear partitioning of the data and, as a consequence, it appears to be little sensitive against model misspecification (Biernacki et al. 2000).

For the standard LCM, using the noninformative conjugate priors defined in Section 9.5, namely Dirichlet priors with hyperparameter b for the mixing proportions and c for the α parameters, the integrated complete-data likelihood (9.4) can be derived in closed form (Biernacki et al. 2010). Replacing the missing labels z by $\hat{z}$:

$$\text{ICL} = \ln \Gamma(bg) - g \ln \Gamma(b) + \sum_{k=1}^g \ln \Gamma(\hat{n}_k + b) - \ln \Gamma(n + bg)$$

$$+ g \sum_{j=1}^d \left\{ \ln \Gamma(cm_j) - m_j \ln \Gamma(c) \right\}$$

$$+ \sum_{k=1}^g \sum_{j=1}^d \left\{ \sum_{h=1}^{m_j} \ln \Gamma \left(\hat{u}_k^{jh} + c \right) - \ln \Gamma(\hat{n}_k + cm_j) \right\}, \tag{9.8}$$

where $\hat{n}_k = \#\{i : \hat{z}_{ik} = 1\}$ and $\hat{u}_k^{jh} = \#\{i : \hat{z}_{ik} = 1, x_i^{jh} = 1\}$.

For the parsimonious models, ICL can also be derived in a closed form with the noninformative prior distributions given in Section 9.5, but it involves ever more combinatorial sums to be calculated as the number of scatter parameters to be estimated decreases.

Finally, it can be noted that although the AIC criterion

$$\text{AIC} = \ln f(x; \hat{\theta}) - \nu \tag{9.9}$$

is known to select too complex mixture models, see McLachlan and Peel (2000) or Frühwirth-Schnatter (2006), a slight modification of this criterion

$$\text{AIC3} = \ln f(x; \hat{\theta}) - \frac{3}{2}\nu \tag{9.10}$$

can, somewhat surprisingly, outperform BIC for the LCM, see Nadif and Govaert (1998) or Nadolski and Viele (2004). All these criteria are compared for example data in Section 9.9.

9.7 A Note on Model-Based Clustering for Ordinal and Mixed Type Data

Often, users are facing data sets where the categorical data are ordinal or gathering categorical, ordinal and continuous variables.

9.7.1 Ordinal Data

A simple way to deal with ordinal data is to consider them as categorical data without taking into account the order. But ignoring the order can lead to an important loss of information and produce poor results. However, few specific latent class models for the clustering of ordinal data have been proposed. A fully developed model is proposed and implemented in the Latent Gold software (Vermunt and Magidson 2005). It is related to the adjacent category logit model (Agresti 2002), which has been conceived to model an ordinal variable x with m levels when a covariable y is available. It is defined by the equation for $h = 1, \ldots, m - 1$:

$$\log \frac{x_h}{x_{h+1}} = \beta_h + \gamma y. \tag{9.11}$$

The extension of this model to a latent class model for ordinal variables proposed in Vermunt and Magidson (2005) under the local independence assumption consists of applying the adjacent category logit model where the latent class indicator is considered as a covariate. With the notation used in this chapter, it leads to the following conditional model for $j = 1, \ldots, d, h = 1, \ldots, m_j$ and $k = 1, \ldots, g$

$$\alpha_k^{jh} = \beta_h^j + \gamma_k. \tag{9.12}$$

9.7.2 Mixed-Type Data

When facing mixed-type data, the main problem is to deal with nominal and continuous data in the same exercise. In theory, it seems that there is no difficulty to deal with such data sets in the model-based clustering framework. Assuming that the continuous and the categorical variables are conditionally independent knowing the clusters, the continuous data could be assumed to arise from a multivariate Gaussian distribution while the categorical variables are assumed to arise from an LCM. In these conditions, there is no difficulty to derive the formulas for the EM algorithm and the parameters of this mixture model can be estimated without particular numerical difficulty. In this setting, a possible and sensible strategy when concerned with a similar number of continuous and categorical variables is to use a mixed mixture model imposing a local independence assumption for all the variables in order to deal with both types of variables in a consistent way. Thus, it leads to the assumption that the Gaussian mixture components for the continuous variables have diagonal variance matrices, while the categorical variables are assumed to arise from a mixture of locally independent multinomial variables.

When there are many more categorical variables than continuous variables, a possibility could be to transform the few continuous variable to categorical data and then to use an LCM on the resulting data set. Transforming continuous data to categorical data could be thought of as a natural and relatively easy way to get a homogeneous resulting data set, and it is certainly a much employed way of dealing with mixed type data. But we do not advocate this way of dealing with mixed type data. As a matter of fact, an important drawback of this strategy is that coding continuous variables in categorical variables could lead to a serious loss of information in many situations (see for instance Celeux and Robert (1993) for a case study illustrating this issue).

All the above-mentioned strategies could be preferred if one wants a clustering that takes all variables into account. The reader is referred to Hennig and Liao (2013) for an interesting case study comparing different clustering approaches from this point of view. Otherwise, other strategies could be more beneficial in some situations.

As a matter of fact, the difficulty with dealing with mixed type data is not numerical, it rather lies in a balanced consideration of both types of variables. When there are many more continuous variables than categorical variables, it could be recommended to use a Gaussian mixture model to cluster the continuous data and then to use the categorical data as illustrative variables to interpret the resulting clusters. On the contrary, when there are many more categorical variables than continuous variables, it could be recommended to use an LCM to cluster the categorical data and then to use the continuous data as illustrative variables to interpret the resulting clusters through a classification procedure for instance.

When the continuous and categorical variables are in similar proportions, an alternative strategy could be to treat the continuous and the categorical variables in two separate clustering analyses (derived from an unrestricted Gaussian mixture model for the continuous variables and from an LCM for the categorical variables). And, finally, one can obtain a final clustering by intersecting the two resulting clusterings, or by producing a consensus clustering (Chapter 22). This way of separating the two cluster analyses is often relevant because in many situations the continuous variables and the categorical variables do not give the same view on the objects to be analyzed.

9.8 Software

There are two kinds of software-proposing model-based clustering methods for qualitative data.

- The first category of software is concerned with latent structure analysis methodology and is not restricted to latent class methodology and cluster analysis. Readers will find a good description of these programs, which are commercial for the most part, on the John Uebersax page (Uebersax 2012). Among them, the Latent Gold software of Vermunt and Magidson (2005) is by far the most complete and elaborate one. In the Uebersax list, an other remarkable software is LEM, a free latent class analysis software proposed by Vermunt.

- The second category of software is model-based clustering software. As a matter of fact, most of these programs are essentially concerned with Gaussian mixture and are not proposing specific programs to deal with categorical data. Notable exceptions are the SNOB software developed by Dowe and the Mixmod software, which has an R interface (RMixmod). This software allows to analyze all the latent class models presented in this chapter through maximum likelihood or classification maximum likelihood.

9.9 An Illustration

In order to illustrate the latent class model in the clustering context, the Carcinoma data set, presented in Agresti (2002), is considered. This data set consists of seven dichotomous variables providing the ratings by seven pathologists of 118 slides on the presence or absence of carcinoma in the uterine cervix. The purpose of Agresti (2002) when studying these data was to model interobserver agreement. Here, the context is different. It is to analyze how subjects could be divided into groups depending upon the consistency of their diagnoses. But the aim of this small study is merely illustrative. It is to exemplify the behavior of the models presented in Section 9.3 and the model selection criteria presented in Section 9.6.

In addition to the four models of Section 9.3 (for binary data, the models $[\varepsilon_k^{jh}]$ and $[\varepsilon_k^j]$ are equivalent), the mixing proportions are assumed to be free or equal. Therefore, eight models are considered from the most complex model with free proportions and free scatter parameters $[p_k, \varepsilon_k^j]$ to the most parsimonious model with equal proportions and scatter parameters $[p, \varepsilon]$. The maximum-likelihood (ML) and regularized maximum-likelihood (RML) estimates have been computed using the EM and EM-Bayes algorithms, respectively, for the eight models with the number of classes ranging from 1 to 10. For the RML estimates, the Bayesian hyperparameters have been set to $b = c = 2$. To select one of the estimated models, the criteria BIC, ICLbic, ICL, AIC, and AIC3, described in Section 9.6, have been computed. The results obtained with the criteria BIC, ICL, ICLbic, and AIC are reported in Figure 9.1 for the ML estimations and in Figure 9.2 for the RML estimations. The results with AIC3 are not reported since this criterion has a behavior quite similar to AIC here.

9.9.1 Running Conditions of the Algorithms

In practical situations, finding the maximum likelihood estimation with an EM-like algorithm is not an easy task. There is no gold standard way to get optimal solutions, but using sensible strategies to get honest estimate parameters is generally highly useful. For these experiments, the so-called em-EM strategy advocated in Biernacki et al. (2003) has been used. This strategy consists of repeating x times the following procedure: (i) r rapid runs of the EM-algorithm without waiting to its convergence by using loose criteria of convergence (the "em" pass); (ii) choose as initial position the solution providing the highest likelihood value among these r rapid runs; (iii) run the EM-algorithm from this initial position with tight criteria of convergence (the "EM" pass). With this kind of strategy, choosing a proper criterion to stop the EM algorithm is crucial. Since EM may have a painfully slow convergence, we do not recommend to base a stopping rule on the loglikelihood. In these experiments, we stop the algorithm as soon as one of the criteria $||\varepsilon^{(r+1)} - \varepsilon^{(c)}|| < s$ or "number of iterations greater than ITEMAX" is fulfilled. The following values have been considered for the em-EM strategy described above: $x = 20$, $r = 50$, $s = 10^{-2}$ (resp. $s = 10^{-6}$) and ITEMAX = 100 (ITEMAX = 2000) for the em pass (for the EM pass). Finally,

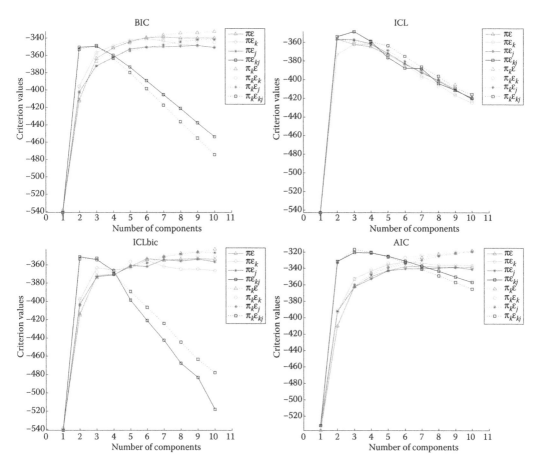

FIGURE 9.1
BIC, ICL, ICLbic, and AIC values derived from the EM-algorithm for the eight latent class models as functions of the number of classes.

although we do not claim that this em-EM strategy is optimal, it is important to stress that this strategy leads to a dramatically improved behavior of the EM algorithm. For instance, using a standard random initialization for EM often leads to a non-increasing likelihood when the complexity of the mixture model increases (numerical experiments not reported here).

The main conclusions are as follows:

- BIC does not work so well since it chooses the model $[p_k, \varepsilon]$ with 10 clusters for ML and with six clusters for RML, and these solutions do not seem attractive options for this data set. On the other hand, ICL and AIC select the model $[p_k, \varepsilon_k^j]$ with three clusters described in Table 9.1.

- Fixing the model to be $[p_k, \varepsilon_k^j]$, all criteria select three clusters.

- The results obtained by model $[p, \varepsilon_k]$ with three clusters (Table 9.2), even if not selected by any selection criteria, are interesting because it provides two very homogeneous classes (slightly more than those provided by the most general model). Furthermore, the values of ε_k (0.06, 0.16, 0.08) confirm the results generally

TABLE 9.1

ML Estimates of a_k^j and ε_k^j for Model $[p_k, \varepsilon_k^j]$

1	1	1	1	1	1	1	.06	.14	.00	.00	.06	.00	.00
2	2	1	1	2	1	2	.49	.00	.00	.06	.25	.00	.37
2	2	2	2	2	1	2	.00	.02	.14	.41	.00	.48	.00

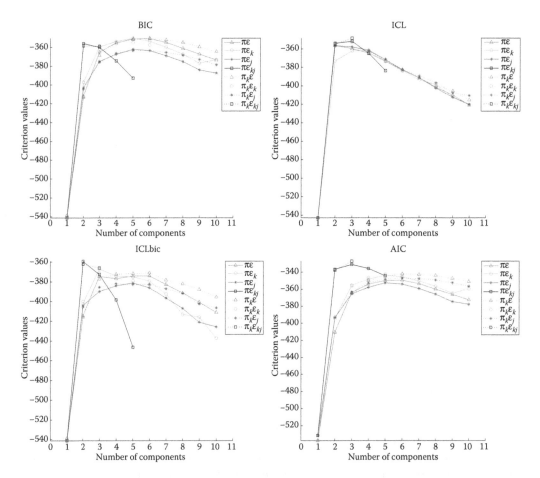

FIGURE 9.2
BIC, ICL, ICLbic, and AIC values derived from the EM–Bayes algorithm for the eight latent class models as functions of the number of classes.

obtained on this data set with two homogeneous classes and a more diffuse class. The results obtained with the simplest model $[p, \varepsilon]$ confirm such a latent structure (same a_k^j's and $\varepsilon = 0.089$).

- All these results have been obtained with the ML methodology. The results with the RML approach are quite similar (Table 9.3 to be compared to Table 9.1). But it must be noticed that for the most complex models $[p, \varepsilon_k^j]$ and $[p_k, \varepsilon_k^j]$, the regularization tends to produce some identical mixture components. For the model $[p, \varepsilon_k^j]$ (resp. $[p_k, \varepsilon_k^j]$), it was not possible to get more than five (three) different classes.

TABLE 9.2

ML Estimates of a_k^j and ε_k for Model $[p, \varepsilon_k]$

1	1	1	1	1	1	1	.03
2	2	1	1	2	1	2	.16
2	2	2	2	2	2	2	.08

TABLE 9.3

RML Estimates of a_k^j and ε_k^j for Model $[p_k, \varepsilon_k^j]$

1	1	1	1	1	1	1	.08	.18	.02	.02	.08	.02	.02
2	2	1	1	2	1	2	.48	.05	.07	.09	.24	.05	.36
2	2	2	2	2	1	2	.02	.04	.17	.41	.03	.47	.02

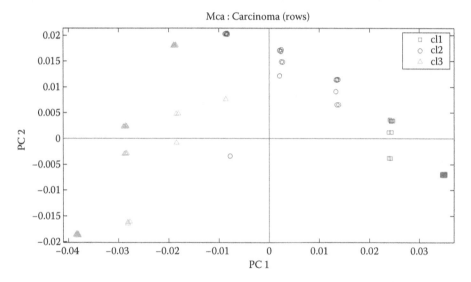

FIGURE 9.3
Representation of the optimal three latent classes derived from the model $[p_k, \varepsilon_k^j]$ in the first plane of a multiple correspondence analysis.

Finally, to illustrate the best estimation (model $[p_k, \varepsilon_k^j]$ with three clusters), Figure 9.3 represents the projection of the data and the partition associated with the estimated model on the plane defined by the first two components obtained by multiple correspondence analysis, a factor analysis technique for nominal categorical data, which is an extension of simple correspondence analysis (Greenacre and Blasius 2006).

References

Agresti, A. 2002. *Categorical Data Analysis, Second Edition*. New York: Wiley.
Aitchison, J. and Aitken, C.G.G. 1976. Multivariate binary discrimination by the kernel method. *Biometrika*, 63, 413–420.

Allman, E., Matias, C., and Rhodes, J. 2009. Identifiability of parameters in latent structure models with many observed variables. *The Annals of Statistics*, 37(6A), 3099–3132.

Anderlucci, L. 2012. *Comparing Different Approaches for Clustering Categorical Data*. Ph.D. thesis, Università di Bologna.

Anderlucci, L. and Hennig, C. 2014. Clustering of categorical data: A comparison of a model- based and a distance-based approach. *Communications in Statistics Theory and Methods*, 43, 704–721.

Benzecri, J.P. 1973. *L'analyse des données*. Paris: Dunod.

Biernacki, C., Celeux, G., and Govaert, G. 2000. Assessing a mixture model for clustering with the integrated completed likelihood. *IEEE Transactions on Pattern Analysis and Machine Intelligence*, 22(7), 719–725.

Biernacki, C., Celeux, G., and Govaert, G. 2003. Choosing starting values for the EM algorithm for getting the highest likelihood in multivariate gaussian mixture models. *Computational Statistics and Data Analysis*, 41, 561–575.

Biernacki, C., Celeux, G., and Govaert, G. 2010. Exact and Monte Carlo calculations of integrated likelihoods for the latent class model. *Journal of Statistical Planning and Inference*, 140(11), 2991–3002.

Bock, H.H. 1986. Loglinear models and entropy clustering methods for qualitative data. In *Classification as a tool of research. Proc. 9th Annual Conference of the Gesellschaft fr Klassifikation*, pp. 18–26. North Holland.

Carreira-Perpiñán, M.Á. and Renals, S. 2000. Practical identifiability of finite mixtures of multivariate Bernoulli distributions. *Neural Computation*, 12(1), 141–152.

Celeux, G. and Govaert, G. 1991. Clustering criteria for discrete data and latent class models. *Journal of Classification*, 8(2), 157–176.

Celeux, G. and Govaert, G. 1995. Gaussian parsimonious clustering models. *Pattern Recognition*, 28(5), 781–793.

Celeux, G., Hurn, M., and Robert, C.P. 2000. Computational and inferential difficulties with mixture posterior distributions. *Journal of American Statistical Association*, 95(3), 957–970.

Celeux, G. and Robert, C. 1993. Une histoire de discrétisation. *Revue de Modulad*, 11, 7–42. (with discussion).

Dowe, D. Snob page. http://www.csse.monash.edu.au/dld/otherSnob.html.

Frühwirth-Schnatter, S. 2006. *Finite mixture and Markov Switching Models*. New York: Springer Verlag.

Frühwirth-Schnatter, S. 2011. Dealing with label switching under model uncertainty. In Mengersen, K.L., Robert, C., and Titterington, D.M. (Eds.), *Mixtures: Estimation and Applications*, pp. 213–240. Wiley.

Gollini, I. and Murphy, T.B. 2014. Mixture of latent trait analyzers for model-based clustering of categorical data. *Statistics and Computing*, 24, 569–588.

Goodman, L.A. 1974. Exploratory latent structure models using both identifiable and unidentifiable models. *Biometrika*, 61, 215–231.

Govaert, G. 1983. *Classification croisée*. Thèse d'état, Université Paris 6, France.

Greenacre, M. and Blasius, J. (Ed.) 2006. *Multiple Correspondence Analysis and Related Methods*. London: Chapman & Hall/CRC.

Gyllenberg, M., Koski, T., Reilink, E., and Verlaan, M. 1994. Nonuniqueness in probabilistic numerical identification of bacteria. *Journal of Applied Probability*, 31, 542–548.

Hennig, C. and Liao, T.F. 2013. Comparing latent class and dissimilarity based clustering for mixed type variables with application to social stratification (with discussion). *Journal of the Royal Statistical Society, Series C*, 62, 309–369.

Hubert, L.J. and Arabie, P. 1985. Comparing partitions. *Journal of Classification*, 2, 193–198.

Langrognet, F. Mixmod. http://www.mixmod.org.

Lazarsfeld, P. 1950. The logical and mathematical foundations of latent structure analysis. In S.A. Stouffer (Ed.), *Measurement and Prediction*, pp. 362–412. Princeton: Princeton University Press.

McLachlan, G.J. and Peel, D. 2000. *Finite Mixture Models*. New York: Wiley.

Nadif, M. and Govaert, G. 1998. Clustering for binary data and mixture models: Choice of the model. *Applied Stochastic Models and Data Analysis*, 13, 269–278.

Nadolski, J. and Viele, K. 2004. The Role of Latent Variables in Model Selection Accuracy. In *International Federation of Classification Societies Meeting*.

Papastamoulis, P. and Iliopoulos, G. 2010. An artificial allocations based solution to the label switching problem in bayesian analysis of mixtures of distributions. *Journal of Computational and Graphical Statistics*, 19, 313–331.

Schwarz, G. 1978. Estimating the number of components in a finite mixture model. *Annals of Statistics*, 6, 461–464.

Stephens, M. 2000. Dealing with label switching in mixture models. *Journal of the Royal Statistical Society. Series B, Statistical Methodology*, 62, 795–809.

Uebersax, J.S. 2010. Latent structure analysis. http://www.john-uebersax.com/stat/index.html

Uebersax, J.S. 2012. Lca software. http://john-uebersax.com/stat/soft.html

Uebersax, J.S. and Grove, W.M. 1993. A latent trait finite mixture model for the analysis of rating agreement. *Biometrics*, 49, 823–835.

Vandewalle, V. 2009. *Estimation et sélection en classification semi-supervisée*. Ph.D. thesis, Université de Lille 1.

Vermunt, J.K. and Magidson, J. 2005. Technical guide for latent gold 4.0: Basic and advanced. http://www.statisticalinnovations.com

10

Dirichlet Process Mixtures and Nonparametric Bayesian Approaches to Clustering

Vinayak Rao

CONTENTS

Abstract

Interest in nonparametric Bayesian methods has grown rapidly as statisticians and machine-learning researchers work with increasingly complex datasets, and as computation grows faster and cheaper. Central to developments in this field has been the Dirichlet process. In this chapter, we introduce the Dirichlet process and review its properties and its different representations. We describe the Dirichlet process mixture model that is fundamental to clustering problems and review some of its applications. We discuss various approaches to posterior inference and conclude with a short review of extensions beyond the Dirichlet process.

10.1 Introduction

The Bayesian framework provides an elegant model-based approach to bringing prior information to statistical problems, as well as a simple calculus for inference. Prior beliefs are represented by probabilistic models of data, often structured by incorporating latent variables. Given a prior distribution and observations, the laws of probability (in particular, Bayes' rule) can be used to calculate posterior distributions over unobserved quantities, and thus to update beliefs about future observations. This provides a conceptually simple approach to handling uncertainty, dealing with missing data, and combining information from different sources. A well-known issue is that complex models usually result in posterior distributions that are intractable to exact analysis; however, the availability of cheap computation, coupled with the development of sophisticated Markov chain Monte Carlo sampling methods and deterministic approximation algorithms, has resulted in a wide application of Bayesian methods. The field of clustering is no exception, (Fraley and Raftery 1998, Richardson and Green 1997), and the references therein.

There still remain a number of areas of active research, and in this chapter, we consider the problem of model selection. For clustering, model selection in its simplest form boils down to choosing the number of clusters underlying a dataset. Conceptually, at least, the Bayesian approach can handle this easily by mixing over different models with different numbers of clusters. An additional latent variable now identifies which of the smaller submodels actually generated the data, and a prior belief on model complexity is specified by a probability distribution over this variable. Given observations, Bayes' rule can be used to update the posterior distribution over models.

However, with the increasing application of Bayesian models to complex data sets from fields such as biostatistics, computer vision and natural language processing, modeling data as a realization of one of a set of finite-complexity parametric models is inadequate. Such data sets raise the need for prior distributions that capture the notion of a world of unbounded complexity, so that, *a priori*, one expects larger datasets to require more complicated explanations (e.g., larger numbers of clusters). A mixture of finite cluster models does not really capture this idea. A realization of such a model would first sample the number of clusters, and this would remain fixed no matter how many observations are produced subsequently. Under a truly nonparametric solution, the number of clusters would be infinite, with any finite dataset "uncovering" only a finite number of active clusters. As more observations are generated, one would expect more and more clusters to become active. The Bayesian approach of maintaining full posterior distributions (rather than fitting point estimates) allows the use of such an approach without concerns about overfitting. Such a solution is also more elegant and principled than the empirical Bayes approach of adjusting the prior over the number of clusters after seeing the data (Casella 1985).

The Dirichlet process (DP), introduced by Ferguson (1973), is the most popular example of a nonparametric prior for clustering. Ferguson introduced the DP in a slightly different context, as a probability measure on the space of probability distributions. He noted two desiderata for such a nonparametric prior: the need for the prior to have large support on the space of probability distributions, and the need for the resulting posterior distribution to be tractable. Ferguson described these two properties as antagonistic, and the Dirichlet process reconciles them at the cost of a peculiar property: a probability measure sampled from a DP is discrete almost surely (Figure 10.1). While there has since been considerable work on constructing priors without this limitation, the discreteness of the DP was

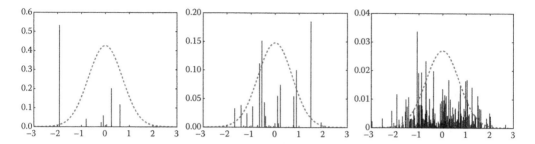

FIGURE 10.1
Samples from a DP whose base measure G_0 (the dashed curve) is the standard normal $\mathcal{N}(0,1)$. From left to right, the concentration parameter α equals 0.1, 1 and 10. G_0 is not normalized in the figure.

seized upon as ideal for clustering appications, providing a prior over mixing proportions in models with an infinite number of components.

The next section defines the DP and introduces various representations useful in applications and for studying its properties. We introduce the DP mixture model that is typically used in clustering applications, and in Section 10.3, describe some applications. We follow that with a description of various approaches to posterior inference (Section 10.4), and end by outlining some approaches to extending the DP.

10.2 The Dirichlet Process

Consider a positive real number α, as well as a probability measure G_0 on some space Θ. For instance, Θ could be the real line, and G_0 could be the normal distribution with mean 0 and variance σ^2. A Dirichlet process, parametrized by G_0 and α, is written as DP(α, G_0), and is a stochastic process whose realizations (which we call G) are probability measures on Θ. G_0 is called the base measure, while α is called the concentration parameter. Often, the DP is parametrized by a single finite measure $\alpha(\cdot)$, in which case, $\alpha = \alpha(\Theta)$ and $G_0(\cdot) = \alpha(\cdot)/\alpha(\Theta)$. We will adopt the former convention.

Observe that for any probability measure G, and for any finite partition $(A_1, \ldots, A_n)$ of Θ, the vector $(G(A_1), \ldots, G(A_n))$ is a probability vector, that is, it is a point on the $(n-1)$-simplex, with non-negative components that add up to one. We call this vector the projection of G onto the partition. Then, the Dirichlet process is defined as follows: for any finite partition of Θ, the projected probability vector has a Dirichlet distribution as specified below:

$$(G(A_1), \ldots, G(A_n)) \sim \text{Dirichlet}\,(\alpha G_0(A_1), \ldots, \alpha G_0(A_n)) \tag{10.1}$$

In words, the marginal distribution induced by projecting the Dirichlet process onto a finite partition is an appropriately parametrized Dirichlet distribution (in particular, the parameter vector of the Dirichlet distribution is the projection of the base measure G_0 onto the partition, multiplied by α). A technical question now arises: does there exist a stochastic process whose projections simultaneously satisfy this condition for all finite partitions of Θ? To answer this, note that equation (10.1) and the properties of the

Dirichlet distribution imply

$$(G(A_1), \ldots, G(A_i) + G(A_{i+1}), \ldots, G(A_n))$$

$$\sim \text{Dirichlet} \ (\alpha G_0(A_1), \ldots, \alpha G_0(A_i) + \alpha G_0(A_{i+1}), \ldots, \alpha G_0(A_n)) \qquad (10.2)$$

This distribution of the projection on the coarsened partition $(A_1, \ldots, A_i \cup A_{i+1}, \ldots, A_n)$, implied indirectly via the distribution on the finer partition $(A_1, \ldots, A_i, A_{i+1}, \ldots, A_n)$, agrees with the distribution that follows directly from the definition of the DP. This consistency is sufficient to imply the existence of the DP via Kolmogorov's consistency theorem* (Kallenberg 2002).

We shall see a more constructive definition of the DP in Section 10.2.2, however there are a number of properties that follow from this definition of the DP. First, note that from the properties of the Dirichlet distribution, for any set A, $E[G(A)] = G_0(A)$. This being true for all sets A, we have that the mean of a draw from a DP is the base probability measure G_0. On the other hand, $\text{var}[G(A)] = \frac{G_0(A)(1-G_0(A))}{1+\alpha}$, so that α controls how concentrated probability mass is around the mean G_0. As $\alpha \to \infty$, $G(A) \to G_0(A)$ for all A, so that a draw from the DP equals the base measure G_0. We will see in Section 10.2.2 that as α tends to 0, G approaches a single Dirac measure, whose location is drawn from G_0.

To appreciate why the DP has large support on the space of probability distributions, observe that any probability measure can be approximated arbitrarily well as a piecewise-constant probability measure on a sufficiently fine, finite partition of Θ. Assuming the partition has m components, and calling the probability vector $p \equiv (p_1, \ldots, p_m)$, note that p lies in the support of an m-dimensional Dirichlet distribution $\text{Dirichlet}(\gamma_1, \ldots, \gamma_m)$ only if for all i such that $p_i > 0$, we have $\gamma_i > 0$. When $\text{Dirichlet}(\gamma_1, \ldots, \gamma_m)$ is the projection of DP with base measure G_0, this is always true if the support of G_0 includes all of Θ (e.g., when Θ is $\mathbb{R}^n$, and G_0 is a multivariate normal on Θ: here, for any open set $A_i, \gamma_i = \int_{A_i} G_0(dx) > 0$). In this case, the support† of the DP includes *all* probability meaures on Θ. More generally, the weak support of $\text{DP}(\alpha, G_0)$ includes all probability measures whose support is included in the support of G_0. Somewhat confusingly, in spite of this large support, we will see that any probability measure sampled from a DP is discrete almost surely.

We next characterize the DP posterior by considering the following hierarchical model:

$$G \sim \text{DP}(\alpha, G_0) \qquad (10.3)$$

$$\theta_i \sim G \quad \text{for } i \text{ in 1 to } N \qquad (10.4)$$

Thus, we sample N observations independently from the random probability measure G (itself drawn from a DP). Let N_A represent the number of elements of the sequence $\theta \equiv (\theta_1, \ldots, \theta_N)$ lying in a set A, that is $N_A = \sum_{i=1}^{N} \delta_A(\theta_i)$, where $\delta_A(\cdot)$ is the indicator function for the set A. For some partition $(A_1, \ldots, A_m)$, the vector of counts $(N_{A_1}, \ldots, N_{A_m})$ is multinomially distributed with a Dirichlet prior on the probability vector (equation 10.1). From the conjugacy of the Dirichlet distribution, we have the posterior distribution

$$(G(A_1), \ldots, G(A_m))|(\theta_1, \ldots, \theta_N) \sim \text{Dirichlet} \ (\alpha G_0(A_1) + N_{A_1}, \ldots, \alpha G_0(A_m) + N_{A_m}) \quad (10.5)$$

---

* While Kolmogorov's consistency theorem guarantees the existence of a stochastic process with specified finite marginals, a subtle issue concerns whether realizations of this stochastic process are probability measures. This actually requires mild additional assumptions on the space Θ, (Ghosal 2010) or (Orbanz 2011) for more detailed accounts of this.

† Weak support, to be precise.

This must be true for any partition of Θ, so that the posterior is a stochastic process, all of whose marginals are Dirichlet distributed. It follows that the posterior is again a DP, now with concentration parameter $\alpha + N$, and base measure $\left(\frac{\alpha}{\alpha+N} G_0 + \frac{1}{\alpha+N} \sum_{i=1}^{N} \delta_{\theta_i} \right)$:

$$G | (\theta_1, \ldots, \theta_N) \sim \mathrm{DP} \left(\alpha + N, \frac{\alpha}{\alpha + N} G_0 + \frac{1}{\alpha + N} \sum_{i=1}^{N} \delta_{\theta_i} \right) \tag{10.6}$$

This is the conjugacy property of the DP. The fact that the posterior is again a DP allows one to integrate out the infinite-dimensional random measure G, obtaining a remarkable characterization of the marginal distribution over observations. We discuss this next.

10.2.1 The Pólya Urn Scheme and the Chinese Restaurant Process

Consider a single observation θ drawn from the DP-distributed probability measure G. The definition of the DP in the previous section shows that the probability that θ lies in a set A, marginalizing out $G(A)$, is just $E[G(A)] = G_0(A)$. Since this is true for all A, it follows that $\theta \sim G_0$. Since the posterior given N observations is also DP-distributed (equation 10.6), the predictive distribution of observation $N + 1$ is just the posterior base measure:

$$\theta_{N+1} | \theta_1, \ldots, \theta_N \sim \frac{1}{\alpha + N} \left(\alpha G_0(\cdot) + \sum_{i=1}^{N} \delta_{\theta_i}(\cdot) \right) \tag{10.7}$$

We see that the predictive distribution of a new observation is a mixture of the DP base measure G_0 (weighted by the concentration parameter), and the empirical distribution of the previous observations (weighted by the number of observations). The result above allows us to sequentially generate observations from a DP-distributed random probability measure G, without having to explicitly represent the the infinite-dimensional variable G. This corresponds to a sequential process knows as a Pólya urn scheme (Blackwell and MacQueen 1973). Here, at stage N, we have an urn containing N balls, the ith 'coloured' with the associated value θ_i. With probability $\alpha/(\alpha + N)$, the $(N + 1)$st ball is assigned a colour θ_{N+1} drawn independently from the base measure G_0, after which it is added to the urn. Otherwise, we uniformly pick a ball from the urn, and set θ_{N+1} equal to its colour, and return *both* balls to the urn. Observe that there is nonzero probability that multiple balls have the same colour and that the more balls share a particular colour, the more likely a new ball will be assigned that colour. This rich-get-richer scheme is key to the clustering properties of the DP.

Let $\theta^* \equiv (\theta_1^*, \ldots, \theta_{K_N}^*)$ be the sequence of unique values in θ, K_N being the number of such values. Let π_i^N index the elements of θ that equal θ_i^*, and let $\pi^N = \{\pi_1^N, \ldots, \pi_{K_N}^N\}$. Clearly, (θ^*, π^N) is an equivalent representation of θ. π^N is a partition of the integers 1 to N, while θ^* represents the parameter assigned to each element of the partition. Define n_i as the size of the ith cluster, so that $n_i = |\pi_i^N|$. Equation (10.7) can now be rewritten as

$$\theta_{N+1} \sim \frac{1}{\alpha + N} \left(\alpha G_0 + \sum_{c=1}^{K_N} n_c \delta_{\theta_c^*} \right) \tag{10.8}$$

The preceding equation is characterized by a different metaphor called the Chinese restaurant process (CRP) (Pitman 2002), one that is slightly more relevant to clustering applications. Here, a partition of N observations is represented by the seating arrangement of N "customers" in a "restaurant." All customers seated at a table form a cluster, and the dish served at that table corresponds to its associated parameter, θ^*. When a new customer (observation $N+1$) enters the restaurant, with probability proportional to α, she decides to sit by herself at a new table, ordering a dish drawn from G_0. Otherwise, she joins one of the existing K_N tables with probability proportional to the number of customers seated there. We can write down the marginal distribution over the observations $(\theta_1, \ldots, \theta_n)$:

$$P(\pi, \theta^*) = \left(\frac{\alpha^{K_N-1}}{[\alpha+1]_1^N} \prod_{c=1}^{K_N} (n_c - 1)! \right) \left(\prod_{c=1}^{K_N} G_0(\theta_c^*) \right) \tag{10.9}$$

Here, $[x]_a^n = \prod_{i=0}^{n-1} (x + ia)$ is the rising factorial. The equation above makes it clear that the partitioning of observations into clusters, and the assignment of parameters to each cluster are independent processes. The former is controlled by the concentration parameter α, while all clusters are assigned parameters drawn independently from the base measure. As α tends to infinity, each customer is assigned to her own table, and has her own parameter (agreeing with the idea that the θ's are drawn i.i.d. from a smooth probability measure). When α equals 0, all customers are assigned to a single cluster, showing that the random measure G they were drawn from was a distribution degenerate at the cluster parameter.

Marginalizing out the cluster parameters θ^*, consider the distribution over partitions specified by the CRP:

$$P(\pi^N) = \frac{\alpha^{K_N-1}}{[\alpha+1]_1^N} \prod_{c=1}^{K_N} (n_c - 1)! \tag{10.10}$$

This can be viewed as a distribution over partitions of the integers 1 to N and is called the Ewens' sampling formula (Ewens 1972). Ewens' sampling formula characterizes the clustering structure induced by the CRP (and thus the DP); for a number of properties of this distribution over partitions (Pitman 2002). We discuss a few below.

First, observe that the probability of a partition depends only on the number of the blocks of the partition and their sizes and is independent of the identity of the elements that constitute the partition. In other words, the probability of a partition of the integers 1 to N is invariant to permutations of the numbers 1 to N. Thus, despite its sequential construction, the CRP defines an *exchangeable* distribution over partitions. Exchangeability has important consequences which we discuss later in Section 10.5.1.

Next, under the CRP, the probability that customer i creates a new table is $\alpha/(\alpha + i - 1)$. Thus, K_N, the number of clusters that N observations will be partitioned into, is distributed as the sum of N independent Bernoulli variables, the ith having probability $\alpha/(\alpha + i - 1)$. Letting $\mathcal{N}(0, 1)$ be the standard normal distribution, and letting $\xrightarrow{d}$ indicate convergence in distribution, one can show that as $N \to \infty$,

$$K_N/\log(N) \to \alpha, \quad (K_N - \alpha \log(N))/\sqrt{\alpha \log(N)} \xrightarrow{d} \mathcal{N}(0, 1) \tag{10.11}$$

Finally, let $C_k^{(N)}$ represent the number of clusters with k customers. As $N \to \infty$,

$$\left(C_1^{(N)}, C_2^{(N)}, C_3^{(N)}, \ldots \right) \xrightarrow{d} (Z_1, Z_2, Z_3, \ldots) \tag{10.12}$$

where the $Z_i, i = 1, 2, \ldots$ are independent Poisson-distributed random variables with $E[Z_i] = \alpha/i$. Thus, even as the number of observations N tends to infinity, one expects that under the CRP, the number of clusters with, say, just 1 observation remains $O(1)$. Pitman (2002) also includes distributions over these quantities for finite N, but they are more complicated. Results like these are useful to understand the modelling assumptions involved in using a Dirichlet process for clustering. Despite its nonparametric nature, these assumptions can be quite strong, and an active area of research is the construction and study of more flexibile alternatives. We review a few in Section 10.5.

10.2.2 The Stick-Breaking Construction

The previous section showed that observations drawn from a DP-distributed probability measure G have non-zero probability of being identical. This suggests that G has an atomic component. In fact, G is purely atomic (Blackwell 1973), this follows from the fact any sample drawn from G has nonzero probability of being repeated later on. Moreover, the previous section showed that the number of components is unbounded, growing as $\log(N)$, and implying that G has an infinite number of atoms. In fact, G can be written as

$$G = \sum_{i=1}^{\infty} w_i \delta_{\theta_i} \tag{10.13}$$

As equation (10.9) suggests, under a DP, the sequence of weights (w_i) and the sequence of locations (θ_i) are independent of each other, with the latter drawn i.i.d. from the base distribution G_0. The weights, which must sum to 1, are clearly not independent, however (Sethuraman 1994) provided a remarkably simple construction of these weights.

Consider a reordering the weights obtained by iteratively sampling without replacement from the set of weights $\{w_i\}$. At any stage, a weight w_i selected from the infinite set of remaining weights with probability proportional to w_i. This constitutes a "size-biased" reordering of the weights.[*]

Consider a second sequence of weights, now obtained by repeatedly breaking a "stick" of length 1. With an initial length equal to 1, repeatedly break off a Beta$(1, \alpha)$-distributed fraction of the remaining stick-length. Letting $V_i \sim \text{Beta}(1, \alpha)$, the sequence of weights is

$$(V_1, V_2(1 - V_1), \quad V_3(1 - V_2)(1 - V_1), \ldots) \tag{10.14}$$

This countable sequence of weights (which adds up to 1) has what is called a GEM$(\alpha, 0)$ distribution (after Griffiths, Engen, and McCloskey). Importantly, this sequence of weights has the same distribution as the size-biased reordering of the DP weights (Pitman 2002). The sequence of weights, along with an infinite (and independent) sequence of θ_i's, defines a sample from a DP via equation (10.13).

The stick-breaking representation provides a simple constructive definition of the DP. Since the sequence of weights returned by the stick-breaking construction is stochastically decreasing, we can construct truncated approximations to the DP (e.g., by setting $V_k = 1$); such truncations are useful for posterior computation. The stick-breaking construction can also be generalized to construct other nonparametric priors (Ishwaran and James 2001).

---

[*] As a side note, when the weights are placed in the order their corresponding clusters were created under the CRP, one obtains a size-biased reordering.

10.2.3 Dirichlet Process Mixtures and Clustering

Since a sample G drawn from a DP is discrete, it is common to smooth it by convolving with a kernel. The resulting model is a nonparametric prior over smooth probability densities and is called a Dirichlet process mixture model (DPMM) (Lo 1984). Let $f(x, \theta)$ be a nonnegative kernel on $\mathcal{X} \times \Theta$, with $\int_{\mathcal{X}} f(x, \theta) dx = 1$ for all θ. Thus, the function f defines a family of smooth probability densities on the space $\mathcal{X}$ (the observation space), indexed by elements of Θ (the parameter space). Observations then come from the following hierarchical model:

$$G \sim \text{DP}(\alpha, G_0) \tag{10.15}$$

$$\theta_i \sim G, \quad x_i \sim f(x, \theta_i) \tag{10.16}$$

In the context of clustering, the DPMM corresponds to an infinite mixture model (Rasmussen 2000). The DP induces a clustering among observations, with observations in the cth cluster having parameter θ_c^*. These cluster parameters are drawn i.i.d. from the base measure G_0, which characterizes the spread of clusters in parameter space. The cluster parameter determines properties like location, spread, and skewness of the smoothing kernel f.

Figure 10.2 shows the distribution of 1,000 data points from two DPMMs. In both cases, the concentration parameter α was set to 1. Both models used Gaussian kernels (and so are DP mixtures of Gaussians); however, for the first plot, all kernels had a fixed, isotropic covariance. Here, the parameter of each cluster was its mean, so that both Θ and $\mathcal{X}$ were the two-dimensional Euclidean space $\mathcal{R}^2$. We set the base measure G_0 as a conjugate normal with mean zero and a standard deviation equal to 10. In the second case, the covariance of the Gaussian kernels also varied across clusters, so that $\Theta = \mathcal{R}^2 \times \mathcal{S}_2^+$ ($\mathcal{S}_+^2$ being the space of positive-definite, two-dimensional symmetric matrices). In this case, we set the base measure as the conjugate normal-inverse-Wishart distribution.

In the examples above, we set parameters to clearly demonstrate the clustering structure of DPMM. Real data are usually more ambigious with multiple explanations that are all plausible. It is in such situations that the Bayesian approach of maintaining a posterior distribution over all possible configurations is most useful. At the same time, it is also in such situations that inferences are most sensitive to aspects of the prior, and one needs to

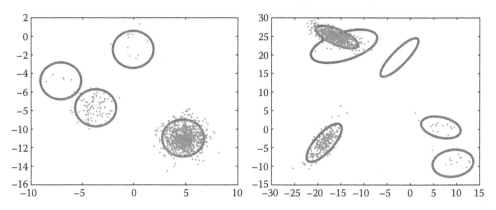

FIGURE 10.2
Realizations of two DP mixture models, with the DP determining cluster means (left), and means and covariance matrices (right). Ellipses are two standard deviations.

be careful about a cavalier use of conjugate priors. Often, the base-measure G_0 is chosen to have heavy tails, and typically, hyperpriors are placed on parameters of G_0. The clustering structure is sensitive to the parameter α, and one often places priors on this as well.

The Chinese restaurant process of Section 10.2.1 to sample parameters from the DP is easily extended to sample observations from the DPMM. Now, after customer $N + 1$ chooses her table (and thus her parameter θ_{N+1}), she samples a value x_{N+1} from the probability density $f(\cdot, \theta_{N+1})$. When the G_0 is conjugate to f, one need not even represent the table parameters θ^*, and can directly sample a new observation given the previous observations associated with that table.

Given observations from a DPMM, the posterior distribution over clusterings is much more complicated than with the DP. Why this is so is easy to see: given the set of parameters θ, we already have the clustering stucture, this is no longer the case given X. Where previously the posterior over the random measure G was also a Dirichlet process, it now is a mixture of Dirichlet processes (with a combinatorial number of components, one for each possible clustering of the observations). Thus, one has to resort to approximate inference techniques like MCMC or variational approximations. We describe a few such techniques later in Section 10.4, first, we concretize ideas by discussing some modelling applications.

10.3 Example Applications

The Dirichlet process has found wide application as a model of clustered data. Often, applications involve more complicated nonparametric models extending the DP, some of which we discuss in Section 10.5. Applications include modeling text (clustering words into topics (Teh et al. 2006)), modelling images (clustering pixels into segments (Sudderth and Jordan 2008)), genetics (clustering haplotypes (Xing et al. 2007)), biostatistics (clustering functional response trajectories of patients (Bigelow and Dunson 2009)), neuroscience (clustering spikes (Wood et al. 2006)), as well as fields like cognitive science (Griffiths et al. 2007) and econometrics (Griffin 2011). Even in density modelling applications, using a DPMM as a prior involves an implicit clustering of observations, this can be useful in interpreting results. Below, we look at three examples.

10.3.1 Clustering Microarray Expression Data

A microarray experiment returns expression levels y_{rgt} of a group of genes $g \in \{1, \ldots, G\}$ across treatment conditions $t \in \{1, \ldots, T\}$ over repetitions $r \in \{1, \ldots, R_t\}$. In Dahl (2006), the expression levels over repetitions are modelled as i.i.d. draws from a Gaussian with mean and variance determined by the gene and treatment: $y_{rgt} \sim \mathcal{N}(\mu_g + \tau_{gt}, \lambda_g^{-1})$. Here, (μ_g, λ_g) are gene-specific mean and precision parameters, while τ_{gt} represent gene-specific treatment effects. To account for the highly correlated nature of this data, genes are clustered as co-regulated genes, with elements in the same cluster having the same parameters $(\mu^*, \tau^*, \lambda^*)$. In Dahl (2006), this clustering is modelled using a Dirichlet process, with a convenient conjugate base distribution $G_0(\mu^*, \tau^*, \lambda^*)$ governing the distribution of parameters across clusters. Besides allowing for uncertainty in the number of clusters and the cluster assignments, Dahl (2006) shows how such a model-based approach allows one to deal with nuisance parameters: in their situation, they were not interested in the effects of the gene-specific means μ_g. The model (including all hyperparameters) was fitted using an MCMC algorithm, and the authors were able to demonstrate superiority over a number of other clustering methods.

10.3.2 Bayesian Haplotype Inference

Most differences in the genomes of two individuals are variations in single nucleotides at specific sites in the DNA sequence. These variations are called single-nucleotide polymorphisms (SNPs), and in most cases, individuals in the population have one of two base-pairs at any SNP site. This results in one of two alleles at each SNP site, labelled either "0" or "1". A sequence of contiguous alleles in a local region of a single chromosome of an individual is called a haplotype. Animals such as human beings are diploid with two chromosomes, and the genotype is made up of two haplotypes. Importantly, with present-day assaying technology, the genotype obtained is usually unphased, and does not indicate which haplotype each of the two base pairs at each site belongs to. Thus, the genotype is a string of "00"s, "11"s and "01"s, with the last pair ambiguous about the underlying haplotypes. Given a collection of length M genotypes from a population, Xing et al. (2007) considers the problem of disentangling the two binary strings (haplotypes) that compose each unphased genotype. To allow statistical sharing across individuals, they treat each haplotype as belonging to one of an infinite number of clusters, with weights assigned via a Dirichlet process. The parameter of each cluster is drawn from a product of M independent Bernoulli random variables (this is the base measure G_0), and can be thought of as the cluster prototype. The individual haplotypes in the cluster differ from the cluster prototype at any location with some small probability of mutation (this forms the kernel f). The two haplotypes of each individual are two independent samples from the resulting DP mixture model. Given the observed set of genotypes, the authors infer the latent variables in the model by running an MCMC sampling algorithm.

10.3.3 Spike Sorting

Given a sequence of action potentials or spikes recorded by an electrode implanted in an animal, a neuroscientist has to contend with the fact that a single recording includes activity from multiple neurons. Spike sorting is essentially the process of clustering these action potentials, assigning each spike in a spike train to an appropriate neuron. The number of neurons is unknown, and Wood et al. (2006) assumes an infinite number using a Dirichlet process. Each neuron has its own stereotypical spike shape, which can be modeled using a Gaussian process as the base measure (Wood et al. (2006), as is typical, model lower-dimensional principal component analysis (PCA) projections of the spike shape as multivariate Gaussians). Adding a Gaussian smoothing kernel f to model measurement noise, the resulting model is a DP mixture of Gaussians. Again, given observations, one can perform posterior inference using techniques we describe next.

10.4 Posterior Inference

Given observations X from a Dirichlet process mixture model, inference involves characterizing the posterior distribution over cluster assignments of data points, as well as the cluster parameters. Implicit in this distribution are quantities like the distribution over the number of clusters underlying the observed dataset, the probability that two observations belong to the same cluster and so on. A straightforward expression for this distribution follows from the CRP representation (recall that π^N is a partition of the integers 1 to N):

$$p(\pi^N, \{\theta_1^*, \ldots, \theta_{|\pi^N|}^*\} | X) \propto \left(\alpha^{|\pi^N|-1} \prod_{i=1}^{|\pi^N|} (|\pi_i^N| - 1)! \prod_{i=1}^{|\pi^N|} G_0(\theta_i^*) \right) \prod_{i=1}^{N} F(x_i | \theta_{c_i}^*) \qquad (10.17)$$

Unfortunately, hidden in the expression about is an intractible normalization constant which involves summing over the combinatorial number of partitions of N data points. This makes calculating exact posterior expectations computationally intractable even when all distributions are chosen to be conjugate, and one has to resort to approximate inference techniques. Below, we look at two dominant approaches, sampling and deterministic approximations.

10.4.1 Markov Chain Monte Carlo

By far the most widespread approach to posterior inference for DP mixture models is Markov chain Monte Carlo (MCMC). The idea here is to set up a Markov chain whose state at any iteration instantiates all latent variables of interest. At each iteration, the state of the chain is updated via a Markov transition kernel whose stationary distribution is the desired posterior distribution over the latent variables. By running the chain for a large number of iterations, we obtain a sequence of samples of the unobserved variables whose empirical distribution converges to the posterior distribution. After discarding initial "burn-in" samples corrupted by the arbitrary initialization of the chain, the remaining samples can be used to calculate posterior expectations and to make predictions. Typical quantities of interest include the probability two observations are co-clustered (often represented by a co-occurence matrix), the posterior over the number and sizes of clusters, and the posterior over cluster parameters. Note that each MCMC sample describes a hard clustering of the observations, and very often a single sample is used to obtain a "typical" clustering of observations.

The different representations of the DP outlined in the earlier sections can be exploited to construct different samplers with different properties. The simplest class of samplers are marginal samplers that integrate out the infinite dimensional probability measure G, and directly represent the partition structure of the data (Escobar and West 1995). Both the Chinese restaurant process and the Pólya urn scheme provide such representations, moreover, they provide straightforward cluster assignment rules for the last observation. Exploiting the exchangeability of these processes, one cycles through all observations, and treating each as the last, assigns it to a cluster given the cluster assignments of all other observations, and the cluster parameters.

Algorithm 10.4.1 outlines the steps involved, the most important being step 4. In words, the probability an observation is assigned to an existing cluster is proportional to the number of the remaining observations assigned to that cluster, and to how well the cluster parameter explains the observation value. The observation is assigned to a new cluster with probability proportional to the product of α and its marginal probability integrating θ out. When the base measure G_0 is conjugate, the latter integration is easy. The nonconjugate case needs some care, and we refer the reader to (Neal 2000) for an authoritative account of marginal Gibbs sampling methods for the DP.

While the Gibbs sampler described above is intuitive and easy to implement, it makes very local moves, making it difficult for the chain to explore multiple posterior modes.

Algorithm AN ITERATION OF MCMC USING THE CRP REPRESENTATION

Input: The observations $(x_1, \ldots, x_N)$
A partition π of the observations, and the cluster parameters θ^*

Output: A new partition $\tilde{\pi}$, and new cluster parameters $\tilde{\theta}^*$

1. **for** i from 1 to N **do**:
2. Discard cluster assignment c_i of observation i, and call the updated partition $\tilde{\pi}^{\backslash i}$.
3. If i belonged to its own cluster, discard θ_{c_i} from θ^*.
4. Update $\tilde{\pi}$ from $\tilde{\pi}^{\backslash i}$ by assigning i to a cluster with probability

$$p(c_i = k | \tilde{\pi}^{\backslash i}, \theta^*) \propto \begin{cases} |\tilde{\pi}_k^{\backslash i}| f(x_i, \theta_k^*) & k \leq |\tilde{\pi}^{\backslash i}| \\ \alpha \int_\Theta f(x_i, \theta) G_0(\mathrm{d}\theta) & k = |\tilde{\pi}^{\backslash i}| + 1 \end{cases}$$

5. If we assign i to a new cluster, sample a cluster parameter from

$$p(\tilde{\theta}_{c_i}^* | \tilde{\pi}, x_1, \ldots, x_N) \propto G_0(\tilde{\theta}_{c_i}^*) f(x_i, \tilde{\theta}_{c_i}^*)$$

6. **end for**
7. Resample new cluster parameters $\tilde{\theta}_c^*$ (with $c \in \{1, \ldots, \tilde{\pi}|\}$) from the posterior

$$p(\tilde{\theta}_c^* | \tilde{\pi}, x_1, \ldots, x_N) \propto G_0(\tilde{\theta}_c^*) \prod_{i \text{ s.t. } c_i = c} f(x_i, \tilde{\theta}_c^*)$$

Consider a fairly common situation where two clusterings are equally plausible under the posterior, one where a set of observations are all allocated to a single cluster, and one where they are split into two nearby clusters. One would hope that the MCMC sampler spends an equal amount of time in both configurations, moving from one to the other. However, splitting a cluster into two requires the Gibbs sampler to sequentially detach observations from a large cluster, and assign them to a new cluster. The rich get richer of the CRP, which encourages parsimony by penalizing fragmented clusterings, makes these intermediate states unlikely, and results in a low probability valley separating these two states. This can lead to the sampler mixing poorly. To overcome this, one needs to interleave the Gibbs updates with more complex Metropolis–Hastings proposals that attempt to split or merge clusters. Marginal samplers that attempt more global moves include Jain and Neal (2004) and Liang et al. (2007).

A second class of samplers are called blocked or conditional Gibbs samplers, and explicitly represent the latent mixing probability measure G. Since the observations are drawn i.i.d. from G, conditioned on it, the assignment of observations to the components of G can be jointly updated. Thus, unlike the marginal Gibbs sampler, conditioned on G, the new partition structure of the observations is independent of the old. A complication is that the measure G is infinite-dimensional, and cannot be represented exactly in a computer simulation. A common approach is to follow (Ishwaran and James 2001) and maintain an approximation to G by truncating the the stick-breaking construction to a finite number of components.

Since the DP weights are stochastically ordered under the stick-breaking construction, one would expect only a small error for a sufficiently large truncation. In fact, Ishwaran and James (2001) shows that if the stick-breaking process is truncated after K steps, then the error decreases exponentially with K. Letting N be the number of observations, and $\|G - \tilde{G}_K\|_1 = \int_\Theta |G(d\theta) - \tilde{G}_K(d\theta)|$ be the L_1-distance between G and its truncated version $\tilde{G}_K$, we have

$$\|G - \tilde{G}_K\|_1 \sim 4N \quad \exp\left(-(K-1)\alpha\right) \tag{10.18}$$

As Ishwaran and James (2001) points out, for $N = 150$ and $\alpha = 5$, a truncation level of $K = 150$ results in an error bound of 4.57×10^{-8}, making it effectively indistinguishable from the true model. Algorithm 10.4.1 outlines a conditional sampler with truncation.

Algorithm AN ITERATION OF MCMC USING THE STICK-BREAKING REPRESENTATION

Input: The observations $(x_1, \ldots, x_N)$, and a truncation level K
        Stick-breaking proportions $\boldsymbol{V} = (V_1, \ldots, V_{K-1})$ (with $V_K = 1$)
        Component parameters $\theta^* = (\theta_1, \ldots, \theta_K)$
        Component indicator variables $\boldsymbol{Z} = (z_1, \ldots, z_N)$
Output: New values of $\tilde{V}, \tilde{\theta}$, and $\tilde{Z}$

1. Sample new component assignments $\tilde{z}_i$, (with $i \in \{1, \ldots, N\}$) with

$$p(\tilde{z}_i = k) \propto V_k \prod_{j=1}^{k-1} (1 - V_j) f(x_i, \theta_k), \quad k \le K$$

2. Resample new component parameters $\tilde{\theta}_k^*$ (with $c \in \{1, \ldots, K\}$) from the posterior

$$p(\tilde{\theta}_k^* | \tilde{\pi}, x_1, \ldots, x_N) \propto G_0(\tilde{\theta}_k^*) \prod_{i \text{ s.t. } \tilde{z}_i = k} f(x_i, \tilde{\theta}_k^*)$$

3. Resample new stick-breaking proportions $\tilde{V}_k$ (with $k \in \{1, \ldots, K-1\}$) with

$$\tilde{V}_k \sim \text{Beta}\left(1 + m_k, \alpha + \sum_{j=k+1}^{K} m_j\right), \quad \text{with } m_k = \sum_{j=1}^{N} \delta_k(\tilde{z}_j)$$

Often, there are situations when one does not wish to introduce a truncation error. This is particularly true when the infinite-dimensional G has some prior other than the Dirichlet process, leading to error bounds that decay much more slowly. A solution is to have a random truncation level, where the data is allowed to guide the truncation level. Walker (2007) or Papaspiliopoulos and Roberts (2008) was referred for descriptions of such samplers; these remain asymptotically unbiased despite working with finite truncations of infinite-dimensional objects.

10.4.2 Variational Inference

The idea behind variational inference is to approximate the intractable posterior with a simpler distribution and use this to approximate quantities like cluster parameters and assignment probabilities. Variational methods have the added advantage of providing approximations to the marginal likelihood of the observed data. For concreteness, we assume the smoothing kernel in the DPMM belongs to an exponential family distribution parametrized by θ^*: $f(x, \theta^*) = h(x) \exp\left(x^T \theta^* - a(\theta^*)\right)$ ($a(\theta^*)$ is the log normalization constant). We also assume the base-measure G_0 belongs to the conjugate exponential family with sufficient statistics $(\theta^*, -a(\theta^*))$ and natural parameters (λ_1, λ_2): $G_0(\theta^*) \propto \exp((\theta^*)^T \lambda_1 - a(\theta^*)\lambda_2)$.

Recall the DPMM posterior is a complicated joint distribution over variables like a probability measure with infinite atoms and indicator variables assigning observations to atoms. These variables are dependent; for instance, the distribution over atom locations depends on the assigned observations, and the weights depend on the locations of the atoms and assigned observations. In the mean-field approximation of Blei and Jordan (2006), these dependencies are discarded, and the posterior is approximated as a product of two independent distributions, one over probability measures, and the other over cluster assignments. The distribution over probability measures is restricted to measures with K atoms (K is a parameter of the algorithm) of the form

$$\tilde{G} = \sum_{i=1}^{K} w_i \delta_{\theta_i^*} \tag{10.19}$$

Further, under the posterior approximation, the locations and the weights are assumed independent. The posterior over the ith location θ_i^* is approximated as a member of the same exponential family distribution as the base-measure, with natural parameters (τ_{i1}, τ_{i2}). Write it as $q(\cdot \mid \tau_i)$. The set of weights $(w_1, \ldots, w_K)$ are distributed according a generalized truncated stick-breaking construction, with the ith stick-breaking proportion drawn from a Beta$(\gamma_{i1}, \gamma_{i2})$ distribution. In equations,

$$w_i = V_i \prod_{j < i}(1 - V_j); \quad V_K = 1, \quad V_i \sim \text{Beta}(\gamma_{i1}, \gamma_{i2}), \quad i < K; \quad \theta_i^* \sim q(\cdot \mid \tau_i) \tag{10.20}$$

Finally, the posterior assignment probability of the ith observation is independent of the other quantities and is specified by a K-dimensional probability vector ϕ_i. The set of parameters $\tau_i, \gamma_{i1}, \gamma_{i2}, \phi_{ij}$ specify the approximate posterior distribution, and one optimizes these to minimize the Kullback-Leibler divergence from the true posterior.

Algorithm 10.4.2 from Blei and Jordan (2006) describes an iteration of a coordinate-descent algorithm reducing the KL-divergence to a local minimum. Variational algorithms are fast and relatively simpler to debug than the stochastic MCMC algorithms. They, however, introduce bias that is hard to quantify, and are not as modular as MCMC which is quite easily extended to more complex hierarchical models. Other deterministic inference schemes include Kurihara et al. (2007), Minka and Ghahramani (2003), Wang and Dunson (2010).

10.4.3 Comments on Posterior Computation

In general, beyond a little additional bookkeeping, these algorithms for posterior inference are not much less efficient than those for the corresponding finite mixture models.

Algorithm AN ITERATION OF THE VARIATIONAL BAYES ALGORITHM OF BLEI AND JORDAN (2006)

Input: The observations $(x_1, \ldots, x_N)$, and a truncation level K
Stick-breaking proportions $V = (V_1, \ldots, V_{K-1})$ (with $V_K = 1$)
Component parameters $\theta^* = (\theta_1, \ldots, \theta_K)$
Cluster assignment probabilities $\Phi = (\phi_1, \ldots, \phi_N)$
Output: New values of V, θ, and Φ

1. Update the Beta parameters for the stick-breaking proportions V as

$$\gamma_{k1} = 1 + \sum_{i=1}^{N} \phi_{ik}, \quad \gamma_{k2} = 1 + \sum_{i=1}^{N} \sum_{j=k+1}^{K} \phi_{ij}, \quad k \in \{1, \ldots, N\}$$

2. Update τ

$$\tau_{k1} = \lambda_1 + \sum_{i=1}^{N} \phi_{ik} x_i, \quad \tau_{k1} = \lambda_2 + \sum_{i=1}^{N} \phi_{ik} \quad \{k = 1, \ldots, K\}$$

3. Update the component assignment probability vectors ϕ_i as

$$\phi_{ik} \propto \exp(S_i), \quad S_i = E_q[\log V_i] + \sum_{j=1}^{i-1} E_q[\log(1 - V_i)] + E_q[(\theta_i^*)^T X_i] - E_q[a(\theta_t^*)]$$

For MCMC sampling algorithms (especially without truncation), the number of clusters will vary from iteration to iteration, and a balance must be struck between reallocating/deallocating memory for the various data structures, and maintaining large and fragmented data structures.

As far as MCMC sampling is concerned, marginal samplers are generally acknowledged to have better mixing properties, though they usually require some form of split-merge to get out of local optima. Conditional samplers can suffer from correlations between cluster assignments and Dirichlet process weights, though the fact that observations can be assigned to clusters independently offers promise for parallelized algorithms. Samplers are typically run for 5,000 to 10,000 iterations, and mixing is assessed using standard MCMC diagnostics (Brooks and Gelman 1998) on statistics like the number of clusters, the probability two observations are co-clustered or the size of the cluster an observation is assigned to.

While variational algorithms offer a number of advantages, these often get trapped in local optima and might require multiple reruns. Additionally, the mean-field nature of these algorithms results in updates that are no longer sparse: unlike an MCMC update which conditions on the cluster assignments of a set of observations, a variational update typically depends on the probabilities of each observation being assigned to all clusters. Thus, even if a variational algorithm requires fewer iterations than an MCMC sampler, each update can require more computation that an MCMC update.

10.5 Extensions

10.5.1 Exchangeability and Consistency

In this chapter, our concern was the study of Bayesian nonparametric approaches to clustering. Abstractly, this can be viewed as the study of flexible probability distributions over partitions of integers. Via the Dirichlet process, we arrived at Ewens' sampling formula (equation 10.10) giving probabilities of partitions of the integers 1 to N. Though constructed sequentially via the Chinese restaurant process, we saw that the resulting probability is exchangeable: it is independent of the order in which the customers arrive and is thus invariant to permutations of the integers 1 to N. Exchangeability is important in many clustering applications where we do not want the order of observations to affect inferences.

Ewens' sampling formula also has a consistency (or projectivity) property: the probability of π^N is equal to the sum of the probabilities of all partitions π^{N+1} of 1 to $N+1$ that are consistent with π^N. Consistency follows directly from the sequential construction of the CRP and is an important property as well: we do not want observations that were not seen to affect our inferences, these should remain irrelevant. A sequence of consistent partitions for all natural numbers implies a distribution over partitions of the natural numbers $\mathcal{N}$ (Pitman 2002), and when each finite distribution is exchangeable as well, the resulting distribution is called an infinitely exchangeable partition function (EPPF).

There is another way to see why Ewens' sampling formula is infinitely exchangeable: this is a consequence of the fact that the observations are drawn i.i.d. from the DP-distributed probability measure G. Conditioned on G, the order of the observations is irrelevant, and will remain so with G integrated out. Consistency is an easy consequence of the i.i.d. construction as well. According to de Finetti's theorem (Kallenberg 2002), for any sequence of infinitely exchangeable observations, there exists a latent variable (possible infinite dimensional), conditioned on which, the observations are i.i.d. For the CRP, this latent variable is the DP-distributed probability G. Drawing observations i.i.d. from G induces a partition of N with observations with the same value in the same cluster. Treating these values as colours drawn i.i.d. from a 'paintbox' G, Kingman's paintbox construction (Kingman 1975) specializes de Finetti's theorem to infinitely exchangeable partitions: any infinite EPPF has a mixture of paintboxes respresentation, and can be induced by sampling from a random probability measure G.

Even if one is only interested in the clustering of observations, a consistent, exchangeable distribution over clusterings corresponds to an underlying random measure G. This viewpoint can facilitate the study of asymptotics of clustering processes, the development of efficient inference techniques, the construction of new clustering models as well as extensions to more structured data where exchangeability is relaxed. We consider two extensions below.

10.5.2 Pitman-Yor Processes

Under the CRP, the number of clusters grows logarithmically with the number of observations. Conditioned on the the number of clusters, the distribution over partitions is independent of the concentration parameter, suggesting a somewhat limited control in the prior specification. The Pitman-Yor process (Pitman and Yor 1997) (also called the two-parameter Poisson–Dirichlet process) is a popular extension of the DP that remedies some of these limitations. Like the DP, this too is a prior over discrete probability measures, and is parametrized as $\mathcal{PY}(\alpha, d, G_0)$. The extra parameter d, called the discount parameter, takes values in $[0, 1)$, while α (the concentration parameter) satisfies $\alpha > -d$. The Pitman-Yor process also has a stick-breaking construction; in this case, rather than all stick-breaking proportions being i.i.d. distributed, the ith breaking proportion V_i has a Beta$(1 - d, \alpha + id)$ distribution. Again, the random measure can be integrated out, resulting in a sequential clustering process that generalizes the Chinese restaurant process (now called the two-parameter CRP). Here, when the $(N + 1)$st customer enters the restaurant, she joins a table with n_c customers with probability proportional to $n_c - d$. On the other hand, she creates a new table with probability proportional to $\alpha + K_N d$, where as before, K_N is number of tables. When the discount parameter equals 0, the Pitman-Yor process reduces to the Dirichlet process. Setting d greater than 0 allows the probability of a new cluster to increase with the number of existing clusters, and results in a power-law behavior:

$$\frac{K_N}{N^d} \to S_d \tag{10.21}$$

Here, S_d is a strictly positive random variable (having the so-called polynomially tilted Mittag–Leffler distribution Pitman (2002)). Power law behavior has been observed in models of text and images, and applications of the $\mathcal{PY}$ process in these domains (Sudderth and Jordan 2008, Teh 2006) have been shown to perform significantly better than the DP (or simpler parametric models). The effect of α and d on the number of clusters is illustrated in Figure 10.3.

The two-parameter CRP representation of the $\mathcal{PY}$-process allows us to write down an EPPF that generalizes Ewens' sampling formula:

$$P(\pi^N) = \frac{[\alpha + d]_d^{K_N - 1}}{[\alpha + 1]_1^N} \prod_{c=1}^{K_N} [1 - d]_1^{n_c} \tag{10.22}$$

The EPPF above belongs to what is known as a Gibbs-type prior: there exist two sequences of nonnegative reals $\mathbf{v} = (v_0, v_1, \dots)$ and $\mathbf{w} = (w_1, w_2, \dots)$ such that the probability of a partition π^N is

$$P(\pi^N) \propto w_{|\pi^N|} \prod_{c=1}^{|\pi^N|} v_{|\pi_c^N|} \tag{10.23}$$

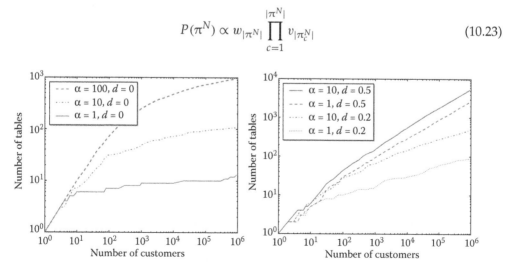

FIGURE 10.3

Number of clusters vs. number of observations for the DP (left) and the Pitman-Yor process (right) simulation.

This results in a simple sequential clustering process: given a partition π^N, the probability that customer $N+1$ joins cluster $c \leq |\pi^N|$ is proportional to $v_{|\pi_c^N|+1}/v_{|\pi_c^N|}$, while the probability of creating a new cluster is $v_1 w_{|\pi^N|+1}/w_{|\pi^N|}$. Pitman (2002) shows that any exchangeable and consistent Gibbs partition must result from a Pitman-Yor process or an m-dimensional Dirichlet distribution (or various limits of these). The CRP for the finite Dirichlet distribution has $\alpha = -\kappa < 0$ and $d = m\kappa$ for some $m = 1, 2, \dots$. Any other consistent and exchangeable distribution over partitions will result in a more complicated EPPF (and thus, for example, a more complicated Gibbs sampler). For more details, refer De Blasi et al. (2013).

10.5.3 Dependent Random Measures

The assumption of exchangeability is sometimes a simplification that disregards structure in data. Observations might come labeled with ordinates like time, position or category, and lumping all data points into one homogeneous collection can be undesirable. The other extreme of assigning each group an independent clustering is also a simplification, it is important to share statistical information across groups. For instance, a large cluster in one region of space in one group might *a priori* suggest a similar cluster in other groups as well. A popular extension of the DP that achieves this is the hierarchical Dirichlet process (HDP) (Teh et al. 2006). Here, given a set of groups $\mathcal{T}$, any group $t \in \mathcal{T}$ has its own DP-distributed random measure G_t (so that observations within each group are exchangeable). The random measures are coupled via a shared, random base measure G which itself is distributed as a DP. The resulting hierarchical model corresponds to the following generative process:

$$G \sim \mathrm{DP}(\alpha_0, G_0) \tag{10.24}$$

$$G_t \sim \mathrm{DP}(\alpha, G) \quad \text{for all } t \in \mathcal{T} \tag{10.25}$$

$$\theta_{it} \sim G_t \quad \text{for } i \text{ in 1 to } N_t \tag{10.26}$$

In our discussion so far, we have implicitly assumed the DP base measure to be smooth. For the group-specific probability measures G_t of the HDP, this is not the case; now, the base measure G (which itself is DP-distributed) is purely atomic. A consequence is that the parameters of clusters across groups (which are drawn from G) now have nonzero probability of being identical. In fact, these parameters themselves are clustered according to a CRP and Teh et al. (2006) show how all random measures can be marginalized out to give a sequential clustering process they call the Chinese restaurant franchise. The clustering of parameters means that the more often a parameter is present, the more likely it is to appear in a new group. Additionally, a large cluster in one group implies (on average) large clusters (with the same parameter) in other groups.

A very popular application of the HDP is the construction of infinite topic models for document modeling. Given a fixed vocabulary of words, a topic is a multinomial distribution over words. A document, on the other hand, is characterized by a distribution over topics, and the clustering problem is to infer a set of topics, as well as an assignment of each word of each document to a topic. A nonparametric approach assumes an infinite number of topics, with a random DP-distributed measure G that characterizes the average distribution over topics across *all* documents. Any particular document has its own distribution over topics, centered around G by drawing it from a DP with base measure G. The discreteness of G ensures the same infinite set of topics is shared across all documents, and allows inferences from one document to propagate to other documents. Figure 10.4, adapted from

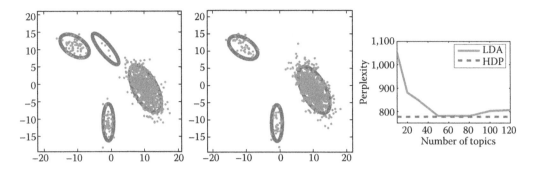

FIGURE 10.4
Thousand samples from an HDP with two groups (left) and (middle). Both groups have three clusters in common, with an additional cluster in the first. Perplexity on test documents for HDP and LDA with different number of topics (right).

Teh et al. (2006), shows that this nonparametric model automatically achieves a performance comparable to choosing the optimal number of topics in the parametric version of the model. Here, the performance measure is "perplexity," corresponding to how surprised different models are upon seeing a held-out test document.

The HDP assumes that the groups themselves are exchangeable, again, one might wish to refine this modelling assumption. An obvious approach is to hierarchically organize the groups themselves. Teh (2006) considers n-level hierarchies, while Wood et al. (2011) allow these hierarchies to be infinitely deep. Alternatively, the groups could be indexed by elements of a space $\mathcal{T}$ with some topological structure ($\mathcal{T}$ could be space or time), and one might wish to construct a *measure-valued stochastic process* G_t for all $t \in \mathcal{T}$. Here, the measure G_t at any point has an appropriate distribution (say, the Dirichlet process), and the similarity between measures varies gradually with Δt. Note that under the HDP, the group-specific measures G_t are DP-distributed conditioned on G, this no longer is true when G is marginalized out. Consequently, the clustering assumptions one makes at the group-level are different from those specified under a DP. An active area of research is the construction of dependent random probability measures with specified marginal distributions over probability measures, as well as flexible correlation structures (MacEachern 1999).

10.6 Discussion

In this chapter, we introduced the Dirichlet process, and described various theoretical and practical aspects of its use in clustering applications. There is a vast literature on this topic, and a number of excellent tutorials. Among these, we especially recommend (Ghosal 2010, Teh 2010, Teh and Jordan 2010). In spite of its nonparametric nature, the modeling assumptions underlying the DP can be quite strong. While we briefly discussed a few extensions, there is much we have had to leave uncovered (see Lijoi and Pruenster (2010) for a nice overview). Many generalizations build on the stick-breaking representation or the CRP representation of the DP. Another approach extends the DP's construction as a normalized Gamma *process* (Ferguson 1973) to construct random probability measures by normalizing completely random measures (Kingman 1967); these are atomic measures whose weights

w_i are essentially all independent. The resulting class of normalized random measures (James et al. 2005) form a very flexible class of nonparametric priors that can be used to construct different extensions of the DP (Rao and Teh 2009).

A different class of nonparametric priors are based on the Beta process (Hjort 1990) and are used to construct infinite feature models (Griffiths et al. 2007). Here, rather than assigning each observation to one of a infinite number of clusters, each observation can have a finite subset of an infinite number of features. Such a distributed representation allows a more refined representation for sharing statistical information. Infinite feature models also come with notions of exchangeability and consistency. We recommend Broderick et al. (2012) for a nice overview of these ideas, and their relation to ideas discussed here.

A major challenge facing the more widespread use of Bayesian nonparametric methods is the development of techniques for efficient inference. While we described a few MCMC and variational approaches, there are many more. More recent areas of research include the development of algorithms for online inference and parallel inference.

Finally, there is scope for a more widespread application of nonparametric methods to practical problems. As our understanding of the properties of these models develops, it is important that they are applied thoughtfully. At the end, these priors represent rich and sophisticated modelling assumptions, and should be treated as such, rather than as convenient ways of bypassing the question "how many clusters?"

Acknowledgments

We thank Marina Meila and Ricardo Silva for reading and suggesting improvements to earlier versions of this chapter. We also thank Jan Gasthaus and Yee Whye Teh for helpful discussions.

References

Bigelow, J.L. and Dunson, D.B. 2009. Bayesian semiparametric joint models for functional predictors. *Journal of the American Statistical Association*, 104(485), 26–36.

Blackwell, D. 1973. Discreteness of Ferguson selections. *The Annals of Statistics*, 1(2), 356–358.

Blackwell, D. and MacQueen, J.B. 1973. Ferguson distributions via Pólya urn schemes. *Annals of Statistics*, 1, 353–355.

Blei, D.M. and Jordan, M.I. 2006. Variational inference for Dirichlet process mixtures. *Bayesian Analysis*, 1(1), 121–144.

Broderick, T., Jordan, M.I., and Pitman, J. 2012. Clusters and features from combinatorial stochastic processes. *pre-print* 1206.5862.

Brooks, S.P. and Gelman, A. 1998. General methods for monitoring convergence of iterative simulations. *Journal of Computational and Graphical Statistics*, 7(4), 434–455.

Casella, G. 1985. An introduction to empirical Bayes data analysis. *The American Statistician*, 39(2), 83–87.

Dahl, D.B. 2006. Model-based clustering for expression data via a Dirichlet process mixture model. In Do, K.-A., Müller, P., and Vannucci, M., Editors, *Bayesian Inference for Gene Expression and Proteomics*. Cambridge University Press, Cambridge.

De Blasi, P., Favaro, S., Lijoi, A., Mena, R.H., Prunster, I., and Ruggiero, M. 2013. Are Gibbs-type priors the most natural generalization of the Dirichlet process? DEM Working Papers Series 054, University of Pavia, Department of Economics and Management.

Escobar, M.D. and West, M. 1995. Bayesian density estimation and inference using mixtures. *Journal of the American Statistical Association*, 90, 577–588.

Ewens, W.J. 1972. The sampling theory of selectively neutral alleles. *Theoretical Population Biology*, 3, 87–112.

Ferguson, T.S. 1973. A Bayesian analysis of some nonparametric problems. *Annals of Statistics*, 1(2), 209–230.

Fraley, C. and Raftery, A.E. 1998. How many clusters? Which clustering method?—Answers via model-based cluster analysis. *Computer Journal*, 41, 578–588.

Ghosal, S. 2010. The Dirichlet process, related priors, and posterior asymptotics. In Hjort, N.L., Holmes, C., Müller, P., and Walker, S.G., Editors, *Bayesian Nonparametrics*. Cambridge University Press, Cambridge.

Griffin, J.E. 2011. Inference in infinite superpositions of non-Gaussian Ornstein–Uhlenbeck Processes using Bayesian nonparametic methods. *Journal of Financial Econometrics*, 9(3), 519–549.

Griffiths, T.L., Canini, K.R., Sanborn, A.N., and Navarro, D.J. 2007. Unifying rational models of categorization via the hierarchical Dirichlet process. In *Proceedings of the Annual Conference of the Cognitive Science Society*, V. 29.

Griffiths, T.L., Ghahramani, Z., and Sollich, P. 2007. Bayesian nonparametric latent feature models (with discussion and rejoinder). In *Bayesian Statistics*, V. 8.

Hjort, N.L. 1990. Nonparametric Bayes estimators based on Beta processes in models for life history data. *Annals of Statistics*, 18(3), 1259–1294.

Ishwaran, H. and James, L.F. 2001. Gibbs sampling methods for stick-breaking priors. *Journal of the American Statistical Association*, 96(453), 161–173.

Jain, S. and Neal, R.M. 2004. A split-merge Markov chain Monte Carlo procedure for the Dirichlet process mixture model. Technical report, Department of Statistics, University of Toronto.

James, L.F., Lijoi, A., and Pruenster, I. 2005. Bayesian inference via classes of normalized random measures. ICER Working Papers—Applied Mathematics Series 5-2005, ICER—International Centre for Economic Research.

Kallenberg, O. 2002. *Foundations of Modern Probability*. Probability and its Applications. Springer-Verlag, New York, Second Edition.

Kingman, J.F.C. 1967. Completely random measures. *Pacific Journal of Mathematics*, 21(1), 59–78.

Kingman, J.F.C. 1975. Random discrete distributions. *Journal of the Royal Statistical Society*, 37(1), 1–22.

Kurihara, K., Welling, M., and Teh, Y.W. 2007. Collapsed variational Dirichlet process mixture models. In *Proceedings of the International Joint Conference on Artificial Intelligence*, V. 20.

Kurihara, K., Welling, M., and Vlassis, N. 2007. Accelerated variational DP mixture models. In *Advances in Neural Information Processing Systems*, V. 19.

Liang, P., Jordan, M.I., and Taskar, B. 2007. A permutation-augmented sampler for Dirichlet process mixture models. In *Proceedings of the International Conference on Machine Learning*.

Lijoi, A. and Pruenster, I. 2010. Models beyond the Dirichlet process. In Hjort, N., Holmes, C., Müller, P., and Walker, S., Editors, *Bayesian Nonparametrics: Principles and Practice*. Cambridge Series in Statistical and Probabilistic Mathematics, No. 28, pp. 80–136, Cambridge University Press, Cambridge.

Lo, A.Y. 1984. On a class of Bayesian nonparametric estimates: I. density estimates. *Annals of Statistics*, 12(1), 351–357.

MacEachern, S. 1999. Dependent nonparametric processes. In *Proceedings of the Section on Bayesian Statistical Science*. American Statistical Association, Alexandria, Virginia.

Minka, T.P. and Ghahramani, Z. 2003. Expectation propagation for infinite mixtures. Presented at NIPS2003 Workshop on Nonparametric Bayesian Methods and Infinite Models.

Neal, R.M. 2000. Markov chain sampling methods for Dirichlet process mixture models. *Journal of Computational and Graphical Statistics*, 9, 249–265.

Orbanz, P. 2011. Projective limit random probabilities on Polish spaces. *Electronic Journal Statistics*, 5, 1354–1373.

Papaspiliopoulos, O. and Roberts, G.O. 2008. Retrospective Markov chain Monte Carlo methods for Dirichlet process hierarchical models. *Biometrika*, 95(1), 169–186.

Pitman, J. Combinatorial stochastic processes. Technical Report 621, Department of Statistics, University of California at Berkeley, 2002. Lecture notes for St. Flour Summer School.

Pitman, J. and Yor, M. 1997. The two-parameter Poisson-Dirichlet distribution derived from a stable subordinator. *Annals of Probability*, 25, 855–900.

Rao, V. and Teh, Y.W. 2009. Spatial normalized gamma processes. In *Advances in Neural Information Processing Systems*, vol. 22, MIT Press, Vancouver, Canada.

Rasmussen, C.E. 2000. The infinite Gaussian mixture model. In *Advances in Neural Information Processing Systems*, vol. 12, MIT Press, Vancouver, Canada.

Richardson, S. and Green, P.J. 1997. On Bayesian analysis of mixtures with an unknown number of components. *Journal of the Royal Statistical Society*, 59(4), 731–792.

Sethuraman, J. 1994. A constructive definition of Dirichlet priors. *Statistica Sinica*, 4, 639–650.

Sudderth, E.B. and Jordan, M.I. 2008. Shared segmentation of natural scenes using dependent Pitman-Yor processes. In *Advances in Neural Information Processing Systems*, Vol. 21, 1585–1592, MIT Press, Vancouver, Canada.

Teh, Y.W. 2006. A hierarchical Bayesian language model based on Pitman-Yor processes. In *Proceedings of the 21st International Conference on Comp. Linguistics and 44th Annual Meeting of the Association for Computational Linguistics*, 985–992, Sydney, Australia.

Teh, Y.W. 2010. Dirichlet processes. In *Encyclopedia of Machine Learning*. Springer, US.

Teh, Y.W. and Jordan, M.I. 2010. Hierarchical Bayesian nonparametric models with applications. In Hjort, N., Holmes, C., Müller, P., and Walker, S., Editors, *Bayesian Nonparametrics: Principles and Practice*. Cambridge University Press, Cambridge.

Teh, Y.W., Jordan, M.I., Beal, M.J., and Blei, D.M. 2006. Hierarchical Dirichlet processes. *Journal of the American Statistical Association*, 101(476), 1566–1581.

Walker, S.G. 2007. Sampling the Dirichlet mixture model with slices. *Communications in Statistics - Simulation and Computation*, 36, 45.

Wang, L. and Dunson, D.B. 2010. Fast Bayesian inference in Dirichlet process mixture models. *Journal of Computational and Graphical Statistics*, 20, 196–216.

Wood, F., Gasthaus, J., Archambeau, C., James, L., and Teh, Y.W. 2011. The sequence memoizer. *Communications of the Association for Computing Machines*, 54(2), 91–98.

Wood, F., Goldwater, S., and Black, M.J. 2006. A non-parametric Bayesian approach to spike sorting. In *Proceedings of the IEEE Conference on Engineering in Medicine and Biologicial Systems*, V. 28, IEEE publishing, Washington, DC.

Xing, E.P., Jordan, M.I., and Sharan, R. 2007. Bayesian haplotype inference via the Dirichlet process. *Journal of Computational Biology*, 14, 267–284.

11

Finite Mixtures of Structured Models

Marco Alfó and Sara Viviani

CONTENTS

Abstract

In this chapter, we describe developments in finite mixture models for structured data, with a particular focus on clustered, multilevel, longitudinal, and multivariate data. Literature on finite mixtures in generalized linear models is now quite extensive and entails application areas such as marketing (Wedel and DeSarbo 1995), biostatistics (Wang et al. 1996), econometrics (Deb and Trivedi 1997), machine-learning (Jacobs et al. 1991), just to mention a few. After a brief introduction, we will discuss some examples of the use of finite mixtures in heterogeneous generalized linear models, with a particular emphasis on model definition. We will also provide a brief review of available software and some suggestions on potential research areas.

11.1 Introduction

In statistics, several applications concern the relation between a response variable Y_i observed on $i = 1, \ldots, n$ units and a set of explanatory variables $x_i = (x_{i1}, \ldots, x_{ip})'$, e.g.

217

through a regression model for the conditional mean of Y_i

$$\mathbf{E}(Y_i \mid x_i) = x_i'\beta, \tag{11.1}$$

where $x_{i1} \equiv 1$ and β represents a p-dimensional vector of regression parameters. In applied contexts, the assumption that the regression coefficient vector is fixed across units may not be adequate, and taking into account the variability across units in model parameters can be of great practical importance. In such circumstances, we may assume that a part of the regression coefficient vector is fixed across observations, while the remaining may take individual-specific values, leading to the general regression model:

$$g(\mathbf{E}(Y_i \mid x_i, b_i)) = x_i'\beta + w_i'b_i. \tag{11.2}$$

Here, $g(\cdot)$ represents the link function and ensures that the expected value of the response is always well defined for any $\beta \in \mathbb{R}^p$. In linear regression, $g(\cdot)$ is the identity function. In the previous equation, w_i is a subset of x_i containing explanatory variables whose effects vary over individuals. The terms b_i, $i = 1, \ldots, n$ may be appropriately scaled to represent individual variation with respect to β, and to make estimation of fixed regression coefficients feasible. It could be of interest to analyze the reasons why the regression coefficients could vary; for example, in the case of cross-section data, the population may contain subgroups characterized by different model structures. When longitudinal observations are at hand, time may influence individual behavior and, therefore, the adopted model structure. In both cases, we may have outlying observations, omitted covariates, questionable assumptions about the response distribution and the link function; furthermore, the observed covariates may be measured with error. In all these cases, the standard homogeneous regression model in Equation 11.1 should be extended to provide reliable model parameter estimates. When defining the model structure, we should take into account all the information we have on the possible reasons for parameter heterogeneity.

While finite mixtures of structured models can be considered a reasonable choice in a model-based cluster analysis perspective, to provide a partition of the sample units according to a criterion of homogeneity of model parameters within clusters, their scope is broader. In fact, finite mixtures may represent an efficient tool to handle empirical situations where model parameters vary across sample units. Even when this variability is modeled by using a parametric, continuous distribution, as in parametric mixed effect models, the information upon the latent heterogeneity distribution is often limited and approximation is anyhow necessary to perform parameter estimation. In this context, common practice is to approximate, in a nonparametric fashion, the possibly continuous distribution by using a discrete distribution defined on a number of locations which is bounded, from above, by the number of units in the sample (indeed by the number of distinct units in the sample). The corresponding likelihood function turns out to be that one of a finite mixture model, where components can be seen as clusters of units characterized by homogeneous values of model parameters. Also in this case, as in the cluster analysis context, posterior probabilities of component membership can be used to allocate each unit to a component; assuming, as it is typical, that each component corresponds to a cluster, the resulting partition of the sample units can be an important by-product of the adopted model, and give further insight into potential causes of individual-specific heterogeneity. For all these reasons, we have decided to discuss the use of finite mixtures in the context of varying coefficient generalized linear models, highlighting the strong similarities between the clustering and the modeling perspectives and stressing that, even when the obtained partition is not the major product of the analysis, it is usually of great importance in describing the heterogeneous behavior of sample units, when compared to an alternative, homogeneous model.

The aim of this chapter is to reflect our experience in the field. In the following section, we will focus on finite mixtures of generalized linear regression models in a wide variety of cases where heterogeneity, measurement error and dependence may arise. Obviously, it cannot be intended as an exhaustive review of the whole amount of proposals appeared in the last years. Some relevant reviews in this field are McLachlan and Peel (2000), who give a comprehensive review of general-purpose finite mixture models, while Frühwirth-Schnatter (2006) provides a more focused discussion of finite mixtures of regression models. Böhning (2000) and Schlattmann (2009) discuss finite mixture models and their applications in the bio-medical sciences. Early references are Everitt and Hand (1981), and Titterington et al. (1981).

11.2 Existing Work

Even if the seminal work in survival analysis by Berkson and Gage (1952) contains a basic mixing of different distributions, the first attempt to explicitly consider varying coefficient models is probably due to Quandt (1958), where a two-state (single shift) regression model is introduced to handle the potential breakpoint in a consumption function, see also Quandt (1972), and Quandt and Ramsey (1978). This scheme is still far from a formal probability representation of the unobserved heterogeneity leading to individual variation in the regression model parameters. A similar approach with a constant regression parameter vector and unequal variances was previously proposed by Box and Tiao (1968); they extend the variance inflation model, with a formal treatment of uncertainty through a probability distribution on the two states. Abraham and Box (1978) discuss a location-shift structure to allow for outliers in a Gaussian regression model, where the switching entails only the intercept; this approach has been further extended by Guttman et al. (1978), where outliers are associated with individual-specific shifts, described by random effects. Similar approaches in the non-Gaussian case have been discussed, among others, by Pregibon (1981) and Verdinelli and Wasserman (1991). Finite mixtures of generalized linear models have experienced a flowering interest in the last 20 years. While earlier proposals were already present in the literature Heckman and Singer (1984); Hinde and Wood (1987); Jansen (1993); Wedel and DeSarbo (1995), the seminal papers by Aitkin (1996, 1999) have the important merit to establish a clear and transparent connection between mixed effect models and finite mixtures, by exploiting the theory of nonparametric maximum likelihood (hereafter NPML) estimation of a mixing distribution (Laird 1978).

11.3 Overdispersed Data

The problem of overdispersion in GLMs has been quite extensively treated. Given a standard exponential family with a specified variance/mean relationship, the observed response shows, conditional on observed information, a variance which is substantially greater than the one predicted by the mean; this may be represented by unusually high Pearson or deviance residuals, see for example McCullagh and Nelder (1989). To deal with this additional source of variation, random effects can be added to the linear predictor. In the following, we assume that the observed responses y_i, $i = 1, \ldots, n$ are realizations from a distribution in the exponential family with canonical parameter θ_i. We will adopt this quite general assumption throughout the chapter. Our interest is focused upon the parameter vector $\theta = (\theta_1, \ldots, \theta_n)'$, which is usually modeled by defining a generalized linear model for the analyzed response, as function of a set of p covariates $x_i = (x_{i1}, \ldots, x_{ip})'$

$$g(m(\theta_i)) = \beta_0 + \sum_{l=1}^{p} x_{il}\beta_l = x_i'\beta. \qquad (11.3)$$

Here, $m(\cdot)$ represents the mean function, and $\beta = (\beta_1, \ldots, \beta_p)'$ denotes a p-dimensional vector of fixed regression parameters. Failure of the adopted model to fit the data could be due to misspecification of any element defining the GLM: a simple way to unify these possibilities is through omitted variables, for a detailed discussion of this topic see Aitkin et al. (2005). According to this approach, we assume that some fundamental covariates were not considered in model specification; should these be orthogonal to the observed ones, we may summarize their joint effect by adding a set of unobserved, individual-specific variables $b_i, i = 1, \ldots, n$, to the linear predictor

$$g(m(\theta_i)) = x_i'\beta + b_i. \qquad (11.4)$$

The latent terms b_i appear additively in the model, due to the orthogonality condition mentioned above. This assumption can be easily relaxed by associating random parameters to some elements of the adopted covariates set. Let us denote by w_i the vector including the variables whose effects are assumed to vary across subjects. The random effect model in Equation 11.4 can be easily generalized to a *random coefficient* model

$$g(\theta_i) = x_i'\beta + w_i'b_i. \qquad (11.5)$$

A random coefficient can be associated to any of the p covariates: while the term β_l represents the overall effect of variable x_{il}, the random component (b_{il}) represents a mean-zero random deviation from β_l, $l = 1, \ldots, p$, due to the effect of omitted covariates. It is worth noticing that expression (11.5) reduces to expression (11.4) when $w_i \equiv 1$. Let $P(\cdot)$ denote the distribution of the random coefficients; treating the b_i's as nuisance parameters and integrating them out, we obtain for the (marginal) likelihood function the following expression

$$L(\Psi) = \prod_{i=1}^{n} \left\{ \int_{\mathcal{B}} f(y_i \mid x_i, b_i) dP(b_i) \right\}, \qquad (11.6)$$

where Ψ represents the global set of model parameters, and $\mathcal{B}$ is the support for $P(b_i)$. Several alternative parametric specifications have been proposed for the random terms; but only in specific cases the above integral has an analytical solution, leading to a well-known marginal distribution. In all other cases, approximate methods should be used to estimate the parameter vector β. The first possibility is to turn to numerical quadrature techniques, such as ordinary (Abramowitz and Stegun 1964; Press et al. 2007), or adaptive

Gaussian quadrature (AGQ). The latter represents an importance sampling version of the ordinary Gaussian quadrature rule, where the grid of the b_i domain is centered (and scaled) around the conditional mode $\hat{b}_i$ (Liu and Pierce 1994; Pinheiro and Bates 1995). A disadvantage of this approach lies in the required computational effort, which is exponentially increasing with the dimension of the random parameter vector, and needs the calculation of posterior modes and curvatures for each unit in the sample. A further alternative could be represented by pseudo-adaptive quadrature (Rizopoulos 2012); this technique significantly reduces computational complexity, and has been shown to perform reasonably well in a series of empirical applications (Viviani et al. 2014). Monte Carlo and simulated ML approaches are potential alternatives as well (Lerman and Manski 1981; McFadden 1989; Geyer and Thompson 1992; McCulloch 1994; Munkin and Trivedi 1999; Chib and Winkelmann 2001). Other alternatives are based on series expansion of the response or the random effect distribution as in the studies of Gurmu and Elder (1997) and Cameron and Johansonn (1997). The reader may also refer to penalized and/or marginal quasi-likelihood methods described, among others, by Green (1987) and Breslow and Clayton (1993). Parametric specifications of the mixing distribution can be, however, overly restrictive; as noted by Knorr-Held and Raßer (2000), a fully parametric specification of the random term distribution could result in oversmoothing and masking local discontinuities especially when the true distribution is characterized by a finite number of locations. Any parametric assumption on the random term distribution may not be tested adequately, raising doubts on the sensitivity of parameter estimates to such assumptions; for a discussion of sensitivity issues in this field, with a particular focus on standard error estimation see (Molenberghs and Verbeke 2005; Rizopoulos et al. 2008; McCulloch and Neuhaus 2011). A flexible specification of the mixing distribution is therefore to be preferred, as suggested by Heckman and Singer (1984). Their results are consistent with Brannas and Rosenqvist (1994), and Davidian and Gallant (1993). A detailed discussion of this topic can be found in Lindsay and Lesperance (1995). As proposed by Aitkin (1996, 1999), we may leave $P(\cdot)$ completely unspecified, drop restrictive distributional assumptions upon the b_i's and provide a nonparametric ML estimate for $P(\cdot)$. The problem of estimating the fixed parameter vector in presence of individual-specific random coefficients may be considered as a particular version of the nuisance parameter problem which dates back at least to Neymann and Scott (1948). They considered models of the form $f(y_i \mid \phi_i, \beta)$, and showed that maximum likelihood estimates of structural parameters β can be inconsistent or inefficient as $n \to \infty$. Kiefer and Wolfowitz (1956) suggested to model ϕ_i as i.i.d. random variables with a common, unknown, distribution function $P(\cdot)$; they proposed the method of maximum likelihood to estimate both β and $P(\cdot)$ and showed that, under mild conditions, the corresponding MLEs are strongly consistent. For the computation of the maximum-likelihood estimator, the results of Laird (1978) and Lindsay (1983a, b) are of great interest. In particular, the latter has shown that finding the MLE of $P(\cdot)$ involves a standard problem of convex optimization: maximizing a concave function over a convex set. As long as the likelihood is bounded, the MLE of $P(\cdot)$ is concentrated on a support of cardinality at most that of the number of distinct points in the analyzed sample, that is, for fixed β, the likelihood is maximized with respect to $P(\cdot)$ by at least one discrete distribution $\widehat{P}_K(\cdot)$ with $K \leq n$ support points. The maximum-likelihood estimator $\hat{\beta}$ can be calculated by maximizing the profile likelihood $L(\beta, \widehat{P}_K(\cdot))$ or, preferably, by including estimation of β into the maximization step of a general EM algorithm (Dempster et al. 1977). Since the NPML estimate of the random parameter/coefficient distribution is a discrete distribution on a finite number of locations, say K, the likelihood function can be expressed as

$$L(\Psi) = \prod_{i=1}^{n} \left\{ \sum_{k=1}^{K} f(y_i \mid x_i, \boldsymbol{b_k})\pi_k \right\} = \prod_{i=1}^{n} \left\{ \sum_{k=1}^{K} [f_{ik}\pi_k] \right\}, \tag{11.7}$$

which clearly resembles the likelihood function for a finite mixture model with K support points, locations $\boldsymbol{b_k}$ and associated masses π_k, $k = 1, \ldots, K$. Here, $\Psi = \{\beta, \boldsymbol{b_1}, \ldots, \boldsymbol{b_k}, \pi_1, \ldots, \pi_K\}$ is the global set of model parameters, and $f_{ik} = f(y_i \mid x_i, \boldsymbol{b_k})$ denotes the response distribution in the k-th component of the finite mixture, indexed by the canonical parameter

$$\theta_{ik} = x_i'\beta + w_i'\boldsymbol{b_k}. \tag{11.8}$$

As noticed by McCulloch and Neuhaus (2011), the use of a discrete distribution for the random coefficients has been criticized as unrealistic, see e.g., Magder and Zeger (1996), who proposed to fit a smooth version of the nonparametric mixing distribution. Verbeke and Lesaffre (1996) suggested to use mixtures of Gaussian distributions. All these proposals trade computational complexity for a more flexible and smooth distributional model for the random effects, and can be quite easily generalized to random coefficient models as well. While we understand that the NPML estimate of the mixing distribution may be less attractive than a parametric continuous mixing distribution, though consistent, our point of view is that the random coefficients represent nuisance parameters and the primary focus is on the fixed effect parameter vector. If one needs to have a more refined estimate of the mixing distribution (e.g., to predict individual values of random effects/coefficients), the use of a simple estimate as the one provided by a discrete distribution could not be attractive. However, the empirical findings discussed by Neuhaus and McCulloch (2011a, b) show that empirical Bayes estimates obtained by using a discrete distribution for the random coefficients are quite accurate. We may notice that the number of proposals based on finite mixtures have experienced an increasing impact in the literature as well as in the available software tools. The reason is probably that the use of finite mixtures has some advantages over parametric mixture models. First, since the locations and the corresponding probabilities are completely free to vary over the corresponding support, the proposed approach can readily accommodate extreme and/or strongly asymmetric departures from the basic, homogeneous, regression model. Secondly, the discrete nature of the estimate allows us to classify subjects in clusters characterized by homogeneous values of regression parameters, which can represent either the main objective of the analysis (in a cluster analysis perspective) or an important by-product of the primary model. From this perspective, rather than invoking an NPML estimate of the random coefficient distribution, we may simply think at the target population as composed by K different sub-populations (components), with component-specific densities from the exponential family, and canonical parameters that are modeled via a generalized linear model with component-specific parameter vectors. From a computational perspective, the EM algorithm is quite simple to implement, see Section 11.5.

While not directly connected to overdispersed data, measurement error may produce unobserved variation in some of the regression coefficients as well as a missfit for the regression model, see Carroll et al. (2006). In this area, we will briefly review the work by Aitkin and Rocci (2002).

Let y_i and w_i be the observed response and an explanatory variable measured with error for the i-th unit, $i = 1, \ldots, n$, with x_i being the expected, *true*, value for w_i. Modeling X_i

as a normal random variable leads to a nonidentifiable model, while working with non-Gaussian assumptions produces identified but analytically tough expression for the model likelihood. Aitkin and Rocci (2002) discuss an efficient estimation method for both Gaussian and non-Gaussian true covariates. Gaussian quadrature- and finite mixture-based approaches may be carried out efficiently using the finite mixture maximum-likelihood method described by Aitkin (1996). To obtain maximum likelihood estimates, the true scores x_i are treated as missing data, as in random effect models. Under Gaussianity for all the involved processes, and considering an additional, error free vector of observed covariates z_i, we may specify the modeling approach as follows:

$$Y_i \mid w_i, x_i, z_i \sim N(\alpha + x_i\beta + z_i'\gamma, \sigma^2)$$

$$W_i \mid x_i, z_i \sim N(x_i, \sigma_w^2), \quad X_i \mid z_i \sim N(\mu + z_i'\lambda, \sigma_x^2).$$

The distribution of z_i is not needed, as it is fully observed; for this reason, we suppress the conditioning on z_i. For the sake of simplicity, we may consider the transform

$$X_i^* = X_i - z_i'\lambda, \quad \gamma^* = \gamma + \beta\lambda,$$

so that, dropping the stars, we obtain

$$Y_i \mid w_i, x_i \sim N(\alpha + x_i\beta + z_i'\gamma, \sigma^2)$$

$$W_i \mid x_i \sim N(x_i + z_i'\lambda, \sigma_w^2), \quad X_i \sim N(\mu, \sigma_x^2).$$

In this case, by fixing σ_w^2, ML estimation can be obtained through an EM algorithm based on the following complete log-likelihood

$$\ell(\Psi) = \sum_{i=1}^{n}\left[-\frac{1}{2}\log(2\pi) - \log(\sigma) - \frac{1}{2\sigma^2}(y_i - \alpha - x_i\beta - z_i'\gamma)^2\right.$$

$$-\frac{1}{2}\log(2\pi) - \log(\sigma_w) - \frac{1}{2\sigma_w^2}\left(w_i - x_i - z_i'\lambda\right)^2$$

$$\left.-\frac{1}{2}\log(2\pi) - \log(\sigma_x) - \frac{1}{2\sigma_x^2}(x_i - \mu)^2\right].$$

In the case of generalized linear models, the assumption of Gaussian true scores leads to numerical integration of the likelihood over the true score distribution; the integral may be replaced by a finite sum over Gaussian quadrature locations X_k, with masses π_k, as in Equation 11.7. Given the observed data $(y_i, w_i, z_i), i = 1, \ldots, n$, the marginal likelihood

$$L(\Psi) = \prod_{i=1}^{n}\int f(y_i \mid x_i, \theta)h(w_i \mid x_i, \theta)p(x_i)dx_i \tag{11.9}$$

can be approximated by

$$L(\Psi) \approx \prod_{i=1}^{n}\sum_{k=1}^{K} f(y_i \mid x_k, \theta)h(w_i \mid x_k, \theta)\pi_k, \tag{11.10}$$

where x_k and π_k are Gaussian quadrature locations and masses. The maximization of (11.9) over the global parameter vector Ψ is equivalent to a finite mixture maximum-likelihood problem. The same procedure can be followed if $p(X)$ is not a Gaussian density. The mixture proportions π_k are estimated as $\sum_i \tau_{ik}/n$, where τ_{ik} represents the posterior probability that the i-th unit belongs to the k-th component, while x_k is estimated as a weighted average of residuals from the two regressions and the previous estimate. The interested reader is also referred to Rabe-Hesketh et al. (2003).

11.4 Clustered Data

The general model in Equation 11.2 can be used to model clustered data, as shown in Aitkin (1999). Let us consider the simple two-level variance component model, where the upper and the lower-level sampling units are indexed by $j = 1, \ldots, r$ and $i = 1, \ldots, n_j$, respectively. The objective is to represent the distribution of the response Y through explanatory variables x observed at both levels. The nested structure of the response y_{ij} induces an intra-class correlation between the responses corresponding to the same upper-level unit; the addition of an unobserved random effect to the linear predictor, for each unit in the same upper-level unit, can be a proper representation of this common variation. As pointed out in Section 11.3, the use of a discrete distribution for b_i through a finite mixture model could be a proper alternative. Hinde and Wood (1987) discuss the computational issues in the framework of two-level variance component models, by showing that both the locations and the masses could be simply estimated by EM-type algorithms. This approach has been extended to general random coefficient models by Aitkin (1999).

Let us denote by θ_{ij} and μ_{ij}, $i = 1, \ldots, n_j$, $j = 1, \ldots, r$ and $\sum n_j = n$, the canonical and mean parameters for Y_{ij}. The explanatory variables x_{ij} are related to μ_{ij} through the link function $g(\mu_{ij}) = g[m(\theta_{ij})] = x'_{ij}\beta$. If we consider upper-level explanatory variables only, then $x_{ij} = x_j$, that is, the upper-level vector x_j is replicated n_j times for the lower-level units. In a random coefficient model, we assume that a common, unobserved, random term b_j is shared by lower-level units within the j-th upper-level unit. Assuming conditional independence given the random coefficients, and using a finite mixture representation to avoid unverifiable parametric assumptions on the random coefficient distribution, the observed log-likelihood is given by

$$\ell(\Psi) = \sum_{j=1}^{r} \log \sum_{k=1}^{K} f_{jk}\pi_k, \tag{11.11}$$

where $f_{jk} = \prod_{i=1}^{n_j} f_{ijk}$, $f_{ijk} = f(y_{ij} \mid x_{ij}, b_k)$. As before, the terms π_k, $k = 1, \ldots, K$, denote proportions of the components in the mixture associated to the locations b_k. Also, in this case, EM-type algorithms are a popular choice.

11.4.1 Longitudinal Data

During the past few decades, longitudinal data have raised great interest in the statistical literature; they may be considered as a particular case of clustered data, where upper-level units have been recorded in a series of time occasions, representing (ordered) lower-level units. As with clustered data, the presence of association between repeated measures recorded on the same subject may be of concern. Let us start assuming that a response variable Y_{it} has been recorded on n individuals, $i = 1, \ldots, n$, at time occasions $t = 1, \ldots, T_i$ with a set of p explanatory variables, $x_{it} = (x_{it1}, \ldots, x_{itp})'$. To deal with such correlated outcomes, we may mimic the path followed in the clustered data case, and define a variance components generalized linear model. The observed responses are assumed to be independent, conditional on the individual-specific random coefficients b_i. Adopting a canonical link, the following regression model holds

$$\theta_{it} = x'_{it}\beta + w'_{it}b_i. \tag{11.12}$$

Individual-specific random coefficients b_i are *shared* by lower-level units (time occasions) within the same i-th upper-level unit (individual) and represent time-invariant sources of heterogeneity. The vector β represents a p-dimensional vector of fixed regression parameters, while the explanatory variables with varying effects are collected in the design vector $w_{it} = (w_{it1}, \ldots, w_{itq})'$. The distribution of b_i may be parametrically specified, usually $b_i \sim \text{MVN}_q(0, \Sigma)$, or may be left unspecified and indicated, without any loss of generality, by $P(b_i)$. Treating the b_i's as nuisance parameters and integrating them out, we obtain for the marginal likelihood the following expression

$$L(\Psi) = \prod_{i=1}^{n} \int \prod_{t=1}^{T_i} f(y_{it} \mid x_{it}, b_i) \mathrm{d}P(b_i), \tag{11.13}$$

where Ψ represents the vector of model parameters. For a Gaussian assumption on the random coefficients, numerical integration techniques are needed. The parametrization $b_i = \Sigma^{1/2} b_i^*$ is usually adopted, where $b_i^* \sim \text{MVN}_q(0, I)$. For the sake of simplicity, we drop the superscript * and write $b_i = b_i^*$ without any loss of generality. The regression model can be modified accordingly

$$\theta_{it} = x'_{it}\beta + w'_{it}\Sigma^{1/2}b_i = x'_{it}\beta + (w_{it} \otimes b_i)'\sigma, \tag{11.14}$$

where $\sigma = vec(\Sigma^{1/2})$. Using standard Gaussian Quadrature formulas, the likelihood function can be approximated by

$$L(\beta) = \prod_{i=1}^{n} \sum_{k=1}^{K} \prod_{t=1}^{T_i} f(y_{it} \mid x_{it}, b_k)\pi_k. \tag{11.15}$$

In the case of random effect models, $b_k = b_k$ represents one of the K standard quadrature points with associated mass equal to π_k. When the random coefficient dimension is $q > 1$, $b_k = (b_{k1}, \ldots, b_{ks})'$ is a quadrature vector with associated mass equal to $\pi_k = \prod_{j=1}^{s} \pi_{kj}$. In this case, $k = (k_1, \ldots, k_s)$ represents a multiple index with $k_j \in \{1, \ldots, K\}$, where K quadrature points are used in each dimension. Parameter estimation can be carried out by implementing a finite mixture EM algorithm, with fixed locations and masses. If a finite mixture representation is employed, estimation of the locations and corresponding masses can be achieved by following a path similar to the one described for overdispersed and/or clustered data. Although it has been widely used to model longitudinal responses, the variance component approach may not be the appropriate choice, since the corresponding intra-class correlation structure is too simple and does not allow for serial dependence. The basic structure may be extended by including in the linear predictor some form of *true* contagion, see Heckman (1981); such an approach may be based either on observables (1998), or on unobservables, see Maruotti and Ryden (2009), and Bartolucci et al. (2011). In the following, we restrict our attention to AR(1) models and consider a common time scale $t = 1, \ldots, T$ for each individual. In this case, independence holds conditional on the explanatory variables x_{it}, the past history summarized by $y_{i,t-1}$ and the random coefficient vector b_i. The heterogeneous AR(1) model can be stated as

$$\theta_{it} = x'_{it}\beta + w'_{it}b_i + \alpha h(y_{i,t-1}). \tag{11.16}$$

Model (11.16) may be considered an extension of transitional GLMs, see Brumback et al. (2000), to the analysis of longitudinal responses in the presence of unobserved heterogeneity; the transform $h(y_{i,t-1})$ is used to avoid potential inconsistencies in the linear predictor formulation (Zeger and Qaqish 1988). Using the properties of Markov chains, the joint distribution of the observed individual sequence is

$$f(y_i) = \int \prod_{t=2}^{T_i} f(y_{it} \mid y_{i,t-1}, b_i) f(y_{i1} \mid b_i) dG(b_i) \tag{11.17}$$

The term $f(y_{i1} \mid b_i)$ is not specified by the adopted modeling assumptions: hence, the joint distribution $f(y_i)$ is not determined and the full likelihood is unavailable. Several proposals have been introduced to overcome this problem. For a full treatment of this topic, the reader is referred to Skrondal and Rabe-Hesketh (2014). According to Aitkin and Alfò (1998), we may estimate the parameter vector by maximizing the likelihood conditional on the first observed outcome y_{i1}. The conditional density can be written as

$$f(y_{iT}, \ldots, y_{i2} \mid y_{i1}) = \int \prod_{t=2}^{T_i} f(y_{it} \mid y_{i,t-1}, b_i) dP(b_i \mid y_{i1})$$

with conditional log-likelihood given by

$$\ell(\Psi \mid y_{i1}) = \sum_{i=1}^{n} \log \left\{ \int \prod_{t=2}^{T_i} f(y_{it} \mid y_{i,t-1}, b_i) dP(b_i \mid y_{i1}) \right\}$$

Discarding the conditioning on y_{i1} would imply considering the first outcome as an exogeneous variable, and it may lead to endogeneity bias, see Fotouhi and Davies (1997) and Davidson and MacKinnon (1993). Following Follmann and Lambert (1989), Aitkin and Alfò (1998) propose to handle the conditioning through a simple linear location change in the b_i's induced by Y_{i1}, and discuss parameter estimation using a finite mixture EM algorithm. A similar approach is discussed in Wooldridge (2005).

However, in the case of short individual time series, this conditional approach may produce inefficient parameter estimates. Chan (2000) provides an alternative way to handle the problem, by specifying a parametric assumption for $f(y_{i1} \mid b_i)$ which is similar in spirit to what is usually done in Hidden Markov Models (MacDonald and Zucchini 1997), for the initial states distribution; in this way, the marginal distribution in Equation 11.17 can be handled and parameter estimates obtained by using standard algorithms. A further proposal in short longitudinal studies is discussed in Aitkin and Alfó (2003), who propose to jointly model the first and subsequent outcomes. The model structure is, for $t > 1$,

$$\theta_{it} = x'_{it}\beta + (w_{it} \otimes b_i)'\sigma + \alpha y_{i,t-1}, \tag{11.18}$$

while, for $t = 1$, a modified model is obtained by marginalizing over the unobserved Y_{i0}:

$$\theta_{i1} = x'_{i1}\beta_0 + (w_{i1} \otimes b_i)'\sigma_0. \tag{11.19}$$

In this case, β_0 and σ_0 represent the regression parameter vector and the vectorized covariance matrix of the random coefficients for the first occasion. The two submodels are connected by the common (but for a scale change) random coefficient structure. For parameter estimation, we may simply define a dummy variable d_{it}, $d_{it} = 0$ if $t = 1$, $d_{it} = 1$ otherwise, and consider its interaction with appropriate terms. Expressions (11.18) and (11.19) allow simultaneous modeling of the first and the subsequent outcomes. The model structure captures the difference between the first and the subsequent occasions and provides reliable estimates for the regression parameter vector. If a finite mixture representation is used, ML estimates can be obtained using a standard (random coefficient) EM algorithm. Therefore, no new computational tool is needed.

11.4.2 Multivariate Data

A further example of clustered data is given by genuine multivariate data; recently, these data have been the subject of an increasing number of proposals in the econometric literature, in particular for count data, see Jung and Winkelmann (1993), Munkin and Trivedi (1999), Chib and Winkelmann (2001), Karlis and Meligkotsidou (2003; 2007), and for general multivariate data, Jedidi et al. (1996) and Rabe-Hesketh et al. (2005). Chib and Winkelmann (2001) propose to model multivariate jointly determined counts using a set of Gaussian latent effects (in general coefficients); in this section, we describe how to employ a finite mixture approach to either relax unverifiable assumptions upon the latent effects distribution or to classify subjects in clusters characterized by homogeneous values of model parameters. The treatment is consistent with the proposal of Karlis and Meligkotsidou (2003), who suggest to use finite mixtures of multivariate Poisson's distributions to allow for general covariance structures between counts. Nevertheless, this argument may be extended to other random variables in the exponential family as well. We assume that the analyzed sample is composed by n individuals: responses y_{ij} and possibly outcome-specific p_j-dimensional covariate vectors $x'_{ij} = (x_{ij1}, \ldots, x_{ijp_j})$ are recorded for $i = 1, \ldots, n$ units and $j = 1, \ldots, s$ outcomes. Adopting the usual notation for multivariate data, let $y_i = (y_{i1}, \ldots, y_{is})'$ denote the vector of observed reponses for the i-th unit, $i = 1, \ldots, n$. When we face multivariate outcomes, the univariate approach we have discussed so far needs to be extended. In a sense, multivariate data can be seen as a particular case of clustered data, where outcomes are nested within individuals, but they may be defined on a different scale since they may not refer to the same outcome. Omitted covariates may affect more than one outcome, leading to outcome-specific unobserved heterogeneity; on the modeling side, the association among observed outcomes and the nature of the stochastic dependence among the analyzed phenomena could be of interest. When responses are correlated, the univariate approach is less efficient than the multivariate one: the latter takes into account of potential zero restrictions on outcome-specific parameters, see Zellner

(1962), Davidson and Mackinnon (1993), for a discussion of this topic. A standard approach to account for profile-specific heterogeneity and for dependence among outcomes is to assume they share common (but unobservable) features. These features may be represented through a convolution-type scheme, Karlis (2002), Karlis and Meligkotsidou (2003), and/or latent effects, Munkin and Trivedi (1999), and Chib and Winkelmann (2001). The first approach has essentially been adopted in the multivariate count data case: the association arises since observed counts are convolutions of unobservable Poisson-distributed random variables with at least one common term, Y_0. Overdispersion is introduced marginally for the single variable level, but only non-negative correlations are allowed. In the latent effect case, the standard approach is, as discussed by Alfó and Trovato (2004), to define s conditional univariate models connected through a common structure, represented by a set of (possibly outcome-specific) random coefficients, which account for heterogeneity between subjects and dependence among outcomes at the marginal level. In this context, a first rational choice is to apply directly, as they are, random coefficient models developed for the longitudinal case to the multivariate context. This implies the assumption that the Y_{ij}'s represent conditionally independent responses, given a set of subject-specific random coefficients, say b_i, which are supposed to be common to all outcomes. These models are sometimes referred to as one-factor, as in Winkelmann (2000) or uni-factor, as in Munkin and Trivedi (1999), models. The multivariate regression model can be specified as

$$\theta_{ij} = x'_{ij}\beta_j + w'_{ij}b_i, \tag{11.20}$$

where, as before, a canonical link has been adopted. The log-likelihood function is obtained by integrating out the random coefficient vector

$$\ell(\Psi) = \sum_{i=1}^{n} \log\left\{ \int_{\mathcal{B}} f(\boldsymbol{y}_i \mid X_i, \boldsymbol{b}_i)\mathrm{d}P(\boldsymbol{b}_i) \right\} = \sum_{i=1}^{n} \log\left\{ \int_{\mathcal{B}} \prod_{j=1}^{s} f(y_{ij} \mid x_{ij}, \boldsymbol{b}_i)\mathrm{d}P(\boldsymbol{b}_i) \right\} \tag{11.21}$$

according to the standard hypothesis of conditional independence. Also in this case, we may use a particular parametric distribution, for example multivariate Gaussian, for the random coefficients b_i or use a discrete distribution on a set b_k, $k = 1, \dots, K$ of locations with corresponding masses π_k. In this case, the previous integral becomes

$$\ell(\Psi) = \sum_{i=1}^{n} \log\left\{ \sum_{k=1}^{K} f(\boldsymbol{y}_i \mid X_i, \boldsymbol{b}_k)\pi_k \right\}. \tag{11.22}$$

Parameter estimation can be carried out extending the multilevel (Goldstein 1995) model described for the clustered data case. We pool data from each outcome into a single two-level model, where each outcome is treated as a lower-level unit nested within subjects, which represent upper-level units. Parameter estimates for the outcome-specific regression parameters β_j are obtained by interacting the indicator variables d_{ij}, $d_{ij} = 1$ if we model the j-th outcome and $d_{ij} = 0$ otherwise, and the covariates vector x_{ij}. The locations b_k are estimated by introducing in the linear predictor the interaction between a component factor with K levels and the design vector w_{ij}. The same EM algorithm for the clustered data case applies. While simple to be implemented, the previous approach lacks generality: the same heterogeneity sources affect all outcomes, but for a scale change. A direct implication of this assumption, as noted by Winkelman (2000), is that we consider only non-negative correlations between the outcomes. A straightforward generalization can be based on so-called *correlated* random coefficients. Parametric examples of such an approach are provided, in the multivariate count data context by Munkin and Trivedi (1999), Chib and Winkelmann (2001), and van Ophem (1999). To explain the fundamental features of this modeling approach, let us denote by $\underline{b}_i = (b_{i1}, \ldots, b_{is})$ a set of unit- and outcome-specific random coefficients. The basic hypothesis is that the Y_{ij} represent independent random variables conditional on the set b_{ij} which vary over outcomes and account for individual-specific unobserved heterogeneity and dependence among outcomes. These models are referred to as multi-factor models, Winkelman (2000). The model could be further generalized by separating sources of unobserved heterogeneity at the univariate levels and sources of dependence among outcomes as in the convolution model introduced for spatial data by Molliè (1996). However, in this case, we should choose which of the two sources has to be represented parametrically, since a full nonparametric specification seems to be unfeasible. Now, given these assumptions, the generic regression model can be written as follows:

$$\theta_{ij} = x'_{ij}\beta_j + w'_{ij}b_{ij}, \tag{11.23}$$

where, as mentioned before, b_{ij} represent unit- and outcome-specific heterogeneity in the regression parameters. The expression for the log-likelihood function follows:

$$\ell(\Psi) = \sum_{i=1}^{n} \log \left\{ \int_B \left[\prod_{j=1}^{s} f(y_{ij} \mid x_{ij}, b_{ij}) \right] dP(\underline{b}_i) \right\}. \tag{11.24}$$

This multiple integral cannot be evaluated in closed form, even if simplifications are possible for certain choices of random coefficient covariance structure. To deal with the general case, we should turn to numerical integration or simulation methods (Leyland et al. 2000; Chib and Winkelmann 2001; Gueorguieva 2001), or use a finite mixture representation. In this last case, we can rewrite the log-likelihood function as follows:

$$\ell(\Psi) = \sum_{i=1}^{n} \log \left\{ \sum_{k=1}^{K} f(y_{ij} \mid x_{ij}, b_{jk}) \pi_k \right\}, \tag{11.25}$$

where $\pi_k = Pr(\underline{b}_k) = Pr(b_{1k}, \ldots, b_{sk})$, $k = 1, \ldots, K$ represents the joint probability of locations $\underline{b}_k$. For purpose of estimation, the previous model can be summarized adopting a standard multilevel notation. Locations $\underline{b}_k$ and corresponding masses π_k can be estimated by introducing in the linear predictor the interaction between the design vector z_{ij}, the K level factor and the indicator variables d_{ij}. The same EM algorithm defined for the univariate case can be used, but with modified weights.

11.4.3 Further Issues

As it has been remarked above, the treatment in this chapter cannot be considered exhaustive and several interesting proposals have not found their space in the text. Nevertheless, while space is limited, we would like to mention a few additional issues from both modeling and theoretical perspectives. Several authors have proposed to model prior probabilities of component membership by using a set of covariates, see e.g., Dayton and MacReady (1988); in this case, let us begin by denoting by $\zeta_i = (\zeta_{i1}, \ldots, \zeta_{iK})$ the unobservable vector of component-indicator variables, where

$$\zeta_{ik} = \begin{cases} 1 & \text{if unit } i \text{ comes from component } k \\ 0 & \text{otherwise} \end{cases}.$$

$k = 1, \ldots, K$. These are independent multinomial random variables; corresponding probabilities can be modeled as:

$$\pi_{k|z} = \frac{\exp(\phi_{0k} + z'_i \phi_k)}{\sum \exp(\phi_{0k} + z'_i \phi_k)},$$

where, to ensure identifiability, $\phi_{0K} = 0$, and $\phi_K = 0$. Concomitant variable regression models taking this form have been proposed by Kamakura et al. (1994) and Wang et al. (1998), by using a nominal logistic regression model for priors. Recently, ordered logit representation has been used by Roy (2003) and, in a mixed HMM context, by Maruotti and Rocci . The former discusses, in particular, a pattern mixture model for longitudinal data subject to potentially informative dropout, where the latent components are ordered classes of dropout with probabilities depending on the dropout time, used as an exogeneous covariate. That is, increases in the dropout time are assumed to progressively increase the probability of being in the last dropout classes. Pattern mixture models of this kind have been discussed, among others, by Wu and Bailey (1989), Wu and Carroll (1988), and Alfó and Atkin (2000). A mixed effect model for Gaussian longitudinal responses and an event process in continuos time have been described by Lin et al. (2002). A random coefficient-based dropout selection model for longitudinal binary data has been proposed in a finite mixture framework by Alfó and Maruotti (2009), extended by Tsonaka et al. (2009) to nonmonotone missingness, and by Tsonaka et al. (2010) to marginalized models for longitudinal categorical responses. Further extensions of random coefficient generalized linear models have been also proposed in the context of three-way data, Vermunt (2007), and multivariate categorical responses with potentially nonignorable missingness, Formann (2007); Rigby and Stasinopoulos (2005) introduced generalized additive models for location, scale, and shape (GAMLSS), while Ingrassia et al. (2014) have discussed simultaneous clustering of responses and covariates in a so-called clusterwise regression context.

11.5 ML Estimation

In this section, we briefly sketch maximum-likelihood estimation of model parameters for the multivariate regression model with correlated random coefficient discussed in the previous section. Estimation through the EM algorithm for all other cases that have been mentioned can be simply derived by this general specification. We briefly review some questions related to inference in such models; namely, the general structure of the EM algorithm, the strategy for efficient initialization of the algorithm, the methods to choose the number of components in the finite mixture, the provision of standard errors, and the issue of identifiability.

11.5.1 The EM Algorithm

In this section, we briefly propose a general EM algorithm for maximum-likelihood estimation of model parameters in the finite mixture multivariate regression models discussed above. The univariate algorithm is well known and has been presented in quite a general form by Aitkin (1996) and Wang et al. (1996). Let us consider a multivariate response variable $(Y_{i1}, \ldots, Y_{is})$ with observed values $y_i = (y_{i1}, \ldots, y_{is})'$, $j = 1, \ldots, s$, $i = 1, \ldots, n$. As discussed before, we use a finite mixture representation where the number of components, K, is unknown and should be estimated along with other model parameters. To describe the algorithm in its general form, let us start assuming that each unit belongs to one of K distinct components, each characterized by a possibly different parameter vector. We denote by π_k the prior probability that a generic unit belongs to the k-th component of the mixture. The algorithm is run for fixed K and reaches a solution which can be successively used to estimate model parameters as the number of components is increased to $K + 1$. Several authors have discussed algorithms for joint estimation of K and model parameters, such as VEM or VDM (Böhning 2000); a computationally feasible and efficient solution is to update estimates for fixed K improving step by step as in EMGFU (Böhning 2003). We follow the latter approach and make the assumptions that:

1. responses Y_{ij}, $i = 1, \ldots, n$; $j = 1, \ldots, J$ are conditional on the random coefficients b_{ij} independent random variables with density in the exponential family and canonical parameter θ_{ij};

2. for each unit, the unobserved multivariate random variable $\underline{b}_i = (b_{i1}, \ldots, b_{is})'$ has a discrete distribution on K support points with associated masses π_k, where $\pi_k = Pr(\underline{b}_i = \underline{b}_k)$ and $\sum_{k=1}^{K} \pi_k = 1$.

Let us denote by $\zeta_i = (\zeta_{i1}, \ldots, \zeta_{iK})$ the unobservable vector of component-indicator variables. Should these indicator variables be known, this problem would reduce to a simple multivariate regression model with group-specific parameter vectors. Since component memberships are unobservable, they can be considered as missing data. The complete data are $(y_i, \underline{x}_i, \zeta_i)$. Using a multinomial distribution for the unobservable vector ζ_i, the log-likelihood for the complete data can be written as

$$\ell_c(\cdot) = \sum_{i=1}^{n} \sum_{k=1}^{K} \zeta_{ik} \{\log(\pi_k) + \log f_{ik}\}, \tag{11.26}$$

where $f_{ik} = \prod_{j=1}^{s} f(y_{ij} \mid x_{ij}, b_{kj})$ according to (local) conditional independence and $\pi_k = Pr(\underline{b}_k)$. As usual, within the E-step, we define the log-likelihood for *observed* data by taking the expectation of the log-likelihood for *complete* data over the unobservable component indicator vector ζ_i conditional on the observed data y_i and the current ML parameter estimates, say $\Psi^{(r)}$. Roughly speaking, we replace ζ_{ik} by its conditional expectation:

$$\hat{\zeta}_{ik}(\Psi^{(r)}) = \tau_{ik}^{(r)} = \frac{\pi_k^{(r)} \prod_{j=1}^{s} f(y_{ij} \mid x_{ij}, \underline{b}_k^{(r)})}{\sum_{l=1}^{K} \pi_l^{(r)} \left\{ \prod_{j=1}^{s} f(y_{ij} \mid x_{ij}, \underline{b}_l^{(r)}) \right\}}, \tag{11.27}$$

where $\hat{\zeta}_{ik}(\Psi^{(r)}) = \tau_{ik}^{(r)}$ is the posterior probability that the i-th unit belongs to the k-th component of the finite mixture. The conditional expectation of the complete log-likelihood given the vector of the observed responses y_i and the current parameter estimates is

$$Q(\Psi \mid \Psi^{(r)}) = E_{\Psi^{(r)}}\{\ell_c(\Psi) \mid y_i\} = \sum_{i=1}^{n}\sum_{k=1}^{K} \tau_{ik}^{(r)}\{\log(\pi_k^{(r)}) + \log f_{ik}^{(r)}\}. \tag{11.28}$$

Maximizing $Q(\Psi \mid \Psi^{(r)})$ with respect to Ψ, we obtain the ML parameter estimates $\hat{\Psi}^{(r+1)}$ based on the posterior probabilities $\hat{\tau}_{ik}^{(r)} = \hat{\zeta}_{ik}(\Psi^{(r)})$. The estimated parameters are the solution of the following M-step equations

$$\frac{\partial Q}{\partial \pi_k} = \sum_{i=1}^{n}\left\{\frac{\tau_{ik}}{\hat{\pi}_k} - \frac{\tau_{iK}}{\hat{\pi}_K}\right\} = 0, \tag{11.29}$$

$$\frac{\partial Q}{\partial \Psi} = \sum_{i=1}^{n}\frac{\partial}{\partial \Psi}\sum_{k=1}^{K}\tau_{ik}\log(f_{ik}), \tag{11.30}$$

To obtain updated estimates of the unconditional probability π_k, we solve Equation 11.29 and obtain

$$\hat{\pi}_k^{(r)} = \sum_{i=1}^{n}\frac{\tau_{ik}^{(r)}}{n}, \tag{11.31}$$

which represents a well-known result from ML in finite mixtures. Since closed form solutions of Equation 11.30 are usually unavailable, a standard Newton–Raphson algorithm can be used to obtain regression parameter estimates. The E- and M-steps are alternated repeatedly until the relative difference

$$\frac{\ell^{(r+1)} - \ell^{(r)}}{|\ell^{(r)}|} < \epsilon, \quad \epsilon > 0 \tag{11.32}$$

changes by an arbitrarily small amount if the adopted criterium is based on the sequence of likelihood values $\ell^{(r)}$, $r = 1, \dots$. Since $\ell^{(r+1)} \geq \ell^{(r)}$, convergence is obtained with a sequence of likelihood values which are bounded from above. As it is often remarked, results obtained using the EM algorithm tend to be quite sensitive to the choice of starting values. To avoid local maxima and to study the sensitivity of the parameters estimates to different initializations, we may adopt simple or complex strategies. Among the former, we may start by perturbing the standard Gaussian quadrature locations using either a constant scale parameter (Alfó and Trovato 2004), or a series of varying scale coefficients drawn from an appropriate distribution (e.g., uniform). A further alternative is to run a short-length CEM (Classification EM) algorithm as suggested by Biernacki et al. (2003). It is simple, performs well in a lot of situations regardless of the particular form of the mixing distribution, and seems little sensitive to noisy data. In any case, empirical experience seems to suggest that a few different starting values are enough to obtain a refined solution for fixed K; among those fitted for a given K, the model with the best maximized log-likelihood value is retained.

11.5.2 The Choice of the Number of Components

Once the algorithm has reached its end for a given K, we can proceed to increase the number of components to $K + 1$ and estimate corresponding model parameters. After that, a formal comparison should be done using either penalized likelihood criteria or a formal likelihood ratio test. In the first case, the choice of the critierium is largely subjective, even if a quite general result due to Keribin (2000) shows that the BIC (Schwarz 1978), consistently estimates the *true* number of components, while the AIC (Akaike 1978), in the limit does not underestimate it, see also Leroux (1992). However, as shown in a different setting by Karlis and Meligkotsidou (2007), the AIC may produce a more refined estimate of the mixing distribution when compared to BIC, CAIC (Bozdogan 1987), or ICL (Biernacki et al. 2000). If a formal test is to be employed, the K value could be selected by bootstrapping the LRT statistic for different number of components as proposed by McLachlan (1987), since standard regularity conditions for the asymptotic distribution of the LRT statistic do not hold. To illustrate, let us suppose we wish to test the null hypothesis $H_0 : K = K_1$ versus the alternative hypothesis $H_1 : K = K_2$, $K_1 < K_2$. A bootstrap sample is generated from the mixture density $\sum_{k=1}^{K_1} f(y_i | \widehat{\Psi}_k)\widehat{\pi}_k$, where, $\widehat{\Psi}_k = \Psi(\widehat{\beta}, \underline{\widehat{b}}_k, \widehat{\pi}_k | K = K_1)$. The value of $-2\log(\xi)$ is computed for the bootstrap sample after fitting mixture models for $K = K_1$ and $K = K_2$. This process is repeated a number of times B, and the replicated values are used to assess the null distribution of $-2\log(\xi)$. The choice of the number of components can be done by comparing the observed value $-2\log(\widehat{\xi})$ to the estimate of the $(1 - \alpha)$th quantile of the simulated null distribution $(\widehat{\xi}_{1-\alpha})$, where α is the chosen size. Obviously, this procedure is substantially influenced by the observed sample size.

11.5.3 The Provision of Standard Errors

The EM algorithm does not provide a direct evaluation of the standard errors for parameter estimates, since it is not based on the calculation of either the observed or the expected information matrix. In this section, we discuss some possible approaches for standard error estimation. Louis (1982) derives a procedure to extract the observed information matrix when an EM algorithm is adopted. The key idea of this work is to build up the observed information matrix by using the complete-data gradient vector and Hessian matrix, rather than the corresponding quantities associated with the incomplete data likelihood. According to Ordchard and Woodbury (1972), the observed information matrix is written as the difference between two matrices corresponding to the complete information, which is usually known, and the missing information. The approach of Louis (1982) presents some computational issues; to overcome these difficulties, Oakes (1999) developed an explicit formula for the second-order derivative matrix of the log-likelihood based on the first derivative of the conditional expectation of the score of the complete data log-likelihood, given the observed data

$$H(\Psi) = -\frac{\partial^2 Q(\Psi | \Psi^{(r)})}{\partial \Psi \Psi'}\bigg|_{\Psi^{(r)} = \Psi} + \frac{\partial^2 Q(\Psi | \Psi^{(r)})}{\partial \Psi^{(r)} \Psi^{(r)'}}\bigg|_{\Psi^{(r)} = \Psi}. \tag{11.33}$$

In this identity, the first part of the sum is the second derivative of the conditional expected value of the complete data log-likelihood given the observed data, and it is simple to obtain when an EM algorithm is used; the second part is the second derivative of the same expected log-likelihood with respect to the current values of the parameters. The Oakes's identity has seen to behave well in different model frameworks, such as hidden Markov models, Bartolucci et al. (2011). The observed information matrix can also be approximated by calculating the second-order partial derivatives or the gradient of the log-likelihood function through numerical differentiation (1989). Jamshidian and Jennrich (2000) discuss various issues concerning approximation of the observed information matrix by numerical differentiation. A further alternative is to use a boostrap approach to standard error approximation, see e.g. McLachlan and Basford (2008). A parametric bootstrap scheme similar to the one in the previous section can be used to approximate the covariance matrix of the parameter estimates through the sample covariance matrix:

$$cov(\widehat{\Psi}) \simeq \sum_{b=1}^{B} (\widehat{\Psi}_b - \overline{\widehat{\Psi}})(\widehat{\Psi}_b - \overline{\widehat{\Psi}})'/(B-1),$$

where $\overline{\widehat{\Psi}} = \sum_{b=1}^{B} \widehat{\Psi}_b / B$. Usually, based on Efron and Tibshirani (1993), 50–100 boostrap replications are sufficient for this purpose, even if the behaviour of this estimate clearly depends on the shape of the log-likelihood function.

11.5.4 Identifiability

As a prerequisite for a well-behaved estimation, we should assume that the mixture is identifiable, that is, that two sets of parameters which do not agree after permutation cannot yield the same mixture distribution. Despite the close relationship between these two model classes, generic identifiability of linear regression models does not, in general, overlap with identifiability of Gaussian mixtures, as claimed by DeSarbo and Cron (1988). Wang et al. (1996) point out that a sufficient condition for identifiability in the context of Poisson regression models is that the covariate matrix X is full rank. However, this is rather a necessary condition, since, as pointed out by Hennig (2000), nonidentifiability may occur also when the full rank condition is fulfilled. He showed that the regression parameters are identifiable if the number of components K is lower than the number of distinct $(p-1)$-dimensional hyperplanes spanned by the covariates (but the intercept). This result is closely related to a similar result in mixed effect logistic regression due to Follmann and Lambert (1989), where it is claimed that the number of identifiable components is not higher than a function of the number of distinct $(p-1)$-dimensional points in the sample. Roughly speaking, nonidentifiability may occur when several categorical explanatory variables are used. Grün and Leisch (2004) extend the results of Hennig (2000) to finite mixtures of multinomial logit models with mixed effects. For a thoughtful discussion on this point see (Frühwirth-Schnatter 2006).

11.6 Software

When talking about available software implementations for fitting finite mixtures of generalized linear models, we should say that any software platform with basic programming features can be used to write an EM code. However, as any user-written code, it could contain errors, may be incomplete or may not offer as elegant features as the ones available in pre-reviewed libraries. So, we may give a few suggestions on programs that include routines which can be used to this purpose. Starting from the R environment, we cannot forget the library NPMLREG by Einbeck et al. (2007) which is, probably, the more direct descendant of the original approach to model overdispersed or clustered data via random effects, with either parametric Gaussian or nonparametric distributions. The package, however, allows to fit random effect models only, since the random slopes are defined to be equal to the correponding random intercept, but for a scale change. A powerful set of functions to fit even very complex structures, with concomitant variables, general random coefficient structure and many other features, is FLEXMIX (Grün and Leisch 2007). A further library that contains some programs to fit finite mixtures of (flexible) regression models is GAMLSS (Stasinopoulos and Rigby 2007); the library has been designed for a broader aim than simply fitting finite mixture of regression models, but contains some tools to do it. SAS does contain some procedures that may be used to analyze overdispersed, clustered or lngitudinal data; among others, we may cite proc NLMIXED for fitting random coefficient models with parametric distribution, and proc FMM for fitting finite mixtures of simple generalized linear models. STATA program GLLAMM (Rabe-Hesketh et al. 2003) can be used to fit complex latent variable models, including random coefficient models for overdispersed, clustered, longitudinal and multivarate data via ordinary/adaptive Gaussian quadrature and finite mixtures. FMM can be used to fit finite mixtures of generalized linear models with concomitant variables (Deb 2007). MPLUS (Muthén and Muthén 2008) and Latent GOLD (Vermunt and Magidson 2005) are powerful programs that may be used for fitting latent class and finite mixture models, with a view toward structural equation models.

11.7 Conclusion

While the theory and practical implementation of finite mixture regression models is now quite widespread, still there are questions that remain open, at least in our perspective. First, the choice of the number of components; while, in some applications, it may not be a concern, it may play a major role if the interest is on predicting the individual-specific random effect (coefficient) values. In this context, interesting perspectives have been discussed in recent research on the effects of misspecifying the random coefficient distribution (McCulloch and Neuhaus 2011a, b). Whatever the criterion to choose the number of components, a question arises on whether the maximum-lkelihood estimate of the mixing distribution should be calculated by looking at the log-likelihood values only, or by using penalization-based criteria. This is somewhat a philosophical question and the answer lies on whether we are looking for the ML estimate of a possibily continuous mixing distribution (as when we invoke the NPML theory), or just try to approximate the unknown $P(\cdot)$ with a discrete distribution, and consider a clustering perspective. This point still deserves further research. When defining random coefficient models for multivariate responses, we should bear in mind that random coefficients account for the potential overdispersion in the univariate profiles and the association between the profiles; the implicit unidimensionality of the latent coefficient distribution, where all the random terms are constrained to share a common distribution with the same number of components in each profile, may represent an unnecessary limitation of the method. Unlike parametric mixing approaches, where ovedispersion in the univariate profiles and association between profiles are kept separate, the approach based on finite mixtures suffers from some rigidity in this respect. Multivariate-mixed responses are a further challenge for this reason.

References

Abraham, B. and Box, G.E.P. 1978. Liear models and spurious observations. *Journal of the Royal Statistical Society, Series C*, 27, 131–138.

Abramowitz, M. and Stegun, I. 1964. *Handbook of Mathematical Functions*. Washington, DC: National Bureau of Standards.

Aitkin, M. 1996. A general maximum likelihood analysis of overdispersion in generalized linear models. *Statistics and Computing*, 6, 127–130.

Aitkin, M. 1999. A general maximum likelihood analysis of variance components in generalized linear models. *Biometrics*, 55, 117–128.

Aitkin, M.A. and Alfó, M. 2003. Random effect ar models for longitudinal binary responses. *Statistical Modelling*, 3, 291–303.

Aitkin, M.A., Francis, B., and Hinde, J. 2005. *Statistical Modelling in GLIM*. Oxford: Oxford University Press, Second Edition.

Aitkin, M. and Rocci, R. 2002. A general maximum likelihood analysis of measurement error in generalized linear models. *Annals of Mathematical Statistics*, 12, 163–174.

Akaike, H. 1973. Information theory as an extension of the maximum likelihood principle. In Petrov, B.N. and Csaki, F., Editors, *Second International Symposium on Information Theory*. Budapest: Akademiai Kiado.

Alfó, M. and Aitkin, M. 1998. Regression models for binary longitudinal responses. *Statistics and Computing*, 8, 289–307.

Alfó, M. and Aitkin, M. 2000. Random coefficient models for binary longitudinal responses with attrition. *Statistics and Computing*, 10, 275–283.

Alfó, M. and Maruotti, A. 2009. A selection model for longitudinal binary responses subject to non-ignorable attrition. *Statistics in Medicine*, 28, 2435–2450.

Alfó, M. and Trovato, G. 2004. Semiparametric mixture models for multivariate count data, with application. *Econometrics Journal*, 7, 426–454.

Bartolucci, F., Bacci, S., and Pennoni, F. 2011. Mixture latent autoregressive models for longitudinal data. *arXiv:1108.1498*.

Bartolucci, F., Farcomeni, A., and Pennoni, F. 2011. A note on the application of the oakes' identity to obtain the observed information matrix in hidden markov models. *Statistics and Computing*, 21(2), 275–288.

Berkson, J. and Gage, R.P. 1952. Survival curve for cancer patients following treatment. *Journal of the American Satistical Association*, 47, 501–515.

Biernacki, C., Celeux, G., and Govaert, G. 2000. Assessing a mixture model for clustering with the integrated completed likelihood. *IEEE Transactions on Pattern Analysis and Machine Learning*, 22, 719–725.

Biernacki, C., Celeux, G., and Govaert, G. 2003. Choosing starting values for the em algorithm for getting the highest likelihood in multivariate gaussian mixture models. *Computational Statistics and Data Analysis*, 41, 561–575.

Böhning, D. 2000. *Computer-Assisted Analysis of Mixtures and Applications: Meta-Analysis, Disease Mapping and Others*. New York: Chapman & Hall/CRC.

Böhning, D. 2003. The em algorithm with gradient function update for discrete mixtures with known (fixed) number of components. *Statistics and Computing*, 13, 257–265.

Box, G.E.P. and Tiao, G.C. 1968. A Bayesian approach to some outlier problems. *Biometrika*, 55, 119–129.

Bozdogan, H. 1987. Model selection and akaike's information criterion (aic): The general theory and its analytical extensions. *Psychometrika*, 52, 345–370.

Brannas, K. and Rosenqvist, G. 1994. Semiparametric estimation of heterogeneous count data models. *European Journal of Operational Research*, 76, 247–258.

Breslow, N.E. and Clayton, D.G. 1993. Approximate inference in generalized linear mixed models. *Journal of the American Statistical Association*, 88, 9–25.

Brian Leroux. 1992. Consistent estimation of a mixing distribution. *Ann. Stat.*, 20, 1350–1360.

Brumback, B.A., Ryan, L.M., Schwartz, J.D., Neas, L.M., Stark, P.C., and Burge, H.A. 2000. Transitional regression models, with application to environmental time series. *Journal of the American Statistical Association*, 95, 16–27.

Cameron, A.C. and Johansson, P. 1997. Count data regression using series expansion: With applications. *Journal of Applied Econometrics*, 12, 203–233.

Carroll, R.J. Ruppert, D., Stefanski, L.A., and Crainiceanu, C. 2006. *Measurement Error in Nonlinear Models: A Modern Perspective*. Boca Raton, FL: Chapman and Hall/CRC.

Chan, J.S.K. 2000. Initial stage problem in autoregressive binary regression. *The Statistician*, 49, 495–502.

Chib, S. and Winkelmann, R. 2001. Markov chain monte carlo analysis of correlated count data. *Journal of Business and Economic Statistics*, 19, 428–435.

Davidian, M. and Gallant, A.R. 1993. The nonlinear mixed effects model with a smooth random effects density. *Biometrika*, 80, 475–488.

Davidson, R. and Mackinnon, J.G. 1993. *Estimation and Inference in Econometrics*. Oxford: Oxford University Press.

Dayton, C.M. and MacReady, G.B. 1988. Concomitant-variable latent-class models. *Journal of the American Statistical Association*, 83, 173–178.

Deb, P. 2007. Fmm: Stata module to estimate finite mixture models. Statistical Software Components, Boston College Department of Economics.

Deb, P. and Trivedi, P.K. 1997. Demand for medical care by the elderly: A finite mixture approach. *Journal of Applied Econometrics*, 12, 313–336.

Dempster, A.P., Laird, N.M. and Rubin, D.B. 1977. Maximum likelihood from incomplete data via the em algorithm. *Journal of the Royal Statistical Society, Series B*, 39, 1–38.

DeSarbo, W.S. and Cron, W.L. 1988. A maximum likelihood methodology for clusterwise regression. *Journal of Classification*, 5, 249–282.

Efron, B. and Tibshirani R.J. 1993. *An Introduction to the Bootstrap*. New York: Chapman & Hall.

Einbeck, J., ad Hinde, J., and Darnell, R. 2007. A new package for fitting random effect models. *R News*, 7, 26–30.

Everitt, B.S. and Hand D.J. 1981. *Finite Mixture Distributions*. London: Chapman & Hall.

Follmann, D.A. and Lambert, D. 1989. Generalizing logistic regression by nonparametric mixing. *Journal of the American Statistical Association*, 84, 295–300.

Formann, A.K. 2007. Mixture analysis of multivariate categorical data with covariates and missing entries. *Computational Statistics and Data Analysis*, 51, 5236–5246.

Fotouhi, A.R. and Davies, R.B. 1997. Modelling repeated durations: the initial conditions problem. In Forcina, A., Marchetti, G.M., Hatzinger, R., and Galmacci, G., Editors, *Statistical Modelling (Proceedings of the 11th International Workshop on Statistical Modelling;* Orvieto, Italy, 15–19 July 1996). Città di Castello: Graphos.

Frühwirth-Schnatter, S. 2006. *Finite Mixture and Markov Switching Models*. New York: Springer.

Geyer, C.J. and Thompson, E.A. 1992. Constrained monte carlo maximum likelihood for dependent data. *Journal of the Royal Statistical Society B*, 54, 657–699.

Goldstein, H. 1995. *Multilevel Statistical Models*. London: Edward Arnold.

Green, P.J. 1987. Penalized likelihood for general semi-parametric regression models. *International Statistical Review*, 55, 245–260.

Grün, B. and Leisch, F. 2004. Bootstrapping finite mixture models. In Jaromir Antoch, Editor, *Compstat 2004—Proceedings in Computational Statistics*, pages 1115–1122. Heidelberg: Physica Verlag.

Grün, B. and Leisch, F. 2007. FlexMix: An R package for finite mixture modelling. *R News*, 7, 8–13.

Gueorguieva, R. 2001. A multivariate generalized linear mixed model for joint modelling of clustered outcomes in the exponential family. *Statistical Modelling*, 1, 177–193.

Gurmu, S. and Elder, J. 1997. A simple bivariate count data regression model. *Technometrics*, 20, 187–194.

Guttman, I., Dutter, R., and Freeman, P. 1978. Care and handling of univariate outliers in the general linear model to detect spuriosity—a Bayesian approach. *Technometrics*, 20, 187–194.

Heckman, J. 1981. Heterogeneity and state dependence. In Sherwin Rosen, Editor, *Studies in Labor Markets*. Chicago: University of Chicago Press.

Heckman, J. and Singer, B. 1984. A method for minimizing the impact of distributional assumptions in econometric models for duration data. *Econometrica*, 52, 271–320.

Hennig, C. 2000. Identifiability of models for clusterwise linear regression. *Journal of Classification*, 17, 273–296.

Hinde, J.P. and Wood, A.T.A. 1987. Binomial variance component models with a non-parametric assumption concerning random effects. In Crouchley, R., Editor, *Longitudinal Data Analysis*. Aldershot, Hants: Avebury.

Ingrassia, S., Minotti, S.C., and Punzo, A. 2014. Model-based clustering via linear cluster-weighted models. *Computational Statistics and Data Analysis*, 71, 159–182.

Jacobs, R.A. Jordan, M.I. Nowln, S.J., and Hinton, G.E. 1991. Adaptive mixtures of local experts. *Neural Computation*, 3, 79–87.

Jamshidian, M. and Jennrich, R.I. 2000. Standard errors for em algorithm. *Journal of he Royal Statistical Society, Series B*, 62, 257–270.

Jansen, R.C. 1993. Maximum likelihood in a generalized linear finite mixture model by using the em algorithm. *Biometrics*, 49, 227–231.

Jedidi, K., Ramaswamyb, V., DeSarbo, W.S., and Wedel, R. 1996. On estimating finite mixtures of multivariate regression and simultaneous equation models. *Structural Equation Modeling: A Multidisciplinary Journal*, 3, 266–289.

Jung, R.C. and Winkelmann, R. 1993. Two aspects of labor mobility: A bivariate poisson regression approach. *Empirical Economics*, 18, 543–556.

Kamakura, W.A. Wedel, M., and Agrawal, J. 1994. Concomitant variable latent class models for conjoint analysis. *International Journal for Research in Marketing*, 11, 451–464.

Karlis, D. 2002. An em algorithm for multivariate poisson distribution and related models. *Journal of Applied Statistics*, 30, 63–77.

Karlis, D. and Meligkotsidou, L. 2003. Model based clustering for multivariate count data. *Proceedings of the 18th International Workshop on Statistical Modelling*, Leuven, Belgium.

Karlis, D. and Meligkotsidou, L. 2007. Finite multivariate poisson mixtures with applications. *Journal of Statistical Planning and Inference*, 137, 1942–1960.

Keribin, C. 2000. Consistent estimation of the order of mixture models. *Sankhyā: Indian Journal of Statistics*, 62, 49–66.

Kiefer, J. and Wolfowitz, J. 1956. Consistency of the maximum likelihood estimator in presence of infinitely many incidental parameters. *Annals of Mathematical statistics*, 27, 886–906.

Knorr-Held, L. and Raßer, S. 2000. Bayesian detection of clusters and discontinuities in disease maps. *Biometrics*, 56, 13–21.

Laird, N.M. 1978. Nonparametric maximum likelihood estimation of a mixing distribution. *Journal of the American Statistical Association*, 73, 805–811.

Lerman, S. and Manski, C. 1981. On the use of simulated frequencies to approximate choice probabilities. In Manski, C. and McFadden, D., Editors, *Structural Analysis of Discrete Data with Econometric Applications*. MIT Press.

Leyland, A., Langford, I., Rasbash, J., and Goldstein, H. 2000. Multivariate spatial models for event data. *Statistics in Medicine*, 19, 2469–2478.

Lin, H.Q. Turnbull, B.W. McCulloch, C.E., and Slate, E.H. 2002. Latent class models for joint analysis of longitudinal biomarker and event process data. *Journal of the American Statistical Association*, 97, 53–65.

Lindsay, B.G. 1983. The geometry of mixture likelihoods: a general theory. *Annals of Statistics*, 11, 86–94.

Lindsay, B.G. 1983. The geometry of mixture likelihoods, part ii: The exponential family. *Annals of Statistics*, 11, 783–792.

Lindsay, B.G. and Lesperance, M.L. 1995. A review of semiparametric mixture models. *Journal of Statistical Planning and Inference*, 47, 29–39.

Liu, Q. and Pierce, D.A. 1994. A note on gaussian-hermite quadrature. *Biometrika*, 81, 624–629.

Louis, T.A. 1982. Finding the observed information matrix when using the em algorithm. *Journal of the Royal Statistical Society, B*, 44, 226–233.

MacDonald, I.L. and Zucchini, W. 1997. *Hidden Markov and Oter Models for Discrete-Valued Time Series*. London: Chapman & Hall.

Magder, L.S. and Zeger, S.L. 1996. A smooth nonparametric estimate of a mixing distribution using mixtures of gaussians. *Journal of the American Statistical Association*, 91, 1141–1151.

Maruotti, A. and Rocci, R. 2012. A mixed non-homogeneous hidden markov model for categorical data, with application to alcohol consumption. *Statistics in Medicine*, 31, 871–886.

Maruotti, A. and Rydén, T. 2009. A semiparametric approach to hidden markov models under longitudinal observations. *Statistics and Computing*, 19, 381–393.

McCullagh, P. and Nelder, J. 1989. *Generalized Linear Models*. London: Chapman & Hall, 2nd Edition.

McCulloch, C.E. 1994. Maximum likelihood estimation of variance components for binary data. *Journal of the American Statistical Association*, 89, 330–335.

McCulloch, C.E. and Neuhaus, J.M. 2011. Misspecifying the shape of a random effects distribution: Why getting it wrong may not matter. *Statistical Science*, 26, 388–402.

McCulloch, C.E. and Neuhaus, J.M. 2011. Prediction of random effects in linear and generalized linear models under model misspecification. *Biometrics*, 67, 270–279.

McFadden, D. 1989. A method of simulated moments for estimation of discrete response models without numerical integration. *Econometrica*, 57, 995–1026.

McLachlan, G.J. 1987. On bootstrapping the likelihood ratio test statistic for the number of components in a normal mixture. *Journal of the Royal Statistical Society C*, 36, 318–324.

McLachlan, G.J. and Basfor, T. 2008. *The EM Algorithm and Extensions*. New York: John Wiley and Sons.

McLachlan, G.J. and Peel, D. 2000. *Finite Mixture Models*. New York: John Wiley and Sons.

Meilijson, I. 1989. A fast improvement to the em algorithm on its own terms. *Journal of the Royal Statistical Society, series B*, 51, 127–138.

Molenberghs, G. and Verbeke, G. 2005. *Models for Discrete Longitudinal Data*. New York: Springer.

Molliè, A. 1996. Bayesian mapping of disease. In Gilks, W.R. Richardson, S., and Spiegelhalter, D.J. Editors, *Markov Chain Monte Carlo in Practice*. London: Chapman & Hall.

Munkin, M.K. and Trivedi, P.K. 1999. Simulated maximum likelihood estimation of multivariate mixed-poisson regression models, with application. *The Econometrics Journal*, 2, 29–48.

Muthén, B. and Muthén L. 2008. *Mplus User's Guide*. LosAngeles: Muthén and Muthén.

Neymann, J. and Scott, B. 1948. Consistent estimates based on partially consistent observations. *Econometrika*, 16, 1–32.

Oakes, D. 1999. Direct calculation of the information matrix via the em algorithm. *Journal of the Royal Statistical Society, series B*, 61, 479–482.

Ophem van, H. 1999. A general method to estimate correlated discrete random variables. *Econometric Theory*, 15, 228–237.

Ordchard, T. and Woodbury, M.A. 1972. In Proceedings of the Sixth Berkeley Symposium on Mathematical Statistics and Volume 1: Theory of Statistics Probability, Editors, *A Missing Information Principle: Theory and Applications*. University of California Press.

Pinheiro, J. and Bates, D.M. 1995. Approximations to the log-likelihood function in the nonlinear mixed-effects model. *Journal of Computational and Graphical Statistics*, 4, 12–35.

Pregibon, D. 1981. Logistic regression diagnostics. *Annals of Statistics*, 9, 705–724.

Press, W.H. Teukolsky, S.A. Vetterling, W.T., and Flannery, B.P. 2007. *Numerical Recipes: The Art of Scientific Computing*. New York: Cambridge University Press.

Quandt, R.E. 1958. The estimation of the parameters of a linear regression system obeying two separate regimes. *Journal of the American Statistical Association*, 53, 873–880.

Quandt, R.E. 1972. A new approach to estimating switching regressions. *Journal of the American Statistical Association*, 67, 306–310.

Quandt, R.E. and Ramsey, J.B. 1978. Estimating mixtures of normal distributions and switching regressions (with discussion). *Journal of the American Statistical Association*, 73, 730–752.

Rabe-Hesket, S., Skrondal, A., and Pickles, A. 2005. Maximum likelihood estimation of limited and discrete dependent variable models with nested random effects. *Journal of Econometrics*, 128, 301–323.

Rabe-Hesketh, S., Pickles, A., and Skrondal, A. 2003. Correcting for covariate measurement error in logistic regression using nonparametric maximum likelihood estimation. *Statistical Modelling*, 3, 215–232.

Rigby, R.A. and Stasinopoulos, D.M. 2005. Generalized additive models for location, scale and shape. *Journal of the Royal Statistical Society, Series C, Applied Statistics*, 54, 507–554.

Rizopoulos, D. 2012. Fast fitting of joint models for longitudinal and event time data using a pseudo-adaptive gaussian quadrature rule. *Computational Statistics and Data Analysis*, 56, 491–501.

Rizopoulos, D., Molenberghs, G., and Verbeke, G. 2008. Shared parameter models under random-effects misspecification. *Biometrika*, 95, 63–74.

Roy, J. 2003. Modelling longitudinal data with nonignorable dropouts using a latent dropout class model. *Biometrics*, 59, 829–836.

Schlattmann, P. 2009. *Medical Applications of Finite Mixture Models*. Berlin: Springer.

Schwarz, G. 1978. Estimating the dimension of a model. *The Annals of Statistics*, 6, 461–464.

Skrondal, A. and Rabe-Hesket, S. 2014. Handling initial conditions and endogenous covariates in dynamic/transition models for binary data with unobserved heterogeneity. *Journal of the Royal Statistical Society, Series C, Applied Statistics*, 63, 211–237.

Stasinopoulos, D.M. and Rigby, R.A. 2007. Generalized additive models for location scale and shape (gamlss) in r. *Journal of Statistical Software*, 23(7), 1–46.

Titterington, D.M., Smith, A.F.M., and Makov, U.E. 1981. *Finite Mixture Distributions*. New York: John Wiley and Sons.

Tsonaka, R., Rizopoulos, D., Verbecke, G., and Lesaffre, E. 2010. Nonignorable models for intermittently missing categorical longitudinal responses. *Biometrics*, 66, 834–844.

Tsonaka, R., Verbecke, G., and Lesaffre, E. 2009. A semi-parametric shared parameter model to handle nonmonotone nonignorable missingness. *Biometrics*, 65, 81–87.

Verbecke, G. and Lesaffre, E. 1996. A linear mixed-effects model with heterogeneity in the random-effects population. *Journal of the American Statistical Association*, 91, 217–221.

Verdinelli, I. and Wasserman, L. 1991. Bayesian analysis of outlier problems using the gibbs sampler. *Statistics and Computing*, 1, 105–117.

Vermunt, J.K. 2007. A hierarchical mixture model for clustering three-way data sets. *Computational Statistics and Data Analysis*, 51, 5368–5376.

Vermunt, J.K. and Magidson, J. 2005. *Latent GOLD 4.0 User's Guide*. Belmont, MA: Statistical Innovations Inc.

Viviani, S., Alfó, M., and Rizopoulos, D. 2014. Generalized linear mixed joint model for longitudinal and survival outcomes. *Statistics and Computing*, 24, 417–427.

Wang, P., Cockburn, I., and Puterman, M.L. 1998. Analysis of patent data a mixed poisson regression model approach. *Journal of Business and Economic Statistics*, 16, 17–41.

Wang, P., Puterman, M.L. Cockburn, I., and Le, N. 1996. Mixed poisson regression models with covariate dependent rates. *Biometrics*, 52, 381–400.

Wedel, M. and DeSarbo, W.S. 1995. A mixture likelihood approch for generalized linear models. *Journal of Classification*, 12, 21–55.

Winkelmann, R. 2000. *Econometric Analysis of Count Data*. Berlin: Springer-Verlag.

Winkelmann, R. 2000. Seemengly unrelated negative binomial regression. *Oxford Bullettin of Economics and Statistics*, 62, 553–560.

Wooldridge, J.M. 2005. Simple solutions to the initial conditions problem in dynamic, nonlinear panel data models with unobserved heterogeneity. *Journal of Applied Econometrics*, 20, 39–54.

Wu, M.C. and Bailey, K.R. 1989. Estimation and comparison of changes in the presence of informative right censoring: Conditional linear model. *Biometrics*, 45, 939–955.

Wu, M.C. and Carroll, R.J. 1988. Estimation and comparison of changes in the presence of informative right censoring by modeling the censoring process. *Biometrics*, 44, 175–188.

Zeger, S.L. and Qaqish, B. 1988. Markov regression models for time series: A quasilikelihood approach. *Biometrics*, 44, 1019–1031.

Zellner, A. 1962. Estimators for seemingly unrelated regression equations: Some exact finite sample results. *Journal of the American Statistical Association*, 58, 977–992.

12

Time-Series Clustering

Jorge Caiado, Elizabeth Ann Maharaj, and Pierpaolo D'Urso

CONTENTS

Abstract

The literature on time-series clustering methods has increased considerably over the last two decades with a wide range of applications in many different fields, including geology, environmental sciences, finance, economics, and biomedical sciences. Clustering of time series can technically be put into three groups, viz., methods based on the actual observations, on features derived for the time series and on parameters estimates of fitted models. Selections of work from each of these three groups are presented together with some examples and applications.

12.1 Introduction

The general problem of time-series clustering is concerned with the separation of a set of time-series data into groups, or clusters, with the property that series in the same group have a similar structure and series in other groups are quite distinct. A fundamental problem in clustering and classification analysis is the choice of a relevant metric. Some studies use nonparametric approaches for splitting a set of time series into clusters by looking to the Euclidean distance in the space of points. As it is known already, the classical unweighted Euclidean distance between raw time series is not a good metric for classifying time series since it is invariant to permutation of the coordinates, and hence, we are not using the information about the dynamic dependence structure. Then, the Euclidean metric requires the time series being compared to be of the same length (Liao 2005).

Standard methods for clustering analysis use a distance or quasi-distance matrix between any two possible groups or clusters as an input. For-time series clustering, each series $x_i = (x_{i1}, x_{i2}, \ldots, x_{iT})$ for $i = 1, 2, \ldots, M$ is considered a point in the space R^T, where T is the length of the series x_i. A key problem is to decide whether two time series should be in the same group. It is known that looking for vectors that are close in the Euclidean metric is not a good idea.

12.2 Existing Work

Various algorithms that are relevant for static data, or other especially developed algorithms have been used to cluster time-series data. Some existing traditional and fuzzy clustering methods are presented. Traditional clustering methods assign time series to mutually exclusive groups, while fuzzy clustering methods did not assign time series exclusively to only one cluster, but allow the time series to different clusters with various membership degrees. The various methods are based on parameters estimates of models fitted to the time series, or on the actual observations, on features derived for the time series (Liao 2005).

12.2.1 Model-Based Clustering

Model-based methods for clustering time series are among the earliest works on this topic. For these methods, it is assumed that a set of times series generated from the same model would most likely have similar patterns. The time series are clustered by means of parameter estimates or by means of the residuals of the fitted models.

Piccolo (1990) was one of the first authors to consider model-based clustering of time series. For every pair of time series under consideration, he fitted autoregressive integrated moving averages (ARIMA) models and used the Euclidean distance between their corresponding autoregressive expansions as the metric. Then, he applied hierarchical clustering. Tong and Dabas (1990) fitted linear or nonlinear models to their time series under consideration and then clustered the residuals of the time series using both multidimensional scaling and hierarchical methods. The main idea of the classification method of Tong and Dabas is that if two residual series have a high degree of affinity, then models generating them must also have a high degree of affinity. As in the Piccolo's metric, a limitation of the residual-based metrics of Tong and Dabas is the need for *ad hoc* modelling of time series.

Maharaj (1999, 2000) extended the model-based clustering idea of Piccolo (1990) by testing for significant differences between underlying models of each pair of time series and used the *p-values* of each test to cluster the time series. Time series, which are assumed to be stationary or could be rendered stationary by differencing, are fitted with AR(k) models. For the null hypothesis,

$$H_0 : \pi_x = \pi_y,$$

where π_x and π_y are vectors of AR(k) parameters associated with each for two series, a Wald test statistic is derived. Two series are grouped together if the associated p-value of the test is greater than some prespecified level of significance. Using simulated series, the clustering results were assessed by means of a measure of discrepancy that was defined as the difference between the true number of cluster and the actual number of corrected cluster generated. Maharaj (1999) extended this approach to multivariate time series by using the estimated parameters of vector autoregressive models.

Otranto (2008) used Wald tests and the autoregressive metrics to measure the distance between the generalized autoregressive conditional heteroskedasticity (GARCH) processes and developed a clustering algorithm, which can provide three types of classifications with increasing degree of depth based on the heteroscedastic patterns of the time series. Otranto (2010) proposed the detection of groups of homogeneous time series in terms of correlation dynamics for a widely used model in financial econometrics, viz., the dynamic conditional correlation (DCC) model. The clustering algorithm is based on the distance between dynamic conditional correlations and the classical Wald test, to compare the coefficients of two groups of dynamic conditional correlations.

Kalpakis et al. (2001) investigated the clustering of time series generated from ARIMA models using the cepstrals coefficients derived from AR(k) models fitted to the series as the clustering variables. It is noted that the cepstrum of a time series is the spectrum of the logarithm of the spectrum, while the spectrum or spectral density function of a stationary time series is the Fourier transform of the autocovariance function of the time series (Childers et al. 1975). Kalpakis et al. (2001) used the k-medoids method with various similarity measures for the purpose of clustering the time series.

Boets et al. (2005) proposed a time-series clustering approach based on a cepstral distance between stochastic models. Savvides et al. (2008) considered an approximate nonparametric form of the cepstrum and use the derived cepstral coefficients to cluster biological time series. Since as the number of coefficients increases the cepstrum converges in mean square to the theoretical cepstrum (Hart, 1997), it is sufficient to choose only a small number of cepstral coefficients to describe the second-order characteristics of a time series.

Ramoni et al. (2002) proposed a Bayesian algorithm for clustering discrete-valued time series in an attempt to discover the most probable set of generating processes. The task of clustering is regarded as a Bayesian model selection problem. The similarity between two estimated Markov chain transition matrices was measured as an average of the symmetrized Kullback–Liebler distance between corresponding rows in the matrices. They also presented a Bayesian clustering algorithm for multivariate time series.

Xiong and Yeung (2002) proposed an approach for clustering of time series using mixtures of autoregressive moving average (ARMA) models. They derived an expectation-maximization (EM) algorithm for learning the mixing coefficients as well as the parameters of the component models. The model selection problem was addressed using the Bayesian information criterion (BIC) to determine the number of clusters in the data.

Caiado and Crato (2010) introduced a volatility-based metric for cluster analysis of stock returns using the information about the estimated parameters in the threshold GARCH (or TGARCH) equation. They computed a Mahalanobis-like distance between the stock returns that takes into account the information about the stochastic dynamic structure of the time series volatilities and allows for unequal length time series. Clusters are formed by looking to the hierarchical structure tree (or dendrogram) and the computed principal coordinates. They employ these techniques to investigate the similarities between the "blue-chip" stocks used to compute the Dow Jones industrial average (DJIA) index.

12.2.2 Observation-Based Clustering

A "direct" approach for clustering time series can be based on a comparison of the observed time series or a suitable transformation of the observed time series. This clustering approach is particularly useful when the (univariate or multivariate) time series are not very long. Various distance measures based on the time observations or their suitable transformations have been proposed in the literature (Carlier 1986; D'Urso 2000; Liao 2005; Jeong et al. 2011).

In a multivariate framework, some distances based on the time observations (position) and/or the so-called velocity (slope) and/or acceleration (concavity/convexity) of the time series have been proposed by D'Urso (2000). By denoting two generic multivariate time series by $\mathbf{x}_{it} = (x_{i1t}, \ldots, x_{ijt}, \ldots, x_{iJt})'$ and $\mathbf{x}_{lt} = (x_{l1t}, \ldots, x_{ljt}, \ldots, x_{lJt})'$—e.g., x_{ijt} represents the j-th variable observed in the i-th unit at time t—a simple distance measure is

$$d_{OBS}(x_i, x_l) = \sqrt{\sum_{t=1}^{T} \|\mathbf{x}_{it} - \mathbf{x}_{lt}\|^2 w_t} \tag{12.1}$$

where w_t is a suitable weight at time t.

Distance measures based on suitable transformations of the observed time series, i.e., on the slope (velocity), concavity/convexity (acceleration), that have been proposed by D'Urso (2000) are as follows:

$$d_{VEL}(x_i, x_l) = \sqrt{\sum_{t=2}^{T} \|\mathbf{v}_{it} - \mathbf{v}_{lt}\|^2 w_t} \tag{12.2}$$

$$d_{ACC}(x_i, x_l) = \sqrt{\sum_{t=3}^{T} \|\mathbf{a}_{it} - \mathbf{a}_{lt}\|^2 \tilde{w}_t} \tag{12.3}$$

where w_t and $\tilde{w}_t$ are suitable weights, respectively, for the time intervals $[t-1, t]$ and $[t-2, t]$; $\mathbf{v}_{it} = (\mathbf{x}_{it} - \mathbf{x}_{it-1})$ and $\mathbf{v}_{lt} = (\mathbf{x}_{lt} - \mathbf{x}_{lt-1})$ are the *velocities* of i-th and l-th time series; $\mathbf{a}_{it} = 1/2(\mathbf{v}_{it} - \mathbf{v}_{it-1})$ and $\mathbf{a}_{lt} = 1/2(\mathbf{v}_{lt} - \mathbf{v}_{lt-1})$ are the *accelerations* of i-th and l-th time series. In $\mathbb{R}^2$, the velocity of each segment of the time series is the *slope* of the straight line passing through it: if the velocity is negative (positive) the slope will be negative (positive) and the angle made by each segment of the time series with the positive direction of the t-axis will be obtuse (acute); the acceleration of each pair of segments of time series represents its convexity or concavity: if the acceleration is positive (negative), the trajectory of the two segments is convex (concave).

The distances (12.1)–(12.3) can be utilized separately in a clustering procedure, e.g., in a hierarchical clustering. In order to simultaneously take into account in the time series clustering process, the features captured by the distances (12.1)–(12.3), different mixed distance measures, based on suitable combinations of the distances (12.1)–(12.3), have been proposed (D'Urso 2000; Coppi and D'Urso 2001). To this purpose, different criteria have been suggested for objectively computing the weights of the linear combination (Coppi and D'Urso 2001).

For a lagged time versions of the above-mentioned distances, see D'Urso and De Giovanni (2008). Furthermore, Coppi and D'Urso (2000) have extended the distances (12.1)–(12.3) and the mixed version for comparing fuzzy time series.

A distance similar to (12.1) (in the sense that it captures the same kind of information) is based on the computation, e.g., for univariate time series, of the area between each pair of time series (D'Urso 2000).

D'Urso (2000) proposed a distance measure for clustering observed time series based on the following *polygonal representation* of a (univariate) time series:

$$x_{it'} = \sum_{t=1}^{T} b_{it}|t' - t| \qquad t' = 1, T \tag{12.4}$$

where $b_{it}(i = 1, I; t = 1, T)$ are the *polygonal coefficients*, obtained by solving (12.4); i.e., setting $A = x_{i1} + x_{iT}/T - 1$, we have $b_{i1} = 1/2(x_{i2} - x_{i1} + A)$, $b_{it} = 1/2[(x_{it+1} - x_{it}) - (x_{it} - x_{it-1})](t = 1, T - 2)$ and $b_{it} = 1/2(A - x_{it} + x_{iT-1})$. The polygonal coefficients b_{it} represent the semi-difference of the velocities (acceleration) of the (univariate) time series in the different time intervals; i.e., they indicate the "weights" of some time circumstances that determinate the vertices of the polygonal line representing each time series. Moreover, the coefficients b_{it} supply indications of the "oscillations" of each time series. For instance, in a time series, there is an oscillation if for a set of consecutive values $x_{it-1}, x_{it}, x_{it+1}, x_{it-1} < x_{it}$ and $x_{it} > x_{it+1}$ or $x_{it-1} > x_{it}$ and $x_{it} < x_{it+1}$; then, the existence of vertices in the polygonal line (which depends on the non-negative values of the polygonal coefficients b_{i1}, b_{it+1} and b_{iT}) is the necessary condition for the existence of the oscillations, and the oscillations of the time series depend on polygonal coefficients.

Thus, a distance for (multivariate) observed time series based on the polygonal coefficients of each component of the multivariate time series is:

$$d_{POL}(x_i, x_l) = \sqrt{\sum_{t=1}^{T} \|\mathbf{b}_{it} - \mathbf{b}_{lt}\|^2} \tag{12.5}$$

where $\mathbf{b}_{it}, \mathbf{b}_{lt}$ represent the J-dimensional polygonal coefficients vectors of the i-th and l-th multivariate time series. Notice that $\mathbf{b}_{it} = \mathbf{a}_{it+1}$.

In addition to the above-mentioned distance measures, there are various distances for time series based on the Dynamic Time Warping (DTW) which is a well-known technique for finding an optimal alignment between two given (time-dependent) sequences under certain restrictions (Jeong et al. 2011). All these distances can be suitably utilized for classifying time series by means of hierarchical and non-hierarchical clustering.

Following a nonhierarchical scheme, of particular interest is the fuzzy clustering approach. This clustering approach shows great sensitivity in capturing the details characterising the time series. In fact, often the dynamics are drifting or switching, hence, the standard clustering approaches are likely to fail to find and represent underlying structure in given data related to time. Thus, the switches from one time state to another, which are usually vague and not focused on any particular time point, can be naturally treated by means of fuzzy clustering (D'Urso 2005).

By considering a distance for observed time series, D'Urso (2004, 2005) proposed the following *fuzzy c-means* clustering model for observed time series:

$$\min \sum_{i=1}^{I} \sum_{c=1}^{C} {}_1 u_{ic}^m \sum_{t=1}^{T} ({}_1 w_t \|x_{it} - {}_1 h_{ct}\|)^2 \tag{12.6}$$

with the constraints

$$\sum_{c=1}^{C} {}_1u_{ic} = 1,\; {}_1u_{ic} \geq 0, \tag{12.7}$$

$$\sum_{t=1}^{T} {}_1w_t = 1,\; {}_1w_t \geq 0, \tag{12.8}$$

where the minimization is with respect to ${}_1u_{ic}$, ${}_1w_t$, and ${}_1\mathbf{h}_{ct}$.

${}_1u_{ic}$ indicates the membership degree of the i-th time series to the c-th cluster; ${}_1w_t$ is the t-th weight associated with $\|\mathbf{x}_{it} - {}_1\mathbf{h}_{ct}\|^2$, that is, the squared Euclidean distance between the i-th observation $\mathbf{x}_{it}$ and the c-th centroid ${}_1\mathbf{h}_{ct}$ at time t; $m > 1$ is a weighting exponent that controls the fuzziness of the obtained partition (refer to Chapter 5.5: Fuzzy Clustering).

The weight ${}_1w_t$ is intrinsically associated with the squared distance $({}_1w_t\|\mathbf{x}_{it} - {}_1\mathbf{h}_{ct}\|)^2$ at time t. Hence, the influence of the various times when computing the dissimilarity between time series can be appropriately tuned. The weight ${}_1w_t$ constitutes a specific parameter to be estimated within the clustering model. D'Urso (2004, 2005) derived iterative solutions for (12.6)–(12.8).

As remarked by D'Urso (2004, 2005), in the model (12.6)–(12.8), the evolutive features of the time series are not taken into account but only their instantaneous positions are. In fact the clustering model (12.6)–(12.8) is based on the instantaneous differences between the time series.

Since, in a time series, the successive observations are dependent, i.e., the set of observed time points holds the strict monotonic ordering property, and any permutation of the time points destroy this natural dependence, it is useful to build a (squared) distance that should take into account the ordering property of the time points and should be sensitive to any permutation of the time points.

When the interest is in capturing the differences concerning the *variational* pattern of the time series, a clustering model based on the comparison of the so-called velocity vectors associated with the time series can be considered (D'Urso 2004, 2005). For this purpose, in order to take into account the variational information of each observed time series in the clustering process, D'Urso (2004, 2005) proposed clustering time series by utilising the same modelling structure as (12.6)–(12.8) but with a new comparison measure, viz., the velocity-based (squared) distance which compares the dynamic features in each time interval of the time series. In particular, in the interval $[t - 1, t]$, the velocities of the time series are compared by considering the following velocity-based (squared) distance between the i-th observed time series and the c-th centroid time series:

$$\sum_{t=2}^{T} ({}_2w_t\|\mathbf{v}_{it} - {}_2\mathbf{h}_{ct}\|)^2,$$

where ${}_2w_t$ is a weight in $[t - 1, t]$ (*"velocity" weight*); $\|\mathbf{v}_{it} - {}_2\mathbf{h}_{ct}\| = \|(\mathbf{x}_{it} - \mathbf{x}_{it-1}) - ({}_1\mathbf{h}_{ct} - {}_1\mathbf{h}_{ct-1})\|$ is the Euclidean distance between the velocities of the i-th observed time series and the c-th centroid time series in the time interval $[t - 1, t]$, in which ${}_2\mathbf{h}_{ct} = ({}_1\mathbf{h}_{ct} - {}_1\mathbf{h}_{ct-1})$ is the velocity vector of the c-th centroid time series in $[t - 1, t]$ and $\mathbf{v}_{it}$ is the velocity vector of the i-th time series.

In this case, the weight ${}_2w_t$ is intrinsically associated with the squared distance $({}_2w_t\|\mathbf{v}_{it} - {}_2\mathbf{h}_{ct}\|)^2$ at time interval $[t - 1, t]$. Analogous observations, indicated for instantaneous weights $_1w_t$, instantaneous (position) squared distance and clustering model (12.6)–(12.8), can be considered for the task of the velocity weights in the velocity (squared) distance and then in the velocity (squared) distance-based clustering model.

The squared Euclidean distance $\|\mathbf{v}_{it} - {}_2\mathbf{h}_{ct}\|$ compares the slopes (velocities) in each time interval $[t-1, t]$ of the segments of each time series concerning the i-th time series with the corresponding slopes of the c-th centroid time series.

Summing up, this distance is able to capture and so to measure similarity of shapes of the time series, which are formed by the relative change of amplitude and corresponding temporal information. In this way, the problem is approached by taking into account the time series as piecewise linear functions and measuring the difference of slopes (velocities) between them.

Thus, by taking the previous velocity-based (squared) distance, we can obtain a *velocity-based fuzzy C-means clustering model* following a similar mathematically formalization to (12.6)–(12.8).

Refer to D'Urso (2000, 2005) for the fuzzy clustering model based on a mixed (squared) distance, in which the instantaneous and slope (velocity) features of the time series are simultaneously taken into account.

Several variants of the previous fuzzy clustering models have been proposed in the literature. Fuzzy clustering models with entropy regularization and spatial additional information have been suggested, respectively, by Coppi and D'Urso (2006) and Coppi et al. (2010).

Around medoids-based fuzzy clustering procedures have been proposed by Coppi and D'Urso (2006). Self-organizing maps based on the distance (12.1) and/or (12.2) and their lagged time-extensions have been proposed by D'Urso and De Giovanni (2008). Finally, different fuzzy clustering models for fuzzy time series had been earlier on developed by Coppi and D'Urso (2002, 2003).

12.2.3 Feature-Based Clustering

If a time series consists of a large number of observations, clustering time series based on these observations is not a desirable option because of the noise that is present, and the fact that the autocorrelation structure of the time series is ignored. Many featured-based methods have been developed to address the problem of clustering noisy raw time series data. Methods based on features extracted in the time domain, frequency domain, and from wavelet decomposition of the time series are presented. The features are then clustered using either traditional clustering methods (hierarchical and non-hierarchical) or fuzzy clustering methods.

12.2.3.1 Time-Domain Features

The autocorrelation function (ACF) of a time series describes its dependence structure and hence it is a good representation of the dynamics of actual time series. While the estimated autocorrelation function technically has one less value than the actual time series, in practice, the number of estimated autocorrelations of no more than a quarter of the length of the actual time series is sufficient to describe the dynamics of this time series, see Box and Jenkins (1976).

Galeano and Peña (2000) introduced a metric for time series based on the estimated ACF. Let $X = (x_{1t}, \ldots, x_{kt})'$ be a vector time series and $\widehat{\rho}_i = (\widehat{\rho}_{i1}, \ldots, \widehat{\rho}_{im})$ be a vector of the estimated autocorrelation coefficients of the time series i for some m such that $\widehat{\rho}_k \cong 0$ for $k > m$.

A distance between two time series x and y can be defined by

$$d_{ACF}(x, y) = \sqrt{(\widehat{\rho}_x - \widehat{\rho}_y)'\Omega(\widehat{\rho}_x - \widehat{\rho}_y)},$$

where Ω is some matrix of weights (Galeano and Peña 2000). Caiado et al. (2006) proposed three possible ways of computing a distance by using the sample autocorrelation function (ACF). The first uses a uniform weighting (ACFU), that is, $\Omega = I$, where I is an identity matrix, and is equivalent to the unweighted Euclidean distance between the autocorrelation coefficients. The second uses a geometric decay (ACFG), that is, $\Omega = D$, where D is a diagonal matrix with the geometric weights on the main diagonal. The third uses the Mahalanobis distance between the autocorrelations (ACFM), that is, $\Omega = M^{-1}$, where M is the sample covariance matrix of the autocorrelation coefficients given by the truncated Bartlett's formula (Brockwell and Davis 1991). It is straightforward to show that the proposed ACF-based metrics fulfil the usual properties of a distance. In a Monte Carlo simulation study, Caiado et al. (2006) show that the metrics based on the autocorrelation coefficients can perform quite well in time series classification. Alonso and Maharaj (2006) introduced a hypothesis testing procedure for the comparison of stationary time series based on the Euclidean distance between the autocorrelations. An illustrative example with financial time series will be presented in the next section.

Another autocorrelation distance measure, based on the Kullback–Leibler information (KLD), is defined by

$$d_{KLD}(x, y) = tr(R_x R_y^{-1}) - \log\frac{|R_x|}{|R_y|} - n, \tag{12.9}$$

where R_x and R_y are the $L \times L$ autocorrelation matrices of time series x and y, respectively, made at L successive times. Since $d_{KLD}(x, y) \neq d_{KLD}(y, x)$, a symmetric distance or quasi-distance can be defined as

$$d_{KLJ}(x, y) = \frac{1}{2}d_{KLD}(x, y) + \frac{1}{2}d_{KLD}(y, x), \tag{12.10}$$

which also satisfies all the usual properties of a metric except the triangle inequality.

Caiado et al. (2006) also introduced distance measures based on the partial autocorrelation function (PACF) and on the inverse autocorrelation function (IACF). A PACF metric between the time series x_t and y_t is defined by

$$d_{PACF}(x, y) = \sqrt{(\widehat{\phi}_x - \widehat{\phi}_y)'\Omega(\widehat{\phi}_x - \widehat{\phi}_y)}, \tag{12.11}$$

where $\widehat{\phi}$ is a vector of the sample partial autocorrelation coefficients and Ω is also a matrix of weights. A distance between the inverse autocorrelations of the time series x and y is defined by

$$d_{IACF}(x, y) = \sqrt{(\widehat{\rho}_x^{(I)} - \widehat{\rho}_y^{(I)})'\Omega(\widehat{\rho}_x^{(I)} - \widehat{\rho}_y^{(I)})}, \tag{12.12}$$

where $\widehat{\rho}_x^{(I)}$ and $\widehat{\rho}_y^{(I)}$ are the sample inverse autocorrelation functions of time series x_t and y_t, respectively. As in the ACF-based distances, uniform weights of PACF and IACF coefficients or weights that decrease with the autocorrelation lag can be used.

D'Urso and Maharaj (2009) proposed an Autocorrelation-based Fuzzy C-means Clustering (A-FCM) model for clustering stationary time series. The model is formalized as follows:

$$min : J_m = \sum_{k=1}^{K} \sum_{c=1}^{C} \mu_{kc}^m \, d_{kc}^2 = \sum_{k=1}^{K} \sum_{c=1}^{C} \mu_{kc}^m \sum_{r=1}^{R} (\hat{\rho}_{kr} - \hat{\rho}_{cr})^2, \quad (12.13)$$

with the constraints

$$\sum_{c=1}^{C} \mu_{kc} = 1 \qquad \mu_{kc} \geq 0 \qquad -1 \leq \hat{\rho}_{cr} \leq 1.$$

$d_{kc}^2 = \sum_{r=1}^{R} (\hat{\rho}_{kr} - \hat{\rho}_{cr})^2$ is the squared Euclidean distance between the k-th time series and the c-th prototype or centroid time series, based on the autocorrelation function at R lags. $m > 1$ is a parameter that controls the fuzziness of the partition (Bezdek 1981), μ_{kc} is the membership degree of the k-th time series to the c-th cluster, K is the number of time series and C is the number of clusters. Refer to D'Urso and Maharaj (2009) for the iterative solutions to Equation 12.13 which satisfy the autocorrelation constraints, viz., $-1 \leq \hat{\rho}_{cr} \leq 1$.

D'Urso and Maharaj (2009) have shown through simulation studies that when clusters are well separated, the performance of the (A-FCM) model is similar to that of the k-means and hierarchical methods. However, for clusters that are not well separated, the A-FCM model has the advantage of identifying time series that are members of more than one cluster simultaneously.

Other time domain features that have been used to cluster time series are correlation and cross correlations. Clustering of financial and economic time series with correlation or cross-correlation coefficients include those by Dose and Cincotti (2005), Basalto et al. (2007), Takayuki et al. (2006), Miskiewicz and Ausloos (2008), and Ausloos and Lambiotte (2007) while Goutte et al. (1999) clustered functional magnetic resonance image (fMRI) time series using cross-correlations.

12.2.3.2 Frequency-Domain Features

Spectral analysis provides useful information about the time series behavior in terms of cyclic patterns and periodicity. In this sense, nonparametric methods based on spectral analysis can be useful for classification and clustering of time series data. For these purposes, some distance-based methods in the frequency domain that have been proposed in the literature are presented.

Caiado et al. (2006) proposed a spectral domain method for clustering time series of equal length. Let $P_x(\omega_j) = n^{-1} |\sum_{t=1}^{n} x_t e^{-it\omega_j}|^2$ be the periodogram of time series x at frequencies $\omega_j = 2\pi j/n$, $j = 1, \ldots, [n/2]$, with $[n/2]$ the largest integer less or equal to $n/2$ (similar expression applies to series y). The Euclidean distance between the periodograms of x and y is given by

$$d_P(x,y) = \sqrt{\sum_{j=1}^{[n/2]} [P_x(\omega_j) - P_y(\omega_j)]^2}. \quad (12.14)$$

Since the variance of the periodogram ordinates is proportional to the spectrum at the corresponding Fourier frequencies, Caiado et al. (2006) suggest the use of sample variances to normalize the periodograms and then take logarithms to attain homoscedasticity. Thus the distance becomes

$$d_{LNP}(x,y) = \sqrt{\sum_{j=1}^{[n/2]} \left[\log \frac{P_x(\omega_j)}{\widehat{\sigma}_x^2} - \log \frac{P_y(\omega_j)}{\widehat{\sigma}_y^2} \right]^2}, \qquad (12.15)$$

where $\widehat{\sigma}_x^2$ and $\widehat{\sigma}_y^2$ are the sample variances of x and y, respectively. These metrics satisfy the usual properties of a metric: positivity, symmetry, identity, and triangle inequality. A simulation study comparing this metric with the ACF-based distance defined in (12.9) and the classical Euclidean distance is provided in the next section.

The Kullback–Leibler information discrepancy in the frequency domain (Kakizawa et al. 1998), which is asymptotically equivalent to (12.9) but much easier to compute, is given by

$$d_{KLF}(x,y) = \sum_{j=1}^{[n/2]} \left(\frac{P_x(\omega_j)}{P_y(\omega_j)} - \log \frac{P_x(\omega_j)}{P_y(\omega_j)} - 1 \right). \qquad (12.16)$$

This measure is greater or equal to zero, with equality if and only if $P_x(\omega_j) = P_y(\omega_j)$ almost everywhere. In a simulation study, Caiado et al. (2006) show that (12.15) and the normalized periodogram version of (12.16) for high frequency components can perform quite well in distinguishing between nonstationary and near-nonstationary time series.

In many real applications, it is often the case that the data sets have different lengths and the sets of frequencies at which the periodogram ordinates are usually computed are different. A solution to the problem is to extend the length of the shorter series to the length of the longer series by adding zeros to the shorter series and computing the periodogram. This matches the frequencies of the longer series and produces a smoothed periodogram. This approach, known as the "zero-padding" procedure, is widely used in the pattern recognition and signal processing literature (Wang and Blostein 2004) due to its simplicity and applicability.

Let x_t and y_t be two time series with different sample sizes, $n_x > n_y$. Let $P_x(\omega_j)$ with $\omega_j = 2\pi j/n_x$, $j = 1, \ldots, m_x = [n_x/2]$ and $P_y(\omega_p)$ with $\omega_p = 2\pi p/n_y$, $j = 1, \ldots, m_y = [n_y/2]$ be the periodograms of series x_t and y_t, respectively. Caiado et al. (2009) define a zero-padding periodogram discrepancy statistic between x_t and y_t by

$$d_{ZP}(x,y) = \sqrt{\frac{1}{m_x} \sum_{j=1}^{m_x} \left[P_x(\omega_j) - P_{y'}(\omega_j) \right]^2}, \qquad (12.17)$$

where $P_{y'}(\omega_j)$ is the periodogram of $y_t' = \begin{cases} y_t, & t = 1, \ldots, n_y \\ 0, & t = n_y + 1, \ldots, n_x, \end{cases}$. This metric performs

reasonably well for the comparison of stationary time series with similar autocorrelation structure. However, it is not able to distinguish between longer stationary and shorter nonstationary time series due to distortion of the zero-padding approach for very unbalanced sample sizes.

As put forward by Caiado et al. (2009), a second solution to the problem of clustering of time series of unequal lengths is to compute both periodograms at a common frequency. For instance, the periodogram ordinates of the longer series x_t at the frequencies of the shorter series y_t may be calculated and then the following reduced periodogram discrepancy statistic can be used.

$$d_{RP}(x, y) = \sqrt{\frac{1}{m_y} \sum_{p=1}^{m_y} [P_x^{RP}(\omega_p) - P_y(\omega_p)]^2}, \tag{12.18}$$

where $P_x^{RP}(\omega_p) = \frac{1}{n_x} \left| \sum_{t=1}^{n_x} x_t e^{-it\omega_p} \right|^2$ with $\omega_p = 2\pi p/n_y$, $p = 1, \ldots, m_y < m_x$ is the reduced periodogram. From the results of a simulation study, Caiado et al. (2009) concluded that the distance (12.18) performs poorly for classifying stationary and nonstationary time series of short length.

A third solution proposed by Caiado et al. (2009) for handling series of unequal lengths in the frequency domain is to construct an interpolated periodogram ordinates for the longer series at the frequencies defined by the shorter series. Without loss of generality, let $r = [p(m_x/m_y)]$ be the largest integer less or equal to $p(m_x/m_y)$ for $p = 1, \ldots, m_y$, and $m_y < m_x$. The periodogram ordinates of x_t can be estimated as

$$P_x^{IP}(\omega_p) = P_x(\omega_r) + (P_x(\omega_{r+1}) - P_x(\omega_r)) \times \frac{\omega_{p,y} - \omega_{r,x}}{\omega_{r+1,x} - \omega_{r,x}}$$

$$= P_x(\omega_r) \left(1 - \frac{\omega_{p,y} - \omega_{r,x}}{\omega_{r+1,x} - \omega_{r,x}} \right) + P_x(\omega_{r+1}) \left(\frac{\omega_{p,y} - \omega_{r,x}}{\omega_{r+1,x} - \omega_{r,x}} \right). \tag{12.19}$$

This procedure will yield an interpolated periodogram with the same Fourier frequencies of the shorter periodogram $P_y(\omega_p)$. If one is interested only in the dependence structure and not in the process scale, then the periodograms can be standardized dividing them by the sample variances, and then taking the logarithms: $\log NP_x^{IP}(\omega_p) = \log \left(P_x^{IP}(\omega_p)/\widehat{\sigma}_x \right)$ and $\log NP_y(\omega_p) = \log \left(P_y(\omega_p)/\widehat{\sigma}_y \right)$. Caiado et al. (2011) extended this procedure for hypothesis testing purposes.

They define the interpolated periodogram discrepancy statistic by

$$d_{IP}(x, y) = \sqrt{\frac{1}{m_y} \sum_{p=1}^{m_y} [P_x^{IP}(\omega_p) - P_y(\omega_p)]^2}, \tag{12.20}$$

or, by using the log normalized periodogram,

$$d_{ILNP}(x, y) = \sqrt{\frac{1}{m_y} \sum_{p=1}^{m_y} \left[\log \frac{P^{IP}(\omega_p)}{\widehat{\sigma}_x^2} - \log \frac{P_y(\omega_p)}{\widehat{\sigma}_y^2} \right]^2}. \tag{12.21}$$

A simulation study carried out by them indicated that the statistic (12.21) has an exceptional performance for time series clustering for a wide type of comparisons: stationary time series with similar autocorrelation functions, near nonstationary versus nonstationary time series, long-memory versus short-memory time series, and deterministic trend versus stochastic trend series.

Maharaj and D'Urso (2011) proposed a class of frequency domain-based fuzzy clustering models which can be formalized as follows:

$$min : J_m = \sum_{k=1}^{K} \sum_{c=1}^{C} \mu_{kc}^m \ d_{kc}^2 = \sum_{k=1}^{K} \sum_{c=1}^{C} \mu_{kc}^m \sum_{r=1}^{R} (\tilde{\rho}_{kr} - \tilde{\rho}_{cr})^2, \tag{12.22}$$

with the constraints

$$\sum_{c=1}^{C}\mu_{kc}=1 \qquad \mu_{kc}\geq 0 \qquad \tilde{\rho}_L < \tilde{\rho}_{kc} < \tilde{\rho}_U.$$

$d_{kc}^2 = \sum_{r=1}^{R}(\tilde{\rho}_{kr} - \tilde{\rho}_{cr})^2$ is now based on the relevant frequency domain features, with m, μ_{kc}, k, and c defined as in (12.13). $\tilde{\rho}_L$ and $\tilde{\rho}_U$ are the upper and lower limits of the frequency domain feature if they exist. Maharaj and D'Urso (2011) derived iterative solutions to (12.22) which satisfy the constraints pertinent to the specific frequency domain features, viz., normalised periodogram ordinates, log normalized periodogram ordinates, and cepstral coefficients.

For the Normalized Periodogram-based Fuzzy Clustering Model, $\tilde{\rho}_{kr} \equiv n\rho_{kr}$, the normalized periodogram ordinate of the k-th time series at the r-th frequency and $\tilde{\rho}_{cr} \equiv n\rho_{cr}$, the normalized periodogram ordinate of the centroid time series at the r-th frequency. $d_{kc}^2 \equiv \sum_{r=1}^{R}(n\rho_{kr} - n\rho_{cr})^2$ the squared Euclidean distance measure between the k-th time series and the centroid time series of the c-th cluster based on the normalized periodogram at r-th frequency, $R = [(T-1)/2]$ and T is the length of the time series. $0 < n\rho_{kr} < 1$ is the constraint on the normalized periodogram ordinates.

For the log normalized periodogram-based fuzzy clustering model, $\tilde{\rho}_{kr} \equiv ln\rho_{kr}$, the log normalized periodogram ordinate of the k-th time series at the r-th frequency and $\tilde{\rho}_{cr} \equiv ln\rho_{cr}$, the lognormalized periodogram ordinate of the centroid time series at the r-th frequency. $d_{kc}^2 \equiv \sum_{r=1}^{R}(ln\rho_{kr} - ln\rho_{cr})^2$ the squared Euclidean distance measure between the k-th time series and the centroid time series of the c-th cluster based on the log normalized periodogram at r-th frequency, $K = [(T-1)/2]$ and T is the length of the time series. $-\infty < ln\rho_{kr} < 0$ is the constraint on the log normalized periodogram ordinates.

For the cepstral-based fuzzy clustering model, $\tilde{\rho}_{kr} \equiv c\rho_{kr}$, the normalized periodogram ordinate of the k-th time series at the r-th frequency and $\tilde{\rho}_{cr} \equiv c\rho_{cr}$, the normalized periodogram ordinate of the centroid time series at the r-th frequency. $d_{kc}^2 \equiv \sum_{r=1}^{R}(c\rho_{kr} - c\rho_{cr})^2$ the squared Euclidean distance measure between the k-th time series and the centroid time series of the c-th cluster based on the cepstrum at r-th frequency, and in this case K is just a small fraction of $R = [(T-1)/2]$ and T is the length of the time series. There are no constraints on $c\rho_{kr}$ since $-\infty < c\rho_{kr} < \infty$.

Maharaj and D'Urso (2011) conducted a series for simulation studies using time series generated from both linear and non-linear models and they found that cepstral-based fuzzy clustering model generally performs better than the normalized and log normalized periodogram-based fuzzy clustering models, the autocorrelation-based fuzzy C-means clustering from D'Urso and Maharaj (2009), and a Fuzzy Clustering model based or discrete wavelet transform (DWT) coefficients. The DWT decomposes the time series into different frequency bands (scales) and hence the total energy of the time series is decomposed into frequency bands (from high frequency to low). Some of the results of the application to electroencephalogram (EEG) time series to differentiate between patterns associated with a sample of healthy volunteers and a sample of epileptic patients during seizure activity are presented in the next section.

Amongst other authors who investigated clustering time series in the frequency domain are Kakizwa et al. (1998) and Shumway (2003) and Maharaj and D'Urso (2010). Kakizawa et al. (1998) proposed clustering and classification methods for stationary multivariate time series. They measured the disparity between two multivariate time series based on their estimated spectral matrices, e.g., for a two dimensional time series X, the spectral matrix is 4-dimensional with estimated spectral density functions of each component of the series on the right diagonal and the estimated cross spectrum elements on the left diagonal. This measure of disparity between two multivariate time series was used as a quasi-distance measure for clustering multivariate time series. Approximations for distance measures were computed from the J-divergence (see Equation 12.10 with correlation matrices replaced by spectral matrices) and that Chernoff information divergence to which they applied hierarchical and k-means clustering. Refer to Kakizwa et al. (1998), p. 332 for details of the Chernoff measure. They applied their clustering methods to differentiate between waveforms of earthquake and explosion time series.

By applying locally stationary versions of the Kullback–Leibler discriminant information measures with respect to estimated spectral densities, Shumway (2003) proposed the clustering non-stationary time series and also applied his method to differentiate between waveforms of earthquake and explosion time series. Maharaj and D'Urso (2010) investigated clustering stationary time series using the squared coherence between every pair of time series under consideration. The squared coherence is a useful measure derived for the cross-spectral density function between two time series. These measures are analogous to time domain measure of cross-correlations, and measures the strength of the linear relationship between then two time series. Their motivation for using squared coherence coefficients instead of cross correlation coefficients in the context of clustering time series is that fewer coherence coefficients than the cross-correlation coefficients are required. They are determined at just positive frequencies whereas the cross-correlation coefficients are determined at both positive and negative lags. In practice, the number of frequencies is equal to half the length of a time series under consideration, while the number of positive and negative lags totals twice the length of the time series minus one. Hence, using cross-correlations as the clustering variables for a large number of time series will result in a high-dimensionality space. The use of high-dimensionality spaces in cluster analysis can make it increasingly difficult to detect clusters of objects that can be meaningfully interpreted (Beyen et al. 1999). Hence, the performance of the clustering process can be quite poor. They clustered GDP series using squared coherence as the input into the k-medoids algorithm.

12.2.3.3 Wavelet Features

Among the authors who have used features from the wavelet decomposition of time series in clustering are Zhang et al. (2005), Maharaj et al. (2010), and D'Urso and Maharaj (2012). Zhang et al. (2005) used discrete wavelet transform (DWT) coefficients to cluster univariate time series. They considered the DWT in a k-means clustering process, keeping, as features, the wavelet coefficients of only those frequency bands that have low energy. To reduce the noise effect, they remove the coefficients of the high-frequency bands.

Maharaj et al. (2010) considered the modified discrete wavelet transform (MODWT) variances at the all frequency bands as the extracted features in the clustering process, using hierarchical, non-hierarchical, and fuzzy cluster analysis. See Percival and Walden (2000) for more details about the DWT and MODWT. Note that under the MODWT, the number of wavelet coefficients created will be the same as the number of observations in the original time series, whereas the DWT does not. Because the MODWT decomposition retains all possible times at each time scale, the MODWT has the advantage of retaining the time invariant property of the original time series.

Since the wavelet coefficients are not used as they are by Zhang et al. (2005), but the wavelet variances associated with each frequency band are used, the dimensionality of the data set is reduced to just the number of frequency bands. Maharaj et al. (2010) proposed a fuzzy c-means clustering approach with the wavelet variances to enable the identification of time series that may simultaneously belong to more than one cluster based on their patterns of variability. As in (12.13), they formalised the fuzzy model with the wavelet variances and derived the iterative solutions subject to constraints on the wavelet variances. They applied their approach to cluster market returns of developed and developing countries.

D'Urso and Maharaj (2012) used MODWT wavelet variance and wavelet correlations to cluster multivariate time series. These features were input into the k-means, k-medoids, fuzzy c-means, and fuzzy relational algorithms. Comparison of their approach to some other methods for clustering multivariate time series, in particular, those by Maharaj (1999), Singhal and Seborg (2005), and Wang et al. (2007), reveals a favorable performance of their approach.

12.3 Examples and Applications

A synthetic example from D'Urso (2005), a simulation example from Caiado et al. (2009) and an application from Caiado and Crato (2010), and from Maharaj and D'Urso (2011) are presented as follows.

12.3.1 Example 1—Data with Switching Time Series

In order to show the usefulness and performance of the fuzzy clustering model (12.6)–(12.8), D'Urso (2005) illustrated an applicative example of the model to a set of synthetic short time series (Figure 12.1) consisting of three well-separated clusters of time series (respectively with 4, 2, and 2 time series each) and one switching time series (the 7-th time series).

The switching time series presents, in the initial times, instantaneous position and slope similar to the time series belonging to cluster 2 (time series 5 and 6), while, in the final times, it has instantaneous position and slope similar to the time series belonging to cluster 3 (time series 8 and 9). It can be observed from Table 12.1 that by applying the fuzzy clustering model (12.6)–(12.8) (with $C = 3$ and $m = 2.3$) the real situation is well captured.

12.3.2 Example 2—Simulation Study

To illustrate the performance of the Euclidean autocorrelation and periodogram-based distances defined in (12.2.3.1) and (12.15), respectively, Caiado et al. (2009) performed a Monte Carlo simulation study for a wide type of comparisons:

1. AR(1), $\phi = 0.9$ versus AR(1), $\phi = 0.5$;
2. AR(1), $\phi = 0.9$ versus ARIMA(0,1,0);
3. AR(2), $\phi_1 = 0.6$, $\phi_2 = -0.3$ versus MA(2), $\theta_1 = -0.6$, $\theta_2 = 0.3$;
4. ARFIMA(0,0.45,0) versus white noise;
5. ARFIMA(0,0.45,0) versus AR(1), $\phi = 0.95$;
6. ARFIMA(0,0.45,0) versus IMA(1,1), $\theta = 0.4$;

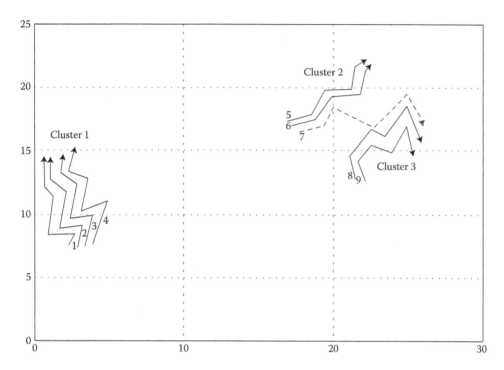

FIGURE 12.1
Plot of a set of short time series (with switching time series).

TABLE 12.1

Membership Degrees of Each Time Series within Each Cluster

	Cluster 1	Cluster 2	Cluster 3
1	0.973	0.013	0.014
2	0.991	0.005	0.004
3	0.995	0.003	0.002
4	0.961	0.024	0.015
5	0.003	0.977	0.002
6	0.001	0.997	0.002
7	0.084	0.497	0.419
8	0.004	0.027	0.969
9	0.001	0.002	0.997

Time-weighting system
(0.196; 0.198; 0.171; 0.136; 0.119; 0.18)

7. ARMA(1,1), $\phi = 0.95$, $\theta = 0.74$ versus IMA(1,1), $\theta = 0.8$.

Two series of length $n = 50, 100, 200, 500, 1000$ were generated from each process. So four different time series were simulated for each replication. For each case, the four generated series were grouped into two clusters by the complete linkage method. Each comparison is defined as a success when the two time series generated by the same process are classified in the same group. One thousands replications were conducted in the experiment. For reference, we also considered the classic Euclidean distance, $d_{EUCL}(x, y) = \sqrt{\sum_{t=1}^{n}(x_t - y_t)^2}$.

The classic Euclidean distance is unable to distinguish successfully similar processes for time series of any length. The log-normalized periodogram discrepancy statistic showed a remarkable good performance on the comparison between different stationary processes with similar sample properties, short memory and long memory processes, and nonstationary and near nonstationary processes. The autocorrelation discrepancy statistic revealed to be competitive for comparing stationary and nonstationary processes with short length.

12.3.3 Application 1—Autocorrelation-Based Clustering

The autocorrelations of the squared returns provide useful information about the time-series behavior in terms of the presence of nonlinear dependence and possible autoregressive heteroskedasticity (ARCH) effects. In contrast to the autocorrelation of returns, which are typically zero or very close to zero, the autocorrelations of squared returns or absolute returns are generally positive and significance for a substantial number of lags. Caiado and Crato (2010) used the discrepancy statistic (12.2.3.1), based on the estimated autocorrelations of the squared returns, to identify similarities between the Dow Jones Industrial Average (DJIA) stocks (Table 12.2). The data correspond to closing prices adjusted for dividends and splits and cover the period from June 11, 1990 to September 12, 2006 (4100 daily observations).

In order to better visualize and interpret similarities among stocks, the outlier INTC was dropped from the clustering analysis. Figure 12.2 shows the complete linkage dendrogram, which minimizes the maximum distance between stocks in the same group. Three groups of corporations were identified. One group is composed of basic materials (Alcoa,

TABLE 12.2

Stocks Used to Compute the Dow Jones Industrial Average (DJIA) Index

Stock	Code	Sector	Stock	Code	Sector
Alcoa Inc.	AA	Basic materials	Johnson & Johnson	JNJ	Healthcare
American Int. Group	AIG	Financial	JP Morgan Chase	JPM	Financial
American Express	AXP	Financial	Coca-Cola	KO	Consumer goods
Boeing Co.	BA	Industrial goods	McDonalds	MCD	Services
Caterpillar Inc.	CAT	Financial	3M Co.	MMM	Conglomerates
Citigroup Inc.	CIT	Industrial goods	Altria Group	MO	Consumer goods
El Dupont	DD	Basic materials	Merck & Co.	MRK	Healthcare
Walt Disney	DIS	Services	Microsoft Corp.	MSFT	Technology
General Electric	GE	Industrial goods	Pfizer Inc.	PFE	Healthcare
General Motors	GM	Consumer goods	Procter & Gamble	PG	Consumer goods
Home Depot	HD	Services	AT&T Inc.	T	Technology
Honeywell	HON	Industrial goods	United Technol.	UTX	Conglomerates
Hewlett-Packard	HPQ	Technology	Verizon Communic.	VZ	Technology
Int. Bus. Machines	IBM	Technology	Wal-Mart Stores	WMT	Services
Inter-tel Inc.	INTC	Technology	Exxon Mobile CP	XOM	Basic materials

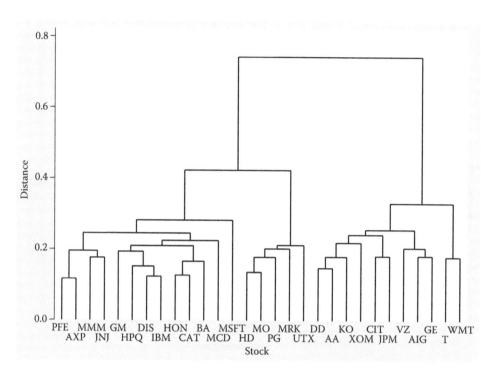

FIGURE 12.2
Complete linkage dendrogram for DJIA stocks using the ACF-based distance for squared returns observations.

El Dupont and Exxon Mobile), communications (AT&T and Verizon), industrial goods (Citigroup and General Electric), financial (AIG and JP Morgan Chase), and consumer goods and services (Coca-Cola and Wal-Mart Stores) corporations. The second group is composed of technology (IBM, Microsoft and Hewlett-Packard), healthcare (Johnson & Johnson and Pfizer), financial (American Express and Caterpiller), industrial goods and consumer goods (Boeing, Honeywell, and General Motors), services (Walt-Disney and McDonalds), and conglomerates (3M) corporations. The third group is composed of consumer goods (Altria and Procter & Gamble) and miscellaneous sector (Home Depot, Merck, and United Technologies) corporations.

Figure 12.3 shows the corresponding two-dimensional map of DJIA stocks by metric multidimensional scaling (Johnson and Wichern 1992). The first dimension accounts for 68.36% of the total variance of the data and the second dimension accounts for 5.23% of the total variance. The map tends to group the basic materials and the communications corporations in a distinct cluster and most technology, healthcare, financial, services, industrial goods, and consumer goods corporations in another distinct cluster.

12.3.4 Application 2—Cepstral-Based Fuzzy Clustering

Maharaj and D'Urso (2011) considered a subset of 200 electroencephalogram (EEG) time series from a suite studied by Andrzejak et al. (2001). These 200 series are divided into two sets denoted by A and E, each containing 100 EEG segments of 23.6 seconds duration (4096 observations): Set A EEG recordings are of healthy volunteers, while Set E consists of EEG recordings of epileptic patients during seizure activity. Figure 12.4 shows a typical EEG

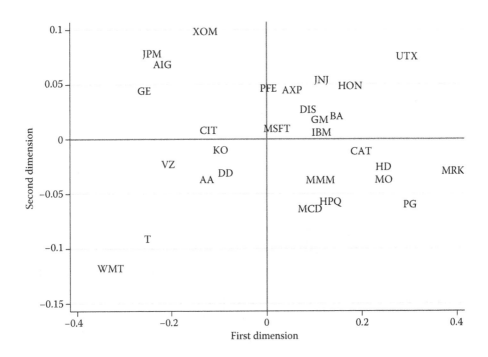

FIGURE 12.3

Two-dimensional scaling map of DJIA stocks using the ACF-based distance for squared returns observations.

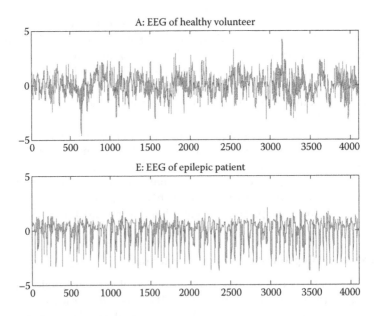

FIGURE 12.4

Standardized EEG recordings from each of sets A and E.

TABLE 12.3

Mean Membership Degrees

p		$m = 1.8$		$m = 2.0$		$m = 2.2$	
		C1	C2	C1	C2	C1	C2
64	A	0.83	0.17	0.78	0.22	0.74	0.26
	E	0.28	0.72	0.32	0.68	0.35	0.65
128	A	0.80	0.20	0.74	0.26	0.70	0.30
	E	0.30	0.70	0.34	0.66	0.37	0.63
256	A	0.73	0.27	0.67	0.33	0.62	0.38
	E	0.33	0.67	0.38	0.62	0.41	0.59

record from each of Sets A and E from where it can be observed that there does appear to be an overall difference in patterns. However, since there is much noise present in both types of records, misclassifications are likely to occur.

Using adaptive neuro fuzzy inference system networks to discriminate among the patterns A and E, Kannathal et al. (2005) achieved a classification error rate of 7.8%, while Nigam and Graupe (2004) used a large memory storage and retrieval neural networks to discriminate among the patterns A and E and achieved a classification error rate of 2.8%. On the other hand, Maharaj and Alonso (2007) achieved a 0% classification error rate using time-dependent wavelet variances in a quadratic discriminant analysis procedure.

While the aim of cluster analysis is to find groups in data when usually the groupings are not known a priori, Maharaj and D'Urso (2011) demonstrate that clustering with cepstral coefficients using the k-means algorithm can discriminate between the EEG patterns of sets A and E with error rates comparable to Kannathal et al. (2005), while fuzzy c-means clustering can explain some of the misclassifications.

Since each EEG record is of length $T = 2^{12} = 4096$, using the guidelines they put forward by for the minimum number of cepstral coefficients p to achieve successful classification, Maharaj and D'Urso (2011) performed the analysis for $p = 64, 128$, and 256 and for fuzzy parameter values of $m = 1.8, 2$, and 2.2. For all (m, p) combinations, the same five EEG recordings from Set A were misclassified as having patterns of the Set E records, and the same 10 EEG recordings from Set E were misclassified as having patterns of the Set A records. For the 200 EEG records, k-means clustering achieved a classification error rate of 7.5%.

Table 12.3 shows the mean membership degrees across the 100 cases of Set A EEG records in Cluster 1 (C1) and in Cluster 2 (C2), and the 100 cases of Set E EEG records in Cluster 1 and in Cluster 2. Tables 12.4 shows the membership degrees from fuzzy c-means clustering of the Set A and E EEG records that were misclassified into Clusters 2 and 1, respectively, by k-means clustering for $p = 64$.

On comparing the mean membership degrees (Table 12.3) of the

1. Set A EEG records in Cluster 1 with that of the five misclassified Set A records (Table 12.4), it can be observed that in all cases the membership degrees are smaller than the mean and are fairly fuzzy; hence these five records have fuzzy membership in both clusters.

2. Set E EEG record in Cluster 2 with that of the 10 misclassified Set E records (Table 12.4), it can be observed that of the 10 records, only 117 and 190 have greater

TABLE 12.4

Membership Degrees of the Set A and E EEG Records That Were
Misclassified by k-Means Clustering for $p = 64$

$p = 64$		k-Means		Fuzzy-c-means				
			$m = 1.8$		$m = 2.0$		$m = 2.2$	
EEG Record	Set	Cluster	C1	C2	C1	C2	C1	C2
1	A	C2	0.49	0.51	0.49	0.51	0.49	0.51
27	A	C2	0.45	0.55	0.46	0.54	0.46	0.54
52	A	C2	0.47	0.53	0.47	0.53	0.47	0.53
78	A	C2	0.45	0.55	0.46	0.54	0.46	0.54
99	A	C2	0.33	0.67	0.36	0.64	0.38	0.62
107	E	C1	0.58	0.42	0.56	0.44	0.55	0.45
117	E	C1	**0.75**	0.25	**0.70**	0.30	**0.66**	0.34
131	E	C1	0.62	0.38	0.60	0.40	0.58	0.42
132	E	C1	0.58	0.42	0.56	0.44	0.54	0.46
134	E	C1	0.56	0.44	0.55	0.45	0.54	0.46
135	E	C1	0.69	0.31	0.65	0.35	0.62	0.38
143	E	C1	0.69	0.31	0.65	0.35	0.62	0.38
172	E	C1	0.61	0.39	0.58	0.42	0.56	0.44
181	E	C1	0.63	0.37	0.60	0.40	0.58	0.42
190	E	C1	**0.79**	0.21	**0.74**	0.26	**0.69**	0.31

membership degrees than the mean membership degrees in Table 12.3, while the
remaining eight records have fuzzy membership in both clusters.

Similar observations were made for $p = 128$ and 256. It can be concluded from this that of
the 200 EEG records, only two EEG records are clearly misclassified, while the other 13 have
fuzzy membership in both clusters. The validation of these results is based on the study of
Maharaj and D'Urso (2011). From the results of this application, they concluded that while
classification error rates for discriminating between the patterns of the EEG records in Sets
A and E, achieved by k-means clustering are not always comparable to those achieved by
other methods, fuzzy c-means clustering adds a new dimension to the analysis of EEG
records that are misclassified, namely, the advantage of fuzzy clustering in a situation such
that those time records that are potentially misclassified are identified and hence could be
rechecked using another method such as a human reader.

12.4 Conclusion

As mentioned in Section 12.2.3, a number of new parametric and nonparametric methods
have recently been developed to address the limitations of the classical Euclidean distance.
Many of these approaches use distance measures based on the structural characteristics
of time series. It seems reasonable to apply classification and clustering methods that best

capture the specific features of time series, such as autocorrelation, nonstationarity, periodicity and cyclic fluctuations, nonlinearity, and volatility. Also, for these purposes, it is often important to provide statistical methods for handling time series of unequal lengths.

A parametric approach to do this would be to associate with each series x_i of length n_i a vector of parameters or features, $\beta_i, i = 1, \ldots, M$, corresponding to M time series and then search for distances among these vectors of parameters or features. The problem is how to define a relevant measure of distance between these vectors or how to define a clustering procedure for making groups associating with each series these vectors of parameters or features. Assuming that the parameters or features β_i are generated by a mixture of normal distributions, $\beta_i \to \sum_{i=1}^{K} \alpha_i N(\mu_i, \sigma_i^2)$, the objective is to find the number of distributions, K, and the probability of each series coming from each distribution. This means the vector of parameters or features will have to be defined and an approach for fitting mixtures of normal variables will have to be put forward. The projection pursuit approach (Peña and Prieto 2001) by projecting the points (autoregressive estimates, autocorrelations, periodogram ordinates or other features extracted from time series) onto certain directions according to some optimally criterion might be a good way to find these groups.

As far as fuzzy clustering methods for time series are concerned, in future, from a theoretical point of view, it could be interesting to investigate the performance of new distance measures or alternative fuzzy clustering approaches (e.g., kernel-based clustering, support vector machines) for time series. As well, it will be useful to further investigate the usefulness of the various fuzzy clustering techniques for classifying nonstationary and nonlinear time series.

References

Alonso, A.M. and Maharaj, E.A. 2006. Comparison of time series using subsampling, *Computational Statistics & Data Analysis*, 50, 2589–2599.

Andrzejak, R.G., Lehnertz, K., Rieke, C., Mormann, F., David, P., and Elger, C.E. 2001. Indications of nonlinear deterministic and finite dimensional structures in time series of brain electrical activity: Dependence on recording region and brain state, *Physical Review E*, 64, 061907.

Ausloos, M.A. and Lambiotte, R. 2007. Clusters or networks of economies? A macroeconomy study through gross domestic product, *Physica A*, 382, 16–21.

Basalto, N., Bellotti, R., De Carlo, F., Facchi, P., Pantaleo, E., and Pascazio, S. 2007. Hausdorff clustering of financial time series, *Physica A*, 379, 635–644.

Beyen, K., Goldstein, J., Ramakrishnan, R., and Shaft, U. 1999. When is the nearest neighbor meaningful? *Proceeding of the 7th International Conference on Database Theory*, 217–235.

Bezdek, J.C. 1981 *Pattern Recognition with Fuzzy Objective Function Algorithms*, Plenum Press, New York.

Boets, J., De Cock, K., Espinoza, M., and De Moor, B. 2005. Clustering time series, subspace identification and cepstral distances, *Communication in Information and Systems*, 5(1), 69–96.

Box, G.E.P. and Jenkins, G.M. 1976. *Time Series Analysis. Forecasting and Control*, Holden Day, San Francisco.

Brockwell, P.J. and Davis, R.A. 1991. *Time Series: Theory and Methods*. 2nd Edition, Springer, New York.

Caiado, J. and Crato, N. 2010. Identifying common dynamic features in stock returns, *Quant. Finance*, 10, 797–807.

Caiado, J., Crato, N., and Peña, D. 2006. A periodogram-based metric for time series classification, *Computational Statistics & Data Analysis*, 50, 2668–2684.

Caiado, J., Crato, N., and Peña, D. 2009. Comparison of time series with unequal length in the frequency domain, *Communications in Statistics—Simulation and Computation*, 38, 527–540.

Caiado, J., Crato, N., and Peña, D. 2011. Tests for comparing time series of unequal lengths, *Journal of Statistical Computing and Simulation*, 82, 1715–1725, DOI: 10.1080/00949655.2011.592985.

Carlier, A. 1986. *Factor Analysis of Evolution and Cluster Methods on Trajectories*, COMPSTAT, Physica-Verlag, Heidelberg, 140–145.

Childers, D.G., Skinner D.P., and Kemerait, R.C. 1975. The cepstrum: A guide to processing, *Proceedings of the IEEE*, 65(10), 1428–1443.

Coppi, R. and D'Urso, P. 2000. Fuzzy time arrays and dissimilarity measures for fuzzy time trajectories, in *Data Analysis, Classification, and Related Methods* (eds. H.A.L Kiers, J.P. Rasson, P.J.F. Groenen, and M. Schader), 273–278, Springer-Verlag, Berlin.

Coppi, R. and D'Urso, P. 2001. The geometric approach to the comparison of multivariate time trajectories, in *Advances in Data Science and Classification* (eds. S. Borra, R. Rocci, M. Vichi, and M. Schader), 93–100, Springer-Verlag, Heidelberg.

Coppi, R. and D'Urso, P. 2002. Fuzzy K-means clustering models for triangular fuzzy time trajectories, *Statistical Methods and Applications*, 11, 21–40.

Coppi, R. and D'Urso, P. 2003. Three-way fuzzy clustering models for LR fuzzy time trajectories, *Computational Statistics & Data Analysis*, 43, 149–177.

Coppi, R. and D'Urso, P. 2006. Fuzzy unsupervised classification of multivariate time trajectories with the shannon entropy regularization, *Computational Statistics & Data Analysis*, 50(6), 1452–1477.

Coppi, R., D'Urso P., and Giordani, P. 2010. A fuzzy clustering model for multivariate spatial time series, *Journal of Classification*, 27, 54–88.

Dose, C. and Cincotti, S. 2005. Clustering of financial time series with application to index and enhanced-index tracking portfolio, *Physica A*, 355, 145–151.

D'Urso, P. 2000. Dissimilarity measures for time Trajectories, *Statistical Methods and Applications*, 1–3, 53–83.

D'Urso, P. 2004. Fuzzy C-means clustering models for multivariate time-varying data: Different approaches, *International Journal of Uncertainty, Fuzziness and Knowledge-Based Systems*, 12(3), 287–326.

D'Urso, P. 2005. Fuzzy clustering for data time arrays with inlier and outlier time trajectories, *IEEE Transactions on Fuzzy Systems*, 13(5), 583–604.

D'Urso, P. and De Giovanni, L. 2008. Temporal self-organizing maps for telecommunications market segmentation, *Neurocomputing*, 71, 2880–2892.

D'Urso, P. and Maharaj, E.A. 2009. Autocorrelation-based fuzzy clustering of time series, *Fuzzy Sets and Systems*, 160, 3565–3589.

D'Urso, P. and Maharaj, E.A. 2012. Wavelets-based clustering of multivariate time series, *Fuzzy Sets and Systems*, 196, 33–61.

Galeano, P. and Peña, D. 2000. Multivariate analysis in vector time series, *Resenhas*, 4, 383–404.

Goutte, C., Toft, P., Rostrup, E., Nielsen, F., and Hansen, L.K. 1999. On clustering fMRI time series, *NeuroImage*, 9(3), 298–310.

Hart, J.D. 1997. *Nonparametric Smoothing and Lack-of-Fit Tests, Springer Series in Statistics*, Springer, New York.

Jeong, Y.-S., Jeong, M.K., and Omitaomu, O.A. 2011. Weighted dynamic time warping for time series classification, *Pattern Recognition*, 44, 2231–2240.

Johnson, R.A. and Wichern, D.W. 1992. *Applied Multivariate Statistical Analysis*, 3rd Edition, Prentice-Hall, Englewood Cliffs, NJ, V. 2007, 706–715.

Kakizawa, Y., Shumway, R.H., and Taniguchi, M. 1998. Discrimination and clustering for multivariate time series, *Journal of the American Statistical Association*, 93, 328–340.

Kalpakis, K., Gada, D., and Puttagunta, V. 2001. Distance measures for the effective clustering of ARIMA time-series, *Proceeding of the IEEE International Conference on Data Mining*, San Jose, 273–280.

Kannathal, N., Choo, M.L., Acharya, U.R., and Sadasivan, P.K. 2005. Entropies in the detection of epilepsy in EEG. *Computer Methods and Programs in Biomedicine*, 80(3), 187–194.

Liao, T.W. 2005. Clustering of time series data: A Survey, *Pattern Recognition*, 38, 1857–1874.

Maharaj, E.A. 1996. A significance test for classifying ARMA models *Journal of Statistical Computation and Simulation*, 54, 305–331.

Maharaj, E.A. 1999. The comparison and classification of stationary multivariate time series, *Pattern Recognition*, 32(7), 1129–1138.

Maharaj, E.A. 2000. Clusters of time series, *Journal of Classification*, 17, 297–314.

Maharaj, E.A. and Alonso, A.M. 2007. Discrimination of locally stationary time series using wavelets, *Computational Statistics and Data Analysis*, 52, 879–895.

Maharaj, E.A. and D'Urso, P. 2010. A coherence-based approach for the pattern recognition of time series, *Physica A., Statistical Mechanics*, 389(17), 3516–3537.

Maharaj, E.A. and D'Urso, P. 2011. Fuzzy clustering of time series in the frequency domain, *Information Sciences*, 181, 1187–1211.

Maharaj, E.A., D'Urso, P., and Galagedera, D.U.A. 2010. Wavelets-based fuzzy clustering of time series, *Journal of Classification*, 27(2), 231–275.

Miskiewicz, J. and Ausloos, M. 2008. Correlation measure to detect time series distances, whence economy globalization, *Physica A* 387, 6584–6594.

Nigam, V.P. and Graupe, D. 2004. A neural-network-based detection of epilepsy, *Neurological Research*, 26(1), 55–60.

Otranto, E. 2008. Clustering heteroskedastic time series by model-based procedures, *Computational Statistics & Data Analysis*, 52, 4685–4698.

Otranto, E. 2010. Identifying financial time series with similar dynamic conditional correlation. *Computational Statistics & Data Analysis*, 54(1), 1–15.

Peña, D. and Prieto, F.J. 2001. Cluster identification using projections. *Journal of the American Statistical Association*, 96, 1433–1445.

Percival, D.B. and Walden, A.T. 2000. *Wavelet Methods for Time Series Analysis*, Cambridge University Press. Cambridge.

Piccolo, D. 1990. A distance measure for classifying ARIMA models, *Journal of Time Series Analysis*, 11(2), 153–164.

Ramoni, M., Sebastiani, P., and Cohen P. 2002. Bayesian clustering by dynamics, *Machine Learning*, 47(1), 91–121.

Savvides, A., Promponas, V.J., and Fokianos, K. 2008. Clustering of biological time series by cepstral coefficients based distances, *Pattern Recognition*, 41, 2398–2412.

Shumway, R.H. 2003. Time-frequency clustering and discriminant analysis. *Statistics Probability Letters*, 63(3), 307–314.

Singhal, A. and Seborg, D. 2005. Clustering multivariate time series data, *Journal of Chemometrics*, 19, 427–438.

Takayuki, M., Takayasu, H., and Takayasu, M. 2006. Correlation networks among curriencies, *Physica A*, 364, 336–342.

Tong, H. and Dabas, P. 1990. Clusters of time series models: An example, *Journal of Applied Statistics*, 17, 187–198.

Wang, N. and Blostein, S. 2004. Adaptive zero-padding OFDM over frequency-selective multipath channels, *Journal on Applied Signal Processing*, 10, 1478–1488.

Wang, X., Wirth, A., and Wang, L. 2007. Structure-based statistical features and multivariate time series clustering. *Seventh IEEE International Conference on Data Mining*, pp. 351–360.

Xiong, Y. and Yeung, D.Y. 2002. Mixtures of ARMA models for model-based time series clustering, *Proceedings of the IEEE International Conference on Data Mining*, Maebaghi City, Japan.

Zhang, H., Ho, T.B., Zhang, Y., and Lin, M. 2005. Unsupervised feature extraction for time series clustering using orthogonal wavelet transform, *Informatica*, 30, 305–319.

13

Clustering Functional Data

David B. Hitchcock and Mark C. Greenwood

CONTENTS

Abstract

Functional data are data that arise as curves. These curves may be functions of time, space, or some other independent variable(s). Clustering of functional data may be based on characteristics of the curves such as positions, shapes, or derivatives. Clustering algorithms may be based on specialized dissimilarity measures, may be based on clustering the coefficients of basis function expansions of the data, or may cluster the functions directly. We review the ever-growing literature of methods for clustering functional data. Early methods emphasized clustering basis coefficients, while later research put forth model-based methods and incorporated more complex dependency structures. We discuss available software for functional clustering and give two illustrative examples on real and simulated data. We discuss open problems for future research in clustering functional data.

13.1 Introduction

Functional data are data that can be thought of as curves, or functions of one or several independent variables. A simple and classical example is the set of growth curves for children in the Berkeley Growth Study (Tuddenham and Snyder 1954). The response variable (denoted by y), height, can be understood to vary continuously across values of the independent variable (or charting variable), age (denoted by t). In practice, if response values are measured on a grid $t_1, \ldots, t_{n_i}$ of t-values for a sample of N children, then a smooth function approximating $y_i(t)$ may be fit for each child, and the data analysis is undertaken on the sample of these curves $y_1(t), \ldots, y_N(t)$.

The term "functional data analysis" was coined by Ramsay and Dalzell (1991) and popularized by the book of Ramsay and Silverman (1997, 2005), although the approach dates back at least to the work of Rao (1958) and Tucker (1958). Functional data are similar in nature to longitudinal data or time-series data, but there are distinguishing qualities that particularly characterize functional data. For a variable to be considered functional, one must be able to conceive of the quantity existing at any value of time (or other charting variable) along an interval T; in practice, the response is measured at typically many (or occasionally a few) snapshots within T. (When the measurements represent aggregates or averages over a subinterval of time, these do not strictly qualify as snapshots, but if they are averages over some relatively short time period, they may practically be treated as such.) The correlation structure between adjacently measured response points $y_i(t_j), y_i(t_{j+1})$ is important and is usually modeled, but it is probably even more important to model the overall (mean) structure of the entire curve $y_i(t)$. In addition, functional data analysis is typically designed to examine patterns of variation for a whole set of curves $y_1(t), \dots, y_N(t)$ taken to exist on a fixed interval T, as opposed to a time series forecasting problem in which the goal is to model the structure of a single response function in order to predict future measurements (extending beyond the observed interval). While the internal dynamics of the multivariate time series can be useful in estimating the functional observations, they are not the main interest in functional data analysis.

Often in analyzing functional data, information about the derivatives of the function is more important than the function itself. In the growth curve example, variation among subjects is typically more apparent from the rate of change of height (first derivative) or even the height acceleration (second derivative) than the height function itself. Fine details that are imperceptible on examining the observed functions become clear on examining their derivatives. Modeling the observed responses as smooth functions allows quick access to that derivative information—and consistent estimates of derivative functions (Hall et al. 2009)—a key difference from multivariate data analysis. In terms of clustering based on derivative information, various approaches could be taken. One might determine smooth functions to represent the observed data, calculate the corresponding derivatives, and then cluster the derivative functions using the type of ordinary clustering methods for functional data that will be described in this chapter. Or one could employ a clustering algorithm that intrinsically groups the observed curves based on rate-of-change variation rather than on variation in the original curves.

Most typically clustering is performed in an L_2-metric space (where the squared L_2 distance between curves i and i' is defined as $\int_T [y_i(t) - y_{i'}(t)]^2 dt$) leading to the standard k-means or hierarchical clustering methods discussed later. The transition to an L_2 comparison of the derivatives of the functions, $\int_T [y_i'(t) - y_{i'}'(t)]^2 dt$, moves to semimetric comparisons of functions (Ferraty and Vieu 2006). Researchers are able to select dissimilarity measures to highlight aspects of the functions of interest and identify groups of functions that share characteristics that are possibly not obvious in initial inspection of the observations.

The methods are generally organized into two-step methods based on calculating a dissimilarity matrix and applying a conventional clustering algorithm, and algorithms that cluster directly using features of functional data, some of which are model-based. Also, some methods in each group are limited in application to functional observations observed over a fine grid of timepoints, while others have the potential to be applied when observations are sparsely observed over a coarse time grid. Sparse functional data are observations that are assumed to vary smoothly over time but are irregularly measured over the time interval of interest.

The overarching goal of investigation patterns of variation in a set of curves makes the cluster analysis of functional data a popular technique. Numerous applications across a variety of fields will be given in Section 13.2.2. As a common example, biologists may measure expression ratios across time for various genes. For each object (gene), the expression ratio as a function of time is the functional datum. Cluster analysis allows biologists to identify similar genes and meaningful groups of genes and may increase scientific knowledge.

13.2 Existing Work

13.2.1 Origins

In the early development of functional data analysis, much work consisted of taking classical multivariate data analyses and introducing analogous methodology in the realm of functional data. For example, important early work in functional principal components analysis (PCA) included that of: Besse and Ramsay (1986); Ramsay and Dalzell (1991); Kneip (1994); Ramsay et al. (1995); and Besse et al. (1997). Meanwhile, Hastie et al. (1995) presented an early treatment of discriminant analysis and supervised classification in the functional context. Soon after these seminal works on FDA, methods of cluster analysis for functional data would follow.

The number of methods for clustering vector-valued data is seemingly myriad, and so a natural approach to take when clustering functional data are to represent each observed function as a vector of numbers and then cluster those using a standard multivariate technique. This could be performed on the initial measurements. Another example of this is to smooth each observed curve via nonparametric regression and take the N vectors of fitted values, say $\hat{\mu}_i, i = 1, \ldots, N$, along some grid of t-values as the object to be input into the clustering algorithm. One drawback of this approach, sometimes known as regularization, is that the number of values on the grid may need to be quite large to capture well the peaks and valleys of the curves, and it may be inappropriate to treat all the many fitted values equally for the cluster analysis. It is important to remember that each functional datum is assumed to be continuous and thus the resolution of the discretization of the process for this type of clustering is to some degree arbitrary.

A more parsimonious way to represent the curve is to model it with a set of basis functions, i.e.,

$$y_i(t) = \beta_{i0} + \sum_{j=1}^{p} \beta_{ij} \phi_j(t).$$

The basis coefficients $\beta_i = \beta_{i0}, \beta_{i1}, \ldots, \beta_{ip}$, which characterize the ith curve, can be estimated via least squares. Then the objects are clustered based on the estimated coefficients $\hat{\beta}_i, i = 1, \ldots, N$. This is sometimes known as filtering (James and Sugar 2003). Abraham et al. (2003) used a B-spline basis to represent each curve and used k-means on the estimated coefficients to cluster the objects. This reduces each continuous functional datum to $m + d - 1$ coefficients, where m is the number of knots used in the spline basis and d is the degree of the B-splines (often 3 which provides cubic B-splines). Abraham et al. (2003) prove a consistency result: When using their method, as the number of observed

curves $N \to \infty$ and the number of measurement points $n \to \infty$, the cluster centers converge to a unique set that defines the clusters. The theorem assumes the signal functions
$\mu_i(t), i = 1, \ldots, n$ come from a (rather general) class of bounded functions, and that the random measurement errors follow an i.i.d. mean-zero, constant variance distribution. (Only
the assumption of independent measurement errors might be unrealistic in certain applications.) The theorem holds for other bases, but B-splines are often preferred because they
are highly flexible even with a small number of coefficients, each of which plays a meaningful role in characterizing the curve. If the observations are periodic, Fourier bases can
be useful and if the observations are "rough," e.g., containing sharp peaks, wavelets can
provide efficient representations. Wavelets and Fourier bases can be inefficient for smooth
but non-periodic functional observations.

More theory about the practice of using k-means on basis coefficients was developed by
Tarpey and Kinateder (2003). They defined the principal points of a random function as a
set of k functions that essentially approximated the function optimally in a squared-error-
loss sense. They connected these principal points to the cluster means from the k-means
clustering of regression coefficients from an orthonormal basis function expansion. If the
random function can be assumed to be Gaussian, the k principal points are shown to lie in
the subspace spanned by the first few eigenfunctions. This implies that clustering Gaussian
random functions always reduce to a finite-dimensional problem. Tarpey and Kinateder
(2003) give one example that uses a Fourier basis. In their second example, clusters of
the raw data curves are simple vertical shifts of each other, but clustering the derivative
functions yields three intriguingly distinct groups of trends in stock prices.

Serban and Wasserman (2005) also propose a method based on k-means on a set of coefficient estimates that includes a variety of interesting other aspects. They use a Fourier
basis, choosing the number of basis functions to minimize a total regret measure (a sum of
marginal risks). To improve the clustering, they remove all curves that are deemed "flat" by
N simultaneous hypothesis tests on the curves, each testing whether all basis coefficients
are 0 for the respective curve. This removal may not make sense for all applications. The
most novel aspect of this approach is that they estimate the clustering error rate that arises
from using estimated curve coefficients instead of the unknown true coefficients.

Tarpey (2007), also considering k-means on estimated coefficients, found that different
choices of basis that produce nearly identical fitted curves can yield markedly different
clustering structures. The k-means algorithm is based on minimizing a sum of squared
Euclidean distances from objects to their cluster means. For functional data, the analogous
distance measure is the squared L_2 distance between a function $y_i(t)$ and a cluster mean
curve (say $\xi(t)$):

$$\int_{\mathcal{T}} [y_i(t) - \xi(t)]^2 \, dt.$$

Tarpey (2007) showed that for k-means clustering on the coefficients to correspond to
this L_2 distance measure, the coefficients must be transformed (or else the basis must be
orthonormal on $\mathcal{T}$). The optimal transformation is one that maximizes between-cluster
variability; that is, the transformation from $\mathbf{y}_i$ to (the transformed) $\hat{\beta}_i$ should stretch the
data along the direction corresponding to the differences in clusters. Unfortunately, this
direction is unlikely to be known at the time of the cluster analysis, so the appropriate
transformation (or equivalently, the best choice of basis) is far from obvious.

One drawback to the classical k-means approach is the fact that the L_2-based objective
function leads to an algorithm that is not robust to outlying objects. In the multivariate

clustering setting, alternative methods have been proposed, such as the k-medoids method (Kaufman and Rousseeuw 1987), implemented by the Partitioning Around Medoids (PAM) algorithm of Kaufman and Rousseeuw (1990). Cuesta-Albertos et al. (1997) proposed trimmed k-means, another robust alternative to k-means. Trimmed k-means calculates the cluster means based on only $N(1 - \alpha)$ observations (rather than all N) for some trimming amount $\alpha \in (0, 1)$. García-Escudero and Gordaliza (2005) adapted the trimmed k-means method for functional data clustering. Using least squares, they fit each observed curve via a cubic B-spline basis and perform trimmed k-means clustering on the estimated coefficients. This yields a connection with functional data depth (Cuesta-Albertos and Fraiman 2006); the larger α must be for a particular functional object to be "trimmed," the deeper within the sample that curve is. The joint choice of α and the number of clusters k is determined using trimmed k-variance functionals, which are based on the rate of change of the minimized trimmed k-means objective function.

Traditional hierarchical clustering methods (and some partitioning algorithms) accept as their input a dissimilarity matrix rather than a raw data matrix (as k-means does). A viable approach to clustering functions would then be to calculate dissimilarities between every pair of observed curves and input these dissimilarities into, say, a hierarchical algorithm (like average linkage or Ward's method). A natural measure of dissimilarity between two functions $y_i(t)$ and $y_{i'}(t)$ is the squared L_2 distance

$$d_{ii'} = \int_T [y_i(t) - y_{i'}(t)]^2 \, dt,$$

which is usually evaluated using numerical integration by Simpson's rule. However, Heckman and Zamar (2000) suggested a different dissimilarity measure based on the *rank correlation* between two functions. Consider a particular value $t^* \in T$. They define the rank of a point on a curve $y(t^*)$ (relative to a measure μ) to be a function of t^* as follows:

$$r^y(t^*) = \mu[s : y(s) < y(t^*)] + 0.5\mu[s : y(s) = y(t^*)].$$

If μ is the counting measure on a set of values $t_1, \ldots, t_n$, then the rank of $y(t_i)$ relative to $y(t_1), \ldots, y(t_n)$ is essentially the number of $y(t_j)$'s less than $y(t_i)$, plus a midrank adjustment for tied values. Then the rank correlation between, say, curves $x(t)$ and $y(t)$ is

$$\rho(x, y) = \frac{\int [r^x(t) - R^x][r^y(t) - R^y] \, d\mu(t)}{\{\int [r^x(t) - R^x]^2 \, d\mu(t) \int [r^y(t) - R^y]^2 \, d\mu(t)\}^{1/2}},$$

where $R^y = \int r^y(t) \, d\mu(t)$. A major benefit of this definition is that the measure μ is highly general, which allows the user to choose μ to compare the two curves on whichever subset of T is of particular interest. In this way this meaning of "similarity" is allowed to be highly malleable. Heckman and Zamar (2000) present a consistent estimator of this rank correlation based on two observed random functions; they use it to cluster Canadian weather stations, based on temperature curves discussed in Ramsay and Silverman (2005), using a hierarchical method.

Other possible dissimilarity measures include one based on a generalization of Spearman correlation:

$$d_{ii'} = 1 - \frac{\int_T [y_i(t) - \bar{y}_i][y_{i'}(t) - \bar{y}_i] \, dt}{\sigma_{y_i} \sigma_{y_{i'}}},$$

where, for example, $\bar{y}_i = \int y_i(t)\,dt$ and $\sigma_{y_i}^2 = \int [y_i(t) - \bar{y}_i]^2\,dt$. Heckman and Zamar (2000) argued that this measure behaves similarly (with similar deficiencies) as squared L_2 distance. In addition, Marron and Tsybakov (1995) suggested "error criteria" designed to give a better visual representation of the discrepancies between curves; these could also be used as dissimilarities in a cluster analysis.

We have seen that classical hierarchical clustering algorithms can be applied to functional data by either (1) reducing the functional observations to multivariate ones, either by working with basis coefficients or with vectorized responses over a grid; or by (2) calculating dissimilarities between pairs of functional observations and inputting the dissimilarities into the algorithm rather than the raw functional data. A hierarchical clustering method that directly handles the data as functional (rather than reducing them to an easier form) was presented by Ferraty and Vieu (2006) and applied to radar waveform data by Dabo-Niang et al. (2007). Ferraty and Vieu (2006), in the context of distributions of random functions, discussed some measures of centrality for such distributions. The sample mean is straightforward: $\bar{y}(t) = N^{-1} \sum_{i=1}^{N} y_i(t)$ estimates $E[y(t)]$. The functional median is defined, for any particular distance or "semi-metric" $d(\cdot, \cdot)$, as

$$\inf_{m(t)} E[d(m(t), y(t))].$$

Given a sample $S = \{y_1(t), \ldots, y_N(t)\}$ of curves, a sample median is the sampled function that produces the smallest sum of distances to other curves in the sample:

$$\inf_{m(t) \in S} \sum_{i=1}^{N} d(m(t), y_i(t)).$$

This semimetric d could be based on an integrated second derivative, a horizontal shift, a PCA, or other choices. Their definition of a mode requires a density f for the functional variable(s). The function $\chi(t)$ (of a certain class) that maximizes the density is the mode. A sample mode requires (1) estimating the density f, which is done via kernel methods, and (2) estimating the mode of f given a sample of curves, which is done by taking the sampled curve with the highest density value $f(y_i(t))$. Finally, a heterogeneity index is proposed that essentially measures the discrepancy between the functional mode of a data set and another measure of centrality (either the mean or median), for example:

$$HI(S) = \frac{d(m(t), \chi(t))}{d(m(t), 0) + d(\chi(t), 0)}.$$

A subsampled $SHI(S)$ is also presented for which the HI is averaged over a large set of randomly generated subsamples of S. The hierarchical clustering method comes from considering splitting the sample S into G subgroups and comparing $SHI(S)$ for the whole sample to

$$SHI(S; S_1, \ldots, S_G) = \frac{1}{Card(S)} \sum_{k=1}^{G} Card(S_k) SHI(S_k),$$

(where $SHI(S_k)$ is the subsampled heterogeneity index applied to the kth subgroup) and proceeding with the division if the *splitting score*

$$SC = \frac{SHI(S) - SHI(S; S_1, \ldots, S_G)}{SHI(S)}$$

is less than some criterion. The number of subgroups, and the division of functions into subgroups, is based on a density estimate for a set of "small ball concentration curves" (Ferraty and Vieu 2006). This approach has similarities to traditional divisive hierarchical clustering methods (such as linkage methods), but in its treatment of distributions of functional data, it is fundamentally different from simply defining a distance metric for functional data and then applying a traditional hierarchical algorithm to the resulting dissimilarity matrix.

Model-based cluster analysis has been a highly common approach to clustering, especially since the work of McLachlan and Basford (1988), Banfield and Raftery (1993), and Fraley and Raftery (2002) popularized it. While the multivariate normal model has been employed for the bulk of model-based clustering applications, formal clustering models for other data types—including functional data—have been proposed.

James and Sugar (2003) proposed a model-based method designed for functional observations (potentially) measured at relatively few values of the charting variable, motivated by the poor performance of regularization and filtering approaches for clustering such *sparse* functional *data*. They modeled a functional observation belonging to cluster k, in its vectorized form $\mathbf{y}_i = (y_i(t_1), \ldots, y_i(t_n))$, as:

$$\mathbf{y} \sim N(\boldsymbol{\mu}_k, \boldsymbol{\Omega}_k + \sigma^2 \mathbf{I}),$$

where $\boldsymbol{\mu}_k = [\mu_k(t_1), \ldots, \mu_k(t_n)]'$ is the (vectorized) mean curve for cluster k and $\boldsymbol{\Omega}_k$ contains the covariances between measurements across timepoints (also specific to cluster k). The structure of the curves is modeled by letting $\mu_k(t) = \mathbf{s}(t)'(\boldsymbol{\lambda}_0 + \boldsymbol{\Lambda}\boldsymbol{\alpha}_k)$, where $\mathbf{s}(t)$ is a spline basis vector and $\sum \boldsymbol{\alpha}_k = \mathbf{0}$ so that $\mathbf{s}(t)'\boldsymbol{\lambda}_0$ represents an overall mean curve, and the covariance elements $\omega_k(t, t') = \mathbf{s}(t)'\boldsymbol{\Gamma}\mathbf{s}(t')$. In this formulation, the spline coefficients are random effects (with covariance matrix $\boldsymbol{\Gamma}$) that are estimated from data across all individuals, rather than fixed effects estimated from data on each individual separately. This provision for "borrowing strength" is convenient when each individual may only have a few measurement points. Sugar and James (2003) gave a method related to this model-based approach for choosing the correct number of clusters. Defining the "distortion" d_K for a K-cluster solution as an average Mahalanobis distance between each vector of spline coefficients and its nearest cluster center, they suggested letting the largest jump between d_K^{-1} and d_{K-1}^{-1} determine the number of clusters K.

Also described as a model-based approach is the method of Ma et al. (2006). Instead of choosing a fixed set of basis functions to approximate the observed curves, they chose a curve estimator based on minimizing a penalized sum-of-squared-errors term

$$\sum_{j=1}^{n} [y_j - f(t_j)]^2 + \lambda \int [f''(t)]^2 \, dt,$$

which results in a cubic smoothing spline estimator being chosen. To improve computational efficiency, a Rejection Control EM algorithm was used to estimate simultaneously the signal curves and cluster membership labels for the curves. The algorithm searches over solutions positing a varied number of clusters; the best model is chosen using BIC, which tends to penalize solutions with a large number of clusters. The method was applied by Ma et al. (2006) to two genetic data sets using sampled flies and nemotodes.

This model was extended by Ma et al. (2008) via a more overtly Bayesian approach involving a mixed-effect smoothing spline model for each function, a Dirichlet prior on the cluster probabilities, and a Gibbs sampler to sample from the posterior distribution for the various parameters. Finally, Ma and Zhong (2008) proposed a method for functional clustering in the presence of covariates. They proposed a general model for the ith (discretized) random function y_i measured at timepoints t_i:

$$\mathbf{y}_i = \mu(t_i) + Z_i b_i + \epsilon_i.$$

The Z_i portion of the model can account for a variety of covariate types, such as treatment groups or regressors. The authors chose $\mu(t)$ to minimize a penalized criterion over a reproducing kernel Hilbert space. Putting priors on the elements of a decomposition of μ, they employed Bayesian inference, which yields posterior probabilities of cluster membership for each functional object.

Approaches to clustering based on finite mixture models often result in cluster membership probabilities for each functional object; each observed function is not necessarily assigned to a single cluster with certainty, but rather could straddle two or more clusters with certain probabilities of belonging to each. This is reminiscent of the well-known *fuzzy clustering* of multivariate data. Tokushige et al. (2007) adapted the fuzzy k-means algorithm of Ruspini (1960) to the functional case. The membership probabilities from this fuzzy method may not arise as organically as those generated within a mixture-model framework, but an interesting aspect of the method of Tokushige et al. (2007) is that the cluster membership probabilities vary with the functional argument t. In other words, a functional object may be likely to belong to, say, cluster 1 at time t_1 but be more likely to belong to cluster 2 at a later time t_2. This seems to make interpretations of the partitions of objects more reasonable when considering a cross-section of the curves (at one fixed time) than when considering the curves in their entireties.

In cluster analysis of multivariate data, it is common that only a few of the (potentially many) observed variables contribute to the clustering structure. The unimportant (from a clustering perspective) variables, often known as masking variables, should be identified and their influence on the cluster analysis downweighted or eliminated. Gattone and Rocci (2012) clustered functional data within a reduced-dimensional subspace, with a similar goal. They proposed a functional k-means algorithm in which cluster centroids are constrained to lie in the lowest-dimensional subspace that still contains much of the cluster-discrimination information in the original data. The constraint is incorporated through an alteration of the ordinary k-means maximization criterion. They represent the curves via basis functions and regularize them with a roughness penalty. It is notable that if the number of clusters is set equal to the number of functional observations, this method reduces to a functional principal components analysis (FPCA).

Peng and Müller (2008) developed methods for calculating a distance matrix between sparsely observed functional data that relies on an FPCA and the assumption that each functional observation is a smoothly varying stochastic process. They deal with the missing information by sharing information across different curves via an estimated variance-covariance matrix. One of their primary tools is to perform an FPCA, estimating the overall mean and variance-covariance matrix of the functions, and truncating the eigenvectors and eigenvalues to provide a lower dimensional representation of the functional data. They suggest an alternative to the standard L_2-metric that is conditional on the observed times of (sparse) observations: the distance between two complete curves that were not observed $(y_i(t), y_{i'}(t))$ based on the observed (sparse) curves $(y_i^*(t), y_{i'}^*(t))$ using $\tilde{d}_{ii'} = \{E(d_{ii'}^2|y_i^*(t), y_{i'}^*(t))\}^{1/2}$. An unbiased estimator of this quantity is developed by Karhunen–Loève expansion of each function

$$y_i(t) = \mu(t) + \sum_{j=1}^{\infty} \xi_{ij} \rho_j(t),$$

where ξ_{ij} is the coefficient for the ith curve and jth eigenfunction $\rho_j(t)$. The L_2 distance between the completely observed functions is equivalent to the distance between the coefficients if the curves were completely observed. After truncating for the first K eigenvectors, the distance conditional on the actually observed functions is defined as

$$\tilde{d}_{ii'}^K = \left\{ E \left(\sum_{k=1}^{K} (\xi_{ik} - \xi_{i'k})^2 | y_i^*(t), y_{i'}^*(t) \right) \right\}^{1/2}.$$

The derivation of an estimator of this distance is nontrivial, but the result is based on various quantities from the FPCA, including estimates of the mean and variance-covariance of the observations. After obtaining the estimated conditional distance matrix, the researcher is free to choose any applicable clustering algorithm. They utilized k-means in their paper for an application to online bid trajectories.

A drawback to the basis-function-based approach to functional clustering is that the same basis must be used for all N observed functions. This can be practically feasible if the basis is sufficiently flexible (e.g., by containing a large number of basis functions), but it is also sensible that the same basis will not be the ideal representation of many functions. Chiou and Li (2007) viewed each observed curve as a realization from a general distribution of random functions. Specifically, the observations come from a Hilbert space $L_2(\mathcal{T})$ of square-integrable functions on interval $\mathcal{T}$; the curves are further identified with subpopulations defined by a clustering variable C. They consider Karhunen–Loève expansion of each function for the different clusters; the discrepancy (measured via an L_2 distance) between the observed function and the (truncated) expansion associated with cluster c, say, indicates the likelihood that the function belongs in the cth cluster. Each function is assigned to that cluster for which this discrepancy is minimized. The functions are iteratively partitioned in a manner not unlike the k-means algorithm, until an integrated-squared error criterion is met. An advantage of k-centres is that since the eigenfunctions capture information about the covariance structure of the functional data, the clusters are based on differences in covariance as well as mean structure among the observed curves. Some drawbacks include the strong sensitivity of the results to the initial partition (which is based on functional principal component scores) and the fact that the algorithm is computationally slower than competitors that involve simpler bases. The estimation of different mean and covariance structures for each cluster distinguishes this method from Peng and Müller (2008), which uses a common estimate for all the observations, making it more applicable to sparse data situations than this method.

In some applications, a nonpositive semidefinite variance-covariance matrix is observed leading to possible negative distances between observations. More recently, Peng and Paul (2009) proposed an improved estimation technique for FPCA that avoids these issues. The newer estimation algorithm should alleviate the issues found in the original conditional distance estimator. When functional observations are completely observed on a regular grid (not sparse), the conditional distance is equivalent to the standard L_2 distance. Also note that in the original implementations, observations are allowed to be sparse but are required to have more than one measurement time per functional datum.

Like any random quantity, observed functions contain both signal and noise components. In many functional data analyses, the observed curves are smoothed before the analysis itself begins. The idea is—as much as possible—to perform the method (e.g., clustering) on the signal curves rather than on the noise. The effect of such pre-smoothing is usually beneficial, but could be detrimental if the data are, say, oversmoothed. A general data-driven James-Stein-type smoothing method was presented by Hitchcock et al. (2007). Let $\mathbf{y}_i = \big(y_i(t_1), \ldots, y_i(t_n)\big)$ denote the discretized version of observed function $y_i(t)$. Let $\hat{\mu}_i = \mathbf{S}y_i$ be a smoothed version of y_i, where $\mathbf{S}$ is the smoothing matrix of some linear smoother. Then

$$\hat{\mu}_i^{(JS)} = \mathbf{S}y_i + \left(1 - \frac{n - r - 2}{||y_i - \mathbf{S}y_i||^2}\right)_+ (y_i - \mathbf{S}y_i),$$

where $(\cdot)_+$ is the positive part, and $r = \text{rank}(\mathbf{S})$. This is essentially a weighted average of the observed discretized curve and the smoothed version, with data-driven weights. The authors chose a cubic B-spline smoother, and found that a large degree of smoothing within $\mathbf{S}$—compensated by the contribution of $\mathbf{y}_i$ which protects against oversmoothing—produced the most accurate results in terms of recovering clustering structure.

13.2.2 More Existing Work: Applications

Interesting applications of functional data clustering have appeared in step with methodological developments, and in many cases the applications have spurred theoretical advances.

In certain functional data applications, information taken from derivatives is at least as valuable as information in the observed curves. The price profiles of eBay online auctions, in which the current price listing varies continuously (in a nondecreasing manner) over time, were studied by Jank and Schmueli (2009). They used smoothing splines (smooth enough to estimate the first two derivative functions) to represent the observed curves and clustered with the k-medoids algorithm on the estimated spline coefficients. They used the information-theoretic approach of Sugar and James (2003) to decide upon three clusters: One cluster is characterized by near-constant price acceleration over time, a second cluster shows steeply increasing price acceleration near the end of the auction, and the third shows a highly variable acceleration profile over the auction.

In clinical trials with placebos, respondents are often classified into clinically meaningful groups (such as nonresponders, placebo responders, true drug responders, and various mixtures thereof). Tarpey et al. (2003) clustered subjects in a longitudinal clinical trial into five specific groups based on responses measured at seven specific times. They used orthogonal polynomial basis functions to model the response curves for the subjects and based the clustering on two of the estimated coefficients. In particular, they viewed principal points and self-consistent points (Flury 1990, 1993) as means of the clusters of curves; these principal points were estimated via a semiparametric approach which performed better than k-means clustering of the coefficients in this case.

Some functional data sets reveal intracurve variation that is highly localized and may contain strong seasonal aspects. The French electricity power demand curves analyzed by Antoniadis et al. (2013) exemplify such data; they used a k-means approach heavily reliant on wavelet-basis representations of the observed curves.

The method of Ferraty and Vieu (2006) was applied by Dabo-Niang et al. (2007) to cluster a set of radar waveform curves generated by a satellite, originally studied by Dabo-Niang et al. (2004). The resulting groups of waveforms were interpreted as reflecting various types of ground (rocky, river, vegetation, lake, etc.) over which the satellite was passing.

While correlation among measurement points on a curve is often modeled, the majority of functional data analyses have treated the whole curves as independent functional observations. In other words, the observed functions could be considered random selections from a distribution of curves. In some applications, however, the curves themselves are correlated. For example, Jiang and Serban (2012) examined records (across time) of service accessibility for census tracts in Georgia. They treated these curves as spatially dependent since accessibility trends in adjacent tracts, say, would be more likely to be similar than trends in geographically disparate tracts. They modeled spatial dependence among clusters and among curves within a cluster. This was accomplished by (1) treating the vector of cluster membership labels as a realization from a locally dependent Markov random field and (2) allowing spatially correlated random errors in the model for a functional datum within a cluster. The number of clusters was chosen to minimize a model selection criterion. Jiang and Serban (2012) addressed the computational challenges that naturally arise when such an intricate model is specified. In the data example, Georgia tracts were divided into five clusters having mean curves of varying shapes. The spatial correlation was clearly seen in a map of Georgia with tracts color-coded by cluster.

For spatially correlated functional data, a hierarchical clustering approach was presented by Giraldo et al. (2012). Their weighted dissimilarity measure for the ith and jth curves, which accounted for spatial dependency, was

$$d_{ij}^w = d_{ij}\hat{\gamma}(h),$$

where d_{ij} is a distance like those considered in Tarpey (2007), the squared L_2 distance between basis coefficients (possibly weighted by a function of the basis functions unless an orthonormal basis is chosen). The spatial weight function $\hat{\gamma}(h)$ is the estimated trace-variogram function

$$\hat{\gamma}(h) = \frac{1}{2|N(h)|} \sum_{i,j \in N(h)} \int [x_i(t) - x_j(t)]^2 \, dt,$$

where $N(h)$ is a set of pairs of functions that are sufficiently close in space. This weighting in d_{ij}^w implies that functions that are far apart spatially are penalized in terms of their pairwise dissimilarity and are thus less likely to be grouped into the same cluster. Their method was applied to the Canadian temperature data of Ramsay and Silverman (2005).

In any type of cluster analysis, one should consider whether transformations of the data before clustering will improve the eventual result. Alignment of the observed curves is a preprocessing step of particular interest with functional data. The variation among functional data $y_i(t), i = 1, \ldots, N$ is often decomposed into phase variation and amplitude variation. Amplitude variation describes the different "vertical behaviors" of the curves—distinctions among the curves in terms of the response variable y—and it is this variation that any functional data analysis seeks to reveal or characterize. On the other hand, phase variation is the phenomenon that causes interesting characteristics of the curves to appear at different t locations across curves. For example, in the Berkeley growth data (Tuddenham and Snyder 1954; Ramsay and Silverman 2005), the inevitable pubertal growth spurt occurs at different times on different children's growth curves. To compare curves appropriately, this phase variation should be removed whenever possible to better reveal the amplitude patterns across curves. This removal is called alignment, or registration, of the functional data. It is often done by subjecting the t-axis of each curve to a warping operation denoted by a function $h(t)$; if done well, the resulting warped data set will have little phase variation and the remaining variability among curves will be amplitude variation.

Sangalli et al. (2010) presented a method that simultaneously aligned and clustered functional data. The method estimated k-template curves (one for each cluster) to which each of the curves assigned to that cluster was aligned. A cycle of (1) template curve identification, (2) assignment to clusters and alignment of curves, and (3) normalization of cluster locations was carried out in an iterative manner. Tang and Müller (2000) suggested a cluster-specific time warping method that removes the time-axis variation among curves, after which the warped curves can be clustered with traditional algorithms. They applied the method to clustering gene expression curves. Liu and Yang (2009) proposed a shape-invariant model for the ith observed function

$$y_{ij}(t) = d_i + m(b_i + t_{ij}) + \epsilon_{ij}, i = 1, \ldots, N, j = 1, \ldots, n_i$$

and modeled $m(t)$ using a cubic B-spline basis. They used an indicator vector Z_i for cluster membership and incorporated Z_i into a mixture model. Treating Z_i and the two "shift parameters" d_i and b_i as missing data, they used an EM algorithm to fit the model. An adjusted BIC measure was used to select to number of clusters and the number of knots in the B-spline basis. Liu and Yang (2009) derived some theoretical properties of their method and applied it to the Berkeley growth data.

Greenwood et al. (2011) applied functional clustering to high time frequency time series measurements of wetland hydraulic gradients. The analysis first considered the decomposition of the original time series into long-term trend functions and diurnal functions using generalized additive models (Wood 2006). The long-term trends were subject to occasional missing values due to equipment malfunction. A modified semimetric was suggested to compare the functional responses when both pairs were observed, weighting the comparisons based on the length of time both were available. Both the low frequency and diurnal functions were clustered using the partitioning around medoids (PAM) algorithm of Kaufman and Rousseeuw (1990), showing that group memberships differed across different years and whether the long-term trend or diurnal variation was clustered. The paper involved a translation from initial time series into nonparametric model estimated components to isolate different aspects of the initially observed time series; those estimated components are continuous and thus suitable for functional clustering.

Huzurbazar and Humphrey (2008) applied hierarchical functional clustering using a derivative-based semi-metric (Ferraty and Vieu 2006) to highlight connections in water pressure in different boreholes drilled in a glacier. Their data set presented two challenges that made functional clustering quite useful. First, the time of observations for different boreholes was not aligned although measurements were being made over the same time periods. Second, the water pressure was a function of the depth of the borehole; connectivity between locations (the reason for clustering the observations) was demonstrated through common variation in the derivatives of the records. Indications of the evolution of glacial hydrological systems were provided by studying changes in the cluster solutions for different time periods of the study.

13.3 Examples, Applications, Software

13.3.1 Software for Functional Data Clustering

Any implementation of cluster analysis with real data, especially functional data, will necessarily be computationally intensive and will require some type of statistical software. For the general practitioner (as opposed to the methodological researcher), methods for statistical clustering may be divided into several classes:

- Methods with available software that is straightforward to implement (with few and simple required choices of tuning parameters)
- Methods for which software is available but which is difficult for the non-expert to use due to less extensive documentation and/or the need for complicated tuning parameter choices
- Methods described in the literature but for which a practitioner must write his or her own code to implement (which could be difficult or easy depending on the method)

The simplest methods to implement tend to be those that are variants of well-known algorithms for clustering multivariate data. For example, one may employ the method of Abraham et al. (2003) in R (R Development Core Team, 2012) by fitting a B-spline curve estimate to each observed function using (for instance) the bs function in R, extracting the estimated coefficients, and inputting the coefficients into the kmeans function. There are still modeling specifications to be made, such as knot number and placement, spline degree, and number of clusters, but these can typically be done at least adequately by most practitioners. For dissimilarity-based methods, the user may need to write code to calculate the defined distance between a pair of functions, but once that is done, a standard R function such as hclust (or pam or agnes in the cluster package) can accept the dissimilarity matrix as input. Similarly, PROC CLUSTER within SAS software (SAS Institute Inc., 2008) can apply standard algorithms to cluster functional data in this manner.

Other algorithms—especially model-based methods—are designed specifically for functional data. The web site of G. James (at http://www-bcf.usc.edu/~gareth/) provides R functions to implement the methods of James and Sugar; these are particularly helpful for users whose data are sparse or contain different numbers of measurements across observed functions. There is documentation explaining the required inputs for the functions. Users with "nice" functional data in the form of an $N \times n$ matrix will need to preprocess the data (converting it into a long strung-out vector) before using the main clustering function.

The website (at http://www.math.univ-toulouse.fr/staph/npfda/) connected with the book of Ferraty and Vieu (2006) contains R functions to implement their hierarchical-type method. The provided documentation is a bit terse and opaque, however, and the main clustering function requires several other functions to be run (and occasionally adjusted) beforehand. These functions are provided "as is" and so may not be as useful to the non-expert practitioner.

At least two R packages contain functions to implement functional data clustering. The MFclust function in the MFDA package (Zhong and Ma 2012) (currently only available in the CRAN archive) implements the model-based method of Ma et al. (2006). The function works easily if the data are contained in a matrix where the rows represent (functional) observations and the columns represent (common) measurement points. The MFclust function will search for the best cluster solution (according to BIC) across a variety of numbers of clusters. The `fda.usc` package (Delaigle and Hall 2010) contains a `kmeans.fd` function that performs basic k-means clustering, for a fixed number of clusters, on "functional data objects" (which can be created by the `fdata` function from this package). The novelty is that `kmeans.fd` can base the cluster analysis on any of several semi-metrics defined by Ferraty and Vieu (2006).

In addition, the `km_PCA` function in the `modelcf` package, in its default setting, runs a simplified version of the k-centres method of Chiou and Li (2007).

13.3.2 Illustrative Real Data Examples and Applications

To illustrate the implementation and interpretation of cluster analyses for functional data using available software, we now present examples with two real data sets.

Hitchcock et al. (2007) analyzed wind speed data gathered at buoys in the Atlantic Ocean and Gulf of Mexico. The data are available at the National Buoy Data Center (NBDC) historical data page http://www.ndbc.noaa.gov/hmd.shtml at the National Oceanic and Atmospheric Administration (NOAA) web site. Wind speeds at 18 sites were measured each hour (in fact the average wind speed for each hour is recorded) during the first week of 2005, making 168 measurements at each site. Each site is located in one of four different regions (Northeast, Southeast, Florida/Eastern Gulf, and Western Gulf), as classified on the NBDC web site (Table 13.1). Initially, to remove any vertical shift variation among the curves, each functional observation was centered by subtracting off the mean of its 168 measurements, so that each curve was vertically centered at zero.

In the data set analyzed, curves 1 through 6 belonged to the Northeast region, curves 7–11 to the Southeast region, curves 12–14 to Florida/Eastern Gulf, and curves 15–18 to Western Gulf. An interesting question is whether a clustering solution tends to recover this regional partition of sites. We clustered these data using three different methods.

The model-based method of Ma et al. (2006) selects (via BIC) the number of clusters to be four. This approach was implemented via the `MFclust` function in the `MFDA` package in R; the resulting partition is given in the first column of Table 13.2. Using a method analogous to that of Abraham et al. (2003), we estimated the 18 signal curves using a cubic B-spline basis (with 4 knots equally spaced across the domain). We then used k-means clustering on the estimated coefficients to place the sites into four groups. The resulting partition is given in the second column of Table 13.2. Finally, we used the `kmeans.fd` function in the `fda.usc` package in R, specifying four clusters and an L_2 semi-metric. The resulting partition is given in the third column of Table 13.2.

As seen in Table 13.2, the Ma et al. (2006) (`MFclust`) and the Abraham et al. (2003) methods produce partitions fairly similar to the regional partition. Notably, site 18, despite being a Western Gulf station, is deemed closer to the "Southeast cluster" by all three algorithms.

TABLE 13.1

The 18 Sampled Sites, Categorized According to Their Regions

Region	NBDC Site Number
Northeast	44004, 44005, 44007, 44008, 44013, 44018
Southeast	40014, 41001, 41002, 41004, 41008
Florida/Eastern Gulf	41009, 41010, 42003
Western Gulf	42001, 42019, 42038, 42041

TABLE 13.2

The Classification of the 18 Sites into Four Clusters, Based on Three Different Clustering Approaches

Cluster	MFclust	k-means on B-spline Coefficients	k-means.fd	Regional Partition
1	1,2,3,4,5	2,3,4,5	2,3,4	1,2,3,4,5,6
2	6,7,8,9,18	8,9,10,18	7,8,9,10,11,18	7,8,9,10,11
3	10,11,12,13	1,6,7,11,12,13	1,5,6	12,13,14
4	14,15,16,17	14,15,16,17	12,13,14,15,16,17	15,16,17,18

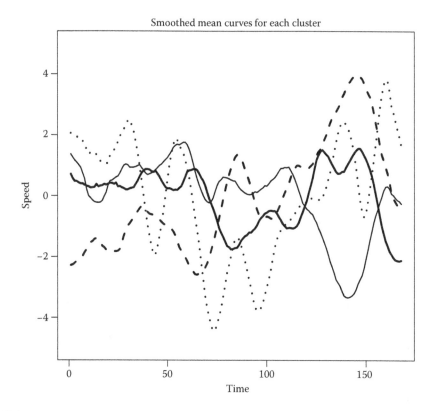

FIGURE 13.1

Smoothed cluster mean curves. Cluster 1 (dotted); cluster 2 (thick solid); cluster 3 (thin solid); cluster 4 (dashed).

We plot the smoothed cluster mean curves for the MFclust partition in Figure 13.1. The plot indicates that on average: Cluster 1 sites tend to have wildly oscillating windspeeds; cluster 2 sites have stable windspeeds except for a severe drop toward the end of the week; cluster 3 sites' windspeeds are stable, with a dip and then a rise near the end of the week, and cluster 4 sites have windspeeds that gradually rise over the week.

A functional PCA was conducted using the fdata2pc function in the fda.usc package; a plot of the first two principal component scores with separate symbols by cluster is given in Figure 13.2. We see clusters 2 and 4 are farthest apart on the first component of variation,

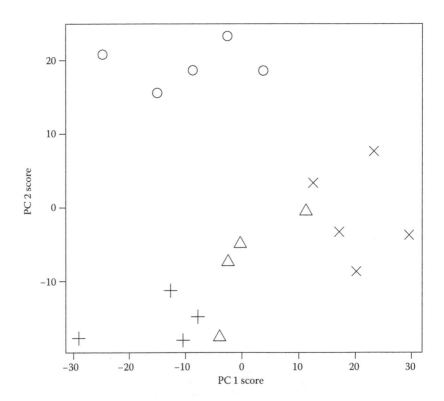

FIGURE 13.2
Principal component scores for the 18 sites, by cluster. Cluster 1 (circle); cluster 2 (X); cluster 3 (triangle); cluster 4 (plus).

while cluster 1 is well-separated from the other clusters in terms of the second component of variation.

To illustrate methods for clustering sparse functional data, we more formally introduce the famous Berkeley growth data set mentioned previously. The data set contains heights of 54 girls and 39 boys measured at 31 stages from 1 to 18 years with four measurements in the first year and bi-yearly measurements in the last 10 years of the study. These data are available in the `fda` package (Ramsay et al. 2012), are displayed in Figure 13.3a and have been analyzed in many papers, including being clustered in Chiou and Li (2007). Because this data set contains two potentially distinct groups, it is often of interest to compare a two-cluster solution to the sexes of the children. The children's growth trajectories are not necessarily completely distinct between the sexes, so reconstructing these groups based on their trajectories is a challenging problem.

The standard version of the data set belies the challenges of many longitudinal studies where observations, especially of human subjects, are difficult to consistently obtain. In order to make this data set more realistic, a random sample of only five of the 31 time points for each child is retained in the data set (Figure 13.3b). There are two methods discussed previously that are explicitly designed for sparse data situations such as these: James and Sugar's (2003) model-based clustering approach and Peng and Müller's (2008) conditional distance metric that is clustered using k-means. In order to assess the performance of the methods, the cluster solutions for the completely observed data sets are compared to the

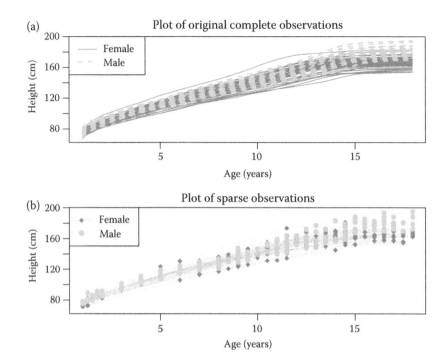

FIGURE 13.3
Plot of original Berkeley Growth Curve data (a) and sparse version of data set (b).

results from the sparse data set that contains only 16% of the original observations for each child. The adjusted Rand index (Hubert and Arabie 1985) provides a measure for comparing the similarity of different cluster solutions, adjusting the amount of agreement in the cluster solutions for chance. The larger the adjusted Rand index, the more agreement in the cluster solutions, with a maximum attainable value of one.

In the conditional distance implementation, the FPCA using five eigenvalues explains 95.5% of the variation in the responses. The two-cluster k-means result for the conditional distance for the original data set provides a 0.088 index with the sex of the children; the sparse data actually provides an index of 0.101, slightly better than for the original data set. The increased performance from the sparse data is likely an artifact of the particular random sample of times selected, with the agreement between the sparse and full data solutions of 0.578. It does show that these methods can provide reasonable clustering performance even when much of the individual trajectories is unobserved. Also note that the conditional distance is equivalent to the L_2 distance when the functional observations are completely observed. Interestingly, clustering using the PAM algorithm on the conditional distance matrix provides better adjusted Rand index performance both between the sparse and original data sets and relative to the sex of the children.

Another feature of the FPCA approach is the possibility of exploring estimated reconstructions of the complete trajectories based on the sum of the product of the first K coefficients and eigenvectors. The eigenvectors for the sparse data set are displayed in Figure 13.4a and show that the first vector is linear, the second provides the curvature in the middle of the time of the study and the third vector accounts for differences at the end

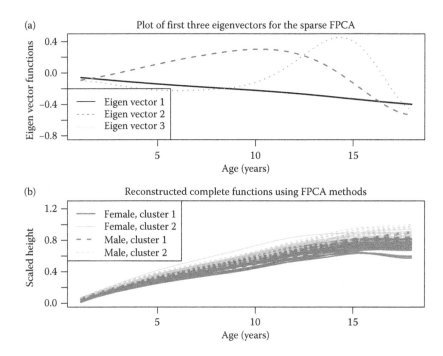

FIGURE 13.4
Plot of first three eigenvectors from the sparse FPCA (a) and reconstructed curves based on the sparse observations with sex and clustering results (b).

of the growth trajectories. In Figure 13.4b, the reconstructed functions are provided with shading based on the cluster solution and line-type for the sex of the children. Compared to Figure 13.3a, it is clear that the general patterns of the original complete observations are reconstructed well. The cluster analysis struggles to separate the boys who are shorter at the beginning of the study from the girls who start as tall or taller. For those subjects who happened to not be observed in the last few years of the study, it is difficult to separate the groups. Note that it is recommended that the responses are rescaled to be between 0 and 1 to enhance the performance of the algorithm.

James and Sugar (2003) provide another approach for clustering sparse functional observations, using a model-based algorithm. It shares the common feature of using an esti-mated variance-covariance matrix to share information between functional observations but considers this estimation iteratively with identifying the cluster members. Because of the iterative nature of the method, it has the potential to adjust the estimated covariance structure based on the current cluster members and thus potentially outperform the con-ditional distance approach. The two-cluster solution on the sparse data set provides one cluster that generally has higher mean heights than the other cluster. For the same sparse data set, the adjusted Rand index of the two-cluster solution with the sex of the children is 0.165 and the full data cluster solution had an index of 0.510. This suggests that this method provides clusters that are more aligned with the sex of the children than the conditional dis-tance approach but that there is slightly less agreement between the full and sparse data applications of the same methods. The model provides estimated mean functions for each cluster (Figure 13.5) and estimates of the probability of the subject belonging to each group. An R function of James and Sugar allows the prediction for each subject of complete curves

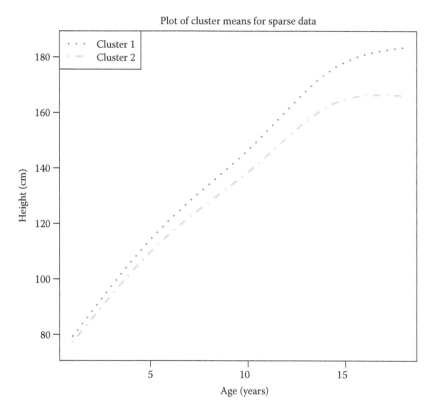

FIGURE 13.5
Plot of cluster means for the sparse Berkeley Growth Curve data based on James and Sugar's (2003) model-based cluster analysis.

from the observed sparse data; the reconstruction of the complete functions is not done in the same manner as in the FPCA approach.

In both analyses, the gender groups are not obvious, and the cluster analysis is correctly not directly tied to genders based on this version of the observations. Clustering the first derivative of growth (the velocity) could more clearly show the differences between the groups. With sparse data, the amount of information available might make derivative estimation unreliable and is currently not a feature of any of the sparse data methods.

13.4 Conclusion

The area of functional clustering is still relatively new, with most serious research about the problem being done in the past decade or so. A discussion of open problems is always somewhat tricky, since the direction of future research in this area is relatively unpredictable.

One of the great achievements of modern research on cluster analysis for multivariate data is the development of model-based clustering. This model-based approach conveniently allows for formal inference and probability statements about the clustering structure; it is often placed into a Bayesian framework as well. To a smaller extent, model-based methods have been developed for functional data. This area will likely see continued development. Mixture models in which one (or more) of the mixture components is a distribution for functions are a natural approach in this vein. In addition, cluster analysis of mixed data in which the variables measured on each individual are of different types (continuous, categorical, functional, directional, etc.) has been a past area of research. The nature of modern data indicates that many data sets of the future will have not one but several functions measured on each individual. Other data sets (having mixed variable types) may have some categorical, some continuous, and some functional variables measured on each individual. How to appropriately cluster such data sets poses an interesting challenge.

Recently, Chen et al. (2011) proposed methods for performing functional data analysis of unordered responses, sorting them using one-dimensional multidimensional scaling in a method that they call stringing. Essentially, they create the interval T and a charting variable using the responses. This makes FDA possible in situations where many response variables are measured simultaneously or where the original order of measurements is arbitrary or even randomized such as in surveys. This could allow functional clustering methods to directly compete with some non-functional methods.

In the history of cluster analysis, theoretical findings have lagged behind the introduction of new methodology, and functional clustering is no exception. With the wide range of clustering methods for functional data—many of which are described in this article—that have been proposed, there is certainly a great need for studies of their theoretical properties. This could allow an easier comparison among methods and perhaps better frame and unify the research area. Potentially, the work on distributions of functional variables by Delaigle and Hall (2010) and Ferraty and Vieu (2006) could be progenitors for more theoretical research on functional clustering.

Another important issue is finding the most appropriate visualization of cluster solutions, once a partition has been found. When the total number of curves in the functional data set is not excessively large, a rainbow plot of the curves, with each curve plotted in a different color according to its assigned cluster, can be informative. However, when there are many curves to be plotted, this picture may be too dense to show any meaningful information. In this case, simply plotting (possibly smoothed) mean curves for each cluster is more useful. Alternatively, projecting the data onto the space of the first two (functional) principal components, and then plotting the PC scores (as is done in Figure 13.2) can provide a quick picture of the cluster separation, although the functional nature of the data is lost in this type of plot.

The choice of the correct number of clusters is a perpetual concern of the clustering practitioner. In the functional clustering context, Sugar and James (2003) discuss a possible method for this problem, and a variety of model-based methods include the number of clusters as a parameter to be estimated in the clustering model. More work in this area will surely be welcomed in the future.

In the future, we expect research on functional data clustering to continue to be driven by the increasingly common applications that feature vast numbers of measurements obtained on a fine scale. Perhaps future applied research will become ever more specialized to the type of study being conducted (e.g., proteomic, genomic, geostatistical, environmental, etc.). To guard against the field of study becoming too splintered, we hope qualified researchers will study properties of the best algorithms with the goal of providing a unified vision of functional clustering.

References

Abraham, C., Cornillon, P.A., Matzner-Løber, E., and Molinari, N. 2003. Unsupervised curve clustering using B-splines. *Scandinavian Journal of Statistics*, 30(3), 581–595.

Antoniadis, A., Brossat, X., Cugliari, J., and Poggi, J.M. Clustering functional data using wavelets. *International Journal of Multiresolution and Information Process*, 11:1350003+, May 2013. arXiv: 1101.4744.

Banfield, J. and Raftery, A.E. 1993. Model-based Gaussian and non-Gaussian clustering. *Biometrics*, 49(3), 803–821.

Besse, P., Cardot, H., and Ferraty, F. 1997. Simultaneous non-parametric regressions of unbalanced longitudinal data. *Computational Statistics and Data Analysis*, 24(3), 255–270.

Besse, P. and Ramsay, J.O. 1986. Principal component analysis of sampled curves. *Psychometrika*, 51, 285–311.

Chen, K., Chen, K., Müller, H.G., and Wang, J.-L. 2011. Stringing high-dimensional data for functional analysis. *Journal of American Statistical Association*, 106(493), 275–284.

Chiou, J.M. and Li, P.L. 2007. Functional clustering and identifying substructures of longitudinal data. *Journal of Royal Statistical Society Services B Statistical Methodology*, 69(4), 679–699.

Cuesta-Albertos, J.A. and Fraiman, R. 2006. Impartial trimmed means for functional data. In *Data Depth: Robust Multivariate Analysis, Computational Geometry and Applications*, volume 72 of *DIMACS Ser. Discrete Mathematics and Theoretical Computer Science*, pages 121–145. American Mathematical Society, Providence, RI.

Cuesta-Albertos, J.A., Gordaliza, A., and Matrán, C. 1997. Trimmed *k*-means: An attempt to robustify quantizers. *Annals of Statistics*, 25(2), 553–576.

Dabo-Niang, S., Ferraty, F., and Vieu, P. 2004. Nonparametric Unsupervised Classification of Satellite Wave Altimeter Forms. In *COMPSTAT 2004—Proceedings in Computational Statistics*, pages 879–886. Physica, Heidelberg.

Dabo-Niang, S., Ferraty, F., and Vieu, P. 2007. On the using of modal curves for radar waveforms classification. *Computational and Statistical Data Analysis*, 51(10), 4878–4890.

Delaigle, A. and Hall, P. 2010. Defining probability density for a distribution of random functions. *Annals of Statistics*, 38(2), 1171–1193.

Febrero-Bande, M. and Oviedo de la Fuente, M. *fda.usc: Functional Data Analysis and Utilities for Statistical Computing(fda.usc)*, 2012. R package version 0.9.7.

Ferraty, F. and Vieu, P. 2006. *Nonparametric Functional Data Analysis. Theory and Practice*. Springer Series in Statistics. Springer, New York.

Flury, B.A. 1990. Principal points. *Biometrika*, 77(1), 33–41.

Flury, B.D. 1993. Estimation of principal points. *Journal of Royal Statistical Society Services C*, 42(1), 139–151.

Fraley, C. and Raftery, A.E. 2002. Model-based clustering, discriminant analysis, and density estimation. *Journal of American Statistical Association*, 97(458), 611–631.

García-Escudero, L.A. and Gordaliza, A. 2005. A proposal for robust curve clustering. *Journal Classification*, 22(2), 185–201.

Gattone, S.A. and Rocci, R. 2012. Clustering curves on a reduced subspace. *Journal of Computational and Graphical Statistics*, 21, 361–379.

Giraldo, R., Delicado, P., and Mateu, J. 2012. Hierarchical clustering of spatially correlated functional data. *Statistica Neerlandica*, 66(4), 403–421.

Greenwood, M., Sojda, R., Sharp, J., Peck, R., and Rosenberry, D.O. 2011. Multi-scale clustering of functional data with application to hydraulic gradients in wetlands. *Journal of Data Science*, 9(3), 399–426.

Hall, P., Müller, H.G., and Yao, F. 2009. Estimation of functional derivatives. *The Annals of Statistics*, 37(6A), 3307–3329.

Hastie, T., Buja, A., and Tibshirani, R. 1995. Penalized discriminant analysis. *The Annals of Statistics*, 23(1), 73–102.

Heckman, N.E. and Zamar, R.H. 2000. Comparing the shapes of regression functions. *Biometrika*, 87(1), 135–144.

Hitchcock, D.B., Booth, J.G., and Casella, G. 2007. The effect of pre-smoothing functional data on cluster analysis. *Journal of Statistical Computation ajd Simulation*, 77(11–12), 1089–1101.

Hubert, L. and Arabie, P. 1985. Comparing partitions. *Journal of Classification*, 2, 193–218.

Huzurbazar, S. and Humphrey, N.F. 2008. Functional clustering of time series: An insight into length scales in subglacial water flow. *Water Resources Research*, 44(W11420), 1–9.

James, G.M. and Sugar, C.A. 2003. Clustering for sparsely sampled functional data. *Journal of American Statistical Association*, 98(462), 397–408.

Jank, W. and Schmueli, G. 2009. Studying Heterogeneity of Price Evolution in Ebay Auctions via Functional Clustering. In *Handbook on Information Series: Business Computing*, pages 237–261. Elsevier.

Jiang, H. and Serban, N. 2012. Clustering random curves under spatial interdependence with application to service accessibility. *Technometrics*, 54(2), 108–119.

Kaufman, L. and Rousseeuw, P.J. 1987. Clustering by means of medoids. In *Statistical Data Analysis Based on the L_1 Norm*, pages 405–416. Elsevier.

Kaufman, L. and Rousseeuw, P.J. 1990. *Finding Groups in Data. An Introduction to Cluster Analysis*. Wiley Series in Probability and Mathematical Statistics: Applied Probability and Statistics. John Wiley & Sons Inc., New York. A Wiley-Interscience Publication.

Kneip, A. 1994. Nonparametric estimation of common regressors for similar curve data. *The Annals of Statistics*, 22(3), 1386–1427.

Liu, X. and Yang, M.C.K. 2009. Simultaneous curve registration and clustering for functional data. *Computional Statistics and Data Analysis*, 53(4), 1361–1376.

Ma, P., Castillo-Davis, C.I., Zhong, W., and Liu, J.S. 2006. A data-driven clustering method for time course gene expression data. *Nucleic Acids Research*, 34, 1261–1269.

Ma, P. and Zhong, W. 2008. Penalized clustering of large-scale functional data with multiple covariates. *Journal of American Statistical Association*, 103(482), 625–636.

Ma, P., Zhong, W., Feng, Y., and Liu, J.S. 2008. Bayesian functional data clustering for temporal microarray data. *International Journal of Plant Genomics*, 2008, 1–4.

Marron, J.S. and Tsybakov, A.B. 1995. Visual error criteria for qualitative smoothing. *Journal of American Statistical Association*, 90(430), 499–507.

McLachlan, G.J. and Basford, K.E. 1988. *Mixture Models*, Volume 84 of *Statistics: Textbooks and Monographs*. Inference and Applications to Clustering. Marcel Dekker Inc., New York.

Peng, J. and Müller, H.G. 2008. Distance-based clustering of sparsely observed stochastic processes, with applications to online auctions. *Annals of Applied Statistics*, 2(3), 1056–1077.

Peng, J. and Paul, D. 2009. A geometric approach to maximum likelihood estimation of the functional principal components from sparse longitudinal data. *Journal of Computational and Graphical Statistics*, 18(4), 995–1015.

R Development Core Team. 2012. *R: A Language and Environment for Statistical Computing*. R Foundation for Statistical Computing, Vienna, Austria. ISBN 3-900051-07-0.

Ramsay, J.O. and Dalzell, C.J. 1991. Some tools for functional data analysis. *Journal of Royal Statistical Society Series B*, 53(3), 539–572. With discussion and a reply by the authors.

Ramsay, J.O. and Silverman, B.W. 1997. *Functional Data Analysis*. Springer Series in Statistics. Springer, New York, 1st Edition.

Ramsay, J.O. and Silverman, B.W. 2005. *Functional Data Analysis*. Springer Series in Statistics. Springer, New York, 2nd Edition.

Ramsay, J.O., Wang, X., and Flanagan, R. 1995. A functional data analysis of the pinch force of human fingers. *Applied Statistics*, 44, 17–30.

Ramsay, J.O., Wickham, H., Graves, S., and Hooker, G. 2012. *fda: Functional Data Analysis*, R package version 2.2.8.

Rao, C.R. 1958. Some statistical methods for comparison of growth curves. *Biometrics*, 14, 1–17.

Ruspini, E.H. 1960. A new approach to clustering. *Information Control*, 15, 22–32.

Sangalli, L.M., Secchi, P., Vantini, S., and Vitelli, V. 2010. *k*-mean alignment for curve clustering. *Computational Statistics and Data Analysis*, 54(5), 1219–1233.

SAS Institute Inc. *SAS/STAT Software, Version 9.2*. Cary, NC, 2008.

Serban, N. and Wasserman, L. 2005. CATS: Clustering after transformation and smoothing. *Journal of American Statistical Association*, 100(471), 990–999.

Sugar, C.A. and James, G.M. 2003. Finding the number of clusters in a dataset: An information-theoretic approach. *Journal of American Statistical Association*, 98(463), 750–763.

Tang, R. and Müller, H.G. 2000. Time-synchronized clustering of gene expression trajectories. *Biostatistics*, 10, 32–45.

Tarpey, T. 2007. Linear transformations and the *k*-means clustering algorithm: Applications to clustering curves. *American Statistics*, 61(1), 34–40.

Tarpey, T. and Kinateder, K.K.J. 2003. Clustering functional data. *Journal of Classification*, 20(1), 93–114.

Tarpey, T., Petkova, E., and Ogden, R.T. 2003. Profiling placebo responders by self-consistent partitioning of functional data. *Journal of American Statistical Association*, 98(464), 850–858.

Tokushige, S., Yadohisa, H., and Inada, K. 2007. Crisp and fuzzy *k*-means clustering algorithms for multivariate functional data. *Computational Statistics*, 22(1), 1–16.

Tucker, L.R. 1958. Determination of parameters of a functional relationship by factor analysis. *Psychometrika*, 23, 19–23.

Tuddenham, R.D. and Snyder, M.M. 1954. Physical growth of California boys and girls from birth to eighteen years. *University of California Publications in Child Development*, 1, 183–364.

Wood, S.N. 2006. *Generalized Additive Models: An Introduction with R*. Texts in Statistical Science Series. Chapman & Hall/CRC, Boca Raton.

Zhong, W. and Ma, P. *MFDA: Model Based Functional Data Analysis*, 2012. R package version 1.1-4.

14

Methods Based on Spatial Processes

Lisa Handl, Christian Hirsch, and Volker Schmidt

CONTENTS

Abstract

This chapter shows how cluster methods that are based on spatial processes can be applied to a classification problem in biomedical science. First, we give a formal definition of a certain generalization of the classical single-linkage method that provides more flexibility when working with clusters of varying densities. Second, we provide an overview of the random-forest method, which is a popular classification tool from machine learning. Finally, we combine these two techniques to analyze a biomedical data set describing knee-cartilage patterns.

14.1 Introduction

In this chapter, we present methods of cluster analysis to identify and classify patterns that are formed by stationary point processes and used, for example, in the modeling of human cartilage cells. In particular, we consider a generalization of the single-linkage approach adapted to deal with inhomogeneous data and describe how the random-forest methodology can be applied to the classification of the previously identified clusters. Note that the problem of clustering point patterns occurs not only at microscopic scales in fields like biology, physics, and computational materials science, but also in geographic applications at macroscopic scales. Often point-process models play an important role in these applications and the development of special cluster algorithms adapted to the problem at hand is imperative.

Let us first discuss a macroscopical example that builds on a point-process model developed in Baccelli and Zuyev (1999). As explained in Bandyopadhyay and Coyle (2004) due to recent technological advances, it is now possible to cost-efficiently produce small sensors that can be used to gather information on their environment. Due to their small size, such sensors can carry only very limited battery capacities and devising energy-efficient computation schemes is crucial. As the cost of transmitting information over large distances is usually higher than the cost of computation itself, it makes sense to group these sensors into hierarchical clusters so that only one node in each cluster has to transmit the information to the next hierarchy level. In particular, it is desirable to have a cluster algorithm that is capable of creating a large number of moderately sized clusters. The classical single-linkage algorithm is inappropriate for this task. Indeed, by percolation theory, we can either observe the occurrence of an infinite component or we obtain a large number of clusters consisting of only very few sensors (Meester and Roy 1996; Penrose 2003).

Next, let us discuss an application at a microscopic scale. Recently, it was observed (Rolauffs et al. 2008, 2010; Meinhardt et al. 2012) that the arrangement of chondrocyte cells in the superficial zone of human cartilage form distinct patterns according to the health status of the cartilage (Figure 14.1). By being able to analyze these patterns and to understand their connection to the health status of the examined cartilage, it might become possible to see already on the scale of cells, where degenerative diseases such as osteoarthritis might develop soon. In order to create a diagnostic tool based on this observation, one important step is to develop point-process models for such cartilage patterns (Meinhardt et al.

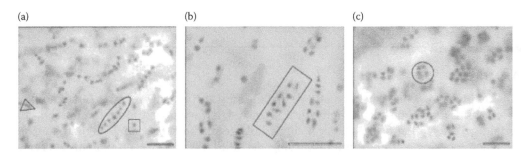

FIGURE 14.1
Chondrocyte patterns in the superficial zone of the condyle of the human knee joint showing: (a) single chondrocytes, pairs and strings, (b) doublestrings, and (c) round clusters. (Figure taken from Rolauffs, B. et al. 2010. *Arthritis and Rheumatism*, 62, 489–498.)

2012) for early results in this direction. Using this type of stochastic model makes diagnoses based on cell patterns amenable to rigorous statistical testing.

Motivated by these applications, the goal of this chapter is to describe how statistical techniques for spatial point processes can be used in the validation of cluster identification and classification steps, where the chapter is structured as follows. In Section 14.2, we present a generalization of the classical single-linkage algorithm which may be useful in situations where substantial inhomogeneities of cluster densities can be observed. To tackle the problem of cluster classification, we recall in Section 14.3 the probabilistic method based on random forests which was introduced in a seminal paper of Breiman (2001). In Section 14.4, we show how the introduced methods can be applied to point patterns formed by cell nuclei in human cartilage tissue. Finally, in Section 14.5 we show how statistical techniques for spatial point processes can be used in the validation of the identification and classification steps, where we will first describe the model used as test scenario and then explain our results.

14.2 Generalized Single-Linkage Algorithm

We first review some techniques to reconstruct clusters and correctly determine their shape from a single realization of the model. Naturally, this task splits up into two parts. In the current section, we discuss the issue of cluster identification while Section 14.3 is devoted to the discussion of the classification step. For generalities on the single-linkage algorithm and other classical hierarchical clustering algorithms, we refer the reader to Chapter 6.

14.2.1 Description of the Algorithm

When considering the identification of clusters generated by a point-process model, the classical method of the single-linkage algorithm seems a reasonable choice. Indeed this algorithm is capable of identifying, for example, long string-like clusters generated inside elongated ellipsoids, see Section 14.5. However, the presence of inhomogeneities of cluster densities makes it difficult to apply this algorithm directly. Recall from Chapter 6 that the classical single-linkage algorithm has only one parameter corresponding to a global threshold value. Therefore, it is difficult to choose a single global threshold value appropriate for both dense as well as sparse clusters. We return to this point in Section 14.5.4 where we show numerical evidence for this observation. We also refer the reader to the survey paper (Murtagh 1985) for related generalizations of classical single-linkage (Rohlf 1973; Gaiselmann et al. 2013).

In order to describe our generalization of the classical single-linkage algorithm, we recall some basic notions from graph theory. A graph G is called locally finite if every vertex has finite degree. Furthermore, a rooted graph is a graph $G = (\varphi, E)$ together with a distinguished vertex $v_0 \in \varphi$. Also, recall that a minimum spanning tree on a finite set $\varphi \subset \mathbb{R}^3$ is a tree of minimal total length among all trees whose vertices are given by φ. Recall from Chapter 6 that the single-linkage algorithm defines the clustering corresponding to the connected components of the graph obtained from deleting all edges of length larger than the global threshold value from the minimum spanning tree. To deal with the problem of varying cluster densities described above, we propose to consider a generalization of the single-linkage algorithm capable of taking into account not only the length of a specified edge but also the local geometry of the minimum spanning tree close to this edge. This approach can be seen as a mathematical formalization of the graph-theoretical methods introduced in Zahn (1971) for the detection of clusters of different shapes which also make use of local characteristics of minimum spanning trees.

Definition 14.1

Write $\mathcal{G}^*$ for the set of all locally finite rooted graphs in $\mathbb{R}^3$. A splitting rule is a function $f : \mathcal{G}^* \times \mathcal{G}^* \times [0, \infty) \to \{0, 1\}$ satisfying $f(g_1, g_2, x) = f(g_2, g_1, x)$ for all $g_1, g_2 \in \mathcal{G}^*$ and $x \in [0, \infty)$.

A splitting rule can be used to define a clustering in the following way.

Definition 14.2

Let $\varphi \subset \mathbb{R}^3$ be locally finite and let $f : \mathcal{G}^* \times \mathcal{G}^* \times [0, \infty) \to \{0, 1\}$ be a splitting rule. Write $T = (\varphi, E_T)$ for the (Euclidean) minimum spanning forest on φ, that is, $\{x, y\} \in E_T$ if and only if there do not exist an integer $n \geq 1$ and a sequence of vertices $x = x_0, x_1, \ldots, x_n = y \in \varphi$ such that $|x_i - x_{i+1}| < |x - y|$ for all $0 \leq i \leq n - 1$. For $\{x_1, x_2\} \in E_T$ write T_{x_1}, T_{x_2} for the connected components of $(\varphi, E_T \setminus \{\{x_1, x_2\}\})$ containing x_1 and x_2, respectively (and rooted at x_1 and x_2, respectively). Then, denote by φ_f the family of connected components of the graph $(\varphi, \{\{x_1, x_2\} \in E_T : f(T_{x_1}, T_{x_2}, |x_1 - x_2|) = 1\})$ and call φ_f the clustering of φ induced by f.

14.2.2 Examples of Splitting Rules

To illustrate the usage of the generalized single-linkage algorithm introduced in Section 14.2.1, let us consider several examples of splitting rules.

Example 14.1

For $f(g_1, g_2, \ell) = \mathbb{1}_{\{x : x < \alpha\}}(\ell)$, we recover the usual definition of single-linkage clustering at level $\alpha > 0$, where $\mathbb{1}_A$ is the indicator of the set A. See Figure 14.2 for an illustration of this clustering rule.

Example 14.2

For a finite set A denote by $|A|$, the number of elements in A. For $n \geq 1$ we define $f(g_1, g_2, \ell) = 1 - \mathbb{1}_{\{x : x > n\}}(|g_{1,\ell}|)\mathbb{1}_{\{x : x > n\}}(|g_{2,\ell}|)$, where $g_{i,\ell}$ denotes the connected component containing the root of g_i of the subgraph of g_i obtained by retaining only those edges of length not exceeding ℓ. See Figure 14.3 for an illustration of this splitting rule. Observe that as $n \to \infty$ the clustering induced by f approaches the minimum spanning forest from below. Furthermore, this splitting rule has the property of being equivariant under scaling, that is, for all $c > 0$ we have $(c\varphi)_f = c(\varphi_f)$.

Although the scaling property of these graphs makes them interesting candidates for the inhomogeneous clustering problem, we still need to add further modifications. Indeed, it can be shown (Hirsch submitted) that on a large class of stationary point processes these graphs almost surely (a.s.) do not percolate (i.e., all connected components are finite). This

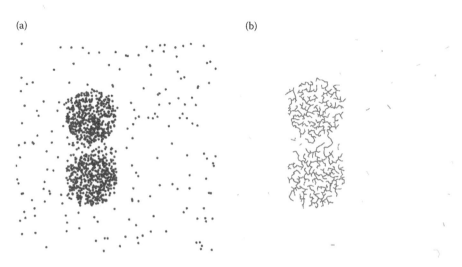

FIGURE 14.2
Clustering induced by the splitting rule of Example 14.1.

property is beneficial, for example, for clustering of large-scale sensor networks, where one is faced with the task of separating a diffuse set of points into a large number of connected components of similar and rather moderate sizes (Foss and Zuyev 1996; Baccelli and Zuyev 1999; Bandyopadhyay and Coyle 2004). In Bandyopadhyay and Coyle (2004), a Voronoi-based hierarchical network is constructed and the energy-efficiency of this network is analyzed in a simulation study. One simplifying assumption in Bandyopadhyay and Coyle (2004) consists in the hypothesis that each sensor can communicate with all other sensors in their radio range using the same amount of energy. However, in reality, the energy cost of transmissions increases super-linearly in Euclidean distance (Dasgupta and Raftery 1998) which suggests that minimum-spanning-tree based approaches, where short connections are preferred, could lead to a further increase in energy efficiency. By

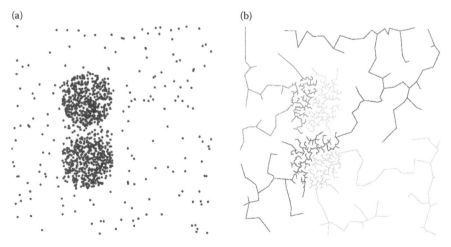

FIGURE 14.3
Clustering induced by the splitting rule of Example 14.2.

making use of the truncation parameter n in Example 14.2, it is guaranteed that the determination of the edge set at a given node does not involve the complete graph but can be achieved in a neighborhood of this graph and additionally allows for the introduction of clusters at different hierarchy level.

However, in the current setting, where clusters contain a large number of points, this property is rather detrimental since it tends to destroy connectivity inside clusters, even if mutually well-separated.

Example 14.3

Taking the discussion following Example 14.2 into account, we propose to use the following splitting rule $f : \mathcal{G}^* \times \mathcal{G}^* \times [0, \infty) \to \{0, 1\}$. For $0 < \alpha_1 < \alpha_2$ and $n_0, n_1 \geq 1$ consider

$$f(g_1, g_2, \ell) = \mathbb{1}_{\{x:x<\alpha_2\}}(x)\left(1 - \mathbb{1}_{\{x:\alpha_1<x\}}(\ell)\mathbb{1}_{\{x:x>n_0\}}(|g_{1,\ell}|)\mathbb{1}_{\{x:x>n_0\}}(|g_{2,\ell}|)\right.$$

$$\left.\times \left(1 - \mathbb{1}_{\{x:x\leq n_1\}}(|g_{1,\ell}|)\mathbb{1}_{\{x:x\leq n_1\}}(|g_{2,\ell}|)\right)\right).$$

See Figure 14.4 for an illustration of this clustering rule.

Let us explain the intuition behind the various components of f.

1. The first factor $\mathbb{1}_{\{x:x<\alpha_2\}}(\ell)$ ensures that it is possible to distinguish between clusters having the property that distances between points from one cluster to points from the other cluster are rather large (i.e., having at least distance α_2 from each other).

2. The term $\mathbb{1}_{\{x:\alpha_1<x\}}(\ell)$ makes sure that close points (i.e., whose distance is at most α_1) always end up in the same cluster.

3. By including the term $\mathbb{1}_{\{x:x>n_0\}}(|g_{1,\ell}|)\mathbb{1}_{\{x:x>n_0\}}(|g_{2,\ell}|)$, it is granted that the algorithm does not create unrealistically small clusters (unless they are already separated due to item 1).

4. Finally, the term $1 - \mathbb{1}_{\{x:x\leq n_1\}}(|g_{1,\ell}|)\mathbb{1}_{\{x:x\leq n_1\}}(|g_{2,\ell}|)$ is included to prevent that the algorithm creates unrealistically large clusters.

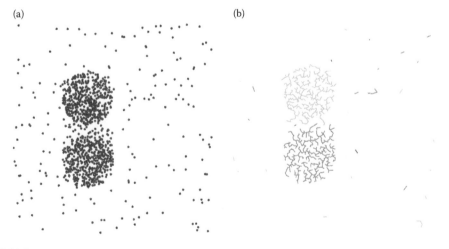

(a) (b)

FIGURE 14.4
Clustering induced by the splitting rule of Example 14.3.

14.2.3 Further Methods

The generalized single-linkage algorithms proposed in Sections 14.2.1 and 14.2.2 are mainly designed for cluster identification in off-grid data. Further, very interesting methods using spatial processes for cluster analysis of pixel-based data are discussed in Murtagh et al. (2002a, b), where a Bayesian approach is considered in conjunction with techniques of Markov random fields to reconstruct clusters of objects from image data arising in meteorology and astrophysics. We also remark that probabilistic methods for cluster analysis of pixel-based data such as the stochastic watershed algorithm play an important role in computational materials science, where a crucial task is the extraction of voxel clusters from (noisy) image data, see Faessel and Jeulin (2010); Brereton et al. (2012); Thiedmann et al. (2012); Gaiselmann et al. (2013). Furthermore, model-based clustering could prove to be an alternative approach for our problem to identify and classify patterns which are formed by stationary point processes (Section 14.5 for a more detailed discussion).

14.3 Classification Methodology

After having discussed the task of cluster identification in Section 14.2, we would like to classify these entities according to their shapes. To present the main ideas of the principal methodology clearly, we consider a stochastic point-process model which is capable of creating two types of cluster shapes, for example, balls and elongated ellipsoids, Section 14.5. For this classification step, we use the random-forest approach proposed by Breiman (2001). To provide the reader with a gentle introduction to this topic, we recall some basic definitions and results from classification theory based on resampling methods, Chapter 28.

14.3.1 Preliminaries on Decision Trees

To understand the methodology of random forests, let us first review the notion of deterministic decision trees, a classical tool in classification theory (Duda et al. 2001; Hastie et al. 2009). Informally speaking, the idea of decision trees is to classify an object by answering a sequence of questions concerning its properties. The question posed at a given point does not need to be independent of the answer to its preceding questions, that is, each answer can influence further questions. Such a sequence of questions and possible answers can be represented by a directed graph, more specifically by a tree, in which each inner node corresponds to a question and each edge corresponds to a possible answer or a set of possible answers to the question of the node it is emanating from Figure 14.5. The decision tree is always rooted, where the root node corresponds to the question from which we start classification. Furthermore, each leaf node, that is, each node with no further leading edges, needs to be associated with a category label. To classify a given object, one starts with the question at the root node and consecutively follows the edges with the answers applying to the currently classified object until a leaf node is reached. The label of this leaf node is the category our object is assigned to.

It is important to make sure that the sets of answers corresponding to the edges departing from a node should be pairwise disjoint and exhaustive. This means, for each possible answer there should be one and only one departing edge that fits this answer and that can

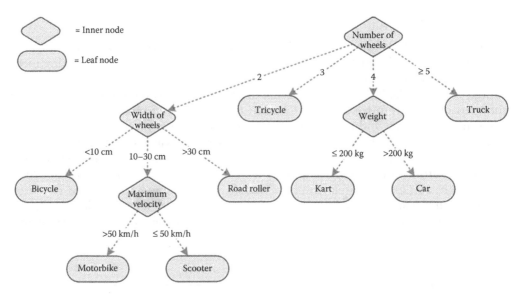

FIGURE 14.5
Example of a decision tree classifying vehicles.

be followed in this case. The category labels at the leaf nodes do not need to be unique, so there can be more than one path through the tree leading to the same category. In our case, we only consider binary decision trees, which means that at each inner node there are exactly two descending nodes. This is reasonable because any decision tree can be represented by a binary one and because it strongly simplifies the construction principle to be explained in Section 14.3.2.

14.3.2 Measures of Impurity and Construction Principle

Next, we need a way to measure the quality of the classification determined by a fixed decision tree. We have a labeled training dataset $\widetilde{M} \subset M \times C$, where M is the space of objects we want to construct a classifier for (in our case the set of possible clusters in $\mathbb{R}^3$, that is, $M = \{\varphi \in \mathbb{R}^3 : 3 \leq |\varphi| < \infty\}$), and C is the finite set of categories to consider (in our case C contains the two elements ball and elongated ellipsoid). One can interpret a decision tree as a successive splitting of our training data set $\widetilde{M}$ into subsets: At each node, a subset of the dataset $\widetilde{M}$ is split into one part going to the left and one part going to the right descending node. Building a decision tree can therefore be done by finding a sequence of splits of $\widetilde{M}$ that is optimal in some way. In order to measure the quality of a split quantitatively, the standard approach is to make use of an impurity measure (Duda et al. 2001; Hastie et al. 2009). If V denotes the vertex set of the graph representing the decision tree at hand, an impurity measure is a function $i \colon V \to [0, \infty)$ satisfying the following conditions:

1. The impurity $i(v)$ at a node $v \in V$ should be 0 if all patterns reaching v are from the same category.
2. The impurity at a node should be large if all categories are equally present.

There are three popular impurity measures we want to mention here, which all fulfill the conditions given above. If v is any node of a decision tree, we write $\widetilde{M}_v$ for the subset of all patterns in $\tilde{M}$ that reach v and we denote by $P_v(j)$ the fraction of patterns in $\tilde{M}_v$ that belong to category $j \in C$. Given this notation, we define

- *Gini impurity* $i_G : V \to [0, \infty)$ by $i_G(v) = (1/2)\left(1 - \sum_{j \in C} P_v^2(j)\right)$,
- *Entropy impurity* $i_e : V \to [0, \infty)$ by $i_e(v) = -\sum_{j \in C} P_v(j) \log_2(P_v(j))$,
- *Misclassification impurity* $i_m : V \to [0, \infty)$ by $i_m(v) = 1 - \max_{j \in C} P_v(j)$.

In general, none of these three impurity measures is superior to the others and it depends on the specific application as to which one is to be preferred. An illustration of these functions in the two-category case can be found in Figure 14.6.

To define which split is the best according to a fixed impurity measure $i : V \to [0, \infty)$, we compare the impurity $i(v)$ of the node v to the impurities $i(v_L)$ and $i(v_R)$ of the two descending nodes v_L (to the left) and v_R (to the right) that result from the split. It is desirable to maximize the impurity decrease from v to v_L and v_R. We therefore define the *drop of impurity* as $\Delta i(v) = i(v) - P_v(v_L)i(v_L) - (1 - P_v(v_L))i(v_R)$, where $P_v(v_L)$ is the fraction of patterns at node v that go to the left descendent node v_L and where $i(v_L)$ and $i(v_R)$ are the impurities at the nodes v_L and v_R. The best split at the node v should maximize $\Delta i(v)$.

After having introduced means to measure the quality of a split, it is desirable to construct decision trees that are optimal with respect to the above measure. In our case, we only want to use one property for each decision, because a simple structure of the decision tree is essential for fast computations as they are needed in random forests. This means that if we consider k properties of our objects in M, we have to find k best splits of $\tilde{M}_v$ at a given node v—one according to each property. This yields k values $\Delta_1 i(v), \ldots, \Delta_k i(v)$ for the respective drops of impurity. The best property to split on can then be chosen as the one corresponding to the maximal value of $\Delta_1 i(v), \ldots, \Delta_k i(v)$.

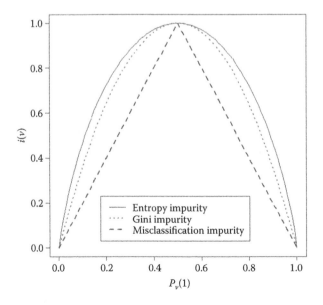

FIGURE 14.6
Illustration of (appropriately scaled) impurity measures in the two-category case, i.e. $C = \{1, 2\}$.

In order to find the best split according to a given (numerically valued) property $\ell \in \{1, \ldots, k\}$ for numerical data as in our case, Section 14.5.3 below, we confine ourselves to splittings of $\tilde{M}_v$ according to a certain value of threshold. That is, one subset consists of all objects in $\tilde{M}_v$ for which the value of the considered property is below this threshold, the other one to the objects for with the value is greater or equal. The best value of threshold is then the solution of a one-dimensional optimization problem, which maximizes $\Delta_\ell i(v)$ and can be solved numerically.

In the simplest case, the above-explained procedure is started with the full training data set $\tilde{M}$ and continued at each descending node, until either the descendent node v is pure, which means that $i(v) = 0$, or until none of the possible splits leads to a further drop of impurity, e.g. if two objects from different categories $i \neq j$ in C have exactly the same properties. Then, the node $v \in V$ is declared a leaf node and associated with the category $j \in C$ such that the majority of objects present at node v belongs to category j. Note that it is not assured that by consecutively choosing the locally optimal split also a globally optimal decision tree is built. There are several approaches for further optimizing this procedure, but for the construction of random forests considered in Section 14.3.3 below, only this simple form of decision trees is needed. We refer the reader to Duda et al. (2001) for more detailed information on this topic.

14.3.3 Random Forests

One major problem of optimal deterministic decision trees discussed in Section 14.3.2 is that they are optimized with respect to minimizing the training error in contrast to the desired minimization of the classification error. In other words, these deterministic trees often lead to a considerable overfitting to the special training set at hand, that is, the training error is much smaller than the classification error. The basic idea of the random-forest approach due to Breiman (2001) is to resolve this problem by using a suitable randomization. The random-forest approach is very flexible and yields robust results also for problems with missing data or where the number of considered properties (so-called predictors) is huge when compared to the size of the training set. There are many approaches for randomization, the most popular is random feature selection, where only a randomly chosen part of the available properties is considered for each splitting in a decision tree, and bagging, where each decision tree is built on the basis of a new training data set which is drawn with replacement from the original training dataset. Since, in our application, the number of predictors is small, we concentrate on the description of randomization by bagging. See Breiman (2001) for a detailed description of these and further techniques.

Let $r \geq 1$ be the size of the training set $\tilde{M} = \{\tilde{m}_1, \ldots, \tilde{m}_r\}$. The basic idea of bagging is to generate new random training sets by drawing r elements of $\tilde{M}$ uniformly with replacement. Thus, we consider a random vector $\Theta = (\Theta^{(1)}, \ldots, \Theta^{(r)})$ and the random set $\tilde{M}_\Theta = \{\tilde{m}_{\Theta^{(1)}}, \ldots, \tilde{m}_{\Theta^{(r)}}\}$, where $\Theta^{(1)}, \ldots, \Theta^{(r)}$ are independent random variables with uniform distribution on $\{1, \ldots, r\}$. Typically, drawing with replacement has the effect of leaving out approximately $(1 - 1/r)^r \approx 37\%$ of the elements of the original training set $\tilde{M}$ which are not involved in the construction of the best-split decision tree and therefore may be used e.g. for cross-validation purposes. These elements are called out-of-bag samples.

By repeating this random generation of training sets independently $n \geq 1$ times, i.e. by considering n independent copies $\Theta_1, \ldots, \Theta_n$ of Θ, we obtain a sequence of independent random training sets $\tilde{M}_{\Theta_1}, \ldots, \tilde{M}_{\Theta_n}$ and may construct decision trees $T_{\Theta_1}, \ldots, T_{\Theta_n}$ as described in Section 14.3.2, where each decision tree T_{Θ_i} is based on the random training set $\tilde{M}_{\Theta_i}$. Hence, the randomized decision trees T_{Θ_i} are random objects as well and as the construction of a decision tree T_{θ_i} is deterministic once we consider a fixed realization θ_i of Θ_i, the decision trees $T_{\Theta_1}, \ldots, T_{\Theta_n}$ are independent copies of the random classifier T_{Θ}.

The collection $\{T_{\Theta_1}, \ldots, T_{\Theta_n}\}$ of random decision trees is called random forest. For realizations $\theta_1, \ldots, \theta_n$ of $\Theta_1, \ldots, \Theta_n$, respectively, we call $\{T_{\theta_1}, \ldots, T_{\theta_n}\}$ a realization of a random forest. We can construct a classifier based on $T_{\theta_1}, \ldots, T_{\theta_n}$ as follows. If $x \in M$ is an element to be classified, we first consider the category that was assigned to x by the decision tree T_{θ_i} for each $i \in \{1, \ldots, n\}$ and afterwards assign x to the most frequent category in this list. We define the margin function $mg: M \times C \times (\mathbb{R}^r)^n \rightarrow [-1, 1]$ of a random forest by

$$mg(x, y, \theta_1, \ldots, \theta_n) = \frac{1}{n} \sum_{i=1}^{n} \mathbb{1}_{\{h(x, \theta_i) = y\}} - \max_{j \neq y} \left(\frac{1}{n} \sum_{i=1}^{n} \mathbb{1}_{\{h(x, \theta_i) = j\}} \right), \quad (14.1)$$

where $\mathbb{1}_A$ is the indicator of the set A and $h(x, \theta_i)$ denotes the category assigned to x by T_{θ_i}. This function measures how far the average number of votes for the right category exceeds the average numbers of votes for all other categories. In particular, $mg(x, y, \theta_1, \ldots, \theta_n) < 0$ if and only if the object $x \in M$ with category $y \in C$ is misclassified by the (deterministic) forest $\{T_{\theta_1}, \ldots, T_{\theta_n}\}$.

If in (14.1), we insert the random vectors $\Theta_1, \ldots, \Theta_n$ instead of their realizations $\theta_1, \ldots, \theta_n$ and a random object X together with its random category Y instead of (x, y), we may consider the probability of misclassification. Note that (X, Y) and $(\Theta_1, \ldots, \Theta_n)$ are assumed to be independent random vectors on the same probability space. The probability for misclassification conditioned on $\Theta_1, \ldots, \Theta_n$ is given by

$$PE_n^* = P(mg(X, Y, \Theta_1, \ldots, \Theta_n) < 0 \mid \Theta_1, \ldots, \Theta_n)$$

called the generalization error, where

$$P(mg(X, Y, \Theta_1, \ldots, \Theta_n) < 0 \mid \Theta_1, \ldots, \Theta_n) = \mathbb{E}(\mathbb{1}_{\{mg(X, Y, \Theta_1, \ldots, \Theta_n) < 0\}} \mid \Theta_1, \ldots, \Theta_n)$$

that is, the generalization error PE_n^* is a random variable. Using the strong law of large numbers, one can show (Breiman 2001) that for unboundedly growing n the generalization error PE_n^* converges to a fixed value, which is given by

$$\lim_{n \to \infty} PE_n^* = P\left(P(h(X, \Theta) = Y \mid X, Y) - \max_{j \neq Y}(P(h(X, \Theta) = j \mid X, Y)) \right) < 0, \quad (14.2)$$

where $h(X, \Theta)$ is the category which is assigned to X by the random classifier T_{Θ}. This is the reason why random forests do not overfit.

Apart from (at least partially) resolving the problem of overfitting, the bagging technique has the additional advantage of removing the need to manually leave out a part of the training set for the purpose of cross-validation. For each labeled object (x, y), we can average the votes of only those decision trees T_{θ_i} that did not receive (x, y) in their training set $\tilde{M}_{\Theta_i}$. We assign x to the category which gets most votes in this modified procedure and call this out-of-bag classifier. For a realization $\{\theta_1, \ldots, \theta_n\}$ of the random forest $\{\Theta_1, \ldots, \Theta_n\}$, we can then define the out-of-bag classification error of a specified class $j \in C$ as the fraction of elements in $\tilde{M}$ from class j that were misclassified by the out-of-bag classifier. By repeatedly generating realizations of the considered random forest and averaging the out-of-bag errors over all iterations, we obtain an estimator for the classification error of the random forest classifier for this problem. In contrast to the naive error rate, this estimator avoids the bias arising if objects are classified which were already included in the training set. However, as each out-of-bag predictor only uses about one-third of all decision trees of the forest, it should be assured that a sufficiently high number of decision trees is generated for each random forest.

Finally, the randomization approach also provides us with a tool to single out the most relevant predictors. Indeed, let $j \in C$ be a category and $\pi \in \{1, \dots, k\}$ be a predictor. In each iteration we may consider the effect of applying a random permutation on the values of predictor π among all out-of-bag elements of class j. Then, we may compare the out-of-bag classification error of class j before and after this permutation and compute its relative increase. This value measures the extent to which disturbing the predictor π decreases the classification accuracy of elements of class j. Averaging these quantities over all iterations, we obtain a measure of how essential predictor π is for the classification in category j. This value is called importance score. A more formal approach to select the most important variables with the aim of creating a parsimonious model is described in Meinshausen (2000). To achieve this goal, a convex penalty for trees is used which is related to the so-called nonnegative Garrote (Breiman 1995).

An implementation of random forests based on the original Fortran code by Breiman can be found in the R package random forest, see Liaw and Wiener (2002).

14.4 Application to Patterns of Cell Nuclei in Cartilage Tissue

After having introduced our methodology, we give a biomedical example for its application. As already mentioned in the introduction of this chapter, the cell nuclei of chondrocytes in human cartilage form distinct patterns according to the type, localization and health state of the cartilage (Rolauffs et al. 2008; Meinhardt et al. 2012). We now apply our methods to the point patterns formed by these cell nuclei in order to detect different types of clusters. First, we describe our data together with some biological background and apply cluster analysis to the originate point patterns. Later, in this section, we assign the obtained clusters to different types by use of the random-forest approach and explain our results.

14.4.1 Data Description

Cartilage tissue contains neither nerves nor veins, it consists of chondrocytes—the only cells that are found in it—and an extracellular matrix that is produced by these chondrocytes and encloses them. The matrix mainly consists of proteoglycans and collagens, where the exact composition varies according to the specific type of cartilage. The chondrocytes are organized in super-imposed layers, a superficial zone at about 0%–10%, a middle zone at 10%–40% and a deep zone at 40%–100% tissue depth. In the superficial zone, the cell density is very high and chondrocytes are organized in horizontal patterns as already illustrated in Figure 14.1, while in the deeper zones, the cell density is much lower and chondrocytes form vertical columns.

All our investigated cartilage samples were taken from the human knee joint of patients of the BG Unfallklinik Tübingen who were treated with ACI (autologous chondrocyte implantation), only the superficial zone of the cartilage was examined. ACI is a biomedical treatment method that tries to repair local cartilage damages as they occur, for example, as a consequence of an accident. Therefore, intact cartilage is taken from an area of the joint which is not meant to bear much weight. The cells from this healthy cartilage are isolated, stimulated in vitro to make them proliferate and applied to some carrier material, for example a biomembrane. When there are enough chondrocytes grown, this carrier material containing the cells is reimplanted into the damaged joint. The pieces of cartilage we examined are the ones that are taken out from around the lesion during this surgery in order to clear enough space for the implant.

We distinguished three different health states of these cartilage samples: intact cartilage, cartilage with early osteoarthritis and already severely damaged cartilage. The respective health state was determined by visual inspection according to the following criteria: the sample was considered intact if no or nearly no damage was visually recognizable, that is, if its surface was smooth and glossy or only slightly roughened. It was considered to be cartilage with early osteoarthritis if fibrillations, fissuring or major surface roughening could be detected and it was considered severely damaged if full defects could be found already.

In total, we investigated 20 different cartilage samples, which were taken from five different patients. In each of these samples, the cell nuclei were stained with a fluorescent dye and three-dimensional images were taken by use of a fluorescence microscope and the technique of structured illumination. This technique makes it possible to record two-dimensional sections through the specimen separately without or with only very few noise from above or below the plane in focus. A stack of these two-dimensional sections can later be combined to the final three-dimensional binary image (Felka submitted). Note that 10 of the 20 images we obtained like this were taken with 10-fold, 10 with 20-fold magnification.

After some preprocessing steps, we obtained the three-dimensional coordinates of all cell nuclei contained in each binary image, given in μm relative to the observation window.

As the cells found in the type of cartilage, we examined, form clusters with characteristic shapes according to its health state, our goal was to use the methods introduced in Sections 14.2 and 14.3 to first extract and then classify the clusters contained in our samples.

14.4.2 Results

In order to extract the clusters from the point patterns, we applied the single-linkage algorithm as already discussed in Section 14.2, see also Chapter 6 using the Euclidean distance and a threshold value of 22.5 μm. This value led to optically good results and is close to the values that have already been used earlier in Meinhardt et al. (2012) for the analysis of similar data. An example of a clustered point pattern can be found in Figure 14.7. Cell nuclei which belong to a given cluster are plotted using the same symbol.

Next, we classify the resulting clusters according to their shapes. We therefore shortly discuss the choice of properties, which were considered for classification and the choice of categories the clusters shall be assigned to. We excluded clusters containing less than three points from the further analysis, as pairs and singletons do not show any relevant geometric substructure and thus form their own trivial categories. So the space of objects

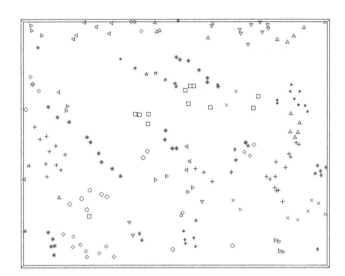

FIGURE 14.7
3D pattern of cell nuclei with detected clusters (indicated by different symbols).

we want to classify consists of the set of all finite subsets of $\mathbb{R}^3$ containing at least three points, that is, $M = \{\varphi \subset \mathbb{R}^3 : 3 \leq |\varphi| < \infty\}$.

The set of possible categories E the clusters in M should be assigned to by the classifier we are going to construct are the specific patterns that are "typical" for the health states of human cartilage according to earlier studies (Rolauffs et al. 2010). Three categories will be distinguished in the following:

1. *String:* A group of points that are one after another lying preferably regularly on a line. This pattern is typical for intact cartilage samples from the considered region of the human knee joint, a microscopic picture of a typical string can be found in Figure 14.1, enclosed by an ellipse.

2. *Doublestring:* A string in which at least one cell has divided once more, leading to two cells lying next to each other against the orientation of the original string. In the extreme case, this results in two close parallel strings. This is typical for early osteoarthritis and the beginnings of degeneration in the cartilage. In Figure 14.1, a typical doublestring is enclosed by a rectangle.

3. *Round cluster:* A round or ellipsoid-shaped aggregation of points without relevant substructure. This pattern is typical for already severely damaged cartilage and severe forms of osteoarthritis. A typical round cluster is enclosed by a circle in Figure 14.1.

We labeled the clusters according to the health state of the cartilage sample they had been extracted from, hence clusters from intact cartilage were labeled "strings," clusters from cartilage samples with beginning osteoarthritis were labeled "doublestring" and clusters from severely damaged cartilage were labeled "round cluster." Of course this method leads to some errors because not *only* the specific "typical" pattern is contained in each sample of cartilage but also patterns from the other two types and inarticulate patterns that do not fit any of the three categories do occur. But at least most of the patterns should be assigned to the right category by this procedure and labeling by visual inspection would be very subjective and need a huge amount of time. Using this method, we obtained 332 clusters labeled as strings, 811 clusters labeled as doublestrings and 98 round clusters.

As we want to classify the clusters according to their shapes, we mainly chose geometric features as properties for classification. In detail, we considered:

- The ratio of length and width, where length width and depth of a cluster were defined as the maximum extension in the direction of the coefficient vector of the first, second and third principal component of the cluster,
- The mean distance to the best line fitted through the cluster (by orthogonal distance regression),
- The mean distance to the best plane fitted through the cluster (by orthogonal distance regression),
- The coefficient of determination and the regression coefficient of the following regression: let dk be the distance from an extremal point in the cluster (chosen as one of the two points with the biggest distance) to its kth-nearest neighbor, a line through the origin was fitted to the points (k, dk) by ordinary regression,
- The mean angle between the connection vectors from one extremal point to all other points,
- The mean angle between the connection vectors from each point (the both extremal points excluded) to its first and second nearest neighbor,
- The ratio of the mean distance from the centroid and the number of points in the cluster and
- The ratio of length and number of points.

In Breiman (2001), evidence is provided that random forests are quite robust with respect to weak inputs. Therefore, we put no additional effort in selecting the best among these features but considered all of them simultaneously.

We balanced the data set by a bootstrapping technique. We took about 50% of the number of clusters from the biggest category, that is, 408 as we have 811 doublestrings, and randomly drew that many clusters from each category with replacement. Note that this means that any cluster can be contained more than just once in the training data set.

As test set, we only considered clusters which were not selected at all for the training dataset. This was necessary because after resampling the out-of-bag estimators are most likely positively biased. Many clusters occur more than once in the training data set, and hence, the idea of out-of-bag estimation does not work properly anymore. It may happen that although a cluster is not in the training data set of a decision tree itself and is therefore considered out-of-bag for this decision tree, one or more identical copies of the same cluster are. A second cause of bias in favor of the classification is that patterns that are in the training data set more than once are more likely to be classified correctly. Replicating a cluster has an effect similar as weighting it higher than other clusters, concerning both the classification itself and the computation of the error rates. A pattern with many replications in the training data set is more likely to be classified correctly and the classification outcome is taken into account once for each replication. This bias is the reason why we decided earlier to retain a test set additional to the out-of-bag estimation of error rates.

The error rates we obtained seem quite high compared to other classification problems, see Table 14.1. However, one can see by optical inspection that the majority of clusters in

TABLE 14.1

Mean Values and Standard Deviations of the Error Rates of 100 Realizations of
Random Forests with Balanced Training Data Set

	OOB Estimates of Error Rates		Test Set Error Rates	
	Mean Value	**Standard Deviation**	**Mean Value**	**Standard Deviation**
Total	0.1249	0.0090	0.3948	0.0187
String	0.1435	0.0156	0.4946	0.0551
Doublestring	0.2242	0.0200	0.3757	0.0255
Round cluster	0.0069	0.0039	0.1918	0.2263

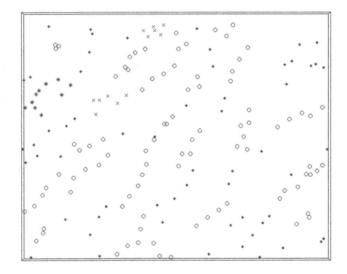

FIGURE 14.8
3D pattern of cell nuclei from cartilage with beginning osteoarthritis. Clusters containing at least 3 points are
shown according to whether they were classified as strings (symbol ○), doublestrings (symbol ∗), or round clusters
(symbol ×).

each cartilage sample was classified correctly, Figures 14.7 and 14.8. As the classification
of single clusters can be seen as a preceding step to the classification of cartilage samples
themselves, which is even more closely related to the problem of diagnosis, this might be
sufficient. Furthermore, it has to be taken into account that the biological data we considered is highly variable, which led to a high variability of cluster shapes. Knowing that there
is in fact quite a high amount of noise in our data, we consider the error rates acceptable.

14.5 Validation of Clustering Methodology

In this section, we validate the methodology for cluster identification and cluster classification stated in Sections 14.2 and 14.3, respectively, by performing a Monte-Carlo analysis
of a synthetic point-process example. Similar to the biomedical problem discussed in
Section 14.4, our goal will be first to identify different clusters in a noisy environment and

after this provide a correct classification of the type of the observed cluster shape. In particular, the main goal of our approach is not necessarily to associate every single data point with its true cluster, but to ensure that the numbers of identified clusters of a given shape are in accordance with those corresponding to the true clusters. Note that the point-process model discussed in this section would be a prototypical example for model-based clustering as discussed in Banfield and Raftery (1993), Allard and Fraley (1997), Byers and Raftery (1998), Dasgupta and Raftery (1998), Fraley and Raftery (1998), Mateu et al. (2007, 2010), and also in Chapter 16. In particular, since, for simplicity, we consider a synthetical example where the cluster shapes are given by ellipsoids, it would be particularly well-suited for this classical method. Nevertheless, as explained in Section 14.4, in practice one often has to deal with non-ellipsoidal shapes and it usually takes some effort to accommodate model-based clustering to this framework (see Allard and Fraley (1997), Mateu et al. (2007, 2010) for possible approaches to resolving this problem).

14.5.1 Description of the Test Scenario

As input data for the validation of the clustering and classification algorithms, we consider synthetic patterns in $\mathbb{R}^3$ formed by realizations of a simple point-process model, see Illian et al. (2008) for a detailed introduction to this class of spatial processes. Note that for simplicity of exposition we consider a three-dimensional example like in Section 14.4, although the methods described in this chapter can be applied in other dimensions as well. Let us first give an informal description of our model. We begin with a primary point process $X^{(1)}$ of possible locations of cluster centers. At each of these centers, we place cluster areas with one of two possible random shapes—either a ball or an elongated ellipsoid, Figure 14.9a. Afterwards we remove some of these cluster areas so that the remaining ones are pairwise disjoint. We then decide for each of the remaining areas independently whether they are supposed to be filled with a dense or sparse cluster and place an appropriate number of points uniformly and independently inside this area, see Figure 14.9b. Finally, some background noise is added, Figure 14.9c.

We now provide a formal mathematical description of this model. For simplicity of exposition, we consider a basic Poisson-type scenario exhibiting the properties of the informal description above. It would be interesting to conduct an extensive analysis on far more complicated (and more realistic) point-process models, but this would be beyond the scope of this expository chapter.

Write $\mathbb{N}$ for the family of all locally finite subsets $\varphi \subset \mathbb{R}^3$ and $\mathcal{C}$ for the family of all convex compact bodies of $\mathbb{R}^3$. Furthermore denote by SO_3 the group of orthogonal matrices in $\mathbb{R}^{3\times3}$ with determinant 1 which can be identified with the group of three-dimensional rotations around the origin. The point process we are interested in is hierarchically built up as follows. Let $\mathbb{Z}_+ = \{1, 2, \ldots\}$ denote the set of positive integers and

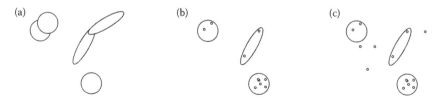

FIGURE 14.9
Construction principle of the point-process model.

consider independent random variables $X^{(1)}, X^{(2)}, X^{(3)}, X^{(4)} \in \mathbb{N}$, $Z^{(5)} = \left\{ Z_n^{(5)} \right\}_{n \geq 1}$, $Z^{(6)} = \left\{ Z_n^{(6)} \right\}_{n \geq 1} \in \{0, 1\}^{\mathbb{Z}_+}$, $U^{(7)} = \left\{ U_n^{(7)} \right\}_{n \geq 1} \in [0, 1]^{\mathbb{Z}_+}$ and $G^{(8)} = \left\{ G_n^{(8)} \right\}_{n \geq 1} \in SO_3^{\mathbb{Z}_+}$ with the following properties. Assume that

1. $X^{(1)}, X^{(2)}, X^{(3)}, X^{(4)}$ are homogeneous Poisson point processes in $\mathbb{R}^3$ with intensities $\lambda_1, \lambda_2, \lambda_3, \lambda_4$, respectively, where $\lambda_3 < \lambda_4$.

2. $Z^{(5)}, Z^{(6)}$ are sequences of independent and identically distributed Bernoulli random variables with $1 - \mathbb{P}\left(Z_1^{(5)} = 0 \right) = \mathbb{P}\left(Z_1^{(5)} = 1 \right) = p_5$ and $1 - \mathbb{P}\left(Z_1^{(6)} = 0 \right) = \mathbb{P}\left(Z_1^{(6)} = 1 \right) = p_6$ for some $p_5, p_6 \in [0, 1]$.

3. $U^{(7)}$ is a sequence of independent and identically distributed random variables uniformly distributed on $[0, 1]$.

4. $G^{(8)}$ is a sequence of independent and identically distributed random rotations uniformly distributed on SO_3.

Let $a > 0$ and let $K_0, K_1 \in \mathcal{C}$ be two fixed compact bodies centered at the origin, where we assume for simplicity that K_0 is an ellipsoid, and K_1 a sphere. Write $X^{(1)} \cap [-a/2, a/2]^3 = \left\{ X_1^{(1)}, \ldots, X_N^{(1)} \right\}$ for the restriction of the Poisson process $X^{(1)}$ to the cube $[-a/2, a/2]^3$, where $N = |X^{(1)} \cap [-a/2, a/2]^3|$ denotes the total number of points of $X^{(1)}$ in $[-a/2, a/2]^3$. For $1 \leq i \leq N$, we say that $X_i^{(1)}$ survives the construction if there does not exist $1 \leq j \leq N$ with $U_i^{(7)} < U_j^{(7)}$ and $\left(X_i^{(1)} + G_i^{(8)} K_{Z_i^{(5)}} \right) \cap \left(X_j^{(1)} + G_j^{(8)} K_{Z_j^{(5)}} \right) \neq \emptyset$. Let us write $\left\{ X_{i_1}^{(1)}, \ldots, X_{i_{N'}}^{(1)} \right\}$ for the set of points of $X^{(1)}$ which survive the construction. The final point configuration X is then obtained as

$$X = \left(X^{(2)} \cap [-a/2, a/2]^3 \right) \cup \bigcup_{j=1}^{N'} \left(X^{\left(3 + Z_{i_j}^{(6)} \right)} \cap \left(X_{i_j}^{(1)} + G_{i_j}^{(8)} K_{Z_{i_j}^{(5)}} \right) \right), \qquad (14.3)$$

where $X^{(2)} \cap [-a/2, a/2]^3$ is some noise.

To summarize, the main features of the point-process model defined in (14.3) are its ability to generate both different cluster shapes as well as different cluster densities. In particular, the Poisson process $X^{(2)}$ generates some background noise, N' denotes the random number of clusters in the cube $[-a/2, a/2]^3$ which survive, and $X^{\left(3 + Z_{i_j}^{(6)} \right)}$ either generates dense clusters (if $Z_{i_j}^{(6)} = 1$) or sparse clusters (if $Z_{i_j}^{(6)} = 0$). Finally, $X_{i_j}^{(1)} + G_{i_j}^{(8)} K_{Z_{i_j}^{(5)}}$ generates the area of a spherical (if $Z_{i_j}^{(5)} = 1$) or ellipsoidal (if $Z_{i_j}^{(5)} = 0$) cluster.

Finally, after having introduced our point-process model and tools of cluster analysis, we now validate both the cluster identification as well as the cluster classification steps by extensive Monte-Carlo simulation based on the underlying spatial point process.

Let us first specify the parameters of the point-process model introduced in (14.3) so that the following properties are satisfied. On the one hand, the number of cluster centers should be not too large (due to computational restrictions) but also not too small to guarantee that for a considerable fraction of cases the distance between two different clusters can be rather small and cluster separation becomes a non-trivial issue. Furthermore, the

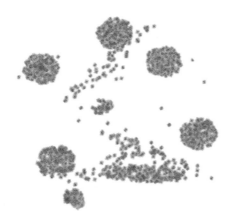

FIGURE 14.10
Sampled 3D point pattern; cutout in a thin slice of the cube $[-a/2, a/2]^3$.

density in dense clusters should be significantly higher than in sparse clusters so that there is a distinct advantage in using the generalized single-linkage algorithm in contrast to the standard single-linkage algorithm. Finally, both cluster shapes should occur with a sufficiently high probability and also the number of background points should not be too small in order to increase the difficulty of cluster identification.

We fix the size of the sampling window $[-a/2, a/2]^3$ by putting $a = 200$ and define suitably chosen values for the intensities $\lambda_1, \ldots, \lambda_4$ introduced above, where we put

- $\lambda_1 = 2.125 \cdot 10^{-6}$ (cluster center intensity).
- $\lambda_2 = 1.250 \cdot 10^{-5}$ (intensity of background noise).
- $\lambda_3 = 0.001705$ (intensity of cluster points in sparse clusters).
- $\lambda_4 = 0.01023$ (intensity of cluster points in dense clusters).

Furthermore, the lengths of the three half-axes of the elongated ellipsoid K_0 are given by $70, 10$ and 10, respectively, while we let K_1 be a ball whose radius is determined by $\nu_3(K_1) = \nu_3(K_0)$, where $\nu_3(K)$ denotes the volume of the set $K \subset \mathbb{R}^3$. Finally, we put $p_5 = p_6 = 1/2$.

Observe that the values of $\lambda_1, \lambda_2, \lambda_3$ and λ_4 have been chosen so that in the sampling window $[-a/2, a/2]^3$ with side length $a = 200$ the expected number of cluster centers is equal to 17, the expected number of noise points is equal to 100 and such that the expected number of points in dense/sparse clusters is equal to 300 and 50, respectively. A realization of the point-process model for the test scenario described above is shown in Figure 14.10.

14.5.2 Choice of Classification Categories

In this section, we show how the general framework of random forests described in Section 14.3 can be applied in the point process setting of Section 14.5.1. Ultimately, our goal is to classify the clusters created in the identification step either as ball-shaped (we call this class *ball*) or elongated ellipsoidal clusters (called *ellipsoid*). However, even when using the generalized single-linkage algorithm some errors occur already at the stage of cluster identification and we have to take this problem into account in the classification step. In general, we cannot say much on the shape of these erratically identified clusters, but we may at least extract two subclasses that are easy to identify. If we have a set of points that was returned as a cluster by the clustering algorithm but that is neither an element of ball nor of ellipsoid then we define its category as follows:

1. If it contains more than a prespecified number of elements κ_{large}, then we define its category to be *errLarge*.
2. If it contains less than a prespecified number of elements κ_{small}, then we define its category to be *errSmall*.
3. If none of the above applies, we define its category to be *errRemainder*.

This subdivision of the error categories makes it possible to identify more easily at least those cases where several dense clusters in the model step are merged into only one cluster in the identification step or if one of the sparse clusters in the model step is subdivided into several smaller clusters in the identification step. In our Monte-Carlo example, we choose $\kappa_{small} = 30$ and $\kappa_{large} = 350$.

14.5.3 Choice of Properties

Finally, let us discuss the set of predictors used to assign clusters to the categories defined in Section 14.5.2. To identify at least some of the obvious misidentifications by the single-linkage algorithms considered in this chapter, a useful predictor is the size (in terms of number of elements) of a cluster. We denote this quantity by *nPoints* in the following. In order to distinguish between the categories *ball* and *ellipsoid*, we need predictors related to the geometry of a cluster. We may make use of the fact that in elongated clusters the variance of the cluster points is particularly pronounced in one direction. Quantitatively, this feature may be captured by principal component analysis on the three spatial coordinates of the points. Indeed, for elongated clusters, the largest eigenvalue of the empirical covariance matrix tends out to be much larger than the second-largest one. On the contrary, for ball-shaped clusters, we would expect the eigenvalues to be of similar size. Therefore, we include predictors η_1 and η_2 which are defined as the quotient of the largest and the second-largest eigenvalue and the quotient of the second-largest and the smallest eigenvalue, respectively.

14.5.4 Validation of Cluster Identification

Naturally, it is desirable that the quality of the cluster analysis is robust with respect to moderate changes of the parameters. We comment on this issue in detail in Section 14.5.5 below, but nevertheless let us summarize the main results at already this point. For all parameter constellations of the point-process model, we have considered, the generalized single-linkage algorithm with the splitting rule given in Example 14.3 of Section 14.2 had a better performance than the standard single-linkage algorithm. Since we only considered performance under optimal parameters of the single-linkage algorithm for cluster identification and since the standard single-linkage algorithm can be seen as a special case of the generalized single-linkage algorithm, this is not surprising. When increasing the ratio between the intensities of dense and sparse clusters, the advantage of the generalized single-linkage algorithm on the standard single-linkage algorithm becomes less significant, but both improve with respect to accuracy. Increasing the noise has a greater impact on the performance of the generalized single-linkage algorithm than on the standard single-linkage algorithm. Finally, increasing the intensity of cluster centers deteriorates the quality of both the standard as well as the generalized single-linkage algorithm. Both cluster identification algorithms were relatively robust with respect to changes in the cluster shape.

The first validation step is to compare the quality of cluster identification of the generalized single-linkage algorithm to that of the standard single-linkage algorithm. This is achieved by first generating $n = 1,000$ realizations of the point-process model and computing the fraction of clusters generated by the point-process model which are detected by each of the algorithms. For both the standard and the modified single-linkage algorithm, the parameters of the algorithm were optimized using the Nelder–Mead method, see Avriel (2003). Note that the Nelder–Mead method is a popular simplex-based method for heuristically finding maxima/minima of multivariate functions. An advantage of the Nelder–Mead method in contrast to other classical optimization techniques such as gradient descent is due to the fact that no knowledge on the derivatives of the objective function is needed. Let us elaborate on the optimization procedure in our specific setting. Denote by $D_\ell = \{D_{\ell,1}, \ldots, D_{\ell,m_\ell}\}$ and $S_\ell = \{S_{\ell,1}, \ldots, S_{\ell,m'_\ell}\}$, the family of dense respectively sparse clusters that appear in the ℓ-th realization of the point process. We say that the algorithm correctly identified a dense cluster $D_{\ell,j}$ if there exists a purported cluster Ξ with $\max(|\Xi \setminus D_{\ell,j}|, |D_{\ell,j} \setminus \Xi|) \leq 30$. Similarly, we say that the algorithm correctly detects a sparse cluster $S_{\ell,j}$ if there is a set of points Ξ which is deemed to be a cluster by the identification algorithm and such that $\max(|\Xi \setminus S_{\ell,j}|, |S_{\ell,j} \setminus \Xi|) \leq 5$ holds. The bounds 30 and 5 correspond to 10% of the expected number of elements in a dense respectively sparse cluster. Observe that we chose different bounds for dense and sparse clusters, since in our application it is appropriate to use a threshold relative to the cluster size. Indeed, our final goal is to determine the shape of various clusters and not to determine the membership of a single point to one of these clusters.

In Figure 14.11, the identification of clusters in a sampled 3D point pattern is visualized using the splitting rule of Example 14.3 in Section 14.2, where we also show an example of a misspecified cluster.

Let ρ_d denote the ratio of the number of correctly identified dense clusters (accumulated over the n iterations) by the overall number of simulated dense clusters. Similarly, we define ρ_s. We want to compare the quality of the two clustering methods under optimal choice of their parameters α and $\alpha_1, \alpha_2, n_0, n_1$, respectively, where we use the Nelder–Mead search for the parameters maximizing the sum $\rho_s + \rho_d$. These parameters were introduced

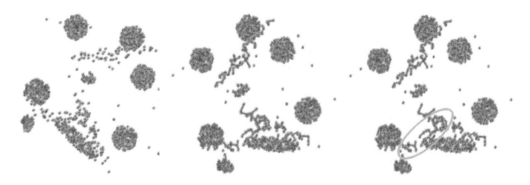

FIGURE 14.11
Identification of clusters in a sampled 3D point pattern; cutout in a thin slice of the cube $[-a/2, a/2]^3$.

TABLE 14.2

Proportion of Clusters Detected in the Identification Step

Algorithm	Dense Clusters	Sparse Clusters	Total
Standard single-linkage	75.7%	52.4%	63.9%
Generalized single-linkage	91.5%	79.5%	85.6%

in Example 14.3 of Section 14.2. For the ordinary single-linkage algorithm, we obtain an optimal threshold of $\alpha = 10.66$, while for the generalized single-linkage algorithm, the optimizer yields the values $\alpha_1 = 6.06$, $\alpha_2 = 16.14$, $n_0 = 10$ and $n_1 = 50$. Table 14.2 shows the fractions of clusters detected in the identification step using either the standard or generalized single-linkage algorithm.

Observe that the α_2-value for the generalized single-linkage algorithm is much larger than the α-value of the standard single-linkage algorithm. This is reasonable, as decreasing the α-value is the only possibility to distinguish between a larger number of clusters, while in the generalized single-linkage algorithm, α_2 can be chosen much higher as we have the additional possibility to delete an edge between two clusters if one of the subgraphs $g_{1,\ell}$ or $g_{2,\ell}$ is sufficiently large. We also note that the generalized single-linkage algorithm considerably improves the proportion both of detected dense and sparse clusters when compared to the standard single-linkage algorithm.

14.5.5 Robustness of Identification Quality

Let us conclude by discussing the robustness of the quality concerning the presented clustering algorithms when changing the parameters of the point-process model. For all configurations of model parameters, the parameters of the single-linkage algorithms were optimized by the Nelder–Mead method. After that $n = 1,000$ realizations of the point-process model were generated and the numbers of correctly identified clusters were computed. As a first step, one may vary the ratio between the intensities of dense and sparse clusters. If we put $p_6 = 0.99$, so that only 1% of all clusters are sparse, then the standard single-linkage algorithm correctly identifies 84.9% of all clusters while the generalized single-linkage algorithm correctly identifies 86.1% of all clusters. On the other hand, if we put $p_6 = 0.01$, so that only 1% of all clusters are dense, then the standard single-linkage algorithm correctly identifies only 68.4% of all clusters while the generalized single-linkage algorithm manages to identify correctly a fraction of 77.4% of all clusters. In a second step, one may change the half-axes of the ellipsoid K_0. If we shrink the size of the largest half-axis by a factor of 2 while increasing the lengths of the other half-axes by a factor of $\sqrt{2}$, then the standard single-linkage algorithm correctly identifies 79.7% of all clusters while the generalized single-linkage algorithm correctly identifies a fraction of 92.1% of all clusters. In a third step, one may change the intensity of the noise, e.g. we may choose $\lambda_2 = 2.5 \cdot 10^{-5}$ (i.e., double the original value). Then the standard single-linkage algorithm correctly identifies 64.0% of all clusters while the generalized single-linkage algorithm correctly identifies a fraction of 82.5% of all clusters. Finally we can increase the intensity of cluster centers, e.g., let us double the value of λ_1 so that $\lambda_1 = 4.5 \cdot 10^{-6}$. Then the standard single-linkage algorithm correctly identifies 54.3% of all clusters while the generalized single-linkage algorithm correctly identifies a fraction of 76.5% of all clusters.

TABLE 14.3

Out-of-Bag Estimators for the Classification Error

	Ball	Ellipsoid	errSmall	errLarge	errRemainder	Total
Oob-estimate	0.54%	2.07%	0.00%	1.36%	28.47%	3.12%

TABLE 14.4

Importance Scores

Predictor	Ball	Ellipsoid	errSmall	errLarge	errRemainder	Total
nPoints	0.11	0.096	0.87	0.85	0.19	0.20
η_1	0.49	0.58	0.10	0.18	0.32	0.46
η_2	0.088	0.054	0.021	0.19	0.027	0.065

14.5.6 Validation of Cluster Classification

We now describe the numerical results of the cluster classification based on random forests. As discussed in Section 14.5.2, each cluster is assigned to precisely one of the five categories *ball, ellipsoid, errLarge, errSmall, errRemainder*. Furthermore, we use the three predictors *nPoints*, η_1 and η_2 introduced in Section 14.5.3. We consider the classification of clusters obtained in 1,000 realizations of the point-process model and use a random forest consisting of 500 decision trees. All clusters consisting of less than four points were discarded, since due to the choice of our intensities the probability that these points constitute the realization of a cluster from the original point process practically vanishes. In this way, we can greatly reduce the number of clusters to be classified. Table 14.3 shows the results obtained for the out-of-bag estimators for the classification error for the five categories under consideration. These estimators were introduced in Section 14.3.

In particular, we see that the classification error for the four classes *ball, ellipsoid, errLarge, errSmall* is very low, while for the category *errRemainder* it is rather large. This is intuitively reasonable. Indeed, the predictor *nPoints* is very well suited to discriminate between the sets of categories {*errSmall*}, {*errLarge*} and {*ball, ellipsoid, errRemainder*}. Similarly, due to the simple geometry of the shapes considered in our point-process model the predictor η_1 is appropriate to distinguish between *ball* and *ellipsoid*. However, as the class *errRemainder* may contain a variety of shapes, it is also not surprising that our simple set of predictors has problems to discriminate between the classes *ball* or *ellipsoid* on the one hand and *errRemainder* on the other hand. These observations are reflected in the importance scores obtained from the out-of-bag sample, see Table 14.4. Furthermore, we see that the importance score for the predictor η_2 is rather low when compared to η_1 or *nPoints* (which is also in accordance with intuition). The notion of the importance score was introduced in Section 14.3.3.

References

Allard, D. and Fraley, C. 1997. Nonparametric maximum likelihood estimation of features in spatial point processes using Voronoi tessellation. *Journal of the American Statistical Association*, 92, 1485–1493.

Avriel, M. 2003. *Nonlinear Programming: Analysis and Methods*. Dover Publishing, New York.

Baccelli, F. and Zuyev, S. 1999. Poisson-Voronoi spanning trees with applications to the optimization of communication networks. *Operations Research*, 47, 619–631.

Bandyopadhyay, S. and Coyle, E.J. 2004. Minimizing communication costs in hierarchically-clustered networks of wireless sensors. *Computer Networks*, 44, 1–16.

Banfield, J.D. and Raftery, A.E. 1993. Model-based Gaussian and non-Gaussian clustering. *Biometrics*, 49, 803–821.

Breiman, L. 1995. Better subset regression using the nonnegative Garrote. *Technometrics*, 37, 373–384.

Breiman, L. 2001. Random forests. *Machine Learning*, 45, 5–32.

Brereton, T.J., Stenzel, O., Baumeier, B., Kroese, D.P., and Schmidt, V. 2012. Efficient simulation of charge transport in deep-trap media. In: C. Laroque, J. Himmelspach, R. Pasupathy, O. Rose, and Uhrmacher, A.M. (eds.), *Proceedings of the Winter Simulation Conference*, IEEE, Piscataway.

Byers, S. and Raftery, A.E. 1998. Nearest-neighbor clutter removal for estimating features in spatial point processes. *Journal of the American Statistical Association*, 93, 577–584.

Ciullo, D., Celik, G.D., and Modiano, E. 2010. Minimizing transmission energy in sensor networks via trajectory control. In *Modeling and Optimization in Mobile, Ad Hoc and Wireless Networks (WiOpt), 2010 Proceedings of the 8th International Symposium on*, pages 132–141. IEEE.

Dasgupta, A. and Raftery, A.E. 1998. Detecting features in spatial point processes with clutter via model-based clustering. *Journal of the American Statistical Association*, 93, 294–302.

Duda, R.O., Hart, P.E., and Stork, D.G. 2001. *Pattern Classification*. Wiley-Interscience, New York, 2nd Edition.

Faessel, M. and Jeulin, D. 2010. Segmentation of 3D microtomographic images of granular materials with the stochastic watershed. *Journal of Microscopy*, 239, 17–31.

Felka, T., Rothdiener, M., Bast, S., Uynuk-Ool, T., Herrmann, S., Fischer, A., Zouhair, S. et al. Chondrocyte spatial organization: A highly promising image-based biomarker with the potential of unprecedented early diagnosis of osteoarthritis. Preprint (submitted).

Foss, S. and Zuyev, S. 1996. On a Voronoi aggregative process related to a bivariate Poisson process. *Advances in Applied Probability*, 28, 965–981.

Fraley, C. and Raftery, A.E. 1998. How many clusters? Which clustering method? Answers via model-based cluster analysis. *The Computer Journal*, 41, 578–588.

Gaiselmann, G., Neumann, M., Spettl, A., Prestat, M., Hocker, T., Holzer, L., and Schmidt, V. 2013. Stochastic 3D modeling of $La_{0.6}Sr_{0.4}CoO_{3-\delta}$ cathodes based on structural segmentation of FIB-SEM images. *Computational Materials Science*, 67, 48–62.

Goldberg, M. and Shlien, S. 1978. A clustering scheme for multispectral images. *IEEE Transactions on Systems, Man and Cybernetics*, 8, 86–92.

Hastie, T., Tibshirani, R., and Friedman, J. 2009. *The Elements of Statistical Learning*. Springer Series in Statistics. Springer, New York, 2nd Edition. Data mining, inference, and prediction.

Hirsch, C., Brereton, T., and Schmidt, V. Percolation and convergence properties of graphs related to minimal spanning forests. Preprint (submitted).

Illian, D.J., Penttinen, P.A., Stoyan, H., and Stoyan, D. 2008. *Statistical Analysis and Modelling of Spatial Point Patterns*. Wiley-Interscience, Chichester.

Liaw, A. and Wiener, M. 2002. Classification and regression by randomforest. *R News*, 2, 18–22.

Mateu, J., Lorenzo, G., and Porcu, E. 2007. Detecting features in spatial point processes with clutter via local indicators of spatial association. *Journal of Computational and Graphical Statistics*, 16, 968–990.

Mateu, J., Lorenzo, G., and Porcu, E. 2010. Features detection in spatial point processes via multivariate techniques. *Environmetrics*, 21, 400–414.

Meester, R. and Roy, R. 1996. *Continuum Percolation*. Cambridge University Press, Cambridge.

Meinhardt, M., Lück, S., Martin, P., Felka, T., Aicher, W., Rolauffs, B., and Schmidt, V. 2012. Modeling chondrocyte patterns by elliptical cluster processes. *Journal of Structural Biology*, 177, 447–458.

Meinshausen, N. 2000. Forest Garrote. *Electronic Journal of Statisitics*, 3, 1288–1304.

Murtagh, F. 1985. A survey of algorithms for contiguity-constrained clustering and related problems. *The Computer Journal*, 28, 82–88.

Murtagh, F., Barreto, D., and Marcello, J. 2002a. Bayesian segmentation and clustering for determining cloud mask images. In *Proceedings of the SPIE*, volume 4877, pages 144–155.

Murtagh, F., Donalek, C., Longo, G., and Tagliaferri, R. 2002b. Bayesian model selection for spatial clustering in 3D surveys. In *Proceedings of the SPIE*, volume 4847, pages 391–401.

Penrose, M. 2003. *Random Geometric Graphs*. Oxford University Press, Oxford.

Rohlf, F.J. 1973. Algorithm 76: hierarchical clustering using the minimum spanning tree. *The Computer Journal*, 16, 93–95.

Rolauffs, B., Williams, J.M., Aurich, M., Grodzinsky, A.J., Kuettner, K.E., and Cole, A.A. 2010. Proliferative remodeling of the spatial organization of human superficial chondrocytes distant from focal early osteoarthritis. *Arthritis and Rheumatism*, 62, 489–498.

Rolauffs, B., Williams, J.M., Grodzinsky, A.J., Kuettner, K.E., and Cole, A.A. 2008. Distinct horizontal patterns in the spatial organization of superficial zone chondrocytes of human joints. *Journal of Structural Biology*, 162, 335–344.

Thiedmann, R., Spettl, A., Stenzel, O., Zeibig, T., Hindson, J.C., Saghi, Z., Greenham, N.C., Midgley, P.A., and Schmidt, V. 2012. Networks of nanoparticles in organic-inorganic composites: Algorithmic extraction and statistical analysis. *Image Analysis and Stereology*, 31, 23–42.

Zahn, C.T. 1971. Graph-theoretical methods for detecting and describing Gestalt clusters. *IEEE Transactions on Computers*, 100, 68–86.

15

Significance Testing in Clustering

Hanwen Huang, Yufeng Liu, David Neil Hayes,
Andrew Nobel, J.S. Marron, and Christian Hennig

CONTENTS

Abstract

In this chapter, we give an overview of principles and ideas for significance testing in cluster analysis. We review test statistics and null models proposed in the literature and discuss issues such as parametric bootstrap, estimating the number of clusters by use of significance tests and p-values for single clusters. Then, we focus on the Statistical Significance of Clustering (SigClust) method which is a recently developed cluster evaluation tool specifically designed for testing clustering results for high-dimensional low sample size data. SigClust assesses the significance of departures from a Gaussian null distribution, using invariance properties to reduce the needed parameter estimation. We illustrate the basic idea and implementation of SigClust and give examples.

15.1 Introduction

Significance testing addresses an important aspect of cluster validation. Many cluster analysis methods (for example hierarchical methods or K means) will deliver clusterings even for homogeneous data. They assume implicitly that a clustering has to be found, regardless of whether this is meaningful or not. Therefore, the question arises how to distinguish between a clustering that reflects meaningful heterogeneity in the data and an artificial clustering of homogeneous data. Significance tests are the standard statistical tools for such distinctions, although there are alternatives such as computing a posterior probability for homogeneity under a Bayesian approach. Testing of a homogeneity model against a clustering alternative ("homogeneity test") is the most straightforward testing task in cluster analysis.

Significance tests are also used for more specific tasks in cluster analysis:

- Testing can be used for estimating the number of clusters.
- In many applications, researchers are interested in interpreting some or all of the individual clusters, so the significance of individual clusters is of interest.

It needs to be noted, however, that these applications of significance tests in most cases violate the original logic of statistical hypothesis testing by testing data dependent hypotheses. More about such use of tests in Section 15.2.4.

The general principle of homogeneity tests is as follows:

- A null hypothesis H_0 is specified, formalizing homogeneity.
- A test statistic T is specified in such a way that it can distinguish homogeneous from clustered data (which should either have systematically larger or smaller values than homogeneous data).
- P_T, the distribution of T assuming H_0, is computed, either theoretically or by Monte Carlo simulations. T may assume that the data are clustered already, for example if it compares within-cluster and overall variability (Section 15.2.1). In such a case, the computation of P_T needs to take into account the clustering method, that is, if data from H_0 is simulated, the clustering method needs to be applied each time before computing T.
- From this, critical values or p-values for given data can be computed so that homogeneity can be significantly rejected if T is in the far tails in the "clustered" direction of its distribution under H_0.
- The test power, that is, how likely it is to reject H_0 significantly under a clustering alternative H_1, is also of interest, although often not easily computable.

Here are some general comments regarding the choice of H_0, T and H_1. Specific methods are explained in Sections 15.2 and 15.3. "Homogeneous data" is an intuitive concept. It may have different meanings in different applications, and often, it may refer to a rather broad class of distributions such as "all unimodal distributions." Computing P_T is straightforward (at least by simulation) if H_0 is a single distribution, or a single family of distributions in which P_T is invariant over the whole parameter space. If H_0 is broader, either it is possible to make a worst case argument (for some choices of T, p-values from the uniform distribution can be shown to be conservative for all unimodal distributions, that is, the uniform is the "least homogeneous distribution among the unimodal ones," or P_T

can be approximated by nonparametric or parametric bootstrap, that is, estimating the distribution from the data.

In order to be optimal according to Neyman and Pearson's classical theory of tests, one would specify H_1 and then choose T so that it achieves optimal power against H_1. This however does not only require the distributions of T under both H_0 and H_1 to be known precisely, but also the existence of a T that is uniformly optimal over H_1. This requirement is hardly fulfilled in cluster analysis, due to the broad nature of the meaning of "homogeneous data" and "clustering." In the literature, T is therefore either chosen as theory-friendly as possible, or in a heuristic manner so that T intuitively distinguishes a certain idea of homogeneity from a certain idea of "clustering." This requires some thought about whether T serves to distinguish the right kind of alternative for the given application from H_0.

A simple example to illustrate this is the test statistic $T_{mnn} =$ "average dissimilarity of points to their m th nearest neighbour," m being a tuning constant, suggested in Chapter 10 of Manly (1997). If the data set is partitioned into clusters of size $\geq m + 1$, it is intuitive that T_{mnn} is rather small, and larger under a homogeneity model assuming that the overall distribution of dissimilarities is (more or less) the same under the clustering alternative as under H_0. Therefore, homogeneity is rejected if T_{mnn} is small enough. This reasoning is rather problematic in many situations, because if H_0 is for example fixed to be an $N(0, \mathbf{I})$-distribution, the statistic does not only distinguish H_0 from a clustering alternative, but also from homogeneous distributions with lower variance. In such situations, T_{mnn} would need to be modified, for example by dividing it by the overall average dissimilarity. In Manly (1997) and Hennig and Hausdorf (2004), T_{mnn} is used together with resampling or parametric bootstrap simulations from H_0 (Section 15.2.3) that preserve the magnitudes of dissimilarities in the data, and therefore, T_{mnn} does not need such an adjustment there. Another aspect of T_{mnn} is the impact of outlying points with large dissimilarities to all (or all but $<m$) other points, which may dominate T_{mnn} and therefore mask clusters. Therefore, T_{mnn} can only reliably detect clusterings in which all points are in clusters of size $\geq m + 1$.

Section 15.2 gives a general overview over significance tests for clustering in the literature. Afterward, the chapter focuses on a specific test, namely, a recently developed Monte Carlo-based cluster evaluation tool called Statistical Significance of Clustering (SigClust) in Section 15.3. Its applications to two real cancer microarray datasets are illustrated in Section 15.4. Some future possible improvements and open questions are discussed in Section 15.5. Section 15.6 provides the conclusion of this chapter.

15.2 Overview of Significance Tests for Clustering

15.2.1 Null Models and Test Statistics for Euclidean Data

Euclidean data have attracted most of the work in cluster analysis, probably because of their connection to geometric intuition. A number of standard null models have been considered for such data, namely

- the multivariate Gaussian distribution,
- the uniform distribution or (often equivalently) a homogeneous Poisson process on a set D,
- the set of all distributions with unimodal density.

Much work is based on the spherical multivariate Gaussian null model and a clustering alternative as implied by the K-means method (Chapter 3), that is, spherical clusters with equal within-cluster variation. Typical test statistics are based on comparing the within- and between-clusters sum of squares, for example

$$T_{ssr} = \frac{\sum_{i=1}^{K} |C_i| \|\bar{\mathbf{x}}_{C_i} - \bar{\mathbf{x}}\|^2}{\sum_{i=1}^{K} \sum_{\mathbf{x}_j \in C_i} \|\mathbf{x}_j - \bar{\mathbf{x}}_{C_i}\|^2},$$

where $C_1, \ldots, C_K$ is the K-means partition of the data set $\{\mathbf{x}_1, \ldots, \mathbf{x}_k\}$ with mean $\bar{x}$ and with $\bar{x}_C$ as the mean within cluster C. An expression for the asymptotic distribution of T_{ssr} under H_0 is available, Bock (1985), but requires simulations in practice. Similar statistics have been proposed elsewhere, for example, Duda et al. (2000), including a handy approximation of the asymptotic distribution under the null. This has been implemented in R's "fpc"-package, and Beale (1969) in probably the first work of this kind. Note that this approach requires the number K of clusters to be specified. One can expect, however, that the test computed with $K = 2$ has a satisfactory power in many situations with $K > 2$ clusters, because the $K = 2$-partition will usually already be significantly better than the homogeneity model. The SigClust test introduced in Section 15.3 is based on the same principle, but for very high-dimensional data. An example where fixing $k = 2$ does not work properly to detect clustering is given in Section 15.5.

Note that the null distributions of these tests are more complicated than the standard analysis of variance (ANOVA) tests for the comparison of groups including the two-sample t-test for one-dimensional data, because the ANOVA tests do not take into account that in cluster analysis the groups are not known in advance, but are arrived at by looking for optimal groupings (Section 15.1).

Surprisingly, Tibshirani et al. (2001) found that the uniform distribution is actually the worst case (in terms of the expected value) of a logarithmized version of the ratio of the within-cluster and overall sum of squares, which they call "gap statistic," among the unimodal distributions for univariate data, suggesting that such test statistics are also suitable for general unimodal distributions.

The term "gap test" has been used by several authors. In Bock (1985), it refers to the maximum of the distances of all points to their nearest neighbour, which can be used to test the uniformity hypothesis.

Yet another "gap test" tests the homogeneous Poisson process hypothesis by looking at the "gap" between two or more clusters (i.e., requiring a clustering method; Rasson and Kubushishi 1994). Other tests of this hypothesis are available, including a likelihood ratio test against an inhomogeneous Poisson process concentrating on subsets interpreted as clusters (Hardy 2011).

A classical example of a test of the unimodality hypothesis is the "dip statistic" (Hartigan and Hartigan 1985) for continuous one-dimensional data, which is the Kolmogorov distance between the empirical distribution of the data and the closest possible unimodal distribution (which can be easily derived). The Kolmogorov distance is the maximum distance between the two cumulative distribution functions, and if there is a clear "dip" in a density, this maximum will occur just where the dip is, hence the name of the statistic. Hartigan and Hartigan (1985) can analytically evaluate the distribution of T under the uniform distribution. They conjecture that the uniform distribution is the "worst unimodal case," so that p-values derived from the uniform are conservative. Alternatively, one could simulate p-values from the unimodal distribution closest to the data (Tantrum et al. 2003), which can be expected to achieve better power, but comes at the price that conservatism (i.e., respecting the level of the test) can no longer be guaranteed and data-dependent simulation is required. If H_1 is taken to be the class of all distributions on the real line that are not unimodal, it is too broad to evaluate power, but it is obvious that T distinguishes unimodal distributions from those with clear "dips," which is therefore the "effective alternative" of the test. The researchers can then decide whether this formalizes their concept of homogeneity and clustering properly. The major drawback of this classical and attractive test, which is implemented in R's "diptest"-package, is that it is difficult to generalize to multivariate data. Bock (1996) lists some generalizations and other multivariate tests of the unimodality hypothesis.

Maitra et al. (1996) consider another H_0 that is more general than a single parametric model, and allows for a more powerful test than testing unimodality, namely general ellipsoidal homogeneous distributions parametrized by a center and a scatter matrix. They suggest a semiparametric bootstrap scheme to simulate from the H_0, which can also be applied for simulating from a K clusters null model in order to test it against a "more clusters"-alternative.

15.2.2 Null Models and Test Statistics for Dissimilarity Data

In many clustering problems, the data are not Euclidean. A number of tests based on general dissimilarity measures have been introduced in the literature. These tests can of course be used for Euclidean data as well, and it is also possible to simulate the distribution of T under H_0 if T is based on dissimilarities but H_0 is a distribution for data on Euclidean space, which may be of interest because other dissimilarity measures may be of interest even for such data in a given application.

Null models have also been proposed for dissimilarity data. Much of this work (Ling 1973; Godehardt and Horsch 1995; Bock 1996) has focused on the distribution of edges in random graphs. For a given dissimilarity between n points in a data set, either a dissimilarity threshold d or a number n_d of edges is fixed, and the graph with all points as vertices and the n_d edges corresponding to the smallest dissimilarities (in case that d is fixed, all similarities below d) is considered. Standard null models assume that n_d edges are randomly drawn out of the $\left(\binom{n}{2}\right)$ possible edges, or that edges are drawn independently with a fixed probability.

Distributional results exist for a number of statistics under these models, such as the number of isolated vertices, the number of connected components (related to single linkage hierarchical clustering), the size of the largest connectivity component and the size of the largest clique (related to complete linkage hierarchical clustering) (Bock 1996). The problem with such null models is that they do not take structural properties of the dissimilarities into account. For example, many dissimilarity measures used in applications are actually metrics and fulfill the triangle inequality, which violates the independence of the occurrence of edges in the graph. See Godehardt and Horsch (1995) for the adjustments of null models for this situation. We are not aware of more recent work in this area, although there are a lot of open problems.

Test statistics can be constructed based on dissimilarities to (m) nearest neighbours and average within-cluster dissimilarities (Bock 1996; Manly 1997; Hennig and Hausdorf 2004) suggest the ratio between the sum of the smallest m and the largest m dissimilarities, for example, $m = n/4$, motivated by the assumption that for clustered data the smallest dissimilarities can be expected within clusters and the largest ones can be expected between clusters, so that this ratio is smaller for clustered data. All these statistics do not require the computation of a clustering and the specification of the number of clusters.

Dissimilarities may also be considered in high-dimensional situations with rather low n in order to avoid the "curse of dimensionality." McShane et al. (2002) use a nearest-neighbor test statistic computed on the first three principal components for Euclidean data under a Gaussian null model for high-dimensional microarray data.

If one is willing to specify a number K of clusters for the alternative ($K = 2$ suggests itself if nothing is known for reasons given above), and to simulate the distribution of T under H_0 by Monte Carlo, in principle any index measuring clustering as discussed in Chapter 26 can be used as a test statistic. Particularly, test statistics for Euclidean data based on sums of squares as introduced in Section 15.2.1 can be generalized to general dissimilarity measures in a straightforward way using

$$\sum_{\mathbf{x}_j \in C} \|\mathbf{x}_j - \bar{\mathbf{x}}_C\|^2 = \frac{1}{2n_C} \sum_{\mathbf{x}_j, \mathbf{x}_k \in C} d(\mathbf{x}_j, \mathbf{x}_k)^2,$$

where the Euclidean distance d can be replaced by any other dissimilarity measure. It is important, though, to choose H_0 and T in such a way that the distribution of T is either invariant for all members of H_0, or that one can simulate from a "worst case" member of H_0 to ensure conservativity.

15.2.3 Nonstandard Tests, Null Models, and Parametric Bootstrap

Apart from the rather general homogeneity models discussed previously, in many applications, it is suitable to use tailor-made null models, because even under homogeneity there is more structure to be expected, or because the data is of a type not covered above.

There is for example a considerable body of work regarding null models for the presence/absence and species abundance data in ecology (Gotelli 2000; Ulrich and Gotelli 2010). Furthermore, time and spatial dependence will affect the distribution of T and make it even in absence of meaningful clustering look significantly different from homogeneity models ignoring such patterns, and models could be chosen that take this into account (Ripley 1981; Manly 1997).

Usually, the distribution of whichever T can then only be obtained by simulation. Some null models are straightforward to simulate, for example, those based on simple permutation of matrix rows and columns (Besag and Clifford 1989; Gotelli 2000), generally referring to null models under which a set of modifications of the data matrix is assumed to be equally likely. However, often the models are characterized by parameters that are not available and need to be estimated from the data. This is in the literature usually referred to as "parametric bootstrap."

The rationale here is to fix or estimate all characteristics of the data that do not contribute to what should be interpreted as a significant clustering. The validity of p-values from parametric bootstrap cannot be guaranteed in general, and for most models, a theoretical justification may not be achievable. Such tests test the significance of a clustering apart from what is reflected in the parameters estimated for the parametric bootstrap. It should therefore be argued, at least heuristically, that these parameters either do not have anything to do with clustering (such as location estimators; scale estimators qualify only if the cluster statistic T is scale invariant), or that it is intended to investigate only those aspects of clustering that cannot be explained by the bootstrap parameters.

Illustrating this, Hennig and Hausdorf (2004) investigated presence/absence data of snail species and were interested in whether the species' distribution ranges were significantly more clustered than what would be expected taking the spatial autocorrelation of presences and absences into account. Their null model, implemented in R's "prabclus"-package, takes from the data the distribution of the number of regions per species, the number of species per region (deviations from uniformity in these distributions should not be interpreted as "significant clustering") and an estimated parameter measuring spatial autocorrelation. Using an appropriate dissimilarity measure between species distribution ranges and a dissimilarity test statistic as introduced previously, a p-value can then be simulated from the null model holding all the data-based parameters listed above fixed. A significant result can be interpreted so that the data set is significantly more clustered than what could be explained by the observed spatial autocorrelation, species size, and region richness distributions alone. The paper also contains power simulations under a biogeographically reasonable alternative distribution. As a general principle, the parametric bootstrap approach is outlined in Hennig and Lin (2015).

Another interesting nonstandard way of testing findings from exploratory graphical data analysis including clusterings is discussed by Buja et al. (2009), where null models are used to generate graphs from which a data analyst is asked to tell apart the dataset of interest.

15.2.4 Testing Clusters and Their Number

Significance tests are often used in the literature in order to estimate the number of clusters, by successively testing $K + 1$ vs. K clusters (Hardy (2011), Maitra et al. (2012); much work in this direction has been done in the area of mixture modeling, McLachlan and Peel (2000) and Chapter 8), or by evaluating a homogeneity test incorporating various numbers of clusters K and picking the one that looks "most significant" (Tibshirani et al. 2001; Hennig and Liao 2013; Hennig and Lin 2015). Although this is an intuitively appealing method for estimating K, it has to be noted that the usual error probabilities on which statistical test theory is based do no longer hold in this case because of multiple testing and the fact that significant test results rejecting too small values of K are mixed up with an insignificant result for the K at which the procedure is stopped. p-values can still be informative, particularly if they are strongly significant or clearly non-significant.

For example, Hennig and Liao (2013) used the average silhouette width index (see Chapter 26) over a range of different values of K for testing clustering against an application-dependent homogeneity model for mixed type continuous/ordinal/nominal data. This revealed that their data set was strongly significantly clustered for $K > 3$ (optimally for $K = 8$), but not for $K = 2$, despite the fact that $K = 2$ yielded the optimal value of the index and would therefore normally have been used as the standard estimator of K.

Sometimes (Tantrum et al. 2003), tests are used in hierarchical clustering to decide whether a cluster should be further split up or not. This adds an additional issue, namely that null hypotheses tested in the process are not fixed in advance, but rather picked by the clustering method in a data dependent way. These null hypotheses have the form "the cluster just about to be considered is homogeneous." This invalidates statistical test theory, and the resulting p-values can only have exploratory value. The resulting methods for estimating K can still be valuable, though.

The same problem applies to all attempts to assign p-values and significance statements to single clusters. Methods with this aim are sought after, because in many applications only some but not all clusters are really meaningful, and researchers would be happy to distinguish them based on statistical significance.

One of the most sophisticated attempts to solve this problem is implemented in the SAS procedure "modeclus" (SAS 2009). The test statistic compares the density in the modal region of a cluster with the density at the boundary to the closest cluster. This is based on the clusters found by "modeclus" before using density or neighborhood-based methods, see SAS (2009). If the latter is significantly smaller than the former, the current cluster is declared to form a cluster without the need to be merged to its neighbour. A heuristic formula is given for the distribution of the test statistic under the H_0 that the considered data subset does not constitute a mode cluster on its own, and the procedure is backed up by extensive simulations.

Most of the literature that applies the terms "p-value" and "significance test" to single clusters is about procedures that do not constitute valid statistical tests, although such procedures can still have exploratory value. Suzuki and Shimodaira (2006) for example consider cluster recovery probabilities under resampling, which in fact explores cluster stability; such approaches are elsewhere suggested without misleading reference to hypothesis testing (Tibshirani and Walther 2005; Hennig 2007). Sometimes p-values are computed comparing single clusters to random samples from a homogeneous distribution (Hautaniemi et al. 2003), ignoring that they are the result of a clustering algorithm.

To our knowledge, there is no proper theory yet for testing the validity of a cluster that was first picked by a clustering method in an isolated manner. One possible strategy would be to define a statistic S that can be computed for a single cluster (as done in some of the cited papers), to assume a homogeneity null model, to apply the same clustering procedure as used for the dataset of interest, and to rank the resulting clusters according to S. Assuming that larger S indicates a "stronger" cluster, a test of the null model can be based on comparing the maximum (over all clusters) of the clusterwise S values in the original dataset and in data simulated from the null model. This test can be seen as a test of the cluster with maximum S. Another test can be based on the second largest S in the same manner and so on, so that every cluster can give rise to a separate valid test of the null model. However, it needs to be kept in mind that the correct interpretation of significance of such tests is "the occurrence of a cluster with the kth best value of S among all clusters is evidence against the homogeneity model" rather than "the cluster with the kth best value of S is significant." Particularly, the cluster with largest S would not necessarily yield the smallest p-value. This could be repaired by comparing every single cluster's S in the original data set with the distribution of the maxima of S from clustering data from the homogeneity model, still yielding valid but more conservative tests.

15.3 The Method of SigClust

Liu et al. (2008) proposed the SigClust method to answer some fundamental statistical questions about clustering evaluation. SigClust was specifically designed to assess the significance of clustering results for "high dimension, low sample size" (HDLSS) data. This is important because HDLSS data are becoming increasingly common in various fields including genetic microarrays, medical imaging and chemometrics. In this section, we introduce the motivation of SigClust as well as its implementation and application. Some discussion about the diagnostics for checking assumptions will also be addressed. In terms of software, the R package for SigClust can be found and freely downloaded on the CRAN website: http://cran.r-project.org/web/packages/sigclust/index.html. Matlab software is also available from: http://www.unc.edu/~marron/marron_software.html

15.3.1 Criterion

The foundation and the first step of almost any clustering evaluation approach is to propose a reasonable definition for a single cluster. SigClust defines a cluster as a subset of data

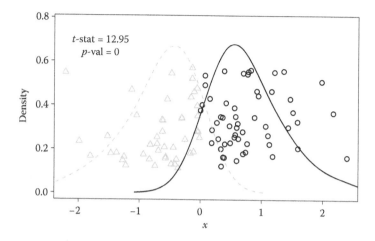

FIGURE 15.1
Plots of a standard Gaussian-simulated example. A random height is added to each point in the y-direction for visualization purposes. If the two extremes are considered as two clusters (distinguished by different shades and symbols), the standard t-test reports a large statistic and significant p-value (shown in the text). However, the data points are sampled from a single Gaussian distribution showing the t-test is inappropriate for finding clusters.

which can be reasonably modeled using a multivariate Gaussian distribution, with an arbitrary covariance matrix. This idea is motivated by studying the difference among different views of a dataset. For illustration, consider a simple example shown in Figure 15.1. The 100 data points are generated from the standard Gaussian distribution. If the two extremes of this dataset are considered as two different clusters, the two sample t-test gives a very significant p-value. However, it is not desirable to claim the existence of two clusters in this data set based on the t-test result. Actually, applying a clustering algorithm to any dataset can divide it into two extreme groups and the p-value based on the t-test will usually be significant. Thus, deeper investigation is needed to determine whether or not there are really two meaningful clusters, for which we will consider the Gaussian distribution as "prototype cluster shape" here. Note that some other unimodal distributions can also serve as the definition for a single cluster. The Gaussian distribution enjoys some advantages over other unimodal distributions, including its simplicity for implementation. Moreover, SigClust is specifically intended for HDLSS data, such as microarray gene expression data. We will see that a Gaussian model for noise is sensible for high dimensional real microarray data in Section 15.4.

Once the cluster definition is established, SigClust formulates the clustering evaluation problem as a hypothesis testing procedure with

H_0: the data are from a single Gaussian distribution versus
H_1: the data are from a mixture of at least two Gaussian distributions.

15.3.2 Invariance Principle

Suppose that the original data set X, of dimension $d \times n$, has d variables and n observations. The null hypothesis of SigClust is that the data are from a single Gaussian distribution $N(\mu, \Sigma)$, where μ is a d-dimensional vector and Σ is a $d \times d$ covariance matrix.

SigClust employs a test statistic called the 2-means cluster index (CI) which is defined as the ratio of the within cluster variation to the total variation. Note that this criterion is optimized in 2-means clustering. This choice substantially simplifies the parameter estimation of the null Gaussian distribution because of location and rotation invariance of the test statistic. Other choices, based on other clustering methods, can also be used here, in a very similar fashion, if they are location and rotation invariant. Location invariance means that without loss of generality it can be assumed that the mean $\mu = 0$. This is because the data can be translated by the sample mean (or any other quantity) and the value of CI stays the same. Rotation invariance implies that instead of estimating $d(d + 1)/2$ covariance parameters, it is enough to work only with a diagonal covariance matrix $\Lambda = \text{diag}(\lambda_1, \ldots, \lambda_d)$. In particular, Λ can be obtained from Σ using the eigenvalue decomposition $\Sigma = U\Lambda U^T$, where U is an orthogonal matrix (essentially a rotation matrix).

The null distribution of the test statistic can be approximated empirically using a direct Monte Carlo simulation from a single Gaussian distribution whose covariance matrix Λ is estimated from the data (parametric bootstrap). The significance of a clustering result can be assessed by computing an appropriate p-value.

15.3.3 Eigenvalue Estimation

In order to simulate the null distribution of the test statistic, SigClust needs to estimate the diagonal null covariance matrix Λ which is the same task as finding the underlying eigenvalues of the covariance matrix. Therefore, the cornerstone of the SigClust analysis is the accurate estimation of the eigenvalues of the covariance matrix of the null multivariate Gaussian distribution.

Although the shift and rotation invariance substantially reduce the number of parameters to be estimated in the SigClust method, the d eigenvalues are still a relatively large number of parameters compared with the sample size n for HDLSS data sets. Further reduction needs to be conducted. Toward this end, a factor analysis model is used to reduce the covariance matrix eigenvalue estimation problem to the problem of estimating a low rank component that models biological effects together with a common background noise level. Specifically, Λ is modeled as

$$\Lambda = \Lambda_B + \sigma_N^2 I, \tag{15.1}$$

where the diagonal matrix Λ_B represents the true underlying biological variation and is typically low-dimensional, and σ_N^2 represents the level of background noise. For this, consider the collection of all $d \times n$ numbers in the full data matrix. For Λ_B of relatively low rank, most of these univariate numbers will be $N(0, \sigma_N^2)$, but a minority will be more variable. We approach this using robust estimate of the standard deviation σ_N, which essentially ignores the more variable components (having originally been designed to eliminate the effect of outliers in the data). In this spirit, σ_N is estimated as

$$\hat{\sigma}_N = \frac{\text{MAD}_{d \times n \text{ data set}}}{\text{MAD}_{N(0,1)}}, \tag{15.2}$$

where MAD stands for the median absolute deviation from the median, and the subscript indicates it is applied to the univariate dataset, which is the collection of all entries of the data matrix. The robust MAD estimate is used here because of its combination of simplicity and its high breakdown properties, see for example Hampel et al. (1986). It will be very effective as long as less than $1/4$ of the $d \times n$ data matrix values contain biological signals. Then Λ is estimated to be

$$\hat{\lambda}_j = \begin{cases} \tilde{\lambda}_j & if \ \tilde{\lambda}_j \geq \hat{\sigma}_N^2 \\ \hat{\sigma}_N^2 & if \ \tilde{\lambda}_j < \hat{\sigma}_N^2, \end{cases} \tag{15.3}$$

where $(\tilde{\lambda}_1, \ldots, \tilde{\lambda}_d)$ are the eigenvalues of the sample covariance matrix.

Once the null Gaussian distribution is estimated, the empirical null distribution of the test statistic can be obtained by simulating from the estimated distribution. The procedure for SigClust can be briefly summarized as follows:

Algorithm SIGCLUST

1. Calculate the CI for the original data set.

2. Obtain estimates $(\hat{\lambda}_1, \ldots, \hat{\lambda}_d)$ of the eigenvalues $(\lambda_1, \ldots, \lambda_d)$ of Σ based on (15.3).

3. Simulate d-dimensional data N_{sim} times, with N_{sim} some large number, from the null distribution, i.e., $(x_1, \ldots, x_d)$ are independent with $x_j \sim N(0, \hat{\lambda}_j)$.

4. Calculate the corresponding CI on each simulated data set from Step 3 to obtain an empirical distribution of the CI based on the null hypothesis.

5. Calculate a p-value for the original dataset and draw a conclusion based on a pre-specified test level.

In some situations, especially when comparing different clusters, it is hard to distinguish between p-values that are $<1/N_{sim}$. A useful approximation is to fit Gaussians to the simulated CI distributions, and use these to obtain p-values or Z-scores. While the Gaussian approximation is often dubious, this approach still provides useful comparisons between settings.

15.3.4 Diagnostics for Assessing Validity of SigClust Assumptions

As illustrated in Section 15.3, the estimation step of the background noise plays an important role in SigClust. To check whether or not the estimation (15.2) can fit the data well, some diagnostic tools exist. Figure 15.2 shows some diagnostics for one realization of a simulated dataset. The data are generated from a d-dimensional multivariate Gaussian distribution with mean 0 and covariance matrix $\Lambda = \mathrm{diag}(\underbrace{v, \ldots, v}_{w}, 1, \ldots, 1)$. Here, $d = 1000$, $v = 10$, $w = 250$, and the sample size is $n = 100$. The left panel of Figure 15.2 displays the distribution of the elements in the $d \times n$ data matrix. Note that 75% of these are standard Gaussian, representing background noise, i.e., have standard deviation $\sigma_N = 1$, and 25% have 10 times the variance, representing important biological structure. The right panel of Figure 15.2 shows the Q-Q plot of the empirical data distribution versus the Gaussian distribution based on the estimated background noise. Note that the central part of the Q-Q plot, between the 0.25 and 0.75 quantiles, follows the diagonal line very well, showing that the MAD will provide an effective estimate of the background noise level, σ_N. In cases where the Q-Q plot trends away from the diagonal line within these two quantiles, then the estimate of σ cannot be trusted, and SigClust will give unreliable inference.

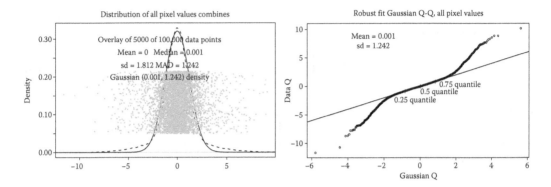

FIGURE 15.2
Summary plot of the elements in the $d \times n$ data matrix for a simulated example. The green points in the left panel, representing randomly chosen 5000 entries from the total $d \times n = 100{,}000$ entries, are plotted with random vertical jitter for better visualization. There is a solid curve which is a kernel density estimate of this distribution and the dashed curve is the Gaussian fit based on the estimated background noise. The right panel is the Q-Q plot for comparing the empirical distribution with the fitted Gaussian distribution.

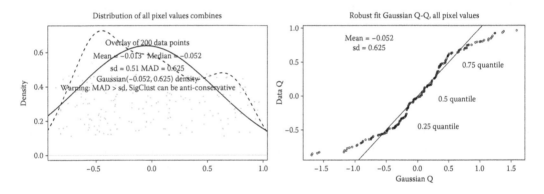

FIGURE 15.3
Diagnostic plots for the simulated uniform on the disk example. This shows how diagnostics are important to flag a hard to interpret SigClust behavior.

Figure 15.3 shows another sense in which these diagnostic plots are very useful. This example is far from Gaussian in a different sense: the data are generated from a uniform distribution on a two dimensional disk. For such data, SigClust reports a significant p-value 0.03, indicating that these data do not come from a single Gaussian distribution. Because one might intuitively view such data as a "single cluster," this example highlights that one must keep in mind that SigClust defines a single cluster by a Gaussian null distribution. For typical applications of SigClust, where there are many background noise values, together with larger structural values, the MAD is smaller than the s.d. In this unusual example, the reverse is true. Hence, the software gives a warning about this hard to interpret behavior when MAD > s.d. In this case, the diagnostics indicate clearly that the data are not a single Gaussian distribution. However, interpretation of the results may require another approach to even the definition of a cluster.

Figure 15.4 plots the elements of the estimated covariance matrix eigenvalues based on the simulated data with $w = 1$ and $v = 1$ in the left panel and $w = 250$ and $v = 10$ in the

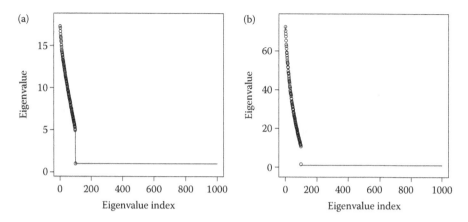

FIGURE 15.4
Plots of the elements of the estimated Λ for the cases of $v = 1, w = 1$ (a) and $v = 10, w = 250$ (b). These show poor performance of the eigenvalue estimation for the low signal case (a), but much better performance for the strong signal case (b).

right panel. The sample size and dimension are the same as the previous example. Note that the data matrix is of size 100×1000, so its rank is upper bounded by 100. As a result, approximately 100 $\hat{\lambda}_j$s that are nonzero. One striking phenomenon is that almost all estimated nonzero $\hat{\lambda}_j$ are larger than the corresponding theoretical λ_j. Especially, for the case of $v = 1$, all the elements in the true Λ are 1, that is, the horizontal line in Figure 15.4(a), in contrast to the first 100 elements in the estimated $\hat{\Lambda}$, which are in the range of 5–17, that is, the points in Figure 15.4(a). This is because the total variation, coming from all 1000 entries in the data vectors, is compressed by PCA into only the first 100 nonzero eigenvalues. For the case of $w = 250$ and $v = 10$ shown in Figure 15.4(b), the similar results can be obtained.

The diagnostic plots in Figures 15.2 and 15.4 show that the current estimation of background noise is reasonable but the estimation of the covariance matrix eigenvalues needs to be improved further. A solution to this problem can be found in Huang et al. (2014).

15.4 Cancer Microarray Data Application

In this section, the method is applied to two real cancer datasets. The first one is the Glioblastoma Multiforme (GBM) brain cancer data from The Cancer Genome Atlas Research Network. The data have been studied in Verhaak et al. (2010), and four clinically relevant subtypes were identified using integrated genomic analysis (meaning they were originally found by a consensus clustering algorithm, but validated by finding important differences between clusters in terms of survival times), which are labeled as proneural (PN), neural (NL), classical (CL), and mesenchymal (MES). After filtering the genes using the ratio of the sample standard deviation and sample mean of each gene, the dataset contained 383 patients with 2727 genes. Among the 383 samples, there are 117 MES samples, 69 NL samples, 96 PL samples, and 101 CL samples.

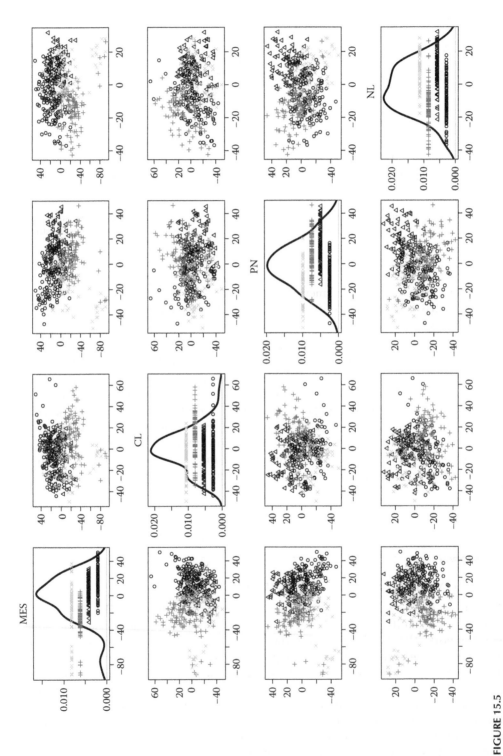

FIGURE 15.5
PCA projection scatter plot view of the GBM data, showing 1D (diagonal) and 2D (off-diagonal) projections of the data onto PC directions. Groupings of symbols and shades indicate biological subtypes.

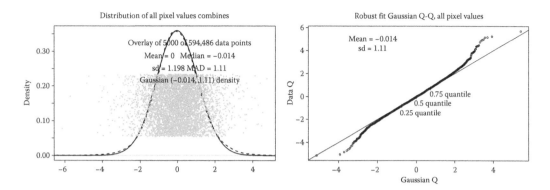

FIGURE 15.6
SigClust Diagnostic test for the pair MES-CL. The view format is similar to that of Figure 15.2. Both panels now show the presence of underlying biological signal, however the Gaussian noise is still sufficiently dominant for the SigClust assumptions to hold.

Figure 15.5 studies the raw GBM data using a scatter plot matrix visualization based on the first four principal component (PC) axes. Observations from different subtypes are distinguished by different colors and symbols. The plots on the diagonal show the one-dimensional projections of the data onto each PC direction vector. A different height is added to each subclass for convenient visual separation. The off-diagonal plots are projections of the data onto two-dimensional planes, determined by the various pairs of the PC directions. Note that none of the four classes appears to be in a separate cluster in this view. However, it is important to keep in mind that only four directions in this 2,727 dimensional space are considered here.

To investigate the structure of these data more deeply, SigClust is applied to every possible pair-wise combination of subclasses and the p-value is calculated based on $N_{sim} = 1000$ Monte Carlo simulation. Here, the cluster index is computed based on the given cluster label. Diagnostic of the Gaussian assumption and validity of the background noise estimation are displayed in Figure 15.6 for the pair MES-CL. Note that in the left plot, the dashed kernel density estimate does not exactly coincide with solid Gaussian fit and has heavier tails, indicating the presence of matrix values with strong biological signal. From the Q-Q plot shown on the right panel, it can be seen that the amount of biological signal is small relative to the noise, that is, the empirical distribution agrees with the Gaussian distribution quite well in the 0.25–0.75 quantile regions, so the MAD-based estimate of the noise standard deviation, σ_N, is expected to be quite accurate. The diagnostic plots for other pairs are not shown here, but the conclusions are quite similar. The SigClust results are listed in the first row of Table 15.1. The p-values for all pairs are highly significant which implies

TABLE 15.1

SigClust p-Values for Each Pair of Subtypes of GBM and BRCA Data. The Known Cluster Labels Are Used to Calculate the Cluster Index. The Gaussian Fit p-Values Which Are Less Than 10^{-100} Are Denoted as 0 Here

GBM	MES-CL	MES-PN	MES-NL	CL-PN	CL-NL	PN-NL
	6.6×10^{-9}	0	0	0	0	0
BRCA	LumA-Basal	LumA-Her2	LumA-LumB	Basal-Her2	Basal-LumB	Her2-LumB
	9.6×10^{-78}	3.9×10^{-13}	0.87	5.7×10^{-26}	1.2×10^{-48}	1.8×10^{-3}

TABLE 15.2

SigClust p-Values for Each Subtype of the GBM Data

MES	CL	PN	NL
6.5×10^{-7}	1.6×10^{-11}	1.4×10^{-6}	3.3×10^{-17}

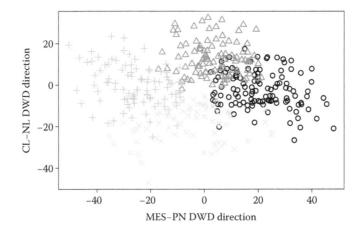

FIGURE 15.7
Scatterplot matrix view of the data, using pairwise DWD directions to determine the projections. The classes MES, CL, PN and NL are denoted by circle, triangle, cross and "x" symbols, respectively.

that they are well separated from each other. Note that because of the multiple comparison problem and the fact that the clusters were determined from the data, these p-values should not be interpreted literally, but are only used in an exploratory way. A number of approaches can be used to make these p-values more formal, including Bonferroni and the false discovery rate.

The SigClust p-value for the whole data set is 5.6×10^{-21}. Table 15.2 shows the SigClust p-values for each subtype of the GBM data.

It is not surprising that the PCA view in Figure 15.5 did not highlight these important clusters in the data. This is because PCA is driven only by variation in the data and ignores class label information. Figure 15.7 provides a much different view of the data (using only different directions in the 2727 dimensional space), where the PC directions are replaced by directions that use class labels in an important way. These directions are computed using Distance Weighted Discrimination (DWD), Marron et al. (2007), trained on MES vs. PN and CL vs. NL, respectively. Note that the four clusters are now visually very apparent. This shows how it is important to couple visualization methods with rigorous statistical validation, as is done by SigClust.

The second real example we used is breast cancer data (BRCA) also from The Cancer Genome Atlas Research Network which include four subtypes: LumA, LumB, Her2 and Basal. The sample size is 348 and the number of genes used in the analysis after filtering is 2043. Among the 348 samples, there are 154 LumA, 81 LumB, 42 Her2, and 66 Basal. The scatter plots from projecting the data points onto the first PC directions are displayed in Figure 15.8. The triangle basal points seem to form a clearly distinct cluster. Others are less clear. The results of applying SigClust to each pair of subclasses are shown in the second row of Table 15.1. All p-values are significant other than for the pair of LumA and

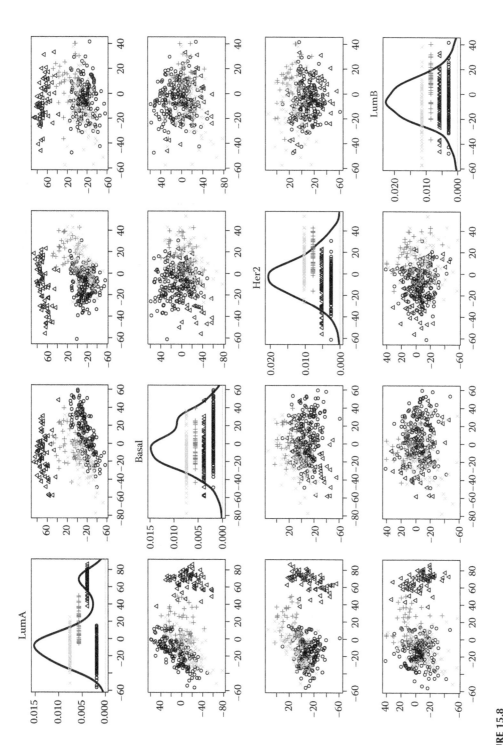

FIGURE 15.8
PCA projection scatter plot view of the BRCA data. Descriptions are similar to Figure 15.5.

LumB. The results are generally consistent with what can be seen in Figure 15.8. The fact that LumA and LumB do not form two significant clusters is consistent with the findings of Parker et al. (2009), who suggest that these are essentially a stretched Gaussian distribution (thus not flagged by SigClust), with an important clinical division within that distribution. Again it is important to keep in mind that these are only meant to be exploratory p-values.

15.5 Open Problems of SigClust

SigClust is based on some approximations, with room for improvement. As discussed in Section 15.3.2, simulation for the null distribution of the SigClust test statistic requires estimation of the eigenvalues of the Gaussian null covariance matrix. It is well known that when the population size is fixed, as the number of samples tends to infinity, the sample covariance matrix is a good approximation of the population covariance matrix. However, for HDLSS situations as often encountered in many contemporary data analysis, the sample covariance matrix is no longer a good approximation to the covariance matrix. As a consequence, it is questionable to use the eigenvalues based on the sample covariance matrix in SigClust. To overcome this problem, currently SigClust employs a factor analysis model to estimate the covariance matrix eigenvalues for HDLSS data. Empirical studies have shown that this method can be improved further, see the example in Figure 15.4.

The current SigClust algorithm is based on splitting a data set into two clusters. In many situations, this still detects heterogeneity if there are more than two clusters in the data set, because the two-cluster solution already deviates significantly from the homogeneity model.

However, the method can fail for some situations such as the toy example shown in Figure 15.9. The two-dimensional example shown in there is sampled from a mixture of

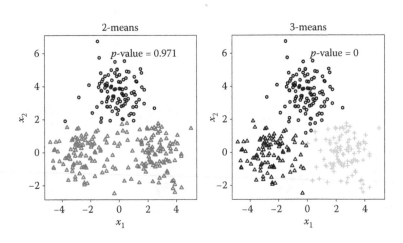

FIGURE 15.9
Application of two-cluster SigClust and three-cluster SigClust to a three-class two-dimensional toy example, with the data generated from a mixture of three Gaussian distributions. Different colors and symbols represent different class labels using 2-means and 3-means clustering algorithms.

three multivariate Gaussian distributions with the same covariance I_2 but different mean vectors $(2.5, 0)$, $(-2.5, 0)$, and $(0, 2.5)$. The left panel in Figure 15.9 shows the result of applying SigClust to the whole data set of this example where the 2-means algorithm is used for the clustering. The p-values are far from statistically significant, so sequential SigClust fails to flag significant clusters in this data set. However, there is a clear visual impression of three clusters. If SigClust is applied to each of the three subsets of the data where one component of the Gaussian mixture has been excluded, we can get highly significant p-values. This indicates that these clusters are important underlying structure that was masked by the third cluster in pairwise SigClust.

A straightforward amendment would be to base tests on cluster indexes not only for $K = 2$ means, but also for $K = 3$ (as shown in the right panel in Figure 15.9) and higher. However, as is the problem for all test statistics assuming implicitly a fixed K, this would require a careful consideration of multiple comparison issues and a stopping rule.

Another possible improvement of SigClust is the incorporation of clustering methods other than K-means, for example hierarchical methods. An important issue to treat here is rotation invariance for covariance matrix estimation, which is no problem if the Euclidean distance is used but may be a problem with other distances such as the L_1 (i.e., Manhattan) and Hamming distance.

15.6 Conclusion

We have given an overview of significance testing in cluster analysis with particular attention to the SigClust method, which is particularly noteworthy, relative to other approaches, for its strong performance in HDLSS contexts. The SigClust method illustrates some of the key issues in significance testing such as the invariance considerations required to make sure that simulations cover the full set of distributions in the null hypothesis, and null model diagnostics. Also, some open problems of SigClust are of interest regarding other testing approaches as well, namely, the interplay between test and cluster analysis method, and the multiple testing issues that arise from testing assuming various numbers of clusters K, or testing on subsets of the data that are picked by first applying a clustering method.

The precise characteristics of using significance tests for determining K or for testing single clusters separately have not been worked out, to our knowledge, for any method, and this remains an important open problem.

Other directions of research are the theoretical derivation of distributions P_T of more test statistics T under more null models, because existing results are rather limited.

Setting up suitable non-standard null models and test statistics for all kinds of further specific situations and data types, e.g., with time series structure, is a fruitful direction for further research as well.

References

Beale, E.M.L. 1969. *Euclidean Cluster Analysis*. In *Bulletin of the IMS*, pp. 92–94. Institute of Mathematical Statistics.

Besag, J. and Clifford, P. 1989. Generalized monte carlo significance tests. *Biometrika*, 76, 633–642.

Bock, H. 1985. On some significance tests in cluster analysis. *Journal of Classification*, 2, 77–108.

Bock, H. 1996. Probabilistic models in cluster analysis. *Computational Statistics & Data Analysis*, 23, 5–28.

Buja, A., Cook, D., Hofmann, H., Lawrence, M., Lee, E.-K., Swayne, D.S., and Wickham, H. 2009. Statistical inference for exploratory data analysis and model diagnostics. *Philosophical Transactions of the Royal Society A*, 367, 4361–4383.

Duda, R.O., Hart, P.E., and Stork, D.G. 2000. *Pattern Classification*. Wiley-Interscience Publication.

Godehardt, E. and Horsch, A. 1995. *Graph-Theoretic Models for Testing the Homogeneity of Data*. In Gaul, W. and Pfeifer, D. (Eds.), *From Data to Knowledge*, pp. 167–176. Springer, Berlin.

Gotelli, N.J. 2000. Null model analysis of species co-occurrence patterns. *Ecology*, 81, 2606–2621.

Hampel, F.R., Ronchetti, E.M., Rousseeuw, P.J., and Stahel, W.A. 1986. *Robust Statistics: The Approach Based on Influence Functions (Wiley Series in Probability and Statistics)*. John Wiley & Sons, New York.

Hardy, A. 2011. *The Poisson Process in Cluster Analysis*. In Fichet, B., Piccolo, D., Verde, R., and Vichi, M. (Eds.), *Classification and Multivariate Analysis for Complex Data Structures*, pp. 51–62.

Hartigan, J.A. and Hartigan, P.M. 1985. The dip test of unimodality. *Ann. Stat.*, 13, 70–84.

Hautaniemi, S., Yli-Harja, O., Astola, J., Kauraniemi, P., Kallioniemi, A., Wolf, M., Ruiz, J., Mousses, S., and Kallioniemi, O.P. 2003. Analysis and visualization of gene expression microarray data in human cancer using self-organizing maps. *Machine Learning*, 52, 45–66.

Hennig, C. 2007. Cluster-wise assessment of cluster stability. *Comput. Stat. Data. Anal.*, 52, 258–271.

Hennig, C. and Hausdorf, B. 2004. Distance-based parametric bootstrap tests for clustering of species ranges. *Computational Statistics & Data Analysis*, 45, 875–895.

Hennig, C. and Liao, T.F. 2013. Comparing latent class and dissimilarity based clustering for mixed type variables with application to social stratification (with discussion). *Journal of the Royal Statistical Society, Series C*, 62, 309–369.

Hennig, C. and Lin, C.J. 2015. Flexible parametric bootstrap for testing homogeneity against clustering and assessing the number of clusters. *Statistics and Computing* 25(4), 821–833.

Huang, Hanwen, Yufeng Liu, Ming Yuan, and J. S. Marron. 2014. Statistical significance of clustering using soft thresholding. *Journal of Computational and Graphical Statistics*, accepted.

Ling, R.F. 1973. A probability theory of clustering. *Journal of the American Statistical Association*, 68, 159–164.

Liu, Y., Hayes, D.N. Nobel, A., and Marron, J.S. 2008. Statistical significance of clustering for high-dimension, low-sample size data. *Journal of the American Statistical Association*, 103(483), 1281–1293.

Maitra, R., Melnykov, V., and Lahiri, S.N. 2012. Bootstrapping for significance of compact clusters in multidimensional datasets. *Journal of the American Statistical Association*, 107, 378–392.

Manly, B.F.J. 1997. *Randomization, Bootstrap and Monte Carlo Methods in Biology*. Chapman & Hall, London.

Marron, J.S., Todd, M., and Ahn, J. 2007. Distance-weighted discrimination. *Journal of the American Statistical Association*, 102, 1267–1271.

McLachlan, G. and Peel, D. 2000. *Finite Mixture Models*. Wiley, New York.

McShane, L.M., Radmacher, M.D., Freidlin, B., Yu, R., Li, M.C., and Simon, R. 2002. Methods for assessing reproducibility of clustering patterns observed in analyses of microarray data. *Bioinformatics*, 18(11), 1462–1469.

Parker, J.S., Mullins, M., Cheang, M.C.U., Leung, S., Voduc, D., Vickery, T., Davies, S., Fauron, C., He, X., Hu, Z. et al. 2009. Supervised risk predictor of breast cancer based on intrinsic subtypes. *Journal of Clinical Oncology*, 27(8), 1160–1167.

Rasson, J.P. and Kubushishi, T. 1994. *The Gap Test: An Optimal Method for Determining the Number of Natural Classes in Cluster Analysis*. In Diday, E., Lechevallier, Y., Schader, M., Bertrand, P., and Butschy, B. (Eds.), *New Approaches in Classification and Data Analysis*, pp. 186–193. Springer, Berlin.

Ripley, B.D. 1981. *Spatial Statistics*. Wiley, Hoboken, NJ.

SAS 2009. Sas/stat(r) 9.2 user's guide, second edition.

Suzuki, R. and Shimodaira, H. 2006. Pvclust: an r package for assessing the uncertainty in hierarchical clustering. *Bioinformatics*, 22(12), 1540–1542.

Tantrum, J., Murua, A., and Stuetzle, W. 2003. Assessment and pruning of hierarchical model based clustering. In *Proceedings of the Ninth ACM SIGKDD International Conference on Knowledge Discovery and Data Mining*, pp. 197–205. ACM, New York.

Tibshirani, R. and Walther, G. 2005. Cluster validation by prediction strength. *Computational & Graphical Statistics*, 14(3), 511–528.

Tibshirani, R., Walther, G., and Hastie, T. 2001. Estimating the number of clusters in a data set via the gap statistic. *Journal of the Royal Statistical Society, Series B*, 63, 411–423.

Ulrich, W. and Gotelli, N.J. 2010. Null model analysis of species associations using abundance data. *Ecology.*, 91, 3384–3397.

Verhaak, R.G., Hoadley, K.A., Purdom, E., Wang, V., Qi, Y., Wilkerson, M.D., Miller, C.R., Ding, L., Golub, T., Mesirov, J.P. et al. Cancer Genome Atlas Research Network 2010. Integrated genomic analysis identifies clinically relevant subtypes of glioblastoma characterized by abnormalities in pdgfra, idh1, egfr, and nf1. *Cancer cell*, 17(1), 98–110.

16

Model-Based Clustering for Network Data

Thomas Brendan Murphy

CONTENTS

Abstract

This chapter reviews some of the most popular statistical model-based methods for clustering network datasets. In particular, the stochastic block model, the mixed membership stochastic block model and the latent position cluster model are described and illustrated using well-known social network data.

16.1 Introduction

A network is any data set that is composed nodes (or actors) upon which we have relational data recorded for each pair of nodes. The relational data can be of any type (categorical

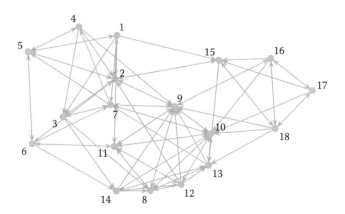

FIGURE 16.1
The social interactions of 18 monks as reported in Sampson (1969) shown as a network. The layout of nodes in the graph is determined using the Fruchterman–Reingold method (Fruchterman and Reingold, 1991).

or numerical), and the relationships can be directed or undirected. Suppose, we have a network consisting of n nodes, then the data are of the form of an $n \times n$ adjacency matrix, Y, where y_{ij} is the recorded relationship between node i and node j. The network is said to be undirected if $y_{ij} = y_{ji}$ for all pairs (i, j); otherwise, the network is said to be directed. The values taken by y_{ij} can be of any type, but are usually binary, count or continuous quantities.

An example of an 18-node network of social interactions between monks is shown as a network in Figure 16.1 and as an adjacency matrix in Table 16.1. In these data, $y_{ij} = 1$ if monk i reported monk j as their friend and $y_{ij} = 0$ otherwise. The graph is directed

TABLE 16.1

The Adjacency Matrix for Social Interactions of 18 Monks as Reported in Sampson (1969)

$$
Y = \begin{pmatrix}
0 & 1 & 1 & 0 & 1 & 0 & 1 & 0 & 0 & 0 & 1 & 0 & 0 & 0 & 1 & 0 & 0 & 0 \\
0 & 0 & 1 & 0 & 1 & 1 & 0 & 0 & 1 & 0 & 0 & 0 & 0 & 0 & 1 & 0 & 0 & 0 \\
0 & 1 & 0 & 0 & 0 & 0 & 1 & 1 & 0 & 0 & 0 & 0 & 0 & 1 & 0 & 0 & 0 & 0 \\
0 & 1 & 1 & 0 & 1 & 0 & 0 & 0 & 1 & 0 & 0 & 0 & 0 & 0 & 0 & 0 & 0 & 0 \\
1 & 1 & 0 & 1 & 0 & 1 & 0 & 0 & 0 & 0 & 0 & 0 & 0 & 0 & 0 & 0 & 0 & 0 \\
0 & 1 & 0 & 0 & 1 & 0 & 1 & 0 & 0 & 0 & 1 & 0 & 0 & 1 & 0 & 0 & 0 & 0 \\
1 & 0 & 1 & 1 & 1 & 0 & 0 & 0 & 1 & 1 & 0 & 0 & 0 & 0 & 0 & 0 & 0 & 0 \\
0 & 0 & 0 & 0 & 0 & 0 & 0 & 0 & 1 & 1 & 1 & 0 & 1 & 0 & 0 & 0 & 0 & 0 \\
0 & 1 & 0 & 0 & 0 & 0 & 1 & 1 & 0 & 1 & 1 & 0 & 0 & 0 & 0 & 1 & 0 & 0 \\
0 & 0 & 0 & 0 & 0 & 0 & 0 & 1 & 1 & 0 & 1 & 1 & 1 & 0 & 0 & 0 & 0 & 0 \\
0 & 0 & 0 & 0 & 0 & 1 & 0 & 1 & 1 & 1 & 0 & 1 & 0 & 0 & 0 & 0 & 0 & 0 \\
0 & 1 & 0 & 0 & 0 & 0 & 0 & 1 & 1 & 1 & 1 & 0 & 1 & 0 & 0 & 0 & 0 & 0 \\
0 & 0 & 0 & 0 & 0 & 0 & 1 & 1 & 1 & 1 & 0 & 0 & 0 & 1 & 0 & 0 & 0 & 0 \\
0 & 0 & 0 & 0 & 0 & 0 & 0 & 1 & 1 & 1 & 0 & 1 & 1 & 0 & 0 & 0 & 0 & 0 \\
0 & 1 & 0 & 0 & 0 & 0 & 0 & 0 & 0 & 0 & 0 & 0 & 1 & 0 & 0 & 0 & 0 & 1 \\
0 & 0 & 0 & 0 & 0 & 0 & 0 & 0 & 1 & 1 & 0 & 0 & 0 & 0 & 1 & 0 & 1 & 1 \\
0 & 0 & 0 & 0 & 0 & 0 & 0 & 0 & 0 & 1 & 0 & 0 & 0 & 0 & 1 & 1 & 0 & 1 \\
0 & 0 & 0 & 0 & 0 & 0 & 0 & 0 & 1 & 1 & 0 & 0 & 1 & 0 & 1 & 1 & 1 & 0
\end{pmatrix}
$$

TABLE 16.2

The In-degree and Out-degree of Each Node in Sampson's Monk Data

	1	2	3	4	5	6	7	8	9	10	11	12	13	14	15	16	17	18
In-degree	6	5	4	4	4	5	6	4	6	5	5	6	5	5	3	5	4	6
Out-degree	2	8	4	2	5	3	5	7	11	10	6	3	6	3	5	3	2	3

and this can be easily seen in the adjacency matrix Y which is non-symmetric. The layout of the nodes in Figure 16.1 is chosen so that connected nodes tend to be adjacent and non-connected nodes are distant. In this case, the Fruchterman–Reingold force-based algorithm (Fruchterman and Reingold, 1991) is used as implemented in the network R package (Butts et al., 2014).

Many network data sets exhibit structure where the subsets nodes have a tendency to form links with each other. A close inspection of Figure 16.1 and Table 16.1 reveals that subsets of nodes $\{1, 2, \ldots, 7\}$, $\{8, 9, \ldots, 14\}$, and $\{15, 16, 17, 18\}$ have a higher tendency to link within subset than across subsets. However, there are some connections across subsets also. In network analysis, such structure is commonly referred to as community structure rather than clustering structure. While, in many applications, the terms cluster and community are identical, we will see in Section 16.3 that clusters can be different to communities.

In this chapter, we introduce a number of model-based clustering methods for network data. Emphasis will be given to clustering binary network data sets, but many of the methods can be extended to other data types. There are many other aspects of network modeling that are not covered in this chapter. Extensive overviews of network analysis and the statistical modeling of network data include Wasserman and Faust (1994), Carrington et al. (2005), Wasserman et al. (2007), Airoldi et al. (2007), Kolaczyk (2009), Goldenberg et al. (2010), and Salter-Townshend et al. (2012).

In the following sections, we will concentrate on the analysis of binary network data, where $y_{ij} = 1$ if node i and node j are related and $y_{ij} = 0$ otherwise. Extending many of the models to more general link types is straightforward.

For any network, a number of summary statistics can be computed. One of interest is the degree of the nodes in the network. For an undirected network, the degree of node i is the number of edges that it is connected to, that is, $d_i = \sum_{j=1}^{n} y_{ij} = \sum_{j=1}^{n} y_{ji}$. For a directed network, a node has an in degree and an out degree which count the number of incoming and outgoing edges, respectively. That is, the in degree is $d_{in}(i) = \sum_{j=1}^{n} y_{ji}$ and the out degree is $d_{out} = \sum_{j=1}^{n} y_{ij}$ (see Table 16.2).

16.2 Example Data

In this section, we introduce example data sets that are used to illustrate the model-based clustering methods for network data.

16.2.1 Sampson's Monk Data

Sampson (1969) recorded the social interactions among a group of 18 novice monks while resident in a monastery. During his stay, a political "crisis in the cloister" resulted in the

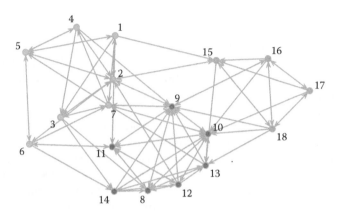

FIGURE 16.2
The social interactions of 18 monks as reported in Sampson (1969). The nodes are shaded according to the grouping given by Sampson.

expulsion of four monks and the voluntary departure of several others. Of particular interest is the data on positive affect relations ("liking"), in which each monk was asked if they had positive relations to each of the other monks.

The data were gathered at three times to capture changes in group sentiment over time. They represent three time points in the period during which a new cohort entered the monastery near the end of the study but before the major conflict began. Each member ranked only his top three choices on "liking." (Some subjects offered tied ranks for their top four choices). A tie from monk A to monk B exists if A nominated B as one of his three best friends at that time point. The most commonly analyzed version of the data aggregates the "liking" data over the three time points. The data are shown in Figure 16.1 and the adjacency matrix of the network is also shown in Table 16.1.

Sampson (1969) grouped the novice monks into three groups which he called "Loyal," "Outcasts," and "Turks"; thus, there are known social clusters in this network. A plot of the data where the nodes are shaded by Sampson's grouping is given in Figure 16.2. This plot indicates that there appears to be higher connectivity within Sampson's groups than between them, thus the nodes in the network exhibit clustering.

16.2.2 Zachary's Karate Club

Zachary's karate club data set (Zachary, 1977) consists of the friendship network of 34 members of a university-based karate club. It was collected following observations over a three-year period in which a factional division caused the members of the club to formally separate into two organizations. While friendships within the club arose organically, a financial dispute regarding the pay of part time instructor Mr. Hi tested these ties, with

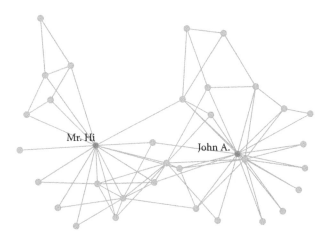

FIGURE 16.3
The friendship network of Zachary's karate club. The two key members in the dispute within the club, Mr. Hi and John A., are labeled and shaded differently.

two political factions developing. Keys to the dispute were two members of the network, Mr. Hi and club President John A. The dispute eventually led to the dismissal of Mr. Hi by John A. and his supporters resigned from the karate club and established a new club, headed by Mr. Hi. The data set exhibits many of the phenomenon observed in sociological studies, in particular community (clustering) structure. The data are shown in Figure 16.3 where the friendship network is shown and the location of Mr. Hi and John A. within the network is highlighted.

16.3 Stochastic Block Model

The most fundamental model for clustering nodes in a network is the stochastic block model (Snijders and Nowicki, 1997; Nowicki and Snijders, 2001); this model is similar in flavor to the latent class model (LCA) (Lazarsfeld and Henry, 1968).

First, we assume that there are G blocks present in the population and that the probability that a node comes from block g is given by τ_g. Let $z_i = (z_{i1}, z_{i2}, \ldots, z_{iG})^T$ be a column indicator vector of block membership for node i, so $z_{ig} = 1$ if node i belongs to block g and $z_{ig} = 0$ otherwise. We also assume that there is a $G \times G$ block-to-block interaction matrix, Θ, where θ_{gh} is the probability that a node in block g is related to a node in block h. Further, we assume that all pairwise interactions are independent conditional on $z_1, z_2, \ldots, z_n$.

Thus,

$$\mathbb{P}\{y_{ij} = 1 \mid \Theta, z_{ig} = 1, z_{jh} = 1\} = \begin{cases} \theta_{gh} & \text{for } y_{ij} = 1 \\ 1 - \theta_{gh} & \text{for } y_{ij} = 0 \end{cases}$$

$$= \begin{cases} z_i \Theta z_j^T & \text{for } y_{ij} = 1 \\ 1 - z_i \Theta z_j^T & \text{for } y_{ij} = 0 \end{cases} . \quad (16.1)$$

The property that the interaction probabilities only depend on the cluster membership of the nodes is called stochastic equivalence (Fienberg and Wasserman, 1981) and this concept has origins in structural equivalence (Lorrain and White, 1971). Further, if the block-to-block interaction matrix, Θ, is diagonally dominant (i.e., $\theta_{gg} > \theta_{gh}$ for $h \neq g$), then the network has evidence of assortative mixing and the blocks will correspond to communities; this is because nodes in the same block are more likely to connect to each other. However, if Θ has the property that $\theta_{gg} < \theta_{gh}$, for $h \neq g$, then disassortative mixing is present; in this case, the blocks do not correspond to communities, but they are still valid clusters because the nodes in the block are stochastically equivalent.

Recent studies have considered the stochastic block model within a broader context. Bickel and Chen (2009) studied the connections between modularity and stochastic block models, while Bickel et al. (2011) discuss the asymptotic properties of a class of models of which stochastic block models are a subset. Rohe et al. (2011) and Choi et al. (2012) also consider the asymptotic properties of the stochastic block models.

Channarond et al. (2012) investigated further properties of the stochastic block model including the degree distribution which can be approximately modeled as a finite mixture of Poisson distributions.

16.3.1 Inference

The complete-data likelihood for the stochastic block model has the form,

$$L_c(\tau, \Theta) = \mathbb{P}\{Y, Z \mid \tau, \Theta\} = \mathbb{P}\{Y \mid Z, \Theta\}\mathbb{P}\{Z \mid \tau\}$$

$$= \prod_{i,j} \left(z_i \Theta z_j^T\right)^{y_{ij}} \left(1 - z_i \Theta z_j^T\right)^{1-y_{ij}} \times \prod_{i=1}^{n} \prod_{g=1}^{G} \tau_g^{z_{ig}}, \qquad (16.2)$$

where $Z = (z_1, z_2, \ldots, z_n)$ and where the product is taken over all pairs $\{(i, j) : i \neq j\}$ if the network is directed and over pairs $\{(i, j) : i < j\}$ if the network is undirected.

The observed likelihood function does not have a tractable form for even moderate values of n because it involves summing (16.2) over all G^n possible values for Z. Thus, maximum-likelihood inference for this model is difficult. However, Snijders and Nowicki (1997) outlines an EM algorithm for fitting stochastic block models when $G = 2$.

Bayesian inference using Markov Chain Monte Carlo (MCMC) has been widely adopted for fitting this model. In order to implement Bayesian inference for the stochastic block model, conjugate priors are used for the model parameters. In particular, the following priors are commonly used $\tau \sim \text{Dirichlet}(\delta)$ and $\theta_{gh} \sim \text{beta}(\alpha, \beta)$ where δ, α, and β are the prior hyperparameters. With this prior specification, a Gibbs sampler (Gelfand and Smith, 1990) can be used to generate samples from the joint posterior, $p(Z, \tau, \Theta \mid Y)$, for the model parameters and cluster memberships. MCMC schemes for fitting the stochastic block model are given in Snijders and Nowicki (1997) for the case when $G = 2$ and in Nowicki and Snijders (2001) for any G.

Recently, McDaid et al. (2012, 2013) developed a collapsed sampler for the stochastic block model by analytically integrating out the model parameters and thus sampling from the posterior $p(Z \mid Y)$. This sampler facilitates fitting stochastic block models to large network datasets.

Further efficient algorithms for fitting stochastic block models have been developed including a CEM algorithm (Zanghi et al., 2008) and a variational Bayesian algorithm

(Daudin et al., 2008; Latouche et al., 2010); these algorithms are implemented in the mixer R package (Ambroise et al., 2013). Finally, Channarond et al. (2012) exploit the degree distribution of the stochastic block model that can be approximately modeled by a finite mixture of Poisson distributions to find a highly efficient algorithm for approximate inference for the stochastic block model; this algorithm uses only the degree of the nodes to estimate the number of blocks, block membership and block parameters, and the resulting estimates are consistent.

16.3.1.1 Sampson's Monks Data

The stochastic block model was fitted to Sampson's monks data from Section 16.2.1 using the Bayesian method of Latouche et al. (2010) and using the mixer R package (Ambroise et al., 2013).

The resulting model is a model with $G = 3$ blocks and the estimated model parameters are

$$\hat{\tau} = \begin{pmatrix} 0.38 \\ 0.38 \\ 0.24 \end{pmatrix} \quad \text{and} \quad \hat{\Theta} = \begin{pmatrix} 0.55 & 0.20 & 0.10 \\ 0.12 & 0.70 & 0.07 \\ 0.07 & 0.27 & 0.79 \end{pmatrix}.$$

The fitted model shows strong within block connection probabilities and low between block connection probabilities. Thus, the model has clustered the monks into subsets with high connectivity. Table 16.3 shows a cross classification of the maximum *a posteriori* block membership for each monk against the groupings given in Sampson (1969). The values show that the model has clustered the monks into the same groupings and thus reveals the structure postulated by Sampson.

16.3.1.2 Zachary's Karate Data

The stochastic block model was fitted to Zachary's karate data from Section 16.2.2 using the Bayesian method of Latouche et al. (2010) as implemented in the mixer R package.

The resulting fitted model is a model with $G = 4$ blocks and the estimated model parameters are

$$\hat{\tau} = \begin{pmatrix} 0.11 \\ 0.36 \\ 0.46 \\ 0.08 \end{pmatrix} \quad \text{and} \quad \hat{\Theta} = \begin{pmatrix} 0.80 & 0.53 & 0.17 & 0.25 \\ 0.53 & 0.13 & 0.00 & 0.10 \\ 0.17 & 0.00 & 0.09 & 0.72 \\ 0.25 & 0.10 & 0.72 & 0.67 \end{pmatrix}. \tag{16.3}$$

TABLE 16.3

Cross Classification of the Maximum *a posteriori* Block Membership for Each Monk against Sampson's Grouping of Monks

	Loyal	Outcasts	Turks
1	7	0	0
2	0	0	7
3	0	4	0

TABLE 16.4

Cross Tabulation of Maximum *a posteriori* Block Membership against Faction Identity and against New Club Membership

	Faction Identity							
	−2	−1	0	1	2		Mr. Hi's	John A's
1	3	0	0	0	0	1	3	0
2	7	5	1	0	0	2	13	0
3	0	0	2	4	10	3	1	15
4	0	0	0	0	2	4	0	2

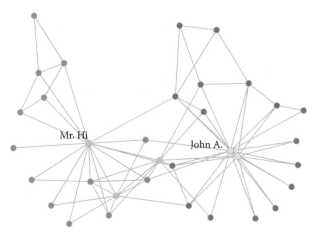

FIGURE 16.4
A plot of Zachary's karate club network with the nodes shaded by maximum *a posteriori* block membership.

It is worth noting that the block-to-block interaction matrix, Θ, does not have a dominant diagonal. Thus, there is some evidence of disassortative mixing in this network.

Table 16.4 shows a cross tabulation of the maximum *a posteriori* block memberships with a faction identity proposed by Zachary (1977); the faction identity score is coded as: −2 (strongly Mr. Hi's), −1 (weakly Mr. Hi's), 0 (neutral), +1 (weakly John's), and +2 (strongly John's). Table 16.4 also shows a cross tabulation of maximum *a posteriori* block membership against the new club membership after the club split. The results show that the stochastic block model captures the divide in the club and the roles within this divide. The results are easily visualized in Figure 16.4, where the nodes are shaded according to maximum *a posteriori* block membership.

16.4 Mixed Membership Stochastic Block Model

The mixed membership stochastic block model (MMSB) (Airoldi et al., 2008) is an extension of the stochastic blockmodel (Section 16.3) that allows nodes to have mixed membership of multiple blocks. Thus, the mixed membership stochastic blockmodel closely related to the latent Dirichlet allocation (Blei et al., 2003) model and the grade of membership model (Erosheva, 2002, 2003; Erosheva et al., 2007) in the same way that the stochastic block model was related to the latent class model.

In this model, as in Sections 16.3, we assume that there are G blocks and any two nodes that belong fully to a block are considered to be stochastically equivalent.

However, in contrast to the stochastic block model, the block membership a node may change depending on which node in the network it is interacting with. Each node may therefore have multiple block memberships as it links with the other nodes in the network.

Let node i have an individual probability τ_{ig} of belonging to block g, and thus a vector of probabilities τ_i, while interacting with any other node in the network. When node i and node j are interacting, they each have an indicator variable $z_{i \to j}$ and $z_{i \leftarrow j}$, then denote sender and receiver cluster membership for each interaction between nodes i and j, respectively. Once cluster membership is accounted for, nodal interaction y_{ij} is modelled, as in Section 16.3, as a Bernoulli random variable with probability $z_{i \to j} \Theta z_{i \leftarrow j}^T$, where Θ is a $G \times G$ matrix of block-to-block interaction probabilities.

Thus, the data generation procedure is as follows:

- For each node $i \in \{1, 2, \ldots, n\}$:
    - Draw a G-dimensional mixed membership vector $\tau_i \sim \text{Dirichlet}(\alpha)$
- For each pair of nodes (i, j):
    - Draw membership indicator for the initiator, $z_{i \to j} \sim \text{Multinomial}(\tau_i)$
    - Draw membership indicator for the receiver, $z_{i \leftarrow j} \sim \text{Multinomial}(\tau_j)$
    - Sample the value of their interaction, $y_{ij} \sim \text{Bernoulli}(z_{i \to j} \Theta z_{i \leftarrow j}^T)$

A number of extensions and alternatives to the MMSB model have recently been developed. In particular, Xing et al. (2010) develop a dynamic version of the mixed membership stochastic blockmodel (dMMSB). Chang and Blei (2010) developed a model based upon the MMSB which jointly model network and nodal covariate data. White and Murphy (2014b) developed version of the MMSB that allows the mixed membership vector to depend on node concomitant variables; the model is called the mixed membership of experts stochastic block model.

16.4.1 Inference

Inference for the mixed membership stochastic block model (MMSB) can be achieved in a Bayesian context using Markov Chain Monte Carlo (MCMC) or variational Bayesian methods. The efficiency of the MCMC algorithm can be improved by analytically integrating out the model parameters in the posterior, $p(\overleftarrow{z}, \overrightarrow{z}, \Theta, \tau \mid y, \alpha)$ and to yield the marginal posterior $p(\overleftarrow{z}, \overrightarrow{z} \mid y, \alpha)$. A collapsed Gibbs sampler (Liu, 1994) can then be used to sample from the posterior for block memberships; this approach is implemented in the lda R package (Chang, 2010). Code for fitting the MMSB is publicly available for Infer.Net (Minka et al., 2010) software and is given in the user guide for the software. Model selection for the MMSB model remains underdeveloped.

16.4.2 Zachary Karate Data

The MMSB model was fitted to Zachary karate data with $G = 4$ blocks to enable comparison with the stochastic block model results of Section 16.3.1. The estimated mixed membership vectors $\hat{\tau}_i$ are shown in a matrix of ternary diagrams in Figure 16.5.

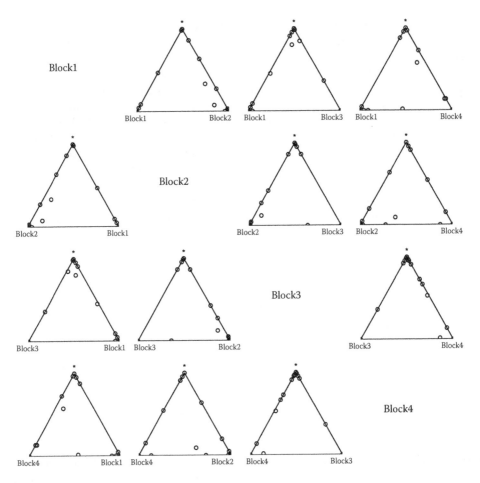

FIGURE 16.5
A matrix of ternary plots which show the mixed membership values for each node in the karate data. Each ternary diagram shows the mixed membership of a pair of blocks versus the remaining blocks.

The resulting block-to-block interaction matrix, Θ, and the mean τ_i value are estimated as

$$
\bar{\tau} = \begin{pmatrix} 0.25 \\ 0.49 \\ 0.06 \\ 0.20 \end{pmatrix} \quad \text{and} \quad \hat{\Theta} = \begin{pmatrix} 0.19 & 0.00 & 0.00 & 0.15 \\ 0.00 & 0.06 & 0.86 & 0.03 \\ 0.00 & 0.97 & 1.00 & 0.22 \\ 0.13 & 0.03 & 0.00 & 0.95 \end{pmatrix}. \tag{16.4}
$$

The fitted model has higher within block connection probabilities than the stochastic block model (16.3). This can be explained as being as a result of the extra flexibility given by the mixed membership structure in the model. This is because nodes are allowed to have mixed membership across blocks, and as a result, the probability of a pairwise interaction can be much higher than the value given by Θ if the pair of nodes being considered have a non-negligible probability of belonging to the same block in their mixed membership vector.

TABLE 16.5

The Mean τ_i Value for Each Fraction Identity in Zachary's Karate Data

	Block1	Block2	Block3	Block4
−2	0.49	0.00	0.02	0.49
−1	0.49	0.17	0.00	0.34
0	0.35	0.65	0.00	0.00
1	0.02	0.89	0.00	0.08
2	0.01	0.84	0.14	0.00

Note: These values show the correspondence between the estimated blocks and the factions.

TABLE 16.6

The Mean τ_i Value for the Two New Clubs Formed after the Break-Up of the Club in Zachary's Karate Data

	Block1	Block2	Block3	Block4
Mr. Hi's	0.50	0.08	0.01	0.41
John A's	0.01	0.89	0.10	0.00

Note: These values show the correspondence between the estimated blocks and the clubs.

White et al. (2012) and White and Murphy (2014a) define the extent of membership (EoM) of a mixed membership vector τ_i as

$$EoM_i = \exp\left(-\sum_{g=1}^{G} \tau_{ig} \log \tau_{ig}\right),$$

which is the exponential of the entropy of τ_i. The EoM value takes values in the range 1 to G, and it is a measure of the effective number of blocks that an observation belongs to; previously, this quantity has been used to measure the effective number of species in ecological applications. A histogram of the EoM_i values for each club member is given in Figure 16.6 and this shows that some members exhibit mixed membership across two or more blocks. Finally, we computed the mean τ_i value for nodes from each faction identity (Section 16.2.2) and by new club membership after the split in the club. The results (Tables 16.5 and 16.6) show that the blocks correspond the split in the club and the role of members within the split and the resulting new clubs that were formed.

16.5 Latent Position Cluster Model

16.5.1 Latent Space Models

Latent space models were introduced by Hoff et al. (2002) under the basic assumption that each node, i, has an unknown position, z_i, in a d-dimensional Euclidean latent space. Network edges are assumed to be conditionally independent given the latent positions, and

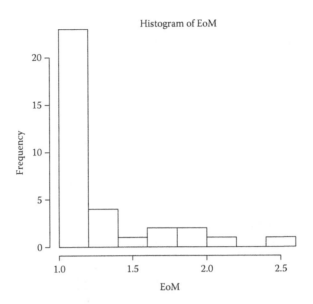

FIGURE 16.6
A histogram showing the extent of membership for each member in Zachary's karate data. The histogram shows that many members have membership in more than one block and one member has mixed membership across more than two blocks.

the probability of an edge/nonedge (y_{ij}) between nodes i and j is modeled as a function of their positions. Generally, in these models, the smaller the distance between two nodes in the latent space, the greater their probability of being connected.

In the case where additional edge covariate information x_{ij} is observed, these models can account for homophily by attributes; this is the propensity for nodes with similar attributes to be more (or less) likely to connect with each other. Node covariate information can be converted into edge covariates using differences in the nodal attributes (and can be either directed or undirected); thus, the covariates can make an edge more (or less) likely to occur between actors that have similar attributes than between those who do not.

The latent position model assumes that

$$\log\left(\frac{\mathbb{P}\{Y_{ij} = 1 \mid z_i, z_j, \alpha\}}{\mathbb{P}\{Y_{ij} = 0 \mid z_i, z_j, \alpha\}}\right) = \eta_{ij} = \alpha - d(z_i, z_j), \tag{16.5}$$

where $d(z_i, z_j)$ is the Euclidean distance between z_i and z_j and η_{ij} is the log-odds of a connection between nodes i and j, and thus

$$\mathbb{P}\{Y_{ij} = 1 \mid z_i, z_j, \alpha\} = \frac{\exp(\eta_{ij})}{1 + \exp(\eta_{ij})} = \frac{\exp[\alpha - d(z_i, z_j)]}{1 + \exp[\alpha - d(z_i, z_j)]}.$$

Further, the latent position model assumes that all edges are independent conditional on the latent positions and thus

$$\mathbb{P}\{Y \mid Z, X, \alpha\} = \prod_{i=1}^{n} \prod_{\substack{j=1 \\ j \neq i}}^{n} P(y_{ij} \mid z_i, z_j, \alpha). \tag{16.6}$$

The inclusion of covariates in the latent space model can be achieved by adding an extra term to (16.5) to yield

$$\log \left(\frac{\mathbb{P}\{Y_{ij} = 1 \mid z_i, z_j, x_{ij}, \alpha, \beta\}}{\mathbb{P}\{Y_{ij} = 0 \mid z_i, z_j, x_{ij}, \alpha, \beta\}} \right) = \eta_{ij}$$

$$= \alpha + \beta^T x_{ij} - d(z_i, z_j),$$

$$= \alpha + \sum_{p=1}^{P} \beta_p x_{ijp} - d(z_i, z_j), \qquad (16.7)$$

where $\beta^T = (\beta_1, \beta_2, \ldots, \beta_P)$ are model coefficients for each edge variable and $x_{ij} = (x_{ij1}, x_{ij2}, \ldots, x_{ijP})$ are the edge covariates for the pair of nodes (i, j).

Finally, a model for the latent positions needs to be specified. The original latent space model of Hoff et al. (2002) proposed modeling the z_i values using a multivariate normal distribution.

16.5.1.1 The Latent Position Cluster Model

The latent position cluster model (LPCM) (Handcock et al., 2007) extended this model by assuming a finite mixture model for the latent positions in order to model network communities. Specifically, the LPCM model assumes a finite mixture of spherical Gaussian distributions for the latent positions:

$$z_i \sim \sum_{g=1}^{G} \tau_g \mathrm{MVN}_d(\mu_g, \sigma_g^2 I), \qquad (16.8)$$

where τ_g is the probability that a node belongs to the gth group, and $\sum_{g=1}^{G} \tau_g = 1$; this structure allows for clusters of highly connected nodes.

The LPCM model with $G = 3$ groups was fitted to Sampson's network data and a plot of the fitted latent positions is shown in Figure 16.7. Model fitting was implemented using the latentnet R package (Krivitsky and Handcock, 2008, 2010).

The model correctly identifies the three groups of monks in the network. Further, the groups are well separated as evidenced by the separate clusters of latent positions and the pie charts which show definitive group membership for each node. Thus, the LPCM clusters the nodes into highly connected clusters and provides a visualization of this clustering. The location of the nodes in Figure 16.7 has a direct interpretation in terms of the model probabilities; this is in contrast to the layout of the same data given in Figure 16.2.

16.5.1.2 The Sender and Receiver Random Effects

Within many networks, some nodes may have a tendency to send and/or receive links more than other nodes. The latent space model and latent position model in their most basic form are unable to account for this network feature. Krivitsky et al. (2009) proposed a model that explicitly considers all this feature by including extra model parameters to account for the tendency of a node to send and/or receive links.

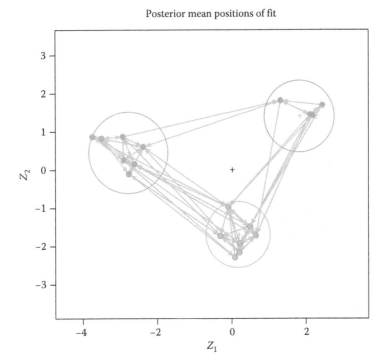

FIGURE 16.7

Plot of the posterior mean latent positions for three groups under the Latent Position Cluster Model for Sampson's network data set. The pie charts depict the posterior probability of group membership for each node, in this case each node belongs almost 100% to one cluster. The empty circles with a plus symbol at the centre show the mean of each group and the 50% contour of the density of latent positions for each group.

In undirected graphs, each node has a single extra parameter, δ_i, called the sociality factor. This parameter denotes the propensity of node i to form edges with other actors. The sociality factor is included in the model to get link probabilities with the following form:

$$\log\left(\frac{\mathbb{P}\{Y_{ij}=1 \mid z_i, z_j, \delta_i, \delta_j, x_{ij}, \alpha, \beta\}}{\mathbb{P}\{Y_{ij}=0 \mid z_i, z_j, \delta_i, \delta_j, x_{ij}, \alpha, \beta\}}\right) = \eta_{ij} = \alpha + \beta^T x_{ij} - |z_i - z_j| + \delta_i + \delta_j, \quad (16.9)$$

where $\delta_i \sim \mathcal{N}(0, \sigma_\delta^2)$ and the variance σ_δ^2 measures the heterogeneity in the propensity to send and receive edges.

In directed graphs, the sociality effect in links depends on two parameters: the sender random effect δ_i for node i and the receiver random effect γ_j for node j. These terms are included in the model to get link probabilities of the form:

$$\log\left(\frac{\mathbb{P}\{Y_{ij}=1 \mid z_i, z_j, \delta_i, \gamma_j, x_{ij}, \alpha, \beta\}}{\mathbb{P}\{Y_{ij}=0 \mid z_i, z_j, \delta_i, \gamma_j, x_{ij}, \alpha, \beta\}}\right) = \eta_{ij} = \alpha + \beta^T x_{ij} - |z_i - z_j| + \delta_i + \gamma_j, \quad (16.10)$$

where $\delta_i \sim \mathcal{N}(0, \sigma_\delta^2)$ and $\gamma_i \sim \mathcal{N}(0, \sigma_\gamma^2)$, and the variances σ_δ^2 and σ_γ^2 measure the heterogeneity in the propensity to send and receive edges, respectively.

16.5.1.3 The Mixture of Experts Latent Position Cluster Model

Gormley and Murphy (2010) proposed the mixture of experts latent position cluster model to extend the latent position cluster model within a mixture of experts framework, assuming that the mixing proportions $(\tau_1, \ldots, \tau_G)$ are node specific and can be modeled as a multinomial logistic function of their covariates $x_i^T = (x_{i1}, \ldots, x_{iP})$ where the probability of belonging to each of $G - 1$ clusters is compared to a baseline cluster, usually $g = 1$. Thus, the distribution of the latent position z_i is assumed to be:

$$z_i \sim \sum_{g=1}^{G} \tau_g(x_i) \text{MVN}_d(\mu_g, \sigma_g^2 I), \tag{16.11}$$

where

$$\tau_g(x_i) = \frac{\exp(\delta_{g0} + \delta_{g1}x_{i1} + \cdots + \delta_{gp}x_{iP})}{\sum_{g'=1}^{G} \exp(\delta_{g'0} + \delta_{g'1}x_{i1} + \cdots + \delta_{g'p}x_{iP})}, \tag{16.12}$$

$(\delta_{10}, \ldots, \delta_{1P}) = (0, 0, \ldots, 0)$ and $\sum_{g=1}^{G} \tau_g(x_i) = 1$.

16.5.1.4 Estimation

For all variants of the latent space models presented above, the complete-data log-likelihood is of the form:

$$\log \mathbb{P}\{Y \mid \eta_{ij}\} = \sum_{i=1}^{n} \sum_{\substack{i=1 \\ i \neq j}}^{n} \{\eta_{ij} y_{ij} - \log(1 + \exp(\eta_{ij}))\}, \tag{16.13}$$

where the form of η_{ij} depends on version of the model used. To estimate the model, the main approaches that have been suggested are maximum likelihood estimation and a Bayesian approach using MCMC or variational approximations.

A fast likelihood-based method can provide point estimates of the distances between the nodes since the complete-data log-likelihood is a convex function of the distances between nodes in the latent space. The drawback of this approach is that the latent positions themselves need to be approximated using multidimensional scaling of the estimated distances because the log-likelihood is not a convex function of the latent positions. Another drawback is that the estimation is done in two steps: in the first step, the likelihood-based estimates of the latent positions are computed without considering any clustering structure, and in the second step, the maximum-likelihood estimates of the mixture model for the latent positions are found conditional on the latent positions estimated at the previous step.

A Bayesian approach to model fitting allows the estimation of all the parameters and the latent positions simultaneously via MCMC sampling; this approach was developed in Handcock et al. (2007) and was further refined in Krivitsky et al. (2009). This approach usually gives better results than the two-stage MLE, but it is more computationally intensive. More recently, Salter-Townshend and Murphy (2009, 2013) developed a variational Bayesian inference routine to approximate the posterior distribution of the parameters and latent positions in the LPCM. Further, Raftery et al. (2012) propose a likelihood approximation using case–control sampling to improve the efficiency of fitting the LPCM model by approximating the likelihood using $O(n)$ terms instead of $O(n^2)$ terms.

The latentnet R package (Krivitsky and Handcock, 2010, 2008) provides two-stage maximum likelihood estimation, MCMC-based inference and minimum Kullback–Leibler (Shortreed and Handcock, 2006) estimation for the LPCM. VBLPCM (Salter-Townshend, 2012) is an R package that performs the variational Bayesian inference of the LPCM for Euclidean latent spaces as developed in Salter-Townshend and Murphy (2013).

16.5.2 Zachary's Karate Data Set

The latent position cluster model was fitted to Zachary's karate data with $G = 4$ clusters. A plot of the model fit is given in Figure 16.8 where the posterior mean latent position for each node is shown, the posterior cluster membership is shown using a pie chart, the cluster means and 50% contours for each cluster are also shown. Many of the nodes have high posterior probability of belonging to two components, so it suggests that a model with a lower value of G could be sufficient for these data.

A cross tabulation of the cluster membership versus faction identity and club membership is given in Table 16.7, and it shows that the model has successfully found the division in the karate club, which two components being used to model each side of the divide.

The latent position cluster model was fitted to Zachary's karate data with $G = 2$ clusters but with each actor having a sociality effect. A plot of the model fit is given in Figure 16.9 where the posterior mean latent position for each node is shown, the posterior cluster membership is shown using a pie chart, the size of the plotting symbol shows the magnitude

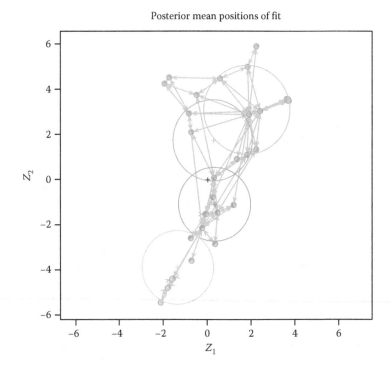

Posterior mean positions of fit

FIGURE 16.8
The estimated latent positions for each node in Zachary's karate club data for the LPCM model with $G = 4$. Each node is represented by a pie chart which represents the posterior probability of the node belonging to each cluster.

TABLE 16.7

A Cross Tabulation of the Cluster Membership Versus Faction Identity and New Club Membership for the LPCM Model with $G = 4$

	-2	-1	0	1	2		Mr. Hi's	John A's
1	1	1	0	2	6	1	3	7
2	0	0	2	2	6	2	0	10
3	4	3	0	0	0	3	7	0
4	5	1	1	0	0	4	7	0

Posterior mean positions of fit

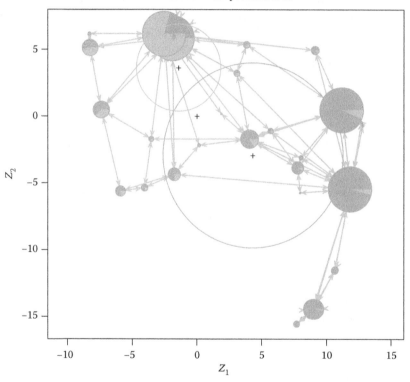

FIGURE 16.9
The estimated latent positions for each node in Zachary's karate club data for the LPCM model with $G = 2$. Each node is represented by a pie chart which represents the posterior probability of the node belonging to each cluster. The size of the pie chart represents the magnitude of the sociality effect.

of the sociality effect, the cluster means and 50% contours for each cluster are also shown. The two key actors in the network (Mr Hi and John A.) have the largest sociality effects.

A cross tabulation of the cluster membership versus faction identity and new club membership shows that the model clusters the nodes into the correct groups, and further, it identifies the most prominent actors in the network (see Table 16.8).

TABLE 16.8

A Cross Tabulation of the Cluster Membership Versus Faction
Identity and New Club Membership for the LPCM Model with
Sender and Receiver Effects with $G = 2$

	−2	−1	0	1	2		Mr. Hi's	John A.'s
1	9	5	1	0	1	1	15	1
2	1	0	2	4	11	2	2	16

16.6 Modeling Issues

Two important issues arise when clustering network data using a model-based approach. The first of these is that the likelihood function for the models described herein involves the product over $O(n^2)$ terms, where n is the number of nodes in the network. Thus, any likelihood-based method is computationally demanding. Raftery et al. (2012) proposed a method to approximate the computationally expensive likelihood term with an approximation by utilizing case–control sampling. Their approach still used a once-off $O(n^2)$ calculation to match cases and controls but involves $O(n)$ calculations thereafter. This approach has only been exploited in the case of the latent position cluster model. There is scope for extending this idea to other network models to increase how well they scale to large datasets.

A second issue with model-based clustering of network data is that methods for model selection, particularly the number of clusters, are underdeveloped. The calculation of the likelihood function with any latent variables marginalized out is computationally difficult or intractable for these models, and thus, likelihood-based methods for model selection are not feasible. Further, the asymptotic properties of network models are difficult to establish, so properties of criteria like the Bayesian information criterion (BIC) are not fully understood.

This issue aside, a number of methods for model selection have been proposed for the models described herein. Côme and Latouche (2013) derived the integrated completed likelihood (ICL) criterion for the stochastic block model and this facilitates selecting the number of blocks. Handcock et al. (2007) propose a BIC-like criterion for selecting the number of components in the latent position cluster model. Airoldi et al. (2008) also used a BIC-like criterion for model selection in the mixed membership stochastic block model. These criteria have been shown to work well in a number of empirical settings but their properties in a general setting are less well established.

Recently, Friel et al. (2013) proposed a collapsed MCMC sampler to estimate the posterior distribution of the number of clusters in the latent position cluster model. There is much scope for further work along these lines to estimate the posterior model probabilities for network models in a more accurate manner.

16.7 Other Models

Other models have been developed for model-based clustering of network data. These models are closely related to the other models described in this chapter.

Latouche et al. (2011) and McDaid and Hurley (2010) developed a model called the overlapping stochastic block model that allows nodes to have full membership of more than one block, thus providing an alternative version of mixed membership, and this model was further extended in McDaid et al. (2014).

Salter-Townshend and Murphy (2015) developed a finite mixture of exponential random graph models (ERGM) to cluster networks with similar structure. The model is fitted using a pseudo-likelihood EM algorithm. They applied the method to cluster ego-networks of nodes in a network to find nodes with similar local structure.

Acknowledgments

This study was supported by the Insight Centre for Data Analytics, which is supported by Science Foundation Ireland under grant number SFI/12/RC/2289.

References

Airoldi, E.M., Blei, D.M., Fienberg, S.E., Goldberg, A., Xing, E.P., and Zheng, A.X. 2007. *Statistical Network Analysis: Models, Issues and New Directions*, vol. 4503 of *Lecture Notes in Computer Science*. Springer, Berlin.

Airoldi, E.M., Blei, D.M., Fienberg, S.E., and Xing, E.P. 2008. Mixed-membership stochastic blockmodels. *Journal of Machine Learning Research*, **9**, 1981–2014.

Ambroise, C., Grasseau, G., Hoebeke, M., Latouche, P., Miele, V., Picard, F., and LAPACK authors 2013. *Mixer: Random Graph Clustering*. R package version 1.7, http://ssbgroup.fr/mixnet/mixer.html

Bickel, P.J. and Chen, A. 2009. A nonparametric view of network models and Newman-Girvan and other modularities. *Proceedings of the National Academy of Sciences*, **106**, 21068–21073.

Bickel, P.J., Chen, A., and Levina, E. 2011. The method of moments and degree distributions for network models. *Annals of Statistics*, **39**, 2280–2301.

Blei, D.M., Ng, A.Y., and Jordan, M.I. 2003. Latent Dirichlet allocation. *Journal of Machine Learning Research*, **3**, 993–1022.

Butts, C.T., Handcock, M.S., and Hunter, D.R. 2014. *Network: Classes for Relational Data*. Irvine, CA. R package version 1.10.2, http://statnet.org/

Carrington, P.J., Scott, J., and Wasserman, S. 2005. *Models and Methods in Social Network Analysis*. Cambridge University Press, Cambridge.

Chang, J. 2010. *lda: Collapsed Gibbs Sampling Methods for Topic Models*. R package version 1.2.1, http://CRAN.R-project.org/package=lda

Chang, J. and Blei, D.M. 2010. Hierarchical relational models for document networks. *Annals of Applied Statistics*, **4**, 124–150.

Channarond, A., Daudin, J.-J., and Robin, S. 2012. Classification and estimation in the stochastic blockmodel based on the empirical degrees. *Electronic Journal of Statistics*, **6**, 2574–2601.

Choi, D.S., Wolfe, P.J., and Airoldi, E.M. 2012. Stochastic blockmodels with growing number of classes. *Biometrika*, **99**, 273–284.

Côme, E. and Latouche, P. 2013. Model selection and clustering in stochastic block models with the exact integrated complete data likelihood. Tech. rep., Université Paris 1.

Daudin, J.-J., Picard, F., and Robin, S. 2008. A mixture model for random graphs. *Statistics and Computing*, **18**, 173–183.

Erosheva, E. 2002. *Grade of Membership and Latent Structure Models with Application to Disability Survey Data*. Ph.D. thesis, Department of Statistics, Carnegie Mellon University.

Erosheva, E.A. 2003. Bayesian estimation of the grade of membership model. In *Bayesian Statistics, 7* (Bernardo, J., Bayarri, M., Berger, J., Dawid, A., Heckerman, D., Smith, A., and West, M. eds.). Oxford University Press, Oxford, UK, 501–510.

Erosheva, E.A., Fienberg, S.E., and Joutard, C. 2007. Describing disability through individual-level mixture models for multivariate binary data. *The Annals of Applied Statistics, 1*, 502–537.

Fienberg, S.E. and Wasserman, S. 1981. Discussion of "An exponential family of probability distributions for directed graphs" by Holland and Leinhardt. *Journal of the American Statistical Association, 76*, 54–57.

Friel, N., Ryan, C., and Wyse, J. 2013. Bayesian model selection for the latent position cluster model for social networks. Tech. rep., University College Dublin.

Fruchterman, T.M.J. and Reingold, E.M. 1991. Graph drawing by force-directed placement. *Software—Practice and Experience, 21*, 1129–1164.

Gelfand, A.E. and Smith, A. F.M. 1990. Sampling-based approaches to calculating marginal densities. *Journal of the American Statistical Association, 85*, 398–409.

Goldenberg, A., Zheng, A.X., Fienberg, S.E., and Airoldi, E.M. 2010. A survey of statistical network models. *Foundations and Trends in Machine Learning, 2*, 129–233.

Gormley, I.C. and Murphy, T.B. 2010. A mixture of experts latent position cluster model for social network data. *Statistical Methodology, 7*, 385–405.

Handcock, M.S., Raftery, A.E., and Tantrum, J.M. 2007. Model-based clustering for social networks. *Journal of the Royal Statistical Society: Series A, 170*, 1–22.

Hoff, P., Raftery, A., and Handcock, M.S. 2002. Latent space approaches to social network analysis. *Journal of the American Statistical Association, 97*, 1090–1098.

Kolaczyk, E.D. 2009. *Statistical Analysis of Network Data: Methods and Models.* Springer, New York.

Krivitsky, P.N. and Handcock, M.S. 2008. Fitting latent cluster models for networks with latentnet. *Journal of Statistical Software, 24*, 1–23.

Krivitsky, P.N. and Handcock, M.S. 2010. *Latentnet: Latent Position and Cluster Models for Statistical Networks.* R Package Version 2.4-4, http://CRAN.R-project.org/package=latentnet

Krivitsky, P.N., Handcock, M.S., Raftery, A.E., and Hoff, P.D. 2009. Representing degree distributions, clustering, and homophily in social networks with latent cluster random effects models. *Social Networks, 31*, 204–213.

Latouche, P., Birmelé, E., and Ambroise, C. 2010. Bayesian methods for graph clustering. In *Advances in Data Analysis, Data Handling and Business Intelligence* (Fink, A., Lausen, B., Seidel, W., and Ultsch, A. eds.). Studies in Classification, Data Analysis, and Knowledge Organization, Springer, Berlin, Heidelberg, 229–239.

Latouche, P., Birmelé, E., and Ambroise, C. 2011. Overlapping stochastic block models with application to the French political blogosphere. *Annals of Applied Statistics, 5*, 309–336.

Lazarsfeld, P.F. and Henry, N.W. 1968. *Latent Structure Analysis.* Houghton Mifflin, Boston.

Liu, J.S. 1994. The collapsed Gibbs sampler in Bayesian computations with applications to a gene regulation problem. *Journal of the American Statistical Association, 89*, 958–966.

Lorrain, F. and White, H.C. 1971. Structural equivalence of individuals in social networks. *Journal of the Mathematical Sociology, 1*, 49–80.

McDaid, A. and Hurley, N.J. 2010. Detecting highly overlapping communities with model-based overlapping seed expansion. In *International Conference on Advances in Social Networks Analysis and Mining (ASONAM)* (Memon, N. and Alhajj, R. eds.). IEEE Computer Society, Piscataway, NJ, 112–119.

McDaid, A., Hurley, N., and Murphy, B. 2014. Overlapping stochastic community finding. In *Advances in Social Networks Analysis and Mining (ASONAM), 2014 IEEE/ACM International Conference on* (Wu, X., Ester, M., and Xu, G. eds.). IEEE, Piscataway, NJ, 17–20.

McDaid, A.F., Murphy, T.B., Friel, N., and Hurley, N. 2012. Model-based clustering in networks with Stochastic Community Finding. In *Proceedings of COMPSTAT 2012: 20th International Conference on Computational Statistics* (Colubi, A., Fokianos, K., Kontoghiorghes, E.J., and Gonzáles-Rodríguez, G. eds.). ISI-IASC, 549–560.

McDaid, A.F., Murphy, T.B., Friel, N., and Hurley, N.J. 2013. Improved bayesian inference for the stochastic block model with application to large networks. *Computational Statistics & Data Analysis*, **60**, 12–31.

Minka, T., Winn, J., Guiver, J., and Knowles, D. 2010. Infer.NET. Version 2.4, http://research. microsoft.com/en-us/um/cambridge/projects/infernet/?

Nowicki, K. and Snijders, T.A.B. 2001. Estimation and prediction of stochastic blockstructures. *Journal of the American Statistical Association*, **96**, 1077–1087.

Raftery, A.E., Niu, X., Hoff, P.D., and Yeung, K.Y. 2012. Fast inference for the latent space network model using a case-control approximate likelihood. *Journal of Computational and Graphical Statistics*, **21**, 901–919.

Rohe, K., Chatterjee, S., and Yu, B. 2011. Spectral clustering and the high-dimensional stochastic blockmodel. *Annals of Statistics*, **39**, 1878–1915.

Salter-Townshend, M. 2012. *VBLPCM: Variational Bayes Latent Position Cluster Model for Networks*. R package version 2.0, http://CRAN.R-project.org/package=VBLPCM

Salter-Townshend, M. and Murphy, T.B. 2009. Variational Bayesian inference for the latent position cluster model. In *NIPS Workshop on Analyzing Networks and Learning with Graphs*.

Salter-Townshend, M. and Murphy, T.B. 2013. Variational Bayesian inference for the latent position cluster model for network data. *Computational Statistics & Data Analysis*, **57**, 661–671.

Salter-Townshend, M. and Murphy, T.B. 2015. Role analysis in networks using mixtures of exponential random graph models. *Journal of Computational and Graphical Statistics*, **24**(2), 520–538.

Salter-Townshend, M., White, A., Gollini, I., and Murphy, T.B. 2012. Review of statistical network analysis: Models, algorithms, and software. *Statistical Analysis and Data Mining*, **5**, 243–264.

Sampson, S.F. 1969. *Crisis in a Cloister*. Ph.D. thesis, Cornell University.

Shortreed, S. and Handcock, M. 2006. Positional estimation within a latent space model for networks. *Methodology: European Journal of Research Methods for the Behavioral and Social Sciences*, **2**, 24–33.

Snijders, T.A.B. and Nowicki, K. 1997. Estimation and prediction for stochastic bockmodels for graphs with latent block structure. *Journal of Classification*, **14**, 75–100.

Wasserman, S. and Faust, K. 1994. *Social Network Analysis: Methods and Applications*. Cambridge University Press, Cambridge.

Wasserman, S., Robins, G., and Steinley, D. 2007. Statistical models for networks: A brief review of some recent research. In *Statistical Network Analysis: Models, Issues, and New Directions* (Airoldi, E., Blei, D.M., Fienberg, S.E., Goldenberg, A., Xing, E.P., and Zheng, A.X. eds.), vol. 4503 of *Lecture Notes in Computer Science*. Springer, Berlin, Heidelberg, 45–56.

White, A., Chan, J., Hayes, C., and Murphy, T.B. 2012. Mixed membership models for exploring user roles in online fora. In *Proceedings of the Sixth International AAAI Conference on Weblogs and Social Media (ICWSM 2012)* (Breslin, J., Ellison, N., Shanahan, J., and Tufekci, Z. eds.). AAAI Press, Palo Alto, CA, 599–602.

White, A. and Murphy, T.B. 2014a. Exponential family mixed membership models for soft clustering of multivariate data. Tech. rep., School of Mathematical Sciences, University College Dublin, Ireland.

White, A. and Murphy, T.B. 2014b. Mixed membership of experts stochastic block model. Tech. rep., School of Mathematical Sciences, University College Dublin, Ireland.

Xing, E.P., Fu, W., and Song, L. 2010. A state-space mixed membership blockmodel for dynamic network tomography. *Annals of Applied Statistics*, **4**, 535–566.

Zachary, W. 1977. An information flow model for conflict and fission in small groups. *Journal of Anthropological Research*, **33**, 452–473.

Zanghi, H., Ambroise, C., and Miele, V. 2008. Fast online graph clustering via erdös-rényi mixture. *Pattern Recognition*, **41**, 3592–3599.

Section IV

Methods Based on Density Modes and Level Sets

17

A Formulation in Modal Clustering Based on Upper Level Sets

Adelchi Azzalini

CONTENTS

Abstract

In the so-called modal approach, clusters are associated with regions of high density of the underlying population density function. In practice, the density must be estimated, typically in a nonparametric approach. This basic idea can be translated into an operational procedure via a few alternative routes. In the direction undertaken here, the key concept is represented by the upper level set, formed by connected subsets of the Euclidean space with density exceeding a certain threshold; these subsets are associated with clusters. As the density threshold spans the range of density values, a tree of clusters is generated. The actual development of this logical scheme, which builds on concepts of spatial tessellation, is presented in detail. A simple illustrative example is examined, and more substantial practical cases are summarized.

17.1 Background and General Aspects

17.1.1 Clustering via Density Estimation

While the classical literature on clustering has focused largely on the concept of dissimilarity between objects, much work of the relatively more recent literature, at least the one stemming from the statistical community, has been dedicated to the "model-based approach" which involves the introduction of a probability model of mixture type with K components, say. It is not our purpose to discuss in any detail this approach (widely documented in the literature and also in the present volume), only to recall its essential ingredients. In this logic, each component of the mixture model is the parent distribution of the observations comprising a given cluster. In the case of d observations of continuous type, these are interpreted as independent realizations of a d-dimensional random variable whose density function is assumed to be of the form

$$f(x) = \sum_{k=1}^{K} \pi_k \, f_0(x; \theta_k), \quad x \in \mathbb{R}^d, \tag{17.1}$$

where π_k is the weight of the k-th subpopulation having density function $f_0(x; \theta_k)$, identified by the set of parameters θ_k within the parametric family f_0. Clearly, the conditions $\pi_k > 0$ and $\pi_1 + \cdots + \pi_K = 1$ must hold. Once the parameters θ_k, π_k (for $k = 1, \ldots, K$) have been estimated from the available data, allocation of any given point $\tilde{x} \in \mathbb{R}^d$ to some subpopulation is based on the highest posterior probability, starting from the marginal probabilities π_k's.

The model-based approach to clustering has, with respect to many other constructions, the conceptual advantage of framing the problem in a probability context and of stating unambiguously what a cluster is. It has, however, the intrinsic limitation of postulating a parametric family $f_0(x; \theta)$ for each component of the mixture. As a consequence, as noted by Stuetzle (2003) and Chaudhuri et al. (2014), "the conceptual problem with this approach is that the optimal number of mixture components is not an intrinsic property of the data density but instead depends on the family of distributions that are being mixed." Recently, moving from the Gaussian assumption to more flexible parametric families, this problem has been alleviated, but conceptually, and partly operationally, it remains there.

The development to be presented below shares with the model-based approach the adoption of a probability formulation and an unambiguous definition of what a cluster is, but it uses a different definition. At variance with equation (17.1), it works in a nonparametric formulation; this allows cluster shapes to be unconstrained. It will also turn out that the number of clusters does not need to be specified at the beginning and it is estimated by the method.

17.1.2 Modal Clustering

The conversion of (17.1) into a nonparametric formulation leads to the mixture density

$$f(x) = \sum_{m=1}^{M} \pi_m \, f_m(x), \quad x \in \mathbb{R}^d, \tag{17.2}$$

where now each f_m is essentially an unconstrained density, up to some weak conditions to be discussed later. Clearly, even if $f(x)$ was known exactly, there is no way to disentangle the components $f_m(x)$'s without the introduction of some criterion to separate them. The commonly adopted criterion is to associate each component f_m with a mode of $f(x)$, whence the name "modal clustering" given to this approach. Therefore, we link the modes of $f(x)$ to the components f_m's of the mixture (17.2), and, when actual data are observed, we form a cluster of data points attributed to f_m.

While it is conceptually useful to retain equation (17.2) as a background concept, the actual development of the modal approach, in its various formulations, is not so explicitly linked to this representation. However, at certain steps (17.2) will re-emerge.

In a large number of cases, decomposition of a density $f(x)$ via (17.1) and (17.2) leads to the same number of components, hence of clusters, but not always; this is why the number of summands in (17.2) has been denoted M instead of K. To perceive one such situation, consider the case where the non-parametric nature of the f_m's in (17.2) allows one of them to encompass in a single cluster a set of observed points whose shape is not necessarily accommodated by a single component of (17.1), since f_0 implies some shape constraint on them.

Another situation of divergence between the model-based and the modal approach is as follows. Consider a mixture of two strongly nonspherical Gaussian distributions, located such that their tails overlap giving rise to a third mode, in addition to the main ones at the centres of the Gaussian components. This case, illustrated graphically in Ray and Lindsay (2005), is regarded as having two subpopulations in the model-based approach, three in the modal approach.

When we move from the probability formulation to statistical work, $f(x)$ must be estimated nonparametrically from a set of n data points in $\mathbb{R}^d$, assumed to be independently sampled from $f(x)$. To ensure consistency and other formal property of the estimate, denoted $\hat{f}(x)$, some regularity conditions on $f(x)$ must be introduced, such as differentiability and alike. The specific assumptions vary somewhat depending on the chosen estimation methodology and the required formal properties, so we cannot be more detailed at this point. One inherent assumption is, however, that the number M of modes is finite. Clearly, for any finite n, we cannot identify infinite modes and even for diverging n the task is beyond hope. This requirement appears quite reasonable on general grounds and is qualitatively similar to the assumption that K in (17.1) is finite.

The rest of the present contribution focuses in one specific direction within this approach, without attempting a full coverage. A broad discussion of modal clustering, inclusive of directions not discussed here, is provided by Menardi (2015). Consequently, we mention only some selected contributions within the modal clustering approach, partly here and partly in Section 17.2.2 below, and refer the reader to Menardi (2015) for a more extensive discussion.

Another form of modal clustering goes under the name mode-hunting and is linked to mean-shift algorithm (Chapter 18). Another formulation which has some connection with the present one is the stream of literature originating with the DBSCAN technique (Ester et al. 1996), followed by various subsequent works. The points of similarity with us are to start from density estimation as a basic notion and consideration of whether the density at a given observation point exceed a certain threshold. In the original paper (Ester et al. 1996), the concept of "density" employed was not in the sense of density of a probability distribution, but simply referred to the count of points within a given distance; however, in more recent accounts, the adopted notion of density has evolved toward a probabilistic meaning (Kriegel et al. 2011); where the links with Wishart (1969) and Hartigan (1975) are highlighted. Another point of difference from us originates from the rule adopted to decide about connectivity of points, which both in Ester et al. (1996) and in Kriegel et al. (2011) depends on the distance between points, not only on the density function.

17.2 High-Density Clusters

17.2.1 Upper Level Sets and the Cluster Tree

The core idea of the formulation to be described in the next pages has appeared a long time ago. Some early occurrences are not always very specific, but Wishart (1969) already puts forward an operational procedure for "hierarchical mode analysis." A more lucid formulation is presented in Hartigan (1975, 11.13), starting from the following criterion:

> "Clusters may be thought of as regions of high density separated from other such regions by regions of low density".

In the ensuing discussion, Hartigan introduced the concept of "density-contour clusters" identified by connected regions with sufficiently high density and establishing their hierarchical tree structure, denoted "density-contour tree." This formulation has, however, not really been followed up in Hartigan's book and only to a limited extent in the subsequent literature. A plausible explanation lies on the computational burden involved, not easily accomplished by the computing resources at that time.

However, gradually over the years, the exploration of this construction has evolved and, thanks to the much improved computational resources, it is now easily accessible to the practitioner. What follows is an account of a specific development in this direction.

For a d-dimensional density function $f(\cdot)$ whose support is a connected set, denote by

$$R(c) = \{x : x \in \mathbb{R}^d, f(x) \geq c\}, \tag{17.3}$$

$$p_c = \mathbb{P}(R(c)) = \int_{R(c)} f(x)\,dx, \tag{17.4}$$

the region with density not less than c and its probability, for any $c \geq 0$. We call $R(c)$ an "upper level set"; similar names are also in use in the literature.

If $f(x)$ is unimodal, $R(c)$ is a connected set for all c. Otherwise, there exists some c whose $R(c)$ is not connected and is formed by a number of (maximal) connected sets, such that any path going from one of these sets to another one inevitably crosses a region of lower density. These, according to the above-stated criterion, constitute distinct clusters. As c varies, a family of clusters is generated and this family forms a tree in the sense that, for any two such clusters, A and B, either $A \subset B$, $B \subset A$ or $A \cap B = \emptyset$, the empty set. If c decreases, separate clusters eventually coalesce, until at the bottom level, where $c = 0$, there will be a single cluster.

The above concepts are more easily perceived with the aid of an illustrative example. Consider the density $f(x)$ in $\mathbb{R}^2$ depicted in Figure 17.1 using a contour plot with the areas between the contours filled in solid color; this is sometimes called a level plot. Of this function, we consider sets $R(c)$, in principle for all possible c, although Figure 17.2 visualizes

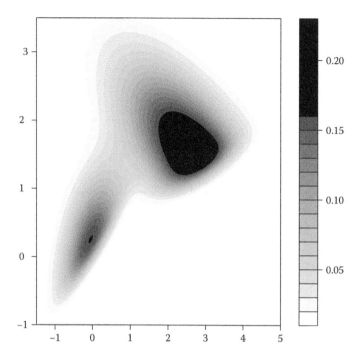

FIGURE 17.1
Level plot of a bivariate density function.

c :	0.20	0.15	0.10	0.0576	0.05
p_c :	0.0755	0.2475	0.49	0.723	0.77
$m(p_c)$:	1	2	2	2→1	1

only a few of them. Specifically, the panels of Figure 17.2 refer, moving left to right and top to bottom, to the following values of c, the corresponding probabilities p_c and the number of modes $m(p_c)$, where the last two columns are both represented in the bottom-right plot, with the dashed lines denoting the transition moment from two clusters to one.

For the distribution of Figure 17.2, the correct number of modes, two, is obtained with some choices of c but not with others. In more complex situations, with several modes, it can happen that there exists no single value c indicating the correct number of modes. We must really examine a range of c value.

The left plot of Figure 17.3 displays the cluster tree for the distribution of Figure 17.1, for the full range of p_c values. Since c and p_c are monotonically linked, either of them could be used to index the vertical axis, but the probability values p_c are somewhat more meaningful. Notice that the top value $p_c = 1$ corresponds to the bottom level of the density and the top mode, marked "1," is at the bottom of the tree. To ease comparison with the earlier plots, the p_c values of the four shaded areas in Figure 17.2 are marked by ticks along the vertical axis of the left plot and along the horizontal axis of the right plot.

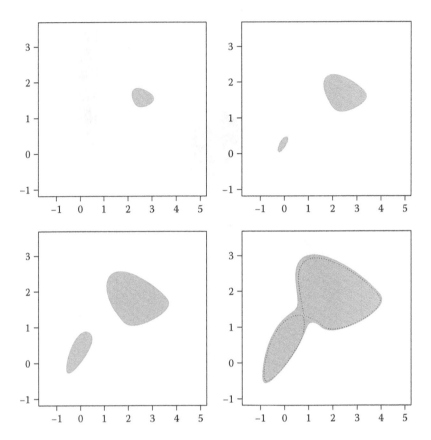

FIGURE 17.2
High-density regions $R(c)$ of the density in Figure 17.1 for a few values of c.

Another diagram associated to f is the "mode function," introduced in Azzalini and Torelli (2007), which for any value of p_c indicates the number of modes $m(p_c)$ identified by the upper level set $R(c)$, with the conventional position that $m(0) = m(1) = 0$. It turns out that the total number of increments of the mode function, counted with their multiplicity, is equal to the total number of modes; a similar fact holds for the number of decrements. Additional information about $f(x)$ can be extracted from the mode function (Azzalini and Torelli 2007).

We have already used the term "mode" without introducing an explicit definition. In common cases, one can simply live with the intuitive idea of a local maximum of $f(x)$ and this is adequate enough for following the present construction. However, for mathematical completeness, it seems unavoidable to state a formal definition of mode, given its role in this formulation. Since a comprehensive definition has to deal with discontinuities, discard high and low values at isolated points, allow for flat regions, poles and so on, it tends to be intricate.

In light of the key role of upper level sets in the present formulation, we introduce a definition of mode which hinges on this notion. The first step is to restate the concept of upper level set in a form more suitable for what follows. We shall denote by $\text{cl}(A)$ the closure of a set A and by $\text{int}(A)$ its interior.

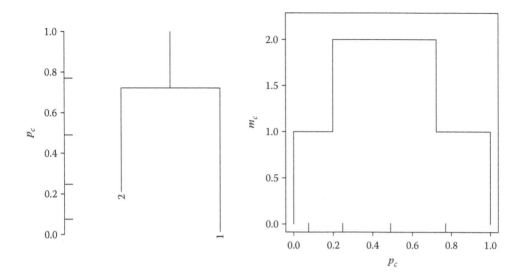

FIGURE 17.3
Cluster tree and mode function of the density in Figure 17.1.

Definition (Upper level set). For a density function $f(x)$ on $\mathbb{R}^d$ and any $c \geq 0$, the set

$$R(c) = \text{cl}(\text{int}(\{x : x \in \mathbb{R}^d, f(x) \geq c\}))$$

is an upper level set (with level c), provided $\mathbb{P}(R(c)) > 0$.

In the simples cases where $f(x)$ is continuous, this concept of upper level set coincides with the earlier expression (17.3), except for the additional requirement of positive probability to avoid the empty set and other trivialities.

Definition (Mode). For a fixed upper level set $R(c')$ of density $f(x)$, consider a fixed maximal connected subset of $R(c')$, say $\bar{R}$. For any other upper level set $R(c)$ of $f(x)$, write $U(c, c') = \text{cl}(R(c) \cap \bar{R})$ and let C represent the set of c values such that $\mathbb{P}(U(c, c')) > 0$. Then a mode is any maximal connected set of

$$\bar{M} = \bigcap_{c \in C} U(c, c') \,.$$

Usually, $\bar{M}$ is represented by a single connected set; it can be formed by a collection of disconnected sets only in unusual cases where each of these sets has the same density level. Also, usually, a mode is constituted by a single point, but it can be a set having positive probability if $f(x)$ is constant over $\bar{M}$ and takes lower values in a neighborhood of $\bar{M}$.

In this definition, isolated points with high or low values are effectively discarded and the mode corresponds to the maximum of the regularized version of $f(x)$, since density values at isolated points do not affect the probability distribution. However, poles (single points or sets of contiguous points) are allowed to be a mode if $f(x)$ is continuous in their neighborhood.

Conceptually, the above definition also provides a procedure to find a mode. For any choice of c' and $\bar{R}$, it lends one mode, except for the special case of several modes having equal density level. To find all modes of $f(x)$, the procedure must be applied starting from different values of c' and different selections of $\bar{R}$.

17.2.2 Density Estimation and Spatial Tessellations

Consider now a set $\mathcal{D} = \{x_1, \ldots, x_n\}$ of d-dimensional observations, independently sampled from a continuous random variable with density function $f(x)$ on $\mathbb{R}^d$. It is assumed that f has connected support and a finite number of modes, as already indicated.

From $\mathcal{D}$, a nonparametric estimate $\hat{f}(x)$ of $f(x)$ is computed. At the very least, $\hat{f}(x)$ must be finite and a consistent estimate, but somewhat more stringent requirements will be introduced in a moment. Correspondingly, we denote by $\hat{R}(c)$ the sample version of (17.3) obtained replacing f with $\hat{f}$.

The natural question arises of consistency of the $\hat{R}(c)$ sets and of the whole cluster tree when n diverges. For a detailed discussion, we refer the reader to Wong and Lane (1983) and references therein, but the essential fact is that the consistency property is ensured if a uniformly consistent estimate $\hat{f}$ is employed. Under uniform continuity of f, uniform consistency of $\hat{f}$ holds for the kth nearest neighbor (k-NN) estimate on which Wong and Lane (1983) is focused, but other estimates can also be employed. Uniform consistency of kernel density estimates is examined by Devroye and Wagner (1980); again, uniform continuity of f is the key condition to achieve uniform convergence of $\hat{f}$.

Since, at least in its standard form, the clustering problem concerns allocation of the sample points rather than construction of connected subsets of $\mathbb{R}^d$, we extract from $\hat{R}(c)$ the elements belonging to $\mathcal{D}$ and define

$$S(c) = \{x_i : x_i \in \mathcal{D}, \hat{f}(x_i) \geq c\}, \quad \hat{p}_c = |S(c)|/n, \tag{17.5}$$

where $|\cdot|$ denotes the cardinality of a set. Correspondingly, for a given $p \in (0,1)$, $\hat{c}_p$ is the value c such that $\hat{p}_c$ is closest to p.

As noted in Stuetzle (2003), "an obvious way of estimating the cluster tree of a density p from a sample is to first compute a density estimate $\hat{p}$ and then use the cluster tree of $\hat{p}$ as an estimate for the cluster tree of p. A difficulty with this approach is that for most density estimates computing the cluster tree seems computationally intractable." Because of this complexity, Stuetzle (2003) develops a method based on the k-NN density estimate, which allows to build a cluster tree by exploiting a connection between this estimate and the minimum spanning tree of the data. At the end of this process, the tree must be pruned (see also Stuetzle and Nugent 2010).

Finding connected subsets of $\hat{R}(c)$ with other estimates $\hat{f}$, for instance kernel estimates, is more complex. Therefore some authors choose to examine clusters for a fixed value of c; see for instance (Cuevas et al. 2001). Alternatively, Chaudhuri et al. (2014) considers the full range of c values, but only in connection with the use of k-NN density estimate and single linkage algorithm from hierarchical clustering.

A route to finding connected subsets of $S(c)$ for any c while working essentially with any non-parametric estimate $\hat{f}$ has been put forward by Azzalini and Torelli (2007). In this construction, the number M of clusters, equal to the number of modes of $\hat{f}$, is selected automatically by the method; no pruning is necessary. This is the direction on which we shall focus on for the rest of the present contribution.

The proposed formulation involves some concepts of spatial tessellation, which we now recall briefly; a comprehensive account of these themes is provided by Okabe et al. (1992). Given a set $\mathcal{D}$ of points $x_1, x_2, \ldots, x_n$ of $\mathbb{R}^d$, the Voronoi tessellation is defined as the partition of $\mathbb{R}^d$ formed by n sets $V(x_1), \ldots, V(x_n)$ such that, for a generic point x of $\mathbb{R}^d$, $x \in V(x_i)$ if x_i is the closest element of $\mathcal{D}$. The $V(x_i)$ sets turn out to be polyhedra, possibly unbounded. Here, "closest" depends on the adopted definition of distance, usually the Euclidean distance, which is the one we employ here. From the Voronoi tessellation, a second tessellation is defined as follows: Any two elements x_i and x_j of $\mathcal{D}$ are connected by a line segment if the corresponding polyhedra $V(x_i)$ and $V(x_j)$ of the Voronoi tessellation share a portion of their boundary facets. These segments form a new set of polyhedra, usually simplices, that is, they are identified by $d + 1$ vertices. Since for $d = 2$, these simplices are triangles, they form a triangulation of the relevant portion of the space, called Delaunay triangulation. Computationally, the Delaunay triangulation can be obtained directly, without generating the Voronoi tessellation first.

These concepts are illustrated graphically in the left panel of Figure 17.4 for the case $d = 2$. For a set of points $x_1, \ldots, x_n$ sampled from the distribution of Figure 17.1, the Voronoi tessellation is represented by dashed segments, some unbounded; the Delaunay triangulation is represented by continuous segments.

In the clustering methodology of Azzalini and Torelli (2007), the Delaunay triangulation is applied to points $x_1, \ldots, x_n \in \mathcal{D}$ which represent observations. Operationally, this step can be accomplished using the Quickhull algorithm of Bradford Barber et al. (1996), whose implementation is publicly available at http://www.qhull.org

After computing the estimate $\hat{f}$, we can identify $S(c)$ and the connected components of the Delaunay triangulation after removing the edges connecting points which do not belong to $S(c)$, that is, those with $\hat{f} < c$. In the right plot of Figure 17.4, there are two

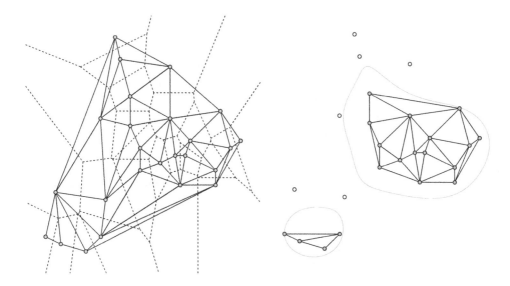

FIGURE 17.4
In the left plot, the small circles denote a set of points sampled from the distribution of Figure 17.1; their Voronoi tessellation is formed by the dashed lines and the Delaunay triangulation by the continuous lines. In the right plot, some edges of the Delaunay triangulation have been removed, those connecting points with $\hat{f} < c$ for some given c.

connected components comprising $S(c)$, enclosed by the two curves which jointly represent the level set $\{x : \hat{f}(x) = c\}$.

17.2.3 Building Clusters

The extraction of connected components of $S(c)$ in the Delaunay triangulation, as illustrated in the right plot of Figure 17.4, must be performed for each value of c or, equivalently, of $\hat{p}_c$. When c decreases (and p_c increases), the connected sets increase in size until they coalesce. This merging is analogous to the transition stage denoted by the dashed lines in the bottom-right plot of Figure 17.2, except that we are now working with clusters of $S(c)$ sets instead of $R(c)$ sets. Just before the two clusters merge, we identify them as *cluster cores*, leaving outside a fraction of data points not yet allocated. Obviously, in cases different from the one displayed in Figure 17.2, there could be other cluster cores appearing as c moves further down and then disappearing as well.

As p_c scans the interval $(0, 1)$, a cluster tree is constructed. Computationally, we consider p_c values over an adequately fine grid of points in $(0, 1)$. When we move from one value of p_c to another in this grid, the membership of the elements of $\mathcal{D}$ belonging to an identified cluster core must be kept track of, to avoid that a given sample element is assigned a different group label. At the end of this stage, we are left with M, say, cluster cores formed by sets of allocated sample points and a number of unallocated points.

Allocating the remaining points to the M cluster cores is essentially a classification problem, although somewhat peculiar because these points are not a random sample from $f(x)$, but inevitably are located in the outskirts of the cluster cores. Given the adopted logic, a coherent route is to form an estimate $\hat{f}_m$ based on the data allocated to the mth cluster core (for $m = 1, \ldots, M$) and allocate a given unallocated point x' to the subpopulation with highest value of

$$r_m(x') = \log \frac{q_m \, \hat{f}_m(x')}{\max_{j \neq m} q_j \, \hat{f}_j(x')}, \quad m = 1, \ldots, M, \tag{17.6}$$

where the q_j can be simply taken to be 1 or could be a weight based on the precision of the estimate $\hat{f}_j$ or on the estimated relative frequency $\hat{\pi}_j$ of the jth subpopulation. The use of logratios instead of simple ratios in (17.6) is merely conventional. Note that here the components f_m of the mixture representation (17.2) come explicitly into play.

An additional point to decide is whether to (a) keep the estimates $\hat{f}_1, \ldots, \hat{f}_M$ fixed across the allocation of all possible x_u points, (b) update sequentially $\hat{f}_m$ when a new point is allocated to the mth cluster, (c) for computational convenience, take an intermediate policy and update the $\hat{f}_m$ in a block-sequential manner, allocating a block of new points before updating the density estimates. For options (b) and (c), the $r_m(x_j)$ values of unallocated points must be sorted in decreasing order and allocation of points with highest $r_m(x_j)$ values takes precedence, since these are the more clear-cut cases to allocate. In the subsequent numerical work, (c) is the option adopted.

The sequential strategy, in both its variants (b) and (c), offers the advantage that it can be stopped when the allocation of additional points becomes too little reliable, which will typically occur for some points located about halfway between two (or possibly more) identified clusters. This policy allows to leave a portion of points unallocated; these can be described as "noise" in the terminology of many authors (Wishart 1969; Ester et al. 1996) or as "fluff" in Stuetzle and Nugent (2010).

We summarize the main steps of the procedure in the following scheme.

Algorithm CLUSTERING VIA UPPER LEVEL SETS

1. Given data $\mathcal{D}$, compute density estimate $\hat{f}(x_i)$ at each $x_i \in \mathcal{D}$.
2. Obtain the Delaunay triangulation of the $\mathcal{D}$ points.
3. For each p belonging to a grid of points in $(0, 1)$:
   (a) remove from the Delaunay triangulation the elements of $\mathcal{D}$ with $\hat{f} < \hat{c}_p$,
   (b) determine the connected sets of the points retained.
4. Build the cluster tree and form the cluster cores.
5. Allocate the remaining points to the cluster cores.

17.3 Practical Aspects and a Variant of the Method

17.3.1 Kernel Density Estimation

The methodology just introduced is not linked to a specific choice of the nonparametric estimate $\hat{f}$. Arguably, the most popular technique for nonparametric density estimation is the kernel method; this is the one adopted by Azzalini and Torelli (2007) and also here in the illustrative numerical work below. For an introduction to this method, see for instance, Silverman (1986) and Bowman and Azzalini (1997).

In the case of a multivariate kernel constructed as the d-fold product of a univariate kernel K, the estimate at point $z = (z_1, \ldots, z_d)$ is

$$\hat{f}(z) = \frac{1}{n} \sum_{i=1}^{n} \prod_{j=1}^{d} \frac{1}{h_{i,j}} K\left(\frac{z_j - x_{i,j}}{h_{i,j}}\right), \tag{17.7}$$

where the smoothing parameters $h_{i,j}$ vary simultaneously with the observations and the variables $(h_{i,j} > 0)$; $x_{i,j}$ denote the jth component of x_i. A common option is to keep the smoothing parameters constant across the observations, so only a vector $(h_1, \ldots, h_d)$ needs to be specified. If K is chosen to be a proper density function, so will be $\hat{f}$. For the purpose of the methodology presented here, $\hat{f}$ must be evaluated only at points $z \in \mathcal{D}$.

It is well-known that the choice of K is not crucial and the simple option of using the $N(0, 1)$ density is often adopted. A far more critical aspect is the selection of the smoothing parameters, for at least two reasons: (i) different choices of these parameters can affect greatly the final outcome, (ii) their optimal choice depends on the unknown density f, making their selection a tricky problem. In the present context, the issue is further complicated by the (conjectured) presence of a (still unknown) number M of subpopulation densities, for which different smoothing parameters may be appropriate locally.

In spite of all of this, the numerical work of Azzalini and Torelli (2007) indicates that the following simple course of action works satisfactorily, at least for a number of dimensions d not larger than 5 or 6, which is what they have examined. For the case of a fixed smoothing vector across all observations, select the jth smoothing parameter, h_j, which is asymptotically optimal in the integrated mean square error sense under the assumption of multivariate normality. This is known to be

$$h_j = \sigma_j \left(\frac{4}{(d+2)\,n} \right)^{1/(d+4)}, \quad (j = 1, \ldots, d), \tag{17.8}$$

where the standard deviation σ_j of the jth variable must be estimated from the observations; see for instance (Bowman and Azzalini 1997, p. 32). The empirical findings of Azzalini and Torelli (2007) indicate that it is convenient to apply a shrinkage factor, denoted h_{mult} here, to the above expression of h_j, which has been derived for the case of a single Gaussian population. A choice of about $h_{mult} = 3/4$ has appeared to be an overall reasonable compromise. However, since there is nothing magic about this choice, it is sensible to explore the outcome obtained with other values of h_{mult}.

We defer numerical examples to Section 17.4, but some comments based at least on the numerical findings just quoted are appropriate. It may be appear surprising that good empirical outcomes are reported with (a) such an apparently naïve choice of h_j, (b) a number of dimensions beyond what is commonly considered in density estimation, unless n is extremely large. The key consideration is that here we are not concerned with estimation of f as such. We are instead concerned with correct identification of the initial cluster cores and allocation of the unlabelled units to the cluster core; these steps may largely work even if the fine details of the densities are not estimated accurately. Numerical experimentation indicates that we can obtain stable results across a reasonably wide range of values of the smoothing parameters. Some numerical evidence in this direction will be provided in Section 17.4.1 (an additional illustration is provided by Figure 7 of Azzalini and Torelli (2007)).

For higher values of d, the data cloud tends to be very sparse in the Euclidean space, even for large values of n, and the use of a constant vector of smoothing parameters becomes progressively less satisfactory. In Menardi and Azzalini (2014), a variable smoothing parameter is employed, to adapt to the local sparseness of the data. Since this route is adopted in connection with a variant of the formulation to be presented in Section 17.3.2 below, further discussion will take place there.

17.3.2 Computational Issues and an Alternative Neighborhood Graph

From a computational viewpoint, the methodology discussed so far scales well for increasing n, both in the computation of $\hat{f}$ and of the Delaunay triangulation, for small to moderate values of d. Even storage is not an issue. Problems arise with large d, since the computational burden of the Delaunay triangulation grows exponentially for large d. For the currently available computational resources, data sets with dimension $d > 6$ are very demanding or just intractable.

To overcome this limitation, in Menardi and Azzalini (2014) an alternative method to Delaunay triangulation is proposed for extracting connected components of $S(c)$. The essence of the underlying idea is as follows. For any two sample points, $x_1, x_2 \in \mathcal{D}$, examine $\hat{f}(x)$ along the segment joining x_1 and x_2. If x_1 and x_2 belong to different sets of high density, $\hat{f}(x)$ must start at some relatively high value when $x = x_1$, descend to lower values when x crosses the area between the two clusters which is called a "valley" and then rise again when x approaches x_2. If instead x_1 and x_2 belong to the same connected component with high density, the path of $\hat{f}(x)$ does not cross a valley and the function will be either monotonic or possibly bell-shaped, when x_1 and x_2 are placed on opposite sides of the mode of that region.

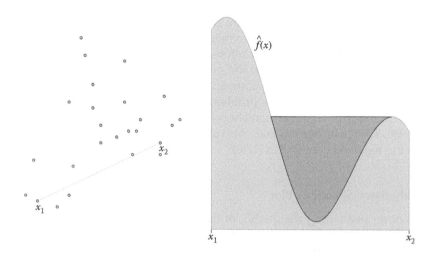

FIGURE 17.5
In the left plot, a segment joins two points, x_1 and x_2, of the sample of Figure 17.4. The right plot displays the restriction of $\hat{f}(x)$ along the selected segment and the dark-shaded area required to fill the valley of the curve.

The left plot of Figure 17.5 presents the same sample values of Figure 17.4 and the segment which joins two specific points, denoted x_1, x_2. The oscillating curve labelled $\hat{f}(x)$ in the right plot is actually the section of the full density estimate along the chosen segment.

The aim is to build a graph similar to the one in the left panel of Figure 17.4 but with edges created by a different criterion, based instead on the restriction of $\hat{f}(x)$ along the segments joining pairs of sample points in $\mathcal{D}$. If the section of $\hat{f}(x)$ along the segment (x_i, x_j) has no valley, then an edge (x_i, x_j) is created on the graph. If there is a valley, an edge is not created, unless the valley is so "small" that it can be regarded as an effect of the inevitable sample variability in estimating $f(x)$. The decision whether a valley is "small" is taken by considering the smallest function $\tilde{f}(x) \geq \hat{f}(x)$, which has no valleys and computing the ratio V of the integral of $\tilde{f}(x) - \hat{f}(x)$ over the integral of $\tilde{f}(x)$. For the right plot of Figure 17.5, this corresponds to

$$V = \frac{\text{(size of dark shaded area)}}{\text{(size of dark and light shaded areas)}}.$$

Operationally, the integrals are approximated numerically; given the context, one does not need a highly accurate evaluation, so computation can be fairly fast. If V is smaller than a certain threshold, an edge between the sample points is inserted, otherwise it is not. In the numerical work of Menardi and Azzalini (2014) a threshold of 0.1 is indicated, but this choice is generally not critical and results remain fairly stable with other values; however, a threshold above about 0.30 is not recommended, and it makes little sense anyway.

After replicating this process for all possible pairs (x_i, x_j) of distinct sample points, we obtain a graph which we use in place of the Delaunay triangulation. The remaining steps of the procedure remain as described in Section 17.2. The connection graph so obtained does not coincide with the one of the Delaunay triangulation, but the final outcomes with the original and the variant procedure are largely similar.

Since the scan of $\hat{f}(x)$ along segments (x_i, x_j) is a one-dimensional process irrespectively of the value of d, this variant of the method allows to consider higher values of d.

In the numerical work of Menardi and Azzalini (2014), high-dimensional cases have been considered, with d up to 22, still leading to meaningful results.

Working with larger values of d than before, one faces more frequently the problem of sparseness of the data, especially so in the outskirt of the data cloud, even for sizeable values of n. The use of variable bandwidths in (17.7), adapting $h_{i,j}$ to the local sparseness of the data, is a way to deal with this problem. Specifically, Menardi and Azzalini (2014) adopt the square-root rule of Silverman choosing a value of $h_{i,j}$ proportional to $f^{-1/2}(x_i)$ (Silverman 1986, 5.3). To be operational, this rule requires a pilot estimate of $f(x)$, which is computed using the fixed bandwidth (17.8).

17.3.3 Density-Based Silhouette Plot

A classical tool for evaluating the quality of the partition generated by a clustering technique is the silhouette plot, introduced by Rousseeuw (1987). The method is based on the average distance from one given object to those of the same cluster as that object, compared with the similar average distance from the best alternative cluster.

In the context of density-based clustering, distances are naturally replaced by density ratios; this is the idea explored by Menardi (2011). If object x_i has been assigned to cluster $m \in \{1, \ldots, M\}$, then the density-based silhouette is computed as

$$\mathrm{dbs}(x_i) = \frac{r_m(x_i)}{\max_{x_i} |r_m(x_i)|}, \qquad (17.9)$$

where $r_m(x_i)$ is computed from (17.6) with $q_m = \pi_m$, the group marginal probability.

The graphical appearance and the interpretation of density-based silhouette are completely analogous to those of the classical silhouette. An example illustration is displayed in the bottom-right panel of Figure 17.7. The technique applies equally to other density-based clustering methods, including those model-based.

17.3.4 Noncontinuous Variables

An intrinsic limitation of density-based clustering, whether parametric or nonparametric, is that it works with continuous variables only. Various methods have been examined in the literature to overcome this limitation, typically in connection with model-based clustering. Some of these proposals deal with binary data, other allows more general categorical data; some consider also discrete data. The general principle is to introduce a set of continuous latent variables, in some form. Corresponding integration with respect to the distribution of the latent variables to obtain the distribution of the observables, combined with a mixture formulation, tends to be a nontrivial computational task, especially so when the formulation allows mixed-type variables.

References to above-cited formulations are provided by Azzalini and Menardi (2015), where an alternative route is adopted; this route is less refined than those in existing work, but far more easily applicable. It also seems also to be the first exploration of the problem of noncontinuous variables in the modal clustering approach.

The logical scheme of Azzalini and Menardi (2015) is very simple: Start from the dissimilarity matrix of the observables and use multidimensional scaling (MDS) to extract a small number of constructed continuous variables, which are then suitable for use with the clustering technique described earlier.

The dissimilarity matrix is the basic ingredient of the classical approach to clustering recalled in the introductory section of this contribution. By the very nature of MDS, the constructed variables so produced preserve, at least approximately, distances between objects; hence, in essence, the MDS step does not alter the clustering structure of the data.

The basic idea just sketched can be employed in a number of variants. One of these concerns the number of MDS variables to extract; another one is whether to treat the continuous variables originally observed as any other variable or to keep them separate from the MDS stage and merge them with the MDS variables only at the end, for the clustering stage. These and other issues are examined in Azzalini and Menardi (2015), to which the reader is referred to.

17.4 Software, Examples, and Applications

17.4.1 The R Package pdfCluster and an Example in Detail

An R implementation of the proposed methodology is provided by package `pdfCluster` available at `http://cran.R-project.org/package=pdfCluster`. What follows is an introduction of the basic commands and options of the package, combined with an illustration of the practical working of the methodology. A detailed description of the package is provided by Azzalini and Menardi (2014).

For illustration purpose, we consider a classical example in the clustering literature represented by the wine data, introduced by Forina et al. (1986). The data set comprises a large number of chemical and other measurements from $n = 178$ specimens of wine from three types produced in the Piedmont region of Italy; the wines are Barbera, Barolo, and Grignolino. The data themselves, possibly for only a subset of the original variables, are publicly available from various sources, including the package `pdfCluster`.

We start by considering variables (alcohol, alcalinity, phenols, flavanoids), hence $d = 4$, and see whether a clustering of the specimens based on these variables identifies groups corresponding to the actual wine type, without of course using this information in the clustering exercise. The kernel estimate of the joint density obtained by

```
library(pdfCluster)
data(wine)
subset <- c(2,5,7,8)
f <- kepdf(wine[,subset], h=0.75*h.norm(wine[,subset]))
plot(f)
```

is displayed in Figure 17.6 in the form of a set contour level plots, one for each pair of variables. The smoothing parameter h is the one of (17.8) multiplied by the shrinkage factor $h_{mult} = 0.75$. This sort of plot is not required for the clustering operation and has been inserted here only to provide a perception of the data pattern.

The main function of the package, called `pdfCluster` itself, performs all stages of the methodology and creates an object for subsequent manipulation; this object encapsulates the clustering structure and other information. The actual clustering operation and subsequent steps for the above data are accomplished by the commands

```
cl <- pdfCluster(wine[,subset])
plot(cl)
table(wine$Type, groups(cl))
```

which do the following: (a) create a cluster object `cl` on which various "methods" can be applied, in R terminology, (b) display some plots from this object, (c) compute a cross-classification table of the true and the constructed groups. Among the operation performed by function `pdfCluster`, there is kernel density estimation which, in a moderate dimensionality like here, is employed in the form of constant bandwidth with a $h_{mult} = 0.75$ automatically applied to (17.8). If a different shrinkage factor is required, the optional parameter `hmult` allows to change that value.

The graphical outcome produced by `plot(cl)` is displayed in the four panels, Figure 17.7, which represent from left to right, top to bottom: the mode function, the cluster tree, the scatter matrix of the data with numeric codes of the constructed groups and the density-based silhouette plot described in Section 17.3.3. The cross-classification table of the true and the constructed groups is reported in the right-most portion of Table 17.1, under $d = 4$. Below that table, the ARI value refers to the "adjusted Rand index" which quantifies the agreement between two partitions of a data set, up to a permutation of the labels (Hubert and Arabie 1985).

Table 17.1 also includes cross-classifications and ARI values for two other groups of variables. Since Figure 17.6 indicates a strong association for (phenols, flavanoids), one of them can safely be dropped. The case $d = 3$ refers then to (alcohol, alcalinity, flavanoids),

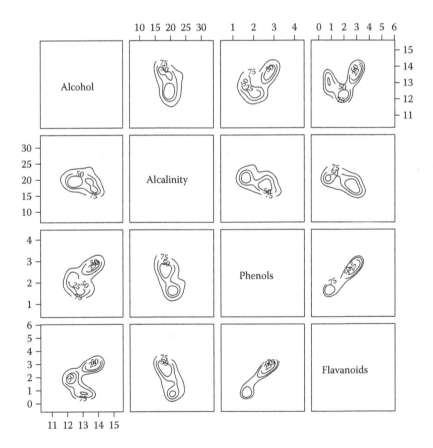

FIGURE 17.6
Wine data: nonparametric density estimate for $d = 4$ variables.

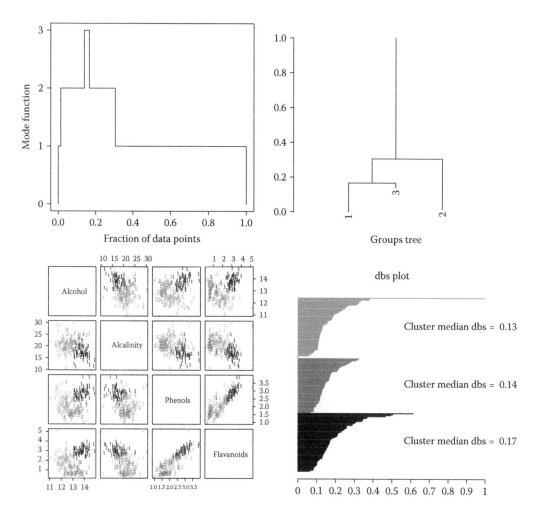

FIGURE 17.7
Wine data: graphical outcome for clusters formed from $d = 4$ variables.

TABLE 17.1

Wine Data: Cross-Classification Tables of the True and the Selected Groups Using $d = 2, 3, 4$ Variables and Their Corresponding ARI Values

	$d = 2$			$d = 3$			$d = 4$		
	1	2	3	1	2	3	1	2	3
Barolo	59	0	0	58	1	0	56	0	3
Grignolino	7	55	9	4	62	5	4	9	58
Barbera	0	0	48	0	0	48	0	47	1
ARI	0.745			0.833			0.731		

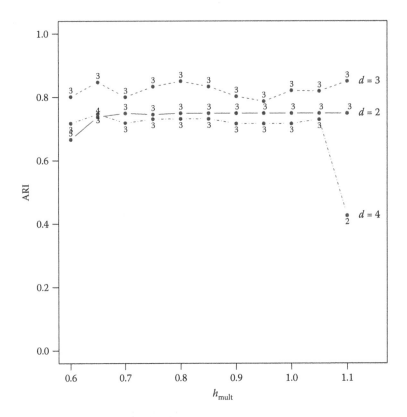

FIGURE 17.8
Wine data: plot of ARI values versus shrinkage factor h_{mult} for three groups of variables; the digit next to each point indicates the number of detected clusters.

which produces a somewhat better ARI value, presumably thanks to an improvement in density estimation working with fewer variables. The case $d = 2$ refers to (alcohol, flavanoids).

In Section 17.3.1, we have claimed a good degree of stability of the results with respect to the smoothing parameter and promised to provide supporting numerical evidence in this section. For each of the three groups of variables used in Table 17.1, we replicated the clustering operations across a range of h_{mult} factors applied to (17.8). Figure 17.8 displays the outcome as a plot of the ARI versus the shrinkage factor h_{mult}, ranging from 0.6 to 1.1; the number printed next to each points indicates the number M of detected clusters. The outcome, both as ARI and M values, is essentially constant over a comfortable range, from about 0.65 to 1.05, which corresponds to a wide degree of modification of the density estimate. These reassuring indications do not, however, must be interpreted in the sense that in practical data analysis one should forget about the general cautionary practice of exploring the effect of adjusting the tuning parameters of a method, h_{mult} in this case.

17.4.2 Some Applications

The numerical example considered in Section 17.4.1 was deliberately very simple, given its purely illustrative role. A wider ranges of cases, some numerically more extensive, have been considered by Azzalini and Menardi (2014) and Azzalini and Torelli (2007), partly using simulated data, partly with real data where again the true groups labels were known but not used for clustering. For higher values of d, the variant type of association graph of Section 17.3.2 was employed and kernel density estimation was carried out using variable bandwidth.

In parallel, the outcome of various existing methods has been recorded, for comparative purposes. When a competing method required the number of clusters as an input information, it was given the true one, while `pdfCluster` had to select it autonomously. The overall indication from this numerical work is encouraging: In all cases considered, the upper level set methodology performed either comparably or, fairly often, better than existing methods.

A more substantial application of the `pdfCluster` methodology is represented by the work of De Bin and Risso (2011) for clustering gene expression data arising in microarray experiments. Their method consists in an initial step to reduce dimensionality of the problem via principal component analysis and a subsequent step based on clustering, for which the authors take into consideration the k-means method, a model-based method using mixtures of Gaussians and `pdfCluster`. These methods are tested on data which represent either real observations and simulated data "with structure similar to that of real microarray experiments." The overall indication of their comparisons is a overall superiority of `pdfCluster`, partly in terms of classification errors, partly and more markedly in terms of dimensionality reduction, that is, number of genes selected.

In the analysis of EEG and EMG signals to evaluate sleep/wake status in animals (with distinction of two phases of sleep), an important goal is to develop automated methods which can avoid time-consuming and partly subjective human inspection of the signal records. Such a fully automatic procedure is proposed by Sunagawa et al. (2013) which is composed of three main steps: character extraction, clustering, and annotation. The first step consists in extracting some variables of a principal component analysis of the EEG/EMG power spectra. Subsequently, in the authors' words, "In the clustering step, we have adopted the clustering algorithm based on non-parametric estimation of probability density (Azzalini and Torelli 2007) to avoid subjectiveness from holding up a model before clustering".

17.5 Conclusion

The approach to clustering based on high-density sets of a density f provides a logical framework, which is both conceptually clear and intuitive. The formulation produces a cluster tree whose number of leaves represents unambiguously the number of clusters of f, without involving a subjective choice of this number or introduction of some additional criterion; no pruning is required.

When we translate this logical framework into operational work, there are two main aspects which require careful consideration: (i) finding the connected subsets of the upper level sets and (ii) non-parametric estimation of f.

The first of these has a convincing solution via the use of Delaunay triangulation. Since this runs into computational difficulties with high-dimensional data, the variant form of connection graph described in Section 17.3.2 provides a way to overcome the problem. Alternative ways of building a connection graph, possibly more theoretically motivated, may be worth exploring.

As for the problem of nonparametric estimation of $\hat{f}$, the kernel method is the more popular choice, but again there could be other methods worth considering. However, for (nearly) any estimation method adopted at this stage, there is a smoothing parameter which represents the critical choice for regulating the behaviour of $\hat{f}$. For kernel estimation, the simple rule of using the asymptotically optimal bandwidth (17.8), modified by shrinkage factor h_{mult}, appears to work well in practice. Surely, it would be welcome to have a better justified procedure, more so if specifically targeted to the case of mixture distributions. The problem is even more challenging in the high-dimensional case, where the use of variable bandwidths becomes more compelling.

In Section 17.3.1, we have indicated a possible explanation of the empirically observed fact that the present methodology appears to work reasonably even for a number d of dimensions higher than those for which density estimation is usually considered satisfactory. However, if d is increased more and more, eventually the method cannot be expected to work. In these cases, a preliminary stage must be considered so to reduce the number of variables. Careful removal of variables on the basis of subject-matter considerations and exploratory data analysis has much to recommend it in this process. When an automatic procedure is required instead, principal component analysis has been employed, for instance in the applied work recalled in Section 17.4.2. An alternative option is to apply multidimensional scaling to the distance matrix, similarly to Section 17.3.4, even when the variables are continuous. This route exploits the MDS property of preserving distances among objects, at least approximately; hence, it avoids the risk of destroying the clustering structure of the data, which is the main objection to a so-called "tandem analysis."

Acknowledgment

The definitions of upper level set and of mode in Section 17.2.1 have been elaborated jointly with Giuliana Regoli.

References

Azzalini, A. and Menardi, G. 2014. Clustering via nonparametric density estimation: The R package pdfCluster. *Journal of Statistical Software*, 57(11).

Azzalini, A. and Menardi, G. 2015. Density-based clustering with non-continuous data. Submitted, under revision.

Azzalini, A. and Torelli, N. 2007. Clustering via nonparametric density estimation. *Statistics and Computing*, 17, 71–80.

Bowman, A.W. and Azzalini, A. 1997. *Applied Smoothing Techniques for Data Analysis: The Kernel Approach with S-Plus Illustrations*. Oxford University Press, Oxford.

Bradford Barber, C., Dobkin, D.P., and Huhdanpaa, H. 1996. The Quickhull algorithm for convex hulls. *ACM Transactions on Mathematical Software*, 22(4), 469–483.

Chaudhuri, K., Dasgupta, S., Kpotufe, S., and von Luxburg, U. 2014. Consistent procedures for cluster tree estimation and pruning. *IEEE Transactions on Information Theory*, 60(12), 7900–7912.

Cuevas, A., Febrero, M., and Fraiman, R. 2001. Cluster analysis: A further approach based on density estimation. *Computational Statistics and Data Analysis*, 36, 441–459.

De Bin, R. and Risso, D. 2011. A novel approach to the clustering of microarray data via nonparametric density estimation. *BMC Bioinformatics*, 12, 49.

Devroye, L.P. and Wagner, T.J. 1980. The strong uniform consistency of kernel density estimates. In P.R. Krishnaiah, editor, *Multivariate Analysis, Volume 5*, pages 59–77. North-Holland, Amsterdam.

Ester, M., Kriegel, H.P., Sander, J., and Xu, X. 1996. A density-based algorithm for discovering clusters in large spatial databases with noise. In *Proceedings of the 2nd International Conference on Knowledge Discovery and Data Mining (KDD-96)*, pages 226–231.

Forina, M., Armanino, C., Castino, M., and Ubigli, M. 1986. Multivariate data analysis as a discriminating method of the origin of wines. *Vitis*, 25, 189–201.

Hartigan, J.A. 1975. *Clustering Algorithms*. NY: Wiley & Sons, New York.

Hubert, L. and Arabie, P. 1985. Comparing partitions. *Journal of Classification*, 2, 193–218.

Kriegel, H.-P., Kröger, P., Sander, J., and Zimek, A. 2011. Density-based clustering. *WIREs: Data Mining and Knowledge Discovery*, 1(3), 231–240.

Menardi, G. 2011. Density-based Silhouette diagnostics for clustering methods. *Statistics and Computing*, 21, 295–308.

Menardi, G. 2015. A review of modal clustering. To appear in Int. Stat. Review, DOI: 10.1111/insr.12109.

Menardi, G. and Azzalini, A. 2014. An advancement in clustering via nonparametric density estimation. *Statistics and Computing*, 24, 753–767.

Okabe, A., Boots, B.N., and Sugihara, K. 1992. *Spatial Tessellations: Concepts and Applications of Voronoi Diagrams*. Wiley & Sons, New York: NY.

Ray, S. and Lindsay, B.G. 2005. The topography of multivariate normal mixtures. *Annals of Statistics*, 33, 2042–2065.

Rousseeuw, P.J. 1987. Silhouettes: A graphical aid to the interpretation and validation of cluster analysis. *Journal of Computational and Applied Mathematics*, 20, 53–65.

Silverman, B.W. 1986. *Density Estimation for Statistics and Data Analysis*. Chapman & Hall, London.

Stuetzle, W. 2003. Estimating the cluster tree of a density by analyzing the minimal spanning tree of a sample. *Journal of Classification*, 20, 25–47.

Stuetzle, W. and Nugent, R. 2010. A generalized single linkage method for estimating the cluster tree of a density. *Journal of Computational and Graphical Statistics*, 19, 397–418.

Sunagawa, G.A., Sei, H., Shimba, S., Urade, Y., and Ueda, H.R. 2013. FASTER: An unsupervised fully automated sleep staging method for mice. *Genes to Cells*, 18(6), 502–518.

Wishart, D. 1969. Mode analysis: A generalization of nearest neighbor which reduces chaining effects. In A.J. Cole, editor, *Numerical Taxonomy*, pages 282–308. Academic Press, London.

Wong, A.M. and Lane, T. 1983. The *k*th nearest neighbour clustering procedure. *Journal of the Royal Statistical Society, series B*, 45, 362–368.

18

Clustering Methods Based on Kernel Density Estimators: Mean-Shift Algorithms

Miguel Á. Carreira-Perpiñán

CONTENTS

Abstract

A natural way to characterize the cluster structure of a dataset is by finding regions containing a high density of data. This can be done in a nonparametric way with a kernel density estimate, whose modes and hence clusters can be found using mean-shift algorithms. We describe the theory and practice behind clustering based on kernel density estimates and mean-shift algorithms. We discuss the blurring and nonblurring versions of mean-shift; theoretical results about mean-shift algorithms and Gaussian mixtures; relations with scale-space theory, spectral clustering and other algorithms; extensions to tracking, to manifold and graph data, and to manifold denoising; K-modes and Laplacian K-modes algorithms; acceleration strategies for large datasets; and applications to image segmentation, manifold denoising and multivalued regression.

18.1 Introduction

One intuitive way of defining clusters is to assume that the data points are a sample of a probability density function, and then to define the clusters through this density. For example, Figure 18.1 shows a 2D dataset and a density estimate for it, whose contours clearly suggest that there are two clusters of a complex shape. The first step, then, is to learn an estimate of the density for the data points. This can be done with a parametric model, such as a Gaussian mixture, typically trained with an EM algorithm to maximize the likelihood (McLachlan and Krishnan 2008) (see Chapter 3). Such an approach is often computationally efficient and can give good results with clusters of elliptical shape, but it has several disadvantages. The likelihood function will typically have local optima, and finding a global optimum is, in general, very difficult; thus, the result is dependent on the initialization, and in practice a user will try different initializations (usually random restarts). The selection of the model (what kernel and how many components) is left to the user, as well as the number of clusters to find. And when the clusters have complex shapes, as for example in image segmentation, many components will be required to approximate them well, increasing the training time and the number of local optima.

We focus on *nonparametric, kernel density estimates (KDE)*. A KDE is a generalization of histograms to define density estimates in any dimension that are smooth. They simplify the mathematical and computational treatment of densities and, crucially, enable one to use continuous optimization to find maxima of the density. With a kernel such as the Gaussian kernel, a KDE requires a single user parameter, the *bandwidth* (also referred to

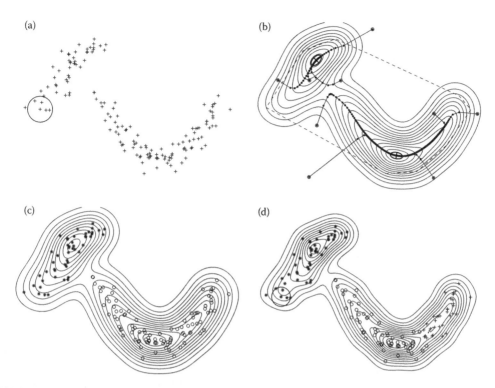

FIGURE 18.1
Illustration in 2D of complex-shaped clusters, the kernel density estimate (KDE) and the mean-shift results. (a): The dataset. The circle has a radius equal to the bandwidth σ used in the KDE. (b): A contour plot of the Gaussian KDE $p(\mathbf{x})$ with bandwidth σ. The KDE has two modes, located at the center of the small ellipses. Each ellipse indicates the eigenvectors (rescaled to improve visibility) of the Jacobian $\mathbf{J}(\mathbf{x}^*)$ of eq. (18.5) at a mode $\mathbf{x}^*$. The dotted-line polygon is the convex hull of the data points. The paths followed by Gaussian MS for various starting points are shown. (c): The resulting clustering using mean-shift, with each point using a marker according to the cluster it belongs to. (d): The resulting clustering and KDE contours using a smaller bandwidth (indicated by the radius of the circle on the left). Now, the KDE defines three clusters.

as *scale*). Given the bandwidth, the KDE is uniquely determined and, as seen below, so will be its clusters, which can take complex nonconvex shapes. Hence, the user need not select the number of clusters or try random restarts. We will focus on clusters defined by the modes of the KDE (although this is not the only way to define the clusters). A *mode* is a local maximum of the density. A natural algorithm to find modes of a KDE is the *mean-shift* iteration, essentially a local average, described in Section 18.2. The basic idea in mean-shift clustering is to run a mean-shift iteration initialized at every data point and then to have each mode define one cluster, with all the points that converged to the same mode belonging to the same cluster. Section 18.2.3 reviews theoretical results regarding the number and location of modes of a KDE, the convergence of mean-shift algorithms and the character of the cluster domains. Section 18.2.4 discusses relations of mean-shift algorithms with spectral clustering and other algorithms. Sections 18.2.5 and 18.2.6 describe extensions of mean-shift for clustering and manifold denoising, respectively. One disadvantage of mean-shift algorithms is their computational cost, and Section 18.2.7 describes several accelerations. Section 18.3 describes another family of KDE-based clustering algorithms which are a hybrid of K-means and mean-shift, the K-modes and Laplacian

K-modes algorithms, which find exactly K clusters and a mode in each, and work better with high-dimensional data. Section 18.4 shows applications in image segmentation, inverse problems, denoising, and other areas. We assume multivariate, continuous (i.e., not categorical) data throughout.

18.2 Problem Formulation and Mean-Shift Algorithms

In mean-shift clustering, the input to the algorithm are the data points (multivariate, continuous feature vectors) and the bandwidth or scale. Call $\{\mathbf{x}_n\}_{n=1}^N \subset \mathbb{R}^D$ the data points to be clustered. We define a kernel density estimate (Wand and Jones 1994)

$$p(\mathbf{x}) = \frac{1}{N} \sum_{n=1}^N K\left(\left\|\frac{\mathbf{x} - \mathbf{x}_n}{\sigma}\right\|^2\right) \quad \mathbf{x} \in \mathbb{R}^D \tag{18.1}$$

with bandwidth $\sigma > 0$ and kernel $K(t)$, for example, $K(t) = e^{-t/2}$ for the Gaussian kernel or $K(t) = 1 - t$ if $t \in [0,1)$ and 0 if $t \geq 1$ for the Epanechnikov kernel. Many of the results below carry over to kernels where each point has its own weight and its own bandwidth, which can be an isotropic, diagonal, or full covariance matrix. To simplify the presentation, we focus on the case where all points have the same, scalar bandwidth σ (the isotropic, homoscedastic case) and the same weight $1/N$ unless otherwise noted. This is the case found most commonly in practice. Also, we mostly focus on Gaussian kernels, which are easier to analyze and give rise to simpler formulas.

We can derive a simple iterative scheme $\mathbf{x}^{(\tau+1)} = \mathbf{f}(\mathbf{x}^{(\tau)})$ for $\tau = 0, 1, 2 \ldots$ to find a mode of p by equating its gradient to zero and rearranging terms (Section 18.2.7 discusses other ways to find modes). We obtain

$$\mathbf{f}(\mathbf{x}) = \sum_{n=1}^N \frac{K'\left(\|(\mathbf{x} - \mathbf{x}_n)/\sigma\|^2\right)}{\sum_{n'=1}^N K'\left(\|(\mathbf{x} - \mathbf{x}_{n'})/\sigma\|^2\right)} \mathbf{x}_n, \tag{18.2}$$

where $K' = dK/dt$ and the vector $\mathbf{f}(\mathbf{x}) - \mathbf{x}$ is the *mean shift*, since it averages the individual shifts $\mathbf{x}_n - \mathbf{x}$ with weights as above. For a Gaussian kernel, $K' \propto K$ and this simplifies to the following, elegant form (where, by Bayes' theorem, $p(n \mid \mathbf{x}) = p(\mathbf{x} \mid n)p(n)/p(\mathbf{x})$ is the posterior probability of the component centered at $\mathbf{x}_n$ given point $\mathbf{x}$) (Carreira-Perpiñán 2000):

$$p(n \mid \mathbf{x}) = \frac{\exp\left((-1/2)\|(\mathbf{x} - \mathbf{x}_n)/\sigma\|^2\right)}{\sum_{n'=1}^N \exp\left((-1/2)\|(\mathbf{x} - \mathbf{x}_{n'})/\sigma\|^2\right)} \quad \mathbf{f}(\mathbf{x}) = \sum_{n=1}^N p(n \mid \mathbf{x})\mathbf{x}_n. \tag{18.3}$$

As discussed below, under mild conditions this scheme converges to modes of p from nearly any initial $\mathbf{x}^{(0)}$. Intuitively, each step moves the iterate $\mathbf{x}^{(\tau)}$ to a local average of the data, in that data points closer to $\mathbf{x}^{(\tau)}$ have larger weight, and this increases the density. Equation 18.2 is called the *mean-shift iteration*, and it can be used in two distinct ways: to find modes, and to filter (or smooth) a dataset. This gives rise to two different clustering algorithms, as follows. We will refer to them as Mean Shift (MS) (where modes are found) and Blurring Mean Shift (BMS) (where the dataset is filtered).

Algorithm mean shift

(a) Gaussian mean-shift (MS) algorithm

for $n \in \{1, ..., N\}$
   $x \leftarrow x_n$
   repeat
       $\forall n:\ p(n|x) \leftarrow \dfrac{\exp\left(-\frac{1}{2}\|(x-x_n)/\sigma\|^2\right)}{\sum_{n'=1}^{N}\exp\left(-\frac{1}{2}\|(x-x_{n'})/\sigma\|^2\right)}$
       $x \leftarrow \sum_{n=1}^{N} p(n|x)x_n$
   until stop
   $z_n \leftarrow x$
end
Connected-Components $(\{z_n\}_{n=1}^{N}, \epsilon)$

(b) Gaussian blurring mean-shift (BMS) algorithm

repeat
   for $m \in \{1, ..., N\}$
       $\forall n:\ p(n|x) \leftarrow \dfrac{\exp\left(-\frac{1}{2}\|(x_m-x_n)/\sigma\|^2\right)}{\sum_{n'=1}^{N}\exp\left(-\frac{1}{2}\|(x_m-x_{n'})/\sigma\|^2\right)}$
       $y_m \leftarrow \sum_{n=1}^{N} p(n|x_m)x_n$
   end
   $\forall m:\ x_m \leftarrow y_m$
until stop
Connected-Components $(\{x_n\}_{n=1}^{N}, \epsilon)$

(c) Gaussian MS algorithm in matrix form

$Z = X$
repeat
   $W = \left(\exp\left(-\frac{1}{2}\|(z_m - x_n)/\sigma\|^2\right)\right)_{nm}$
   $D = \mathrm{diag}\left(\sum_{n=1}^{N} w_{nm}\right)$
   $Q = WD^{-1}$
   $Z = XQ$
until stop
Connected-Components $(\{z_n\}_{n=1}^{N}, \epsilon)$

(d) Gaussian BMS algorithm in matrix form

repeat
   $W = \left(\exp\left(-\frac{1}{2}\|(x_m - x_n)/\sigma\|^2\right)\right)_{nm}$
   $D = \mathrm{diag}\left(\sum_{n=1}^{N} w_{nm}\right)$
   $P = WD^{-1}$
   $X = XP$
until stop
Connected-Components $(\{x_n\}_{n=1}^{N}, \epsilon)$

FIGURE 18.2
Pseudocode for Mean Shift (MS) (left) and Blurring Mean Shift (BMS) (right), in both loop and matrix forms, for the Gaussian kernel. In all cases, the input is a dataset $x_1, ..., x_N \in \mathbb{R}^D$ and a bandwidth $\sigma > 0$. See Section 18.2.1 for the stopping criterion and the Connected-Components threshold distance between points ϵ.

18.2.1 Two Basic Types of Mean-Shift Algorithms: MS and BMS

18.2.1.1 Clustering by Mean-Shift (MS): Find Modes

Here, we declare each mode of p as representative of one cluster, and assign data point x_n to the mode it converges to under the mean-shift iteration, $f^\infty(x_n)$. The algorithm is given in Figure 18.2a for the Gaussian kernel. We can also estimate error bars for each mode from the local Hessian (Carreira-Perpiñán 2000, 2001), given in Equation 18.5, which is related to the local covariance.

Some practical problems need to be solved. Firstly, some points (minima and saddle points) do not converge to modes. It is unlikely that this will happen with a finite sample, but if so such points can be detected by examining the Hessian or by a postprocessing step that checks for small clusters.

Second, the mean-shift iteration is stopped after a finite number of steps, for example when the relative change in the value of x is smaller than a set tolerance `tol` > 0. This means that data points that in theory would converge to the same mode actually stop at numerically slightly different points. A postprocessing step is necessary to merge these into a unique mode. This can be done by finding the connected components* of a graph that has N vertices, one for every convergence point, and has an edge between any pair of vertices lying within a small distance $\epsilon > 0$. The graph need not be explicitly constructed. The user should set `tol` small enough to converge to the modes with good accuracy, while

---

* The Connected-Components algorithm is described in Appendix 1.

limiting the computational cost incurred; and ϵ should be set quite larger than `tol`, but smaller than the distance between different true modes.

18.2.1.2 Clustering by Blurring Mean-Shift (BMS): Smooth the Data

Here, each point x_m of the dataset actually moves to the point $f(x_m)$ given by Equation (18.2). That is, given the dataset $X = \{x_1, \ldots, x_N\}$, for each $x_m \in X$ we obtain a new point $\tilde{x}_m$ by applying one step of the mean-shift algorithm: $\tilde{x}_m = f(x_m)$. Thus, one iteration of blurring mean-shift results in a new dataset $\tilde{X}$ which is a blurred (smoothed) version of X. By iterating this process we obtain a sequence of datasets $X^{(0)}, X^{(1)}, \ldots$ (and a sequence of kernel density estimates $p^{(0)}(x)$, $p^{(1)}(x), \ldots$) where $X^{(0)}$ is the original dataset and $X^{(\tau)}$ is obtained by blurring $X^{(\tau-1)}$ with one mean-shift step (see Figure 18.6).

As will be shown below, Gaussian BMS can be seen as an iterated filtering (in the signal processing sense) that eventually leads to a dataset with all points coincident for any starting dataset and bandwidth. However, before that happens, the dataset quickly collapses into meaningful, tight clusters which depend on σ (see Figure 18.6), and then these point-like clusters continue to move towards each other relatively slowly. A stopping criterion that detects this situation quite reliably is based on whether the entropy of the dataset changes (Carreira-Perpiñán 2006) (a simpler criterion would be to stop when the update to X is small, but this does not always give good clusters). As with MS clustering, a connected-components postprocessing step merges the points into actual clusters. The BMS algorithm is given in Figure 18.2b.

18.2.1.3 Similarities and Differences between MS and BMS

Although both MS and BMS are based on the same mean-shift iteration, they are different algorithms and can produce different clustering results. Specifically, given a value of the bandwidth, the number of clusters resulting from MS and BMS is usually different. However, the collection of clusterings produced over a range of bandwidths can be quite similar.

BMS is quite faster than MS in number of iterations and in runtime, particularly if using the accelerated BMS algorithm (Section 18.2.7), which introduces essentially no approximation error. However, MS (and also BMS) can be considerably accelerated if a small clustering error is tolerated (Section 18.2.7).

In MS, the N optimizations (one for each data point) proceed independently, but they could be done synchronously, as in BMS, without altering the result. However, practically this is wasteful, because the number of iterations required varies considerably among points, and a synchronous scheme would have to run the largest number of iterations. Conversely, it is possible to run BMS with asynchronous iterations, for example, moving points as soon as their update is computed. However, this makes the result dependent on the order in which points are picked, and is unlikely to be faster than the accelerated algorithm described below.

18.2.1.4 Choice of Bandwidth

The fundamental parameter in mean-shift algorithms is the bandwidth σ, which determines the number of clusters. The statistics literature has developed various ways to estimate the bandwidth of a KDE (Silverman 1986; Wand and Jones 1994), mostly in the

1D setting, for example based on minimizing a suitable loss of function (such as the mean integrated squared error), or more heuristic rules (such as making the bandwidth proportional to the average distance of each point to its kth nearest neighbor). While these bandwidth estimators are useful and can give reasonable results, they should be used with caution, because the bandwidth that gives the best density estimate (in a certain sense) need not give the best clustering—clustering and density estimation are, after all, different problems. Besides, clustering is by nature exploratory, and it is best to explore a range of bandwidths. Computationally, this is particularly easy to do in MS (see the scale-space discussion below).

It is also possible to use a different bandwidth σ_n for each point (called *adaptive KDE*), which can help with areas where points are sparse, for example. A good way to do this is with *entropic affinities* (Hinton and Roweis 2003; Vladymyrov and Carreira-Perpiñán 2013), where the user sets a global number of neighbors k and then, for each data point $n = 1, \ldots, N$, the bandwidth σ_n is computed so that point n has a distribution over neighbors with a perplexity (log-entropy) k, that is, each point sets its own bandwidth to have k effective neighbors. One could then vary k to achieve different clusterings. Other ways to construct adaptive KDEs for MS have been proposed (Comaniciu 2003). The mean-shift update with adaptive bandwidths $\sigma_1, \ldots, \sigma_N$ has the form Carreira-Perpiñán (2007):

$$q(n \mid \mathbf{x}) = \frac{p(n \mid \mathbf{x})\sigma_n^{-2}}{\sum_{n'=1}^{N} p(n' \mid \mathbf{x})\sigma_{n'}^{-2}} \quad \mathbf{f}(\mathbf{x}) = \sum_{n=1}^{N} q(n \mid \mathbf{x})\mathbf{x}_n, \tag{18.4}$$

where the $q(m \mid \mathbf{x})$ values are the posterior probabilities $p(m \mid \mathbf{x})$ reweighted by the inverse variance and renormalized (compare with the single-bandwidth Equation (18.3)).

Mean-shift has also been used to track moving objects ("blobs") through a sequence of images (Comaniciu et al. 2003). Since the blobs can change size, using a fixed bandwidth (scale) becomes problematic. Collins (2003) used ideas from scale-space theory (see below) to select a scale that adapts to the blob size. An "image feature" is defined as a point in scale-space where a certain differential operator achieves a local maximum (i.e., a mode) with respect to both space and scale. This mode is then tracked by a two-step mean-shift procedure that convolves in image space with a filterbank of spatial difference-of-Gaussian filters, and convolves in scale space with an Epanechnikov kernel, in an efficient way.

18.2.1.5 Choice of Kernel

Different kernels give rise to different versions of the mean-shift and blurring mean-shift algorithms. Much previous work (including Fukunaga and Hostetler 1975; Cheng 1995) uses the Epanechnikov kernel for computational efficiency, since the kernel evaluations involve only pairs of neighboring points (at distance $< \sigma$) rather than all pairs of points (though the neighbors must still be found at each iteration), and convergence occurs in a finite number of iterations. However, in practice, the Gaussian kernel produces better results than the Epanechnikov kernel (Comaniciu and Meer 2002), which generates KDEs that are only piecewise differentiable and can contain spurious modes (see below).

18.2.1.6 Hierarchical Mean-Shift and Scale-Space Clustering

The behavior of modes and other critical points (minima, saddles) as a function of the bandwidth (or scale) is the basis of *scale-space theory* in computer vision (Witkin 1983;

Koenderink 1984; Lindeberg 1994). Here, one studies the evolution of an image under Gaussian convolution (blurring). If we represent the image as a sum of delta functions centered at the pixel locations and with value equal to the grayscale, convolving this with an isotropic Gaussian kernel of scale σ gives a KDE (with weighted components). As the scale increases, the image blurs, structures in the image such as edges or objects lose detail, and for large scales the image tends to uniformly gray. Some important structures, such as objects, can be associated with modes of the KDE, and the lifetime of a mode—defined by the scale interval between the creation and destruction (merging with another mode) of that mode—is taken as an indication of its importance in the image.

An ideal convolution kernel should blur structure in the image but never create new structure (which would then reflect properties of the kernel rather than of the image). The result would then be a tree of modes, where there are N modes for $\sigma = 0$ (one mode at each component) and modes merge as σ increases until there is a single mode. Unfortunately, this is only true for the Gaussian kernel and only in dimension one (Section 18.2.3). In dimension two, that is, with images, new modes can be created as the scale decreases. However, in practice these creation events seem rare and short-lived, in that the mode created usually merges with another mode or critical point at a slightly larger scale. Thus, they are generally ignored. Computationally, one starts with a mode at each pixel (or data point) for $\sigma = 0$ and then tracks the location of the modes as in a numerical continuation method (Nocedal and Wright 2006), by running mean-shift at a new, larger scale using as initial points the modes at the previous scale. By construction, this results in a tree— although, unlike in agglomerative clustering (see Chapter 2), the resulting set of clusterings at each scale need not be nested.

Another notion of lifetime, topological persistence, has been explored in computational geometry (Edelsbrunner et al. 2002), and used to define a hierarchical mean-shift image segmentation (Paris and Durand 2007; Chazal 2013). The tree of modes has also been proposed in the statistical literature as a tool for visualization of KDEs (Minnotte and Scott 1993), but this is practical only for small datasets in 1D or 2D. The mode tree is sensitive to small changes in the data and gives no way to differentiate between important modes and those caused by, for example, outliers, so it can help to combine several trees constructed by jittering or resampling the original dataset (Minnotte et al. 1998).

18.2.1.7 Matrix Formulation of BMS and Generalizations of BMS

As indicated in Figure 18.2d, we can equivalently write each Gaussian BMS iteration in matrix form (Carreira-Perpiñán 2006) as $\mathbf{X} \leftarrow \mathbf{XP}$ in terms of the random-walk matrix $\mathbf{P} = \mathbf{WD}^{-1}$, an $N \times N$ stochastic matrix with elements $p_{nm} = p(n \mid \mathbf{x}_m) \in (0, 1)$ and $\sum_{n=1}^{N} p_{nm} = 1$. Here, $\mathbf{X} = (\mathbf{x}_1, \ldots, \mathbf{x}_N)$ is the $D \times N$ matrix of data points, $\mathbf{W} = \left(\exp\left((-1/2) \|(\mathbf{x}_m - \mathbf{x}_n)/\sigma\|^2\right)\right)_{nm}$ is a Gaussian affinity matrix, which defines a weighted graph where each $\mathbf{x}_n$ is a vertex, and $\mathbf{D} = \text{diag}\left(\sum_{n=1}^{N} w_{nm}\right)$ is the degree matrix of that graph. This establishes a connection with spectral clustering, which we describe in Section 18.2.4.

Also, we can regard the iteration $\mathbf{X} \leftarrow \mathbf{XP}$ as a smoothing, or more generally a filtering (Taubin 1995, 2000; Desbrun et al. 1999; Carreira-Perpiñán 2008), of (each dimension of) the data $\mathbf{X}$ with an inhomogeneous filter $\mathbf{P}$, where the filter depends on the data and is updated at each iteration as well, hence resulting in a nonlinear filtering. This in turn suggests that one could use other filters constructed as a function $\phi(\mathbf{P})$, where ϕ is a scalar function, so it modifies the spectrum of $\mathbf{P}$. Carreira-Perpiñán (2008) studied several such filters, including explicit, implicit, power, and exponential ones, depending on a step size parameter η, resulting in generalized Gaussian blurring mean-shift algorithms. He gave convergence conditions on ϕ and η and found that the different filters tend to find similar clusters, that is, over a range of bandwidths they can obtain the same clustering (at possibly different σ values). However, their runtime varies widely. Implicit filters (which involve solving a linear system) or power filters (which involve iterating a matrix product) have a strong clustering effect in each iteration, but their iterations themselves are more costly. When one considers both the number of iterations and the cost of each iteration, the method found fastest was a slightly overrelaxed explicit function of the form $\phi(\mathbf{P}) = (1 - \eta)\mathbf{I} + \eta\mathbf{P}$ with $\eta \approx 1.25$. However, its runtime was very close to that of the standard BMS ($\eta = 1$). An interesting extension of this work would be to be able to design the function ϕ so that the resulting generalized BMS algorithm is optimal (in some sense) for clustering.

It is also possible to write MS in matrix form (Figure 18.2c), where we write the random-walk matrix with the symbol $\mathbf{Q}$ to differentiate it from the standard random-walk matrix $\mathbf{P}$, since $q_{nm} = p(\mathbf{x}_n \mid \mathbf{z}_m)$ is defined on two sets of points ($\mathbf{X}$ and $\mathbf{Z}$) while $\mathbf{P}$ is defined on one ($\mathbf{X} = \mathbf{Z}$). However, the matrix form implies that the MS updates for all points are synchronous, which, as mentioned before, is slow.

18.2.1.8 Mapping New Points to Clusters (Out-of-Sample Mapping)

Having run MS or BMS on a dataset, how should we deal with new data points not in the original dataset? The purist option (and the only one for BMS) is to run the clustering algorithm from scratch on the entire dataset (old and new points), but this is computationally costly. A faster option with MS is to use the original KDE and simply run MS on each of the new points, assigning them to the mode they converge to. This reflects the fact that MS clusters not just the points in the dataset, but (implicitly) the whole space. However, this point of view implies no new clusters are created when new points arrive.

18.2.1.9 Advantages and Disadvantages of Mean-Shift Algorithms

The advantages of mean-shift algorithms stem from the nonparametric nature of the KDE: (1) It makes no model assumptions (other than using a specific kernel), unlike Gaussian mixture models or K-means, for example. (2) It is able to model complex clusters having nonconvex shape (although this does not imply that all shapes can be modeled well), unlike K-means (Chapter 1.1). (3) The user need only set one parameter, the bandwidth, which has an intuitive physical meaning of local scale, and this determines automatically the number of clusters. This is often more convenient than having to select the number of clusters explicitly. (4) It has no local minima, thus the clustering it defines is uniquely determined by the bandwidth, without the need to run the algorithm with different initializations. (5) Outliers, which can be very problematic for Gaussian mixtures and K-means, do not overly affect the KDE (other than creating singleton clusters).

Using KDEs and equating modes with clusters as in mean-shift also has some disadvantages. The most important one is that KDEs break down in high dimensions, where the number of clusters changes abruptly from one for large σ to many, with only a minute decrease in σ. Indeed, most successful applications of mean-shift have been in low-dimensional problems, in particular image segmentation (using a few features per pixel, such as color in LAB space). A way of using modes in high-dimensional spaces is the K-modes algorithm described in Section 18.3.

Other possible disadvantages, depending on the case, are as follows. (1) In some applications (e.g., figure-ground or medical image segmentation), the user may seek a specific number of clusters. However, in mean-shift clustering we have no direct control over the number of clusters: to obtain K clusters, one has to search over σ. This is computationally costly (and sometimes not well defined, since the number of clusters might not be a monotonic function of σ). (2) We do not differentiate between meaningful and non-meaningful modes. For example, outliers will typically create their own mode; or, the density in a cluster may genuinely contain multiple modes (especially with clusters that have a nonconvex or manifold structure, as in Figure 18.1d). Some of these problems may be partially corrected by postprocessing the results from mean-shift (e.g., to remove low-density modes and clusters with few points, which are likely outliers), or by the K-modes algorithm. Finally, mean-shift is slow computationally, and this is addressed in Section 18.2.7.

18.2.2 Origins of the Mean-Shift Algorithm

The mean-shift algorithm is so simple that it has probably been discovered many times. In 1975, Fukunaga and Hostetler (1975) were perhaps the first to propose its idea and also introduced the term "mean shift." They derived the blurring version of the algorithm (BMS) for a KDE with the Epanechnikov kernel as gradient ascent on $\log p(\mathbf{x})$ with a variable step size, without proving convergence or giving a stopping criterion. They observed that it could be used for clustering and dimensionality reduction (or denoising), since points converge locally to cluster centroids or medial lines for appropriate values of the bandwidth. Since 1981, the algorithm was also independently known in the statistics literature as "mean update algorithm" (see Thompson and Tapia (1990) pp. 167ff and references therein). The term "blurring process" is due to Cheng (1995), who discussed the convergence of both the blurring (BMS) and the nonblurring (MS) versions of mean-shift. Carreira-Perpiñán (2000), motivated by the problem of finding all the modes of a Gaussian mixture for multivalued regression (Carreira-Perpiñán 2000, 2001), independently rediscovered the algorithm for the Gaussian kernel and proved its convergence for arbitrary covariance matrices (Carreira-Perpiñán 2001). Since the early 2000s, the nonblurring mean-shift received much attention thanks to the work of Comaniciu and Meer (2002), who demonstrated its success in image filtering, image segmentation and later in tracking (Comaniciu et al. 2003). This was followed by work by many researchers in theoretical, computational, and application issues. Algorithms similar to mean-shift have appeared in scale-space clustering (Wilson and Spann 1990; Wong 1993; Chakravarthy and Ghosh 1996; Roberts 1997), in clustering by deterministic annealing (Rose 1998) and in pre-image finding in kernel-based methods (Schölkopf et al. 1999).

18.2.3 Theoretical Results about Mean-Shift Algorithms and Gaussian Mixtures

Although MS and BMS are defined by very simple iterative schemes, they exhibit some remarkable and somewhat counterintuitive properties regarding whether they converge at all, how fast they converge, and the character of the convergence domains. The geometry of the modes of a KDE is also surprising. The relevant literature is scattered over different areas, including computer vision, statistics, and machine learning. We give a summary of results, with particular attention to Gaussian kernels, without proof. More details can be found in the references cited.

18.2.3.1 Geometry of the Modes of a Gaussian Mixture

Intuitively, one might expect that a sum of N unimodal kernels with bandwidth σ as in Equation (18.1) would have at most N modes, and that the number of modes should decrease monotonically as σ increases from zero and the different components coalesce. In general, this is only true for the Gaussian kernel in dimension one. Motivated by scale-space theory, several articles (Babaud et al. 1986; Yuille and Poggio 1986; Koenderink 1984) showed that, in dimension one, the Gaussian kernel is the only kernel that does not create modes as the scale increases. It is easy to see that, in KDEs with non-Gaussian kernels (Epanechnikov, Cauchy, etc.), modes can appear as σ increases (Carreira-Perpiñán and Williams 2003). The creation of modes need not occur often, though, and some kernels (such as the Epanechnikov kernel) are more likely than others to create modes. It is less easy to see, but nonetheless true, that modes can also appear with the Gaussian kernel in dimension two, that is, with images, and thus in any larger dimension, as shown by an example in Lifshitz and Pizer (1990).

The scale-space theory results were restricted to a single bandwidth σ, and also the creation of modes does not necessarily imply that a mixture with N components may have more than N modes. The results for the general case of Gaussian mixtures (GMs) are as follows. Again, there is a qualitative difference between 1D and 2D or more. A GM (with possibly different bandwidths σ_n) can have at most N modes in 1D (Carreira-Perpiñán and Williams 2003) but more than N modes in 2D or above, even if the components are isotropic (Carreira-Perpiñán and Williams 2003). In 2D, if the components have diagonal or full covariance matrices, it is easy to construct examples with more modes than components (Carreira-Perpiñán 2001; Carreira-Perpiñán and Williams 2003). It is far harder to achieve this if all the components are isotropic with equal bandwidth σ, but still possible. This was first shown by a construction suggested by Duistermaat and studied by Carreira-Perpiñán and Williams (2003), consisting of 3 Gaussians in the vertices of an equilateral triangle. For a narrow interval of scales (lifetime), an extremely shallow fourth mode appears at the triangle barycenter. They generalized this construction to dimension D as a regular simplex with a Gaussian in each vertex, that is, a GM with $D+1$ components, one at each vertex of the simplex (Carreira-Perpiñán and Williams 2003). They showed that the number of modes is either $D+1$ (for small scale, modes near the vertices), 1 (for large scale, at the simplex barycenter) or $D+2$ (for a narrow range of intermediate scales, modes at the barycenter, and near the vertices). Interestingly, the lifetime of the mode that is created in the barycenter peaks at $D=698$ and then slowly decreases towards zero as D increases. Small perturbations (but not vanishingly small) of the construction prevent the creation of the extra mode, which suggests that it may be unlikely for isotropic GMs to have more modes than components. However, apart from a few isolated studies, the geometry of the modes of GMs in high dimensions is poorly understood.

As for the location of the modes of a GM, they are in the interior of the convex hull of the data points if the components are isotropic (with possibly different bandwidths σ_n), for any dimension (Carreira-Perpiñán 2001; Carreira-Perpiñán and Williams 2003), as seen in Figure 18.1b. With KDEs, it is easy to see this from the fact that Equation (18.2) is a convex sum if $K' \geq 0$. If the components have diagonal or full covariance matrices, the modes can be outside the convex hull (Carreira-Perpiñán 2001; Carreira-Perpiñán and Williams 2003a, b).

Most work on mean shift and scale space theory tacitly assumes that the modes of a GM are finite in number or at least are isolated. Although sometimes this is claimed to be a consequence of Morse theory, no proof seems to exist for an arbitrary dimension.* Indeed, kernels such as the uniform or triangular kernel do create continuous ridges of modes even in 1D.

18.2.3.2 Convergence of MS

An attractive feature of MS is that it is defined without step sizes, which makes it deterministic given the dataset and bandwidth. However, this is only useful if it converges at all, and whether this is the case depends on the kernel used. Not all kernels K give rise to convergent MS updates, even if they are valid kernels (i.e., they are nonnegative and integrate to one); an example where MS diverges appears in Comaniciu and Meer (2002). For kernels where MS diverges, it is of course possible to devise optimization algorithms that will converge to the modes, for example, by introducing a line search, but we lose the simplicity of the MS iteration. For kernels where MS converges, the convergence is to a mode for most initial points, but in general to a stationary point (minima and saddle points in addition to modes). Here, we review convergence results for MS, with a focus on the most common kernels (Gaussian and Epanechnikov).

With the Epanechnikov kernel, convergence occurs in a finite number of iterations (Comaniciu and Meer 2002). The intuitive reason for this is that the KDE, which is the sum of N kernels, is piecewise quadratic. There is a finite number of "pieces" (each corresponding to a possible subset of the N kernels), and, within one piece, a single Newton iteration would find the maximum (indeed, the MS update coincides with a Newton step for the Epanechnikov kernel (Fashing and Tomasi 2005)).

With the Gaussian kernel, MS also converges, but in an infinite number of iterations, and the convergence rate is generally linear (sublinear or superlinear in limit cases) (Carreira-Perpiñán 2001, 2007). One convenient way to see this is by relating MS with the EM algorithm (McLachlan and Krishnan 2008) (see Chapter 3), as follows.

In general for mixtures of Gaussians (using arbitrary covariance matrices) or other kernels (and thus in particular for KDEs), Gaussian MS is an expectation-maximization (EM) algorithm and non-Gaussian MS is a generalized EM algorithm (Carreira-Perpiñán 2001, 2007). This can be seen by defining a certain dataset and a certain probabilistic model with hidden variables (missing data), deriving the corresponding EM algorithm to maximize the log-likelihood for it, and verifying it coincides with the MS algorithm. For the GM case, the model consists of a constrained mixture of Gaussians, where each component has the given weight, mean, and covariance matrix (constant, data point, and bandwidth for KDEs, respectively), but the whole mixture can be freely (but rigidly) translated. The dataset consists solely of one point located at the origin. Thus, maxima of the likelihood occur whenever the translation vector is such that a mode of the mixture coincides with the origin. The missing data is the index of the mixture component that generated the origin. The resulting E step computes the posterior probabilities $p(n \mid \mathbf{x})$ of Equation (18.3), and the M step maximizes a lower bound on the log-likelihood in closed form giving the MS update. With non-Gaussian kernels, the M step cannot be solved in closed form, and

---

* In dimension one, this is simple to prove assuming infinite differentiabilty (as for the Gaussian kernel), by setting f to the gradient of the mixture in the following result. Let $f(x)$ be an infinitely differentiable real function of $x \in \mathbb{R}$. Then either f is identically zero or its zero crossings are isolated. Indeed, if f is zero in a nonempty interval (a, b), then $\forall x \in (a, b)$ we have $\lim_{h \to 0} (f(x + h) - f(x))/h = 0 = f'(x)$. Repeating this argument for f', f'', etc., we obtain that all derivatives at any $x \in (a, b)$ are zero, hence by Taylor's theorem $f(x) = 0 \ \forall x \in \mathbb{R}$.

the MS update corresponds to a single iteration to solve this M step. (Whether this iteration actually increases the likelihood, leading to convergence, depends on the kernel.)

Viewing MS as an EM algorithm has several consequences. Convergence for Gaussian MS is assured by the EM convergence theorems (McLachlan and Krishnan 2008) (which apply because the Q function in the EM algorithm is continuous), and each iterate increases or leaves unchanged the density. General results for EM algorithms also indicate that the convergence order will typically be linear (with a rate dependent on the ratio of information matrices). Finally, MS can be seen as a bound optimization (see also Fashing and Tomasi (2005)).

The convergence order can be studied in detail (Carreira-Perpiñán 2007) by linearizing the MS update $x \leftarrow f(x)$ of Equation 18.3 around a mode x^*, so the update can be written as $x^{(\tau+1)} - x^* \approx J(x^*)(x^{(\tau)} - x^*)$, where $J(x^*)$ is the $D \times D$ Jacobian of f. The Jacobian of f and the Hessian of p are related as follows:

$$J(x^*) = \frac{1}{\sigma^2}\Sigma(x^*) \quad \nabla^2 p(x^*) = \frac{p(x^*)}{\sigma^2}(J(x^*) - I) \quad \Sigma(x^*) = \sum_{n=1}^{N} p(n \mid x^*)(x_n - x^*)(x_n - x^*)^T$$

(18.5)

where $\Sigma(x^*)$ is the local covariance matrix at x^*, since (from $x^* = f(x^*)$ and the definition of f) the local mean is x^* itself. The eigenvalues of $J(x^*)$ are in $(0, 1)$ and the convergence rate is given by the largest one, with the iterates approaching x^* along the eigenvector associated with the largest eigenvalue (assuming distinct eigenvalues). This shows that the convergence order depends on the bandwidth: it is nearly always linear, approaches superlinear convergence when $\sigma \to 0$ or $\sigma \to \infty$, and converges sublinearly at mode merges. The practically useful cases of MS use an intermediate σ value, for which the rate r of linear convergence (i.e., the ratio of distances to the mode after and before the update) can be close to 1, thus convergence will be slow. The MS iterates smoothly approach the mode along the principal component of the local covariance matrix of the data points, from within the convex hull of the data points (see Figure 18.1b).

18.2.3.3 Other Properties of MS

We focus on Gaussian MS on a KDE, that is, with isotropic covariances of bandwidth σ. As seen in Figure 18.1b, if initialized at points far from a mode, the first MS steps are often large and make considerable progress towards the mode. This is an advatageous property generally observed in alternating optimization algorithms (such as EM). After that, the steps become small, in agreement with the linear convergence rate.

The path followed by the MS iterates has the following properties (illustrated in Figure 18.1b). (1) Near convergence, the path follows the direction of the principal eigenvector of the local covariance at the mode (Carreira-Perpiñán 2007). (2) Each iterate is a convex linear combination of the data points, so the path lies in the interior of the convex hull of the data points (Carreira-Perpiñán 2000, 2007). This is also true of non-Gaussian kernels if they satisfy $K' \geq 0$. (3) The path is smooth in the sense that consecutive steps (consecutive mean-shift vectors $f(x) - x$) always make an angle in $(-\pi/2, \pi/2)$ (Comaniciu and Meer 2002). (4) The mean shift vector $f(x) - x$ is proportional to the gradient of p, so the MS iteration is a gradient step, but the step size is not the best one (i.e., does not maximize p) along the gradient.

In K-means (see Chapter 1), the centroids partition the space into Voronoi cells, which are convex polytopes. In MS, it is harder to characterize the regions that the modes partition the space into, that is, their domains of convergence, which can in fact have surprising properties (Carreira-Perpiñán 2007). In general, they can be curved, nonconvex, and disconnected (for example, one domain can be completely surrounded by another). This allows MS to represent complex-shaped clusters and is an advantage over K-means. A less desirable aspect is that the domain boundaries can show fractal behavior, although this seems confined to cluster boundaries, and could be removed if necessary by postprocessing the clusters. This fractal behavior is due to the iterated MS mapping $\mathbf{f}(\mathbf{x})$, and would not occur if we defined the clusters purely based on flow lines of the gradient.

18.2.3.4 Convergence of BMS

Cheng (1995) proved convergence of blurring mean-shift, as follows. (1) For kernels broad enough to cover the dataset $\mathbf{X}$ (e.g., infinite-support kernels such as the Gaussian) convergence is to a dataset $\mathbf{X}^{(\infty)}$ with all points coincident ($\mathbf{x}_1^{(\infty)} = \cdots = \mathbf{x}_N^{(\infty)}$), regardless of the value of σ. This can be seen by noting that the diameter of the data set decreases at least geometrically. (2) For finite-support kernels and small enough σ, convergence is to several clusters with all points coincident in each of them; the clusters depend on the value of σ.

Another proof can be obtained from the matrix formulation of BMS (Carreira-Perpiñán 2006), since at each iteration the dataset $\mathbf{X}$ is multiplied times a stochastic matrix $\mathbf{P}(\mathbf{X})$. By the Perron-Frobenius theorem [Horn and Johnson (1986), Chapter 8], with broad kernels this will have a single eigenvalue equal to 1 and all other eigenvalues with magnitude less than 1. Since a fixed point verifies $\mathbf{X} = \mathbf{XP}$ then $\mathbf{X} = \mathbf{x}\mathbf{1}^T$ for some $\mathbf{x} \in \mathbb{R}^D$, that is, all points coincide. For nonbroad kernels, the unit eigenvalue is multiple, resulting in multiple clusters where all points coincide.

While, for Gaussian BMS, the dataset converges to a size (therefore variance) zero, its variance need not decrease monotonically, in fact it is easy to construct examples in 1D where it increases at some steps.

Carreira-Perpiñán (2006) completely characterized the behavior with the Gaussian kernel (Gaussian BMS), assuming that the dataset is infinite with a Gaussian distribution. In this case, one can work with distributions rather than finite samples, and the iteration $\mathbf{X} \leftarrow \mathbf{XP}$ can be written as a Gaussian integral and solved in closed form. The result is that the data distribution $p(\mathbf{x})$ is Gaussian after each iteration, with the same mean, and it shrinks towards its mean independently along each principal axis and converges to it with cubic order. Specifically, the standard deviation s along a given principal axis evolves as $s \leftarrow r(s)s$ with $r(s) = 1/(1 + \sigma/s)^2 \in (0, 1)$, where σ is the BMS bandwidth, and s converges to 0 cubically. The reason for this extremely fast convergence is that, since σ is kept constant but the dataset shrinks, effectively σ increases. Thus, at each iteration both s and $1/(1 + \sigma/s)^2$ decrease. Note that the smaller the initial s is, the faster the convergence and so the direction of largest variance (principal component) collapses much more slowly (in relative terms) than all other directions.

This explains the practical behavior shown by Gaussian BMS (see Figure 18.6): (1) clusters collapse extremely fast (in a handful of iterations, for a suitable bandwidth); (2) after a few iterations only the local principal component survives, resulting in temporary linearly shaped clusters (that quickly straighten). These two behaviors make BMS useful for clustering and denoising, respectively.

The same proof technique applies to the generalized Gaussian BMS algorithms that use an update of the form $X \leftarrow X \phi(P(X))$. With a Gaussian dataset, each iteration produces a new Gaussian with a standard deviation (separately along each principal axis) $s \leftarrow |\phi(r(s))| s$. This allows a complete characterization of the conditions and order of convergence in terms of the real function $\phi(r)$, $r \in (0,1)$, instead of a matrix function. Convergence occurs if $|\phi(r)| < 1$ for $r \in (0,1)$ and $\phi(1) = 1$. Depending on ϕ, the convergence order can vary from linear to cubic and beyond (Carreira-Perpiñán 2008).

18.2.4 Relations with Other Algorithms

18.2.4.1 Spectral Clustering

As noted earlier, putting Gaussian BMS in matrix form in terms of a Gaussian affinity matrix W uncovers an intimate relation with spectral clustering (see Chapter 2.2). Each BMS iteration is a product $X \leftarrow XP$ of the data times $P = (p(n \mid x_m))_{nm}$, the stochastic matrix of the random walk in a graph (Chung 1997; Meilă and Shi 2001), which in BMS represents the posterior probabilities of each point under the kernel density estimate (18.1). P is closely related to the matrix $N = D^{-\frac{1}{2}} W D^{-\frac{1}{2}}$ (equivalent to the normalized graph Laplacian) commonly used in spectral clustering, for example, in the normalized cut (Shi and Malik 2000). The eigenvalue/eigenvector pairs (μ_n, u_n) and (λ_n, v_n) of P and N satisfy $\mu_n = \lambda_n$ and $u_n = D^{-\frac{1}{2}} v_n$. In spectral clustering, given σ and X one computes the eigenvectors associated with the top K eigenvalues of N (if K clusters are desired); in this spectral space the clustering structure of the data is considerably enhanced and so a simple algorithm such as K-means can often find the clusters. In BMS, we iterate the product $X \leftarrow XP$. If P were kept constant, this would be the power method (Golub and van Loan 1996) and each column of X would converge to the leading left eigenvector of P (the vector of ones, i.e., a single cluster), with a rate of convergence given by the second eigenvalue $\mu_2 < 1$ (the Fiedler eigenvalue in spectral clustering). However, the dynamics of BMS is more complex because P also changes after each iteration. In practice, P and X quickly reach a quasistable state where points have collapsed in clusters which slowly approach each other and P remains almost constant (at which point BMS is stopped). Thus, BMS can be seen as refining the original affinities into a matrix consisting of blocks of (nearly) constant value and then (trivially) extracting piecewise-constant eigenvectors for each cluster with the power method. With the generalized BMS algorithm, one uses instead the matrix $\phi(P)$, which has the same eigenvectors $v_1, \ldots, v_N$ as P but eigenvalues $\phi(\lambda_1), \ldots, \phi(\lambda_N)$. However, this manipulation of the spectrum of P is performed implicitly, without actually having to compute the eigenvectors as in spectral clustering.

While both spectral clustering and BMS rely on the random-walk matrix P, they differ in several respects. They do not give the same clustering results. In spectral clustering, the user sets the desired number of clusters K and the bandwidth σ (if using Gaussian affinities), while in BMS the user sets only σ and K is determined by this. Computationally, spectral clustering solves an eigenproblem (and then runs K-means), while BMS (especially if using its accelerated version) performs a small number of matrix products (and then runs Connected-Components), thus it is considerably faster.

18.2.4.2 Bilateral Filtering and Nonlinear Diffusion

Many image processing algorithms (for example, for image denoising) operate on *range variables* (intensity, color) defined on the *space variables* (of the image lattice). This includes bilateral filtering (Paris et al. 2008), nonlinear diffusion (Weickert 1998) and others. Mean-shift is basically an iterated, local averaging that operates jointly on range and space, and bears both similarities and differences with those algorithms, which are described in Barash and Comaniciu (2004). For example, in bilateral filtering the spatial component is fixed during iterations, so only the range variables are updated, and a stopping criterion is necessary to prevent excessive smoothing.

18.2.4.3 Other Mean-Shift-Like Algorithms

Mean-shift algorithms appear whenever we have expressions having the form of a sum over data points of a function of squared distances of the parameter $\mathbf{x}$ (whether the latter is a vector or a matrix or a set thereof). Equating the gradient of this expression to zero, we can solve for $\mathbf{x}$ and obtain a fixed-point iteration with the form of a weighted average of the data points, where the weights depend on $\mathbf{x}$. One example mentioned below are Riemannian centers of mass. Another example are Laplacian objective functions, of the form $\sum_{n,m=1}^{N} w_{nm} \|\mathbf{x}_n - \mathbf{x}_m\|^2 = 2\text{trace}(\mathbf{X L X}^T)$, where $\mathbf{X}$ (of $L \times N$) are coordinates of the N data points in a low-dimensional space, $\mathbf{W} = (w_{nm})$ is an affinity matrix, and $\mathbf{L}$ is its graph Laplacian. Alternating optimization over each $\mathbf{x}_n$ can be done with a mean-shift algorithm. Laplacian objectives appear in algorithms for (among other problems) dimensionality reduction and clustering, such as Laplacian eigenmaps (Belkin and Niyogi 2003), the elastic embedding (Carreira-Perpiñán 2010), or spectral clustering (Shi and Malik 2000). However, as noted in Section 18.2.3, the resulting mean-shift iteration need not converge in general. This is the case, for example, when the weights or the kernels can take negative values, as in the elastic embedding. In this case, one should use a line search or a different optimization algorithm altogether.

18.2.5 Extensions of Mean-Shift for Clustering

18.2.5.1 Tracking: Mode Finding over Time

In some applications, the distribution of the data changes over time, and we want to track clusters and their location, in order to predict their location at a future time (assuming the distribution changes slowly). One example is tracking a nonrigid object over a sequence of images, such as a video of a moving car. A robust way to represent the object is by the color histogram of all the pixels within the object. In its simplest form, one can use a region of fixed shape and size but variable location, initialized by the user in the first image. Then, the most likely location of the object in the next image can be defined as the region (near the current region) having the histogram closest to the current histogram. Comaniciu et al. (2003) noted that we can use a differentiable KDE instead of a histogram, and that any differentiable similarity function of two KDEs also has the form (up to constant factors) of a weighted KDE to first order (a good approximation if the object moves slowly). Hence, we can use mean-shift iterations to maximize the (linearized) similarity over the location, thus finding the location where pixels look most like in the previous image's region. The resulting algorithm is fast enough to track several objects in real-time in video, and is robust to partial occlusion, clutter, distractors, and camera motion.

Hall et al. (2006) define a spatiotemporal KDE $p(\mathbf{x}, t)$ with a product kernel $K_{\mathbf{x}}(\mathbf{x} - \mathbf{x}_n; \sigma_{\mathbf{x}}) K_t(t - t_n; \sigma_t)$. The kernel over the spatial variables $\mathbf{x}$ is typically Gaussian and the kernel over the temporal variable t considers only past observations and gives more importance to recent ones. At each new value of t, one finds modes over $\mathbf{x}$ by starting from the modes of the previous time value (as in scale-space theory, but using time rather than scale).

18.2.5.2 Manifold Data

The original mean-shift algorithm is defined on the Euclidean space $\mathbb{R}^D$. Sometimes, the data to be clustered lies on a low-dimensional manifold, so the mean-shift iterates and the modes should also lie on this manifold. In some applications (such as motion segmentation, diffusion tensor imaging, and other computer vision problems) this is a known Riemannian manifold, such as rotation matrices, Grassmann manifolds, or symmetric positive definite matrices (Berger 2003). Subbarao and Meer (2009) extended mean-shift to Riemannian manifolds by defining the squared distances appropriately with respect to the manifold (although this does not result in a proper KDE on the manifold). However, most times the manifold is not known *a priori*. Shamir et al. (2006) extended mean-shift to 3D point clouds in computer graphics applications, by constraining the mean shift steps to lie on the surfaces of a triangulated mesh of the data. The Laplacian K-modes described later uses a different approach to find modes that are valid patterns with data lying on nonconvex manifolds.

18.2.5.3 Graph Data

The mean-shift algorithm operates on data where each data point $\mathbf{x}_n$ is defined by a feature vector in $\mathbb{R}^D$. In some applications, the data is best represented as a weighted graph, where each vertex is a data point $\mathbf{x}_n$, and an edge $(\mathbf{x}_n, \mathbf{x}_m)$ represents a neighborhood relation between a pair of data points, with a real-valued weight $\delta(\mathbf{x}_n, \mathbf{x}_m)$ representing distance (see Chapter 2.2). This allows one to work with data that need not live in a Euclidean space, and can be used to represent distances along a manifold (e.g., by approximating geodesics through shortest paths over the graph).

Some algorithms have been proposed for graph data that are based on a KDE. They work by assigning to each data point a "parent" in the graph, which is another data point in their neighborhood having a higher density (roughly speaking). This results in a forest of directed trees that spans the graph, whose roots are "modes" in the sense of not having any neighboring point with higher density. Each tree corresponds to a cluster. The parent of a data point is defined as the data point that optimizes a criterion based on the KDE and the distances between data points. The crucial aspect is that this optimization is defined only over the N data points rather than over the space $\mathbb{R}^D$ (as happens with K-medoid algorithms, Chapter 4).

One of the earliest approaches is the algorithm of Koontz et al. (1976). This uses as criterion $(p(\mathbf{x}) - p(\mathbf{x}_n))/\delta(\mathbf{x}, \mathbf{x}_n)$, which is a numerical approximation to the directional gradient of the KDE. This is maximized over the data points $\mathbf{x} \in \{\mathbf{x}_1, \ldots, \mathbf{x}_N\}$ that are in a neighborhood of $\mathbf{x}_n$ (e.g., the k nearest neighbors or the points within a distance ϵ). The output of this algorithm can be improved with cluster-merging based on topological persistence (Chazal et al. 2013).

The medoidshift algorithm of Sheikh et al. (2007) is more directly related to mean-shift. It uses as criterion the Riemannian center of mass [sometimes called Fréchet mean (Pennec et al. 2006; Afsari 2011)] $\mathbf{f}(\mathbf{x}) = \arg\min_{\mathbf{x} \in \{\mathbf{x}_1, \ldots, \mathbf{x}_N\}} \sum_{n=1}^{N} w_n \delta(\mathbf{x}, \mathbf{x}_n)$, where $w_n = p(\mathbf{x}^{(\tau)}, \mathbf{x}_n) \propto K(\|(\mathbf{x}^{(\tau)} - \mathbf{x}_n)/\sigma\|^2)$ is evaluated at the current iterate $\mathbf{x}^{(\tau)}$. The algorithm then alternates computing the Riemannian center of mass (over the data points) with updating the weights. We recover mean-shift by using the squared Euclidean distance for δ and optimizing over $\mathbb{R}^D$. Unlike mean-shift (which finds maxima of the KDE), this does not seem to maximize a global objective function, but rather maximizes a local objective at each iteration and data point. The algorithm gives better clusterings and is faster in its "blurring" version, where at every iteration the entire dataset is updated. In matrix form (using the random-walk matrix $\mathbf{P}$ and the matrix $\mathbf{\Delta} = (\delta(\mathbf{x}_n, \mathbf{x}_m))_{nm}$ of pairwise distances, and updating all N points synchronously), this takes the form $\mathbf{X} \leftarrow \arg\min_{\mathbf{x}_1,\ldots,\mathbf{x}_N} (\mathbf{\Delta P})$, where the minimization is columnwise. Hence, medoidshift can be seen as a discrete filter, while the BMS iteration $\mathbf{X} \leftarrow \mathbf{XP}$ is a continuous filter. As in the accelerated BMS, each iteration contracts the graph, reducing the number of points. It is possible to define other variations by using different types of distances δ, such as the local Tukey median in the medianshift algorithm of Shapira et al. (2009).

18.2.5.4 Other Extensions

The mean-shift update as originally proposed (Fukunaga and Hostetler 1975) results from estimating the gradient of a density in two steps: first, one estimates the density with a KDE, and then one differentiates this. Sasaki et al. (2014) directly estimate the gradient of the log-density by using score matching (Hyvärinen 2005) and derive a mean-shift algorithm based on this (by equating the gradient to zero and obtaining a fixed-point iteration). In score matching, one does a least-squares fit to the true log-density gradient using a model. If the model is a linear combination of basis functions (one per data point), one obtains an update that is identical to the mean-shift update of Equation 18.3 but with weighted kernels (although, since these weights can be negative, this fixed-point iteration is not guaranteed to converge). A faster, parametric method is obtained by using fewer basis functions than data points. They observe better results than the original mean-shift with high-dimensional data. Note that, as seen in Equation (18.4), using an adaptive KDE (e.g., with bandwidths obtained using the entropic affinities) also gives a weighted mean-shift update, but the weights obey a different criterion.

Ranking data consist of permutations of a given set of items, and are discrete. Meilă and Bao (2010) extended blurring mean-shift to ranking data by using the Kendall distance rather than the Euclidean distance, and by rounding the continuous average of Equation 18.3 to the nearest permutation after each iteration (so the algorithm stops in a finite number of steps).

Functional data, such as a collection of curves or surfaces, live in an infinite-dimensional space. Mean-shift clustering can be extended to functional data, where it corresponds to a form of adaptive gradient ascent on an estimated surrogate density (Ciollaro et al. 2014).

18.2.6 Extensions of Mean-Shift beyond Clustering: Manifold Denoising

The mean-shift iteration (18.2) or (18.3) is essentially a *local smoothing*, which provides a smooth or denoised version of a data point as a weighted average of its nearby data points. Clustering is just one specific problem that can make use of this smoothing. Here we describe work based on mean-shift for manifold denoising. This was already pointed out in Fukunaga and Hostetler (1975). Consider data lying on a low-dimensional manifold in $\mathbb{R}^D$, but with noise added (as in Figure 18.3). If we run one mean-shift iteration for each data point in parallel, and then replace each point with its denoised version, we obtain a denoised dataset, where points have moved towards the manifold. Repeating this eventually compresses the dataset into clusters, and this is the basis of the BMS algorithm,

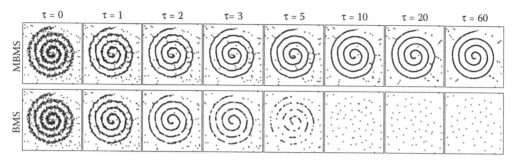

FIGURE 18.3
Denoising a spiral with outliers over iterations ($\tau = 0$ is the original dataset). Each box is the square $[-30, 30]^2$, where 100 outliers were uniformly added to an existing 1 000-point noisy spiral. Algorithms: MBMS and BMS, both with the same bandwidth σ for the Gaussian kernel. MBMS denoises while preserving the spiral structure and ignoring the outliers. BMS locally collapses points onto clusters, destroying the spiral.

although of course it destroys the manifold structure. However, if we stop BMS very early, usually after one or two iterations, we obtain an algorithm that can remove noise from a dataset with manifold structure.

The denoising ability of local smoothing was noted independently in the computer graphics literature for the problem of surface fairing or smoothing. Earlier, the Laplacian had also been used to smooth finite element meshes (Ho-Le 1988). In surface smoothing, 3D point clouds representing the surface of an object can be recorded using LIDAR, but usually contain noise. This noise can be eliminated by Laplacian smoothing, which replaces the 3D location of each point with the average of its neighbors (Taubin 1995, 2000). The neighbors of each point are obtained from a triangulated graph of the cloud, which is usually available from the scanning pattern of LIDAR. Typically, the Laplacian is constructed without the notion of a KDE or a random-walk matrix, and both the values of its entries and the connectivity pattern are kept constant during iterations.

One important problem of BMS (or Laplacian smoothing) is that points lying near the boundary of a manifold or surface will move both orthogonally to and tangentially along the manifold, away from the boundary and thus shrinking the manifold. While this is good for clustering, it is not for surface smoothing because it distorts the object shape, or for manifold learning because it distorts the manifold structure. For example, in the MNIST dataset of handwritten digit images (LeCun et al. 1998), motion along the manifold changes the style of a digit (e.g., slant, thickness). Various approaches, such as rescaling to preserve the object volume, have been proposed in computer graphics (Desbrun et al. 1999). In machine learning, the Manifold Blurring Mean Shift (MBMS) algorithm (Wang and Carreira-Perpiñán 2010) is an extension of BMS to preserve the local manifold structure. Rather than applying to each point the mean-shift vector directly, this vector is corrected to eliminate tangential motion. The mean-shift vector is first projected onto the local tangent space to the manifold at that point (estimated using local PCA on the nearest neighbors of the point), and this projection is removed from the motion. Hence, the motion is constrained to be locally orthogonal to the manifold. Comparing the local PCA eigenvalues between the orthogonal and tangent spaces gives a criterion to stop iterating.

Figure 18.4 shows MNIST images before and after denoising with MBMS. The digits look smoother (as if they had been anti-aliased to reduce pixelation) and easier to read (compare the original 𝒪 1 2 3 4 5 6 7 8 9 vs. the denoised 0 1 2 3 4 5 6 7 8 9),

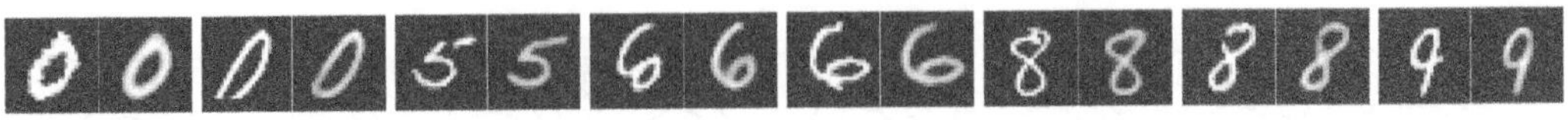

FIGURE 18.4
Sample pairs of (original, MBMS-denoised) images from the MNIST dataset.

and indeed a classifier trained on the denoised data performs better (Wang and Carreira-Perpiñán 2010). While this smoothing homogenizes the digits somewhat, it preserves distinctive style aspects of each digit. MBMS performs a sophisticated denoising (very different from simple averaging or filtering) by intelligently closing loops, removing or shortening spurious strokes, enlarging holes, removing speckle noise and, in general, subtly reshaping the digits while respecting their orientation, slant, and thickness. MBMS has also been applied to the identification of the structure of tectonic faults from seismic data (Grillenzoni 2013).

(Manifold) denoising can be used as a preprocessing stage for several tasks, leading to more robust models, such as dimensionality reduction (Wang and Carreira-Perpiñán 2010), classification (Hein and Maier 2007; Wang and Carreira-Perpiñán 2010), density estimation (Hall and Minnotte 2002), or matrix completion (Wang et al. 2011).

In some algorithms, one can establish a continuum between clustering and dimensionality reduction, with clustering being the extreme case of dimensionality reduction where the manifold dimensionality is zero. This is the case with MBMS, where a tangent space of dimension zero recovers BMS. Another case is spectral clustering and spectral dimensionality reduction, for example, normalized cut (Shi and Malik 2000) versus Laplacian eigenmaps (Belkin and Niyogi 2003): both operate by extracting eigenvectors from the graph Laplacian, but in spectral clustering, the number of eigenvectors is the number of clusters, while in spectral dimensionality reduction, one cluster is assumed and the eigenvectors correspond to degrees of freedom along the manifold defined by the cluster.

18.2.7 Accelerated Mean-Shift Algorithms

Computationally, mean-shift algorithms are slow, because their complexity is quadratic on the number of points (which would limit their applicability to small data sets), and many iterations may be required to converge (for MS). Significant work has focused on faster, approximate mean-shift algorithms, using a variety of techniques such as subsampling, discretization, search data structures (Samet 2006), numerical optimization (Nocedal and Wright 2006), and others, which we review below. Some of them are specialized to image segmentation while others apply to any data. A practical implementation of a fast mean-shift algorithm could combine several of these techniques. In addition, although it does not seem to have been studied in detail, mean-shift algorithms benefit from embarrasing parallelism, since the iterations for each point proceed independently from the rest of the points.

18.2.7.1 Accelerating Mean-Shift (MS)

With the Epanechnikov kernel, mean-shift converges in a finite number of iterations. If the neighborhood structure of the data is known *a priori*, as is the case in image segmentation (Comaniciu and Meer 2002), MS can be very efficient without any approximations, since the iteration for each point involves only its neighbors.

With the Gaussian kernel and with arbitrary data, for which the neighborhood structure is not known and varies over iterations, the runtime of MS can be large, particularly with large datasets, and much work has tried to accelerate Gaussian MS. Computationally, MS has two bottlenecks (Carreira-Perpiñán 2006): (1) Accurately converging to a mode can require many iterations, since its convergence is typically linear and can approach sublinearity in regions where two modes are close. (2) Each iteration is linear on the number of points N, because it is an average of the data. Thus, running MS for the entire dataset is $\mathcal{O}(IDN^2)$, where I is the average number of iterations per point and D the dimension. For example, in image segmentation, typical values are 10 to 100 for I (depending on the accuracy sought and the bandwidth σ), $D \leq 5$ (higher if using texture features) and thousands to millions for N (the number of pixels). Acceleration algorithms should attack either or both bottlenecks, while keeping a low approximation error, that is, producing a clustering (assignment of points to modes found) that is close to that of the naive MS—otherwise one would really be running a different clustering algorithm.

As for accelerating the convergence, the obvious answer is using Newton's method (with a modified Hessian to ensure ascent), because it has quadratic convergence, computing the Hessian of a Gaussian mixture is simple (Carreira-Perpiñán 2000) and Newton iterations are typically not much costlier than those of MS. The reason is that, for the low-dimensional problems for which MS is best suited, computing the Newton direction (which involves solving a $D \times D$ linear system) is not costly compared to computing the gradient or Hessian themselves. However, Newton's method does have other problems with MS: (1) it introduces user parameters such as step sizes or damping coefficients; (2) it need not be effective far from a mode, where the Hessian is undefined; (3) since Newton's method is very different from MS, it can make a point converge to a different mode than under MS. A solution to this is to use an MS–Newton method (Carreira-Perpiñán 2006) that starts with MS iterations (which often suffice to move the iterate relatively close to a mode) and then switches to Newton's method.

As for reducing the cost of one iteration below $\mathcal{O}(N)$, the most obvious strategy is to approximate the average in each iteration with a small subset of the data, namely, the points closest to the current iterate. This can be done by simply ignoring faraway points in the average, which introduces an approximation error in the modes, or by updating faraway points infrequently as in the sparse EM algorithm of Carreira-Perpiñán (2006), which guarantees convergence to a true mode. Unfortunately, either way this requires finding nearest neighbors, which is itself $\mathcal{O}(N)$ unless we know the neighborhood structure *a priori*. And, if the bandwidth is relatively large (which is the case if we want a few clusters), then a significant portion of the data have non-negligible weights in the average. Another approach is to use N-body methods such as the (improved) fast Gauss transform (Greengard and Strain 1991; Yang et al. 2003) or various tree structures (Wang et al. 2007). These can reduce the $\mathcal{O}(N^2)$ cost of one iteration for every data point to $\mathcal{O}(N)$ or $\mathcal{O}(N \log N)$, respectively. However, this implies we run synchronous iterations (all points move at once), and does not scale to high dimensions (because the data structures involved grow exponentially with the dimension), although it is generally appropriate for the low dimensions used in image segmentation.

A final approach consists of eliminating many of the iterations for many of the points altogether, in effect predicting what mode a given point will converge to. This cannot be simply done by running MS for a small subset of the points and then assigning the remaining points to the mode of their closest point in the subset, because mean-shift clusters have complex shapes that significantly differ from a nearest-neighbor clustering, and the clustering error would be large. A very effective approach for image segmentation is the spatial discretization approximation of Carreira-Perpiñán (2006). This is based on the fact that the trajectories followed by different points as they converge towards a given mode collect close together (see Section 18.2.3 and Figure 18.1b). Thus, we can run the naive MS for a subset of the points, keeping track of the trajectories they followed. For a new point, as soon as we detect that its iterate is close to an existing trajectory, we can stop and assign it to that trajectory's mode. With image segmentation (even if the number of features, and thus dimensions, is large), the trajectories can be coded by discretizing the image plane into subpixel cells and marking each cell the first time an iterate visits it with the mode it converges to. Since nearly all these cells end up empty, the memory required is small. This reduces the average number of iterations per point I to nearly 1 with only a very small clustering error (Carreira-Perpiñán 2006). This error can be controlled through the subpixel size.

Yuan et al. (2010) show that if $\mathbf{y} = \mathbf{f}(\mathbf{x})$ is the result of applying a mean-shift step to a point $\mathbf{x}$, then all the points $\mathbf{z}$ within a distance $\|\mathbf{y} - \mathbf{x}\|$ from $\mathbf{y}$ have a KDE value $p(\mathbf{z}) \geq p(\mathbf{x})$. Not all these points $\mathbf{z}$ actually converge to the same mode as $\mathbf{x}$, however many do. A fast mean-shift algorithm results from not running mean-shift iterations for all those points and assigning them to the same mode as $\mathbf{x}$. Hence, as in the discretization approach above, only a few data points actually run mean-shift iterations. The approximation error was not controlled in Yuan et al. (2010), although one could use instead a distance $r\|\mathbf{y} - \mathbf{x}\|$ and use $r \in [0, 1]$ to control it.

Paris and Durand (2007) combine several techniques to accelerate mean-shift image segmentation, including spatial discretization and 1D convolutions (classifying most pixels without iterating, as above), and construct a hierarchy of clusterings based on the notion of topological persistence from computational geometry (Edelsbrunner et al. 2002). The algorithm is fast enough to segment large images and video, although it is not clear how good an approximation it is to the true mean-shift.

As the dimension of the data points increases, searching for nearest neighbors (to compute the mean shift update in Equation 18.2 or 18.3 with a truncated kernel) becomes a bottleneck. One can then use approximate nearest-neighbor algorithms. For example, Locality-Sensitive Hashing (LSH) (Andoni and Indyk 2008) is used in Georgescu et al. 2003.

18.2.7.2 Accelerating Blurring Mean-Shift (BMS)

A simple and very effective acceleration that has essentially zero approximation error was given in Carreira-Perpiñán (2006). It essentially consists of interleaving connected-components and blurring steps. The fact that BMS collapses clusters to a single point suggests that as soon as one cluster collapses we could replace it with a single point with a weight porportional to the cluster's number of points. This will be particularly effective if clusters collapse at different speeds, which happens if they have different sizes, as predicted in Section 18.2.3; for example, see Figure 18.6. The total number of iterations remains the same as for the original BMS but each iteration uses a dataset with fewer points and is thus faster. Specifically, the Gaussian kernel density estimate is now $p(\mathbf{x}) = \sum_{n=1}^{N} \pi_n p(\mathbf{x} \mid n)$ where $p(\mathbf{x} \mid n) = N(\mathbf{x}; \mathbf{x}_n, \sigma^2 \mathbf{I})$ and the posterior probability is $p(n \mid \mathbf{x}_m) = p(\mathbf{x}_m \mid n)\pi_n / p(\mathbf{x}_m)$ as in Carreira-Perpiñán (2000). At the beginning $\pi_n = 1/N \; \forall n$ and when clusters m and n merge then the combined weight is $\pi_m + \pi_n$. Using the matrix notation of Section 18.2.1, we have $w_{nm} \propto p(\mathbf{x}_m \mid n)$, $\mathbf{\Pi} = \text{diag}(\pi_n)$, $d_m = \sum_{n=1}^{N} w_{nm} \pi_n = p(\mathbf{x}_m)$ and $(\mathbf{\Pi} \mathbf{W} \mathbf{D}^{-1})_{nm} = p(n \mid \mathbf{x}_m)$. This can be proven to be equivalent to the original BMS.

The reduction step where coincident points are replaced with a single point can be approximated by a Connected-Components step where points closer than ϵ are considered coincident, where $\epsilon > 0$ takes the same value as in the final Connected-Components step of BMS (Figure 18.2b). Thus, ϵ is the resolution of the method (below which points are indistinguishable), and while BMS applies it only after having stopped iterating, the accelerated version applies it at each iteration. Hence, the accelerated BMS algorithm alternates a Connected-Components graph contraction with a BMS step.

Experimentally with image segmentation, a remarkable result arises: the runtime of this accelerated BMS is almost constant over bandwidth values, unlike for both MS and BMS, corresponding to about 4–5 BMS iterations, and is dominated by the first few iterations, where almost no merging occurs. This means a speedup factor of 2–4 over BMS and 5–60 over MS (Carreira-Perpiñán 2006).

18.3 *K*-Modes and Laplacian *K*-Modes Algorithms

As discussed earlier, equating modes with clusters as in mean-shift implies we lose direct control on the number of clusters and do not differentiate between meaningful and nonmeaningful modes. Recently, a *K*-modes algorithm (Carreira-Perpiñán and Wang 2013) has been proposed that addresses these problems.[*] The (Gaussian) *K*-modes algorithm takes *both K* and σ as user parameters and maximizes the objective function

$$\max_{\mathbf{Z},\mathbf{C}} \frac{1}{N} \sum_{k=1}^{K} \sum_{n=1}^{N} z_{nk} G\left(\left\|\frac{\mathbf{x}_n - \mathbf{c}_k}{\sigma}\right\|^2\right) \quad \text{s.t. } \mathbf{Z}\mathbf{1}_K = \mathbf{1}_N, \ \mathbf{Z} \in \{0,1\}^{NK} \quad (18.6)$$

over the cluster centroids $\mathbf{c}_k \in \mathbb{R}^D$ and the assignments of points to clusters $z_{nk} \in \{0,1\}$, where G is the Gaussian kernel. For a given assignment $\mathbf{Z}$, this can be seen as the sum of a KDE as in Equation 18.1 but separately for each cluster. Thus, a good clustering must move centroids to local modes, but also define K separate KDEs. This naturally combines the idea of clustering through binary assignment variables (as in *K*-means or *K*-medoids) with the idea that, for suitable bandwidth values, high-density points are representative of a cluster (as in mean-shift). As a function of the bandwidth σ, the *K*-modes objective function becomes *K*-means for $\sigma \to \infty$ and a version of *K*-medoids for $\sigma \to 0$. The training algorithm alternates an assignment step as in *K*-means, where each data point is assigned to its closest centroid, with a mode-finding step as in mean-shift, but only for each centroid $\mathbf{c}_k$ in the KDE defined by its current cluster (rather than for each data point). A more robust, homotopy-based algorithm results from starting with $\sigma = \infty$ (i.e., *K*-means) and gradually decreasing σ while optimizing the objective for each σ. The computational complexity of this algorithm is the same as for *K*-means ($\mathcal{O}(DNK)$ per iteration), although it requires more iterations, and is thus much faster than mean-shift ($\mathcal{O}(DN^2)$ per iteration).

K-modes obtains exactly K modes even if the KDE of the data has more or fewer than K modes, because it splits the data into K KDEs. Using the homotopy-based algorithm, it tends to track a major mode in each cluster and avoid outliers. Thus, *K*-modes can work well even in high dimensions, unlike mean-shift.

---

[*] There exists another algorithm called "*K*-modes" (Huang 1998; Chaturvedi et al. 2001), but defined for categorical data, rather than continuous data.

The fundamental disadvantage of K-modes is that the clusters it defines are convex, as with K-means, since they are the Voronoi cells defined by the K modes. This is solved in the Laplacian K-modes algorithm (Wang and Carreira-Perpiñán 2014), which minimizes the objective function

$$\min_{\mathbf{Z},\mathbf{C}} \quad \frac{\lambda}{2} \sum_{n,m=1}^{N} w_{nm} \|\mathbf{z}_n - \mathbf{z}_m\|^2 - \frac{1}{N} \sum_{k=1}^{K} \sum_{n=1}^{N} z_{nk} G\left(\left\| \frac{\mathbf{x}_n - \mathbf{c}_k}{\sigma} \right\|^2 \right) \quad \text{s.t. } \mathbf{Z}\mathbf{1}_K = \mathbf{1}_N, \ \mathbf{Z} \geq \mathbf{0}.$$

(18.7)

This relaxes the hard assignments of K-modes to soft assignments (so each point may belong to different clusters in different proportions) and adds a term to the objective with a weight $\lambda \geq 0$ that encourages neighboring points to have similar soft assignments. This term can be equivalently written as $\lambda \operatorname{trace}(\mathbf{Z}^T \mathbf{L} \mathbf{Z})$ in terms of the graph Laplacian $\mathbf{L} = \mathbf{D} - \mathbf{W}$ constructed from an affinity matrix $\mathbf{W}$, where $\mathbf{D} = \operatorname{diag}(\sum_{m=1}^{N} w_{nm})$ is the degree matrix. The training algorithm is as with K-modes, but the assignment step becomes a convex quadratic program on $\mathbf{Z}$, which can be solved efficiently with a variety of solvers.

Laplacian K-modes becomes the K-modes algorithm with $\lambda = 0$, and a *Laplacian K-means* algorithm with $\sigma = \infty$. The introduction of the Laplacian term allows the clusters to be nonconvex, as happens with other Laplacian-based algorithms such as spectral clustering (see Chapter 2.2). However, here we solve a quadratic program rather than a spectral problem, and obtain a KDE, centroids (= modes) and soft assignments of points to clusters (unlike spectral clustering, which only returns hard assignments).

An out-of-sample mapping to predict soft assignments $\mathbf{z}(\mathbf{x})$ for a test point $\mathbf{x}$ not in the original training set can be obtained by augmenting 18.7 with $\mathbf{x}$ and solving but keeping the centroids and the soft assignments for the training points fixed. The result equals the projection on the simplex of the average of two terms: the average of the assignments of $\mathbf{x}$'s neighbors, and a term dependent on the distances of $\mathbf{x}$ to the centroids (Wang and Carreira-Perpiñán 2014). The assignment $z_k(\mathbf{x})$ can be readily interpreted as a posterior probability $p(k \mid \mathbf{x})$ of $\mathbf{x}$ belonging to cluster k. Hence, Laplacian K-modes can be seen as incorporating *nonparametric posterior probabilities* into mean-shift (or spectral clustering). The usual way of obtaining posterior probabilities in clustering is by using a mixture model, for which an EM algorithm that maximizes the likelihood is derived (see Chapter 3.1). In contrast to this parametric approach, in Laplacian K-modes the assignments optimize an objective function that is designed specifically for clustering, unlike the likelihood; the clusters are not obliged to follow a model (such as Gaussian); and, consequently, the optimization algorithm does not depend on the particular type of clusters (unlike the EM algorithm, which depends on the cluster model). These soft assignments or posterior probabilities are helpful to estimate the uncertainty in the clustering or for other uses. For example, in image segmentation they can be used as a smooth pixel mask for image matting, conditional random fields, or saliency maps.

18.3.1 Cluster Representatives: Means, Modes, and Medoids

In centroid-based clustering algorithms, each cluster is associated with a centroid: the cluster mean in K-Means (Chapter 3), an exemplar (data point) in K-Medoids (Chapter 4), and the KDE mode in mean-shift, K-modes, and Laplacian K-modes. A desirable property of centroids, which makes them interpretable, is that they should be valid patterns and be

TABLE 18.1

Comparison of Properties of Different Clustering Algorithms

	K-Means	K-Medoids	Mean-Shift	Spectral Clustering	K-Modes	Laplacian K-Modes	Gaussian Mixture, EM
Centroids	Likely invalid	"Valid"	"Valid"	N/A	Valid	Valid	Likely invalid
Nonconvex clusters	No	Depends	Yes	Yes	No	Yes	To some extent
Density	No	No	Yes	No	Yes	Yes	Yes
Assignment	Hard	Hard	Hard	Hard	Hard	Soft, nonparam.	Soft, param.

representative of their cluster ("look" like a typical pattern in the cluster). A well-known disadvantage of K-means is that the mean of a nonconvex cluster need not be a valid pattern, since it may lie in a low-density area. This is often the case with clusters with manifold structure. For example, given a sequence of rotated digit-1 images ⁄ ⁄ ∣ ∣ ∖ ∖, its mean ⼁ is not a valid digit-1 image itself. In K-medoids, exemplars (data points) are valid patterns by definition, although when the data is noisy, individual exemplars may look somewhat atypical (see the manifold denoising discussion in Section 18.2.6). Modes can strike an optimal tradeoff, in being typical, valid, and yet denoised representatives, as follows. First, modes are by definition on high-density areas, so in this sense they are representative of the KDE and cluster. Second, a mode can be characterized as a scale-dependent, local average of the data. Indeed, a mode $\mathbf{x}^*$ is a maximum of the density p, so its gradient at $\mathbf{x}^*$ is zero, and this implies that $\mathbf{x}^*$ equals the weighted average of the data in the sense of Equations 18.2 and 18.3 (where $\mathbf{x}^* = \mathbf{f}(\mathbf{x}^*)$), as seen in Section 18.2. As a function of the scale, a mode spans a continuum between equaling the regular mean of the whole data ($\sigma = \infty$) and equaling any individual data point ($\sigma = 0$). With an intermediate bandwidth in the KDE, they are local averages of the data and so can remove noise that affects each individual exemplar, thus being even more prototypical than actual data points. In this sense, (Laplacian) K-modes achieves a form of intelligent denoising similar to that of MBMS (Section 18.2.6).

Table 18.1 compares several clustering algorithms in terms of whether their centroids are valid patterns, whether they can model nonconvex clusters, whether they estimate a density, and whether the cluster assignments are hard or soft.

18.4 Examples, Applications, Software

We illustrate MS and BMS in the clustering application where they have been most effective (image segmentation), and point out clustering applications where they are not so effective (high-dimensional data or data with manifold structure), as well as applications beyond clustering where MS and BMS have been used. Matlab code for most of the algorithms described may be obtained from the author.

18.4.1 Image Segmentation

MS and BMS are most effective with low-dimensional data, and its most successful application is in image segmentation (Comaniciu and Meer 2002). They are applied as follows.

Given an image, we consider each pixel as a data point in a space consisting of two types of features: *spatial features* (the i and j location of a pixel) and *range features* (the intensity I or grayscale value, the color values, texture features, etc., of the pixel). The range features are scaled to span approximately the range of the spatial features. This way, all features and the bandwidth have pixel units. For example, for the image of Figure 18.5, we rescale the original intensity values to the range $[0, 100]$, so a feature vector $(4, 13, 80)$ would correspond to the pixel located at coordinates $(4, 13)$, which has an intensity equal to 80% of the maximum intensity (white). The precise scaling will affect the clustering and should be done carefully. Using spatial features is beneficial because they introduce spatial coherence (nearby pixels tend to belong to the same cluster), although sometimes only the range features are used. One should use a perceptually uniform color space such as LAB rather than the RGB space, so that Euclidean distances approximately match perceptual differences in color (Forsyth and Ponce 2003).

Figure 18.5 shows an example with a grayscale image of 50×50 pixels. Thus, the dataset contains $N = 2\,500$ points in 3D (location i, j, and intensity I). Figure 18.5 shows the result with MS while Figure 18.6 shows the result with BMS. As noted earlier, MS and BMS generally give similar clusterings over a range of bandwidths.

With the Gaussian kernel, reasonable clusterings arise for bandwidths around one-fifth of the image size (vertical or horizontal), although one should explore a range of bandwidths. In some computer vision applications, MS is sometimes used with a very small bandwidth in order to oversegment an image into uniform patches which are then further processed.

18.4.2 Cases When Mean-Shift Clustering Does Not Work Well

Figure 18.7 illustrates two situations where MS (and BMS) do not produce good clusterings. The first one is data having manifold structure (of lower dimension than the feature space), as in the 1D spirals in a 2D space shown in the left panel. The KDE will have either many modes spread over the data manifolds (for smaller bandwidths) or few modes that mix manifolds (for larger bandwidths). No bandwidth value will cluster the spirals correctly. The Laplacian K-modes algorithm is able to cluster the spirals correctly, while estimating a reasonable KDE and mode for each spiral (Wang and Carreira-Perpiñán 2014).

The second situation is high-dimensional data, as in the MNIST handwritten digits (LeCun et al. 1998), where each data point is a grayscale image of 28×28 pixels (i.e.,

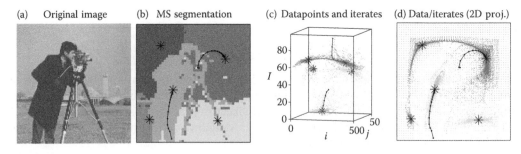

FIGURE 18.5
(a) Test image (cameraman 50×50 grayscale). (b) Gaussian MS segmentation with bandwidth $\sigma = 8$ pixels, resulting in five clusters, with the respective modes marked $*$. (c) The dataset (pixels) and all the MS iterates for all starting points in (i, j, I) space. For example, the cluster near the bottom corresponds to the photographer's body. (d) Projection of the middle plot on (i, j) space. The paths for two starting pixels are shown in plots (b–d).

$D = 784$ feature dimensions). In high-dimensional spaces, MS tends to produce either a single mode (for larger bandwidths), which is uninformative, or many modes (for smaller bandwidths), which often sit on small groups of points or outliers. In the results in Figure 18.7 (right panel), the MS bandwidth was tuned to produce $K = 10$ clusters, and K-means and Laplacian K-modes were run with $K = 10$. Since the data forms nonconvex clusters, K-means produces centroids that average distant data points, of possibly different digit classes, and thus do not look like digits. Again, Laplacian K-modes can partition the data into exactly K clusters, while finding a mode for each that looks like a valid digit and is more representative of its class (Wang and Carreira-Perpiñán 2014).

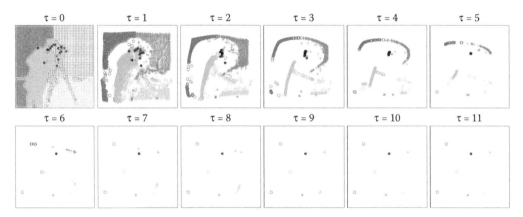

FIGURE 18.6
Sequence of datasets $\mathbf{X}^{(0)}, \dots, \mathbf{X}^{(11)}$ obtained by Gaussian BMS for the `cameraman` image of Figure 18.5 with bandwidth $\sigma = 6$, resulting in seven clusters. We show the 2D projection on the spatial domain of every data point (pixel), colored by cluster, as in Figure 18.5d. Note: (1) points very quickly move toward a centroid and collapse into it (clustering property); (2) for each cluster-to-be, the local direction of maximum variance collapses much more slowly than the lower-variance directions, producing linearly shaped clusters (denoising property) that straighten and shorten. The BMS stopping criterion stopped BMS at iteration $\tau = 11$, where the dataset consists of seven clusters of coincident points. If not stopping, these clusters would keep moving and eventually merge into a single cluster.

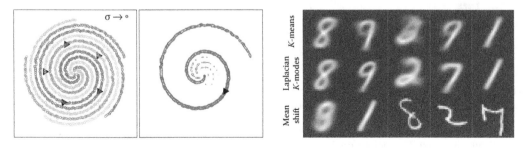

FIGURE 18.7
Left: clustering of five spirals using Laplacian K-modes ($K = 5$, $\sigma = 0.2$, the circle at the top right corner has a radius of σ), with modes denoted by ▶, and contours of the KDE of the one of the clusters. Right: a subset of the ten centroids obtained by K-means, Laplacian K-modes and mean shift for MNIST data.

18.4.3 Multivalued Regression and Inversion

In (univalued) regression, we assume there exists a mapping $f: \mathbf{x} \in \mathbb{R}^d \rightarrow \mathbf{y} \in \mathbb{R}^D$ which assigns a unique vector $\mathbf{y}$ to each possible input $\mathbf{x}$ (i.e., f is a function in the mathematical sense). We estimate f given a training set of pairs $\{(\mathbf{x}_n, \mathbf{y}_n)\}_{n=1}^N$. This is the usual regression setting. In *multivalued regression*, the mapping f assigns possibly multiple vectors $\mathbf{y}_1, \mathbf{y}_2, \ldots$ to each input $\mathbf{x}$. A classical example is learning the inverse of a (univalued) mapping g. For example, the inverse of $g(y) = y^2$ is $f(x) = \pm\sqrt{x}$, which has 0, 1, or 2 outputs depending on the value of x (smaller, equal, or greater than zero, respectively). Carreira-Perpiñán (2000, 2001, 2004) proposed to represent a multivalued mapping $f: \mathbf{x} \in \mathbb{R}^d \rightarrow \mathbf{y} \in \mathbb{R}^D$ by the modes of the conditional distribution $p(\mathbf{y} \mid \mathbf{x})$. Hence, in this case, we apply mean-shift to a *conditional distribution* $p(\mathbf{y} \mid \mathbf{x})$. One advantage of this approach is that the number of inverses is selected automatically for each value of $\mathbf{x}$; essentially, it is the number of "clusters" found by mean-shift in $p(\mathbf{y} \mid \mathbf{x})$. Once the modes have been found, one can also estimate error bars for them from the local Hessian at each mode (Carreira-Perpiñán 2000, 2001). The conditional distribution may be obtained from a Gaussian KDE or Gaussian mixture $p(\mathbf{x}, \mathbf{y})$ for the joint distribution of $\mathbf{x}$ and $\mathbf{y}$ (since the marginal and conditional distributions of any Gaussian mixture are also Gaussian mixtures), but it can also be learned directly, for example, using mixture density networks or particle filters (Bishop 2006).

Figure 18.8 illustrates two examples. On the left panels, we learn the inverse kinematics mapping of a robot arm (Qin and Carreira-Perpiñán 2008a, b). The forward kinematics mapping $\mathbf{x} = \mathbf{g}(\theta)$ is univalued and gives the position $\mathbf{x}$ in workspace of the arm's end-effector given the angles θ at the joints. The inverse mapping $\theta = \mathbf{f}(\mathbf{x})$ is multivalued, as shown by the elbow-up and elbow-down configurations. In this case, the conditional distribution $p(\theta \mid \mathbf{x})$ was learned using a particle filter. On the right panel, we learn to map speech sounds to tongue shapes (articulatory inversion) (Qin and Carreira-Perpiñán 2010). The American English/ɹ/sound (as in "<u>r</u>ag" or "<u>r</u>oll") can be produced by one of two tongue shapes: bunched or retroflex. The figure shows, for a specific/ɹ/sound, three modes

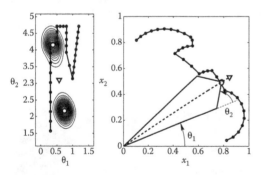
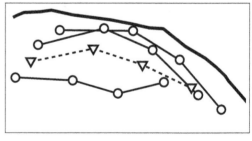

FIGURE 18.8

Left two panels: illustration for a 2D robot arm of using conditional modes to represent multiple inverses. The plots correspond to the joints' angle space $\theta = (\theta_1, \theta_2)$ and the end-effector workspace $\mathbf{x} = (x_1, x_2)$ where we want to reach the point near $(x_1, x_2) = (0.8, 0.5)$ (which is part of the desired workspace trajectory, the solid line with circular markers). That point $\mathbf{x}$ can be reached by two joint angle configurations, elbow up/down, which correspond to the two modes of the conditional distribution $p(\theta \mid \mathbf{x})$ (whose contours are shown). Note that the mean in θ-space (marked by a triangle) does not correspond to a valid inverse (dashed arm). Right panel: profile view of the palate (black outline) and several tongue shapes corresponding to the modes (solid lines) and mean (dashed line) of the conditional distribution for a speech sound /ɹ/ in the utterance "<u>r</u>ag."

in tongue shape space (two bunched and one retroflex). In both inverse kinematics and articulatory inversion, using the conditional mean (shown as a dotted line), or equivalently doing univalued least-squares regression, leads to invalid inverses.

18.4.4 Other Applications

Algorithms based on mean-shift have also been applied to video segmentation (DeMenthon 2002; Wang et al. 2004; Paris and Durand 2007), image denoising and discontinuity preserving smoothing (Comaniciu and Meer 2002), and to object tracking (Collins 2003; Comaniciu et al. 2003), among other problems, as well as for manifold and surface denoising (as described in Section 18.2.6).

18.5 Conclusion and Open Problems

Mean-shift algorithms are based on the general idea that locally averaging data results in moving to higher-density, and therefore more typical, regions. Iterating this can be done in two distinct ways, depending on whether the dataset itself is updated: mode finding (MS), or smoothing (BMS), both of which can be used for clustering, but also for manifold denoising, multivalued regression, and other tasks.

In practice, mean-shift algorithms are very attractive because—being based on nonparametric kernel density estimates—they make few assumptions about the data and can model nonconvex clusters. The user need only specify the desired scale of the clustering but not the number of clusters itself. Although computationally slow, mean-shift algorithms can be accelerated and scale to large datasets. They are very popular with low-dimensional data having nonconvex clusters, such as image segmentation. The (Laplacian) K-modes algorithms remain nonparametric but still perform well with high-dimensional data and outliers.

Mean-shift is intimately related to kernel density estimates and Gaussian mixtures, and to neighborhood graphs. The graph Laplacian or random-walk matrix $\mathbf{P}$ arises as a fundamental object in clustering and dimensionality reduction that encapsulates geometric structure about a dataset. Although not usually seen this way, MS/BMS algorithms are essentially based on $\mathbf{P}$, and they alternate between smoothing the data with $\mathbf{P}$ (a power iteration), with possibly modified eigenvalues, and updating $\mathbf{P}$ itself.

Some directions for future research are as follows. (1) Understanding the geometry of Gaussian mixtures and its modes in high dimensions. (2) Finding a meaningful extension of (blurring) mean-shift to directed graphs. (3) Designing, or learning, random-walk or Laplacian matrices that are optimal for clustering in some sense. (4) Finding a meaningful definition of clusters that is based on "bumps" of a kernel density estimate, that is, distinct regions of high probability rather than modes.

Acknowledgments

I thank Weiran Wang and Marina Meilǎ for comments on this chapter.

Connected-Components Algorithm

Consider an undirected graph $G = (V, E)$ with vertex set V and edge set E. A connected component of G is a maximal subset of V such that every pair of vertices in it can be connected through a path using edges in E (Cormen et al. 2009). Hence, the vertex set V can be partitioned into connected components. The connected component of a given vertex $v \in V$ can be found by depth-first search (DFS), which recursively follows edges adjacent to v until all vertices reachable from v have been found. This can be repeated for the remaining vertices until all connected components have been found, with a total runtime $\mathcal{O}(|V| + |E|)$.

Connected-Components is a clustering algorithm in its own right. To use it, one provides a matrix of distances d_{nm} between every pair of vertices (data points x_n and x_m) and a threshold $\epsilon > 0$, and defines a graph having as vertices the N data points and as edges (x_n, x_m) if $d_{nm} < \epsilon$ for $n, m = 1, \ldots, N$ (this is sometimes called an ϵ-ball graph). The connected components of this graph give the clusters.

Connected-Components gives poor clustering results unless the clusters are tight and clearly separated. However, it can be reliably used as a postprocessing step with mean-shift algorithms, as described in the main text, to merge points that ideally would be identical. For example, with MS, the final iterates for points that in the limit would converge to the same mode are numerically different from each other (by a small amount if using a sufficiently accurate stopping criterion in the mean-shift iteration). Hence, these iterates form a tight cluster around their mode, which is widely separated from the tight clusters corresponding to other modes.

Naively implemented (Figure 18.9 left), Connected-Components would run in $\mathcal{O}(DN^2)$ time, because to construct the graph we have to threshold all pairwise distances, each of which costs $\mathcal{O}(D)$. However, in the case of tight, clearly separated clusters, we need not construct the graph explicitly, and it runs in $\mathcal{O}(DNK)$ if there are K connected components. Let the data satisfy the following "tight clusters" assumption: there exists $\epsilon > 0$ that is larger than the diameter of each component (the largest distance between every pair of points in a component) but smaller than the distance between any two components (the smallest distance between two points belonging to different components). Then the connected components can be found incrementally by connecting each point to the representative of its component (Figure 18.9 right), where the representative of component k is c_k, which is a point in that component. This can be done on the fly, as we process each data point in MS, or given the final set of iterates in BMS.

Algorithm Connected-Components

(a) Naive Connected-Components algorithm	(b) Efficient Connected-Components algorithm

Define an ϵ-ball graph:
 	
• Vertices $x_1, \ldots, x_N$
 	
• Edges $(x_n, x_m) \Leftrightarrow d(x_n, x_m) < \epsilon$,
  $\forall n, m = 1, \ldots, N$.
 	
Apply DFS to this graph.
	$K \leftarrow 1, c_1 \leftarrow x_1$ First component
	for $n = 2$ to N
	  for $k = 1$ to K
	    if $d(x_n, c_k) < \epsilon$ Old component
	      Assign x_n to component k; <u>break</u>
	  if x_n was not assigned
	    $K \leftarrow K + 1, c_k \leftarrow x_n$ New component

FIGURE 18.9
Pseudocode for Connected-Components, implemented naively in $\mathcal{O}(DN^2)$ (left) and under the "tight clusters" assumption in $\mathcal{O}(DNK)$ (right), where K is the number of components. In all cases, the input is a dataset $x_1, \ldots, x_N \in \mathbb{R}^D$, a distance function $d(\cdot, \cdot)$ applicable to any pair of points, and a threshold $\epsilon > 0$.

Some heuristics can further accelerate the computation. (1) The distances can be computed incrementally (dimension by dimension), so that as soon as ϵ is exceeded we exit the distance calculation. In the tight cluster assumption, this could reduce the cost to $\mathcal{O}(N(K + D))$, since for distances above ϵ each individual distance dimension will typically exceed ϵ. (2) In image segmentation, we can scan the pixels in raster order, so that for most pixels their component is the same as for the previous pixel (this will not be true when crossing a cluster boundary, but that happens infrequently). Then, when computing the distance, we always try first the last pixels's component. The average cost for the entire Connected-Components runtime is then $\mathcal{O}(ND)$, because most pixels compute a single distance.

The value of ϵ depends on the problem but it can usually be chosen in a wide interval. For example, in image segmentation with features in pixel units, we can set $\epsilon = 0.5$, since we do not need subpixel accuracy to locate a mode, and modes must be at least one pixel apart to be meaningful. In BMS, we can safely set ϵ to a smaller value (say, 0.01), since its cubic convergence rate produces extremely tight clusters quickly.

References

Aapo Hyvärinen. Estimation of non-normalized statistical models by score matching. *J. Machine Learning Research*, 6:695–708, 2005.

Alan L. Yuille and Tomaso A. Poggio. Scaling theorems for zero crossings. *IEEE Trans. Pattern Analysis and Machine Intelligence*, 8(1):15–25, 1986.

Alexandr Andoni and Piotr Indyk. Near-optimal hashing algorithms for approximate nearest neighbor in high dimensions. *Comm. ACM*, 51(1):117–122, 2008.

Andrew P. Witkin. Scale-space filtering. In *Proc. of the 8th Int. Joint Conf. Artificial Intelligence (IJCAI'83)*, pages 1019–1022, Karlsruhe, Germany, 1983.

Anil Chaturvedi, Paul E. Green, and J. Douglas Caroll. *k*-modes clustering. *Journal of Classification*, 18(1):35–55, 2001.

Ariel Shamir, Lior Shapira, and Daniel Cohen-Or. Mesh analysis using geodesic mean shift. *The Visual Computer*, 22(2):99–108, 2006.

Bernard W. Silverman. *Density Estimation for Statistics and Data Analysis*. Number 26 in Monographs on Statistics and Applied Probability. Chapman & Hall, London, New York, 1986.

Bernhard Schölkopf, Sebastian Mika, Christopher J. C. Burges, Philipp Knirsch, Klaus-Robert Müller, Gunnar Rätsch, and Alexander Smola. Input space vs. feature space in kernel-based methods. *IEEE Trans. Neural Networks*, 10(5):1000–1017, 1999.

Bijan Afsari. Riemannian l^p center of mass: Existence, uniqueness, and convexity. *Proc. Amer. Math. Soc.*, 139:655–673, 2011.

Bogdan Georgescu, Ilan Shimshoni, and Peter Meer. Mean shift based clustering in high dimensions: A texture classification example. In *Proc. 9th Int. Conf. Computer Vision (ICCV'03)*, 2003, pages 456–463.

Carlo Grillenzoni. Detection of tectonic faults by spatial clustering of earthquake hypocenters. *Spatial Statistics*, 7:62–78, 2013.

Changjiang Yang, Ramani Duraiswami, Nail A. Gumerov, and Larry Davis. Improved fast Gauss transform and efficient kernel density estimation. In *Proc. 9th Int. Conf. Computer Vision (ICCV'03)*, 2003, pages 464–471.

Chao Qin and Miguel Á. Carreira-Perpiñán. Trajectory inverse kinematics by conditional density modes. In *Proc. of the 2008 IEEE Int. Conf. Robotics and Automation (ICRA'08)*, pages 1979–1986, Pasadena, California, May 19–23, 2008.

Chao Qin and Miguel Á. Carreira-Perpiñán. Trajectory inverse kinematics by nonlinear, nongaussian tracking. In *Proc. of the IEEE Int. Conf. Acoustics, Speech and Sig. Proc. (ICASSP'08)*, pages 2057–2060, Las Vegas, Nevada, March 31–April 4, 2008.

Chao Qin and Miguel Á. Carreira-Perpiñán. Articulatory inversion of American English /r/ by conditional density modes. In Takao Kobayashi, Keikichi Hirose, and Satoshi Nakamura, editors, *Proc. of Interspeech'10*, pages 1998–2001, Makuhari, Japan, September 26–30, 2010.

Christopher M. Bishop. *Pattern Recognition and Machine Learning*. Springer Series in Information Science and Statistics. Springer-Verlag, Berlin, 2006.

Daniel DeMenthon. Spatio-temporal segmentation of video by hierarchical mean shift analysis. In *Statistical Methods in Video Processing Workshop (SMVP 2002)*, Copenhagen, Denmark, June 1–2, 2002.

Danny Barash and Dorin Comaniciu. A common framework for nonlinear diffusion, adaptive smoothing, bilateral filtering and mean shift. *Image and Vision Computing Journal*, 22(1):73–81, 2004.

David A. Forsyth and Jean Ponce. *Computer Vision. A Modern Approach*. Prentice-Hall, Upper Saddle River, NJ, 2003.

Dorin Comaniciu. An algorithm for data-driven bandwidth selection. *IEEE Trans. Pattern Analysis and Machine Intelligence*, 25(2):281–288, 2003.

Dorin Comaniciu and Peter Meer. Mean shift: A robust approach toward feature space analysis. *IEEE Trans. Pattern Analysis and Machine Intelligence*, 24(5):603–619, 2002.

Dorin Comaniciu, Visvanathan Ramesh, and Peter Meer. Kernel-based object tracking. *IEEE Trans. Pattern Analysis and Machine Intelligence*, 25(5):564–577, 2003.

Fan R. K. Chung. *Spectral Graph Theory*. Number 92 in CBMS Regional Conference Series in Mathematics. American Mathematical Society, Providence, RI, 1997.

Frédéric Chazal, Leonidas J. Guibas, Steve Y. Oudot, and Primoz Skraba. Persistence-based clustering in Riemannian manifolds. *Journal of the ACM*, 60(6):41, 2013.

Gabriel Taubin. A signal processing approach to fair surface design. In Susan G. Mair and Robert Cook, editors, *Proc. of the 22nd Annual Conference on Computer Graphics and Interactive Techniques (SIGGRAPH 1995)*, pages 351–358, Los Angeles, CA, August 6–11, 1995.

Gene H. Golub and Charles F. van Loan. *Matrix Computations*. Johns Hopkins University Press, Baltimore, third edition, 1996.

Geoffrey Hinton and Sam T. Roweis. Stochastic neighbor embedding. In Suzanna Becker, Sebastian Thrun, and Klaus Obermayer, editors, *Advances in Neural Information Processing Systems (NIPS)*, volume 15, pages 857–864. MIT Press, Cambridge, MA, 2003.

Geoffrey J. McLachlan and Thriyambakam Krishnan. *The EM Algorithm and Extensions*. Wiley Series in Probability and Mathematical Statistics. John Wiley & Sons, second edition, 2008.

Hanan Samet. *Foundations of Multidimensional and Metric Data Structures*. Morgan Kaufmann, San Francisco, 2006.

Herbert Edelsbrunner, David Letscher, and Afra Zomorodian. Topological persistence and simplification. *Discrete & Computational Geometry*, 28(4):511–533, 2002.

Hiroaki Sasaki, Aapo Hyvärinen, and Masashi Sugiyama. Clustering via mode seeking by direct estimation of the gradient of a log-density. In Toon Calders, Floriana Esposito, Eyke Hüllermeier, and Rosa Meo, editors, *Proc. of the 25th European Conf. Machine Learning (ECML–14)*, pages 19–34, Nancy, France, September 15–19, 2014.

K. Ho-Le. Finite element mesh generation methods: A review and classification. *Computer-Aided Design*, 20(1):27–38, 1988.

James R. Thompson and Richard A. Tapia. *Nonparametric Function Estimation, Modeling, and Simulation*. Other Titles in Applied Mathematics. SIAM Publ., 1990.

Jan J. Koenderink. The structure of images. *Biol. Cybern.*, 50(5):363–370, 1984.

Jean Babaud, Andrew P. Witkin, Michael Baudin, and Richard O. Duda. Uniqueness of the Gaussian kernel for scale-space filtering. *IEEE Trans. Pattern Analysis and Machine Intelligence*, 8(1):26–33, 1986.

Jianbo Shi and Jitendra Malik. Normalized cuts and image segmentation. *IEEE Trans. Pattern Analysis and Machine Intelligence*, 22(8):888–905, 2000.

Joachim Weickert. *Anisotropic Diffusion in Image Processing*. ECMI Series. Teubner, Stuttgart, Leipzig, 1998.

Jorge Nocedal and Stephen J. Wright. *Numerical Optimization*. Springer Series in Operations Research and Financial Engineering. Springer-Verlag, New York, second edition, 2006.

Jue Wang, Yingqing Xu, Heung-Yeung Shum, and Michael F. Cohen. Video tooning. *ACM Trans. Graphics*, 23(3):574–583, 2004.

Keinosuke Fukunaga and Larry D. Hostetler. The estimation of the gradient of a density function, with application in pattern recognition. *IEEE Trans. Information Theory*, IT–21(1):32–40, 1975.

Kenneth Rose. Deterministic annealing for clustering, compression, classification, regression, and related optimization problems. *Proc. IEEE*, 86(11):2210–2239, 1998.

W. L. G. Koontz, P. M. Narendra, and Keinosuke Fukunaga. Graph-theoretic approach to nonparametric cluster-analysis. *IEEE Trans. Computers*, C–25(9):936–944, 1976.

Lawrence M. Lifshitz and Stephen M. Pizer. A multiresolution hierarchical approach to image segmentation based on intensity extrema. *IEEE Trans. Pattern Analysis and Machine Intelligence*, 12(6):529–540, 1990.

Leslie Greengard and John Strain. The fast Gauss transform. *SIAM J. Sci. Stat. Comput.*, 12(1): 79–94, 1991.

Lior Shapira, Shai Avidan, and Ariel Shamir. Mode-detection via median-shift. In *Proc. 12th Int. Conf. Computer Vision (ICCV'09)*, pages 1909–1916, Kyoto, Japan, September 29–October 2, 2009.

Marcel Berger. *A Panoramic View of Riemannian Geometry*. Springer-Verlag, Berlin, 2003.

Marina Meilă and Jianbo Shi. Learning segmentation by random walks. In Todd K. Leen, Tom G. Dietterich, and Volker Tresp, editors, *Advances in Neural Information Processing Systems (NIPS)*, volume 13, pages 873–879. MIT Press, Cambridge, MA, 2001.

Marina Meilă and Le Bao. An exponential model for infinite rankings. *J. Machine Learning Research*, 11:3481–3518, 2010.

Mark Fashing and Carlo Tomasi. Mean shift is a bound optimization. *IEEE Trans. Pattern Analysis and Machine Intelligence*, 27(3):471–474, 2005.

Mathieu Desbrun, Mark Meyer, Peter Schröder, and Alan H. Barr. Implicit fairing of irregular meshes using diffusion and curvature flow. In Warren Waggenspack, editor, *Proc. of the 26th Annual Conference on Computer Graphics and Interactive Techniques (SIGGRAPH 1999)*, pages 317–324, Los Angeles, CA, August 8–13, 1999.

Matthias Hein and Markus Maier. Manifold denoising. In Bernhard Schölkopf, John Platt, and Thomas Hofmann, editors, *Advances in Neural Information Processing Systems (NIPS)*, volume 19, pages 561–568. MIT Press, Cambridge, MA, 2007.

Mattia Ciollaro, Christopher Genovese, Jing Lei, and Larry Wasserman. The functional mean-shift algorithm for mode hunting and clustering in infinite dimensions. arXiv:1408.1187 [stat.ME], August 6 2014.

Max Vladymyrov and Miguel Á. Carreira-Perpiñán. Entropic affinities: Properties and efficient numerical computation. In Sanjoy Dasgupta and David McAllester, editors, *Proc. of the 30th Int. Conf. Machine Learning (ICML 2013)*, pages 477–485, Atlanta, GA, June 16–21, 2013.

Michael C. Minnotte, David J. Marchette, and Edward J. Wegman. The bumpy road to the mode forest. *Journal of Computational and Graphical Statistics*, 7(2):239–251, 1998.

Michael C. Minnotte and David W. Scott. The mode tree: A tool for visualization of nonparametric density features. *Journal of Computational and Graphical Statistics*, 2(1):51–68, 1993.

Miguel Á. Carreira-Perpiñán. Mode-finding for mixtures of Gaussian distributions. *IEEE Trans. Pattern Analysis and Machine Intelligence*, 22(11):1318–1323, 2000.

Miguel Á. Carreira-Perpiñán. Reconstruction of sequential data with probabilistic models and continuity constraints. In Sara A. Solla, Todd K. Leen, and Klaus-Robert Müller, editors, *Advances in Neural Information Processing Systems (NIPS)*, volume 12, pages 414–420. MIT Press, Cambridge, MA, 2000.

Miguel Á. Carreira-Perpiñán. *Continuous Latent Variable Models for Dimensionality Reduction and Sequential Data Reconstruction*. PhD thesis, Dept. of Computer Science, University of Sheffield, UK, 2001.

Miguel Á. Carreira-Perpiñán. Reconstruction of sequential data with density models. arXiv: 1109.3248 [cs.LG], January 27, 2004.

Miguel Á. Carreira-Perpiñán. Acceleration strategies for Gaussian mean-shift image segmentation. In Cordelia Schmid, Stefano Soatto, and Carlo Tomasi, editors, *Proc. of the 2006 IEEE Computer Society Conf. Computer Vision and Pattern Recognition (CVPR'06)*, pages 1160–1167, New York, NY, June 17–22, 2006.

Miguel Á. Carreira-Perpiñán. Fast nonparametric clustering with Gaussian blurring mean-shift. In William W. Cohen and Andrew Moore, editors, *Proc. of the 23rd Int. Conf. Machine Learning (ICML'06)*, pages 153–160, Pittsburgh, PA, June 25–29, 2006.

Miguel Á. Carreira-Perpiñán. Gaussian mean shift is an EM algorithm. *IEEE Trans. Pattern Analysis and Machine Intelligence*, 29(5):767–776, 2007.

Miguel Á. Carreira-Perpiñán. Generalised blurring mean-shift algorithms for nonparametric clustering. In *Proc. of the 2008 IEEE Computer Society Conf. Computer Vision and Pattern Recognition (CVPR'08)*, Anchorage, AK, June 23–28, 2008.

Miguel Á. Carreira-Perpiñán. The elastic embedding algorithm for dimensionality reduction. In Johannes Fürnkranz and Thorsten Joachims, editors, *Proc. of the 27th Int. Conf. Machine Learning (ICML 2010)*, pages 167–174, Haifa, Israel, June 21–25, 2010.

Miguel Á. Carreira-Perpiñán and Christopher K. I. Williams. An isotropic Gaussian mixture can have more modes than components. Technical Report EDI–INF–RR–0185, School of Informatics, University of Edinburgh, 2003.

Miguel Á. Carreira-Perpiñán and Christopher K. I. Williams. On the number of modes of a Gaussian mixture. In Lewis Griffin and Martin Lillholm, editors, *Scale Space Methods in Computer Vision*, number 2695 in Lecture Notes in Computer Science, pages 625–640. Springer-Verlag, Berlin, 2003.

Miguel Á. Carreira-Perpiñán and Christopher K. I. Williams. On the number of modes of a Gaussian mixture. Technical Report EDI–INF–RR–0159, School of Informatics, University of Edinburgh, 2003.

Miguel Á. Carreira-Perpiñán and Weiran Wang. The K-modes algorithm for clustering. arXiv:1304.6478 [cs.LG], April 23, 2013.

Mikhail Belkin and Partha Niyogi. Laplacian eigenmaps for dimensionality reduction and data representation. *Neural Computation*, 15(6):1373–1396, 2003.

Peter Hall, Hans-Georg Müller, and Ping-Shi Wu. Real-time density and mode estimation with application to time-dynamic mode tracking. *Journal of Computational and Graphical Statistics*, 15(1):82–100, 2006.

Peter Hall and Michael C. Minnotte. High order data sharpening for density estimation. *Journal of the Royal Statistical Society, B*, 64(1):141–157, 2002.

Ping Wang, Dongryeol Lee, Alexander Gray, and James Rehg. Fast mean shift with accurate and stable convergence. In Marina Meilă and Xiaotong Shen, editors, *Proc. of the 11th Int. Conf. Artificial Intelligence and Statistics (AISTATS 2007)*, San Juan, Puerto Rico, March 21–24, 2007.

Raghav Subbarao and Peter Meer. Nonlinear mean shift over Riemannian manifolds. *Int. J. Computer Vision*, 84(1):1–20, 2009.

Robert T. Collins. Mean-shift blob tracking through scale space. In *Proc. of the 2003 IEEE Computer Society Conf. Computer Vision and Pattern Recognition (CVPR'03)*, pages 234–240, Madison, Wisconsin, June 16–22, 2003.

Roger A. Horn and Charles R. Johnson. *Matrix Analysis*. Cambridge University Press, Cambridge, U.K., 1986.

Roland Wilson and Michael Spann. A new approach to clustering. *Pattern Recognition*, 23(12):1413–1425, 1990.

Srinivasa V. Chakravarthy and Joydeep Ghosh. Scale-based clustering using the radial basis function network. *IEEE Trans. Neural Networks*, 7(5):1250–1261, 1996.

Stephen J. Roberts. Parametric and non-parametric unsupervised cluster analysis. *Pattern Recognition*, 30(2):261–272, 1997.

Sylvain Paris and Frédo Durand. A topological approach to hierarchical segmentation using mean shift. In *Proc. of the 2007 IEEE Computer Society Conf. Computer Vision and Pattern Recognition (CVPR'07)*, Minneapolis, MN, June 18–23, 2007.

Sylvain Paris, Pierre Kornprobst, Jack Tumblin, and Frédo Durand. Bilateral filtering: Theory and applications. *Foundations and Trends in Computer Graphics and Vision*, 4(1):1–73, 2008.

G. Taubin. Geometric signal processing on polygonal meshes. In *Eurographics'2000: State of the Art Reports*, 2000.

Thomas H. Cormen, Charles E. Leiserson, Ronald L. Rivest, and Clifford Stein. *Introduction to Algorithms*. MIT Press, Cambridge, MA, third edition, 2009.

Tony Lindeberg. *Scale-Space Theory in Computer Vision*. Kluwer Academic Publishers Group, Dordrecht, The Netherlands, 1994.

M. P. Wand and M. C. Jones. *Kernel Smoothing*. Number 60 in Monographs on Statistics and Applied Probability. Chapman & Hall, London, New York, 1994.

Weiran Wang and Miguel Á. Carreira-Perpiñán. Manifold blurring mean shift algorithms for manifold denoising. In *Proc. of the 2010 IEEE Computer Society Conf. Computer Vision and Pattern Recognition (CVPR'10)*, pages 1759–1766, San Francisco, CA, June 13–18, 2010.

Weiran Wang and Miguel Á. Carreira-Perpiñán. The Laplacian K-modes algorithm for clustering. arXiv:1406.3895 [cs.LG], June 15, 2014.

Weiran Wang, Miguel Á. Carreira-Perpiñán, and Zhengdong Lu. A denoising view of matrix completion. In J. Shawe-Taylor, R. S. Zemel, P. Bartlett, F. Pereira, and K. Q. Weinberger, editors, *Advances in Neural Information Processing Systems (NIPS)*, volume 24, pages 334–342. MIT Press, Cambridge, MA, 2011.

Xavier Pennec, Pierre Fillard, and Nicholas Ayache. A Riemannian framework for tensor computing. *Int. J. Computer Vision*, 66(1):41–66, 2006.

Xiaotong Yuan, Bao-Gang Hu, and Ran He. Agglomerative mean-shift clustering. *IEEE Trans. Knowledge and Data Engineering*, 24(2):209–219, 2010.

Yann LeCun, Léon Bottou, Yoshua Bengio, and Patrick Haffner. Gradient-based learning applied to document recognition. *Proc. IEEE*, 86(11):2278–2324, 1998.

Yaser Ajmal Sheikh, Erum Arif Khan, and Takeo Kanade. Mode-seeking via medoidshifts. In *Proc. 11th Int. Conf. Computer Vision (ICCV'07)*, Rio de Janeiro, Brazil, October 14–21, 2007.

Yiu-fai Wong. Clustering data by melting. *Neural Computation*, 5(1):89–104, 1993.

Yizong Cheng. Mean shift, mode seeking, and clustering. *IEEE Trans. Pattern Analysis and Machine Intelligence*, 17(8):790–799, 1995.

Zhexue Huang. Extensions to the k-means algorithm for clustering large data sets with categorical values. *Data Mining and Knowledge Discovery*, 2(2):283–304, 1998.

19

Nature-Inspired Clustering

Julia Handl and Joshua Knowles

CONTENTS

Abstract

We survey data clustering techniques that are inspired by natural systems. It is instructive to divide this group into two distinct classes: those methods derived directly from clustering phenomena in nature, such as the self-organizing behavior of some ants, or the schooling of fish and flocking of birds; and those methods that view clustering as an optimization problem, and apply nature-inspired general heuristics (or meta-heuristics) to this problem. The first group work largely in a bottom-up fashion, making them potentially powerful for parallelization and for operating in dynamic and noisy clustering applications, such as distributed robotics. The second group has the advantage that the clustering problem can be specified very precisely using one or more mathematical objective functions; then, due to the general-purpose properties of the heuristics, very good performance against these objectives can be achieved. This chapter charts the progress in these two broad classes of nature-inspired clustering algorithms and points out some further prospects in this area.

19.1 Introduction

We consider clustering approaches that take inspiration from the observation of self-organization and emergent behaviors in nature. Goldstein (1999) defines emergence as "the arising of novel and coherent structures, patterns, and properties during the process of self-organization in complex systems." More specifically, emergent behaviors are characterized as complex collective behaviors or patterns that arise from the interaction of simple entities within an environment. Examples of self-organization and emergence can be observed in many areas including biology, physics, and behaviors of man-made collectives such as financial markets and the Internet. Some of the most prominent examples are the seemingly intelligent behaviors exhibited by insect colonies, such as the construction of termite mounds and the optimization of foraging-trails in ant colonies (Corne et al., 1999). Processes such as these have long intrigued natural scientists and much observation and empirical work have gone into unlocking their secrets. From a modern computer science perspective, these systems provide valuable lessons regarding the bottom-up design of distributed systems and give important design principles for the achievement of efficiency, robustness, and adaptability. These aspects can give self-organizing approaches an edge in applications that exhibit dynamic behavior and uncertainty, as well as those involving hard optimization problems. Several of these conditions may be present in practical clustering applications and, for this reason, algorithms inspired by natural processes of self-organization give rise to a promising set of novel, alternative approaches to data clustering.

Concretely, nature-inspired methods have influenced the field of clustering in two different ways, and this is reflected by the structure of this chapter. In the first part, we will highlight a subset of techniques that have been developed as *specialist heuristics* for the problem of clustering. Some of these methods directly build upon computational models of (self-organizing) clustering phenomena that can be observed in nature (e.g., in ant-colonies and bird flocks), but we also include self-organizing algorithms with less direct parallels to nature. The second part of the chapter focuses on those approaches that use general-purpose *meta-heuristics* for optimization. These approaches are traditional in the sense that a mathematical formulation of the clustering problem is assumed. The resulting (typically hard) optimization problem is then tackled using a meta-heuristic optimization technique. This is where a nature-inspired element comes into play, as state-of-the-art meta-heuristics often take their inspiration from nature. We provide an overview of this second approach, as well as a brief description of established meta-heuristics.

19.2 Specialist (Self-Organizing) Clustering Heuristics

Self-organization can be defined as "a dynamical and adaptive process where systems acquire and maintain structure themselves, without external control" (Wolf and Holvoet, 2005). Thus, a core feature of the clustering techniques described in this chapter is the fact that a clustering is obtained as the emergent product of a bottom-up process, that is, there is no external, central control.* Instead, the clustering process is implemented based on a

---

* As clarified in Wolf and Holvoet (2005), the notion of external control does not preclude the use of data inputs (in this case the data set to be clustered), if this input does not provide direct control instructions.

set of local rules, which define individual interactions between elements of the overall system. Typically, some aspect of the system is progressively adjusted by the algorithm and the final state of the system is interpreted as a clustering solution.

Existing self-organizing methods for clustering differ in the nature of the local interactions and the corresponding rules. However, as we will see in the following, the representation of the system may also take a variety of forms: in some of the algorithms it has a concrete and direct physical interpretation as the environment in which data-items reside, or agents navigate. Furthermore, very different choices have been made regarding the part of the system used to represent a clustering solution. Depending on this last aspect, the interpretation of the final system state as a partitioning can be more or less straightforward and it may require additional postprocessing steps.

19.2.1 Self-Organizing Maps

Probably the best-known class of self-organizing clustering techniques are self-organizing maps (SOMs) or Kohonen maps (Kohonen, 2001). Kohonen maps present a special, unsupervised form of artificial neural networks, and, like all artificial neural networks, their inspiration comes from the central nervous systems of organisms (Haykin, 1994). A self-organizing map consists of a set of neurons that are arranged in a (usually two-dimensional) rectangular or hexagonal grid. Each neuronal unit in the grid is associated with a numerical vector of a fixed dimensionality.* The learning process of a self-organizing map involves the adjustment of these vectors to provide a suitable representation of the input data.

Self-organizing maps can be used for the clustering of numerical data in vector format. Their learning algorithm is set out in Algorithm SOM below, and is described in the following. In terms of input parameters, the training of a self-organizing map first requires a decision on the size of the grid to be used. For use in the context of clustering, this is commonly decided based on the number of clusters K expected in the data and, consequently, a $1 \times k$ grid of neurons is used: in this case, each neuron has a straightforward interpretation as the center of a single cluster. An alternative approach is the use of a grid with a higher resolution. This provides greater flexibility, for example, catering for different cluster shapes, but it also complicates the final retrieval and interpretation of clusters.

During the initialization phase, the vectors associated with each neuron are initialized randomly. After this initialization step, competitive learning takes place as follows: All data items in the data set are presented to the network sequentially (this can be either done systematically, in a randomized order or based on sampling). For each data item, a comparison to the vectors associated with each neuronal unit is made in order to find the vector that is closest in terms of the distance measure used. The neuron that this vector is associated with is referred to as the best matching unit (BMU). Once the BMU for a given input i has been determined, the vectors of the BMU and all surrounding neurons v are adjusted using the update rule:

$$\mathbf{w}_v(s+1) = \mathbf{w}_v(s) + \theta(u, v, s)\alpha(s)(\mathbf{d}(i) - \mathbf{w}_v(s)) \tag{19.1}$$

Here, u is the index of the BMU for i, $\mathbf{d}(i)$ is the data vector describing item i, and $\mathbf{w}_v(s)$ is the weight vector currently associated with v. During training, a step counter s is maintained and $\alpha(s)$ is a learning coefficient that decreases monotonically as a function of s

---

* The dimensionality of these vectors derives from the dimensionality of the input data.

(a decrease takes place either after every step or after a fixed number of steps). The neighborhood function $\theta(u, v, s)$ decreases as a function of the grid distance between the BMU (neuron u) and neuron v; most commonly this takes the form of a Gaussian centered around the BMU. Like the learning coefficient, the radius of the neighborhood function also decreases over time, which has the effect of gradually shifting the learning process from a global to a local level.

Algorithm SOM

Input parameter grid size, number of steps λ

   1. Randomly initialize the weight vectors associated with all neurons

   2. $s = 0$

   3. WHILE ($s < \lambda$)

      a. Select an input vector

      b. FOR each neuron

         i. Calculate the dissimilarity between the input vector and the neuron's weight vector

      c. Identify the BMU u for this input vector

      d. FOR each neuron in the neighborhood of u

         i. Apply the SOM update rule in Equation 19.1

      e. $s = s + 1$

Output weight vectors of all neurons

Despite the competitive learning approach adopted by SOMs, there are similarities between SOMs and the traditional K-means algorithm. Specifically, in the final stages of competitive learning, the SOM's neighborhood function includes a single BMU only. This last phase therefore corresponds to the direct minimization of the distance between data items and their corresponding BMU (and is equivalent to the local optimization of K-means' minimum sum-of-squares criterion (Trosset, 2008)). Yet, there are also clear differences in the performance of K-means and SOM. In particular, the SOM algorithm has been shown to be more robust than K-means towards increases in the number of data items and the number of clusters. K-means is known to suffer from problems with local optima and SOMs seem to ameliorate this issue through the information-exchange between neighboring neurons in the early stages of the algorithm (Trosset, 2008).

A further advantage of SOMs is the additional information provided by the SOM output: in addition to information about cluster memberships, the grid topology conveys information about the relationship between clusters. This last property makes them a particularly attractive tool for applications that involve both an element of visualization and of clustering (Murtagh and Hernandez-Pajares, 1995). For example, SOMs have been previously applied to generate "spatializations," a form of visualization based on a landscape metaphor. In spatialization, 3D effects are used to illustrate the density of data in a two-dimensional mapping of a data set. For example, in the context of document mining, a collection of documents can be visualized as a landscape, with individual density peaks labeled to highlight the topics associated with these particular clusters (Fabrikant and Buttenfield, 2000).

Implementations of self-organizing maps are available for the use in many popular computing environments, for example, as a part of the Neural Network Toolbox in Matlab and as the package "som" in R.

19.2.2 Ant-Based Clustering

We now shift our focus to a second self-organizing technique that also produces an embedding of a multidimensional data set into two dimensions (not entirely unlike the output produced by a self-organizing map or multidimensional scaling (Kuntz et al., 1997)). The technique of ant-based clustering takes its inspiration from the observation of emerging behaviors in some types of ant species, such as "cemetery formation" (the aggregation of corpses into piles) in the ant species *Pheidole pallidula* (Corne et al., 1999). The first ant-based clustering method was developed by biologists as a simple mathematical model aimed at re-creating the emergent phenomenon observed in nature, and to explore the use of this model in the context of self-organized robotics (Deneubourg et al., 1991). This original algorithm was capable of differentiating between two classes of data items only, but it was soon extended to account for the segregation of items into an arbitrary number of clusters, using information about pairwise dissimilarities of all data items (Lumer and Faieta, 1994).

Different from self-organizing maps, ant-based clustering does not require numerical data in vector format, but it works directly on a dissimilarity matrix. A basic approach to ant-based clustering is shown in Algorithm ANTS below. Intuitively, this algorithm provides an embedding of a data set within a two-dimensional (usually toroidal) grid space. The algorithm starts with a random projection of the set of all data items onto the grid space; the subsequent sorting process is achieved by means of a set of ant-like agents, who perform a random walk around the grid, picking and dropping the encountered data with appropriate probabilities. Suitable probabilities can be derived from the density as well as the similarity of data items in an ant's local neighborhood. For example, suitable probability functions for data clustering are defined by Lumer and Faieta as follows (Lumer and Faieta, 1994):

$$p_{pick}(i) = \left(\frac{k^+}{k^+ + f(i)} \right)^2 \tag{19.2}$$

$$p_{drop}(i) = \begin{cases} 2f(i), & \text{if } f(i) < k^- \\ 1, & \text{otherwise} \end{cases} \tag{19.3}$$

$$f(i) = \max\left(0, \frac{1}{\sigma^2} \sum_{j \in N(i)} \left(1 - \frac{d(i,j)}{\alpha} \right) \right) \tag{19.4}$$

Here, i is the data item currently considered by an agent, $d(i,j)$ gives the distance between data items i and j, k^+ and k^- are fixed parameters that define the shape of the probability function, and α is a user-defined threshold on data dissimilarities. $N(i)$ is a neighborhood of size $\sigma \times \sigma$, which is centred around the data item's current location on the grid.

Algorithm ANTS

Input parameter grid size, number of ant-like agents, number of steps λ

1. Assign data items and ant-like agents to random grid cells (at most one agent and item per cell)

2. $s = 0$

3. WHILE $(s < \lambda)$

   a. Select an ant-like agent

   b. Make a random move to an adjacent cell on the grid

   c. IF the agent carries a data item AND the grid cell is empty

      i. Set i to be the data item currently carried by the ant

      ii. Use $f(i)$ in Equation 19.3 to compare data item i to items in the surrounding grid cells

      iii. Use $p_{drop}(i)$ in Equation 19.2 to make a decision on dropping the data item i

   d. IF the agent does not carry a data item AND the grid cell contains an item

      i. Set i to be the data item contained in the grid cell

      ii. Use $f(i)$ in Equation 19.3 to compare data item i to items in the surrounding grid cells

      iii. Use $p_{pick}(i)$ in Equation 19.1 to make a decision on picking up the data item i

   e. $s = s + 1$

Output grid positions of all data items

The default end product of ant-based clustering is not an explicit clustering solution, but an arrangement of the data on the grid. A final postprocessing step is then required to move from the (visual) arrangements of items to an explicit description of the clustering implicit to the embedding. Compared to the output of a SOM, the grid layout produced by ant-based clustering tends to be of a more structured nature, with cluster boundaries captured by empty grid regions. This aspect of the output may be exploited during the final postprocessing step (Handl et al., 2006). To obtain an explicit partitioning, a single-link agglomerative clustering can be applied to the *grid-positions* of the data-items, with a generic distance threshold (in grid-space) defined as the algorithm's stopping criterion.

At least implicitly, ant-based clustering provides an automatic estimate of the number of clusters in the data set. This presents an advantage compared to techniques such as K-means, which require the specification of the expected number of clusters. One may argue, however, that the need for the specification of the number of clusters has been side-stepped through the introduction of a different parameter: in the case of Lumer and Faeita's original ant-based clustering model (see above), this crucial parameter is the distance threshold α (see Equation 19.3), which impacts on the number of clusters formed. Fortunately, it has been shown that, for data sets with clear cluster structures, the parameter α can be adjusted automatically during the clustering process (Handl et al., 2006), reducing the reliance of algorithm performance on user input.

For small-to-medium-sized data sets, the results obtained by ant-based clustering have been shown to be comparable in accuracy to those from K-means, SOMs, and hierarchical clustering (Handl et al., 2006). Yet, it is currently unclear to what extent these results carry over to very large sets of data. Compared to self-organizing maps, a noteworthy feature of ant-based clustering is the fact that most implementations do not include any time-dependent learning parameters. This property makes the algorithm attractive for incremental clustering where data items are added to a data stream over time. Finally, while the distributed nature of ant-based clustering makes it immediately attractive for robotics applications (Deneubourg et al., 1991), it is currently unclear to what extent the parallelism intrinsic to the algorithm can be exploited in computational data clustering applications. A limiting factor for parallelization may lie in the fact that information about the occupation of the grid (as the primary medium enabling indirect communication) needs to be available to all ant-like agents.

Demonstrations and code examples for ant-based clustering are available online (see e.g., the authors' version at http://personalpages.manchester.ac.uk/mbs/julia.handl/ants.html), but an integration into standard packages has yet to be achieved.

19.2.3 Flocking Behavior

In the ant-based clustering algorithm discussed above, the clustering process is implemented through a set of agents who re-position data items within a two-dimensional grid. These agents provide the means of manipulating the items' grid positions, but they do not play a direct role in the representation of the final clustering solution. In this subsection, we look at a further type of self-organizing technique in which this differentiation between agents and data-items is not present. The technique of "information flocking" is a visualization and clustering technique that takes its inspiration from spontaneous group formation observed in nature, such as flock formation in birds, schooling of fish, and herding of cattle.

Models of bird flocking were first developed for use in computer animations (Reynolds, 1987). The initial "Boids" model uses a group of simple, local agents, each associated with a position and a velocity component in three-dimensional space. The behavior of these agents (boids) is then determined by the synergy of three simple behavioral rules: (i) Each boid steers to avoid collision with others (Avoidance); (ii) Each boid aims to align its direction of movement with that of other flock members (Alignment); (iii) Each boid steers towards the average location of other flock members (Cohesion). In the original Boids model, these individual rules take into account either the position or direction of flight of local (nearby) flock members, as described by the following equations:

$$\Delta \mathbf{v}_i^{\text{avoidance}}(t) = \frac{1}{|N_1(i,t)|} \sum_{j \in N_3(i,t)} \mathbf{x}_j(t) - \mathbf{x}_i(t) \tag{19.5}$$

$$\Delta \mathbf{v}_i^{\text{alignment}}(t) = \frac{1}{|N_2(i,t)|} \sum_{j \in N_1(i,t)} \mathbf{v}_j(t) \tag{19.6}$$

$$\Delta \mathbf{v}_i^{\text{cohesion}}(t) = \frac{1}{|N_3(i,t)|} \sum_{j \in N_2(i,t)} \mathbf{x}_j(t) - \mathbf{x}_i(t) \tag{19.7}$$

Here, $\mathbf{x}_i(t)$ describes the position of boid i at time t, and $\mathbf{v}_i(t)$ describes its speed and direction. $N_k(i,t)$ describes the neighborhood of boid i at time t, with $k = 1,2,3$ catering for differences in the neighborhood size employed for the characterization of each behavior.

The above three rules feed into the following update rules for a boid's position and velocity:

$$\mathbf{x}_i(t+1) = \mathbf{x}_i(t) + \mathbf{v}_i(t) \tag{19.8}$$

$$\mathbf{v}_i(t+1) = \alpha_0 \times \mathbf{v}_i(t) + \alpha_1 \times \Delta\mathbf{v}_i^{\text{avoidance}}(t) + \alpha_2 \times \Delta\mathbf{v}_i^{\text{alignment}}(t) + \alpha_3 \times \Delta\mathbf{v}_i^{\text{cohesion}}(t) \tag{19.9}$$

where the parameters α_k describe the weighting factors for the boid's original speed and each of the three incremental speed components $\Delta\mathbf{v}_i(t)$, respectively.

To apply this model to data clustering, the approach is modified by associating each boid with an individual item from the data set. Similarity between pairs of data items is employed to introduce additional rules or to adjust the existing behavioral alignment rules. For example, a weight factor proportional to the similarity between data items may be used to obtain preferential alignment of groups of similar items; this results in the formation of distinct groups of boids, which can then be interpreted as distinct clusters (Proctor and Winter, 1998). Similar to ant-based clustering, information flocking can work directly on a dissimilarity matrix. Basic pseudocode for the approach is shown in Algorithm FLOCKING below.

Algorithm FLOCKING

Input number of steps λ

1. Create a boid for each data item and assign a random velocity vector and position in three-dimensional space
2. $s = 0$
3. WHILE $(s < \lambda)$
   a. Select a boid
   b. Use the velocity vector to update the boid's position (Equation 19.8)
   c. Calculate the required update to the velocity vector (Equation 19.9)
   d. $s = s + 1$

Output position vectors of all boids

Information-flocking shares a number of advantages with the technique of ant-based clustering. Like ant-based clustering, it does not require *a priori* information about the number of clusters. Furthermore, the nature of the algorithm allows it to adjust rapidly to changes within the data set. Finally, the approach can be used to generate a dynamic visualization of the data, and it has been argued that this improves the interpretability of results (compared to the static visualization returned by techniques such as SOMs) (Proctor and Winter, 1998). This is because the movement of boids over time provides an additional dimension that may capture valuable information about the validity of individual neighborhood relationships (which may not be as clear from a static low-dimensional embedding).

Regarding public software for information flocking, the authors are unaware of any software that is currently available.

19.2.4 Affinity Propagation

The last type of self-organizing clustering technique that we discuss here is the approach of affinity propagation proposed in Dueck and Frey (2007), which is based on the principle of recursive message-passing. Similar to standard k-medoids clustering, affinity propagation aims to determine a set of "exemplars," that is, data points that are most suitable to represent the clusters within a data set. However, different from the k-medoids method, affinity propagation does not iteratively adjust a set of candidate centroids. Instead, for a data set with n items, the algorithm operates on two $n \times n$ matrices which reflect the pairwise "responsibility" and "availability" between all pairs of data points. In the following, these two matrices are denoted by $\mathbf{a}(,)$ and $\mathbf{r}(,)$, respectively. The algorithm uses repeated iterations of message-passing to estimate the most appropriate entries for these matrices. Specifically, given a pairwise similarity function $\text{sim}(i, j)$, updates for individual entries of the matrices are calculated based on the consideration of pairwise similarities:

$$\mathbf{r}(i, j) = \text{sim}(i, j) - \max_{j' \neq j} \left(\mathbf{a}(i, j') + \text{sim}(i, j') \right) \tag{19.10}$$

$$\mathbf{a}(i, j) = \min \left(0, \mathbf{r}(j, j) + \sum_{i' \neq i} \max(0, \mathbf{r}(i', j)) \right) \quad \text{for } i \neq j \tag{19.11}$$

$$\mathbf{a}(i, j) = \sum_{i' \neq j} \max(0, \mathbf{r}(i', j)) \quad \text{for } i = j \tag{19.12}$$

Intuitively, the responsibility $\mathbf{r}(i, j)$ captures information about how suitable data item j would be to act as a the representative of data item i. This is assessed by assessing the similarity $\text{sim}(i, j)$ between the items j and i and comparing this to the maximum sum of two components: the similarity observed between i and any other potential exemplar (which could be any other data item) and the corresponding availability of that particular exemplar. The availability $\mathbf{a}(i, j)$ is calculated as the self-responsibility of $\mathbf{r}(j, j)$ plus the sum of all positive responsibilities of j for any other data items. Intuitively, the first part of the expression assesses how suitable j would be as an exemplar for itself, and the second part boosts this estimate by the overall amount of support for using j as an exemplar (discounting negative contributions).

Affinity propagation can be used to cluster dissimilarity data, including those that are sparse or nonmetric. Pseudocode for the approach is shown in algorithm AFFINITY below. Here, ω is a damping factor typically set to $\omega = 0.5$ (Dueck and Frey, 2007).

Unlike ant-based clustering and information flocking, an explicit clustering criterion has been proposed for affinity propagation (Dueck and Frey, 2007) and this has been shown to be closely related to the k-medoids criterion (Walter, 2007). Nevertheless, affinity propagation has been reported to outperform significantly the optimization performance of the k-medoids algorithm for large and high-dimensional data sets. This is thought to be due to the fact that affinity propagation does not start from an initial set of medoids, but slowly determines suitable exemplars through its iterative process of message-passing. This gradual information exchange helps reduce preliminary convergence to local optima, somewhat similar to the effects of information exchange in self-organizing maps.

Affinity propagation can be seen as a self-organizing technique in the sense that a global order (clustering) arises from a sequence of simple pair-wise interactions between data points. Where ant-based clustering and information flocking use the spatial location of data items as a mechanism for indirect communication, affinity propagation achieves this through the update of the responsibility and availability matrices.

Algorithm AFFINITY

Input number of steps λ, damping factor ω
   1. $\forall j$: Set sim(j, j) to indicate initial preference for j to be an exemplar
   2. $\forall i, j$: Set sim(i, j) to reflect similarity
   3. $\forall i, j : a(i, j) = 0$
   4. $s = 0$
   5. WHILE $(s < \lambda)$
       a. $r_{old}(i, j) = r(i, j)$
       b. Calculate $r(i, j)$ using Equation 19.10
       c. $r(i, j) \leftarrow \omega \times r_{old}(i, j) + (1 - \omega) \times r(i, j)$
       d. $a_{old}(i, j) = a(i, j)$
       e. Calculate $a(i, j)$ using Equation 19.11
       f. $a(i, j) \leftarrow \omega \times a_{old}(i, j) + (1 - \omega) \times a(i, j)$
       g. $s = s + 1$
Output $\forall i$: Return $\arg\max_j (a(i, j) + r(i, j))$ as the exemplar of i

---

19.2.5 Prospects and Future Developments

Considering all of the algorithms discussed above, the self-organizing nature of these techniques leads to distinct differences compared to traditional algorithms. Importantly, the assumptions made about the desired partitioning are somewhat less stringent than those for traditional algorithms. This is true regarding the number of clusters expected in the data (which does not need to be specified explicitly), but also regarding the clustering criterion to be optimized. To some extent, this property resounds with the spirit of exploratory data analysis and may be seen as an advantage in settings where no such assumptions are available. However, the lack of mechanisms for integrating assumptions may become problematic in situations where explicit prior knowledge is available. Furthermore, it is unclear to what extent there is simply a *lack of understanding* regarding the assumptions implicit to the algorithm as opposed to an actual *lack of assumptions*. It seems to be the latter for affinity propagation, at least, since an explicit clustering criterion has been derived. Some assumptions are clearly implicit to the parameterization of these algorithms. For example, in ant-based clustering the definitions of the probability and neighborhood functions have potential consequences for the types, sizes, and shapes of clusters that can be identified. Similarly, changes to the neighborhood size in information flocking may produce a range of behaviors from fully independent motion to a single big flock (Proctor and Winter, 1998). Compared to an algorithm like K-means, the nature and the implications of these assumptions are not well understood, and this may impact on performance or complicate the interpretation of clustering results.

There also remain other performance aspects of these techniques that are not fully understood. For example, several of the approaches discussed (SOMs, ant-based clustering, and information flocking) use a system representation in two or three dimensions. As discussed above, this enables their combined use as a clustering and visualization technique, and has been a driving factor in their uptake. However, while the effectiveness of the embedding has been established for techniques such as self-organizing maps, this is not the case for some of the other approaches. In the case of information flocking, there has been little quantitative analysis regarding the extent to which distances between or within clusters reflect underlying data dissimilarities. For ant-based clustering, detailed analysis of the results has provided disappointing results, indicating that the positioning of clusters produced by ant-based clustering provides little information on actual inter-cluster relationships (Handl et al., 2006). Similarly, the spatial arrangement of data items within each cluster was found to be poor compared to the layout obtained from established techniques such as multi-dimensional scaling (Handl et al., 2006).

In conclusion, while the idea of self-organizing clustering and visualization techniques is appealing, further (theoretical and empirical) work remains crucial in establishing an accurate picture of the strength and weaknesses of these approaches.

19.3 Meta-Heuristic Clustering Approaches

In this second part of the chapter, we consider techniques that are based on the mathematical formulation of clustering as an optimization problem. The definition of clustering in this form requires the implementation of suitable optimization techniques, and an effective and flexible option is available in the form of meta-heuristic optimizers. Very commonly, the design of novel meta-heuristics has taken inspiration from natural phenomena, and the optimization approach is thus highly relevant to the core theme of this chapter. This section will provide a general overview of cluster-optimization using meta-heuristics and highlight examples of techniques that are commonly used.

19.3.1 Meta-Heuristics

Meta-heuristics are high-level procedures that are designed to find approximate solutions to hard optimization problems in reasonable time. One of the attractive properties of meta-heuristics is the absence of strong mathematical assumptions about the properties of an optimization problem. As a result, meta-heuristics can be flexibly applied to a wide variety of optimization problems, without the need for simplifications that may be unrealistic.

Meta-heuristics work by generating and evaluating new candidate solutions. They attempt to provide an efficient search the space of possible solutions, with the aim of rapidly identifying some of the best possible solutions to the problem. While the overall process that guides the search is generic (i.e., it needs *no* adjustment for the application to different optimization problems), meta-heuristics do contain three components that are problem-specific:

- Encoding: This provides a computational representation of a candidate solution to the optimization problem.
- Variation operators: These introduce perturbations to a given solution. Given one or more candidate solutions as the input, a variation operator will generate novel, alternative solutions to the problem. These new solutions typically share some

properties with the original input solutions (in other words, perturbations are mostly small).

- Objective function: This is the optimization criterion, which does *not* need to take a closed mathematical form. The objective function loosely corresponds to any approach (mathematical, computational, human) that is capable of considering a candidate solution and mapping it to a final objective value—the functional form of this mapping can be completely unknown. This objective function provides the means of assessing the quality of each candidate solution generated during the heuristic search.

The application of a meta-heuristic to a particular optimization problem requires crucial design decisions regarding all three of the above components: suitable specifications for all three components are application-dependent and may impact significantly on a meta-heuristics optimization performance. Possible choices for the problem of data clustering are made explicit in the following subsections.

19.3.2 Solution Representation (and Operators) for Data Clustering

Possibly the most general representation of a clustering solution is the use of a single n-dimensional integer vector, with the ith entry indicating the cluster assignment of the ith data item. While this encoding provides a more flexible representation of a partitioning (no assumption is made on cluster properties and shapes), its main limitation lies in the re-labeling required to enable the comparison of different clustering solutions (and the design of effective variation operators).

A common alternative is, therefore, the use of centroid-based encodings. This assumes knowledge about the desired number of clusters K, but a clustering solution can then be represented by a set of K d-dimensional real vectors, which have a direct interpretation as cluster representatives (or prototypes). This is an *indirect* representation in the sense that cluster assignments are not explicit: an additional interpretation step is required to make the clustering explicit, which is usually done by assigning each data item to its closest centroid vector. This encoding is advantageous in terms of its scalability for large data sets, as the length of this encoding scales as a function of K and d only. Furthermore, as the encoding simply corresponds to a set of continuous numbers, standard variation operators for continuous optimization may be employed for this representation (although more specialized operators for clustering have also been proposed). A crucial limitation of centroid-based encodings lies in the assumptions intrinsic to this representation: the use of centroids assumes the presence of spherically shaped clusters, and may therefore prevent the detection of nonspherical clusters.

More specialized encodings include the linkage-based encoding scheme (Park and Song, 1998) and Falkenauer's grouping GA (Falkenauer, 1994). These encodings are also indirect, and they have been designed to provide effective mechanisms for the exploration of the clustering search space. In particular, Falkenauer's grouping GA comes with a set of operators specifically designed for the clustering problem, whereas linkage-based encoding has been shown to perform well in combination with standard variation operators for integer encodings (such as uniform crossover and standard mutation in the case of an evolutionary algorithm (Handl et al., 2006)). Both encodings are capable of describing any possible partitioning of a data set, thus avoiding the loss of flexibility that a centroid-based representation would incur.

19.3.3 Objective Functions (Clustering Criteria)

One of the key challenges encountered in unsupervised classification is the lack of a clear and unique definition of a good or natural clustering. As a result of this, there is an increasing consensus that the quality of a clustering can only be evaluated objectively in terms of the quality of the final deliverable. For example, in situations where a clustering algorithm is used as a preprocessing step to improve the performance of a supervised classifier, the quality of a clustering should be measured *indirectly* by assessing the classification accuracy of the *final* supervised classification task. This is a setup that meta-heuristics naturally allow for, as a mathematical description of the objective function is not required (see above).

Nevertheless, some clustering applications require us to make assumptions regarding the types of clusters expected in the data. When such clustering tasks are formulated as optimization problems, these assumptions are made explicit in the form of the optimization criterion. The advantage of this explicit approach is a good understanding of the properties of resulting clustering solutions (different from some self-organizing approaches, see above). The clustering literature provides a range of mathematical measures that have been designed to capture aspects of a good partitioning—such measures typically capture the compactness of a cluster, the separation between clusters, or a combination of these two aspects. Any of these may be employed as optimization criteria within a meta-heuristics. For a detailed discussion of possible clustering criteria, the reader is referred to the relevant chapters of this book.

19.3.4 Meta-Heuristics for Clustering

Some of the best-known and most established meta-heuristics are those that draw their inspiration from nature. In the following, we provide a brief introduction to some of these techniques including evolutionary algorithms, simulated annealing, ant colony optimization, and particle swarm optimization. For a detailed survey of meta-heuristics, the interested reader is referred to Blum and Roli (2003).

Implementations of meta-heuristic optimizers are available widely, for example, as a part of the Global Optimization Toolbox for Matlab, which currently includes implementations of genetic algorithms, simulated annealing, and multiobjective genetic algorithms (see below).

Evolutionary algorithms. Evolutionary algorithm (Bäck et al., 1997) is the modern generic term for a group of stochastic computational techniques that encompasses genetic algorithms, genetic programming, evolutionary programming, and evolution strategies. Evolutionary algorithms simulate key processes in natural evolution and Mendelian genetics with the aim of evolving some artifact so that it is adapted to its environment. When used for optimization, this artifact corresponds to a solution to the optimization problem (and is encoded by a set of genes); playing the role of the environment is a (static) objective function, which evaluates the quality (fitness) of each solution. A key feature of evolutionary algorithms is that they are population-based. This means that competition between members of the population can be used to drive progress towards the goal via a series of generations of selection (biased so that fitter individuals survive over less fit ones), and reproduction (involving variation of solutions through mutation and recombination).

Evolutionary algorithm provide effective optimization approaches for continuous or combinatorial optimization. In the context of data clustering, either approach may be employed dependent on the cluster representation of choice (see previous sub-section). Evolutionary algorithms have also been used in combination with local search techniques, most prominently the use of K-means to locally optimize the position of candidate centroids (Hruschka et al., 2009).

Simulated annealing. Simulated annealing (Aarts and Korst, 1988) is a stochastic iterative optimization method based on statistical physics models describing the process of physical annealing, where a system is led to a very low energy configuration by slow cooling. In simulated annealing, the objective function value of a solution to an optimization problem takes on the role of the energy of a physical state. The state evolution of the system is modeled as a Markov chain where the state transition probabilities depend upon three things: a neighborhood function; the energy difference between state pairs (i.e., neighboring solutions); and, the temperature, which is a global variable controlled by the algorithm. In the algorithm, an initial solution is typically generated at random, and a form of neighborhood search then proceeds. Here, the concept of a neighborhood is defined by the variation operator designed for the optimization problem, for example, the immediate neighbors of a solution consist of all those solutions that can be reached in a single perturbation step. During the neighborhood search, a randomly selected neighbor x' of the current solution x is accepted (becomes the new current solution) with a probability given by the Metropolis criterion. Thus, the algorithm implements neighborhood descent, but allows uphill (worsening) moves, dependent on temperature, enabling it to escape from local optima.

Similar to evolutionary algorithms, simulated annealing protocols may be implemented for both continuous and combinatorial optimization. Implementations for data-clustering may therefore be based on any of the representations discussed in the previous sub-section, see, for example, Güngör and Ünler (2007); Selim and Alsultan (1991).

Particle swarm optimization. Particle swarm optimization (Eberhart and Kennedy, 1995) is a population-based meta-heuristic initially designed for continuous optimization problems. Inspired by models of the flocking behavior of birds (and by other social behavior), the method consists of particles that move around the search space, collectively affecting each others' trajectories. Trajectory updates for each particle are most influenced by those particles whose position corresponds to promising candidate solutions. Over time, the communication between particles leads to the convergence of the swarm to optima in the search space.

Consistent with PSO's original design, PSO for clustering has typically been implemented as a continuous optimization problem. Specifically, it is most commonly used in combination with a centroid-based encoding and, frequently, in hybridization with K-means (Handl and Meyer, 2007; Van der Merwe and Engelbrecht, 2003).

Ant colony optimization. Ant colony optimization (ACO) (Dorigo et al., 1996; Dorigo and Stützle, 2003) is inspired by the foraging behavior of ant species that use pheromone-laying to promote efficient collective route-finding. The mechanism used by real Argentine ants was uncovered using a key experiment (Deneubourg et al., 1990) where a double-bridge, connecting an ant nest to a food source, was presented to a colony of foraging ants. The ants quickly established a stronger pheromone trail on the shorter of the two arms of the bridge, causing almost all subsequent ants (attracted to the pheromone) to take this shorter path. A positive feedback effect explains this result: ants initially choose either of the two paths at random, but the shorter path builds up pheromone more quickly, as ants taking that route return in less time.

The ACO meta-heuristic simulates this process. It was initially demonstrated on a route-finding problem, but it can be used on any combinatorial optimization problem (see e.g., Dorigo and Stützle, 2003) provided a suitable encoding of the problem as a construction graph can be designed.

A few implementations of ACO have been proposed for data-clustering, with the construction graph typically employed to directly represent cluster assignments (Handl and Meyer, 2007; Runkler, 2005).

19.3.5 Prospects and Future Developments

Regarding all of the above meta-heuristics, their relative performances in data clustering are not fully understood. Furthermore, there remains significant debate regarding the best design choices for their use in clustering (in terms of objective functions, encoding, and operators used), and the synergies between different types of choices. Further theoretical and empirical work in these areas is crucial to improve the scalability of these clustering approaches and equip them for increases in data volumes.

Traditional clustering algorithms optimize a single criterion of cluster quality and this is also the way in which meta-heuristics have been employed for clustering: techniques such as EAs, PSO, SA, and ACO are typically used in conjunction with a single clustering criterion, and often in combination with local search techniques such as K-means. More recently though, a number of multiobjective approaches to data clustering have been proposed, with promising results in terms of the accuracy of the solutions and the algorithms' robustness to different data properties. Multiobjective clustering is a key area in which meta-heuristics can make a significant contribution to the field of data clustering, and we therefore conclude this section with a brief overview of the theoretical considerations, methods, and results that underpin recent developments in this area.

19.3.5.1 Multiobjective Clustering

The idea of multiobjective clustering was first formalized in the seminal work of Delattre and Hansen (1980) who observed that informal definitions of clustering capture the presence of two conflicting objectives: (i) the grouping of similar data items; and (ii) the separation of dissimilar items. The multicriterion nature of the clustering problem is evident from this, as there is clear reference to two different properties of a partitioning (intra-cluster homogeneity and inter-cluster separation) that need to be measured separately and are, to some extent, conflicting. Delattre and Hansen noted that, despite this implicit multicriterion nature of data clustering, traditional approaches to data-clustering typically address the problem as a single-objective optimization problem. For example, the well-known K-means algorithm (locally) optimizes a single objective that captures intra-cluster homogeneity alone. The second objective to data-clustering (inter-cluster separation) is not considered; furthermore, as the objective of intra-cluster homogeneity naturally improves with an increasing number of clusters, algorithms such as K-means require the analyst to specify the required number of clusters.

In contrast to this, multiobjective approaches to data clustering explicitly acknowledge the multicriterion nature of the problem. This has the advantage that candidate solutions can be assessed in terms of a set of desirable properties and that the conflict between criteria can be explicitly explored. Both are important advantages in a situation of exploratory data analysis, where cluster properties may differ, where a single clustering criteria may not fully capture all aspects of a good solution, and where the importance (and thus prioritization or weighting) of different clustering criteria is notoriously difficult to establish *a priori*. The core premise of the multicriterion approach to data-clustering is further illustrated in Figure 19.1 (from Handl and Knowles, 2007).

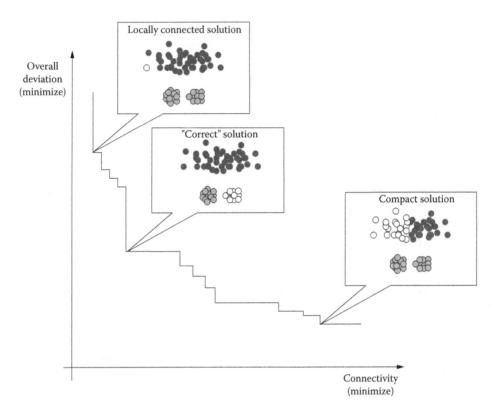

FIGURE 19.1
In this example, three different partitions are shown for a simple three-cluster data set. Their evaluation with respect to two different criteria is indicated by their relative positioning in two-objective space, and the partitioning of the data is indicated through the use of different colors. The solution to the top left is generated by an algorithm like single link agglomerative clustering, which prioritizes inter-cluster separation. The solution to the bottom right is generated by an algorithm like-means, which prioritizes intra-cluster homogeneity. The best solution does well with respect to both criteria, but would not usually be discovered by either method. A multi-criterion approach considering the trade-offs between the two objectives is able to access this solution much more readily.

Given an optimization problem with multiple criteria, a single optimal solution cannot usually be identified and the task becomes one of finding a set of optimal solutions, specifically those that provide the best trade-off between the criteria considered. More formally, this notion is captured by the principle of Pareto optimality. For a multicriterion optimization problem, the set of Pareto optimal solutions is defined as the subset of all those solutions $\mathbf{x}$ for which the following condition is met: no other solution exists that is equal or better than $\mathbf{x}$ with respect to all of the criteria considered. In objective space, this set of Pareto optimal solutions maps to the Pareto front. A variety of approaches can be used to identify the Pareto optimal set of a multicriterion problem, and meta-heuristic approaches have quickly gained in importance in this area. A particularly popular approach is the use of multiobjective evolutionary algorithms (MOEA), due to their adaptability, speed, and competitive performance in practice. Like EAs, MOEAs are not guaranteed to provide exact solutions to an optimization problem, but they are designed to obtain a good approximation to the Pareto front in reasonable time.

The first implementation of a Pareto optimization approach for clustering was described by Delattre and Hansen (1980). This seminal work employed graph-coloring to provide an exact solution to a problem of bi-criterion clustering and served as an excellent proof of concept. The algorithm described is specific to the optimization of two concrete clustering criteria (diameter and split), however, and the computational expense of graph coloring restricts the applicability of the algorithm to small data sets.

It has since been recognized that these limitations of the approach may be overcome through the use of general-purpose meta-heuristic optimizers. In 2006, Handl and Knowles first described a multiobjective clustering approach (Multiobjective clustering with automatic K-determination, MOCK) that employed an MOEA as the optimizer, could be used with different sets of objectives and has been demonstrated to scale to data sets containing several thousand data points (Handl and Knowles, 2007). MOCK explores partitions corresponding to a range of different numbers of clusters, and supports the user in selecting the most appropriate solution, thus integrating aspects of cluster generation, cluster validation, and model selection. Empirical results on synthetic and real-world data sets have illustrated concrete performance advantages of the multicriterion approach when compared to traditional clustering techniques but also in comparison to state-of-the-art ensemble approaches and evidence accumulation (Handl and Knowles, 2007, 2013). In addition to the high accuracy of MOCK's clustering solutions, an advantageous feature of the algorithm is its robustness towards different data properties, a property that can be attributed directly to the use of multiple criteria (see Figure 19.1 above). The empirical performance advantages of multiobjective data clustering have been further confirmed by other researchers, but theoretical aspects of the approach as well as further methodological improvements (e.g., to achieve scalability to very large data sets) present challenging avenues for future research.

An implementation of multiobjective clustering is available from the authors' web page (see http://personalpages.manchester.ac.uk/mbs/julia.handl/mock.html), and an application example is described in the following section.

19.3.5.2 Application Example

As an example of the application of multiobjective clustering, we consider a data set from computational biology. The Leukemia data set (available from http://www.broad .mit.edu/cgi-bin/cancer/datasets.cgi) measures gene expression across 6817 genes for a set of 38 bone marrow samples (Golub et al., 1999). The samples come from patients with different types of leukemia: myeloid precursors (AML), as well as two sub-classes arising from lymphoid precursors (T-lineage ALL and B-lineage ALL). This data set is interesting, as it contains a hierarchical cluster structure (and some degree of overlap between clusters), but it is also small enough to provide an intuitive visualization of clustering results.

Prior to multiobjective clustering, a number of preprocessing steps are applied. Specifically, lower and upper threshold values of 600 and 16,000 are applied to all measurements and the 100 genes with the largest variation across samples are selected. Finally, all expression values are subjected to a log-transformation. This results in a data set of reduced dimensionality (100 genes) that is amenable to cluster analysis. We conduct a cluster analysis across samples, using the Pearson correlation coefficient as the similarity measure between samples.

The correct classifications of samples, and the results from multiobjective clustering, are shown in Figures 19.2 and 19.3. Multiobjective clustering identifies a three-cluster solution that misclassifies a single sample only. An alternative two-cluster solution is also identified

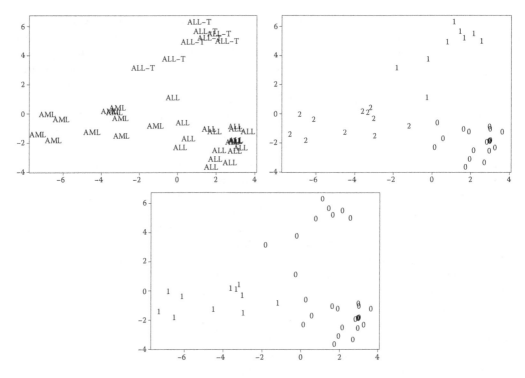

FIGURE 19.2
(Top left) Correct classification of samples; (Top right) Three-cluster solution identified by multiobjective cluster-ing; (Bottom left) Two-cluster solution identified multiobjective clustering. Data points are positioned according to the results from classical multidimensional scaling.

and corresponds to a perfect separation between the AML and ALL types. A visualiza-tion of all optimal solutions in objective space (see Figure 19.2) highlights the three-cluster solution as the most pronounced partition.

19.4 Conclusions

In this chapter, we have surveyed two groups of nature-inspired approaches that comple-ment the traditional clustering literature.

The first of these are self-organizing techniques with a direct inspiration from nature. Some of the interest in these techniques derives from the distributed nature of the algo-rithms, which achieve a complex solution as the consequence of simple local interactions. In practice, the bottom-up design of several of these techniques has been shown to result in tangible benefits regarding the accuracy of the resulting clustering solutions. Furthermore, these algorithms provide powerful approaches to non-standard clustering applications such as distributed robotics or the dynamic clustering of data streams. There remains nevertheless a challenge in assessing the real clustering capability of these techniques, and it is fair to say that really strong performance in large-scale applications is yet to be demonstrated.

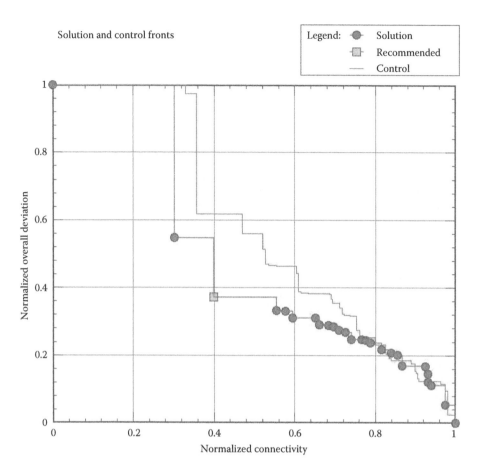

FIGURE 19.3

Results from multiobjective clustering.

The second half of this chapter discussed clustering approaches based on the use of nature-inspired meta-heuristics. We gave an overview of the variety of techniques available for heuristic optimization, and the ways in which these can and have been applied to clustering. In our opinion, the core strength of meta-heuristics is their flexibility; it makes them a particularly effective tool for the integration of knowledge into the clustering process, and enables very specific clustering objectives to be pursued. The advent of multiobjective clustering has begun to show that by combining objectives a better overall performance can be achieved compared with traditional approaches, while model selection is conveniently supported by this approach as well.

References

Aarts, E. and J. Korst. 1988. *Simulated Annealing and Boltzmann Machines*. John Wiley and Sons Inc., New York, NY.

Bäck, T., D. B. Fogel, and Z. Michalewicz. 1997. *Handbook of Evolutionary Computation*. IOP Publishing Ltd, Bristol, UK.

Blum, C. and A. Roli. 2003. Metaheuristics in combinatorial optimization: Overview and conceptual comparison. *ACM Computing Surveys* 35(3), 268–308.

Corne, D., M. Dorigo, and F. Glover (Eds.). 1999. *New Ideas in Optimization*. McGraw-Hill, New York.

Delattre, M. and P. Hansen. 1980. Bicriterion cluster analysis. *IEEE Transactions on Pattern Analysis and Machine Intelligence 4*, 277–291.

Deneubourg, J.-L., S. Aron, S. Goss, and J. M. Pasteels. 1990. The self-organizing exploratory pattern of the Argentine ant. *Journal of Insect Behavior* 3(2), 159–168.

Deneubourg, J.-L., S. Goss, N. Franks, A. Sendova-Franks, C. Detrain, and L. Chrétien. 1991. The dynamics of collective sorting: Robot-like ants and ant-like robots. In *Proceedings of the First International Conference on Simulation of Adaptive Behavior: From Animals to Animats*, pp. 356–363.

Dorigo, M., V. Maniezzo, and A. Colorni. 1996. Ant system: Optimization by a colony of cooperating agents. *IEEE Transactions on Systems, Man, and Cybernetics, Part B* 26(1), 29–41.

Dorigo, M. and T. Stützle. 2003. The ant colony optimization metaheuristic: Algorithms, applications, and advances. In *Handbook of Metaheuristics*, pp. 250–285. Springer, New York.

Dueck, D. and B. J. Frey. 2007. Non-metric affinity propagation for unsupervised image categorization. In *11th IEEE International Conference on Computer Vision (ICCV 2007)*, pp. 1–8. IEEE Press.

Eberhart, R. and J. Kennedy. 1995. A new optimizer using particle swarm theory. In *Proceedings of the Sixth International Symposium on Micro Machines and Human Science*, pp. 39–43. IEEE.

Fabrikant, S. I. and B. P. Buttenfield. 2000. *Spatial metaphors for browsing large data archives*. PhD thesis, University of Colorado.

Falkenauer, E. 1994. A new representation and operators for genetic algorithms applied to grouping problems. *Evolutionary Computation* 2(2), 123–144.

Goldstein, J. 1999. Emergence as a construct: History and issues. *Emergence* 1(1), 49–72.

Golub, T. R., D. K. Slonim, P. Tamayo, C. Huard, M. Gaasenbeek, J. P. Mesirov, H. Coller, M. L. Loh, J. R. Downing, M. A. Caligiuri, et al. 1999. Molecular classification of cancer: Class discovery and class prediction by gene expression monitoring. *Science* 286(5439), 531–537.

Güngör, Z. and A. Ünler. 2007. K-harmonic means data clustering with simulated annealing heuristic. *Applied Mathematics and Computation* 184(2), 199–209.

Handl, J. and J. Knowles. 2007. An evolutionary approach to multiobjective clustering. *IEEE Transactions on Evolutionary Computation* 11(1), 56–76.

Handl, J. and J. Knowles. 2013. Evidence accumulation in multiobjective data clustering. In *Evolutionary Multi-Criterion Optimization*, Volume 7811 of *LNCS*, pp. 543–557. Springer.

Handl, J., J. Knowles, and M. Dorigo. 2006. Ant-based clustering and topographic mapping. *Artificial Life* 12(1), 35–62.

Handl, J. and B. Meyer. 2007. Ant-based and swarm-based clustering. *Swarm Intelligence* 1(2), 95–113.

Haykin, S. 1994. *Neural Networks: A Comprehensive Foundation*. Prentice-Hall, Upper Saddle River, NJ.

Hruschka, E. R., R. J. G. B. Campello, A. A. Freitas, and A. C. P. L. F. de Carvalho. 2009. A survey of evolutionary algorithms for clustering. *IEEE Transactions on Systems, Man, and Cybernetics, Part C: Applications and Reviews* 39(2), 133–155.

Kohonen, T. 2001. *Self-Organizing Maps*. Springer, New York. 3rd edition.

Kuntz, P., P. Layzell, and D. Snyers. 1997. A colony of ant-like agents for partitioning in VLSI technology. In *Proceedings of the Fourth European Conference on Artificial Life*, pp. 417–424. MIT Press, Cambridge, MA.

Lumer, E. D. and B. Faieta. 1994. Diversity and adaptation in populations of clustering ants. In *Proceedings of the Third International Conference on Simulation of Adaptive Behavior: From Animals to Animats 3*, pp. 501–508. MIT Press.

Murtagh, F. and M. Hernandez-Pajares. 1995. The Kohonen self-organizing map method: An assessment. *Journal of Classification 12*, 165–190.

Park, Y. and M. Song. 1998. A genetic algorithm for clustering problems. In *Proceedings of the Third Annual Conference on Genetic Programming*, pp. 568–575. Morgan Kaufmann.

Proctor, G. and C. Winter. 1998. Information flocking: Data visualisation in virtual worlds using emergent behaviours. In *Virtual Worlds*, pp. 168–176. Springer, Berlin, Germany.

Reynolds, C. W. 1987. Flocks, herds and schools: A distributed behavioral model. In *ACM SIGGRAPH Computer Graphics*, Volume 21 (4), pp. 25–34. ACM Press.

Runkler, T. A. 2005. Ant colony optimization of clustering models. *International Journal of Intelligent Systems* 20(12), 1233–1251.

Selim, S. Z. and K. Alsultan. 1991. A simulated annealing algorithm for the clustering problem. *Pattern Recognition* 24(10), 1003–1008.

Trosset, M. W. 2008. Representing clusters: K-means clustering, self-organizing maps, and multidimensional scaling. Technical Report 08-03, Department of Statistics, Indiana University.

Van der Merwe, D. W. and A. P. Engelbrecht. 2003. Data clustering using particle swarm optimization. In *Proceedings of the Congress on Evolutionary Computation, (CEC'03)*, Volume 1, pp. 215–220. IEEE.

Walter, S. F. 2007. Clustering by affinity propagation. Master's thesis, Department of Physics, ETH Zürich.

Wolf, T. D. and T. Holvoet. 2005. Emergence versus self-organisation: Different concepts but promising when combined. In *Engineering Self-Organising Systems*, pp. 1–15. Springer-Verlag, Berlin/Heidelberg, Germany.

Section V

Specific Cluster and Data Formats

20

Semi-Supervised Clustering

Anil Jain, Rong Jin, and Radha Chitta

CONTENTS

Abstract

Clustering is an unsupervised learning problem, whose objective is to find a partition of the given data. However, a major challenge in clustering is to define an appropriate objective function in order to find an *optimal* partition that is useful to the user. To facilitate data clustering, it has been suggested that the user provide some supplementary information about the data (e.g., pairwise relationships between few data points), which, when incorporated in the clustering process, could lead to a better data partition. Semi-supervised clustering algorithms attempt to improve clustering performance by utilizing this supplementary information. In this chapter, we present an overview of semi-supervised clustering techniques and describe some prominent algorithms in the literature. We also present several applications of semi-supervised clustering.

20.1 Introduction

Clustering is an inherently ill-posed problem due to its unsupervised nature. Consider a grouping of a subset of face images from the CMU Face data set (Mitchell 1999),* shown in

---

* This data set is available in the UCI repository (Newman et al. 1998).

FIGURE 20.1
A clustering of a subset of face images from the CMU face data set (Mitchell 1999). It is possible to cluster these images in many ways, all of them equally valid. Faces have been clustered into two clusters based on the facial expression of the subjects in (a), and on the basis of whether or not the subjects are wearing sunglasses in (b). Without additional information from the user, it is not possible to determine which one is the correct or preferred partition.

Figure 20.1. These images have been clustered based on facial expression in Figure 20.1a, and on the basis of presence of sunglasses in Figure 20.1b. Both these partitions are equally valid, illustrating that the given data can be partitioned in many ways depending on user's intent and goal.

Most clustering algorithms seek a data partition that minimizes an objective function defined in terms of the data points and cluster labels. It is often the case that multiple partitions of the same data are equally good in terms of this objective function, making it difficult to determine the optimal data partition. Consider the two-dimensional data shown in Figure 20.2. If an algorithm such as K-means (Jain 2010) is used to cluster the points into two clusters, then both the partitions shown in Figures 20.2c and 20.2d are equally good in terms of the *sum-of-squared-error* criterion.* Hence, additional information (constraints) is required to resolve this ambiguity and determine the partition sought by the user. In many clustering applications, the user is able to provide some *side-information* besides the vector representation of data points (or the pairwise similarity between the data points). This side-information can be used to tune the clustering algorithm towards finding the data partition sought by the user. Semi-supervised clustering deals with mechanisms

---

* Sum-of-squared-error is defined as the sum of the square of the distances between each point and its corresponding cluster center.

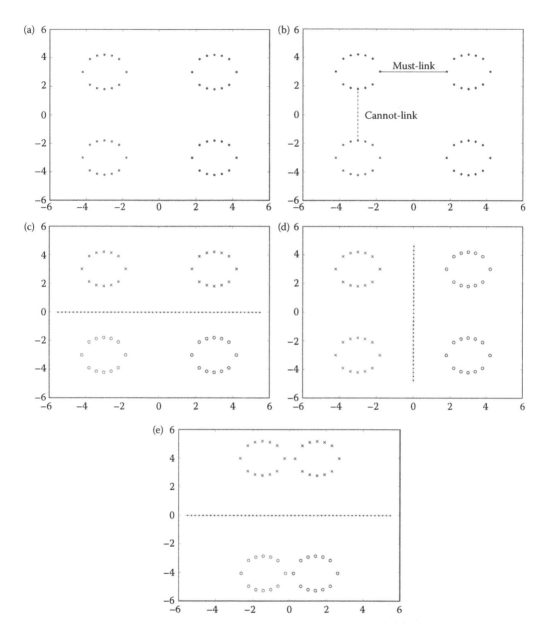

FIGURE 20.2
User-specified pairwise constraints. It is possible to partition the data points in (a) into two clusters in the following two ways: horizontally, as shown in (c), and vertically as shown in (d). While both of them yield the same sum-of-squared-error value, it is not possible to determine the desired partition, without additional information. Given the two pairwise constraints shown in (b), we can ascertain that the correct partition is the one in (c). A distance metric learning technique (Yang and Jin 2006) for incorporating the pairwise constraints is illustrated in (e).

to obtain and incorporate this side-information in the clustering process to attain better clustering performance. Formally, given a data set $\mathcal{D} = (\mathbf{x}_1, \ldots, \mathbf{x}_n)$ containing n points, side-information π, and a clustering algorithm $\mathcal{A}$, the objective of semi-supervised clustering is to augment $\mathcal{A}$ with π, and partition $\mathcal{D}$ into K clusters $\{\mathcal{C}_1, \ldots, \mathcal{C}_K\}$. It is expected that the resulting partition of $\mathcal{D}$ will be better than the partition obtained from $\mathcal{A}$ in the absence of side-information. In this chapter, two measures of clustering performance are employed: F-measure and Normalized Mutual Information. These are *external* measures adapted from information retrieval, which compare the data partition obtained from the clustering algorithm with the true class labels:

F-measure: A pair of points $(\mathbf{x}_i, \mathbf{x}_j)$ are said to be correctly paired if (i) they have the same cluster label, and they have the same class label (represented as true positive TP), or (ii) they have different cluster labels and different class labels (represented as true negative TN). Otherwise they are said to be paired incorrectly: False positive FP if they fall in the same cluster but belong to different classes, and false negative FN if they are assigned to different clusters but belong to the same class. F-measure is defined as

$$\frac{2\#TP}{2\#TP + \#FN + \#FP},$$

where # represents the number of the corresponding quantity.

Normalized Mutual Information (NMI): Let n_i^c represent the number of data points that have been assigned to the ith cluster ($1 \leq i \leq K$), n_j^p the number of data points from the jth class ($1 \leq j \leq P$), and $n_{i,j}^{c,p}$ the number of data points from class j that have been assigned to the ith cluster. NMI is defined as

$$\frac{\displaystyle\sum_{i=1}^{K} \sum_{j=1}^{P} n_{i,j}^{c,p} \log\left(n \frac{n_{i,j}^{c,p}}{n_i^c n_j^p}\right)}{\sqrt{\left(\displaystyle\sum_{i=1}^{K} n_i^c \log \frac{n_i^c}{n}\right)\left(\displaystyle\sum_{j=1}^{P} n_j^p \log \frac{n_j^p}{n}\right)}}.$$

There are two main issues that need to be addressed in semi-supervised clustering: (a) How is the side-information obtained and specified? and (b) How can the side-information be used to improve the clustering performance? These questions are addressed in Sections 20.1.1 and 20.1.2.

20.1.1 Acquisition and Expression of Side-Information

The most common forms of expressing the side-information are pairwise must-link and cannot-link constraints (Basu et al. 2009). A *must-link* constraint between two data points $\mathbf{x}_a$ and $\mathbf{x}_b$, denoted by $\mathcal{ML}(\mathbf{x}_a, \mathbf{x}_b)$, implies that the points $\mathbf{x}_a$ and $\mathbf{x}_b$ must be assigned the same cluster label. A *cannot-link* constraint, denoted by $\mathcal{CL}(\mathbf{x}_a, \mathbf{x}_b)$, implies that the points $\mathbf{x}_a$ and $\mathbf{x}_b$ should be assigned to different clusters. Figure 20.2b illustrates two such constraints applied to partition the data shown in Figure 20.2a. Given the pairwise must-link and cannot-link constraints, we can bias the clustering algorithm towards obtaining the desired partition. Figure 20.3 shows the improvement in the clustering performance, measured in terms of the F-measure (Manning et al. 2008), as a result of providing pairwise constraints, on six benchmark data sets. All the semi-supervised clustering algorithms considered in (Bilenko et al. 2004), namely, supervised-means, PCK-means, MK-means, and MPCK-means, perform better than the K-means clustering algorithm, when provided with a sufficient number of pairwise constraints (Bilenko et al. 2004).

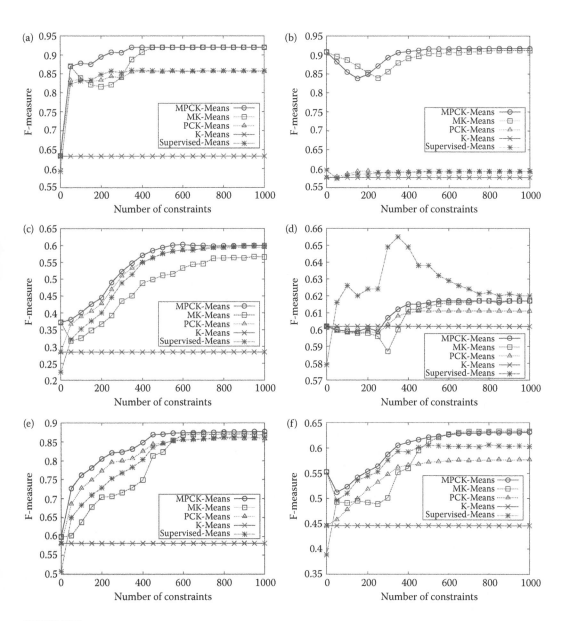

FIGURE 20.3
Improvement in clustering performance of the *K*-means algorithm using pairwise constraints on six benchmark data sets. (a Iris, b Wine, c Protein, d Ionosphere, e Digits-389, f Letters-IJL). (Reproduced from M. Bilenko, S. Basu, and R.J. Mooney. *Proceedings of the International Conference on Machine Learning*, pages 81–88, 2004.)

Pairwise constraints for data clustering occur naturally in many application domains. In applications involving graph clustering, such as social network analysis, the given edges in the graph indicate the pairwise relationships. Some protein data sets* contain information about co-occurring proteins, which can be viewed as must-link constraints during clustering (Kulis et al. 2005). In image segmentation, neighboring pixels are likely to be a

---

* The Database of Interacting Proteins (Xenarios et al. 2000).

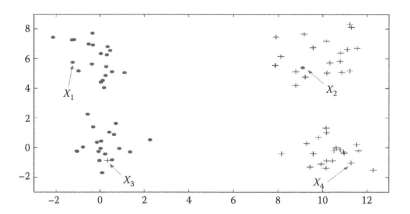

FIGURE 20.4
Pairwise constraints can lead to counter-intuitive solutions. Points pairs (x_1, x_2) and (x_3, x_4) are related by a must-link constraints. x_2 and its neighbors, and x_3 and its neighbors are related by cannot-link constraints. This yields a counter-intuitive clustering solution, where points x_2 and x_3 are assigned cluster labels that are different from those of their neighbors. (From M.H.C. Law, A. Topchy, and A.K. Jain. *Proceedings of SIAM International Conference on Data Mining*, pages 641–645, 2005.)

part of the same homogeneous region in an image, whereas pixels which are far from each other tend to belong to different regions (Law et al. 2007). This fact can be used to generate pairwise constraints. Pairwise constraints can be derived from the domain knowledge and other external sources as well (Huang et al. 2008; Pletschacher et al. 2009). For example, Wikipedia (Allab and Benabdeslem 2011) was used to identify semantic relationships between documents in Huang et al. (2008).

Although the side-information is, in general, expected to improve the clustering performance, inaccurate or conflicting pairwise constraints may actually degrade the clustering performance (Davidson and Ravi 2005a,b, 2006; Davidson et al. 2006). For example, consider the scenario shown in Figure 20.4, where point pairs (x_1, x_2) and (x_3, x_4) are involved in must-link constraints, but x_2 and its neighbors, and x_3 and its neighbors are related by cannot-link constraints. These constraints lead to a counter-intuitive clustering solution, where points x_2 and x_3 are assigned cluster labels that are different from those of their neighbors. This behavior can also be observed while clustering real data sets, as shown in Figure 20.3. Some semi-supervised clustering algorithms exhibit a small dip in the performance, for some number of constraints. In other words, the performance of semi-supervised clustering algorithms is not guaranteed to increase monotonically with the number of constraints. For example, in Figure 20.3d, the accuracy of the MK-means algorithm on the Ionosphere data set is lower than that of the unsupervised K-means algorithm when the number of constraints is less than 400. This suggests that the nature of the constraints, not simply the number of constraints, is crucial for attaining performance gain from semi-supervised clustering.

In order to identify the most informative pairwise constraints, some studies (Basu et al. 2004; Huang and Lam 2007; Wang and Davidson 2010) have focused on active learning (Settles 2009), originally developed for semi-supervised classification (Cohn et al. 1994). Semi-supervised classification techniques based on active learning assume the presence of an oracle which can supply the class labels for points selected from a large pool of unlabeled points. Starting with a small set of labeled points, the oracle is iteratively queried for the labels of the points most useful for determining the classification model. The key idea behind active learning-based semi-supervised clustering is to find the constraints that would be violated if the clustering algorithm was executed without supervision (Allab and Benabdeslem 2011; Zhao et al. 2011). Most active clustering techniques assume an oracle can answer queries involving pairwise constraints among data points. These algorithms differ in the manner in which the queries are made. For instance, the active PCK-means algorithm (Basu et al. 2004) aims at identifying the individual cluster boundaries. It first identifies pairs of data points which are farthest from each other, and queries the oracle until a predefined number of cannot-link constraints are obtained. It then queries the relationship between the points involved in the cannot-link constraints and their nearest neighbors to obtain must-link constraints. The active spectral clustering technique (Wang and Davidson 2010), on the other hand, iteratively refines the data partition by querying the pairwise relationship between the data points which leads to the largest change in the current partition towards the desired partition. Figure 20.5 compares the PCK-means algorithm, which assumes pairwise constraints are available *a priori*, with the active PCK-means algorithm on three data sets. The active PCK-means algorithm achieves better accuracy, measured in terms of the Normalized Mutual Information (Kvalseth 1987) with respect to the true cluster membership, with fewer number of constraints, demonstrating that active clustering is able to identify the most informative constraints (Basu et al. 2004).

Given that the most common form of specifying side-information to clustering algorithms is pairwise constraints, semi-supervised clustering is also referred to as *constrained clustering* to distinguish between semi-supervised clustering and general semi-supervised learning (Chapelle et al. 2006). In its most popular form, semi-supervised learning involves using a large number of unlabeled examples along with the labeled training set to improve the learning efficiency. Figure 20.6 illustrates the spectrum of learning methodologies as we transition from supervised learning to unsupervised learning.

Besides pairwise constraints, other forms of side-information and constraints to obtain the desired clusters have also been studied in the literature. Class labels for a subset of the data set to be clustered can be used as the side-information (Demiriz et al. 1999; Basu et al. 2002; Kamvar et al. 2003). In applications such as document clustering and multimedia retrieval, class labels are easily obtained through crowdsourcing (Barbier et al. 2011) tools such as the Amazon Mechanical turk (https://www.mturk.com/). Class labels of subsets of data can be used to assign temporary labels to the remaining unlabeled data points by learning a classifier. These labels can then be employed to constrain the solution search space (Demiriz et al. 1999; Gao 2006). The class labels can also be employed to initialize the clusters in addition to restricting the possible partitions (Basu et al. 2002).

Triplet constraints deal with the relative distance between sets of three data points, and are less restrictive than pairwise constraints. A triplet constraint (x_a, x_b, x_c) indicates that x_a is closer to x_b than x_a is to x_c, which, in turn, implies that x_a is more likely to form a cluster with x_b than with x_c. These constraints are used to estimate the similarity between data points and thereby enhance the clustering performance (Kumar and Kummamuru 2008; Huang et al. 2010; Bi et al. 2011). The SSSVaD algorithm (Kumar and Kummamuru 2008) uses the triplet constraints to learn the underlying dissimilarity measure. In Bade and Nürnberger (2008), Zhao and Qi (2010), Zheng and Li (2011), the triplet constraints are used to determine the order in which the agglomerative hierarchical clustering algorithm merges the clusters.

When the data set is sufficiently small in size, user feedback can also be employed to obtain the partition sought by the user (Cohn et al. 2003; Gondek 2005; Halkidi et al. 2005; Hu et al. 2011). For instance, in Gondek (2005), the user is iteratively presented with a possible partition of the data, and allowed to choose whether or not it is the desired partition. This feedback is used to eliminate undesired partitions that are similar to the one presented

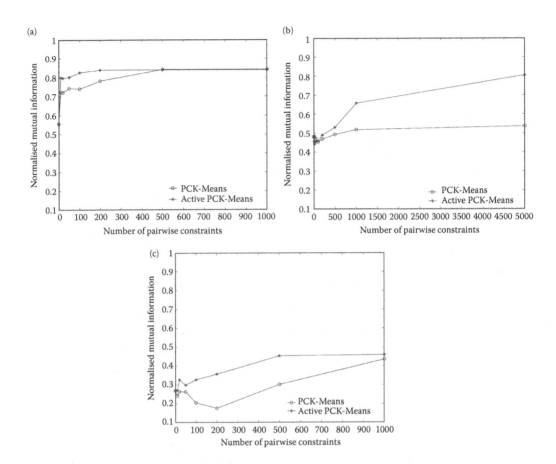

FIGURE 20.5
Better clustering results are achieved by employing active learning to obtain the most informative pairwise constraints. (a Iris, b Newsgroups, c Classic 3). (Reproduced from S. Basu, A. Banerjee, and R.J. Mooney. *Proceedings of the SIAM International Conference on Data Mining*, pages 333–344, 2004.)

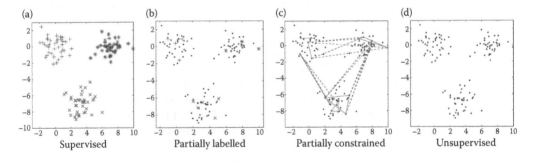

FIGURE 20.6
Spectrum between supervised and unsupervised learning. Dots correspond to points without label information. Points with labels are denoted by pluses, asterisks, and crosses, each representing a different class. In (c), the must-link and cannot-link constraints are denoted by solid and dashed lines, respectively. (From T. Lange et al. *Proceedings of the IEEE Conference on Computer Vision and Pattern Recognition*, volume 1, pages 731–738, 2005.)

to the user. User supervision is employed to select the most informative features that lead to the desired clusters in Hu et al. (2011).

20.1.2 Incorporation of Side-Information

Several mechanisms have been developed to exploit the side-information to achieve better clustering performance. They can be classified into two main categories: (i) methods based on constraining the solution space, and (ii) methods based on distance metric learning (Basu et al. 2009).

20.1.2.1 Solution Space Restriction

Most methods in this category deal with side-information in the form of pairwise constraints. They use the pairwise constraints to restrict the feasible data partitions when deciding the cluster assignment. The COP-Kmeans (Wagstaff et al. 2001) and the SS-SOM (Allab and Benabdeslem 2011) algorithms modify the cluster membership update phase of the K-means and the Self Organizing map algorithms respectively, to ensure that the data partitions are consistent with the given pairwise constraints. In the COP-Kmeans algorithm, the cluster centers are first initialized randomly. Each data point is then assigned to the nearest cluster center ensuring that no constraints are violated. The cluster centers are updated by finding the mean of the points assigned to the cluster, like in the K-means algorithm. In Shental et al. (2004), the generalized Expectation Maximization (EM) algorithm is modified such that only the mixture models that are compliant with the constraints are considered. These approaches treat the side-information as *hard constraints* and ensure that all the constraints are strictly satisfied. As mentioned before, such an approach may lead to counter-intuitive clustering solutions, as shown in Figure 20.4, and may even render the clustering problem infeasible.

A number of studies have used the side-information in the form of *soft constraints*. Instead of trying to satisfy all the constraints, the key idea behind such methods is to satisfy as many constraints as possible, and introduce a penalty term to account for constraints that cannot be satisfied (Kamvar et al. 2003; Basu et al. 2004; Lange et al. 2005; Law et al. 2005; Bekkerman and Sahami 2006). In Law et al. (2005), the authors modified the mixture model for data clustering by redefining the data generation process through the introduction of hidden variables. In Lange et al. (2005), a mean field approximation method was proposed to find appropriate data partition that is consistent with pairwise constraints. The pairwise constraints are enforced in the form of additional penalty terms in the objective function for clustering in spectral learning (Kamvar et al. 2003) and PCK-means (Basu et al. 2004).

20.1.2.2 Distance Metric Learning

The distance measure used to determine the dissimilarity between data points is crucial to the clustering process. Semi-supervised clustering methods which fall under this category attempt to find and apply a transformation to the data such that (a) the data points in must-link constraints are separated by small distances, and (b) data points in cannot-link constraints are separated by larger distances (Yang and Jin 2006). The distance between any two data points x_a and x_b is expressed as

$$d_M(x_a, x_b) = \|x_a - x_b\|_M^2 = (x_a - x_b)^\top M(x_a - x_b),$$

where M is the distance metric cast as a positive semi-definite matrix.

An example of distance metric learning is illustrated in Figure 20.2e. The data points which satisfy must-link constraints move closer to each other, and the data points that satisfy cannot-link constraints move farther away. After learning this distance metric, any conventional clustering algorithm such as K-means or spectral clustering can be applied to the resulting similarity matrix.

Techniques for distance metric learning have been studied extensively in the semi-supervised and unsupervised learning literature (Yang and Jin 2006). *Local* distance metric learning techniques only focus on constraints in local regions, and are typically used in semi-supervised classification (Atkeson et al. 1997; Gunopulos 2001). On the other hand, *global* distance metric learning methods consider all the pairwise constraints simultaneously (Xing et al. 2002; Liu et al. 2007; Cheng et al. 2008; Yan et al. 2009; Guillaumin et al. 2010). For example, a convex optimization problem that minimizes the distance between points that are related by must-link constraints, and maximizes the distance between points that are related by cannot-link constraints, is solved to find the optimal metric in (Xing et al. 2002). Techniques to learn nonlinear distance metrics such as Bregman divergence were proposed in Cohn et al. (2003), Wu et al. (2010). Besides pairwise constraints, triplet constraints have also been employed for distance metric learning.

The idea of distance metric learning for semi-supervised clustering has been extended to learn the kernel representing the pairwise data similarity (Hoi et al. 2007; Domeniconi et al. 2011). Similar to distance metric learning, the kernel similarity function is modified to accommodate the given pairwise constraints, that is, (a) data points in must-link relationships have large similarity, and (b) data points in cannot-link relationships have small similarity. The kernel similarity is modified by incorporating the constraints in the objective function in (Kulis et al. 2005; Basu et al. 2006; Chen et al. 2007). Non-parametric approaches for kernel learning are proposed in Hoi et al. (2007), Baghshah and Shouraki (2010) to learn the pairwise similarity measure.

Methods that combine the two approaches to semi-supervised clustering have also been proposed (Bilenko et al. 2004; Basu et al. 2006). One example is the MPCK-means (Bilenko et al. 2004), which performs both solution space restriction and distance metric learning. It performs better than the methods that employ only distance metric learning, or methods that only constrain the solution space.

Table 20.1 presents a summary of the major semi-supervised clustering techniques proposed in the literature. They are classified based on the form of the available side-information and the manner in which the side-information is acquired and incorporated.

20.2 Semi-Supervised Clustering Algorithms

We describe three semi-supervised clustering algorithms in this section. These three algorithms are representatives of various approaches that have been used for semi-supervised clustering. The semi-supervised kernel K-means (SSKKM) (Kulis et al. 2005) and Boost-Cluster (Liu et al. 2007) algorithms are based on distance metric learning. The SSKKM algorithm modifies the pairwise similarity between the data points using the must-link and cannot-link constraints. The BoostCluster algorithm projects the data into a subspace where points related by must-links are closer to each other, than the points related by cannot-links. In both these algorithms, the pairwise constraints are assumed to be available *a priori*. The active spectral clustering algorithm (Wang and Davidson 2010) obtains the constraints using the active learning mechanism. It then finds the solution by restricting the solution space of the spectral clustering algorithm.

20.2.1 Semi-Supervised Kernel *K*-Means

The semi-supervised kernel *K*-means algorithm, abbreviated as **SSKKM**, is based on the strategy of distance metric learning. It aims to enhance the accuracy of the *K*-means algorithm by constructing a kernel matrix, which incorporates the given pairwise constraints.

The objective of the *K*-means algorithm is to minimize the sum-of-squared-error, expressed as the following optimization problem:

$$\min \sum_{k=1}^{K} \sum_{x_i \in C_k} \|x_i - c_k\|^2, \tag{20.1}$$

where c_k is the cluster center of $C_k, k = 1, 2, \ldots, K$.

TABLE 20.1

Summary of Prominent Semi-Supervised Clustering Techniques

Type of Side-Information	Side-Information Acquisition	Side-Information Incorporation	Examples
Pairwise constraints	Prior knowledge	Constrain the solution space	PCK-Means (Basu et al. 2004), Spectral Learning (Kamvar et al. 2003), COPK-Means (Wagstaff et al. 2001; Shental et al. 2004; Lange et al. 2005; Law et al. 2005; Bekkerman and Sahami 2006; Yan and Domeniconi 2006; Li et al. 2007; Ruiz et al. 2010; Wang and Davidson 2010)
Pairwise constraints	Prior knowledge	Distance metric learning	HMRF-K-means (Basu et al. 2006), MPCK-means (Bilenko et al. 2004), SSKKM (Kulis et al. 2005), BoostCluster (Xing et al. 2002; Cohn et al. 2003; Chen et al. 2007, 2009; Hoi et al. 2007; Liu et al. 2007; Yan et al. 2009; Baghshah and Shouraki 2010; Wu et al. 2010)
Pairwise constraints	Active learning	Constrain the solution space	Active PCK-Means (Basu et al. 2004), Active spectral clustering (Huang and Lam 2007; Huang et al. 2008; Wang and Davidson 2010; Allab and Benabdeslem 2011; Zhao et al. 2011)
Pairwise constraints	Active learning	Distance metric learning	Guillaumin (2010)
Class labels	Prior knowledge, Crowdsourcing	Constrain the solution space	Constrained *K*-means (Demiriz et al. 1999; Basu et al. 2002; Gao et al. 2006)
Triplet constraints	Prior knowledge	Distance metric learning	SSSVaD (Kumar and Kummamuru 2008; Zhao and Qi 2010; Zheng and Li 2011)
	User feedback	Constrain the solution space	Cohn et al. (2003a,b); Gondek 2005; Halkidi et al. (2005); Hu et al. (2011)

Let $w_{a,b}$ be the cost of violating the constraint between data points $\mathbf{x}_a$ and $\mathbf{x}_b$. The set of pairwise must-link and cannot-link constraints, denoted by $\mathcal{ML}$ and $\mathcal{CL}$ respectively, are embedded in the optimization problem (20.1) as follows:

$$\min \sum_{k=1}^{K} \left(\sum_{\mathbf{x}_i \in \mathcal{C}_k} \|\mathbf{x}_i - \mathbf{c}_k\|^2 - \sum_{\substack{(\mathbf{x}_p, \mathbf{x}_q) \in \mathcal{ML} \\ \mathbf{x}_p, \mathbf{x}_q \in \mathcal{C}_k}} \frac{w_{p,q}}{n_k} + \sum_{\substack{(\mathbf{x}_r, \mathbf{x}_s) \in \mathcal{CL} \\ \mathbf{x}_r, \mathbf{x}_s \in \mathcal{C}_k}} \frac{w_{r,s}}{n_k} \right),$$

which is equivalent to

$$\min \sum_{k=1}^{K} \left(\sum_{\mathbf{x}_i, \mathbf{x}_j \in \mathcal{C}_k} \frac{\|\mathbf{x}_i - \mathbf{x}_j\|^2}{n_k} - \sum_{\substack{(\mathbf{x}_p, \mathbf{x}_q) \in \mathcal{ML} \\ \mathbf{x}_p, \mathbf{x}_q \in \mathcal{C}_k}} \frac{2w_{p,q}}{n_k} + \sum_{\substack{(\mathbf{x}_r, \mathbf{x}_s) \in \mathcal{CL} \\ \mathbf{x}_r, \mathbf{x}_s \in \mathcal{C}_k}} \frac{2w_{r,s}}{n_k} \right), \tag{20.2}$$

where n_k is the number of data points assigned to cluster $\mathcal{C}_k$.

Let $U = [u_{k,j}]$ be the $K \times n$ normalized cluster membership matrix defined as

$$u_{k,j} = \begin{cases} 1/\sqrt{n_k} & \text{if } \mathbf{x}_j \in \mathcal{C}_k, \\ 0 & \text{otherwise.} \end{cases}$$

Also, define the Euclidean similarity matrix $E = [E_{a,b}]$ and the constraint matrix $W = [W_{a,b}]$ as

$$E_{a,b} = \|\mathbf{x}_a - \mathbf{x}_b\|^2 = \mathbf{x}_a^\top \mathbf{x}_a + \mathbf{x}_b^\top \mathbf{x}_b - 2\mathbf{x}_a^\top \mathbf{x}_b \tag{20.3}$$

and

$$W_{a,b} = \begin{cases} w_{a,b} & \text{if there is a must-link constraint between } \mathbf{x}_a \text{ and } \mathbf{x}_b, \\ -w_{a,b} & \text{if there is a cannot-link constraint between } \mathbf{x}_a \text{ and } \mathbf{x}_b, \\ 0 & \text{otherwise.} \end{cases}$$

The problem in (20.2) can be re-written as

$$\min \operatorname{trace}\left(U \left(E - 2W \right) U^\top \right),$$

which is equivalent to the trace maximization problem

$$\max \operatorname{trace}\left(U \mathcal{K} U^\top \right), \tag{20.4}$$

where the kernel matrix is given by $\mathcal{K} = S + W$ and the entries of the similarity matrix $S = [S_{a,b}]$ are given by $S_{a,b} = \mathbf{x}_a^\top \mathbf{x}_b$.

It has been shown that the optimization problem (20.4) can be solved by applying the kernel K-means algorithm (Scholkopf et al. 1996; Girolami 2002) on the kernel matrix $\mathcal{K}$. To ensure that $\mathcal{K}$ is positive semi-definite, it is diagonal shifted using a positive parameter σ. The method is illustrated in Figure 20.7.

Algorithm SSKKM

Input

- Input data set $\mathcal{D} = (x_1, \ldots, x_n)$.
- Set of pairwise constraints $\mathcal{ML}$ and $\mathcal{CL}$.
- Constraint penalty matrix W.
- Kernel function $\kappa(\cdot, \cdot)$ to obtain the pairwise similarity matrix.
- Parameter σ for making the kernel matrix positive semi-definite.
- Number of clusters K.

1. Compute the $n \times n$ similarity matrix S using the kernel function $S_{i,j} = \kappa(x_i, x_j)$.
2. Compute the kernel matrix $\mathcal{K} = S + W + \sigma I$.
3. Initialize the cluster membership of all the data points, ensuring no constraints are violated.
4. Repeat until convergence or a pre-defined maximum number of iterations is reached:

   a. For each point x_i and cluster $\mathcal{C}_k$, compute distance

   $$d(x_i, \mathcal{C}_k) = \mathcal{K}_{i,i} - \frac{2\sum_{x_j \in \mathcal{C}_k} \mathcal{K}_{i,j}}{n_k} + \frac{\sum_{x_j, x_l \in \mathcal{C}_k} \mathcal{K}_{j,l}}{n_k^2},$$

   where n_k is the number of points assigned to cluster $\mathcal{C}_k$.

   b. Assign each point x_i to the cluster $\mathcal{C}_k^*$ which minimizes the distance $d(x_i, \mathcal{C}_k)$, resolving ties arbitrarily.

Output Cluster memberships of the data points.

$S_{a,b}$ was replaced by the RBF kernel $\kappa(x_a, x_b) = \exp(-\lambda \|x_a - x_b\|^2)$ to achieve better clustering performance (Kim et al. 2005). An adaptive scheme to estimate the kernel width λ using the pairwise constraints was proposed in (Yan and Domeniconi 2006). This was done by scaling the penalty terms in (20.2) by the kernel distance between the data points involved in the pairwise constraints.

20.2.2 BoostCluster

The BoostCluster algorithm (Liu et al. 2007) follows the general boosting framework employed in data classification. Given a clustering algorithm $\mathcal{A}$, the BoostCluster algorithm iteratively modifies the input data representation, ensuring that the data points related by must-link constraints are more similar to each other than the data points related by cannot-link constraints. This idea is illustrated in Figure 20.8. A two-dimensional representation of a subset of the *Balance scale* data set (http://archive.ics.uci.edu/ml/datasets/Balance Scale) from the UCI repository (Newman et al. 1998) is shown in Figure 20.8a. The three clusters in this data are represented by triangles, crosses, and circles. The must-link and cannot-link constraints are represented using the solid and dotted lines, respectively. The data is iteratively projected into subspaces such that the constraints are satisfied

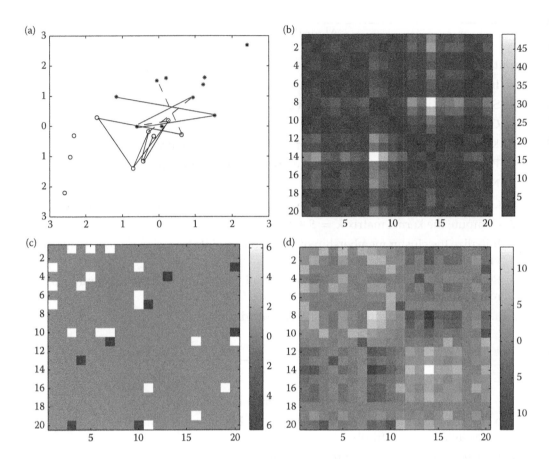

FIGURE 20.7
Illustration of SSKKM algorithm (Kulis et al. 2005) on a toy two-dimensional example. (a) shows the 2-D dataset containing 20 points, along with the must-link (solid lines) and cannot-link constraints (dashed lines). (b)–(d) represent the Euclidean similarity matrix S, the constraint matrix W, and the kernel matrix $\mathcal{K} = S + W$ between the points. The points get clustered perfectly into the two groups indicated in (a) on executing kernel K-means with $\mathcal{K}$ as the input, whereas the points do not get clustered as expected using the Euclidean similarity between them.

(see Figures 20.8(b)–(d)), thereby increasing the separation among the three clusters and enhancing the clustering performance.

The key steps in the BoostCluster algorithm involve (i) identifying the constraints that are not satisfied by the current partition, and (ii) projecting the data into a space where the points linked by must-link constraints are relatively closer to each other than the points linked by cannot-link constraints.

Let kernel matrix $\mathcal{K}$ denote the current similarity between the points. The objective of Boostcluster is to minimize the inconsistency between $\mathcal{K}$ and the pairwise constraints:

$$\mathcal{L}(\mathcal{K}) = \sum_{i,j=1}^{n} \sum_{a,b=1}^{n} \mathcal{ML}(\mathbf{x}_i, \mathbf{x}_j) \mathcal{CL}(\mathbf{x}_a, \mathbf{x}_b) \, \exp(\mathcal{K}_{a,b} - \mathcal{K}_{i,j}), \tag{20.5}$$

where $\mathcal{ML}(\mathbf{x}_i, \mathbf{x}_j) = 1$ if there is a must-link constraint between points $\mathbf{x}_i$ and $\mathbf{x}_j$ and 0 otherwise, and $\mathcal{CL}(\mathbf{x}_i, \mathbf{x}_j) = 1$ if there is a cannot-link constraint between points $\mathbf{x}_i$ and $\mathbf{x}_j$ and 0 otherwise.

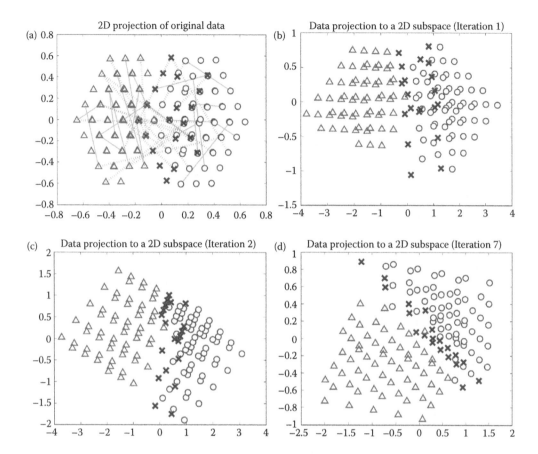

FIGURE 20.8
Illustration of BoostCluster algorithm (Liu et al. 2007) on the Balance Scale data set (http://archive.ics.uci.edu/ml/datasets/Balance Scale). A two-dimensional projection of the Balance Scale data set, obtained using PCA, is shown in Figure (a). Figures (b)–(d) show the derived data representations based on the must-link and cannot-link constraints in iterations 1, 2 and 7 of the BoostCluster algorithm. (a Input data, b Iteration 1, c Iteration 2, d Iteration 7). (From Y. Liu, R. Jin, and A.K. Jain. *Proceedings of the International Conference on Knowledge Discovery and Data Mining*, pages 450–459, 2007.)

Let $\Delta = [\Delta_{i,j}]$ represent the incremental similarity matrix inferred from the current partition of the data. Entry $\Delta_{i,j} = 1$ when points x_i and x_j belong to the same cluster in the current partition, and $\Delta_{i,j} = 0$ otherwise. The kernel is incrementally updated as $\mathcal{K}' = \mathcal{K} + \alpha \Delta$, where α is a weight parameter. Let matrix $T = [T_{i,j}]$, defined by

$$T_{i,j} = \frac{p_{i,j}}{\sum_{a,b=1}^{n} p_{a,b}} - \frac{q_{i,j}}{\sum_{a,b=1}^{n} q_{a,b}},$$

where $p_{i,j} = \mathcal{ML}(x_i, x_j) \exp(-\mathcal{K}_{i,j})$ and $q_{i,j} = \mathcal{CL}(x_i, x_i) \exp(\mathcal{K}_{i,j})$, represent the inconsistency between $\mathcal{K}$ and the constraints. A large positive (negative) value of the entry $T_{a,b}$ indicates that the corresponding entry in the similarity matrix $\mathcal{K}_{a,b}$ does not reflect the must-link (cannot-link) constraint between the data points x_a and x_b. Using Jensen's inequality, the loss $\mathcal{L}(\mathcal{K}')$ can be upper bounded by

$$\mathcal{L}(\mathcal{K}') \leq \mathcal{L}(\mathcal{K}) \times \left(\frac{(\exp(3\alpha) + \exp(-3\alpha) + 1) - (1 - \exp(-3\alpha))\operatorname{trace}(T\Delta)}{3} \right)$$

Algorithm BOOSTCLUSTER

Input
- Input data set $\mathcal{D} = (\mathbf{x}_1, \ldots, \mathbf{x}_n)$.
- Clustering algorithm $\mathcal{A}$.
- Set of pairwise constraints $\mathcal{ML}$ and $\mathcal{CL}$.
  $\mathcal{ML}(\mathbf{x}_i, \mathbf{x}_j) = 1$ if there exists a must-link constraint between points $\mathbf{x}_i$ and $\mathbf{x}_j$, and 0 otherwise.
  $\mathcal{CL}(\mathbf{x}_i, \mathbf{x}_j) = 1$ if there exists a cannot-link constraint between points $\mathbf{x}_i$ and $\mathbf{x}_j$, and 0 otherwise.
- Number of principal eigenvectors s to be used for data projection.
- Number of clusters K.
1. Initialize $\mathcal{K}_{i,j} = 0 \; \forall i, j = 1, 2, \ldots, n$.
2. Repeat until all constraints are satisfied or a predefined maximum number of iterations is reached:
   a. Compute the inconsistency matrix T given by

   $$T_{i,j} = \frac{p_{i,j}}{\sum_{a,b=1}^n p_{a,b}} - \frac{q_{i,j}}{\sum_{a,b=1}^n q_{a,b}},$$

   where $p_{i,j} = \mathcal{ML}(\mathbf{x}_i, \mathbf{x}_j)\exp(-\mathcal{K}_{i,j})$ and $q_{a,b} = \mathcal{CL}(\mathbf{x}_a, \mathbf{x}_b)\exp(\mathcal{K}_{a,b})$.

   b. Construct the projection matrix $P = (\sqrt{\lambda_1}\mathbf{v}_1, \sqrt{\lambda_2}\mathbf{v}_2, \ldots, \sqrt{\lambda_s}\mathbf{v}_s)$, where $\{(\lambda_i, \mathbf{v}_i)\}_{i=1}^s$ represent the top s (non-zero) eigenvalues and the corresponding eigenvectors of $\mathcal{D}T\mathcal{D}^\top$.

   c. Project the data $\mathcal{D}$ into the space spanned by the vectors in P and obtain the new representation of the data set $\widehat{\mathcal{D}} = P^\top \mathcal{D}$.

   d. Run algorithm $\mathcal{A}$ with $\widehat{\mathcal{D}}$ as the input.

   e. Update the similarity matrix $\mathcal{K}$ as $\mathcal{K} = \mathcal{K} + \alpha\Delta$, where

   $$\alpha = \frac{1}{2}\log\left(\frac{\sum_{i,j=1}^n p_{i,j}\delta(\Delta_{i,j}, 1)}{\sum_{i,j=1}^n p_{i,j}\delta(\Delta_{i,j}, 0)} \times \frac{\sum_{i,j=1}^n q_{i,j}\delta(\Delta_{i,j}, 0)}{\sum_{i,j=1}^n q_{i,j}\delta(\Delta_{i,j}, 1)} \right) \quad \text{and}$$

   $$\Delta_{i,j} = \begin{cases} 1 & \text{if } x_i \text{ and } x_j \text{ belong to the same cluster} \\ 0 & \text{otherwise.} \end{cases}$$

3. Run algorithm $\mathcal{A}$ with $\mathcal{K}$ as kernel similarity matrix or with the data representation generated by the top $s + 1$ eigenvectors of $\mathcal{K}$.

Output Cluster memberships of the data points.

In order to ensure that $\mathcal{L}(\mathcal{K}') \leq \mathcal{L}(\mathcal{K})$ in successive iterations of the algorithm, the upper bound is minimized with respect to α, which can be accomplished by maximizing the expression trace $(T\Delta)$. The incremental kernel matrix Δ is approximated as $(P^\top \mathcal{D})^\top (P^\top \mathcal{D})$, where P is a projection matrix that specifies the direction along which the data should be projected to obtain the new data representation. The optimal projection matrix is obtianed as $P = (\sqrt{\lambda_1}\mathbf{v}_1, \sqrt{\lambda_2}\mathbf{v}_2, \ldots, \sqrt{\lambda_s}\mathbf{v}_s)$, where $\{(\lambda_i, \mathbf{v}_i)\}_{i=1}^s$ represent the top s (non-zero) eigenvalues and the corresponding eigenvectors of $\mathcal{D}T\mathcal{D}^\top$.

The BoostCluster algorithm falls into the same category of semi-supervised clustering algorithms as the SSKKM algorithm discussed in Section 20.2.1. It also modifies the similarity between the data points based on the given constraints, and hence yields similar clustering performance enhancement as the SSKKM algorithm. The advantage of Boost-Cluster over other semi-supervised clustering algorithms is that it can serve as a wrapper around any clustering algorithm. Hence, given the pairwise constraints, BoostCluster is able to boost the performance of any clustering algorithm. BoostCluster was used to enhance the performance of K-Means, single-link hierarchical clustering, and spectral clustering algorithms on several data sets (Liu et al. 2007).

20.2.3 Active Spectral Clustering

The spectral clustering algorithm (Shi and Malik 2002) poses data clustering as a graph partitioning problem. The data points are represented as nodes in a graph and the pairwise similarities are represented as the weights on the edges connecting the vertices. The algorithm then finds the minimum weight normalized cut of the graph, and the resulting components of the graph form the clusters. The active spectral clustering algorithm (Wang and Davidson 2010) employs pairwise constraints to enhance the performance of spectral clustering. Instead of using the pairwise relationships between randomly sampled data point pairs, it employs the active learning mechanism to identify a subset of the most informative pairwise constraints.

Let S represent the $n \times n$ pairwise dissimilarity matrix corresponding to the given set of n data points. The objective of spectral clustering, expressed as a graph *bi-partition* problem, is to find the solution to the following optimization problem:

$$\arg \min_{\mathbf{u} \in \mathbb{R}^n} \mathbf{u}^\top L\mathbf{u}$$

$$\text{s.t. } \mathbf{u}^\top D\mathbf{u} = \mathbf{1}^\top D\mathbf{1}$$

$$\mathbf{u}^\top D\mathbf{1} = 0, \tag{20.6}$$

where $L = D - S$ is the graph Laplacian, D is the degree matrix given by

$$D_{i,j} = \begin{cases} \sum_{l=1}^n S_{i,l} & \text{if } i = j, \\ 0 & \text{otherwise} \end{cases}$$

and $\mathbf{u}$ is the relaxed cluster membership vector. If the data is perfectly separable into two clusters, then $\mathbf{u} \in \{-1, 1\}^n$. Its ith element $\mathbf{u}_i = 1$ if the point $\mathbf{x}_i$ belongs to the first cluster, and $\mathbf{u}_i = -1$ if it belongs to the second cluster.

The solution is given by the eigenvector associated with the second smallest eigenvalue of the normalized Laplacian $D^{-1/2}LD^{-1/2}$. A data partition containing K clusters is obtained through recursive bi-partitioning (Shi and Malik 2002).

The active spectral clustering algorithm embeds the pairwise constraints in the form of a constraint matrix W in the above optimization problem as follows:

$$\arg\min_{\mathbf{u}\in\mathbb{R}^n} \mathbf{u}^\top L \mathbf{u}$$

$$\text{s.t. } \mathbf{u}^\top D \mathbf{u} = \mathbf{1}^\top D \mathbf{1}$$

$$\mathbf{u}^\top W \mathbf{u} \geq \alpha, \tag{20.7}$$

Algorithm ACTIVE SPECTRAL CLUSTERING

Input

- Input data set $\mathcal{D} = (\mathbf{x}_1, \ldots, \mathbf{x}_n)$.
- Oracle which has access to the true pairwise relationships W^*.
- Number of clusters K.
- Parameter α.

1. Compute the dissimilarity matrix $S = [d(\mathbf{x}_i, \mathbf{x}_j)]_{n\times n}$.
2. Compute the graph Laplacian $L = D - S$, where

$$D_{i,j} = \begin{cases} \displaystyle\sum_{l=1}^{n} S_{i,l} & \text{if } i = j, \\ 0 & \text{otherwise.} \end{cases}$$

3. Initialize the constraint matrix $W = \mathbf{0}$.
4. Repeat until convergence
   a. Find the eigenvector $\mathbf{u}$ of $L\mathbf{u} = \lambda(W - \alpha I)\mathbf{u}$, which is associated with a positive eigenvalue and minimizes $\mathbf{u}^\top L \mathbf{u}$.
   b. Find the singular vector $\bar{\mathbf{u}}$ corresponding to the largest singular value of W.
   c. Compute the rank one approximation of W as $\bar{W} = \bar{\mathbf{u}}\bar{\mathbf{u}}^\top$.
   d. Calculate the probability $p_{i,j}$ that the data points $\mathbf{x}_i$ and $\mathbf{x}_j$ are related by a must-link constraint, given by

$$p_{i,j} = \frac{1 + \min\{1, \max\{-1, \bar{W}_{i,j}\}\}}{2}.$$

   e. Solve

$$(r,s) = \arg\max_{(i,j)|W_{i,j}=0} p_{i,j}(\mathbf{u}_i\mathbf{u}_j^\top - 1)^2 + (1 - p_{i,j})(\mathbf{u}_i\mathbf{u}_j^\top + 1)^2.$$

   f. Query the oracle for the entry $W_{r,s}^*$, and set $W_{r,s}$ and $W_{s,r}$ equal to $W_{r,s}^*$.

Output Cluster memberships of the data points.

where α is a user-defined parameter, indicating how well the constraints in W are satisfied. A heuristic for selecting a suitable value for α is described in Wang and Davidson (2010). To obtain the constraint matrix, the active spectral clustering algorithm uses the active learning scheme. It assumes the presence of an oracle which has access to the true pairwise relationship matrix $W^* = \mathbf{u}^*\mathbf{u}^{*\top}$, where $\mathbf{u}^*$ is the desired data partition. The objective is to minimize the difference between the solution $\mathbf{u}$ obtained from the spectral clustering algorithm and $\mathbf{u}^*$ by querying the oracle for entries from W^*. The problem (20.7) is solved using the algorithm proposed in Wang and Davidson (2010).

This active clustering mechanism is shown to perform better than methods which assume that the pairwise constraints are available *a priori*. Its only drawback is the high computational complexity of solving the eigenvalue problem. Fast approximate eigende-composition techniques (Sleijpen and Van der Vorst 2000; Tisseur and Meerbergen 2001) can be used to mitigate this issue.

20.3 Applications

Semi-supervised clustering has been successfully applied in several fields including bioin-formatics, medical diagnosis, marketing, social network analysis, and web mining. Prior knowledge has been used to generate the side-information to enhance the clustering performance. Some of the applications are described below:

- **Character recognition:** Semi-supervised clustering was employed to decipher heavily degraded characters in historical typewritten documents in (Pletschacher et al. 2009). Due to various problems such as discoloration, aging, and disinte-gration of portions of documents, commercial OCR systems fail to recognize a majority of the characters. The documents were segmented down to the glyph* level, and each glyph was represented by a set of features such as width, height, and the ratio of the number of black pixels to the number of white pixels. The glyphs were then clustered using the MPCK-means semi-supervised clus-tering algorithm (Bilenko et al. 2004). Pairwise constraints were generated from typography-related domain knowledge. For example, characters with very differ-ent aspect ratios were related by cannot-link constraints. Constraints were also obtained through pre-clustering of the glyph images. Figure 20.9 shows a plot of the clustering performance (F-measure (Manning 2008)) of K-means, FineReader OCR engine, and semi-supervised MPCK-means as a function of the number of constraints. The plots show that the availability of pairwise constraints improves the character recognition performance. The best performance was achieved using a set of constraints containing 731 must-link constraints and 2091 cannot-link constraints.

- **Image segmentation:** Image segmentation is an important problem in computer vision. The goal is to identify homogeneous regions in an image whose pix-els share similar visual patterns (e.g., color and texture). Figure 20.10b shows a segmentation of a Mondrian image, consisting of five textured regions, that was obtained using the mean field approximation technique (Lange et al. 2005).

---

* A glyph represents a character or a symbolic figure.

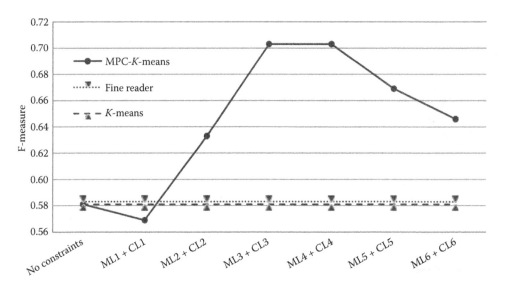

FIGURE 20.9
Character recognition using semi-supervised clustering (Pletschacher et al. 2009). ML1-ML6 and CL1-CL6 represent sets of must-link and cannot-link constraints, respectively with increasing number of constraints. Best performance is achieved using the set ML3 + CL3, containing 731 must-link constraints and 2091 cannot-link constraints.

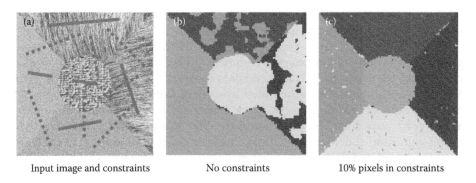

| Input image and constraints | No constraints | 10% pixels in constraints |

FIGURE 20.10
Image segmentation using semi-supervised clustering. (Adapted from T. Lange et al. *Proceedings of the IEEE Conference on Computer Vision and Pattern Recognition*, volume 1, pages 731–738, 2005.)

The segmented image in Figure 20.10b does not capture the true textured regions in Figure 20.10a very well. In order to generate the pairwise constraints, the image was divided into grids; segment labels for the grids were obtained from the ground truth and converted into pairwise constraints. With 10% of the grids in pairwise constraints, the true textured regions were very well identified through semi-supervised clustering, as shown in Figure 20.10c.

Image segmentation algorithms based on semi-supervised clustering have been employed in various applications such as medical imaging (Filipovych et al. 2011; Ribbens et al. 2011), and remote sensing (Torres et al. 2010).

- **Document clustering:** Semi-supervised clustering has been very useful in document clustering. Semi-supervised Non-negative Matrix factorization (SS-NMF) (Chen et al. 2007) and its variants supplied with pairwise relationships between the documents were used to cluster documents in Chen et al. (2007) and Chen et al. (2009). In Hu et al. (2011), users were allowed to label discriminative features in addition to specifying pairwise constraints. This is useful in scenarios where the user desires clusters based on specific attributes. For example, a set of articles related to sports may be organized by sport or by country. The users identified words which described the topic of the document, when presented with a subset of documents for labeling. This information was incorporated in the K-Means algorithm to obtain the desired document clusters.

- **Gene expression data analysis:** Semi-supervised clustering techniques have been popular in the domain of gene expression analysis (Pensa et al. 2006, 2008; Pensa and Boulicaut 2008). In addition to pairwise constraints, interval constraints which define the spatial and temporal cluster boundaries are employed. Genes which are known to have the same function in a biological process are related by must-link constraints. Interval constraints relate genes which are close to each other in terms of sampling time and/or spatial position in the DNA sequence. These constraints are easily attainable and aid biologists in capturing interesting patterns in the data. In Schönhuth et al. (2009), the yeast gene expression data was augmented with labels for a subset of the data. These labels were used to generate pairwise constraints, which were then used in the semi-supervised EM algorithm (Lange et al. 2005) to obtain better clusters.

20.4 Conclusions and Open Problems

Semi-supervised clustering is a useful mechanism to integrate side-information or prior knowledge in the clustering algorithm to find desired clusters in a data set. We have described different ways in which side-information can be utilized in clustering. Pairwise must-link and cannot-link constraints are the most common form of specifying the side-information. Between them, experimental results suggest that must-link constraints are more useful than cannot-link constraints (Baghshah and Shouraki 2010). The side-information can be incorporated in two ways: constraining the set of partitions that can be found by the clustering algorithm, and learning a distance metric which takes the constraints into account. Many semi-supervised algorithms have been developed and studied in the literature, and it continues to be a thriving field of study.

We have seen that the accuracy of the constraints is crucial to the semi-supervised clustering performance. A major challenge in semi-supervised clustering is identifying the most useful constraints, while minimizing user effort (Wagstaff 2006). Though active clustering alleviates this issue to some extent, it may not always lead to the desired solution. In some scenarios, only partial information may be available, and it may not be feasible to determine the pairwise relationship accurately. In Gao et al. (2006), class labels are associated with a confidence rating and used as side-information. Similar mechanisms for assigning confidence measures to the pairwise constraints and incorporating them in the semi-supervised clustering algorithm need to be developed.

References

K. Allab and K. Benabdeslem. Constraint selection for semi-supervised topological clustering. *Machine Learning and Knowledge Discovery in Databases*, pages 28–43, 2011.

C.G. Atkeson, A.W. Moore, and S. Schaal. Locally weighted learning. *Artificial Intelligence Review*, 11(1):11–73, 1997.

K. Bade and A. Nürnberger. Creating a cluster hierarchy under constraints of a partially known hierarchy. In *Proceedings of the SIAM International Conference on Data Mining*, pages 13–24, 2008.

M.S. Baghshah and S.B. Shouraki. Kernel-based metric learning for semi-supervised clustering. *Neurocomputing*, 73:1352–1361, 2010.

G. Barbier, R. Zafarani, H. Gao, G. Fung, and H. Liu. Maximizing benefits from crowdsourced data. *Computational and Mathematical Organization Theory*, 18(3):257–279, 2012.

S. Basu, A. Banerjee, and R. Mooney. Semi-supervised clustering by seeding. In *Proceedings of the International Conference on Machine Learning*, pages 19–26, 2002.

S. Basu, A. Banerjee, and R.J. Mooney. Active semi-supervision for pairwise constrained clustering. In *Proceedings of the SIAM International Conference on Data Mining*, pages 333–344, 2004.

S. Basu, M. Bilenko, A. Banerjee, and R.J. Mooney. Probabilistic semi-supervised clustering with constraints. *Semi-supervised Learning*, pages 71–98, 2006.

S. Basu, I. Davidson, and K.L. Wagstaff. *Constrained Clustering: Advances in Algorithms, Theory, and Applications*. Chapman & Hall/CRC, Boca Raton, FL, 2009.

R. Bekkerman and M. Sahami. Semi-supervised clustering using combinatorial MRFs. In *Proceedings of the ICML Workshop on Learning in Structured Output Spaces*, 2006.

J. Bi, D. Wu, L. Lu, M. Liu, Y. Tao, and M. Wolf. Adaboost on low-rank psd matrices for metric learning. In *Proceedings of the IEEE Conference on Computer Vision and Pattern Recognition*, pages 2617–2624, 2011.

M. Bilenko, S. Basu, and R.J. Mooney. Integrating constraints and metric learning in semi-supervised clustering. In *Proceedings of the International Conference on Machine Learning*, pages 81–88, 2004.

O. Chapelle, B. Schölkopf, A. Zien, et al. *Semi-supervised Learning*, volume 2. MIT Press, Cambridge, MA, 2006.

Y. Chen, M. Rege, M. Dong, and J. Hua. Incorporating user provided constraints into document clustering. In *Proceedings of the International Conference on Data Mining*, pages 103–112, 2007.

Y. Chen, L. Wang, and M. Dong. Semi-supervised document clustering with simultaneous text representation and categorization. *Machine Learning and Knowledge Discovery in Databases*, pages 211–226, 2009.

H. Cheng, K.A. Hua, and K. Vu. Constrained locally weighted clustering. In *Proceedings of the VLDB Endowment*, volume 1, pages 90–101, 2008.

D. Cohn, L. Atlas, and R. Ladner. Improving generalization with active learning. *Machine Learning*, 15(2):201–221, 1994.

D. Cohn, R. Caruana, and A. McCallum. Semi-supervised clustering with user feedback. *Constrained Clustering: Advances in Algorithms, Theory, and Applications*, 4(1):17–32, 2003a.

D. Cohn, R. Caruana, and A. Mccallum. Semi-supervised clustering with user feedback. Technical report, University of Texas at Austin, 2003b.

I. Davidson and S. Ravi. Towards efficient and improved hierarchical clustering with instance and cluster level constraints. Technical report, University of Albany, 2005a.

I. Davidson and S.S. Ravi. Clustering with constraints: Feasibility issues and the k-means algorithm. In *Proceedings of the SIAM International Conference on Data Mining*, pages 138–149, 2005b.

I. Davidson and S.S. Ravi. Identifying and generating easy sets of constraints for clustering. In *Proceedings of the National Conference on Artificial Intelligence*, pages 336–341, 2006.

I. Davidson, K. Wagstaff, and S. Basu. Measuring constraint-set utility for partitional clustering algorithms. In *Proceedings of the European Conference on Principles and Practice of Knowledge Discovery in Databases*, pages 115–126, 2006.

A. Demiriz, K.P. Bennett, and M.J. Embrechts. Semi-supervised clustering using genetic algorithms. *Artificial neural networks in Engineering*, pages 809–814, 1999.

C. Domeniconi and D. Gunopulos. Adaptive nearest neighbor classification using support vector machines. *Advances in Neural Information Processing Systems*, 14:665–672, 2001.

C. Domeniconi, J. Peng, and B. Yan. Composite kernels for semi-supervised clustering. *Knowledge and Information Systems*, 28(1):99–116, 2011.

R. Filipovych, S.M. Resnick, and C. Davatzikos. Semi-supervised cluster analysis of imaging data. *NeuroImage*, 54(3):2185–2197, 2011.

J. Gao, P.N. Tan, and H. Cheng. Semi-supervised clustering with partial background information. In *Proceedings of the SIAM International Conference on Data Mining*, pages 487–491, 2006.

M. Girolami. Mercer kernel-based clustering in feature space. *IEEE Transactions on Neural Networks*, 13(3):780–784, 2002.

D. Gondek. Clustering with model-level constraints. In *Proceedings of the SIAM International Conference on Data Mining*, pages 126–137, 2005.

M. Guillaumin, J. Verbeek, and C. Schmid. Multiple instance metric learning from automatically labeled bags of faces. In *Proceedings of the European Conference on Computer Vision*, pages 634–647, 2010.

M. Halkidi, D. Gunopulos, N. Kumar, M. Vazirgiannis, and C. Domeniconi. A framework for semi-supervised learning based on subjective and objective clustering criteria. In *Proceedings of the International Conference on Data Mining*, pages 637–640, 2005.

S.C.H. Hoi, R. Jin, and M.R. Lyu. Learning nonparametric kernel matrices from pairwise constraints. In *Proceedings of the International Conference on Machine Learning*, pages 361–368, 2007.

Y. Hu, E.E. Milios, and J. Blustein. Interactive feature selection for document clustering. In *Proceedings of the ACM Symposium on Applied Computing*, pages 1143–1150, 2011.

K. Huang, R. Jin, Z. Xu, and C.L. Liu. Robust metric learning by smooth optimization. In *Proceeding of the Conferene on Uncertainity in Artificial Intelligence*, pages 244–251, 2010.

R. Huang and W. Lam. Semi-supervised document clustering via active learning with pairwise constraints. In *Proceedings of the International Conference on Data Mining*, pages 517–522, 2007.

A. Huang, D. Milne, E. Frank, and I.H. Witten. Clustering documents with active learning using wikipedia. In *Proceedings of the International Conference on Data Mining*, pages 839–844, 2008.

T. Hume. Balance scale data set. http://archive.ics.uci.edu/ml/datasets/Balance Scale, 1994.

A.K. Jain. Data clustering: 50 years beyond k-means. *Pattern Recognition Letters*, 31(8):651–666, 2010.

K. Kamvar, S. Sepandar, K. Klein, D. Dan, M. Manning, and C. Christopher. Spectral learning. In *Proceedings of the International Joint Conference on Artificial Intelligence*, 2003.

D.W. Kim, K.Y. Lee, D. Lee, and K.H. Lee. Evaluation of the performance of clustering algorithms in kernel-induced feature space. *Pattern Recognition*, 38(4):607–611, 2005.

B. Kulis, S. Basu, I. Dhillon, and R. Mooney. Semi-supervised graph clustering: A kernel approach. In *Proceedings of the International Conference on Machine Learning*, pages 457–464, 2005.

N. Kumar and K. Kummamuru. Semi-supervised clustering with metric learning using relative comparisons. *IEEE Transactions on Knowledge and Data Engineering*, 20(4):496–503, 2008.

T.O. Kvalseth. Entropy and correlation: Some comments. *IEEE Transactions on Systems, Man and Cybernetics*, 17(3):517–519, 1987.

T. Lange, M.H.C. Law, A.K. Jain, and J.M. Buhmann. Learning with constrained and unlabelled data. In *Proceedings of the IEEE Conference on Computer Vision and Pattern Recognition*, volume 1, pages 731–738, 2005.

M.H.C. Law, A. Topchy, and A.K. Jain. Model-based clustering with probabilistic constraints. In *Proceedings of SIAM International Conference on Data Mining*, pages 641–645, 2005.

T. Li, C. Ding, and M.I. Jordan. Solving consensus and semi-supervised clustering problems using nonnegative matrix factorization. In *Proceedings of the International Conference on Data Mining*, pages 577–582, 2007.

Y. Liu, R. Jin, and A.K. Jain. Boostcluster: Boosting clustering by pairwise constraints. In *Proceedings of the International Conference on Knowledge Discovery and Data Mining*, pages 450–459, 2007.

C.D. Manning, P. Raghavan, and H. Schutze. *Introduction to Information Retrieval*, volume 1. 2008.

T.M. Mitchell. CMU face images data set. http://archive.ics.uci.edu/ml/datasets/CMU%20Face%20 Images, 1999.

D.J. Newman, S. Hettich, C.L. Blake, C.J. Merz, and D.W. Aha. UCI repository of machine learning databases. http://archive.ics.uci.edu/ml/datasets.html, 1998.

R.G. Pensa and J.F. Boulicaut. Constrained co-clustering of gene expression data. In *Proceedings SIAM International Conference on Data Mining*, pages 25–36, 2008.

R. Pensa, C. Robardet, and J.F. Boulicaut. Towards constrained co-clustering in ordered 0/1 data sets. *Foundations of Intelligent Systems*, pages 425–434, 2006.

R.G. Pensa, C. Robardet, and J.F. Boulicaut. Constraint-driven co-clustering of 0/1 data. *Constrained Clustering: Advances in Algorithms, Theory and Applications*, pages 145–170, 2008.

S. Pletschacher, J. Hu, and A. Antonacopoulos. A new framework for recognition of heavily degraded characters in historical typewritten documents based on semi-supervised clustering. In *Proceedings of the International Conference on Document Analysis and Recognition*, pages 506–510, 2009.

A. Ribbens, F. Maes, D. Vandermeulen, and P. Suetens. Semisupervised probabilistic clustering of brain MR images including prior clinical information. In *Proceedings of the Medical Computer Vision Workshop on Recognition Techniques and Applications in Medical Imaging*, pages 184–194, 2011.

C. Ruiz, M. Spiliopoulou, and E. Menasalvas. Density-based semi-supervised clustering. *Data Mining and Knowledge Discovery*, 21(3):345–370, 2010.

B. Scholkopf, A. Smola, and K.R. Muller. Nonlinear component analysis as a kernel eigenvalue problem. *Neural Computation*, 10(5):1299–1314, 1996.

A. Schönhuth, I.G. Costa, and A. Schliep. Semi-supervised clustering of yeast gene expression data. *Cooperation in Classification and Data Analysis*, pages 151–159, 2009.

B. Settles. Active learning literature survey. Computer Science Technical Report 1648, University of Wisconsin–Madison, 2009.

N. Shental, A. Bar-Hillel, T. Hertz, and D. Weinshall. Computing gaussian mixture models with em using equivalence constraints. *Advances in Neural Information Processing Systems*, 16:465–472, 2004.

J. Shi and J. Malik. Normalized cuts and image segmentation. *IEEE Transactions on Pattern Analysis and Machine Intelligence*, 22(8):888–905, 2002.

G.L.G. Sleijpen and H.A. Van der Vorst. A jacobi-davidson iteration method for linear eigenvalue problems. *SIAM Review*, pages 267–293, 2000.

F. Tisseur and K. Meerbergen. The quadratic eigenvalue problem. *SIAM Review*, pages 235–286, 2001.

M. Torres, M. Moreno, R. Menchaca-Mendez, R. Quintero, and G. Guzman. Semantic supervised clustering approach to classify land cover in remotely sensed images. *Signal Processing and Multimedia*, 68–77, 2010.

K.L. Wagstaff. Value, cost, and sharing: Open issues in constrained clustering. In *Proceedings of the International Conference on Knowledge Discovery in Inductive Databases*, pages 1–10, 2006.

K. Wagstaff, C. Cardie, S. Rogers, and S. Schrödl. Constrained k-means clustering with background knowledge. In *Proceedings of the International Conference on Machine Learning*, pages 577–584, 2001.

X. Wang and I. Davidson. Active spectral clustering. In *Proceedings of the International Conference on Data Mining*, pages 561–568, 2010.

X. Wang and I. Davidson. Flexible constrained spectral clustering. In *Proceedings of the International Conference on Knowledge Discovery and Data Mining*, pages 563–572, 2010.

L. Wu, S.C.H. Hoi, R. Jin, J. Zhu, and N. Yu. Learning bregman distance functions for semi-supervised clustering. *IEEE Transactions on Knowledge and Data Engineering*, 478–491, 2010.

I. Xenarios, D.W. Rice, L. Salwinski, M.K. Baron, E.M. Marcotte, and D. Eisenberg. DIP: The database of interacting proteins. *Nucleic Acids Research*, 28(1):289–291, 2000.

E.P. Xing, A.Y. Ng, M.I. Jordan, and S. Russell. Distance metric learning with application to clustering with side-information. *Advances in Neural Information Processing Systems*, 15:505–512, 2002.

B. Yan and C. Domeniconi. An adaptive kernel method for semi-supervised clustering. In *Proceedings of the European Conference on Machine Learning*, pages 521–532, 2006.

S. Yan, H. Wang, D. Lee, and C. Giles. Pairwise constrained clustering for sparse and high dimensional feature spaces. *Advances in Knowledge Discovery and Data Mining*, 620–627, 2009.

L. Yang and R. Jin. Distance metric learning: A comprehensive survey. Technical report, Michigan State Universiy, 2006.

W. Zhao, Q. He, H. Ma, and Z. Shi. Effective semi-supervised document clustering via active learning with instance-level constraints. *Knowledge and Information Systems*, 1–19, 2011.

H. Zhao and Z. Qi. Hierarchical agglomerative clustering with ordering constraints. In *Proceedings of the International Conference on Knowledge Discovery and Data Mining*, pages 195–199, 2010.

L. Zheng and T. Li. Semi-supervised hierarchical clustering. In *Proceedings of the International Conference on Data Mining*, pages 982–991, 2011.

21

Clustering of Symbolic Data

Paula Brito

CONTENTS

Abstract

In this chapter, we present clustering methods for symbolic data. We start by recalling that symbolic data is data presenting inherent variability, and the motivations for the introduction of this new paradigm. We then proceed by defining the different types of variables that allow for the representation of symbolic data, and recall some distance measures appropriate for the new data types. Then we present clustering methods for different types of symbolic data, both hierarchical and nonhierarchical. An application illustrates two well-known methods for clustering symbolic data.

21.1 Introduction

Symbolic data analysis (see Bock and Diday 2000; Billard and Diday 2006; Diday and Noirhomme-Fraiture 2008; Noirhomme-Fraiture and Brito 2011; Brito 2014) introduced by E. Diday in the late eighties of the last century (Diday 1988) is concerned with analyzing data presenting intrinsic variability, which is to be explicitly taken into account. In classical Statistics and Multivariate Data Analysis, the elements under analysis are generally individual entities for which a single value is recorded for each variable—for example, individuals, described by their age, salary, education level, marital status, etc.; cars each described by its weight, length, power, engine displacement, etc., students for each of which the marks at different subjects were recorded. But when the elements of interest are classes or groups of some kind—the citizens living in given towns; teams, consisting of individual players; car models, rather than specific vehicles; classes and not individual students—then there is variability inherent to the data. To reduce this variability by taking central tendency measures—mean values, medians or modes—obviously leads to a too important loss of information. Symbolic data analysis provides a framework allowing representing data with variability, using new variable types. Also, methods have been developed which suitably take data variability into account. Symbolic data may be represented using the usual matrix-form data arrays, where each entity is represented in a row and each column corresponds to a different variable—but now the elements of each cell are generally not single real values or categories, as in the classical case, but rather finite sets of values, intervals or, more generally, distributions.

Symbolic data should not be confused with fuzzy data—although their respective data representation may present some similarities. Symbolic data aims at representing variability intrinsic to the entities under analysis, therefore, the variable values take into account the sets of values, or the distributions, that are observed. In fuzzy data, the variable values are in fact unique, but unknown with precision; for that reason, intervals, or more generally, sets, are recorded, as the possible regions where the (unique) variable value lays.

As concerns clustering, since the initial formalization of symbolic data a large number of methods for clustering such data has been proposed and studied, and applied in different domains. We categorize these methods into two distinct groups (Brito 2007): (i) Methods that result from adapting classical clustering methods based on dissimilarities to the new data types, by properly defining dissimilarity measures for symbolic data. In this case, the clustering methodologies and criteria remain basically unchanged, though some necessary adaptations may have to be considered, and are applied to the obtained dissimilarity matrices. (ii) Methods that do not rely on dissimilarities and use the data (i.e., the descriptions of the elements) explicitly in the clustering process. The criterion to form classes is to obtain a "meaningful" class description, and we are in the scope of the so-called conceptual clustering methods. This categorization is not specific to the problem of clustering symbolic data, the same applies in the classical data case. However, given that we now are considering data which present variability, this issue does have consequences in the outcome of the clustering process. Clustering methods of type (i) will tend to cluster together entities with similar descriptions—this similarity being evaluated by one of the proposed measures—irrespective of the intrinsic variability of the underlying descriptions. In other words, however large is the variability inherent to two given descriptions, if they are alike, their dissimilarity will be low—and the corresponding entities will tend to be clustered together. On the other hand, methods of type (ii) will concentrate on the description of each newly formed cluster, so as to minimize its inherent variability. This means that this kind of method may favor the grouping of entities whose descriptions are less alike, if the description of the resulting cluster presents a lower variability. This duality of criteria does not arise if we are in the presence of classical data. In that case, the closer the values of a given variable, the more specific is their generalization—so both dissimilarity and generalization-based methods will tend to elect the same candidate pairs to be aggregated. Not so for symbolic data: the criteria are different and therefore it makes no sense to compare results issued by the two kinds of methods, since they start from a different concept of "what a cluster is."

In this chapter, we present some of the methods that have been proposed for clustering different types of symbolic data. As it will become clear, interval data constitute by far the most studied case, and for which more methods have been developed. *SODAS* was the first software package allowing representing and analyzing symbolic data, and we will be referring to it at some points of the text. Nowadays, other alternatives exist, including some *R* packages.

In the next section, we recall the different types of variables, giving some small illustrative examples in each case. Given that many clustering methods rely on dissimilarities, we proceed by presenting dissimilarity measures appropriate for symbolic data. Then, we present methods for specific variable types, as well as for multiple types of variables, detailing in some cases the measures used and the considered criteria. An application on a set of car models, described by variables of different types, illustrates two methods: a partitional k-means type method and a hierarchical conceptual clustering one. Finally, we conclude by a brief summary on the topic, and referring to new lines of research.

21.2 Symbolic Data

To represent data variability explicitly, new variable types have been introduced, whose realizations are now not real values, in the numerical case, or individual categories, in the qualitative case. Below we define the different considered variable types, including the classical ones, which may be considered special cases of the symbolic types (see also Noirhomme-Fraiture and Brito 2011 or Brito 2014).

As in the classical case, we distinguish numerical and categorical symbolic variables. A numerical (or quantitative) variable may be single-valued (real or integer), as in the classical framework, if it takes one single value of an underlying domain for each entity. It is multi-valued if its values are finite subsets of the domain and it is an interval-valued variable if its values are intervals of $\mathbb{R}$. When a distribution over a set of subintervals is given, the variable is called a histogram-valued variable. A categorical (or qualitative) variable can be single-valued (ordinal or not), as in the classical context, when it takes one category from a given finite domain $O = \{m_1, \ldots, m_k\}$ for each entity; multi-valued, if its values are finite subsets of the domain O. A categorical modal variable Y with a finite domain $O = \{m_1, \ldots, m_k\}$ is a multistate variable where, for each element, we are given a category set and, for each category m_ℓ, a frequency or probability which indicates how frequent or likely that category is for this element. Let $Y_1, \ldots, Y_p$ be the set of variables, O_j the underlying domain of Y_j and B_j the set where Y_j takes its value for each entity, $j = 1, \ldots, p$. A description d is defined as a p-tuple $d = (d_1, \ldots, d_p)$ with $d_j \in B_j$, $j = 1, \ldots, p$. Let $S = \{s_1, \ldots, s_n\}$ be the set of entities under analysis, then $Y_j(s_i) \in B_j$ for $j = 1, \ldots, p$, $i = 1, \ldots, n$; the data array consists of n descriptions, one for each entity $s_i \in S : d_i = (Y_1(s_i), \ldots, Y_p(s_i))$, $i = 1, \ldots, n$.

21.2.1 Quantitative Single-Valued Variables

Given a set of n entities $S = \{s_1, \ldots, s_n\}$, a quantitative single-valued variable Y is defined by an application $Y : S \to O$ such that $s_i \longmapsto Y(s_i) = c$ where $O \subseteq \mathbb{R}$. This is the classical numerical case, and B is identical to the underlying set O, $B \equiv O$.

21.2.2 Quantitative Multi-Valued Variables

Given a set of n entities S, a quantitative multi-valued variable Y is defined by an application $Y : S \to B$ such that $s_i \longmapsto Y(s_i) = \{c_{i1}, \ldots, c_{in_i}\}$ where B is the power set of an underlying set $O \subseteq \mathbb{R}$. The "values" of $Y(s_i)$ are now finite sets of real numbers.

21.2.3 Interval-Valued Variables

Given $S = \{s_1, \ldots, s_n\}$, an interval-valued variable is defined by an application $Y : S \to B$ such that $s_i \longmapsto Y(s_i) = [l_i, u_i]$ where B is the set of intervals of an underlying set $O \subseteq \mathbb{R}$. Let I be an $n \times p$ matrix representing the values of p interval-valued variables on S. Each $s_i \in S$ is represented by a p-tuple of intervals, $I_i = (I_{i1}, \ldots, I_{ip})$, $i = 1, \ldots, n$, with $I_{ij} = [l_{ij}, u_{ij}]$, $j = 1, \ldots, p$ (see Table 21.1).

The value of an interval-valued variable Y_j for each $s_i \in S$ is defined by the bounds l_{ij} and u_{ij} of $I_{ij} = Y_j(s_i)$. For modeling purposes, however, an alternative parametrization consisting in representing $Y_j(s_i)$ by the midpoint $c_{ij} = (l_{ij} + u_{ij})/2$ and range $r_{ij} = u_{ij} - l_{ij}$ of I_{ij} may be useful.

Example: Consider a dataset containing information about customers of a specialized store during a given period; Table 21.2 presents data for three of those stores. In store A,

TABLE 21.1

Matrix I of Interval Data

	Y_1	$\ldots$	Y_j	$\ldots$	Y_p
s_1	$[l_{11}, u_{11}]$	$\ldots$	$[l_{1j}, u_{1j}]$	$\ldots$	$[l_{1p}, u_{1p}]$
$\ldots$	$\ldots$		$\ldots$		$\ldots$
s_i	$[l_{i1}, u_{i1}]$	$\ldots$	$[l_{ij}, u_{ij}]$	$\ldots$	$[l_{ip}, u_{ip}]$
$\ldots$	$\ldots$		$\ldots$		$\ldots$
s_n	$[l_{n1}, u_{n1}]$	$\ldots$	$[l_{nj}, u_{nj}]$	$\ldots$	$[l_{np}, u_{np}]$

TABLE 21.2

Data for Specialized Stores

Store	Age	Nb. Visits	Expense (Euros)
A	[25, 62]	{1, 2}	[100, 400]
B	[40, 75]	{1, 4, 5, 6}	[80, 300]
C	[20, 50]	{2, 3}	[150, 250]

the age of customers ranges from 25 to 62 years old, each customer made 1 or 2 visits and the expense of the customers ranged from 100 to 400 Euros. Here, age and expense are interval-valued variables whereas the number of visits is a multi-valued quantitative variable. A similar description may be obtained for the remaining stores. Notice that in this example the entities under analysis are the stores, for each of which we have aggregated information, and NOT the individual customers of each store.

In Brito and Duarte Silva (2012) parametric models for interval data are proposed, which consider Multivariate Normal or Skew-Normal distributions for the midpoints and log-ranges of the interval-valued variables. The Gaussian model allows for the application of classical inference methods, the Skew-Normal setup provides some more flexibility. In either case, since the midpoint c_{ij} and the range r_{ij} of the value of an interval-valued variable $I_{ij} = Y_j(s_i)$ are both related to one same variable, they must be considered together: therefore, the global covariance matrix should take into account the link that may exist between midpoints and ranges of the same or different variables. Intermediate parameterizations between the nonrestricted and the noncorrelation setup considered for real-valued data are hence relevant for interval data. The following cases are of particular interest, and have been addressed:

1. Nonrestricted case: allowing for nonzero correlations among all midpoints and log-ranges.
2. Interval-valued variables Y_j are uncorrelated, but for each variable, the midpoint may be correlated with its log-range.
3. Midpoints (log-ranges) of different variables may be correlated, but no correlation between midpoints and log-ranges is allowed.
4. All midpoints and log-ranges are uncorrelated, both among themselves and between each other.

This modeling has been implemented in the R-package MAINT.Data (Duarte Silva and Brito 2011), available on CRAN. MAINT.Data introduces a data class for representing interval data and includes functions for modeling and analyzing such data.

21.2.4 Histogram-Valued Variables

When real-valued data are aggregated by means of intervals, the information on the internal variation inside the intervals is lost, and the corresponding distribution is not taken into account. One way to keep more information about this is to define limits between the global lower (LB) and upper (UB) bounds and compute frequencies between these limits. We obtain for each case and this numerical variable a histogram with k classes (and k frequencies) if $k - 1$ is the number of limits between LB and UB. Given

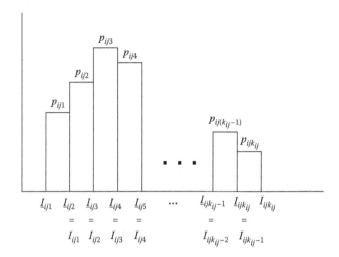

FIGURE 21.1
Histogram representation.

$S = \{s_1, \ldots, s_n\}$, a histogram-valued variable is defined by an application $Y : S \to B$ such that $s_i \longmapsto Y(s_i) = \{[\underline{I}_{i1}, \overline{I}_{i1}], p_{i1}; [\underline{I}_{i2}, \overline{I}_{i2}], p_{i2}; \ldots; [\underline{I}_{ik_i}, \overline{I}_{ik_i}], p_{ik_i}\}$ where $I_{i\ell} = [\underline{I}_{i\ell}, \overline{I}_{i\ell}]$, $\ell = 1, \ldots, k_i$ are the considered subintervals for observation s_i, $p_{i1} + \cdots + p_{ik_i} = 1$ (see Figure 21.1); B is now the set of frequency distributions over $\{I_{i1}, \ldots, I_{ik_i}\}$. It is generally assumed that for each entity s_i values are uniformly distributed within each subinterval. For different observations, the number and length of subintervals of the histograms may naturally be different.

Example: Consider again the specialized stores example, with a new variable which records the waiting time for an ordered product. In this case, information is recorded for three time lengths (0–3 days, 3–7 days, 7–14 days), and the corresponding variable is therefore a histogram-valued variable—see Table 21.3.

The values of a histogram-valued variable may equivalently be represented by an empirical distribution function:

$$
F(x) = \begin{cases}
0 & \text{for} & x < \underline{I}_{i1} \\
p_{i1}(x - \underline{I}_{i1})/(\underline{I}_{i2} - \underline{I}_{i1}) & \text{for} & \underline{I}_{i1} \le x < \underline{I}_{i2} \\
F(\underline{I}_{i2}) + p_{i2}(x - \underline{I}_{i2})/(\underline{I}_{i3} - \underline{I}_{i2}) & \text{for} & \underline{I}_{i2} \le x < \underline{I}_{i3} \\
\vdots & & \\
F(\underline{I}_{ik_i}) + p_{ik_i}(x - \underline{I}_{ik_i})/(\overline{I}_{ik_i} - \underline{I}_{ik_i}) & \text{for} & \underline{I}_{ik_i} \le x < \overline{I}_{ik_i} \\
1 & \text{for} & x \ge \overline{I}_{ik_i}
\end{cases}
$$

TABLE 21.3

Data for Specialized Stores

Store	Age	Nb. Visits	Expense (Euros)	Waiting Time (days)
A	[25, 62]	{1, 2}	[100, 400]	([0, 3 [(0.5), [3, 7 [(0.25), [7, 14] (0.25))
B	[40, 75]	{1, 4, 5, 6}	[80, 300]	([0, 3 [(0.2), [3, 7 [(0.6), [7, 14] (0.2))
C	[20, 50]	{2, 3}	[150, 250]	([0, 3 [(0.4), [3, 7 [(0.3), [7, 14] (0.3))

or by its inverse, the quantile function,

$$\Psi(t) = F^{-1}(t) = \begin{cases} \underline{I}_{i1} + \dfrac{t}{w_{i1}} r_{i1} & \text{if } 0 \le t < w_{i1} \\[2ex] \underline{I}_{I_{i2}} + \dfrac{t - w_{i1}}{w_{i2} - w_{i1}} r_{i2} & \text{if } w_{i1} \le t < w_{i2} \\[2ex] \vdots & \\[2ex] \underline{I}_{ik_i} + \dfrac{t - w_{ik_i-1}}{1 - w_{ik_i-1}} r_{ik_i} & \text{if } w_{in_i-1} \le t \le 1 \end{cases}$$

where $w_{il} = 0$ if $l = 0$ and $w_{il} = \sum_{\ell=1}^{l} p_{i\ell}$ if $l = 1, \ldots, k_i$ and $r_{i\ell} = \overline{I}_{i\ell} - \underline{I}_{i\ell}$ with $\ell \in \{1, \ldots, k_i\}$; k_i is the number of subintervals in $Y(s_i)$.

When $k = 1$ a histogram reduces to an interval. Interval-valued variables may therefore be considered special cases of histogram-valued variables.

21.2.5 Categorical Single-Valued Variables

This is the standard categorical variable. Given $S = \{s_1, \ldots, s_n\}$, a qualitative single-valued variable is defined by an application $Y : S \to O$ such that $s_i \longmapsto Y(s_i) = m$ where O is a finite set of categories, $O = \{m_1, \ldots, m_k\}$ (i.e., in this case, again, $B \equiv O$). In the symbolic context, it means that all the individuals of the concept share the same categorical value. If the categories of O are naturally ordered, the variable is called ordinal, otherwise it is nominal. Often, analysts use such a categorical variable to build new concepts or entities, by aggregating the cases sharing the same category.

21.2.6 Categorical Multi-Valued Variables

A categorical multi-valued variable is defined by an application $Y : S \to B$ where B is the power-set of $O = \{m_1, \ldots, m_k\}$. The "values" of $Y(s_i)$ are now finite sets of categories.

21.2.7 Categorical Modal Variables

A categorical modal variable Y with a finite domain $O = \{m_1, \ldots, m_k\}$ is a variable where, for each element, we are given a category set and, for each category m_ℓ, a weight, frequency, or probability p_ℓ which indicates how frequent or likely that category is for this element, and whose sum adds up to 1. In this case, B is the set of distributions over O, and its elements are denoted $\{m_1(p_1), \ldots, m_k(p_k)\}$.

Example: Consider again the specialized stores example, and consider now that the job of the customers has been recorded. We have then a categorical modal variable job, as in Table 21.4.

The weights may be something else rather than probabilities or frequencies, like capacities (Choquet 1954), necessities, possibilities, and credibilities (Dubois and Prade 1999; Walley 2000). In these cases, their sum does not necessarily add up to 1.

Categorical modal variables are similar to histogram-valued variables for the quantitative case, in that their values are both characterized by classes or categories and weights. Henceforth in this text, by "distributional data" we refer to both types. Nevertheless, from a mathematical point of view, they are clearly of different nature.

TABLE 21.4

Data for Specialized Stores

Store	Age	Nb. Visits	Expense	Job
A	[25, 62]	{1, 2}	[100, 400]	{Manager (0.25), Professor (0.4), Student (0.35)}
B	[40, 75]	{1, 4, 5, 6}	[80, 300]	{Manager (0.10), Clerk (0.15), Lawyer (0.6), Student (0.15)}
C	[20, 50]	{2, 3}	[150, 250]	{Lawyer (0.3), Professor (0.5), Student (0.2)}

21.2.8 Other Types of Symbolic Data

21.2.8.1 Taxonomic Variables

A variable $Y \to O$ is a taxonomic variable if O has a tree structure. Taxonomies may be taken into account in obtaining descriptions of aggregated data: first, values are recorded as in the case of categorical multi-valued variables and then each set of values of O is replaced by the lowest value in the taxonomy covering all the values of the given set. Generally, when at least two successors of a given level h are present, they are replaced by h.

21.2.8.2 Constrained Variables

A variable Y' is hierarchically dependent on a variable Y if its application is constrained by the values taken by Y : Y' cannot be applied if Y takes values within a given set C. In other words, a variable Y' is hierarchically dependent on a variable Y if Y' makes no sense for some values Y may take, and hence becomes "nonapplicable." For instance, if a survey contains an item on the previous job of a person, the variable does not apply when the person's present job is the first one. Descriptions that do not comply with a rule are called "non-coherent." The consideration of hierarchical rules in symbolic data analysis has been widely studied in Vignes (1991), De Carvalho (1994), and Csernel and De Carvalho (1999).

21.3 Dissimilarities and Distances for Symbolic Data

A large number of clustering methods for symbolic data rely on dissimilarities between the entities to be clustered. Given that new variable types are used to represent those entities, appropriate measures must be considered in these cases. Several measures for different types of symbolic data, adopting different points of view on how to measure dissimilarity in this new context, have been proposed and investigated. For an overview and discussion on different alternatives, refer to De Carvalho (1994), Bock and Diday (2000) (Chapter 8), Diday and Esposito (2003) and Esposito et al. (2008). Below, we give some dissimilarity measures for different variable types.

21.3.1 Dissimilarities for Set-Valued Variables

For set-valued variables, that is, quantitative or categorical variables whose values are finite sets (B is the power-set of an underlying domain O), a collection of dissimilarity measures have been proposed and are implemented in the *SODAS* package. Most of these measures rely on set operations between the observed values of the set-valued

TABLE 21.5

Dissimilarity Measures for Set-Valued Data

Name	Variable-Wise Dissimilarity	Global Dissimilarity
U_1	$D(j)(A_{1j}, A_{2j}) = D_\pi(A_{1j}, A_{2j}) +$ $D_s(A_{1j}, A_{2j}) + D_c(A_{1j}, A_{2j})$ $D_\pi(A_{1j}, A_{2j})$ due to position, $D_s(A_{1j}, A_{2j})$ due to spanning, $D_c(A_{1j}, A_{2j})$ due to content.	$\sum_{j=1}^{p} D^{(j)}(A_{1j}, A_{2j})$
U_2	$\phi(A_{1j}, A_{2j}) =$ $|A_{1j} \oplus A_{2j}| - |A_{1j} \otimes A_{2j}| +$ $\gamma(2|A_{1j} \otimes A_{2j}| - |A_{1j}| - |A_{2j}|)$ where meet ($\oplus$) and join ($\otimes$) are Cartesian operators.	$\sqrt[q]{\sum_{j=1}^{p}(\phi(A_{1j}, A_{2j}))^q}$
U_3	$\psi(A_{1j}, A_{2j}) = \frac{\phi(A_{1j}, A_{2j})}{|B_j|}$	$\sqrt[q]{\sum_{j=1}^{p}(\psi(A_{1j}, A_{2j}))^q}$
U_4	$\psi(A_{1j}, A_{2j}) = \frac{\phi(A_{1j}, A_{2j})}{|B_j|}$	$\sqrt[q]{\sum_{j=1}^{p} w_j(\psi(A_{1j}, A_{2j}))^q}$
SO_1	$D^{(1)}(A_{1j}, A_{2j}) =$ $1 - \alpha/(\alpha + \beta + \chi);$ $D^{(2)}(A_{1j}, A_{2j}) =$ $1 - 2\alpha/(2\alpha + \beta + \chi);$ $D^{(3)}(A_{1j}, A_{2j}) =$ $1 - \alpha/(\alpha + 2\beta + 2\chi);$ $D^{(4)}(A_{1j}, A_{2j}) =$ $1 - \frac{1}{2}(\frac{\alpha}{\alpha+\beta} + \frac{\alpha}{\alpha+\chi});$ $D^{(5)}(A_{1j}, A_{2j}) =$ $1 - \alpha/\sqrt{(\alpha+\beta)(\alpha+\chi)}$ with $\alpha = \mu(A_{1j} \cap A_{2j}),$ $\beta = \mu(A_{1j} \cap c(A_{2j})),$ $\chi = \mu(c((A_{1j}) \cap A_{2j}))$	$\sqrt[q]{\sum_{j=1}^{p} w_j(D^{(r)} A_{1j}, A_{2j})^q}$
SO_2	$\psi(A_{1j}, A_{2j}) = \frac{\phi(A_{1j}, A_{2j})}{\mu(A_{1j} \oplus A_{2j})}$	$\sqrt[q]{\sum_{j=1}^{p} \frac{1}{p}(\psi(A_{1j}, A_{2j}))^q}$
SO_3		$\pi(d_{i_1} \oplus d_{i_2}) - \pi(d_{i_1} \otimes d_{i_2}) + \gamma(2\pi(d_{i_1} \otimes d_{i_2}) - \pi(d_{i_1}) - \pi(d_{i_2}))$
SO_4		$\frac{\pi(d_{i_1} \oplus d_{i_2}) - \pi(d_{i_1} \otimes d_{i_2}) + \gamma(2\pi(d_{i_1} \otimes d_{i_2}) - \pi(d_{i_1}) - \pi(d_{i_2}))}{\pi(d_O)}$ where $d_O = (O_1, \ldots, O_p)$
SO_5		$\frac{\pi(d_{i_1} \oplus d_{i_2}) - \pi(d_{i_1} \otimes d_{i_2}) + \gamma(2\pi(d_{i_1} \otimes d_{i_2}) - \pi(d_{i_1}) - \pi(d_{i_2}))}{\pi(d_{i_1} \oplus d_{i_2})}$

variables. Some are defined using a measure on the given set, $\mu(V) = |V|$ if V is a finite set, $\mu(V) = |u - l|$ if the variable is interval-valued and $V = [l, u]$, while some are functions of the "description potential," $\pi(d_i) = \prod_{j=1}^{p} \mu(Y_j(s_i))$, the volume covered by the description $d_i = (Y_1(s_i), \ldots, Y_p(s_i))$. Table 21.5 details the following dissimilarities:

- U_1 : Gowda and Diday's (1991) dissimilarity measure.
- U_2 : Ichino and Yaguchi's (1994) first dissimilarity measure.
- U_3 : Ichino and Yaguchi's (1994) dissimilarity measure normalized w.r.t. domain length.

- U_4 : Ichino and Yaguchi's (1994) normalized and weighted dissimilarity measure.
- SO_1 : De Carvalho's (1994) dissimilarity measure.
- SO_2 : De Carvalho's (1994) extension of Ichino and Yaguchi's dissimilarity.
- SO_3 : De Carvalho's (1998) first dissimilarity measure based on description potential.
- SO_4 : De Carvalho's (1998) second dissimilarity measure based on description potential.
- SO_5 : De Carvalho's (1998) normalized dissimilarity measure based on description potential.

Let $Y_j(s_{i_1}) = A_{1j}$ and $Y_j(s_{i_2}) = A_{2j}$, with $A_{1j}, A_{2j} \in B_j$.

U_1 and U_2 are additive dissimilarity measures. They can be applied to entities described by interval-valued variables, quantitative single-valued variables (as long as domains are restricted by a lower and an upper bound for U_1) and by categorical single- and multi-valued variables. U_1 values are normalized between 0 and $3p$ where p is the number of variables, while U_2 values are always nonnegative. U_3 and U_4 are similar to U_2, the former with the values normalized in the interval $[0, p]$ and the latter in the interval $[0, 1]$. SO_1 and SO_2 are similar to U_4 and their range is $[0, 1]$ too; SO_3 values are nonnegative, SO_4 and SO_5 range in $[0, 1]$ (see Esposito et al. 2008).

21.3.2 Distances for Interval-Valued Variables

When the entities under analysis are described by interval-valued variables, specific distance measures for comparing interval-valued observations may be used. The most common are:

- Minkowski-type distances
  These result from embedding intervals in $\mathbb{R}^2$, where one dimension is used for the lower bound and the other for the upper bound of the intervals. Then these distances are just the Minkowski distances between the corresponding points in the two-dimensional space. So, the Euclidean distance between two intervals $I_1 = [l_1, u_1]$ and $I_2 = [l_2, u_2]$ is defined as $d_E(I_1, I_2) = \sqrt{(l_1 - l_2)^2 + (u_1 - u_2)^2}$; the Euclidean distance between two interval-valued observations becomes $d_2(s_{i_1}, s_{i_2}) = \sqrt{\sum_{j=1}^{p}((l_{i_1 j} - l_{i_2 j})^2 + (u_{i_1 j} - u_{i_2 j})^2)}$. Analogously, the City-Block (or Manhattan) distance is defined as $d_1(s_{i_1}, s_{i_2}) = \sum_{j=1}^{p}(|l_{i_1 j} - l_{i_2 j}| + |u_{i_1 j} - u_{i_2 j}|)$.
- Hausdorff distance
  The Hausdorff distance between two sets is the maximum distance of a set to the nearest point in the other set, that is, two sets are close in terms of the Hausdorff distance if every point of either set is close to some point of the other set. Therefore, the Hausdorff distance between two intervals $I_1 = [l_1, u_1]$ and $I_2 = [l_2, u_2]$ becomes $d_H(I_1, I_2) = \max\{|l_1 - l_2|, |u_1 - u_2|\}$. For multivariate interval-valued observations, these may be combined; often authors choose to combine them in an "Euclidean" way, leading to $d_{H_2}(s_{i_1}, s_{i_2}) = \sqrt{\sum_{j=1}^{p}(\max\{|l_{i_1 j} - l_{i_2 j}|, |u_{i_1 j} - u_{i_2 j}|\})^2}$
- Mahalanobis distance
  The Mahalanobis distance between the two entities s_{i_1}, s_{i_2} described by vectors of intervals is defined in De Souza et al. (2004d, 2004e) on the basis of the vectors of

observed lower bounds $X_{iL} = (l_{i1},\ldots,l_{ip})$ and upper bounds $X_{iU} = (u_{i1},\ldots,u_{ip})$. The distance is then defined as: $d(s_{i_1}, s_{i_2}) = d_M(X_{i_1L}, X_{i_2L}) + d_M(X_{i_1U}, X_{i_2U})$ where $d_M(X_{i_1L}, X_{i_2L}) = (X_{i_1L} - X_{i_2L})^t M_L (X_{i_1L} - X_{i_2L})$ is the Mahalanobis distance between the two vectors X_{i_1L} and X_{i_2L} and, analogously, $d_M(X_{i_1U}, X_{i_2U}) = (X_{i_1U} - X_{i_2U})^t M_U (X_{i_1U} - X_{i_2U})$ is the Mahalanobis distance between the two vectors X_{i_1U} and X_{i_2U}.

21.3.3 Distances for Distributional-Valued Data

Gibbs and Su (2002) provide a study of different measures for comparing probability distributions: Discrepancy, Hellinger distance, Relative entropy (or Kullback–Leibler divergence), Kolmogorov (or Uniform) metric, Lévy metric, Prokhorov metric, Separation distance, Total variation distance, Wasserstein (or Kantorovich) metric, and χ^2 distance. See also Bock (2000) on this topic.

Among these, the Wasserstein distance and its "L_2" counterpart, the Mallows distance, are the most widely used for histogram-valued data, and are defined as follows. Let Ψ_{ij} be the quantile function corresponding to the distribution $Y_j(s_i)$, Y_j a histogram-valued variable, $s_i \in S$.

- Wasserstein distance:

$$D_W(\Psi_{i_1j}, \Psi_{i_2j}) = \int_0^1 |\Psi_{i_1j}(t) - \Psi_{i_2j}(t)| dt$$

- Mallows distance:

$$D_M(\Psi_{i_1j}, \Psi_{i_2j}) = \sqrt{\int_0^1 (\Psi_{i_1j}(t) - \Psi_{i_2j}(t))^2 dt}$$

Assuming uniformity in each subinterval, and that for all variables the weights vector $(p_1,\ldots,p_\ell,\ldots,p_k)$ is the same (which implies that the number of considered subintervals is the same, but they are, of course, in general not equal), Irpino and Verde (2006) rewrote the squared Mallows distance using the centers $c_{ij\ell}$ and the half-ranges $r_{ij\ell}^*$ of the subintervals defining the histograms, obtaining : $d_M^2(\Psi_{i_1j}, \Psi_{i_2j}) :=$
$\sum_{\ell=1}^k p_\ell \left[(c_{i_1j\ell} - c_{i_2j\ell})^2 + \frac{1}{3}(r_{i_1j\ell}^* - r_{i_2j\ell}^*)^2 \right]$.

Irpino and Romano (2007) proved that this distance can be decomposed as:

$$d_M^2 = \underbrace{(\mu_{i_1j} - \mu_{i_2j})^2}_{Location} + \underbrace{(\sigma_{i_1j} - \sigma_{i_2j})^2}_{Size} + \underbrace{2\sigma_{i_1j}\sigma_{i_2j}(1 - \rho_{QQ}(\Psi_{i_1j}, \Psi_{i_2j}))}_{Shape} \qquad (21.1)$$

where

$$\rho_{QQ}(\Psi_{i_1j}, \Psi_{i_2j}) = \frac{\int_0^1 (\Psi_{i_1j}(t) - \mu_{i_1j})(\Psi_{i_2j}(t) - \mu_{i_2j}) dt}{\sigma_{i_1j}\sigma_{i_2j}} = \frac{\int_0^1 \Psi_{i_1j}(t)\Psi_{i_2j}(t)dt - \mu_{i_1j}\mu_{i_2j}}{\sigma_{i_1j}\sigma_{i_2j}}$$

$$(21.2)$$

is the correlation of the quantiles of the two distributions as represented in a classical QQ plot.

From there, Verde and Irpino (2008) obtained the codeviance between two distributions. Let $\bar{F}_j$ be the distribution whose tth quantile is the mean of the tth quantiles of the n distributions $F_{ij}, i = 1, \ldots, n$, then $CODEV(Y_j, Y_{j'}) = \left(\sum_{i=1}^{n} \sigma_{ij}\sigma_{ij'} - n\sigma_j\sigma_{j'}\right) + \left(\sum_{i=1}^{n} \mu_{ij}\mu_{ij'} - n\mu_j\mu_{j'}\right)$. Then, given the vector $\mathbf{F_i} = (F_{i1}, \ldots, F_{ip})$ and the inverse of the $CODEV$ matrix, $\Sigma_F^{-1} = [s_{hk}^{-1}]_{p \times p}$, Verde and Irpino (2008) introduced the Mahalanobis–Wasserstein distance as:

$$d_{MW}(\mathbf{F_{i_1}}, \mathbf{F_{i_2}}) = \sqrt{\sum_{h=1}^{p}\sum_{k=1}^{p}\int_0^1 s_{hk}^{-1}\left(\Psi_{i_1h}(t) - \Psi_{i_2h}(t)\right)\left(\Psi_{i_1k}(t) - \Psi_{i_2k}(t)\right)dt}$$

The Euclidean distance between observations of a histogram-valued variable or a categorical-modal variable with k different categories Y may be used—and thereby it is meant the Euclidean distance between the weights' vectors. In the case of histogram-valued variables, the observations to be compared should then be written with the same partition, that is, the same set of k subintervals,

$$Y(s_{i_1}) = \{I_{i_11}(p_{i_11}); \quad I_{i_12}(p_{i_12}), \ldots, I_{i_1k}, (p_{i_1k})\}$$

$$Y(s_{i_2}) = \{I_{i_21}(p_{i_21}); \quad I_{i_22}(p_{i_22}), \ldots, I_{i_2k}, (p_{i_2k})\}$$

Then, for both variable types, we get $D_2(Y(s_{i_1}), Y(s_{i_2})) = \sqrt{\sum_{\ell=1}^{k}(p_{i_1\ell} - p_{i_2\ell})^2}$.

In all these cases, to obtain a distance between multivariate observations, the distance values between distributions of individual variables must be combined. Generally they are combined additively.

The affinity coefficient (see Bacelar-Nicolau 2000) is also used to compare multivariate observations when a distribution on a finite set of subclasses or categories is observed for each of p variables. It is defined by $aff(s_{i_1}, s_{i_2}) = \sum_{j=1}^{p}\sum_{\ell=1}^{k_j}\sqrt{p_{i_1j\ell} \times p_{i_2j\ell}}$ which is the affinity coefficient (Matusita 1951) between $(p_{i_1j1}, \ldots, p_{i_1jk_j})$ and $(p_{i_2j1}, \ldots, p_{i_2jk_j})$. Notice that this is a similarity measure.

Kim and Billard (2013) introduce various dissimilarity measures for histogram-valued data, in particular, they extend the Gowda–Diday, the Ichino–Yaguchi measures and some De Carvalho measures for interval data. Also, a cumulative distribution measure is developed for histograms. The proposed measures are illustrated in the context of a clustering problem.

Košmelj and Billard (2012) discuss decompositions of the Mallows distance and use them in the interpretation of the results of k-means clustering and multidimensional scaling. An illustration on population pyramids clustering is presented.

21.3.4 Standardization

When the descriptive variables are quantitative, the question of comparability of the measurement scales of the different variables is a major issue for dissimilarity-based clustering. It is well known, and may be easily verified in applications, that dissimilarity values and, therefore, the clustering results are affected by the variables' scales. To obtain an objective or scale-invariant result, some standardization must be performed prior to dissimilarity computations in the clustering process.

21.3.4.1 Standardization of Interval Data

In De Carvalho et al. (2006), three alternative standardization methods for the case of interval data have been proposed; these methods mainly differ in the way dispersion of an interval-valued variable is evaluated. An alternative approach for the standardization of interval data has been proposed in Chavent (2005), when a Hausdorff distance is used. In that paper, it is explained that to compute distances between standardized observations is generally equivalent to using a normalized version of the corresponding distance.

21.3.4.2 Standardization of Histogram Data

Verde and Irpino (2008) define barycenter of a set of distributions $\{F_{ij}, i = 1, \ldots, n\}$ as the distribution obtained by minimizing $I_S = \sum_{i=1}^{n} d_W^2(\Psi_{ij}, \bar{\Psi}_j) = \sum_{i=1}^{n} \int_0^1 \left(\Psi_{ij}^{-1}(t) - \bar{\Psi}_j(t)\right)^2 dt$, by analogy to the real data case.

Following this reasoning, the authors define $VAR_M(Y_j) = \sum_{i_1=1}^{n} \sum_{i_2=1}^{n} d_M^2(\Psi_{i_1 j}, \Psi_{i_2 j})/2n^2$ and $STD_M(Y_j) = \sqrt{VAR_M(Y_j)}$ which allow defining the "standardized deviation–function" as: $SD_{ij}(t) = (\Psi_{ij}(t) - \bar{\Psi}_j(t))/ STD_M(Y_j), 0 \leq t \leq 1$, which verifies $n^{-1} \sum_{i=1}^{n} \int_0^1 (SD_{ij}(t))^2 dt = 1$ as in the classical case. This defines a standardization procedure for histogram-valued data, represented by the respective quantile functions Ψ.

21.4 Clustering Interval Data

21.4.1 Nonhierarchical Clustering

Many methods based on different dissimilarities φ, adaptations of k-means or dynamical clustering have been developed—see Csernel and De Carvalho (1999), De Souza and De Carvalho (2004a), De Souza et al. (2004d, 2004e), De Carvalho et al. (2006a), and Chavent et al. (2006). These use Minkowski-type, the Mahalanobis or the Hausdorff distances. In each case, the method aims at minimizing a criterion that measures the fit between the clusters' members and their prototypes and is, as usual, based on the alternate application of a representation function followed by an assignment function, until convergence is attained (or the prefixed number of iterations is reached). Adapted to interval data, the algorithm for this kind of methods becomes:

The *representation function* $g = g(P_1, \ldots, P_k)$ assigns to each partition $P = (P_1, \ldots, P_k)$ of S the set of centroids $L = L(P) := (\ell_1, \ldots, \ell_k)$ of p-dimensional intervals:

$$\ell_h = ([\bar{a}_{h1}, \bar{b}_{h1}], \ldots, [\bar{a}_{hp}, \bar{b}_{hp}]) \quad \text{for } h = 1, \ldots, k \tag{21.3}$$

whose lower and upper bounds are determined from the data intervals, following an optimization procedure which depends on the chosen distance φ. If φ is the Euclidean distance, as in De Carvalho et al. (2006a), then the lower and upper bounds for the intervals defining the centroid ℓ_h are given by the corresponding averaged bounds of the data intervals $([l_{i1}, u_{i1}], \ldots, [l_{ip}, u_{ip}])$ for elements $s_i \in P_h$.

The *assignment function* $f = f((\ell_1, \ldots, \ell_k)) := (P_1, \ldots, P_k) =: P$ is given by the minimum-distance rule

$$P_h = \{s_i \in S : \varphi(d_i, \ell_h) \leq \varphi(d_i, \ell_m), 1 \leq m \leq k\} \quad h = 1, \ldots, k \qquad (21.4)$$

with φ the selected distance measure. In case of ties the cluster with smallest index is chosen.

Following these works, fuzzy c-means methods have been developed, as well as methods based on adaptive distances, and combinations of both approaches.

Extensions of the k-means algorithm, using adaptive distances—that is, distances that vary from cluster to cluster—have been proposed, using different distance measures to compare the entities under analysis, see now also De Souza and De Carvalho (2004b), De Souza et al. (2004c), De Carvalho et al. (2006b), De Carvalho and Lechevallier (2009), and De Carvalho and Tenorio (2010). Adaptive distances are weighted distances, where the weights vary from cluster to cluster, and may also be different for different variables. In this case, the algorithm has an additional step, consisting in the determination of the weights' vector.

The first fuzzy clustering method for interval data has been proposed by El-Sonbaty and Ismail (1998). Other approaches followed, namely, D'Urso and Giordani (2006), De Carvalho (2007), De Carvalho and Tenorio (2010), and Jeng et al. (2010). Fuzzy k-means methods for interval data generally result from adapting the classical fuzzy c-means algorithm, using appropriate distances, as is done for the crisp algorithms. In De Carvalho (2007) and De Carvalho and Tenorio (2010) fuzzy k-means clustering models for interval-valued data based on adaptive distances are proposed. The models provide a fuzzy partition and a prototype for each cluster by optimizing an adequacy criterion that measures the fit between the fuzzy clusters and their representatives. The adaptive distances change at each algorithm iteration and can be either the same for all clusters or different from one cluster to another. Additional interpretation tools for individual fuzzy clusters of interval-valued data, suitable to these fuzzy-clustering models, are also presented. In D'Urso and Giordani (2006) a fuzzy-clustering model for fuzzy data is proposed, which is based on a weighted dissimilarity measure for comparing pairs of fuzzy data, composed by two distances, the center (mode) distance and the spread distance. The particularity of the proposed model is the objective estimation of suitable weights concerning the distance measures of the center and the spreads of the fuzzy data. Jeng et al. (2010) propose a robust interval competitive agglomeration (RICA) clustering algorithm, meant to deal with the problems of the outliers, the number of clusters and the initialization of prototypes in the fuzzy c-means clustering algorithm for symbolic interval-valued data.

Recently, a model-based approach for clustering interval data has been developed (Brito et al. 2015), applying the Gaussian models proposed in Brito and Duarte Silva (2012) to the model-based clustering context. For this purpose, the EM algorithm has been adapted, for different covariance configurations. Using both synthetic and empirical data sets allowed putting in evidence the well-founding of the proposed methodology, and, in particular, its flexibility to identify heteroscedastic models even in situations with limited information. Moreover, in the reported applications, considering special configurations of the variance–covariance matrix, adapted to specific nature of interval data, proved to be the adequate approach. Clustering and validation of interval data are discussed in Hardy and Baune (2007).

Apart from the above, clustering methods based on Kohonen's self-organizing maps have also been developed for interval data. These methods are oriented towards a graphical visualization of the clusters, representing them in a two-dimensional rectangular lattice (map) of size $a \times b$ (the number of clusters being $m = ab$) such that clusters with a high similarity are grouped together in close locations of the lattice. In the *SODAS* software, Kohonen maps are constructed by the module *SYKSOM* (Bock 2002, 2008). The construction of a Kohonen map is particularly well suited for the analysis of a large number n of entities, since the implicitly involved clustering process reduces then to a smaller number m of clusters (or vertices) whose properties can be easily visualized. Second, in the Kohonen approach both clustering and prototype construction are conducted in a simultaneous process where the data are entered in a sequential way and current results (clusters, prototypes) are iteratively modified and updated by incorporating a new (the subsequent) entity. Pacifico and De Carvalho (2011) introduces a batch self-organizing map algorithm based on adaptive distances, and Dos S. Dantas and De Carvalho (2011) an adaptive batch SOM method for Multiple Dissimilarity Data Tables. Other approaches are investigated by Hajjar and Hamdan (2011a, b) and Yang et al. (2012).

21.4.2 Hierarchical Clustering

Hierarchical clustering specific for interval data has not had the same extensive development witnessed for the partitional methods.

In general, defining a dissimilarity matrix between the entities to be clustered, using a suitable measure (see Section 21.3.2), allows applying the classical hierarchical methods—single-linkage, complete linkage, average linkage—to data described by interval-valued variables.

Rasson et al. (2008) presents a divisive-clustering method based on non-homogeneous Poisson point processes.

Another clustering method based on Poisson point processes has been proposed in Hardy and Kasaro (2009). The first part of the method consists in a monothetic divisive clustering procedure; the cutting rule uses an extension of the hypervolumes criterion to interval data. The pruning step uses two likelihood ratio tests based on the homogeneous Poisson point processes: the Hypervolumes test and the Gap test. A decision tree is hence obtained. The second part of the method uses a merging procedure, which may allow improving the clustering obtained in the first step.

Other methods for hierarchical clustering of symbolic data have been developed for multiple variable types, and will be described in Section 21.6.

21.5 Clustering Distributional Data

21.5.1 Nonhierarchical Clustering of Distributional Data

Clustering methods specific for histogram-valued data are much less numerous than for the interval data case, and most were developed recently. It is therefore to be expected that methodologies for this kind of data will experience a greater development in the years to come.

Verde and Irpino (2007, 2008) used the Mallows distance for partitional clustering of histogram-valued data. To this purpose, and assuming uniformity in each subinterval of each observed histogram, they rewrote it using the center and half-range of the subintervals defining the histograms (see Section 21.3.3). This then allows finding the "barycenter" of a set of histograms. Let $c_{bj\ell} = n^{-1} \sum_{i=1}^{n} c_{ij\ell}; r_{bj\ell}^* = n^{-1} \sum_{i=1}^{n} r_{ij\ell}^*$. The barycenter (*prototype*) of the n histograms may be expressed by the couples: $([c_{bj\ell} - r_{bj\ell}^*; c_{bj\ell} + r_{bj\ell}^*], p_{j\ell})$, $\ell = 1, \ldots, k; j = 1, \ldots, p$ of intervals with associated weights $p_{j\ell}$. It follows that the Mallows distance admits the Huygens decomposition in within and between-class dispersion. A k-means—or "dynamical clustering"—method is then developed from these definitions and results.

In Verde and Irpino (2008), the authors derive the Mahalanobis–Wasserstein distance (see Section 21.3.3) and use it to define a new dynamical-clustering method for histogram-valued data.

Korenjak-Cerne et al. (2011) proposed the "adapted leaders"—a k-means version—clustering method for data described by discrete distributions. The original $n \times p$ data matrix is first transformed in an $n \times (\sum_{j=1}^{p} k_j)$ frequency matrix (somehow in a similar way as in [Hardy and Lallemand 2004], see Section 21.6.2), to which the algorithm is applied, using a classical L_1 or L_2 distance.

Recently, Irpino et al. (2014) proposed a dynamic-clustering algorithm for histogram data with an automatic weighting step of the variables by using adaptive distances. The method uses the Mallows distance and two novel adaptive distance-based clustering schemes.

21.5.2 Hierarchical Clustering of Distributional Data

Together with the so-called "leaders method" (see Section 21.5.1), Korenjak-Cerne et al. (2011) propose an extension of the Ward method. According to the authors, while the leaders method allows to efficiently solve clustering problems with a large number of units, the agglomerative method is applied on the obtained leaders and allows selecting the number of clusters. Both methods were successfully applied in analyses of different datasets, an application on the "Trends in International Mathematics and Science Study (TIMSS)" data set is presented in the referred paper. The descriptions with distributions allowed combining two datasets, answers of teachers and answers of their students, into only one dataset.

21.6 Clustering Methods for Multiple Data Types

21.6.1 Nonhierarchical Clustering

SCLUST (see De Carvalho et al. 2008) is a module of the *SODAS* package that performs nonhierarchical clustering on symbolic data, using a k-means—or dynamical clustering—like method: starting from a partition on a prefixed number of clusters, it alternates an assignment step (based on minimum distance to cluster prototypes) and a representation step (which determines new prototypes in each cluster) until convergence is achieved (or a prefixed number of iterations is reached). The method locally optimizes a criterion that measures the fit between cluster prototypes and cluster members, which is additive, and based on the assignment-distance function. The method allows for all types of variables in the input data; categorical multivalued or distributional data are transformed into frequency tables (assuming uniformity in the first case). Distances for the assigning step are then selected according to the variable type: for quantitative real-valued data the Euclidean distance is used, for interval and quantitative multi-valued data, the Hausdorff distance, for categorical single-valued data, the χ^2 distance is applied, for categorical multi-valued data, De Carvalho distance is used and for distributional data, a classical ϕ^2 distance is used (see De Carvalho et al. 2008, p. 190). The prototypes are obtained according to the variable type: for interval-valued variables, the prototype of each class is defined as the interval that minimizes the Hausdorff distance to all the elements belonging to the class (see De Carvalho et al. 2008); for categorical multivalued or distributional data, average weights are obtained for each category in each class.

The *SCLUST* module also includes functions for the determination of the appropriate number of clusters, based on classical indices—see Hardy (2008).

Yang et al. (2004) present a fuzzy clustering algorithm, c-means like, for mixed variable types, including for fuzzy data. For the symbolic data types, Gowda and Diday's distance is used (see Section 21.3), in a modified form for quantitative data; for fuzzy data, a suitable distance, derived from previous work of the authors, is used. Cluster prototypes are determined separately for symbolic and fuzzy variables.

21.6.2 Hierarchical and Pyramidal Clustering

Gowda and Diday have developed hierarchical clustering methods for symbolic data, where entities may be described by variables of different types, using appropriate comparison measures, based on position, span, and content of the symbolic descriptions (see also Section 21.3.1). In Gowda and Diday (1991) a method is proposed based on a similarity measure. The method is of single linkage type; the single linkage is performed through the identification of a mutual pair having the highest similarity values at various hierarchical levels. On the other hand, in Gowda and Diday (1992) the presented method uses a dissimilarity measure. The algorithm forms composite symbolic descriptions using a Cartesian join operator whenever a mutual pair of symbolic objects is selected for agglomeration based on minimum dissimilarity.

In Gowda and Ravi (1995a) the previous dissimilarity measure is modified, and used in an agglomerative clustering method which uses both similarity and dissimilarity; a composite symbolic description is formed using a Cartesian join operator when two entities are assigned to the same class based on both similarity and dissimilarity. In Gowda and Ravi (1995b), a divisive clustering approach is used. Dissimilarities between all pairs of entities in the dataset are determined according to the variable type and the minimum spanning tree of the dataset is determined. The edge having the highest weight in the minimum spanning tree is removed, leading to two groups; then the Similarity S and Dissimilarity D between the two groups are computed, if the Dissimilarity D is greater than Similarity S the "inconsistent" edge is removed so as to result in two clusters, else the inconsistent edge is placed back. The procedure is repeated until a complete tree is obtained.

Hardy and Lallemand (2004) have developed a module for the *SODAS* package, called *SHICLUST*, which extends four well-known classic hierarchical clustering methods—single linkage, complete linkage, centroid, and Ward methods—to symbolic data, in particular to set-valued data (numerical or categorical), interval-valued, and categorical modal variables (see also Hardy 2008). For interval-valued-variables, the Hausdorff and the Minkowski L_1 and L_2 are used; for multi-valued and categorical modal variables, the original $n \times p$ data matrix is first transformed in an $n \times (\sum_{j=1}^{p} k_j)$ frequency matrix, gathering all frequency values for all categories, and Minkowski L_1 and L_2 distances or De Carvalho distance SO1 are then applied to the extended frequency matrix. Usual aggregation indices may then be applied for hierarchical clustering.

Brito (1991, 1994, 1995) developed a method for "symbolic" hierarchical or pyramidal clustering, which allows clustering entities described by multiple variable types; the method was later extended to distributional variables (Brito 1998), on the one hand, and to allow for the existence of hierarchical rules between categorical variables (Brito and De Carvalho 1999) and between distributional variables (Brito and De Carvalho 2002). The method may be summarized as follows: for each candidate cluster, a description is built, generalizing the descriptions corresponding to the clusters to be merged, and the "candidate cluster" is eligible only if this new description covers all cluster elements and none other. When two given clusters are merged, generalization is performed according to the variable type, for set-valued variables, set-union is performed, for interval-valued variables, the minimum interval that covers those to be merged is considered; for variables whose values are distributions generalization is done by either considering the maximum or the minimum of the probability/frequency values for each category. In this sense, the method comes within the framework of "conceptual clustering" (see Michalski et al. 1981; Michalski and Stepp 1983), since each cluster formed is associated with a conjunction of properties on the descriptive variables, which constitutes a necessary and sufficient condition for cluster membership. An additional criterion is considered to choose among the different aggregations meeting the above condition, following the principle that clusters associated with less general descriptions should be formed first. For set-valued and interval-valued variables, this "generality degree" evaluates the proportion of the description space covered by the considered description; for distributional variables, it evaluates how much the given distribution is close to the Uniform distribution, by computing the affinity between the given distribution and the Uniform (see Brito and De Carvalho 2002, 2008). The generality degree is computed variable-wise; the values for each variable are then combined in a multiplicative way to produce a measure of the variability of the full description. Module *HIPYR* of the *SODAS* package applies this methodology.

More recently, Brito and Polaillon (2011) proposed a common framework for representing and operating with ordinal, numerical single or interval data as well as distributional data, by defining a generalization operator that determines descriptions in the form of intervals. In the case of distributional data, the obtained concepts are more homogeneous and more easily interpretable than those obtained by using the maximum and minimum operators previously proposed. This approach has been applied to hierarchical (or pyramidal) clustering, along the lines of the above presented algorithm; the considered "generality degree" is now additive on the variables, measuring an "average" generality across variables (see Brito and Polaillon 2012).

In Chavent (1998), a divisive hierarchical clustering method is proposed, which produces "monothetic" clusters, that is, each cluster formed is again associated with a conjunction of properties on the descriptive variables, constituting a necessary and sufficient condition for cluster membership. The method may apply both to interval and categorical modal variables (though not simultaneously), and uses a criterion that measures intra-class dispersion, evaluated using distances appropriate to the type of variables. The algorithm successively splits one cluster into two subclusters, according to a condition expressed as a binary question on the values of one variable; the cluster to be split and the condition to be considered at each step are selected so as to minimize intracluster dispersion on the next step. Therefore, each formed cluster is directly interpreted by necessary and sufficient conditions for cluster membership (the conditions that lead to its formation by successive splits). Brito and Chavent (2012) extend the divisive algorithm proposed in Chavent (1998) and Chavent et al. (2007) to data described by interval and/or histogram-valued variables. In both cases, data are represented as distributions, and therefore may be treated together; Malllows or Euclidean distances are now used to evaluate intraclass dispersion.

In Kim and Billard (2012) dissimilarity measures for categorical modal data are proposed, by extending the Gowda and Diday and Ichino and Yaguchi measures. A divisive monothetic algorithm is then developed, along the lines of Chavent's methodology. The authors introduce an association measure, to obtain a representative item for each categorical modal variable; this procedure allows reducing the number of partitions to explore at each step of the algorithm.

Brito and Ichino (2010, 2011a,b) are developing hierarchical-clustering methods based on quantile representations of the data (Ichino 2008, 2011). Having a common representation setup based on quantile-vector representations (for prechosen sets of quantiles) allows for a unified analysis of the dataset, taking variables of different types simultaneously into account. Also, the "full" quantile function may be used, leading to a continuous representation instead of a discrete one. In a numerical clustering context, hierarchical and pyramidal models may be applied and clusters are formed on the basis of quantile proximity. In a conceptual approach, the proposed (agglomerative) hierarchical/pyramidal clustering method merges at each step the two clusters with closest quantile representation; the newly formed cluster is then represented according to the same model, that is, a discrete or continuous quantile representation for the new cluster is determined from the mixture of the respective distributions. Clusters are successively compared on the basis of the current quantile (discrete or continuous) representation. An appropriate dissimilarity is used to compare data units: in the discrete approach this may be the Euclidean distance between standardized quantile vectors, whereas when using a continuous approach the Mallows distance between functions is appropriate. Notice that even if Uniform distributions are assumed for the input data, the formed clusters are generally not Uniform on each variable.

21.7 Application: The *CAR* Dataset

The *CAR* dataset consists of a set of 33 car models described by eight interval-valued and two categorical multi-valued variables. The eight interval-valued variables are Price, Engine Displacement, Maximum Speed, Acceleration, Pace, Length, Width, and Height; the multi-valued categorical variables are Fuel, with category set {Petrol, Diesel} and Traction, with category set {Front-wheel, Rear-wheel, All-wheels}. The 33 models cover four car categories: Supermini (SM), Sedan (SD), Luxury (L) and Sportive (S).

Table 21.6 shows a portion of the dataset.

TABLE 21.6

Car Dataset: Partial View

Model	Price	Engine-Disp.	Fuel	Traction	Max-Speed
Alfa 145	[27,806, 33,596]	[1370, 1910]	{Petrol, Diesel}	{Front-wheel}	[185, 211]
Aston Martin	[260,500, 460,000]	[5935, 5935]	{Petrol}	{Rear-wheel}	[298, 306]
BMW series 3	[45,407, 76,392]	[1796, 2979]	{Petrol}	{Rear-wheel, All-wheels}	[201, 247]

TABLE 21.7

Car Dataset: Global Center and Prototypes of the Four Clusters Obtained by SCLUST

	Whole set	Class 1	Class 2	Class 3	Class 4
Nb. Elements	33	13	7	8	5
Price	[250,269.50, 71,529.50]	[20,402.50, 31,802.50]	[160,888.00, 259,596.00]	[48,769.00, 73,030.00]	[104,892.00, 276,792.00]
Engine-disp.	[2037.00, 2738.00]	[1320.50, 1831.50]	[4292.00, 4494.00]	[1907.00, 2868.00]	[2425.00, 5012.00]
Fuel	{Petrol (0.86), Diesel (0.14)}	{Petrol (0.65), Diesel (0.35)}	{Petrol (1.00), Diesel (0.00)}	{Petrol (1.00), Diesel (0.00)}	{Petrol (1.00), Diesel (0.00)}
Traction	{Front-wheel (0.53), Rear-wheel (0.33), All-wheel (0.14)}	{Front-wheel (0.96), Rear-wheel (0.00), All-wheel (0.04)}	{Front-wheel (0.00), Rear-wheel (0.79), All-wheel (0.21)}	{Front-wheel (0.63), Rear-wheel (0.19), All-wheel (0.19)}	{Front-wheel (0.00), Rear-wheel (0.80), All-wheel (0.20)}
Max-speed	[205.00, 222.00]	[164.50, 180.50]	[287.50, 297.50]	[200.00, 227.00]	[224.00, 248.00]
Acceleration	[8.05, 10.05]	[11.20, 14.10]	[4.50, 5.20]	[7.85, 9.85]	[6.60, 9.00]
Pace	[262.00, 262.00]	[249.00, 249.00]	[260.00, 260.00]	[270.00, 270.00]	[289.00, 289.00]
Length	[447.00, 447.00]	[396.00, 396.00]	[447.00, 447.00]	[453.00, 453.00]	[503.00, 503.00]
Width	[175.00, 175.00]	[169.00, 169.00]	[182.00, 182.00]	[175.00, 175.00]	[186.00, 186.00]
Height	[143.00, 143.00]	[143.00, 143.00]	[129.00, 129.00]	[142.00, 142.00]	[144.00, 144.00]

TABLE 21.8

SCLUST Four Clusters Partition for the Car Dataset

Cluster	Members
Class 1	Alfa 145/SM, Punto/SM, Fiesta/SM, Focus/SD, Lancia Y/SM, Nissan Micra/SM
	Corsa/SM, Vectra/SD, Twingo/SM, Rover 25/SM, Skoda Fabia/SM, Skoda Octavia/SD, Passat/L
Class 2	Aston Martin/S, Ferrari/S, Honda NSK/S, Lamborghini/S
	Maserati GT/S, Mercedes SL/S, Porsche/S
Class 3	Alfa 156/SD, Alfa 166/L, Audi A3/SM, Audi A6/SD
	BMW series 3/SD, Lancia K/L, Mercedes Classe C/SD, Rover 75/SD
Class 4	Audi A8/L, BMW series 5/L, BMW series 7/L
	Mercedes Classe E/L, Mercedes Classe S/L

The *a priori* classification, indicated by the suffix attached to the car model denomination, is as follows:

```
Supermini:
1-Alfa 145/SM      5-Audi A3/SM    12-Punto/SM    13-Fiesta/SM      17-Lancia Y/SM
24-Nissan Micra/SM 25-Corsa/SM     28-Twingo/SM   29-Rover 25/SM    31-Skoda Fabia/SM

Sedan:
2-Alfa 156/SD            6-Audi A6/SD    8-BMW series 3/SD   14-Focus/SD
21-Mercedes Classe C/SD  26-Vectra/SD    30-Rover 75/SD      32-Skoda Octavia/SD

Sportive:
4-Aston Martin/S   11-Ferrari/S        15-Honda NSK/S    16-Lamborghini/S
19-Maserati GT/S   20-Mercedes SL/S    27-Porsche/S

Luxury:
3-Alfa 166/L    7-Audi A8/L          9-BMW series 5/L       10-BMW series 7/L
18-Lancia K/L   22-Mercedes Classe E/L  23-Mercedes Classe S/L  33-Passat/L
```

We have applied two clustering methods to this dataset, which allow clustering sets described by different variable types: *SCLUST* nonhierarchical clustering method (see Section 21.6.1) and *HIPYR* (see Section 21.6.2), a conceptual clustering hierarchical clustering procedure (both available in the *SODAS* package).

Given there are four car categories, we have concentrated on partitions in about four clusters.

As described above, *SCLUST* is a k-means, or dynamical-clustering, type method. The obtained partition in four clusters explains 53.48% of the total dispersion. The four obtained clusters are represented by the prototypes presented in Table 21.7; the composition of the clusters is given in Table 21.8.

As explained above, the prototype of each cluster is defined as follows: for the interval-valued variables, by the interval minimizing the Hausdorff distance to the elements of the cluster; for the categorical multi-valued variables, an average weight is obtained for each category (assuming uniformity of the weights for the categories present in each car model).

From the obtained results, we can see, for instance, that in cluster 2 we find expensive sportive car models, all using petrol as fuel, mainly with rear-wheel traction (but not all), with high engine-displacement and top-speed, and low acceleration values.

The *HIPYR* method produced a conceptual hierarchy, where, as explained above, each cluster is associated with a generalized description which covers all the cluster members

TABLE 21.9

Car Dataset: Descriptions of the Partition into Five Clusters Produced by *HIPYR*

	Class 1	Class 2	Class 3	Class 4	Class 5
Nb. Elements	4	10	3	6	10
Price	[132,800, 262,500]	[16,992, 34,092]	[240,292, 460,000]	[68,216, 394,342]	[27,419, 115,248]
Engine-disp.	[2799, 5987]	[973, 1994]	[3586, 5992]	[1781, 5786]	[1585, 3199]
Fuel	{Petrol}	{Petrol, Diesel}	{Petrol}	{Petrol}	{Petrol, Diesel}
Traction	{Rear-wheel, All-wheel}	{Front-wheel}	{Rear-wheel, All-wheel}	{Front-wheel, Rear-wheel, All-wheel}	{Front-wheel, Rear-wheel, All-wheel}
Max-speed	[232, 305]	[150, 211]	[295, 335]	[210, 250]	[189, 250]
Acceleration	[4.2, 9.7]	[8.3, 17]	[3.9, 5.2]	[5.4, 10.1]	[5.2, 12.7]
Pace	[235, 266]	[235, 262]	[259, 269]	[276, 309]	[250, 275]
Length	[414, 451]	[343, 415]	[447, 476]	[478, 516]	[415, 475]
Width	[175, 183]	[160, 171]	[183, 204]	[180, 188]	[171, 183]
Height	[129, 131]	[132, 148]	[111, 132]	[143, 145]	[142, 146]
Generality	9.04×10^{-6}	9.41×10^{-6}	10.43×10^{-6}	21.93×10^{-6}	41.57×10^{-6}

TABLE 21.10

HIPYR Five Clusters Partition for the Car Dataset

Cluster	Members
Class 1	Mercedes SL/S, Porsche/S, Honda NSK/S, Maserati GT/S
Class 2	Corsa/SM, Skoda Fabia/SM, Punto/SM, Fiesta/SM, Lancia Y/SM
	Nissan Micra/SM, Twingo/SM, Focus/SM, Alfa 145/SM, Rover 25/SM
Class 3	Lamborghini/S, Aston Martin/S, Ferrari/S
Class 4	Mercedes Classe E/L, Audi A6/L, BMW series 5/L
	Mercedes Classe S/L, Audi A8/L, BMW series 7/L
Class 5	Passat/L, Lancia K/L, Alfa 166/L, Rover 75/SD, BMW series 3/SD
	Mercedes Classe C/SD, Audi A3/SM, Skoda Octavia/SD, Alfa 156/SD, Vectra/SD

and none other. If we consider clusters with generality degree above 9×10^{-6}, we obtain a partition into five clusters, whose descriptions are reported in Table 21.9; the composition of the clusters is given in Table 21.10.

21.8 Concluding Remarks

Symbolic data generalize the usual real and categorical single-valued data by allowing for multiple, possibly weighted, values for each variable and each entity under analysis, so as to suitably take variability into account. To this aim, new variable types have been defined, whose realizations are now sets, intervals, or distributions. Dissimilarity measures appropriate to these new data have been proposed and used in different contexts.

Many clustering methods, of different types, have now been designed for symbolic data. It should nevertheless be stressed that there has been considerable greater effort in developing methods for interval-valued data than for any other type of symbolic data. In this chapter, we have tried to give the reader an overview of the issues that are raised when it is wished to apply clustering analysis to symbolic data, that is, when variable values are allowed to assume new forms. Different authors approach the problems in different ways, sometimes by suitably adapting classical methods, otherwise conceiving total new methodologies.

Although having tried to be complete, we are aware that much still remains to be reported; choices had to be done, and, what is more, as we were writing this text, new publications were appearing, making it clear that it is not possible to be exhaustive. Symbolic data analysis is a new and dynamical field of research, where much work is currently being developed.

Many issues are still open, substantive aspects requiring careful thinking. Take as an example, the parametric modeling for some types of symbolic data, which just starts to be studied. This area is certainly a challenge, but also a well of opportunities, for nowadays data analysis researchers.

References

Bacelar-Nicolau, H. 2000. The affinity coefficient. In H.-H. Bock and E. Diday (Eds.), *Analysis of Symbolic Data: Exploratory Methods for Extracting Statistical Information from Complex Data*, Berlin-Heidelberg, pp. 160–165. Springer.

Billard, L. and E. Diday 2006. *Symbolic Data Analysis: Conceptual Statistics and Data Mining*. Chichester: Wiley.

Bock, H.-H. 2000. Dissimilarity measures for probability distributions. In H.-H. Bock and E. Diday (Eds.), *Analysis of Symbolic Data: Exploratory Methods for Extracting Statistical Information from Complex Data*, Berlin-Heidelberg, pp. 153–165. Springer.

Bock, H.-H. 2002. Clustering methods and Kohonen maps for symbolic data. *Journal of the Japanese Society of Computational Statistics* 15(2), 217–229.

Bock, H.-H. 2008. Visualizing symbolic data by Kohonen maps. In E. Diday and M. Noirhomme-Fraiture (Eds.), *Symbolic Data Analysis and the SODAS Software*, Chichester, pp. 205–234. Wiley.

Bock, H.-H. and E. Diday 2000. *Analysis of Symbolic Data: Exploratory Methods for Extracting Statistical Information from Complex Data*. Berlin-Heidelberg: Springer.

Brito, P. 1991. Analyse de Données Symboliques. Pyramides d'Héritage. PhD thesis. Université Paris-IX Dauphine.

Brito, P. 1994. Use of pyramids in Symbolic Data Analysis. In E. Diday, Y. Lechevallier, M. Schader, P. Bertrand, and B. Burtschy (Eds.), *New Approaches in Classification and Data Analysis*, Berlin-Heidelberg, pp. 378–386. Springer.

Brito, P. 1995. Symbolic objects: Order structure and pyramidal clustering. *Annals of Operations Research* 55, 277–297.

Brito, P. 1998. Symbolic clustering of probabilistic data. In A. Rizzi, M. Vichi, and H.-H. Bock (Eds.), *Advances in Data Science and Classification*, Berlin-Heidelberg, pp. 385–389. Springer.

Brito, P. 2007. On the analysis of symbolic data. In P. Brito, P. Bertrand, G. Cucumel, and F. A. T. De Carvalho (Eds.), *Selected Contributions in Data Analysis and Classification*, Heidelberg, pp. 13–22. Springer.

Brito, P. 2014. Symbolic Data Analysis: Another look at the interaction of Data Mining and Statistics. *WIREs Data Mining and Knowledge Discovery* 4(4), 281–295.

Brito, P. and M. Chavent 2012. Divisive monothetic clustering for interval and histogram-valued data. In *Proceedings of ICPRAM 2012—1st International Conference on Pattern Recognition Applications and Methods*.

Brito, P. and F. A. T. De Carvalho 1999. Symbolic clustering in the presence of hierarchical rules. In *Studies and Research Proceedings of the Conference on Knowledge Extraction and Symbolic Data Analysis (KESDA'98)*, Luxembourg, pp. 119–128. Office for Official Publications of the European Communities.

Brito, P. and F. A. T. De Carvalho 2002. Symbolic clustering of constrained probabilistic data. In O. Opitz and M. Schwaiger (Eds.), *Exploratory Data Analysis in Empirical Research*, Heidelberg, pp. 12–21. Springer Verlag.

Brito, P. and F. A. T. De Carvalho 2008. Hierarchical and pyramidal clustering. In E. Diday and M. Noirhomme-Fraiture (Eds.), *Symbolic Data Analysis and the Sodas Software*, Chichester, pp. 181–203. Wiley.

Brito, P. and A. P. Duarte Silva 2012. Modelling interval data with Normal and Skew-Normal distributions. *Journal of Applied Statistics 39*(1), 3–20.

Brito, P., A. P. Duarte Silva, and J. G. Dias 2015. Probabilistic clustering of interval data. *Intelligent Data Analysis, 19*(2), 293–313.

Brito, P. and M. Ichino 2010. Symbolic clustering based on quantile representation. In *Proceedings of COMPSTAT 2010*, Paris.

Brito, P. and M. Ichino 2011a. Clustering symbolic data based on quantile representation. In P. Brito and M. Noirhomme-Fraiture (Eds.), *Proceedings of Workshop in Symbolic Data Analysis*, Namur.

Brito, P. and M. Ichino 2011b. Conceptual clustering of symbolic data using a quantile representation: Discrete and continuous approaches. In *Proceedings of Workshop on Theory and Application of High-dimensional Complex and Symbolic Data Analysis in Economics and Management Science*, Beijing.

Brito, P. and G. Polaillon 2011. Homogeneity and stability in conceptual analysis. In A. Napoli and V. Vychodil (Eds.), *Proceedings of 8th International Conference on Concept Lattices and Their Applications*, Nancy, pp. 251–263. INRIA.

Brito, P. and G. Polaillon 2012. Classification conceptuelle avec généralisation par intervalles. *Revue des Nouvelles Technologies de l'Information E.23*, 35–40.

Chavent, M. 1998. A monothetic clustering method. *Pattern Recognition Letters 19*(11), 989–996.

Chavent, M. 2005. Normalized k-means clustering of hyper-rectangles. In *Proceedings of the XIth International Symposium of Applied Stochastic Models and Data Analysis (ASMDA 2005)*, Brest, pp. 670–677.

Chavent, M., F. A. T. De Carvalho, Y. Lechevallier, and R. Verde 2006. New clustering methods for interval data. *Computational Statistics 21*(2), 211–229.

Chavent, M., Y. Lechevallier, and O. Briant 2007. DIVCLUS-T: A monothetic divisive hierarchical clustering method. *Computational Statistics & Data Analysis 52*(2), 687–701.

Choquet, G. 1954. Theory of capacities. *Annales de l'Institut Fourier 5*, 131–295.

Csernel, M. and F. A. T. De Carvalho 1999. Usual operations with symbolic data under Normal Symbolic Form. *Applied Stochastic Models in Business and Industry 15*, 241–257.

De Carvalho, F. A. T. 1994. Proximity coefficients between boolean symbolic objects. In E. Diday, Y. Lechevallier, M. Schader, P. Bertrand, and B. Burtschy (Eds.), *New Approaches in Classification and Data Analysis*, Berlin-Heidelberg, pp. 387–394. Springer-Verlag.

De Carvalho, F. A. T. 1998. Extension-based proximity coefficients between constrained boolean symbolic objects. In C. Hayashi, K. Yajima, H.-H. Bock, N. Ishumi, Y. Tanaka, and Y. Baba (Eds.), *Data Science, Classification, and Related Methods*, Tokyo, pp. 370–378. Springer-Verlag.

De Carvalho, F. A. T. 2007. Fuzzy c-means clustering methods for symbolic interval data. *Pattern Recognition Letters 28*(4), 423–437.

De Carvalho, F. A. T., P. Brito, and H.-H. Bock 2006a. Dynamic clustering for interval data based on L_2 distance. *Computational Statistics 21*(2), 231–250.

De Carvalho, F. A. T., R. M. C. R. De Souza, M. Chavent, and Y. Lechevallier 2006b. Adaptive Hausdorff distances and dynamic clustering of symbolic interval data. *Pattern Recognition Letters 27*(3), 167–179.

De Carvalho, F. A. T. and Y. Lechevallier 2009. Partitional clustering algorithms for symbolic interval data based on single adaptive distances. *Pattern Recognition 42*(7), 1223–1236.

De Carvalho, F. A. T., Y. Lechevallier, and R. Verde 2008. Clustering methods in symbolic data analysis. In E. Diday and M. Noirhomme-Fraiture (Eds.), *Symbolic Data Analysis and the SODAS Software*, Wiley, Chichester, pp. 182–203.

De Carvalho, F. A. T. and C. P. Tenorio 2010. Fuzzy k-means clustering algorithms for interval-valued data based on adaptive quadratic distances. *Fuzzy Sets and Systems 161*(23), 2978–2999.

De Souza, R. M. C. R. and F. A. T. De Carvalho 2004a. Clustering of interval data based on City-Block distances. *Pattern Recognition Letters 25*(3), 353–365.

De Souza, R. M. C. R. and F. A. T. De Carvalho 2004b. Dynamic clustering of interval data based on adaptive Chebyshev distances. *Electronics Letters 40*(11), 658–659.

De Souza, R. M. C. R., F. A. T. De Carvalho, and F. C. D. Silva 2004c. Clustering of interval-valued data using adaptive squared Euclidean distance. In N. R. Pal, N. Kasabov, R. K. Mudi, S. Pa, and S. K. Paruil (Eds.), *Neural Information Processing—Proceedings of ICONIP*, Berlin-Heidelberg, pp. 775–780. Springer.

De Souza, R. M. C. R., F. A. T. De Carvalho, and C. P. Tenorio 2004d. Two partitional methods for interval-valued data using Mahalanobis distances. In C. Lemaitre, C. A. Reyes, and J. A. Gonzalez (Eds.), *Advances in Artificial Intelligence—Proceedings of IBERAMIA 2004*, Berlin-Heidelberg, pp. 454–463. Springer.

De Souza, R. M. C. R., F. A. T. De Carvalho, C. P. Tenorio, and Y. Lechevallier 2004e. Dynamic cluster methods for interval data based on Mahalanobis Distances. In D. Banks, F. R. McMorris, P. Arabie, and W. Gaul (Eds.), *Classification, Clustering, and Data Mining Applications—Proceedings of IFCS 2004*, Berlin-Heidelberg, pp. 351–360. Springer.

Diday, E. 1988. The symbolic approach in clustering and related methods of data analysis: The basic choices. In H.-H. Bock (Ed.), *Classification and Related Methods of Data Analysis, Proceedings of IFCS'87*, Amsterdam, pp. 673–684. North Holland.

Diday, E. and F. Esposito 2003. An introduction to Symbolic Data Analysis and the SODAS software. *Intelligent Data Analysis 7*, 583–602.

Diday, E. and M. Noirhomme-Fraiture 2008. *Symbolic Data Analysis and the SODAS Software*. Chichester: Wiley.

Dos S. Dantas, A. B. and F. A. T. De Carvalho 2011. Adaptive batch SOM for multiple dissimilarity data tables. In *Proceedings of 23rd IEEE International Conference on Tools with Artificial Intelligence (ICTAI)*, pp. 575–578.

Duarte Silva, A. P. and P. Brito 2011. MAINT.Data: Model and Analyze Interval Data, R package, version 0.2. http://cran.r-project.org/web/packages/MAINT.Data/index.html

Dubois, D. and H. Prade 1999. Properties of measures of information in evidence and possibility theories. *Fuzzy Sets and Systems 100* (supplement), 35–49.

D'Urso, P. and P. Giordani 2006. A weighted fuzzy c-means clustering model for fuzzy data. *Computational Statistics & Data Analysis 50*(6), 1496–1523.

El-Sonbaty, Y. and M. A. Ismail 1998. Fuzzy clustering for symbolic data. *IEEE Transactions on Fuzzy Systems 6*(2), 195–204.

Esposito, F., D. Malerba, and A. Appice 2008. Dissimilarity and matching. In E. Diday and M. Noirhomme-Fraiture (Eds.), *Symbolic Data Analysis and the SODAS Software*, Chichester, pp. 123–148. Wiley.

Gibbs, A. L. and F. E. Su 2002. On choosing and bounding probability metrics. *International Statistical Review 70*, 419–435.

Gowda, K. C. and E. Diday 1991. Symbolic clustering using a new dissimilarity measure. *Pattern Recognition 24*(6), 567–578.

Gowda, K. C. and E. Diday 1992. Symbolic clustering using a new similarity measure. *IEEE Transactions on Systems, Man and Cybernetics 22*(2), 368–378.

Gowda, K. C. and T. V. Ravi 1995a. Agglomerative clustering of symbolic objects using of both similarity and dissimilarity. *Pattern Recognition Letters 16*, 647–652.

Gowda, K. C. and T. V. Ravi 1995b. Divisive clustering of symbolic objects using the concepts of both similarity and dissimilarity. *Pattern Recognition 28*(8), 1277–1282.

Hajjar, C. and H. Hamdan 2011a. Self-organizing map based on Hausdorff distance for interval-valued data. In *Proceedings of IEEE International Conference on Systems, Man, and Cybernetics (SMC)*, pp. 1747–1752.

Hajjar, C. and H. Hamdan 2011b. Self-organizing map based on L_2 distance for interval-valued data. In *Proceedings of 6th IEEE International Symposium on Applied Computational Intelligence and Informatics (SACI)*, pp. 317–322.

Hardy, A. 2008. Validation of clustering structure: determination of the number of clusters. In E. Diday and M. Noirhomme-Fraiture (Eds.), *Symbolic Data Analysis and the SODAS Software*, Chichester, pp. 335–362. Wiley.

Hardy, A. and J. Baune 2007. Clustering and validation of interval data. In P. Brito, P. Bertrand, G. Cucumel, and F. A. T. De Carvalho (Eds.), *Selected Contributions in Data Analysis and Classification*, Heidelberg, pp. 69–82. Springer.

Hardy, A. and N. Kasaro 2009. A new clustering method for interval data. *Mathématiques et Sciences Humaines 187*, 79–91.

Hardy, A. and P. Lallemand 2004. Clustering of symbolic objects described by multi-valued and modal variables. In D. Banks, F. McMorris, P. Arabie, and W. Gaul (Eds.), *Classification, Clustering, and Data Mining Applications*, Heidelberg, pp. 325–332. Springer.

Ichino, M. 2008. Symbolic PCA for histogram-valued data. In *Proceedings of IASC 2008*, Yokohama, p. 123.

Ichino, M. 2011. The quantile method for symbolic principal component analysis. *Statistical Analysis and Data Mining 4*, 184–198.

Ichino, M. and H. Yaguchi 1994. Generalized Minkowski metrics for mixed feature-type data analysis. *IEEE Transactions on Systems, Man, and Cybernetics 24(4)*, 698–708.

Irpino, A. and E. Romano 2007. Optimal histogram representation of large data sets: Fisher vs piecewise linear approximations. *Revue des Nouvelles Technologies de l'Information E-9*, 99–110.

Irpino, A. and R. Verde 2006. A new Wasserstein based distance for the hierarchical clustering of histogram symbolic data. In V. Batagelj, H.-H. Bock, and A. Ferligoj (Eds.), *Proceedings of of the Conference of the International Federation of Classification Societies (IFCS06)*, Heidelberg, pp. 185–192. Springer.

Irpino, A., R. Verde, and F. A. T. De Carvalho 2014. Dynamic clustering of histogram data based on adaptive squared Wasserstein distances. *Expert Systems with Applications 41*, 3351–3366.

Jeng, J.-T., C.-C. Chuan, C.-C. Tseng, and C.-J. Juan 2010. Robust interval competitive agglomeration clustering algorithm with outliers. *International Journal of Fuzzy Systems 12(3)*, 227–236.

Kim, J. and L. Billard 2012. Dissimilarity measures and divisive clustering for symbolic multimodal-valued data. *Computational Statistics & Data Analysis 56(9)*, 2795–2808.

Kim, J. and L. Billard 2013. Dissimilarity measures for histogram-valued observations. *Communications in Statistics—Theory and Methods 42(2)*, 283–303.

Korenjak-Cerne, S., V. Batagelj, and B. Japelj Paveic 2011. Clustering large data sets described with discrete distributions and its application on TIMS data set. *Statistical Analysis and Data Mining 4*, 199–215.

Košmelj, K. and L. Billard 2012. Mallows L_2 distance in some multivariate methods and its application to histogram-type data. *Metodoloski Zvezki 9(2)*, 107–118.

Matusila, K. 1951. Decision rules based on distance for problems of fit, two samples and estimation. *Annals of Mathematical Statistics 3*, 1–30.

Michalski, R. S., E. Diday, and R. Stepp 1981. A recent advance in data analysis: Clustering objects into classes characterized by conjunctive concepts. In L. N. Kanal and A. Rosenfeld (Eds.), *Progress in Pattern Recognition*, Amsterdam, pp. 33–56. North-Holland.

Michalski, R. S. and R. Stepp 1983. Learning from observations: Conceptual clustering. In R. S. Michalsky, J. Carbonell, and T. Mitchell (Eds.), *Machine Learning: An Artificial Intelligence Approach*, Palo Alto, CA, pp. 163–190. Morgan Kaufmann.

Noirhomme-Fraiture, M. and P. Brito 2011. Far beyond the classical data models: Symbolic Data Analysis. *Statistical Analysis and Data Mining 4(2)*, 157–170.

Pacifico, L. D. S. and F. A. T. De Carvalho 2011. A batch self-organizing maps algorithm based on adaptive distances. In *Proceedings of the 2011 International Joint Conference on Neural Networks (IJCNN)*, pp. 2297–2304.

Rasson, J. P., J.-Y. Piron, P. Lallemand, and S. Adans 2008. Unsupervised divisive classification. In E. Diday and M. Noirhomme-Fraiture (Eds.), *Symbolic Data Analysis and the SODAS Software*, Chichester, pp. 149–156. Wiley.

Verde, R. and A. Irpino 2007. Dynamic clustering of histogram data: Using the right metric. In P. Brito, P. Bertrand, G. Cucumel, and F. A. T. De Carvalho (Eds.), *Selected Contributions in Data Analysis and Classification*, Heidelberg, pp. 123–134. Springer.

Verde, R. and A. Irpino 2008. Comparing histogram data using a Mahalanobis-Wasserstein distance. In P. Brito (Ed.), *Proceedings of COMPSTAT'2008*, Heidelberg, pp. 77–89. Springer.

Vignes, R. 1991. *Caractérisation Automatique de Groupes Biologiques*. PhD thesis. Université Paris VI.

Walley, P. 2000. Towards a unified theory of imprecise probability. *International Journal of Approximate Reasoning 24*(2–3), 125–148.

Yang, M.-S., W.-L. Hung, and D.-H. Chen 2012. Self-organizing map for symbolic data. *Fuzzy Sets and Systems 203*, 49–73.

Yang, M.-S., P.-Y. Hwang, and D.-H. Chen 2004. Fuzzy clustering algorithms for mixed feature variables. *Fuzzy Sets and Systems 141*, 301–317.

22

A Survey of Consensus Clustering

Joydeep Ghosh and Ayan Acharya

CONTENTS

Abstract

This chapter describes the problem of combining multiple partitionings of a set of objects into a single consolidated clustering without accessing the features or algorithms that determine these partitionings—popularly known as the problem of "consensus cluster-ing." We illustrate different algorithms for solving the consensus clustering problem. The notion of dissimilarity between a pair of clustering solutions plays a key role in design-ing any cluster ensemble algorithm and a summary of such dissimilarity measures is also provided. We also cover recent efforts on combining classifier and clustering ensembles, leading to new approaches for semisupervised learning and transfer learning. Finally, we describe several applications of consensus clustering.

22.1 Introduction

The design of multiple classifier systems to solve difficult classification problems, using techniques such as bagging, boosting, and output combining (Sharkey 1999; Tumer and Ghosh 2000; Kittler and Roli 2002; Kuncheva 2004), has resulted in some of the most notable advances in classifier design over the past two decades. A popular approach is to train multiple "base" classifiers, whose outputs are combined to form a classifier ensemble. A survey of such ensemble techniques—including applications of them to many difficult real-world problems such as remote sensing, person recognition, one versus all recognition, and medicine—can be found in Oza and Tumer (2008). Concurrently, analytical frame-works have been developed that quantify the improvements in classification results due to combining multiple models (Tumer and Ghosh 1996). The extensive literature on the subject has shown that the ensemble created from independent, diversified classifiers is usually more accurate as well as more reliable than its individual components, that is, the base classifiers.

The demonstrated success of classifier ensembles provides a direct motivation to study effective ways of combining multiple clustering solutions as well. This chapter covers the theory, design and application of *cluster ensembles*, which combine multiple "base clusterings" of the same set of objects into a single consolidated clustering. Each base clus-tering refers to a *grouping* of the same set of objects or its transformed (or perturbed) version using a suitable clustering algorithm. The consolidated clustering is often referred to as the *consensus* solution. At first glance, this problem sounds similar to the problem of designing classifier ensembles. However, combining multiple clusterings poses additional challenges. First, the number of clusters produced may differ across the different *base* solutions (Ayad and Kamel 2008). The appropriate number of clusters in the consensus is also not known in advance and may depend on the scale at which the data is inspected. Moreover, clus-ter labels are symbolic and thus aligning cluster labels across different solutions requires solving a potentially difficult correspondence problem. Also, in the typical formulation,* the original data used to yield the base solutions are not available to the consensus mech-anism, which has only access to the sets of cluster labels. In some schemes, one does have control on how the base clusterings are produced (Fred and Jain 2005), while in others even this is not granted in order to allow applications involving knowledge reuse (Strehl and Ghosh 2002), as described later. Despite these added complications, cluster ensembles are inviting since, typically, the variation in quality across a variety of clustering algo-rithms applied to a specific dataset tend to be larger than the typical variation in accuracies returned by a collection of reasonable classifiers. This suggests that cluster ensembles may

---

* In this chapter, we shall not consider approaches where the feature values of the original data or of the cluster representatives are available to the consensus mechanism, for example, Hore et al. (2009).

achieve greater improvements over the base solutions, when compared with ensembles of classifiers (Ghosh 2002).

In fact, the potential motivations for using cluster ensembles are much broader than those for using classification or regression ensembles, where one is primarily interested in improving predictive accuracy. These reasons include:

22.1.1 Improved Quality of Solution

Just as ensemble learning has been proved to be more useful compared to single-model solutions for classification and regression problems, one may expect that cluster ensembles will improve the quality of results as compared to a single clustering solution. It has been shown that using cluster ensembles leads to more accurate results on average as the ensemble approach takes into account the biases of individual solutions (Hu and Yoo 2004; Kuncheva and Hadjitodorov 2004).

22.1.2 Robust Clustering

It is well known that the popular clustering algorithms often fail spectacularly for certain datasets that do not match well with the modeling assumptions (Karypis et al. 1999). A cluster ensemble approach can provide a "meta" clustering model that is much more robust in the sense of being able to provide good results across a very wide range of datasets. As an example, by using an ensemble that includes approaches such as k-means (Wu et al. 2008), and the Self-Organizing Map (Kohonen 1997), that are typically better suited to low-dimensional metric spaces, as well as base clusterers designed for high-dimensional sparse spaces (spherical k-means, Jaccard-based graph clustering, etc., see Strehl et al. (2000)), one can perform well across a wide range of data dimensionality (Strehl and Ghosh 2002). Sevillano et al. (2006) present several empirical results on the robustness of the results in document clustering by using feature diversity and consensus clustering.

22.1.3 Model Selection

Cluster ensembles provide a novel approach to the model selection problem by considering the match across the base solutions to determine the final number of clusters to be obtained (Ghosh et al. 2002).

22.1.4 Knowledge Re-Use

In certain applications, domain knowledge in the form of a variety of clusterings of the objects under consideration may already exist due to past projects. A consensus solution can integrate such information to get a more consolidated clustering. Several examples are provided in Strehl and Ghosh (2002), where such scenarios formed the main motivation for developing a consensus clustering methodology. As another example, a categorization of web pages based on text analysis can be enhanced by using the knowledge of topical document hierarchies available from Yahoo! or DMOZ.

22.1.5 Multiview Clustering

Often the objects to be clustered have multiple aspects or "views," and base clusterings may be built on distinct views that involve nonidentical sets of features or subsets of data points. In marketing applications, for example, customers may be segmented based on their needs, psychographic or demographic profiles, attitudes, etc. Different views can also be obtained by considering qualitatively different distance measures, an aspect that was exploited in clustering multifaceted proteins to multiple functional groups in Asur et al. (2007). Consensus clustering can be effectively used to combine all such clusterings into a single-consolidated partition. Strehl and Ghosh (2002) illustrated empirically the utility of cluster ensembles in two orthogonal scenarios:

- Feature-distributed clustering: different base clusterings are built by selecting different subsets of the features but utilizing all the data points.
- Object-distributed clustering: base clusterings are constructed by selecting different subsets of the data points but utilizing all the features.

Fern and Brodley (2003) proposed multiple random projections of the data onto subspaces followed by clustering of projected data and subsequent aggregation of clustering results as a strategy for high-dimensional clustering. They showed that such a method performs better than applying PCA and clustering in the reduced feature space.

22.1.6 Distributed Computing

In certain situations, data is inherently distributed and it is not possible to first collect the entire data at a central site due to privacy/ownership issues or computational, bandwidth and storage costs (Merugu and Ghosh 2005). An ensemble can be used in situations where each clusterer has access to only a subset of the features of each object, as well as where each clusterer has access to only a subset of the objects (Ghosh et al. 2002; Strehl and Ghosh 2002).

The problem of combining multiple clusterings can be viewed as a special case of the more general problem of comparison and consensus of data "classifications," studied in the pattern recognition and related application communities in the 1970s and 1980s. In this literature, "classification" was used in a broad sense to include clusterings, unrooted trees, graphs, etc., and problem-specific formulations were made (see Mirkin (1996) for a broad, more conceptual coverage). For example, in the building of phylogenetic trees, it is important to get a strict consensus solution, wherein two objects occur in the same consensus partition if and only if they occur together in all individual clusterings (Day 1986), typically resulting in a consensus solution at a much coarser resolution than the individual solutions. A quick overview with pointers to such literature is given by Ayad and Kamel (2008).

This chapter is organized as follows. In Section 22.2, we formulate the cluster ensemble problem. In Section 22.3, different measures for comparing a pair of clustering solutions are briefly introduced. Details of different cluster ensembles algorithms are presented in Section 22.4. In Section 22.5, a summary of recent works on combining classifier and cluster ensembles is illustrated. Finally, the applications of cluster ensembles are provided in Section 22.6.

22.2 The Cluster Ensemble Problem

We denote a vector by a bold faced letter and a scalar variable or a set in normal font. We start by considering r base clusterings of a dataset $\mathcal{X} = \{x_i\}_{i=1}^{n}$ with the qth clustering containing $k^{(q)}$ clusters. The most straightforward representation of the qth clustering is $C^{(q)} = \{C_\ell \mid \ell = 1, 2, \ldots, k^{(q)} \text{ and } C_\ell \subseteq \mathcal{X}\}$. Here, each clustering is denoted by a collection of subsets (not necessarily disjoint) of the original dataset. For hard partitional clustering (clustering where each object is assigned to a single cluster only), the qth clustering can alternatively be represented by a label vector $C^{(q)} \in \mathbb{Z}_+^n$. In this representation, each object is assigned some cluster label and 0 is used if the corresponding object was not available to that clusterer. The third possible way of representation of an individual clustering is by the binary membership indicator matrix $H^q \in \{0,1\}^{1 \times k^{(q)}}$ which is defined as $H^q = \{h_{i\ell}^q | h_{i\ell}^q \in \{0,1\} \forall x_i, C_\ell, C^{(q)}\}$. For partitional clustering, we additionally have $\sum_{\ell=1}^{k^{(q)}} h_{i\ell}^q = 1 \ \forall x_i \in \mathcal{X}$.

A *consensus function* Γ is defined as a function $\mathbb{Z}_+^{n \times r} \to \mathbb{Z}_+^n$ mapping a set of clusterings to an integrated clustering $\Gamma : \mathcal{C} \to \hat{\mathcal{C}}$, where for conciseness, we shall denote the set of clusterings $\{C^{(q)}\}_{q=1}^r$ that is available to the consensus mechanism by $\mathcal{C}$. Moreover, the results of any hard clustering* of n objects can be represented as a binary, symmetric $n \times n$ *co-association matrix*, with an entry being 1 if the corresponding objects are in the same cluster and 0 otherwise. For the qth base clustering, this matrix is denoted by $S^{(q)}$ and is given by

$$S_{ij}^{(q)} = \begin{cases} 1: & (i,j) \in C_\ell(C^{(q)}) \quad \text{for some } \ell \in \{1,2,\ldots,k^{(q)}\} \\ 0: & \text{otherwise} \end{cases}. \tag{22.1}$$

Broadly speaking, there are two main approaches to obtaining a consensus solution and determining its quality. One can postulate a probability model that determines the labeling of the individual solutions, given the true consensus labels, and then solve a maximum likelihood formulation to return the consensus (Topchy et al. 2004; Wang et al. 2009). Alternately, one can directly seek a consensus clustering that agrees the most with the original clusterings. The second approach requires a way of measuring the similarity between two clusterings, for example, to evaluate how close the consensus solution is to each base solution. These measuring indices will be discussed in more detail in Section 22.3. For now, let $\phi(C^{(a)}, C^{(b)})$ represent a similarity index between two clustering solutions $C^{(a)}$ and $C^{(b)}$. One can express the average similarity measure between a set of r labelings, C, and a single consensus labeling $\hat{C}$, by

$$\phi(C, \hat{C}) = \frac{1}{r} \sum_{q=1}^r \phi(C^{(q)}, \hat{C}). \tag{22.2}$$

This serves as the objective function in certain cluster ensemble formulations, where the goal is to find the combined clustering $\hat{C}$ with $\hat{k}$ clusters such that $\phi(C, \hat{C})$ is maximized. It turns out though that this objective is intractable for the popular similarity measures described in Chapter 27 of the Handbook, so heuristic approaches have to be resorted to.

22.3 Measuring Similarity between Clustering Solutions

Since no ground truth is available for clustering problems, cluster ensemble algorithms instead aim to maximize some similarity measure between the consensus clustering and

---

* This definition is also valid for overlapping clustering.

each of the base clustering solutions. Two of the most desirable properties of such similarity measures are: (i) the index should be normalized for easy interpretation and comparison across solutions with varying number of clusters; and (ii) the expected value of the index between pairs of independent clusterings should be a constant that provides a lower bound on the values obtainable using that index. Often this constant can be made zero by adjusting for "chance overlap," for example using adjusted Rand index instead of Rand index.

A variety of methods have been proposed for measuring the similarity of two clustering solutions (Nguyen et al. 2010). Some of them, such as the classical Rand Index, are based on counting the number of pairs in agreement or in disagreement across the two clusterings. Other indices, such as Variation of Information and Normalized Mutual Information, are founded on concepts from information theory. A detailed exposition of various similarity measures between two clusterings is provided in Chapter 27 of this Handbook.

22.4 Cluster Ensemble Algorithms

Cluster ensemble methods are now presented under three categories: (i) probabilistic approaches, (ii) approaches based on co-association, and (iii) direct and other heuristic methods.

22.4.1 Probabilistic Approaches to Cluster Ensembles

The two basic probabilistic models for solving cluster ensembles are described in this subsection.

22.4.1.1 A Mixture Model for Cluster Ensembles

In a typical mixture model (Bishop 2006) approach to clustering, such as fitting the data using a mixture of Gaussians, there are $\hat{k}$ mixture components, one for each cluster. A component-specific parametric distribution is used to model the distribution of data attributed to a specific component. Such an approach can be applied to form the consensus decision if the number of consensus clusters is specified. This immediately yields the pioneering approach taken in Topchy et al. (2004). We describe it in a bit more detail as this work is essential to build an understanding of later works (Wang et al. 2009, 2010).

In the basic mixture model of cluster ensembles (MMCE) (Topchy et al. 2004), each object x_i is represented by $y_i = \mathcal{C}(x_i)$, that is, the labels provided by the base clusterings. We assume that there are $\hat{k}$ consensus clusters each of which is indexed by $\hat{\ell}$. Corresponding to each consensus cluster $\hat{\ell}$ and each base clustering q, we have a multinomial distribution $\beta_{\hat{\ell}}^{(q)}$ of dimension $k^{(q)}$. Therefore, a sample from this distribution is a cluster label corresponding to the qth base clustering. The underlying generative process is assumed as follows:

For ith data point x_i:

1. Choose $z_i = \mathbf{I}_{\hat{\ell}}$ such that $\hat{\ell} \sim \text{multinomial}(\theta)$. Here $\mathbf{I}_{\hat{\ell}}$ is a probability vector of dimension $k^{(q)}$ with only the $\hat{\ell}$th component being 1, and θ is the parameter of a multinomial distribution of dimension $\hat{k}$.

2. For the qth base clustering of ith data point, choose the base clustering result $y_{iq} = \ell \sim$ multinomial $(\beta_{\ell}^{(q)})$.

These probabilistic assumptions give rise to a simple maximum log-likelihood problem that can be solved using the Expectation Maximization algorithm (see Chapter 9). This model also takes care of the missing labels in a natural way.

22.4.1.2 Bayesian Cluster Ensembles

A Bayesian version of the multinomial mixture model described above was subsequently proposed by Wang et al. (2009). As in the simple mixture model, we assume $\hat{k}$ consensus clusters with $\beta_{\ell}^{(q)}$ being the multinomial distribution corresponding to each consensus cluster $\hat{\ell}$ and each base clustering q. The complete generative process for this model is as follows:

For ith data point x_i:

1. Choose $\theta_i \sim$ Dirichlet(α) where θ_i is the parameter of a multinomial distribution with dimension $\hat{k}$.
2. For the qth base clustering:
   a. Choose $z_{iq} = I_{\hat{\ell}}$ such that $\hat{\ell} \sim$ multinomial (θ_i). $I_{\hat{\ell}}$ is a probability vector of dimension $\hat{k}$ with only $\hat{\ell}$th component being 1.
   b. Choose the base clustering result $y_{iq} = \ell \sim$ multinomial $(\beta_{\ell}^{(q)})$.

So, given the model parameters $(\alpha, \beta = \{\beta_{\ell}^{(q)}\})$, the joint distribution of latent and observed variables $\{y_i, z_i, \theta_i\}$ is given by

$$p(y_i, z_i, \theta_i \mid \alpha, \beta) = p(\theta_i | \alpha) \prod_{q=1, \exists y_{iq}}^{r} p(z_{iq} = I_{\hat{\ell}} \mid \theta_i) p(y_{iq} | \beta_{\ell}^{(q)}), \qquad (22.3)$$

where $\exists y_{iq}$ implies that there exists a qth base clustering result for y_i. The marginals $p(y_i | \alpha, \beta)$ can further be calculated by integrating over the hidden variables $\{z_i, \theta_i\}$. The authors used variational EM and Gibbs sampling for inference and parameter estimation. The graphical model corresponding to this Bayesian version is given in Figure 22.2. To highlight the difference between Bayesian cluster ensembles (BCE) and the mixture model for cluster ensembles (MMCE), the graphical model corresponding to the latter is also shown alongside in Figure 22.1.

22.4.1.3 Nonparametric Bayesian Cluster Ensembles

Recently, a nonparametric version of Bayesian cluster ensemble (NPBCE) has been proposed in Wang et al. (2010), which allows the number of consensus clusters to adapt with data. The stick-breaking construction of the generative process of this model is described below. The authors use a truncated stick breaking construction of the Dirichlet process with truncation enforced at $\hat{k}$. If $\hat{k}$ is made sufficiently large, the resulting process closely approximates a Dirichlet Process.

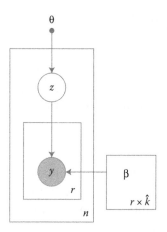

FIGURE 22.1
Graphical model for MMCE.

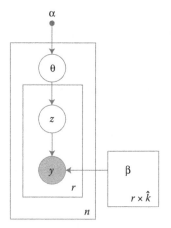

FIGURE 22.2
Graphical model for BCE.

1. Generate $v_{\hat{\ell}} \sim \text{Beta}(1, \alpha) \; \forall \hat{\ell} \in \{1, 2, \ldots, \hat{k}\}$. Let $\theta_{\hat{\ell}} = v_{\hat{\ell}} \prod_{j=1}^{\hat{\ell}-1}(1 - v_j)$.

2. For each base clustering (indexed by q), generate $\boldsymbol{\beta}_{\hat{\ell}}^{(q)} \sim G_0^{(q)} \; \forall \hat{\ell} \in \{1, 2, \ldots, \hat{k}\}$ where $G_0^{(q)}$ is a symmetric Dirichlet distribution of dimension $k^{(q)}$.

3. For ith data point $\mathbf{x}_i$, generate $\mathbf{z}_i \sim$ categorical $(\boldsymbol{\theta})$, where $\mathbf{z}_i$ is an indicator vector of dimension $\hat{k}$ with only one component $\hat{l}$ being unity and others being zero.

4. For the qth base clustering of ith data point, generate the base clustering label $y_{iq} = \ell \sim$ categorical $(\boldsymbol{\beta}_{\hat{\ell}}^{(q)})$.

One should note that NPBCE does not allow multiple base clustering solutions of a given data point to be generated from more than one consensus cluster. Therefore, the model is more restrictive compared to BCE and is really a nonparametric version of MMCE. The graphical model shown in Figure 22.3 illustrates this difference more clearly. It should be noted that although all of the generative models presented above were used only with hard partitional clusterings, they could be used for overlapping clustering as well.

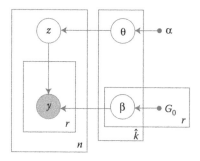

FIGURE 22.3
Graphical model for NPBCE.

22.4.2 Pairwise Similarity-Based Approaches

In pairwise similarity-based approaches, one takes the weighted average of all r co-association matrices to form an *ensemble co-association matrix* $\mathbf{S}$ which is given as follows:

$$\mathbf{S} = \frac{1}{r} \sum_{q=1}^{r} w_q \mathbf{S}^{(q)}. \tag{22.4}$$

Here w_q specifies the weight assigned to the qth base clustering. This ensemble co-association matrix captures the fraction of times a pair of data points is placed in the same cluster across the r base clusterings. The matrix can now be viewed as a similarity matrix (with a corresponding similarity graph) to be used by the consensus mechanism for creating the consensus clusters. This matrix is different from the similarity matrix $\hat{\mathbf{S}}$ that we obtain from the consensus solution $\hat{C}$. We will explain the difference in detail in Section 22.4.2.1.

Note that the co-association matrix size is itself quadratic in n, which thus forms a lower bound on computational complexity as well as memory requirements, inherently handicapping such a technique for applications to very large datasets. However, it is independent of the dimensionality of the data.

22.4.2.1 Methods Based on the Ensemble Co-Association Matrix

The Cluster-based Similarity Partitioning Algorithm (CSPA) (Strehl and Ghosh 2002) used METIS (Karypis and Kumar 1998) to partition the induced consensus similarity graph. METIS was chosen for its scalability and because it tries to enforce comparable sized clusters. This added constraint is desirable in several application domains (Strehl and Ghosh 2000); however, if the data is actually labeled with imbalanced classes, then it can lower the match between cluster and class labels. Assuming quasilinear graph clustering, the worst case complexity for this algorithm is $\mathcal{O}(n^2kr)$. Punera and Ghosh (2007) later proposed a soft version of CSPA, that is, one that works on soft base clusterings. Al-Razgan and Domeniconi (2006) proposed an alternative way of obtaining non-binary co-association matrices when given access to the raw data.

The Evidence Accumulation approach (Fred and Jain 2005) obtains individual co-association matrices by random initializations of the k-means algorithm, causing some variation in the base cluster solutions. This algorithm is used with a much higher value of k than the range finally desired. The ensemble co-association matrix is then formed, each entry of which signifies the relative cooccurrence of two data points in the same cluster. A minimum spanning tree (MST) algorithm (also called the single-linkage or nearest neighbor hierarchical clustering algorithm) is then applied on the ensemble co-association matrix. This allows one to obtain nonconvex-shaped clusters. Essentially, this approach assumes that the designer has access to the raw data, and the consensus mechanism is used to get a more robust solution than what can be achieved by directly applying MST to the raw data. Very recently, probabilistic approaches to evidence accumulation have been proposed (Loureno et al. 2015a,b) that can produce soft clustering assignment as the final solution of a consensus problem.

A related approach was taken by Monti et al. (2003), where the perturbations in the base clustering were achieved by resampling. Any of bootstrapping, data subsampling or feature subsampling can be used as a resampling scheme. If either of the first two options are selected, then it is possible that certain objects will be missing in a given base clustering. Hence when collating the r base co-association matrices, the (i, j)th entry needs to be divided by the number of solutions that included both objects rather than by a fixed r. This work also incorporated a model selection procedure as follows: The consensus co-association matrix is formed multiple times. The number of clusters is kept at k_i for each base clustering during the ith experiment, but this number is changed from one experiment to another. A measurement termed as *consensus distribution* describes how the elements of a consensus matrix are distributed within the 0–1 range. The extent to which the consensus matrix is skewed toward a binary matrix denotes how good the base clusterings match one another. This enables one to choose the most appropriate number of consensus clusters $\hat{k}$. Once $\hat{k}$ is chosen, the corresponding ensemble co-association matrix is fed to a hierarchical clustering algorithm with average linkage. Agglomeration of clusters is stopped when $\hat{k}$ branches are left.

The Iterative Pairwise Consensus algorithm (Nguyen and Caruana 2007) essentially applies model-based k-means (Zhong and Ghosh 2003) to the ensemble co-association matrix $\mathbf{S}$. The consensus clustering solution $\hat{C} = \{C_\ell\}_{\ell=1}^{k}$ is initialized to some solution, after which a reassignment of points is carried out based on the current configuration of $\hat{C}$. The point x_i gets assigned to cluster C_ℓ, if x_i has maximum average similarity with the points belonging to cluster C_ℓ. Then the consensus solution is updated, and the cycle starts again.

However, both Mirkin (1996) and Li et al. (2007) showed that the problem of consensus clustering can be framed in a different way than what has been discussed so far. In these works, the distance $d(C^{(q_1)}, C^{(q_2)})$ between two clusterings $C^{(q_1)}$ and $C^{(q_2)}$ is defined as the number of pairs of objects that are placed in the same cluster in one of $C^{(q_1)}$ or $C^{(q_2)}$ and in different cluster in the other, essentially considering the (unadjusted) Rand Index. Using this definition, the consensus clustering problem is formulated as

$$\arg\min_{\hat{C}} J = \arg\min_{\hat{C}} \frac{1}{r} \sum_{q=1}^{r} d(C^{(q)}, \hat{C}) \tag{22.5}$$

$$= \arg\min_{\hat{S}} \frac{1}{r} \sum_{q=1}^{r} w_q \sum_{i<j} [S_{ij}^{(q)} - \hat{S}_{ij}]^2.$$

Mirkin (1996, Section 22.3.4, p. 260) further proved that the consensus clustering according to criterion (22.5) is equivalent to clustering over the ensemble co-association matrix by subtracting a "soft" and "uniform" threshold from each of the different consensus clusters. This soft threshold, in fact, serves as a tool to balance cluster sizes in the final clustering. The subtracted threshold has also been used in Swift et al. (2004) for consensus clustering of gene-expression data.

In Wang et al. (2009), the consensus clustering is obtained by minimizing a weighted sum of the Bregman divergence (Banerjee et al. 2005) between the consensus partition and the input partitions with respect to their co-association matrices. In addition, the authors also show how to generalize their framework in order to incorporate must-link and cannot-link constraints between objects.

Note that the optimization problem in Equation 22.5 is over the domain of $\hat{S}$. The difference between the matrices S and $\hat{S}$ lies in the way the optimization problem is posed. If optimization is performed with cluster labels only (as illustrated in Section 22.4.3), there is no guarantee of achieving the optimum value $\hat{S} = S$. However, if we are optimizing over the domain of the co-association matrix, we can achieve this optimum value in theory.

22.4.2.2 Relating Consensus Clustering to Other Optimization Formulations

The co-association matrix representation of clustering has been used to relate consensus clustering with two other well-known problems.

22.4.2.2.1 Consensus Clustering as Nonnegative Matrix Factorization (NNMF)

Li and Ding (2008) and Li et al. (2007), using the same objective function as mentioned in Equation 22.5, showed that the problem of consensus clustering can be reduced to an NNMF problem. Assuming $U_{ij} = \hat{S}_{ij}$ to be an integer solution to this optimization problem, we can rewrite Equation 22.5 as

$$\arg\min_{U} \sum_{i,j=1}^{n} (S_{ij} - U_{ij})^2 = \arg\min_{U} \|S - U\|_F^2, \tag{22.6}$$

where the matrix norm is the Frobenius norm. This problem formulation is similar to the NNMF formulation (Lee and Seung 2000) and can be solved using an iterative update procedure. In Goder and Filkov (2008), the cost function J used in Equation 22.5 was further modified via normalization to make it consistent with datasets with different number of data points (n) and different number of base clusterings (r).

22.4.2.2.2 Consensus Clustering as Correlation Clustering

Gionis et al. (2007) showed that a certain formulation of consensus clustering is a special case of correlation clustering. Suppose that we have a dataset $\mathcal{X}$ and some kind of dissimilarity measurement (distance) between every pair of points in $\mathcal{X}$. This dissimilarity measure is denoted by $d_{ij} \in [0,1] \, \forall x_i, x_j \in \mathcal{X}$. The objective of correlation clustering (Bansal et al. 2002) is to find a partition $\hat{C}$ such that:

$$\hat{C} = \arg\min_{C} d(C)$$

$$= \arg\min_{C} \left[\sum_{(i,j):C(x_i)=C(x_j)} d_{ij} + \sum_{(i,j):C(x_i)\neq C(x_j)} (1 - d_{ij}) \right]. \tag{22.7}$$

In Equation 22.7, $\mathcal{C}(\mathbf{x}_i)$ is the cluster label imposed by $\mathcal{C}$ on $\mathbf{x}_i$. The co-association view of the cluster ensemble problem reduces to correlation clustering if the distance d_{ij} is defined as $d_{ij} = 1/r|\{\mathcal{C}^{(q)} : \mathcal{C}^{(q)}(\mathbf{x}_i) \neq \mathcal{C}^{(q)}(\mathbf{x}_j)\}|\ \forall\ i, j$.

22.4.3 Direct Approaches Using Cluster Labels

Several consensus mechanisms take only the cluster labels provided by the base clusterings as input, and try to optimize an objective function such as Equation 22.2, without computing the co-association matrix.

22.4.3.1 Graph Partitioning

In addition to CSPA, Strehl and Ghosh (2002) proposed two direct approaches to cluster ensembles: the Hypergraph Partitioning Algorithm (HGPA) which clusters the objects based on their cluster memberships, and the Meta Clustering Algorithm (MCLA), which groups the clusters based on which objects are contained in them. HGPA considers a graph with each object being a vertex. A cluster in any base clustering is represented by a hyper-edge connecting the member vertices. The hyper-graph clustering package HMETIS (Karypis et al. 1997) was used as it gives quality clusterings and is very scalable. As with CSPA, employing a graph clustering algorithm adds a constraint that favors clusterings of comparable size. Though HGPA is fast with a worst case complexity of $\mathcal{O}(nkr)$, it suffers from an additional problem: if all members of a base cluster are not assigned the same cluster in the consensus solution, the corresponding hyper-edge is broken and incurs a constant penalty; however, it cannot distinguish between a situation where only one object was clustered differently and one where several objects were allocated to other groups. Due to this issue, HGPA is often not competitive in terms of cluster quality.

MCLA first forms a meta-graph with a vertex for each base cluster. The edge weights of this graph are proportional to the similarity between vertices, computed using the binary Jaccard measure (number of elements in common divided by the total number of distinct elements). Since the base clusterings are partitional, this results in an r-partite graph. The meta-graph is then partitioned into k balanced meta-clusters. Each meta-cluster, therefore, contains approximately r vertices. Finally, each object is assigned to its most closely associated meta-cluster. Ties are broken randomly. The worst case complexity is $\mathcal{O}(nk^2r^2)$.

Noting that CSPA and MCLA consider either the similarity of objects or similarity of clusters only, a Hybrid Bipartite Graph Formulation (HBGF) was proposed in Fern and Brodley (2004). A bipartite graph models both data points and clusters as vertices, wherein an edge exists only between a cluster vertex and a object vertex if the latter is a member of the former. Either METIS or other multiway spectral clustering methods are used to partition this bipartite graph. The corresponding soft versions of CSPA, MCLA, and HBGF have also been developed by Punera and Ghosh (2007). It should be noted that all of CSPA, MCLA, and HGPA were compared with one another using the NMI measure in Strehl and Ghosh (2002).

22.4.3.2 Cumulative Voting

The concept of Cumulative Voting was first introduced in Dudoit and Fridlyand (2003), where the authors used bagging to improve the accuracy of clustering. Once clustering is done on a bootstrap sample, the cluster correspondence problem is solved using iterative relabeling via the Hungarian Algorithm (Papadimitriou and Steiglitz 1998). Clustering on each bootstrap sample gives some votes corresponding to each data point and cluster label pair, which, in aggregate, decides the final cluster assignment.

A similar approach was adopted in Ayad and Kamel (2008) wherein each base clustering provides soft or probabilistic scores indicating the fractional assignments of each data point to the different clusters in the consensus solution. These votes are then gathered across the base solutions and thresholded to determine the membership of each object to the consensus clusters. Again, this requires a mapping function from the base clusterings to a stochastic one. An information-theoretic criterion based on the information bottleneck (see Section 1.5.7) principle was used in Ayad and Kamel (2008) for this purpose. The mean of all the stochastic clusterings then yields the consensus partition. This approach is able to accommodate a range of k values in the base clusterings, is fast as it avoids the quadratic time/space complexity of forming a co-association matrix, and has shown good empirical results as well. Noting that the information bottleneck solutions can be obtained as a special case of Bregman clustering (Banerjee et al. 2005), it should be possible to recast this approach as a probabilistic one.

A variety of heuristic search procedures have also been suggested to hunt for a suitable consensus solution. These include a genetic algorithm formulation (Yoon et al. 2006) and one using an ant colony (Yang and Kamel 2006). These approaches tend to be computationally expensive, and the lack of extensive comparisons with the methods covered in this article currently makes it difficult to assess their quality. Also, one can use several heuristics suggested in Fern and Lin (2008) to select only a few clustering solutions from a large ensemble.

22.5 Combination of Classifier and Clustering Ensembles

Based on the success of classifier and cluster ensembles, efforts have been made recently to combine the strength of both types of ensembles (Acharya et al. 2011, 2013; Gao et al. 2011; Xie et al. 2014). Unsupervised models can provide a variety of supplementary constraints which can be useful for improving the generalization capability of a classifier (or a classifier ensemble), especially when labeled data is scarce. Also, they might be useful for designing learning methods that are aware of the possible differences between training and target distributions, thus being particularly interesting for applications in which concept drift might take place (Gao et al. 2008; Acharya et al. 2014). This section focuses on one such algorithm (Acharya et al. 2011, 2014) named Combination of Classification and Clustering Ensembles (C^3E).

The framework described in Acharya et al. (2011, 2014) is depicted in Figure 22.4. A set of classifiers, previously induced from a training dataset, is applied on a new target data $\mathcal{X} = \{x_i\}_{i=1}^n$, thereby generating a set of average class probability estimates $\{\pi_i\}_{i=1}^n$ on the instances in the target data. Suppose that there are k classes denoted by $C = \{C_\ell\}_{\ell=1}^k$. A cluster ensemble is further applied on the target data and a similarity matrix S is obtained from the ensemble outputs. Each entry of this similarity matrix corresponds to the relative co-occurrence of two instances in the same cluster (Strehl and Ghosh 2002). The prediction from the classifier ensemble ($\{\pi_i\}_{i=1}^n$) is then refined with the help of the similarity matrix obtained from the cluster ensemble. The refined class assignment estimate can be

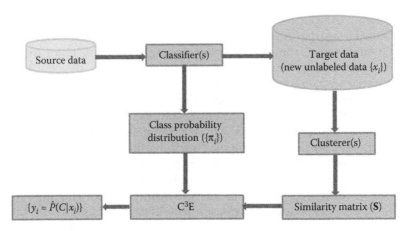

FIGURE 22.4
Overview of C³E.

represented by a set of vectors $\{\mathbf{y}_i\}_{i=1}^{n}$ with $\mathbf{y}_i \in \mathcal{S} \subseteq \mathbb{R}^k \; \forall i$, and $\mathbf{y}_i \propto \hat{P}(C \mid \mathbf{x}_i)$ (estimated posterior class probability assignment).

The problem of combining classifiers and clusterers is then posed as an optimization problem whose objective is to minimize J in Equation 22.8 with respect to the set of vectors $\{\mathbf{y}_i\}_{i=1}^{n}$:

$$J = \sum_{i \in \mathcal{X}} d_\phi(\pi_i, \mathbf{y}_i) + \alpha \sum_{(i,j) \in \mathcal{X}} S_{ij} d_\phi(\mathbf{y}_i, \mathbf{y}_j). \tag{22.8}$$

The term $d_\phi(.,.)$ refers to a Bregman divergence (Banerjee et al. 2005). Informally, the first term in Equation 22.8 captures dissimilarities between the class probabilities provided by the ensemble of classifiers and the output vectors $\{\mathbf{y}_i\}_{i=1}^{n}$. The second term encodes the cumulative weighted dissimilarity between all possible pairs $(\mathbf{y}_i, \mathbf{y}_j)$. The weights to these pairs are assigned in proportion to the similarity values $S_{ij} \in [0, 1]$ of matrix $\mathbf{S}$. The coefficient $\alpha \in \mathbb{R}_+$ controls the relative importance of classifier and cluster ensembles. Therefore, minimizing the objective function over $\{\mathbf{y}_i\}_{i=1}^{n}$ involves combining the evidence provided by the ensembles in order to build a more consolidated classification.

All Bregman divergences have the remarkable property that the single best (in terms of minimizing the net loss) representative of a set of vectors is simply the expectation of this set (!) provided the divergence is computed with this representative as the second argument of $d_\phi(\cdot, \cdot)$—see Theorem 1 in Acharya et al. (2011) (originally from Banerjee et al. (2005)) for further reference. Unfortunately, this simple form of the optimal solution is not valid if the variable to be optimized occurs as the first argument. In that case, however, one can work in the (Legendre) dual space, where the optimal solution has a simple form. Re-examination of Equation 22.8 reveals that the $\mathbf{y}_i$'s to be minimized over occur both as first and second arguments of a Bregman divergence. Hence, optimization over $\{\mathbf{y}_i\}_{i=1}^{n}$ is not available in closed form. This problem is circumvented by creating two copies for each $\mathbf{y}_i$—the left copy, $\mathbf{y}_i^{(l)}$, and the right copy, $\mathbf{y}_i^{(r)}$. The left (right) copies are used whenever the variables are encountered in the first (second) argument of the Bregman divergences. In what follows, it will be clear that the right and left copies are updated iteratively, and an additional soft constraint is used to ensure that the two copies of a variable remain "close enough" during the updates. With this modification, the following objective is minimized:

$$J(\mathbf{y}^{(l)}, \mathbf{y}^{(r)}) = \left[\sum_{i=1}^{n} d_\phi(\pi_i, \mathbf{y}_i^{(r)}) + \alpha \sum_{i,j=1}^{n} S_{ij} d_\phi(\mathbf{y}_i^{(l)}, \mathbf{y}_j^{(r)}) + C \sum_{i=1}^{n} d_\phi(\mathbf{y}_i^{(l)}, \mathbf{y}_i^{(r)}) \right], \qquad (22.9)$$

where $\mathbf{y}^{(l)} = \left(\mathbf{y}_i^{(l)}\right)_{i=1}^{n} \in \mathcal{S}^n$ and $\mathbf{y}^{(r)} = \left(\mathbf{y}_i^{(r)}\right)_{i=1}^{n} \in \mathcal{S}^n$.

To solve the optimization problem in an efficient way, first $\{\mathbf{y}_i^{(l)}\}_{i=1}^{n}$ and $\{\mathbf{y}_i^{(r)}\}_{i=1}^{n}/\{\mathbf{y}_j^{(r)}\}$ are kept fixed, and minimize the objective w.r.t. $\mathbf{y}_j^{(r)}$ only. The associated optimization problem can, therefore, be written as

$$\min_{\mathbf{y}_j^{(r)}} \left[d_\phi(\pi_j^{(r)}, \mathbf{y}_j^{(r)}) + \alpha \sum_{i^{(l)} \in \mathcal{X}} S_{i^{(l)} j^{(r)}} d_\phi(\mathbf{y}_i^{(l)}, \mathbf{y}_j^{(r)}) + C_j^{(r)} d_\phi(\mathbf{y}_j^{(l)}, \mathbf{y}_j^{(r)}) \right], \qquad (22.10)$$

where $C_j^{(r)}$ is the corresponding penalty parameter that is used to keep $\mathbf{y}_j^{(r)}$ and $\mathbf{y}_j^{(l)}$ close to each other.

From the results of Corollary 1 in Acharya et al. (2011), the unique minimizer of the optimization problem in Equation 22.10 is obtained as

$$\mathbf{y}_j^{(r)*} = \frac{\pi_j^{(r)} + \gamma_j^{(r)} \sum_{i^{(l)} \in \mathcal{X}} \delta_{i^{(l)} j^{(r)}} \mathbf{y}_i^{(l)} + C_j^{(r)} \mathbf{y}_j^{(l)}}{1 + \gamma_j^{(r)} + C_j^{(r)}}, \qquad (22.11)$$

where $\gamma_j^{(r)} = \alpha \sum_{i^{(l)} \in \mathcal{X}} S_{i^{(l)} j^{(r)}}$ and $\delta_{i^{(l)} j^{(r)}} = S_{i^{(l)} j^{(r)}}/\left[\sum_{i^{(l)} \in \mathcal{X}} S_{i^{(l)} j^{(r)}} \right]$. The same optimization in Equation 22.10 is repeated over all the $\mathbf{y}_j^{(r)}$. After the right copies are updated, the objective function is (sequentially) optimized with respect to all the $\mathbf{y}_i^{(l)}$. However, one needs to work in the dual space now where one can get a closed form update again for the $\mathbf{y}_i^{(l)}$. The alternating optimization w.r.t. the left and right copies leads to a guaranteed convergence for a jointly convex Bregman divergence (Acharya et al. 2012). Additionally, a linear rate of convergence has been proven for squared loss, KL divergence or I-divergence are used as the loss function. The authors in Acharya et al. (2012) show applications of C³E in semisupervised and transfer learning scenarios using some standard UCI datasets, real word text classification datasets and, remote sensing datasets. The method has been empirically proven to leverage information from cluster ensembles, particularly in the presence of concept drift.

22.6 Applications of Consensus Clustering

The motivation for consensus clustering has already been introduced in Section 22.1. Since it improves the quality of a clustering solution, they can be used for any cluster analysis problem, e.g., image segmentation (Wattuya et al. 2008), bioinformatics, document retrieval (González and Turmo 2008), automatic malware categorization (Ye et al. 2010), clustering of time-series data (Yetere Kurun 2014), just to name a few. Gionis et al. (2007) showed how clustering ensemble algorithms is used to improve the robustness of clustering solution, clustering categorical data (He et al. 2005) and heterogeneous data, identifying the correct number of clusters and detecting outliers (Zimek et al. 2013). Fischer and Joachim (Fischer and Buhmann 2003) showed how resampling the data and subsequent aggregation of the clustering solutions from the sampled sets can improve the quality of clustering solution. The Sawtooth Software (http://www.sawtoothsoftware.com/) has commercialized some of the algorithms in Strehl and Ghosh (2002) for applications in marketing. A package consisting of implementations of all the algorithms in Strehl and Ghosh (2002) is also available on http://strehl.com/soft.html. In this section, we briefly discuss two major application domains.

22.6.1 Gene Expression Data Analysis

Consensus clustering has been applied to microarray data to improve the quality and robustness of the resulting clusters. A resampling-based approach is used by Monti et al. (2003), in which the agreement across the results obtained by executing a base clustering algorithm on several perturbations of the original dataset is used to obtain the final clustering. Swift et al. (2004) use a variety of clustering algorithms on the same dataset to generate different base clustering results and try to find clusters that are consistent across all the base results using simulated annealing. In Filkov and Skiena (2004), the consensus clustering problem is treated as a median partition problem, where the aim is to find a partitioning of the data points that minimizes the distance to all the other partitionings. The authors propose greedy heuristic solutions to find a local optimum. Additionally, Deodhar and Ghosh (2006) used consensus clustering to find overlapping clusters in microarray data. In this work, two different techniques are used to generate the consensus clustering solution from the candidate solutions. The first one is MCLA with some adjustable threshold and the second one is soft kernelized k-means that works on the ensemble co-association matrix. In Chiu et al. (2010), gene expression time series data is clustered at different time intervals and the solutions from different time stamps are merged in a single solution using graph partitioning of the ensemble co-association matrix. Chiu et al. (2010) designs Bayesian networks corresponding to individual partitions of the data and then aggregates the results using a weighted combination of the individual networks.

22.6.2 Image Segmentation

Though there exists several image segmentation algorithms, depending on the application data, some perform better than the others and it is almost impossible to know beforehand which one should be used. The authors in Singh et al. (2010) used the cluster ensemble formulation to aggregate the results of multiple segmentation algorithms like (i) Normalized Cuts, (ii) Energy Minimization by Graph Cuts, and (iii) Curve Evolution to generate image segmentation. However, they found that the ensemble segmentation outperforms any individual segmentation algorithm. One of such results is shown in Figure 22.5. The first result corresponds to Normalized Cuts, the second one is from Graph Cuts, the third and fourth ones are from Curve Evolution, and the last one is due to the ensemble segmentation.

In another similar application, the same authors showed the utility of ensemble methods for better visualization and interpretation of images obtained from the Diffusion Tensor Imaging (DTI) technique. DTI images have become popular within neuroimaging because they are useful to infer the underlying structure and organizational pattern in the body (e.g., neuronal pathways in the brain). To simplify the processing of such images, a number of different measures (or channels) are calculated from the diffusion tensor image. Some of

FIGURE 22.5
Image segmentation results (Singh et al. 2010). The images shown resulted from using (from left to right) Normalized Cuts, Graph cuts, Curve Evolution (two variants) and ensemble segmentation. (Adapted from V. Singh, L. Mukherjee, J. Peng, and J. Xu. Ensemble clustering used semidefinite programming with applications. *Machine Learning*, 79(1–2):177–200, 2010).

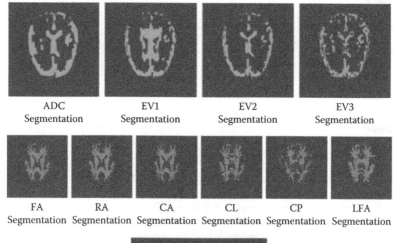

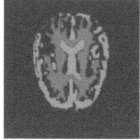

Final segmentation

FIGURE 22.6
Segmentation of tensor images (Singh et al. 2010). The final segmentation is obtained by combining the 10 base clusterings shown above, using a cluster ensemble. See text for details.

these channels are the Apparent Diffusion Coefficient (ADC), Fractional Anisotropy (FA), Mean Diffusivity (MD), Planar Anisotropy (CP), etc.* Segmentations obtained from 10 such channels of a brain are given in the first two rows of Figure 22.6. The last row shows the segmentation obtained using the ensemble strategy.

---

* See Singh et al. (2010) for more details about all of the 10 channels.

22.7 Concluding Remarks

This article showed that cluster ensembles are beneficial in a wide variety of scenarios. It then provided a framework for understanding many of the approaches taken so far to design such ensembles. Even though there seem to be many different algorithms for this problem, we showed that there are several commonalities among these approaches. The design domain, however, is still quite rich, leaving space for more efficient heuristics as well as formulations that place additional domain constraints to yield consensus solutions that are useful and actionable in diverse applications.

Acknowledgment

This work has been supported by NSF Grants (IIS-1421729, SCH-1418511, and IIS-1016614) and ONR Grant (ATL N00014-11-1-0105).

References

A. Acharya, E.R. Hruschka, J. Ghosh, and S. Acharyya. C^{3}E: A framework for combining ensembles of classifiers and clusterers. In *10th International Workshop on MCS*, 2011.

A. Acharya, E.R. Hruschka, J. Ghosh, and S. Acharyya. Transfer learning with cluster ensembles. *JMLR Workshop and Conference Proceedings*, 27:123–132, 2012.

A. Acharya, E.R. Hruschka, J. Ghosh, and S. Acharyya. An optimization framework for combining ensembles of classifiers and clusterers with applications to nontransductive semisupervised learning and transfer learning. *ACM Transactions on Knowledge Discovery from Data*, 9(1):1–35, 2014.

A. Acharya, E.R. Hruschka, J. Ghosh, J.D. Ruvini, and B. Sarwar. *Probabilistic Combination of Classifier and Cluster Ensembles for Non-transductive Learning*, chapter 31, pp. 288–296, 2013.

M. Al-Razgan and C. Domeniconi. Weighted cluster ensemble. In *Proceedings of SIAM International Conference on Data Mining*, pp. 258–269, 2006.

S. Asur, S. Parthasarathy, and D. Ucar. An ensemble framework for clustering protein-protein interaction networks. In *proceedings of 15th Annual International Conference on Intelligent Systems for Molecular Biology (ISMB)*, p. 2007, 2007.

H.G. Ayad and M.S. Kamel. Cumulative voting consensus method for partitions with variable number of clusters. *IEEE Transactions on Pattern Analysis and Machine Intelligence*, 30(1):160–173, 2008.

A. Banerjee, S. Merugu, I. Dhillon, and J. Ghosh. Clustering with Bregman divergences. *Journal of Machine Learning Research*, 6:1705–1749, 2005.

A. Banerjee, S. Merugu, I.S. Dhillon, and J. Ghosh. Clustering with bregman divergences. *Journal of Machine Learning Research*, 6:1705–1749, 2005.

N. Bansal, A.L. Blum, and S. Chawla. Correlation clustering. In *Proceedings of Foundations of Computer Science*, p. 238247, 2002.

C.M. Bishop. *Pattern Recognition and Machine Learning*. Springer, New York, 2006.

T.Y. Chiu, T.C. Hsu, and J.S. Wang. Ap-based consensus clustering for gene expression time series. *Pattern Recognition, International Conference on*, pp. 2512–2515, 2010.

W.H.E. Day. Foreword: Comparison and consensus of classifications. *Journal of Classification*, 3:183–185, 1986.

M. Deodhar and J. Ghosh. Consensus clustering for detection of overlapping clusters in microarray data. In *ICDMW '06: Proceedings of the Sixth IEEE International Conference on Data Mining—Workshops*, pp. 104–108, 2006.

S. Dudoit and J. Fridlyand. Bagging to improve the accuracy of a clustering procedure. *Bioinformatics*, 19(9):1090–1099, 2003.

X.Z. Fern and C.E. Brodley. Random projection for high dimensional data clustering: A cluster ensemble approach. In *Proceedings of 20th International Conference on Machine Learning (International Conference on Machine Learning'03)*, Washington, 2003.

X. Fern and C. Brodley. Solving cluster ensemble problems by bipartite graph partitioning. In *Proceedings of International Conference on Machine Learning*, pp. 281–288, 2004.

X.Z. Fern and W. Lin. Cluster ensemble selection. *Statistical Analysis and Data Mining*, 1(3): 128–141, 2008.

V. Filkov and S. Skiena. Integrating microarray data by consensus clustering. In *International Journal on Artificial Intelligence Tools (IJAIT)*, 4:863–880, 2004.

B. Fischer and J.M. Buhmann. Bagging for path-based clustering. *IEEE Transactions on Pattern Analysis and Machine Intelligence*, 25(11):1411–1415, 2003.

A. Fred and A.K. Jain. Combining multiple clusterings using evidence accumulation. *IEEE Transactions on Pattern Analysis and Machine Intelligence*, 27(6):835–850, 2005.

J. Gao, W. Fan, J. Jiang, and J. Han. Knowledge transfer via multiple model local structure mapping. In *Proceedings of KDD*, pp. 283–291, 2008.

J. Gao, F. Liang, W. Fan, Y. Sun, and J. Han. A graph-based consensus maximization approach for combining multiple supervised and unsupervised models. *IEEE Transactions on Knowledge and Data Engineering*, 25(11):15–28, 2011.

J. Ghosh. Multiclassifier systems: Back to the future (invited paper). In F. Roli and J. Kittler, editors, *Multiple Classifier Systems*, pp. 1–15. LNCS Vol. 2364, Springer, Heidelberg, 2002.

J. Ghosh, A. Strehl, and S. Merugu. A consensus framework for integrating distributed clusterings under limited knowledge sharing. In *Proceedings of NSF Workshop on Next Generation Data Mining*, Baltimore, pp. 99–108, 2002.

A. Gionis, H. Mannila, and P. Tsaparas. Clustering aggregation. *ACM Transactions on Knowledge Discovery from Data*, 1(4):109–117, 2007.

A. Goder and V. Filkov. Consensus clustering algorithms: Comparison and refinement. In *Proceedings of the Tenth Workshop on Algorithm Engineering and Experiments*, pp. 109–117, 2008.

E. González and J. Turmo. Comparing non-parametric ensemble methods for document clustering. In *Proceedings of the 13th International Conference on Natural Language and Information Systems: Applications of Natural Language to Information Systems*, NLDB '08, pp. 245–256, Berlin, Heidelberg, 2008. Springer-Verlag.

Z. He, X. Xu, and S. Deng. A cluster ensemble method for clustering categorical data. *Information Fusion*, 6(2):143–151, 2005.

P. Hore, L.O. Hall, and D.B. Goldgof. A scalable framework for cluster ensembles. *Pattern Recognition*, 42(5):676–688, 2009.

X. Hu and I. Yoo. Cluster ensemble and its applications in gene expression analysis. In *APBC '04: Proceedings of the Second Conference on Asia-Pacific Bioinformatics*, pp. 297–302, Darlinghurst, 2004. Australian Computer Society, Inc.

G. Karypis, R. Aggarwal, V. Kumar, and S. Shekhar. Multilevel hypergraph partitioning: Applications in VLSI domain. In *Proceedings of the Design and Automation Conference*, pp. 526–529, 1997.

G. Karypis, E.-H. Han, and V. Kumar. Chameleon: Hierarchical clustering using dynamic modeling. *IEEE Computer*, 32(8):68–75, 1999.

G. Karypis and V. Kumar. A fast and high quality multilevel scheme for partitioning irregular graphs. *SIAM Journal on Scientific Computing*, 20(1):359–392, 1998.

J. Kittler and F. Roli, editors. *Multiple Classifier Systems*. LNCS Vol. 2634, Springer, Heidelberg, 2002.

T. Kohonen. *Self-Organizing Maps*. Springer, Berlin, Heidelberg, 1995. (Second Extended Edition 1997).

L.I. Kuncheva. *Combining Pattern Classifiers: Methods and Algorithms*. Wiley, Hoboken, NJ, 2004.

L.I. Kuncheva and S.T. Hadjitodorov. Using diversity in cluster ensemble. In *IEEE International Conference on Systems, Man and Cybernetics*, pp. 1214–1219, 2004.

D.D. Lee and H.S. Seung. Algorithms for non-negative matrix factorization. In *In Neural Information Processing Systems*, pp. 556–562. MIT Press, 2000.

T. Li and C. Ding. Weighted consensus clustering. In *Proceedings of Eighth SIAM International Conference on Data Mining*, pp. 798–809, 2008.

T. Li, C. Ding, and M. Jordan. Solving consensus and semi-supervised clustering problems using non-negative matrix factorization. In *Proceedings of Eighth IEEE International Conference on Data Mining*, pp. 577–582, 2007.

A. Loureno, S. Rota Bul, N. Rebagliati, A.N. Fred, M.T. Figueiredo, and M. Pelillo. Probabilistic consensus clustering using evidence accumulation. *Machine Learning*, 98(1–2):331–357, 2015.

A. Loureno, S. Rota Bulo, N. Rebagliati, A. Fred, M. Figueiredo, and M. Pelillo. A map approach to evidence accumulation clustering. In A. Fred and M. De Marsico, editors, *Pattern Recognition Applications and Methods*, volume 318 of *Advances in Intelligent Systems and Computing*, Springer, Heidelberg, pp. 85–100. 2015.

S. Merugu and J. Ghosh. A distributed learning framework for heterogeneous data sources. In *Proceedings of Knowledge Discovery and Data Mining*, pp. 208–217, 2005.

B. Mirkin. *Mathematical Classification and Clustering*. Kluwer, Dordrecht, 1996.

S. Monti, P. Tamayo, J. Mesirov, and T. Golub. Consensus clustering-a resampling-based method for class discovery and visualization of gene expression microarray data. *Journal of Machine Learning*, 52:91–118, 2003.

N. Nguyen and R. Caruana. Consensus clusterings. In *Proceedings of International Conference on Data Mining*, pp. 607–612, 2007.

X.V. Nguyen, J. Epps, and J. Bailey. Information theoretic measures for clusterings comparison: Variants, properties, normalization and correction for chance. *Journal of Machine Learning Research*, 11:2837–2854, 2010.

H. Njah and S. Jamoussi. Weighted ensemble learning of bayesian network for gene regulatory networks. *Neurocomputing*, 150, Part B(0):404–416, 2015.

N.C. Oza and K. Tumer. Classifier ensembles: Select real-world applications. *Information Fusion*, 9:4–20, 2008.

C. Papadimitriou and K. Steiglitz. *Combinatorial Optimization. Algorithms and Complexity*. Dover Publication, Inc., Minneola, NY, 1998.

K. Punera and J. Ghosh. Consensus based ensembles of soft clusterings. In *Proceedings of MLMTA'07— Int'l Conference on Machine Learning: Models, Technologies & Applications*, 2007.

X. Sevillano, G. Cobo, F. Alías, and J.C. Socoró. Feature diversity in cluster ensembles for robust document clustering. In *SIGIR '06: Proceedings of the 29th Annual International ACM SIGIR Conference on Research and Development in Information Retrieval*, pp. 697–698, New York, NY, 2006. ACM.

A. Sharkey. *Combining Artificial Neural Nets*. Springer-Verlag, London, 1999.

V. Singh, L. Mukherjee, J. Peng, and J. Xu. Ensemble clustering using semidefinite programming with applications. *Machine Learning*, 79(1–2):177–200, 2010.

A. Strehl and J. Ghosh. A scalable approach to balanced, high-dimensional clustering of market-baskets. In *Proceedings of HiPC 2000, Bangalore, volume 1970 of LNCS*, pp. 525–536. Springer, 2000.

A. Strehl and J. Ghosh. Cluster ensembles—A knowledge reuse framework for combining multiple partitions. *Journal of Machine Learning Research*, 3:583–617, 2002.

A. Strehl, J. Ghosh, and R. Mooney. Impact of similarity measures on web-page clustering. In *Proceedings of 17th National Conference on AI: Workshop on AI for Web Search (AAAI 2000)*, pp. 58–64. AAAI, 2000.

S. Swift, A. Tucker, V. Vinciotti, and N. Martin. Consensus clustering and functional interpretation of gene-expression data. *Genome Biology* 5:R94, 2004.

S. Swift, A. Tucker, V. Vinciotti, N. Martin, C. Orengo, X. Liu, and P. Kellam. Consensus clustering and functional interpretation of gene-expression data. *Genome Biology*, 5(11):R94, 2004.

A. Topchy, A. Jain, and W. Punch. A mixture model for clustering ensembles. In *Proceedings of SIAM International Conference on Data Mining*, pp. 379–390, 2004.

K. Tumer and J. Ghosh. Analysis of decision boundaries in linearly combined neural classifiers. *Pattern Recognition*, 29(2):341–348, 1996.

K. Tumer and J. Ghosh. Robust order statistics based ensembles for distributed data mining. In H. Kargupta and P. Chan, editors, *Advances in Distributed and Parallel Knowledge Discovery*, pp. 85–110. AAAI Press, Palo Alto, CA, 2000.

P. Wang, C. Domeniconi, and K. Laskey. Nonparametric bayesian clustering ensembles. In *Machine Learning and Knowledge Discovery in Databases*, volume 6323 of *Lecture Notes in Computer Science*, chapter 28, J.L. Balcázar, F. Bonchi, A. Gionis, M. Sebag (eds.). pp. 435–450. Springer, Berlin, 2010.

H. Wang, H. Shan, and A. Banerjee. Bayesian cluster ensembles. In *Proceedings of the Ninth SIAM International Conference on Data Mining*, pp. 211–222, 2009.

F. Wang, X. Wang, and T. Li. Generalized cluster aggregation. In *Proceedings of IJCAI'09*, pp. 1279–1284, San Francisco, CA, 2009. Morgan Kaufmann Publishers Inc.

P. Wattuya, K. Rothaus, J.-S. Prassni, and X. Jiang. A random walker based approach to combining multiple segmentations. In *ICPR'08*, pp. 1–4, 2008.

X. Wu, V. Kumar, J. Ross Quinlan, J. Ghosh, Q. Yang, H. Motoda, G. McLachlan, A. Ng, B. Liu, P. Yu, Z.-H. Zhou, M. Steinbach, D. Hand, and D. Steinberg. Top 10 algorithms in data mining. *Knowledge and Information Systems*, 14(1):1–37, 2008.

S. Xie, J. Gao, W. Fan, D. Turaga, and P.S. Yu. Class-distribution regularized consensus maximization for alleviating overfitting in model combination. In *Proceedings of the 20th ACM SIGKDD International Conference on Knowledge Discovery and Data Mining*, KDD '14, pp. 303–312, 2014.

Y. Yang and M.S. Kamel. An aggregated clustering approach using multi-ant colonies algorithms. *Journal of Pattern Recognition*, 39(7):1278–1289, 2006.

Y. Ye, T. Li, Y. Chen, and Q. Jiang. Automatic malware categorization using cluster ensemble. In *Knowledge Discovery and Data Mining '10: Proceedings of the 16th ACM SIGKnowledge Discovery and Data Mining International Conference on Knowledge discovery and Data Mining*, pp. 95–104, 2010.

A. Yetere Kurun. Consensus clustering of time series data. Msc, Scientific Computing, Institute of Applied Mathematics, Middle East Technical University, 2014.

H.S. Yoon, S.Y. Ahn, S.H. Lee, S.B. Cho, and J.H. Kim. Heterogeneous clustering ensemble method for combining different cluster results. In *Proceedings of BioDM 2006, Lecture Notes in Computer Science*, volume 3916, pp. 82–92, 2006.

S. Zhong and J. Ghosh. A unified framework for model-based clustering. *Journal of Machine Learning Research*, 4:1001–1037, 2003.

A. Zimek, M. Gaudet, R.J.G.B. Campello, and J. Sander. Subsampling for efficient and effective unsupervised outlier detection ensembles. In *Proceedings of the 19th ACM SIGKDD International Conference on Knowledge Discovery and Data Mining*, KDD '13, pp. 428–436, 2013.

23

Two-Mode Partitioning and Multipartitioning

Maurizio Vichi

CONTENTS

Abstract

Two-mode clustering, coclustering, biclustering, and subspace clustering are similar terms used to describe the activity of clustering simultaneously the two-modes, that is, rows and columns, of a rectangular data matrix or a rectangular (dis)similarity matrix. Methodologies in this field of research differ from classical cluster analysis methods, where the assignments of objects to clusters are based on the entire set of variables and, vice versa, the assignments of variables to clusters are based on the total set of objects. The motivation for the increasing use of two-mode clustering methodologies resides in the fact that, frequently groups of objects may be homogeneous only within subsets of variables, while variables may be strongly associated only on subsets of objects. In microarray cluster analysis, for example, the clustering on tissue samples is coregulated within subsets of samples and groups of samples share a common gene expression pattern only for some subsets of genes; in market basket analysis customers have similar preference patterns only on subsets of products and, vice versa, classes of products may more frequently be consumed and preferred by subgroups of customers; on movie recommender systems generally reviewers are homogeneous on subset of movies and vice versa movies are generally chosen by homogeneous typologies of people with similar profiles of preferences. To obtain a two-mode partitioning researchers may think to apply on the data matrix X simply an algorithm for partitioning rows and independently an algorithm for partitioning columns, thus splitting X into rectangular blocks. However, with this simple approach the partitioning of objects would be optimal for the entire profile of variables and the partitioning of variables would be ideal for the entire set of objects, but, this is what we wish to avoid, having hypothesized homogeneous groups of objects only within subset of variables, while variables strongly associated only on subsets of objects. Thus, in general the independent application of a clustering methodology on the rows and columns of a data matrix is not fully appropriate for two-mode clustering and suitable methodologies are needed. In this paper, we wish to overview a large amount of new literature proposed in the last years in two-mode clustering due to the need of data reduction, bioinformatics applications and text-mining analyses.

23.1 Introduction

The term two-mode clustering appeared since the late 1970s (e.g., Hartigan 1972, 1976; Braverman et al. 1974; DeSarbo 1982; DeSarbo and De Soete 1984; Packer 1989; Arabie et al. 1990; Eckes and Orlik 1991, 1993; Eckes 1993; Mirkin et al. 1995) for describing the task of clustering both "modes" (dimensions) of the observed rectangular data matrix X. More specifically, the objective is to cluster both rows and columns to obtain homogeneous subsets of values, frequently representing submatrices with similar values. The dimensions of X may refer to: objects and variables of a classical data matrix; two different sets of objects and/or variables, such as in the case of contingency tables (e.g., occurrences of words for different documents) or preference data tables (e.g., preferences of customers on different products), or, flows tables (e.g., volume of traffic between origins and different destinations), or, in general rectangular (a)symmetric, (dis)similarity matrices. Alternative terms frequently used for the activity of clustering rows and columns of X, simultaneously, are "bi-clustering," "coclustering," and "subspace clustering" (Van Mechelen et al. 2004; Madeira and Oliveira 2004).

There are several key theoretical topics to analyse in formulating a two-mode clustering problem: (i) the choice of appropriate "ideal" two-mode clusters for the data, based on submatrices with desirable characteristics; (ii) the definition and measurement of "heterogeneity" and "isolation," respectively, within and between two-mode clusters, by considering the (dis)similarity between pairs of objects and variables; (iii) the decision of the appropriate two-mode clustering structure (e.g., partition, hierarchy, covering) to fit the observed data; (iv) the selection of the appropriate criterion to measure and quantify the total loss or the discrepancy between the original data matrix and the two-mode clustering structure, assessing in this way the eventual properties of heterogeneity, isolation, partitioning or overlapping, nested-ness of the detected two-mode clusters.

In this work, starting from the relevant and well-structured overviews on two-mode clustering proposed by Van Mechelen et al. (2004) and Madeira and Oliveira (2004), we wish to present some methodologies that have been developed in the last years.

23.2 The Two-Mode Cluster

A *two-mode cluster* $_{OV}C_k$ *of objects and variables* (briefly *2-mc*) is a set of ordered pairs $(o_i, v_j) \in O \times V$, where $O \times V$ is the *Cartesian product* of the sets of objects O and variables V. In this chapter, we will mainly focus on the two-mode "rectangular" clusters. We now give a formal definition. A *two-mode rectangular cluster* $_{OV}C_{pq}$ *of objects and variables* (*2-mrc*) is the collection of ordered pairs (o_i, v_j) obtained by the Cartesian product $_OC_p \times _VC_q$ of two subsets $_OC_p \subseteq O$ and $_VC_q \subseteq V$. Clusters $_OC_p$ and $_VC_q$, respectively, are called *marginal object cluster*, and *marginal variable cluster* of 2-mrc $_{OV}C_{pq}$. Following the definition of *direct cluster* given by Hartigan (1972), the 2-mrc can be seen as a submatrix or *block* $\mathbf{X}_{pq} = [x_{ij} : o_i \in {_OC_p}, v_j \in {_VC_q}]$ of the data matrix $\mathbf{X}$, with dimensions $I_p \times J_q$. $\mathbf{X}_{pq}$ represents a subsample of I_p multivariate observations $\mathbf{x}_i, i = 1, \ldots, I_p$ characterized by J_q variables which are selected to show relevant "homogeneous" parts of statistical information in $\mathbf{X}$. When a two-mode cluster $_{OV}C_k$ is not rectangular, the associated block has not the shape of a rectangular data matrix and generally it cannot be analyzed by using classical tools of multivariate statistics. This peculiarity deserves a specific study and appropriate analyses that will not be fully considered in this work. Examples of rectangular two-mode clusters are shown in Table 23.1 for the data matrix $\mathbf{X}$ of dimension ten objects and five variables.

23.2.1 Models for Two-Mode Clusters

Hartigan (1972), Cheng and Church (2000), Lazzeroni and Owen (2002), and Madeira and Oliveira (2004) propose different models, that is, typologies of expected "ideal" two-mode rectangular clusters $_{OV}C_{pq}$ with associated block $\hat{\mathbf{X}}_{pq}$.

TABLE 23.1

Two-Mode Cluster

						V			
					$_VC_1$			$_VC_2$	
					v_1	v_2	v_3	v_4	v_5
				o_1	5.6	5.9	5.1	1.43	1.44
$\mathbf{X}_{11}$	$\mathbf{X}_{12}$		$_OC_1$	o_2	5.8	5.6	5.6	1.46	1.34
				o_3	5.3	5.7	5.8	1.21	1.38
				o_4	5.1	5.5	5.5	1.02	1.45
$\mathbf{X}=$		$= O$		o_5	5.3	5.0	5.6	1.17	1.15
				o_6	1.23	1.49	1.37	3.30	3.04
				o_7	1.21	1.05	1.19	3.32	3.27
$\mathbf{X}_{21}$	$\mathbf{X}_{22}$		$_OC_2$	o_8	1.06	1.11	1.42	3.39	3.01
				o_9	1.29	1.02	1.27	3.28	3.39
				o_{10}	1.16	1.25	1.09	3.36	3.27

i. *Row and Column Constant 2-mrc Model*, i.e., a two-mode cluster with constant values $\bar{x}_{pq}$:
$$\hat{\mathbf{X}}_{pq} = \bar{x}_{pq} \mathbf{1}_{Ip} \mathbf{1}'_{Jq} \qquad \text{(Table 23.2a)}$$

ii. *Row Constant 2-mrc Model*, i.e., a two-mode clusters with constant row effect,

additive: $\qquad \hat{\mathbf{X}}_{pq} = \bar{x}_{pq} \mathbf{1}_{Ip} \mathbf{1}'_{Jq} + \mathbf{a}_p \mathbf{1}'_{Jq} \qquad$ (Table 23.2b)

or multiplicative: $\quad \hat{\mathbf{X}}_{pq} = \bar{x}_{pq} \mathbf{a}_p \mathbf{1}'_{Jq} \qquad\qquad$ (Table 23.2c)

iii. *Column Constant 2-mrc Model*, i.e., a two-mode cluster with constant column effect,

additive: $\qquad \hat{\mathbf{X}}_{pq} = \bar{x}_{pq} \mathbf{1}_{Ip} \mathbf{1}'_{Jq} + \mathbf{1}_{Ip} \mathbf{b}'_q \qquad$ (Table 23.2d)

or multiplicative: $\quad \hat{\mathbf{X}}_{pq} = \bar{x}_{pq} \mathbf{1}_{Ip} \mathbf{b}'_q \qquad\qquad$ (Table 23.2e)

iv. *Row-Column 2-mrc Model*, i.e., a two-mode cluster with row and column constant effects,

additive: $\qquad \hat{\mathbf{X}}_{pq} = \bar{x}_{pq} \mathbf{1}_{Ip} \mathbf{1}'_{Jq} + \mathbf{a}_p \mathbf{1}'_{Jq} + \mathbf{1}_{Ip} \mathbf{b}'_q \qquad$ (Table 23.2f)

or multiplicative: $\quad \hat{\mathbf{X}}_{pq} = \bar{x}_{pq} \mathbf{a}_p \mathbf{b}'_q \qquad\qquad$ (Table 23.2g)

Vector $\mathbf{a}_p$ is I_p-dimensional and identifies the row effects, $\mathbf{b}_q$ is the J_q-dimensional vector of column effects, $\bar{x}_{pq}$ is the centroid of the 2-mrc $_{OV}C_{pq}$ and $\mathbf{1}_L$ is an L-dimensional vector of ones. In Table 23.2 an example of two mode clusters (i), (ii), (iii), and (iv) is shown, where $\bar{x}_{pq} = 1$, $\mathbf{a} = [0\ 2\ 4\ 3]'$, $\mathbf{b} = [1\ 3\ 0]'$.

From the four models it can be observed that each class $_{OV}C_{pq}$ has a centroid (medoid), symbolizing the center of $_{OV}C_{pq}$ and synthesizing the information of the 2-mrc. The centroid (medoid) is defined as the unobservable (observable) representative value $\bar{x}_{pq}$ (associated with each pair $(o_i, v_j) \in {}_{OV}C_{pq}$), which is the "closest" or the less dissimilar to the values in block $\mathbf{X}_{pq}$. For a binary data matrix $\mathbf{X}$, ideal blocks are (Doreian et al. 1994): (i) *complete* $\hat{\mathbf{X}}_{pq} = \mathbf{1}_{Ip} \mathbf{1}'_{Jq}$; (ii) *null* $\hat{\mathbf{X}}_{pq} = \mathbf{0}_{Ip} \mathbf{0}'_{Jq}$; (iii) *row dominant*, if $\hat{\mathbf{X}}_{pq}$ has at least one row of ones; (iv) *column dominant*, if $\hat{\mathbf{X}}_{pq}$ has at least one column of ones.

TABLE 23.2

Models for Two-Mode Clusters: $\bar{x}_{pq} = 1$, $\mathbf{a} = [0\ 2\ 4\ 3]'$, $\mathbf{b} = [1\ 3\ 0]'$. Two-Mode Cluster (a) with Constant Values, (b) with Rows Constant (Additive Model), (c) with Rows Constant (Multiplicative Model), (d) with Columns Constant (Additive Model), (e) with Columns Constant (Multiplicative Model), (f) with Coherent Values (Additive Model), (g) with Coherent Values (Multiplicative Model)

	v_1	v_2	v_3		v_1	v_2	v_3		v_1	v_2	v_3		v_1	v_2	v_3
o_1	1	1	1	o_1	1	1	1	o_1	0	0	0	o_1	2	4	1
o_2	1	1	1	o_2	3	3	3	o_2	2	2	2	o_2	2	4	1
o_3	1	1	1	o_3	5	5	5	o_3	4	4	4	o_3	2	4	1
o_4	1	1	1	o_4	4	4	4	o_4	3	3	3	o_4	2	4	1
	(a)				(b)				(c)				(d)		

	v_1	v_2	v_3		v_1	v_2	v_3		v_1	v_2	v_3
o_1	1	3	0	o_1	2	4	1	o_1	0	0	0
o_2	1	3	0	o_2	4	6	3	o_2	2	6	0
o_3	1	3	0	o_3	6	8	5	o_3	4	12	0
o_4	1	3	0	o_4	5	7	4	o_4	3	9	0
	(e)				(f)				(g)		

23.3 Two-Mode Partition: Single and Multipartition

A *two-mode partition* $P_{OV} = \{{}_{OV}C_1, \ldots, {}_{OV}C_k, \ldots, {}_{OV}C_K\}$ is a set of K disjoint clusters such that their union is $O \times V$ itself. Note that each pair (o_i, v_j) is completely assigned exactly to one of the K 2-*mrc* and this guarantees that a partition is defined. Two-mode clusters forming a partition have associated blocks $\mathbf{X}_1, \ldots, \mathbf{X}_k, \ldots, \mathbf{X}_K$. A unique observation x_{ij} is a *singleton block*, while matrix $\mathbf{X}$ is the *total block*.

In Figure 23.1, different two-mode partitions of the data matrix $\mathbf{X}$ are shown by means of heatmaps of data matrices of dimension (50×30). An heatmap is a rectangular grid composed of pixels each of which corresponds to a data value in $\mathbf{X}$ and represents a singleton block. The color of a pixel ranges into a scale between black (lowest values) and white (highest values), visualizing the corresponding data value, unless for Figure 23.1c where similarity data are observed and a reverse scale is considered. If the objects and variables in the same clusters are contiguously reordered, the clustering pattern in the data becomes obvious to observe visually.

Two-mode partition (a) and two-mode single-partitions (b), (c)

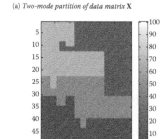

(a) *Two-mode partition of data matrix* $\mathbf{X}$

(b) *Two-mode single-partition of data matrix* $\mathbf{X}$

(c) *Two-mode partition of similarity matrix* $\mathbf{X}$

Data matrix $\mathbf{X}$ (50×30), with values 0–100, split into four nonrectangular blocks of homogeneous values. For each mode there are not partitions that allow to define the two mode partition.

Data matrix $\mathbf{X}$ (50×30), with values 0–100, split into (5×3) rectangular blocks of homogeneous values. In each mode a single-partition is used to define the 2-msp.

Rectangular similarity matrix $\mathbf{X}$ (50×30), split into (4×4) blocks with four diagonal rectangular blocks with high similarity (dark color). Other blocks have low similarities. A single-partition for O and V is defined.

Two-mode multipartitions (d), (e), (f)

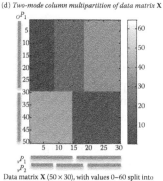

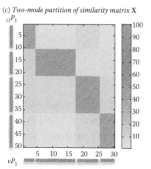

(d) *Two-mode column multipartition of data matrix* $\mathbf{X}$

(e) *Two-mode row multipartition of data matrix* $\mathbf{X}$

(f) *Two-mode row & column multipartition of data* $\mathbf{X}$

Data matrix $\mathbf{X}$ (50×30), with values 0–60 split into five rectangular blocks of homogeneous values. A multipartitioning of columns is obtained for each cluster of rows. Note that multipartitions are not nested.

Data matrix $\mathbf{X}$ (50×30), with values 0–60 split into five rectangular blocks. A multipartitioning of rows is obtained for each cluster of columns. The different partitions are not nested.

Data matrix $\mathbf{X}$ (50×30), with values 0–100 split into (5×3) rectangular blocks. A multipartitioning of rows and columns is obtained. The different partitions are nested.

FIGURE 23.1

Two-mode partition, single-partition and multipartitions of a (50×30) data matrix.

In Figure 23.1a blocks are not rectangular, but may be defined as union of rectangular blocks. Note that for nonrectangular blocks there is not simple way to link the two-mode partition of $\mathbf{X}$ with one or more partitions of the set O of objects and the set V of variables. For rectangular clusters, as those included in cases (b–f), we can show that there is a direct linkage between P_{OV} and P_O, P_V.

We now focus on two-mode partitions linked to partitions of sets O and V. Given marginal partitions $P_O = \{{}_OC_1, \ldots, {}_OC_p, \ldots, {}_OC_P\}$, $P_V = \{{}_VC_1, \ldots, {}_VC_q, \ldots, {}_VC_Q\}$ of O and V, respectively, the Cartesian product $P_O \times P_V$ defines a *two-mode single-partition* (briefly, 2-*msp*) that induces, one-to-one, a *two-mode partition* of $\mathbf{X}$ into blocks $\mathbf{X}_{11}, \ldots, \mathbf{X}_{pq}, \ldots, \mathbf{X}_{PQ}$, after the appropriate reordering. Figure 23.1b shows a 2-*msp* for a data matrix of dimensions (50×30), split into (5×3) blocks. From the Cartesian product $P_O \times P_V$, pairs of marginal clusters $({}_OC_p, {}_VC_q) \; \forall \; p, q$, determine 2-*mrc* ${}_{OV}C_{pq}$, with associated blocks $\mathbf{X}_{pq}$ forming data submatrices after the appropriate reordering. Formally $P_{OV} = \{{}_{OV}C_{11}, \ldots, {}_{OV}C_{pq}, \ldots, {}_{OV}C_{PQ}\}$, such that $\bigcup_{p=1}^{P} \bigcup_{q=1}^{Q} {}_{OV}C_{pq} = O \times V$ and ${}_{OV}C_{pq} \cap {}_{OV}C_{lm} = \varnothing$, for $(p, q) \neq (l, m)$, where

$$ {}_{OV}C_{pq} = \{(o_i, v_j) : o_i \in {}_OC_p, v_j \in {}_VC_q\}, \text{ with } \mathbf{X}_{pq} = [x_{ij} : (o_i, v_j) \in {}_{OV}C_{pq}] \qquad (23.1) $$

Thus the data matrix $\mathbf{X}$ can be rewritten as a block matrix

$$ \mathbf{X} = \begin{bmatrix} \mathbf{X}_{11} & \mathbf{X}_{12} & \cdots & \mathbf{X}_{1Q} \\ \mathbf{X}_{21} & \mathbf{X}_{22} & \cdots & \mathbf{X}_{2Q} \\ \vdots & \vdots & \ddots & \vdots \\ \mathbf{X}_{P1} & \mathbf{X}_{P2} & \cdots & \mathbf{X}_{PQ} \end{bmatrix} $$

where each block represents a 2-*mrc* that can have associated an ideal model, chosen among those described in the previous Section 23.2.1. Therefore, the number of clusters of the two-mode single partition, and their blocks, are PQ. In Table 23.1 the marginal partitions $P_O = \{{}_OC_1, {}_OC_2\}$, $P_V = \{{}_VC_1, {}_VC_2\}$ define a 2-*msp* formed by 2-*mc* ${}_{OV}C_{11}$, ${}_{OV}C_{12}$, ${}_{OV}C_{21}$, and ${}_{OV}C_{22}$. Furthermore, matrix $\mathbf{X}$ is accordingly partitioned into blocks $\mathbf{X}_{11}$, $\mathbf{X}_{12}$, $\mathbf{X}_{21}$, and $\mathbf{X}_{22}$ with centroids 5.5, 1.3, 1.2, and 3.3, respectively. A specific 2-*msp* model related to the data generally representing asymmetric (dis)similarities is connected to the ideal *block diagonal matrix*

$$ \hat{\mathbf{X}} = \text{blockdiag}(\mathbf{X}_1, \ldots, \mathbf{X}_K) \qquad (23.2) $$

Each block $\mathbf{X}_k$ has associated a 2-*mrc*. Here, it is supposed that blocks $\mathbf{X}_k$ $(k = 1, \ldots, K)$, specify high association (similarity) among rows and columns involved, while outside these blocks a low association representing a background is supposed (Lazzeroni and Owen 2002). In this 2-*msp* the same number of clusters K for rows and columns is hypothesized. These data types are illustrated in Figure 23.1c. For microarray analysis every gene in gene-block k is expressed within, and only within, those samples in sample-block k.

The two-mode single-partition is the most parsimonious two-mode partitioning model for rectangular blocks since it has associated a unique partition for rows and a unique partition for columns. This also implies that for each class of rows the same partition of columns is identified and vice versa, for each class of columns the same partition of rows is found.

In some situations this property may be considered too restrictive, because given any cluster of one mode, the partition of the other mode does not change. For example, customers (rows) having different characteristics generally belong to different clusters (segments) and very likely have different preferences that specify different partitions of products (columns).

To overcome this restriction Rocci and Vichi (2008) have proposed the *two-mode multipartition* or *two-mode class-conditional partition model*. Two cases can be defined.

First, the partition P_V can change conditionally to each cluster $_OC_p$ of objects (Figure 23.1d). Thus, *a two-mode multipartition of variables* is specified. Formally, given a partition of objects $P_O = \{_OC_1, \ldots, _OC_p, \ldots, _OC_P\}$, and P partitions of variables $P_{V_p} = \{_{V_p}C_1, \ldots, _{V_p}C_q, \ldots, _{V_p}C_{Q_p}\}$ $p = 1, \ldots, P$, the two-mode multipartition of variables is

$$P_{OV_p} = \{_{OV_p}C_1, \ldots, _{OV_p}C_q, \ldots, _{OV_p}C_{Q_p}\} \tag{23.3}$$

where $_{OV_p}C_q = \{(o_i, v_j) : o_i \in _OC_p, v_j \in _{V_p}C_q\}$ with the associated block $\mathbf{X}_{pq} = [x_{ij} : (o_i, v_j) \in _{OV_p}C_q]$, for $q = 1, \ldots, Q_p$.

Second, the partition P_O can change conditionally to each cluster $_VC_q$ of variables. Hence, *a two-mode multipartition of objects* is specified. Formally, given a partition of variables $P_V = \{_VC_1, \ldots, _VC_q, \ldots, _VC_Q\}$, and Q partitions of variables $P_{O_q} = \{_{O_q}C_1, \ldots, _{O_q}C_p, \ldots, _{O_q}C_{P_q}\}$ $q = 1, \ldots, Q$, the two-mode multipartition of objects is

$$P_{O_q V} = \{_{O_q}VC_1, \ldots, _{O_q}VC_p, \ldots, _{O_q}VC_{P_q}\} \tag{23.4}$$

where $_{O_q}VC_q = \{(o_i, v_j) : o_i \in _{O_q}C_p, v_j \in _VC_q\}$ with the associated block $\mathbf{X}_{pq} = [x_{ij} : (o_i, v_j) \in _{O_q}VC_p]$, for $p = 1, \ldots, P_q$.

In Figure 23.1f, a two-mode multipartition of objects and variables is shown. Values in this data matrix $\mathbf{X}$ increase moving from top left side cell (1,1) to the bottom right side cell (n, J) showing a clear clustering on the columns and two main block clustering for rows. This result is highlighted by the two-mode hierarchical classification shown in Figure 23.2a, where the data in Figure 23.1f are two-mode hierarchically classified. In Figure 23.2b another example of a data matrix where nested clusters for rows and columns are shown by the two-mode hierarchical classification.

23.3.1 Models for Two-Mode Single-Partition: The Double *K*-Means

Any two-mode single-partition can be defined by modeling matrix $\mathbf{X}$ according to

$$\mathbf{X} = \mathbf{U}\bar{\mathbf{X}}\mathbf{V}' + \mathbf{E} \tag{23.5}$$

which has been called *Double K-Means* (DKM) model by Vichi (2001) where unknown partitions P_O and P_V, specified by the membership matrices $\mathbf{U}$ and $\mathbf{V}$, need to be identified in order to best reconstruct matrix $\mathbf{X}$. Matrix $\bar{\mathbf{X}}$ is the centroid matrix, with generic element $\bar{x}_{pq}$ representing the centroid of the 2-mrc $_{OV}C_{pq}$, $p = 1, \ldots, P$ and $q = 1, \ldots, Q$. The centroid matrix represents the relevant information in the data matrix $\mathbf{X}$ and can be seen as the reduced data matrix of dimension $P \times Q$, corresponding to P nonobservable prototype objects described by Q nonobservable prototype variables. The name double K-means model is justified by the property that if matrix $\mathbf{V}$ degenerates into the identity matrix of order J ($\mathbf{V} = \mathbf{I}_J$), Equation 23.5 becomes: $\mathbf{X} = \mathbf{U}\bar{\mathbf{X}} + \mathbf{E}$, that is, the clustering model implicitly associated with the K-means algorithm for partitioning objects (Ball et al. 1967; MacQueen 1967). On the other hand, when $\mathbf{U}$ degenerates into the identity matrix of order I ($\mathbf{U}=\mathbf{I}_I$), Equation 23.5 is $\mathbf{X} = \bar{\mathbf{X}}\mathbf{V}' + \mathbf{E}$, that is, the clustering model implicitly associated

(a)   (b)

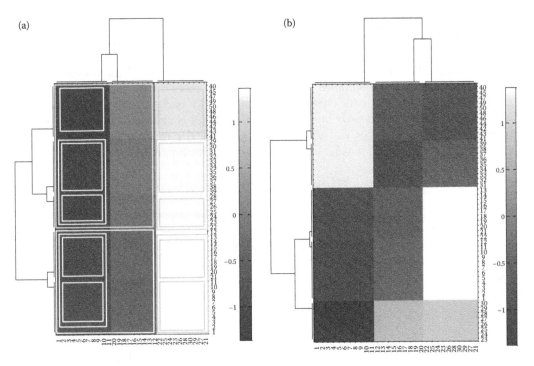

FIGURE 23.2
Two-mode hierarchical clustering.

with the K-means algorithm for partitioning variables. Thus, in general two interconnected K-means problems are considered in the DKM model.

Equation 23.5 for similarity data was introduced by Hartigan (1975) and generalized by De Sarbo (1982), who started from the ADCLUS model: $S = UWU' + C + E$, given by Shepard and Arabie (1979), where S is the observed similarity matrix, W is a diagonal matrix weighting P clusters identified by the membership matrix U and C is a constant matrix. In particular, GENNCLUS model, $S = UWV' + C + E$, was introduced for an asymmetric similarity matrix S, where W (the centroid matrix) is supposed symmetric with zero diagonal elements and represents the association between P derived clusters for rows and Q derived clusters for columns of the asymmetric similarity matrix S. Therefore, GENNCLUS is appropriate for similarity rectangular data, such those described in Figure 23.1c.

Busygin et al. (2008) noted some connection of the DKM model with the *singular value decomposition* (SVD) of a rectangular matrix X which generalizes the spectral decomposition of a square symmetric matrix. SVD is applicable to any rectangular matrix S and specifies orthogonal matrices $\ddot{U}$ and $\ddot{V}$ of *left* and *right singular vectors*, respectively, such that $\ddot{U}'X\ddot{V} = diag(\sigma_1, \ldots, \sigma_K)$, with $K = \min(n, J)$, where $\sigma_1 \geq \cdots \geq \sigma_K$ are the *singular values* of X. Biclustering and SVD may be related when matrix $\bar{X}$ is diagonal; therefore, $\bar{X}$ is constrained to be square of order K and diagonal. The corresponding reconstructed (ideal) matrix is the block diagonal matrix (23.2) which is generally appropriate for (dis)similarity rectangular asymmetric data, as discussed in Section 23.3. Note that any block matrix X_{pq}, associated with 2-*mrc* $_{OV}C_{pq}$ in Equation 23.5, can be reparameterized as an $I \times J$ extended block matrix

$$_e X_{pq} = \mathbf{u}_p \bar{x}_{pq} \mathbf{v}'_q \tag{23.6}$$

that takes the value 0 for all objects and variables that do not belong to the 2-*mrc* $_{OV}C_{pq}$, while that has value $\bar{x}_{pq}$ if the corresponding object and variable belong to $_{OV}C_{pq}$. Since clusters are disjoint, we have

$$\mathbf{U}\bar{\mathbf{X}}\mathbf{V}' = \sum_{p=1}^{P}\sum_{q=1}^{Q} \mathbf{u}_p \bar{x}_{pq} \mathbf{v}'_q \tag{23.7}$$

Long et al. (2005) propose Equation 23.5 for nonnegative $\mathbf{X}$ relaxing the binary form of $\mathbf{U}$ and $\mathbf{V}$, and requiring nonnegative $\mathbf{U}, \mathbf{V}$, and $\bar{\mathbf{X}}$.

Note that DKM model can be extended to include different ideal 2-*mrc* models, considering features discussed in Section 23.2.1. Thus, DKM model can have the form:

i. $\hat{\mathbf{X}} = \mathbf{U}\bar{\mathbf{X}}\mathbf{V}' + \mathbf{a}\mathbf{1}'_J$

including a constant row effect, defined by vector $\mathbf{a}$ $I \times 1$ for different column clusters;

ii. $\hat{\mathbf{X}} = \mathbf{U}\bar{\mathbf{X}}\mathbf{V}' + \mathbf{1}_I \mathbf{b}'$

including a constant column effect, defined by vector $\mathbf{b}$ $J \times 1$ for different row clusters;

iii. $\hat{\mathbf{X}} = \mathbf{U}\bar{\mathbf{X}}\mathbf{V}' + \mathbf{a}\mathbf{1}'_J + \mathbf{1}_I \mathbf{b}'$

including constant row and column effects.

23.4 Isolation of Two-Mode Clusters and Their Measures

To define isolation and heterogeneity of 2-*mrc*, it is useful to specify the notion of object–variable dissimilarity between pairs (o_i, v_j) and (o_l, v_m), within and between two-mode clusters. In general, an object–variable dissimilarity between (o_i, v_j) and (o_l, v_m), written as $d((o_i, v_j), (o_l, v_m))$, or simply as $d_{ij,lm}$, is a mapping d from $(O \times V) \times (O \times V)$ to $\mathfrak{R}^+$ (nonnegative real numbers) satisfying the following properties: (i) $d_{ij,lm} \geq 0, (\forall ij, lm)$; (ii) $d_{ij,ij} = 0$ $(\forall ij)$; (iii) $d_{ij,lm} = d_{lm,ij} (\forall ij, lm)$. The object–variable dissimilarity $d_{ij,lm} = f(d_{il}, d_{jm})$ can be seen as a function f of the dissimilarity between pairs of objects d_{il} and between pairs of variables d_{jm}. For example, $d_{ij,lm} = d_{il} + d_{jm}$, or $d_{ij,lm} = d_{il} \times d_{jm}$. Several objects and variable dissimilarity measures have been proposed in cluster analysis literature, these will not be discussed in this paper (for an extensive discussion of dissimilarities between objects and dissimilarities between variables, e.g., Gower and Legendre 1986; Gordon 1999). Hanish et al. (2002) propose to construct a distance function that combines a graph distance between (objects) with a correlation-based function for variables. On the defined distance matrix, a hierarchical clustering such as the average linkage method is applied.

23.4.1 Heterogeneity within Two-Mode Clusters

Similarly to the case of classical cluster analysis, two features characterize 2-*mrc* $_{OV}C_{pq}$ $(p = 1, \ldots, P; q = 1, \ldots, Q)$: *heterogeneity* and *isolation*. They are related to the *internal cohesion* and *external isolation* described by Cormack (1971) in the formal definition of clusters for multivariate objects.

The *heterogeneity* or *lack of internal cohesion within a two-mode cluster* $_{OV}C_{pq}$, modeled by ideal block, $\hat{\mathbf{X}}_{pq}$, is the extent to which pairs (o_i, v_j), within $_{OV}C_{pq}$ with values x_{ij}, are dissimilar (separate) one from the others; or, values x_{ij} are dissimilar from the expected values $\hat{x}_{ij} \in \hat{\mathbf{X}}_{pq}$. Therefore, there are two ways to formalize a measure of heterogeneity of a two-mode cluster: (i) it can be estimated as a function of the difference between x_{ij} of (o_i, v_j) and x_{lm} of (o_l, v_m), for each (o_i, v_j) and (o_l, v_m) belonging to $_{OV}C_k$; or (ii) it is a function of the difference between x_{ij} and $\hat{x}_{ij}$ for each pair $(o_i, v_j) \in {_{OV}C_{pq}}$.

Measures of heterogeneity can be based on sum of squares and some others on graph theory such as diameter, radius, and cut, similarly to measures of heterogeneity for classical cluster analysis (Gordon 1999). Some measures are defined in the following section.

23.4.2 Measures of Heterogeneity within Two-Mode Clusters

A measure of heterogeneity within a 2-*mrc* $_{OV}C_{pq}$, with observed block $\mathbf{X}_{pq}$ and expected block $\hat{\mathbf{X}}_{pq}$, is the *Sum Square Residual* (SSR) *within* 2-*mrc* $_{OV}C_{pq}$, as described in Cheng and Churh (2000),

$$W_{SSR}(\mathbf{X}_{pq}) = \sum_{i \in {_O}C_p} \sum_{j \in {_V}C_q} (x_{ij} - \hat{x}_{ij})^2 = \|\mathbf{X}_{pq} - \hat{\mathbf{X}}_{pq}\|^2 \qquad (23.8)$$

corresponding to the dissimilarity between observed and expected 2-*mrc* $_{OV}C_{pq}$, where $\|\mathbf{A}\| = \sqrt{\sum_i \sum_j a_{ij}^2}$ is the Euclidean norm of a matrix $\mathbf{A}$. $W_{SSR}(\mathbf{X}_{pq})$ has minimum value equal to zero, if and only if, the observed block is equal to the expected, that is, $\mathbf{X}_{pq} = \hat{\mathbf{X}}_{pq}$.

Obviously, for different expected clusters, a different measure of heterogeneity can be defined. For *Row and Column Constant* 2-*mrc Model* (i) (see Section 23.2.1): $\hat{\mathbf{X}}_{pq} = \hat{x}_{pq}\mathbf{1}_{Ip}\mathbf{1}'_{Jq}$, $W_{SSR}(\mathbf{X}_{pq})$ corresponds to the *Deviance Within* 2-*mrc* $_{OV}C_{pq}$ used by Hartigan (1972) for direct clustering of a data matrix, that is,

$$DEV_W(\mathbf{X}_{pq}) = \sum_{i \in {_O}C_p} \sum_{j \in {_V}C_q} (x_{ij} - \bar{x}_{pq})^2 = \|\mathbf{X}_{pq} - \bar{x}_{pq}\mathbf{1}_{Ip}\mathbf{1}'_{Jq}\|^2 \qquad (23.9)$$

where $x_{ij} \in \mathbf{X}$ is supposed centered and $\bar{x}_{pq}$ is the mean of the values in $\mathbf{X}_{pq}$. For 2-*mrc* in Table 23.1 we have: $_{OV}C_{11}$, $DEV_W(\mathbf{X}_{11}) = 1.0393$; $_{OV}C_{12}$, $DEV_W(\mathbf{X}_{12}) = 0.2287$; $_{OV}C_{21}$, $DEV_W(\mathbf{X}_{21}) = 0.2607$; $_{OV}C_{22}$, $DEV_W(\mathbf{X}_{22}) = 0.1480$. Thus, the most homogeneous two-mode cluster is $_{OV}C_{22}$, while the most heterogeneous is $_{OV}C_{11}$.

It is worthy to note that a robust two-mode clustering methodology can be obtained by substituting, in Equation 23.9 the mean with the median and estimating DKM by considering the L_1 norm. Cheng and Church (2000), Cho et al. (2004) for biclustering of expression data, use the SSR with additive 2-*mrc* model (iv) (see Section 23.2.1), that is,

$$W_{SSRa}(\mathbf{X}_{pq}) = \|\mathbf{X}_{pq} - \bar{x}_{pq}\mathbf{1}_{Ip}\mathbf{1}'_{Jq} - \mathbf{a}_p\mathbf{1}'_{Jq} - \mathbf{1}_{Ip}\mathbf{b}'\|^2 \qquad (23.10)$$

where $\mathbf{a}_p$ and $\mathbf{b}_q$ are the vectors whose elements are rows and columns means of $\mathbf{X}_{pq}$, respectively. The $W_{SSRa}(\mathbf{X}_{pq})$ is zero if all columns of the 2-*mrc* are equal to each other, and vice versa all rows are equal to each other. The sum square residual has been used by Pensa and Boulicau (2008) to propose an iterative algorithm which exploits user-defined constraints. Cheng and Church (2000) define a δ-two-mode (homogeneous) cluster if $W_{SSRa}(\mathbf{X}_{pq}) < \delta$. It is worthy to note that measures (23.8–23.10) are cluster-size dependent and therefore, their normalization, for example, by dividing for the cardinality of the cluster, is needed when it is necessary to compare clusters with different size. The normalized version of measure (23.10), that is, $W_{MSRa}(\mathbf{X}_{pq}) = W_{SSRa}(\mathbf{X}_{pq})/(I_p J_q)$, is called by Cheng and Church (2000) *Mean Square Residual*. They define an algorithm for extracting a list of two-mode clusters all guaranteed to be δ-two-mode clusters. The algorithm computes the score W_{MSRa} for each possible row/column addition/deletion and chooses the action that decreases W_{MSRa} the most. If no action will decrease W_{MSRa}, or if it is less than δ the algorithm stops, thus, extracting one 2-*mrc* at a time. A major option that drastically influences the final solution is the choice of the threshold δ. Authors have provided an empirical threshold $\delta = 300$ for the gene expression data by analyzing 30 datasets well known in literature. Bryan et al. (2005) improved the algorithm of Cheng and Church by applying a simulated annealing technique.

For nonnegative matrices generally corresponding to (dis)similarity measures between elements of O and V, it is useful to consider $\mathbf{X}$ as the adjacency matrix of a bipartite graph $BiG(O, V, E, \mathbf{X})$, where nodes of the graph are split into two sets: O for objects and V for variables. Each edge in the edge set E is allowed between O and V, only, and has a nonnegative weight $x_{ij} \in \mathbf{X}$, corresponding to a frequency, a preference, in general, a (dis)similarity. When $x_{ij} = 0$, there is not an edge between vertex i and j. The heterogeneity within 2-*mrc* $_{OV}C_{pq}$ can be evaluated by

- The *Within Two-mode Cluster Diameter* W_D, that is, the greatest value of $\mathbf{X}_{pq}$
  $$W_D(\mathbf{X}_{pq}) = \max\{x_{ij} : i \in {_O}C_p, j \in {_V}C_q\}$$
- The *Within Two-mode Cluster Radius (or Star)* W_R, that is, the smallest value of $\mathbf{X}_{pq}$
  $$W_R(\mathbf{X}_{pq}) = \min\{x_{ij} : i \in {_O}C_p, j \in {_V}C_q\}$$
- The *Winhin Two-mode Cluster Cut*, that is, twice the sum of the elements of $\mathbf{X}_{pq}$,
  $$W_{Cut}(\mathbf{X}_{pq}) = 2\sum_{i \in {_O}C_p} \sum_{j \in {_V}C_q} x_{ij} = 2\mathbf{u}'_p \mathbf{X} \mathbf{v}_q$$

where $\mathbf{U} = [\mathbf{u}_1, \ldots, \mathbf{u}_p, \ldots, \mathbf{u}_P]$ and $\mathbf{V} = [\mathbf{v}_1, \ldots, \mathbf{v}_q, \ldots, \mathbf{v}_Q]$. The W_{Cut} has been used in graph partitioning, however, this criterion frequently gives skewed cut results, that is, very small subgraphs of isolated nodes. This is not surprising since W_{Cut} increases with the number of edges going across the two classes $_O C_p$ and $_V C_q$. Various normalizations can be introduced to circumvent this problem. The *Cut* can be divided by the cardinality of the marginal clusters of $_{OV}C_{pq}$

$$W_{RatioCut}(\mathbf{X}_{pq}) = \frac{W_{Cut}(\mathbf{X}_{pq})}{I_p + I_q} = \frac{2\mathbf{u}'_p \mathbf{X} \mathbf{v}_q}{\mathbf{u}'_p \mathbf{u}_p + \mathbf{v}'_q \mathbf{v}_q} \tag{23.11}$$

which is called *Within Two-mode cluster Ratio Cut* and has been considered in Hagen and Kahng (1992) for circuit partitioning. Another interesting normalization was introduced by Shi and Malik (2000) for image segmentation

$$W_{NRCut}(\mathbf{X}_{pq}) = \frac{W_{Cut}(\mathbf{X}_{pq})}{\sum_{i \in {_O}C_p} \sum_{j=1}^{J} x_{ij} + \sum_{i=1}^{I} \sum_{j \in {_V}C_q} x_{ij}} = \frac{2\mathbf{u}'_p \mathbf{X} \mathbf{v}_q}{\mathbf{u}'_p \mathbf{X} \mathbf{1}_J + \mathbf{1}'_I \mathbf{X} \mathbf{v}_q} \tag{23.12}$$

which is called *Within Two-mode Cluster Normalized Ration Cut*. It normalizes the Cut by considering the isolation between two classes $_O C_p$ and $_V C_q$, as well as, the heterogeneity within classes. Note that W_{NRCut} is bounded above by 1, which is reached when isolation between classes (see Equation 23.12) is null. For the data in Table 23.1, the Within Diameter of $_{OV}C_{11}$ is $W_D(\mathbf{X}_{11}) = 5.9$, the Within Radius of $_{OV}C_{11}$ is $W_R(\mathbf{X}_{11}) = 5.0$, the Within Cut of $_{OV}C_{11}$ is $W_{Cut}(\mathbf{X}_{11}) = 2 \times 82.4 = 164.8$, the Within Ratio Cut of $_{OV}C_{11}$ is $W_{RatioCut}(\mathbf{X}_{11}) = 2 \times 82.4/8 = 20.6$. Finally, the Normalized Cut of $_{OV}C_{11}$ is $W_{NormalizedCut}(\mathbf{X}_{11}) = 2 \times 82.4/(95.45 + 100.61) = 0.84$, indicating an homogeneous cluster, quite close to the maximum equal to 1.

23.4.3 Within Clusters Criteria for Two-Mode Single-Partitioning

The measures of heterogeneity within two-mode clusters can be used to define a general family of criteria for assessing the heterogeneity of a two-mode single-partition with blocks $\mathbf{X}_{11}, \ldots, \mathbf{X}_{pq}, \ldots, \mathbf{X}_{PQ}$,

$$W_M(\mathbf{X}_{11}, \ldots, \mathbf{X}_{PQ}) = \sum_{p=1}^{P} \sum_{q=1}^{Q} W_M(\mathbf{X}_{pq}) \tag{23.13}$$

by summing the heterogeneity measure of each two mode cluster of the partition $_{OV}C_{pq}$ $(p = 1, \ldots, P, q = 1, \ldots, Q)$, where M is one of the within measures, that is: SSR (Sum Square Residual), D (Diameter), R (Radius), Cut, RatioCut, NRCut.

Thus, considering the SSR, the *Deviance Within clusters of two-mode single-partition* is

$$DEV_W(\mathbf{X}_{11}, \ldots, \mathbf{X}_{PQ}) = \sum_{p=1}^{P} \sum_{q=1}^{Q} \sum_{i \in _OC_p} \sum_{j \in _VC_q} (x_{ij} - \bar{x}_{pq})^2$$

$$= \sum_{p=1}^{P} \sum_{q=1}^{Q} \|\mathbf{X}_{pq} - \bar{x}_{pq} \mathbf{1}_{Ip} \mathbf{1}'_{Jq}\|^2 = \|\mathbf{X} - \mathbf{U}\bar{\mathbf{X}}\mathbf{V}'\|^2 \tag{23.14}$$

For data in Table 23.1, after centering $\mathbf{X}$, $DEV_W = 1.6767$, which is the sum of deviances within clusters $_{OV}C_{11}, _{OV}C_{12}, _{OV}C_{21}$, and $_{OV}C_{22}$.

Especially when $\mathbf{X}$ is a nonnegative matrix, the ideal two-mode single partition considered in several coclustering methodologies (e.g., Ding et al. 2001; Dhillon 2001; Lazzeroni and Owen 2002) would produce a reordering of the data in $\mathbf{X}$, into K rectangular blocks on the diagonal of $\mathbf{X}$. Each block would be expected to have highly similarity values, thus representing the association within 2-*mrc*, while the rest of the matrix would have low similarity values representing the low similarity (high dissimilarity) between clusters. An example of this type of dataset is given in Figure 23.1d.

The family of criteria for assessing heterogeneity in this case is

$$W_M(\mathbf{X}_1, \ldots, \mathbf{X}_K) = \sum_{k=1}^{K} \frac{W_{Cut}(\mathbf{X}_{kk})}{\rho(_{OV}C_k)} \tag{23.15}$$

where for $M = RatioCut$, $\rho(_{OV}C_k) = |_{OV}C_k|$, that is, the cardinality of $_{OV}C_k$; while for $M = NRCut$, $\rho(_{OV}C_k) = \sum_{i \in _OC_k} \sum_{j=1}^{J} x_{ij} + \sum_{i=1}^{I} \sum_{j \in _VC_k} x_{ij}$.

23.4.4 Isolation between Two-Mode Clusters

The *isolation between two-mode clusters* $_{OV}C_{pq}$ and $_{OV}C_{hk}$ with blocks $\mathbf{X}_{pq}$ and $\mathbf{X}_{hk}$ and expected blocks $\hat{\mathbf{X}}_{pq}$ and $\hat{\mathbf{X}}_{hk}$ is the extent to which $_{OV}C_{pq}$ is dissimilar or separate from

$_{OV}C_{hk}$. The isolation can be estimated as a function of differences between values $\hat{x}_{ij} \in \hat{\mathbf{X}}_{pq}$ and a general expected value for all blocks, or, alternatively, as a function of the differences between values $x_{ij} \in \mathbf{X}_{pq}$ and $x_{lm} \in \mathbf{X}_{hk}$ $\forall i,j,l,m$.

In Table 23.1 two-mode clusters $_{OV}C_{11}$ and $_{OV}C_{12}$ are quite isolated from each other, while the isolation between $_{OV}C_{12}$ and $_{OV}C_{21}$ is very low. In fact, elements in $\mathbf{X}_{11}$ are all uniform random values around the value 5.5, while values in $\mathbf{X}_{12}$ are random values around 1.3, and therefore, there is a relevant difference, which in not observed for blocks $\mathbf{X}_{12}$ and $\mathbf{X}_{21}$ having values around 1.3 and 1.2, respectively. We now formalize measures of isolation between two-mode clusters.

23.4.5 Measures of Isolation between Two-Mode Clusters

A measure of isolation between 2-*mrc* $_{OV}C_{pq}$, with block $\mathbf{X}_{pq}$, and the general expected mean, supposed without loss of generality, equal to zero (i.e., matrix $\mathbf{X}$ is centered), is

$$B_{SSR}(\mathbf{X}_{pq}) = \sum_{i \in {_O}C_p} \sum_{j \in {_V}C_q} \hat{x}_{ij}^2 \qquad (23.16)$$

the SSR *Between* 2-*mrc* that for the *Row and Column Constant Model* (i) (see Section 23.3.1) with $\hat{\mathbf{X}} = \bar{x}_{pq} \mathbf{1}_{Ip} \mathbf{1}'_{Jq}$, $SSR_B(\mathbf{X}_{pq})$ is the Deviance Between two-mode clusters $_{OV}C_{pq}$ and $_{OV}C_{hk}$, that is,

$$DEV_B(\mathbf{X}_{pq}) = \left(\sum_{i \in {_O}C_p} \sum_{j \in {_V}C_q} \bar{x}_{pq}^2 \right) = I_p J_q \bar{x}_{pq}^2 \qquad (23.17)$$

For data in Table 23.1, the isolation of clusters $_{OV}C_{11}$, $_{OV}C_{12}$, $_{OV}C_{21}$, $_{OV}C_{22}$ measured through the deviance between 2-*mrc* are respectively: $15 \times (2.1397)^2 = 68.6747$; $10 \times (-0.979)^2 = 9.5844$; $15 \times (2.1397)^2 = 68.6747$; $10 \times (0.979)^2 = 9.5844$.

For nonnegative matrices $\mathbf{X}$, where x_{ij} represents a weight, a frequency, a preference, and a (dis)similarity, the isolation is generally measured between cluster $_{OV}C_{pq}$ and complement cluster $_{OV}\bar{C}_{pq} = \{(o_i, v_j) : o_i \in {_O}C_p, v_j \in V - {_V}C_q\} \cup \{(o_i, v_j) : o_i \in O - {_O}C_p, v_j \in {_V}C_q\}$. Thus, Dhillon (2001) for bipartite spectral cluster partitioning considers the *Between Two-mode Clusters* $_{OV}C_{pq}$, *Cut*, that is,

$$B_{Cut}(\mathbf{X}_{pq}) = \mathbf{u}'_p \mathbf{X} \mathbf{1}_J + \mathbf{1}'_I \mathbf{X} \mathbf{v}_q - 2\mathbf{u}'_p \mathbf{X} \mathbf{v}_q = \sum_{h \neq p}^{P} W_{Cut}(\mathbf{X}_{hq}) + \sum_{k \neq q}^{Q} W_{Cut}(\mathbf{X}_{pk}) \qquad (23.18)$$

The B_{Cut} can be divided by the cardinality of the marginal clusters of $_{OV}C_{pq}$

$$B_{RatioCut}(\mathbf{X}_{pq}) = B_{Cut}(\mathbf{X}_{pq})/(I_p + J_q) \qquad (23.19)$$

which is called *Between Two-mode Clusters Ratio Cut*. Another useful normalization has been proposed by Shi and Malik (2000) for image segmentation

$$B_{NRCut}(\mathbf{X}_{pq}) = \frac{B_{Cut}(\mathbf{X}_{pq})}{\sum_{i \in {_O}C_p} \sum_{j=1}^{J} x_{ij} + \sum_{i=1}^{I} \sum_{j \in {_V}C_q} x_{ij}} = \frac{\mathbf{u}'_p \mathbf{X} \mathbf{1}_J + \mathbf{1}'_I \mathbf{X} \mathbf{v}_q - 2\mathbf{u}'_p \mathbf{X} \mathbf{v}_q}{\mathbf{u}'_p \mathbf{X} \mathbf{1}_J + \mathbf{1}'_I \mathbf{X} \mathbf{v}_q}$$

$$(23.20)$$

which is called the *Between Two-mode clusters Normalized Cut*. Finally, the *Between Two-mode Clusters MinMax cut* (Ding et al. 2001) is defined by

$$B_{MMCut}(\mathbf{X}_{pq}) = \frac{B_{Cut}(\mathbf{X}_{pq})}{W_{Cut}(\mathbf{X}_{pq})} = \frac{\mathbf{u}'_p \mathbf{X} \mathbf{1}_J + \mathbf{1}'_I \mathbf{X} \mathbf{v}_q - 2\mathbf{u}'_p \mathbf{X} \mathbf{v}_q}{2\mathbf{u}'_p \mathbf{X} \mathbf{v}_q} \tag{23.21}$$

For data in Table 23.1, the $B_{Cut}(\mathbf{X}_{11}) = 31.26$; $B_{RatioCut}(\mathbf{X}_{11}) = 3.9075$; $B_{NRCut}(\mathbf{X}_{11}) = 0.1594$; $B_{MMCut}(\mathbf{X}_{11}) = 0.1897$.

23.4.6 Between Clusters Criteria for Two-Mode Single-Partitioning

The Between two-mode clusters SSR can be used to define a criterion for assessing the isolation of a two-mode single-partition with blocks $\mathbf{X}_{11}, \ldots, \mathbf{X}_{PQ}$

$$B_{SSR}(\mathbf{X}_{11}, \ldots, \mathbf{X}_{PQ}) = \sum_{p=1}^{P} \sum_{q=1}^{Q} B_{SSR}(\mathbf{X}_{pq}) \tag{23.22}$$

by summing the SSR of each two mode cluster of the partition $_{OV}C_{pq}$ ($p = 1, \ldots, P$; $q = 1, \ldots, Q$). Thus, by using *the deviance between two-mode clusters, the between deviance of the two-mode single-partition* is

$$DEV_B(\mathbf{X}_{11}, \ldots, \mathbf{X}_{PQ}) = \sum_{p=1}^{P} \sum_{q=1}^{Q} \sum_{i \in {_O}C_p} \sum_{j \in {_V}C_q} \bar{x}_{ij}^2 = \|\mathbf{U}\bar{\mathbf{X}}\mathbf{V}'\|^2 \tag{23.23}$$

For data in Table 23.1, $DEV_B(\mathbf{X}_{11}, \mathbf{X}_{12}, \mathbf{X}_{21}, \mathbf{X}_{22}) = 156.5183$, which is the sum of deviances between clusters. The measures of isolation based on *Cut* have been used for spectral clustering of bipartite graphs (Dhillon 2001; Gu et al. 2001) where K clusters for rows and K clusters for columns are considered for the similarity matrix $\mathbf{X}$, thus, a specific two-mode partition is defined. In this case an ideal two-mode partition would produce a reordering of the data into a number K of rectangular blocks on the diagonal. Each block would have uniform elements (highly similar), and the part of values outside of these diagonal blocks would be of uniform values representing dissimilarities between clusters. Thus,

$$B_M(\mathbf{X}_{11}, \ldots, \mathbf{X}_{KK}) = \sum_{k=1}^{K} \frac{B_{Cut}(\mathbf{X}_{kk})}{\rho(_{OV}C_k)} \tag{23.24}$$

where M is one of the within measures, that is: *RatioCut*, *NRCut*, and *MMCut*. The denominator $\rho(_{OV}C_k)$ is equal to $|_{OV}C_k|$ or $\sum_{i \in {_O}C_p} \sum_{j=1}^{J} x_{ij} + \sum_{i=1}^{I} \sum_{j \in {_V}C_q} x_{ij}$ or $W_{Cut}(\mathbf{X}_{kk})$ if M is equal to *RatioCut* or *NRCut* or *MMCut*, respectively.

Note that Equation 23.24 is minimized when clusters are not too small. In particular the minimum of the normalized RatioCut is achieved if all 2-*mrc* have the same cardinality. So both criteria try to achieve balanced measures by means of the number of vertices, or edge weights, respectively. By using the between MMCut the similarity between clusters is minimized while the similarity within groups is maximized.

For the two-mode partition in Table 23.1:

$$B_{RatioCut}(\mathbf{X}_{11}, \mathbf{X}_{22}) = B_{Cut}(\mathbf{X}_{11})/8 + B_{Cut}(\mathbf{X}_{22})/8 = 3.9075 + 4.4657 = 8.3732;$$

$$B_{NRCut}(\mathbf{X}_{11}, \mathbf{X}_{22}) = B_{Cut}(\mathbf{X}_{11})/196.06 + B_{Cut}(\mathbf{X}_{22})/96.52 = 0.1594 + 0.3239 = 0.4833;$$
$$B_{MMCut}(\mathbf{X}_{11}, \mathbf{X}_{22}) = B_{Cut}(\mathbf{X}_{11})/164.8 + B_{Cut}(\mathbf{X}_{22})/65.26 = 0.1897 + 0.4790 = 0.6687.$$

23.4.7 Links between Partitioning Criteria

23.4.7.1 Decomposition of the Total Deviance of X for a Given Two-Mode Partition

Given a 2-*msp* P_{OV} with marginal partitions P_O and P_V, blocks $\mathbf{X}_{11}, \ldots, \mathbf{X}_{PQ}$, membership matrices $\mathbf{U}$, $\mathbf{V}$, and centroid matrix $\bar{\mathbf{X}}$, the *Total Deviance of the centered matrix* $\mathbf{X}$ is decomposed into the *Deviance Within and Between two-mode clusters* of the two-mode single-partition P_{OV}. In formulas

$$DEV_T(\mathbf{X}) = DEV_W(\mathbf{X}_{11}, \ldots, \mathbf{X}_{PQ}) + DEV_B(\mathbf{X}_{11}, \ldots, \mathbf{X}_{PQ}),$$
$$\|\mathbf{X}\|^2 = \|\mathbf{X} - \mathbf{U}\bar{\mathbf{X}}\mathbf{V}'\|^2 + \|\mathbf{U}\bar{\mathbf{X}}\mathbf{V}'\|^2 \tag{23.25}$$

To prove this result, we observe that

$$\|\mathbf{X}\|^2 = \|\mathbf{X} - \mathbf{U}\bar{\mathbf{X}}\mathbf{V}' + \mathbf{U}\bar{\mathbf{X}}\mathbf{V}'\|^2$$
$$= \|\mathbf{X} - \mathbf{U}\bar{\mathbf{X}}\mathbf{V}'\|^2 + \|\mathbf{U}\bar{\mathbf{X}}\mathbf{V}'\|^2 + 2\mathrm{tr}[(\mathbf{X} - \mathbf{U}\bar{\mathbf{X}}\mathbf{V}')\mathbf{V}\bar{\mathbf{X}}'\mathbf{U}'] \tag{23.26}$$

where $\mathrm{tr}(\mathbf{X}\mathbf{V}\bar{\mathbf{X}}'\mathbf{U}') - \mathrm{tr}(\mathbf{U}\bar{\mathbf{X}}\mathbf{V}'\mathbf{V}\bar{\mathbf{X}}'\mathbf{U}') = 0$ since $\bar{\mathbf{X}} = (\mathbf{U}'\mathbf{U})^{-1}\mathbf{U}'\mathbf{X}\mathbf{V}(\mathbf{V}'\mathbf{V})^{-1}$ implies

$$\mathrm{tr}(\mathbf{U}\bar{\mathbf{X}}\mathbf{V}'\mathbf{V}\bar{\mathbf{X}}'\mathbf{U}') = \mathrm{tr}(\mathbf{U}(\mathbf{U}'\mathbf{U})^{-1}\mathbf{U}'\mathbf{X}\mathbf{V}(\mathbf{V}'\mathbf{V})^{-1}\mathbf{V}'\mathbf{V}\bar{\mathbf{X}}'\mathbf{U}') = \mathrm{tr}(\mathbf{X}\mathbf{V}\bar{\mathbf{X}}'\mathbf{U}') \tag{23.27}$$

23.4.7.2 Decomposition of the Total Cut (Degree or Weight) of X

Given a 2-*msp* P_{OV}, with K rows and columns clusters and associated blocks $\mathbf{X}_{11}, \ldots, \mathbf{X}_{KK}$, the Total Degree (Weight) of $\mathbf{X}$, defined as twice the sum of the elements of $\mathbf{X}$, is decomposed into the *Within and Between Two-mode Clusters Cut* of the 2-*msp* P_{OV}. Thus

$$T_{Cut}(\mathbf{X}) = W_{Cut}(\mathbf{X}_{11}, \ldots, \mathbf{X}_{KK}) + B_{Cut}(\mathbf{X}_{11}, \ldots, \mathbf{X}_{KK})$$
$$2 \times \mathbf{1}'_I \mathbf{X} \mathbf{1}_J = \sum_{k=1}^{K} W_{Cut}(\mathbf{X}_{kk}) + \sum_{k=1}^{K} B_{Cut}(\mathbf{X}_{kk}) \tag{23.28}$$

Since the *Within and Between Two-mode Clusters Cut* are bounded by the total degree, maximizing $W_{Cut}(\mathbf{X}_{11}, \ldots, \mathbf{X}_{KK})$ is equivalent to minimize $B_{Cut}(\mathbf{X}_{11}, \ldots, \mathbf{X}_{KK})$. For the partition in Table 23.1, in two clusters for row and columns with blocks $\mathbf{X}_{11}, \mathbf{X}_{22}$, the $T_{Cut}(\mathbf{X}) = 292.58$ is decomposed into $W_{Cut}(\mathbf{X}_{11}, \mathbf{X}_{22}) = 230.06$ and $B_{Cut}(\mathbf{X}_{11}, \mathbf{X}_{22}) = 62.52$.

23.4.7.3 Decomposition of the Total Normalized Cut (Degree or Weight) of X

Given a 2-*msp* P_{OV}, with K rows and columns clusters and associated blocks $\mathbf{X}_{11}, \ldots, \mathbf{X}_{KK}$, the following decomposition holds for the NRCut,

$$T_{NRCut}(\mathbf{X}) = W_{NRCut}(\mathbf{X}_{11}, \ldots, \mathbf{X}_{KK}) + B_{NRCut}(\mathbf{X}_{11}, \ldots, \mathbf{X}_{KK})$$
$$K = \sum_{k=1}^{K} W_{NRCut}(\mathbf{X}_{kk}) + \sum_{k=1}^{K} B_{NRCut}(\mathbf{X}_{kk}) \tag{23.29}$$

Each 2-*mrc* has a normalized weight equal to 1, thus the *Total Normalized Cut* is equal to K.

For the partition in Table 23.1, with blocks X_{11}, X_{22}, the $T_{NRCut}(X) = 2$ is decomposed into $W_{NRCut}(X_{11}, X_{22}) = 1.5167$ and $B_{NRCut}(X_{11}, X_{22}) = 0.4833$.

For the *RatioCut* the sum of the Within and Between parts is not constant, thus, an equivalent decomposition, as for the *Cut* and the *NRCut*, does not hold.

23.5 Estimation and Algorithm for Double K-Means Model

23.5.1 Least-Squares Estimation of DKM

The least-squares assessment of Equation 23.5 leads to the formulation of the following quadratic optimization problem that has to be solved with respect to variables u_{ip}, v_{jq}, and $\bar{x}_{pq}$ where constraints (23.33) are necessary to specify an objects (rows) partition P_O and a variables (columns) partition P_V (Vichi 2001)

$$J_{DKM}(U, V, \bar{X}) = \sum_{p=1}^{P} \sum_{q=1}^{Q} \sum_{i=1}^{I} \sum_{j=1}^{J} (x_{ij} - \bar{x}_{pq})^2 u_{ip} v_{jq} \tag{23.30}$$

$$= \sum_{p=1}^{P} \sum_{q=1}^{Q} \sum_{i \in {}_O C_p} \sum_{j \in {}_V C_q} (x_{ij} - \bar{x}_{pq})^2 \tag{23.31}$$

$$= \|X - U\bar{X}V'\|^2 \to \min_{U,V,\bar{X}} \tag{23.32}$$

$$\text{s.t. } u_{ip}, v_{jq} \in \{0, 1\}, \sum_{p} u_{ip} = 1, \sum_{q} v_{jq} = 1, \forall i, j, p, q \tag{23.33}$$

Criterion J_{DKM} has been used for proximity data by Hartigan (1975), Bock (1979), De Sarbo (1982) and for nonnegative matrices such as contingency tables by Govaert (1983).

It is interesting to note that the LS solution of DKM can be obtained by solving different mathematical problems with equivalent loss functions. In fact, from the decomposition (23.25) of the total deviance and recalling that J_{DKM} attains a minimum with respect to $\bar{X}$ when $\bar{X} = (U'U)^{-1}U'XV(V'V)^{-1}$, equivalent formulations of the DKM s.t. constraints (23.33) are

$$\|X - H_U X H_V'\|^2 \to \min_{U,V} \quad \text{or} \quad \|H_U X H_V'\|^2 \to \max_{U,V} \tag{23.34}$$

where $H_U = U(U'U)^{-1}U'$ and $H_V = V(V'V)^{-1}V'$ are projector matrices of U and V, respectively.

Another equivalent formulation of DKM is given by considering the normalized membership matrices

$$\ddot{U} = U(U'U)^{-1/2}, \ddot{V} = V(V'V)^{-1/2} \tag{23.35}$$

that are such that $\ddot{\mathbf{U}}'\ddot{\mathbf{U}} = \mathbf{I}_P$ and $\ddot{\mathbf{V}}'\ddot{\mathbf{V}} = \mathbf{I}_Q$. By using (23.35) the *normalized* DKM is defined as

$$\|\mathbf{X} - \ddot{\mathbf{U}}\ddot{\ddot{\mathbf{X}}}\ddot{\mathbf{V}}'\|^2 \to \min_{\ddot{\mathbf{U}},\ddot{\mathbf{X}},\ddot{\mathbf{V}}} \text{ or } \|\mathbf{X} - \ddot{\mathbf{U}}\ddot{\mathbf{U}}'\mathbf{X}\ddot{\mathbf{V}}\ddot{\mathbf{V}}'\|^2 \to \min_{\ddot{\mathbf{U}},\ddot{\mathbf{V}}} \text{ or } \|\ddot{\mathbf{U}}\ddot{\mathbf{U}}'\mathbf{X}\ddot{\mathbf{V}}\ddot{\mathbf{V}}'\|^2 \to \max_{\ddot{\mathbf{U}},\ddot{\mathbf{V}}} \quad (23.36)$$

where $\ddot{\ddot{\mathbf{X}}} = \ddot{\mathbf{U}}'\mathbf{X}\ddot{\mathbf{V}}$ is the centroid matrix for the two-mode single-partition P_{OV}. The normalized DKM is equivalent to the DKM because the loss functions corresponding to the two problems coincide, that is, $\|\mathbf{H}_{\mathbf{U}}\mathbf{X}\mathbf{H}_{\mathbf{V}}\|^2 = \|\ddot{\mathbf{U}}\ddot{\mathbf{U}}'\mathbf{X}\ddot{\mathbf{V}}\ddot{\mathbf{V}}'\|^2$. This is so because $\mathbf{H}_{\mathbf{U}} = \ddot{\mathbf{U}}\ddot{\mathbf{U}}'$ and $\mathbf{H}_{\mathbf{V}} = \ddot{\mathbf{V}}\ddot{\mathbf{V}}'$.

An algorithm for problem (23.36) with objective function $\|\mathbf{X} - \ddot{\mathbf{U}}\ddot{\ddot{\mathbf{X}}}\ddot{\mathbf{V}}'\|^2$ has been given in Cho et al. (2004), who formalize a DKM algorithm by considering the ideal clusters (i) (Section 23.2.1), that is, with $\hat{\mathbf{X}}_k = \bar{\bar{x}}_k \mathbf{1}_{Ik}\mathbf{1}'_{Jk}$. Furthermore, Cho et al. (2004) propose to use the ideal clusters (iv) (Section 23.2.1), that is, with $\hat{\mathbf{X}}_k = \bar{\bar{\bar{x}}}_k \mathbf{1}_{Ik}\mathbf{1}'_{Jk}$.

Note that the loss functions of DKM are not convex in variables $\mathbf{U}$ and $\mathbf{V}$, thus, it is unrealistic to expect that one of these algorithms can always find the global minima solution. In fact, DKM is NP-hard since already each K-means problem included in the DKM is NP-hard (Aloise et al. 2009), and therefore, the global minima solution cannot be guaranteed. Hence, to alleviate the problem of local minima solutions, it is advised to repeat the analysis starting from different random partitions P_{OV}.

Note that a DKM algorithm can be obtained by iteratively applying the steps of the *K-means for objects and K-means for variables*. It fact, the *K-means for objects* can be formalized in one of the following equivalent ways

$$\|\mathbf{X} - \mathbf{U}\bar{\mathbf{X}}_O\|^2 \to \min_{\mathbf{U},\bar{\mathbf{X}}_O} \text{ or } \|\mathbf{X} - \mathbf{H}_{\mathbf{U}}\mathbf{X}\|^2 \to \min_{\mathbf{U}} \text{ or } \|\mathbf{H}_{\mathbf{U}}\mathbf{X}\|^2 \to \max_{\mathbf{U}} \quad (23.37)$$

$$\|\mathbf{X} - \ddot{\mathbf{U}}\ddot{\mathbf{X}}_O\|^2 \to \min_{\ddot{\mathbf{U}},\ddot{\mathbf{X}}_O} \text{ or } \|\mathbf{X} - \ddot{\mathbf{U}}\ddot{\mathbf{U}}'\mathbf{X}\|^2 \to \min_{\ddot{\mathbf{U}}} \text{ or } \|\ddot{\mathbf{U}}\ddot{\mathbf{U}}'\mathbf{X}\|^2 \to \max_{\ddot{\mathbf{U}}} \quad (23.38)$$

where $\bar{\mathbf{X}}_O = (\mathbf{U}'\mathbf{U})^{-1}\mathbf{U}'\mathbf{X}$ is the centroid matrix for the partition P_O and $\ddot{\mathbf{X}}_O = \ddot{\mathbf{U}}'\mathbf{X}$. *K*-means for variables can be equivalently formulated with obvious notation as

$$\|\mathbf{X} - \bar{\mathbf{X}}_V\mathbf{V}'\|^2 \to \min_{\bar{\mathbf{X}}_V,\mathbf{V}} \text{ or } \|\mathbf{X} - \mathbf{X}\mathbf{H}_{\mathbf{V}}\|^2 \to \min_{\mathbf{V}} \text{ or } \|\mathbf{X}\mathbf{H}_{\mathbf{V}}\|^2 \to \max_{\mathbf{V}} \quad (23.39)$$

$$\|\mathbf{X} - \ddot{\mathbf{X}}_V\ddot{\mathbf{V}}\|^2 \to \min_{\ddot{\mathbf{X}}_V,\ddot{\mathbf{V}}} \text{ or } \|\mathbf{X} - \mathbf{X}\ddot{\mathbf{V}}\ddot{\mathbf{V}}'\|^2 \to \min_{\ddot{\mathbf{V}}} \text{ or } \|\mathbf{X}\ddot{\mathbf{V}}\ddot{\mathbf{V}}'\|^2 \to \max_{\ddot{\mathbf{V}}} \quad (23.40)$$

Let us consider the Normalized DKM, thus we have to maximize $\|\ddot{\mathbf{U}}\ddot{\mathbf{U}}'\mathbf{X}\ddot{\mathbf{V}}\ddot{\mathbf{V}}'\|^2$, see (23.36), subject to (23.33) and (23.35). When $\ddot{\mathbf{U}}$ has to be updated, $\ddot{\mathbf{V}}$ is fixed, thus the objective function to maximize is that of K-means for objects applied on the weighted matrix $\mathbf{X}\ddot{\mathbf{V}}\ddot{\mathbf{V}}'$. Analogously, the update of $\ddot{\mathbf{V}}$ given $\ddot{\mathbf{U}}$ can be done by applying a K-means for variables on the weighted matrix $\ddot{\mathbf{U}}\ddot{\mathbf{U}}'\mathbf{X}$. The user can decide if for each update it is better to run a complete K-means or only some iterations. A similar procedure, named Double-conjugated clustering (Busygin et al. 2002), was developed, applying K-means or Self-Organizing Maps (Kohonen 1995). Busygin et al. (2005, 2008) define the solution of this algorithm, consistent, implying separability of the clusters by convex cones. From the conic separability it follows that the convex hulls of the clusters are separated. Thus, Trapp et al. (2010)

propose a new mathematical programming formulation of the DKM model by involving a solution of a fractional 0–1 programming problem. However, the program is not suitable for large data sets so a heuristic algorithm based on multistart procedure and considering the solution obtained assigning each object and each variable to its majority cluster is proposed. We now show an important property that allows to understand why to use the DKM model instead of the simpler independent application of K-means on the rows and K-means for the columns of $\mathbf{X}$. Let us first recall that the independent application on the same matrix of K-means for objects and K-means for variables defines the *Independent Two-Mode Partition* (ITMP) of rows and columns of $\mathbf{X}$ with total loss given by the sum or better the mean of the K-means losses for objects and variables, see (23.37) and (23.39),

$$J_{ITMP}(\mathbf{U}, \mathbf{V}) = \frac{\|\mathbf{X} - \mathbf{H_U X}\|^2 + \|\mathbf{X} - \mathbf{X H_V}\|^2}{2} \tag{23.41}$$

It can be observed that ITMP produces two weighted centroid matrices $\mathbf{H_U X}$ and $\mathbf{X H_V}$ that independently reconstruct the data matrix $\mathbf{X}$, while DKM produces a unique weighted centroid matrix $\mathbf{H_U X H_V}$ to model the observed data. The following equations show links between the solution of DKM and that of each K-means for objects and variables,

$$J_{DKM}(\mathbf{U}, \mathbf{V}) = \|\mathbf{X} - \mathbf{H_U X}\|^2 + \|\mathbf{H_U X} - \mathbf{H_U X H_V}\|^2 \tag{23.42}$$

$$= \|\mathbf{X} - \mathbf{X H_V}\|^2 + \|\mathbf{X H_V} - \mathbf{H_U X H_V}\|^2 \tag{23.43}$$

Thus, the sum of squares of DKM is always larger than the sum of squares of the solutions of the K-means. By summing (23.42) and (23.43), we have

$$J_{DKM}(\mathbf{U}, \mathbf{V}) = J_{ITMP}(\mathbf{U}, \mathbf{V}) + \frac{1}{2}\|\mathbf{H_U X} - \mathbf{H_U X H_V}\|^2 + \frac{1}{2}\|\mathbf{X H_V} - \mathbf{H_U X H_V}\|^2 \tag{23.44}$$

Thus, it can be observed that the solution of DKM is always bounded below by the two solutions of the K-means. It is also interesting to note that

$$\|\mathbf{H_U X} - \mathbf{H_U X H_V}\|^2 + \|\mathbf{X H_V} - \mathbf{H_U X H_V}\|^2 = \|\mathbf{H_U X} - \mathbf{X H_V}\|^2 \tag{23.45}$$

Indeed, DKM has an objective function corresponding to a penalized version of ITMP, where solutions giving very different object and variable centroids are penalized (Rocci and Vichi 2008)

$$J_{DKM}(\mathbf{U}, \mathbf{V}) = J_{ITMP}(\mathbf{U}, \mathbf{V}) + \frac{1}{2}\|\mathbf{H_U X} - \mathbf{X H_V}\|^2 \tag{23.46}$$

Therefore, the DKM solution is a *consensus* between the two solutions identified by the two independent K-means.

The expression (23.45) shows also that centroids of objects should be equal to centroids of variables in order to have the solution of ITMP equal the solution of DKM.

Lee et al. (2010) propose a sparse SVD method for biclustering when the observed data matrix is expected to have the ideal form (23.2). The singular vectors of the SVD are interpreted as regression coefficients of a linear regression subject to penalties to obtain sparse pairs of singular vectors with many elements equal to zero. The algorithm extracts one 2-*mrc* and the corresponding $\mathbf{X}_k$ per time, by solving the following constrained minimization problem

$$\|\mathbf{E}_{k-1} - \bar{x}_k \mathbf{u}_k \mathbf{v}_k'\|^2 + \lambda_1 \sum_{i=1}^{I} w_{ik}|u_{ik}| + \lambda_2 \sum_{j=1}^{J} z_{jk}|v_{jk}| \qquad (23.47)$$

on the residual matrix $\mathbf{E}_{k-1}$ obtained by extracting the first $k-1$ components from $\mathbf{X}$. Penalties functions are of the adaptive lasso type (Zou 2006).

For ideal matrix (23.2), a second model is given by the Plaid Model (PM) (Lazzeroni and Owen 2002) defined according to the following least-squares loss

$$\sum_{i=1}^{I} \sum_{j=1}^{J} \left(x_{ij} - \theta_{ijk} u_{ik} v_{jk}\right)^2 \rightarrow \min_{\theta_{ijk}; u_{ip}, v_{jq} \in \{0,1\}} \qquad (23.48)$$

Note that for $\theta_{ijk} = \mu_k$ the PM is equal to the DKM where $P = Q = K$, that is, the number of clusters in DKM is limited to be K both for rows and columns. Thus, $\bar{\mathbf{X}} = \mathrm{diag}(\mu_1, \ldots, \mu_k)$. In fact, the model is hypothesized for similarity data of the form in Figure 23.1c, where ideal reordering of the data matrix would produce an image with some number K of rectangular blocks on the diagonal. Each block would be nearly uniformly colored, and the part of the image outside of these diagonal blocks would be of a neutral background color (background layer). Note that in the plaid model the constraints $\sum_k u_{ik} = 1, \sum_k v_{jk} = 1$, for all i, j, necessary to obtain a partitions of rows and columns, are removed. This is considered by authors an advantage to obtain a possible cover of the data matrix with possible overlapping two-mode clusters. An algorithm which alternates the estimation of the parameters of the model, cluster by cluster, is given. The PM algorithm has been criticized for producing large heterogeneous clusters (Segal et al. 2003) for this reason Turner et al. (2005a) developed an improved and computationally more efficient algorithm. First it fits the background layer, then searches for one 2-*mrc* at a time until a prespecified number is reached or no more significant 2-*mrc* are found, as determined by a permutation test for fitting the plaid model. Then the algorithm includes a pruning step to remove ill-fitting objects and variables which is repeated until a stable 2-*mrc* is obtained. Turner et al. (2005b) extended the PM to incorporate external grouping information or for two-mode clustering of profiles of repeated measures. Zhu et al. (2010) propose a constrained version of the plaid model to produce clusters of reduced size by penalizing the loss function (23.48) with the L_1 norm of the membership matrices that exploit the sparsity of the two-mode rectangular blocks present in the data.

23.5.2 Maximum Likelihood and Other Methods of Estimation of DKM

Martella and Vichi (2012) estimate the DKM model by considering the maximum likelihood approach. Recall that DKM model can be expressed in row form, that is, by using multivariate observations

$$\mathbf{x}_i = \mathbf{V}\bar{\mathbf{X}}'\mathbf{u}_i + \mathbf{e}_i = \mathbf{V}\bar{\mathbf{x}}_p + \mathbf{e}_i, \quad i = 1, \ldots, I \qquad (23.49)$$

where $\mathbf{x}_i$ is a $J \times 1$ vector representing the i-th multivariate object and $\mathbf{u}_i$ is the i-th object membership vector, while $\mathbf{e}_i$ is the J-dimensional random error specified by the i-th row of E. It is supposed that the population from which the data are observed be composed of P homogeneous subpopulations; and conditionally on class p, $(p = 1, \ldots, P)$, the density of x_i is $x_i | u_{ip} \sim \text{MVN}_J (V\bar{x}_p, \Sigma_p)$, that is, multivariate normal with mean vector $V\bar{x}_p$ and dispersion matrix Σ_p. The DKM model is estimated by using the maximum likelihood clustering (MLC) approach (Scott and Symons, 1971) implicitly supposing that the overlap of the multivariate normal distributions is negligible, because DKM model specifies a 2-*msp* where the two-mode clusters do not overlap,

$$\sum_{i=1}^{I} \sum_{p=1}^{P} u_{ip} \ln \text{MVN}_J (V\bar{x}_p, \Sigma_p) \rightarrow max \qquad (23.50)$$

subject to the usual constraints (23.33). The computation of the maximum likelihood (ML) estimates of the DKM model is based on a coordinate ascent algorithm which alternates the updating of, $\bar{x}_p$, Σ_p, V, and $U = [u_1, \ldots, u_p, \ldots, u_P]$ where

$$\bar{x}_p = (V'V)^{-1} V'X'u_p \left(u'_p u_p\right)^{-1} \qquad (23.51)$$

$$\Sigma_p = \left(u'_p u_p\right)^{-1} \sum_{i=1}^{I} (x_i - V\bar{x}_p)(x_i - V\bar{x}_p)' u_{ip} \qquad (23.52)$$

while the updating of U and V is done for each row by assigning the values 1 in the colums of the current row where the log-likelihood function, that is the loss in (23.50), is maximized. In the estimation of model (23.5), together with the implicit hypothesis of negligible overlap of the multivariate normal distributions, it is implicitly hypothesized, as for the DKM, (but this is true also for the K-means algorithm), that two-mode clusters, and therefore the associated multivariate normal distributions have equal proportions. The specified hypotheses simplify the estimation of DKM and allows us to specify a fast algorithm, however, these should be verified by assessing the overlap of the clusters of the final ML solution, by means of clusters mean vectors distances. If there is a relevant overlap of clusters, the mean vector and dispersion matrices estimators are biased.

The block mixture model (Govaert and Nadif 2003) gives a setting to correctly handle such situations. They propose the mixture model estimation of the two mode-single-partition model (23.5) that they call *block mixture model*. They assume that the population is divided into K blocks X_k associated with two-mode clusters, with a mixture random data generation process that can be described as follows:

1. Generate the membership of I objects by drawing u_i from a multinomial distribution with probabilities $p = (\pi_1, \ldots, \pi_P)'$. This induces a partition P_O of objects.
2. Generate the membership of J variables by drawing v_j from a multinomial distribution with probabilities $q = (\rho_1, \ldots, \rho_Q)'$. This induces a partition P_V of variables.
3. For the generated (u_i, v_j) a random observation is given from the univariate distribution $f(x_{ij}; \theta_{pq})$.

Let $(x_{11}, u_1, v_1), \ldots, (x_{IJ}, u_I, v_J)$ be a i.i.d. sample drawn, under the above mixture sampling scheme. The complete log-likelihood associated to the block mixture model is

$$L_C(\mathbf{U}, \mathbf{V}, \Theta) = \sum_{i=1}^{I} \sum_{j=1}^{J} \sum_{p=1}^{P} \sum_{q=1}^{Q} u_{ip} v_{jq} \ln \left(\pi_p \rho_q f(x_{ij}; \theta_{pq}) \right) \tag{23.53}$$

An extension of the CEM (Celeux and Govaert 1992) algorithm named block CEM (Govaert and Naoif 2003) has been proposed for maximizing (23.53). For binary data, we have

$$f(x_{ij}; \theta_{pq}) = \theta_{pq}^{x_{ij}} (1 - \theta_{pq})^{1-x_{ij}} \tag{23.54}$$

where θ_{pq} is the probability that $x_{ij} = 1$ given class $_OV C_{pq}$.

Govaert and Nadif (2005) propose to estimate model parameters by a Generalized EM algorithm (Dempster et al. 1977). The EM algorithm iteratively maximizes the conditional expectation of the complete log-likelihood given the sample observations and a previous current estimate $\hat{\Theta}$:

$$E_{\hat{\Theta}}(L_C(\mathbf{U}, \mathbf{V}, \Theta) \mid \mathbf{X}) = \sum_{i=1}^{I} \sum_{j=1}^{J} \sum_{p=1}^{P} \sum_{q=1}^{Q} E_{\hat{\Theta}}(u_{ip} v_{jq} \mid \mathbf{X}) \ln \left(\pi_p \rho_q f(x_{ij}; \theta_{pq}) \right) \tag{23.55}$$

However, there are difficulties due to the determination of $E_{\hat{\Theta}}(u_{ip} v_{jq} | \mathbf{X})$ and approximations are required to make the algorithm tractable.

Two versions of the block EM algorithm have been proposed. Govaert and Nadif (2008) compare these two versions together with the block CEM algorithm. The simulation study confirms the theoretical hypothesis that the block EM is more accurate than the block CEM especially in cases when blocks overlap, even if the block CEM is faster and can be applied for larger problems. A general limitation in the block mixture model is connected with the univariate specification of the distribution of each block $\mathbf{X}_k$. In fact, by using the block mixture model the multivariate relationships between variables in each block are lost and this produces also a less flexible possible formulation of the shape of the clusters. A Bayesian Two-mode-Clustering (Bayesian Co-Clustering, BCC) has been proposed by Meeds and Roweis (2007), and Shan and Banerjee (2008). The generative two-mode clustering model specifies Dirichlet priors for the conditional distribution of rows-clusters and column-clusters. These distributions are used to first generate the row- and column-clusters where the current entry should be located, then the entry is generated by the probability density distribution associated with the two-mode cluster. The authors propose a variational Bayesian algorithm to perform inference and estimate the BCC model. A lower bound of the likelihood function is learned and used to estimate model parameters. Wang et al. (2009) propose a collapsed Gibbs sampling and a collapsed variational Bayesian algorithm for it, in order to find more accurate likelihood functions leading to higher predictive performance than those given by the original algorithm of BCC. Meeds and Roweis (2007) propose nonparametric BCC (NBCC) assuming a Pitman-Yor Process (Pitman and Yor 1997) prior, which generalizes the Dirichlet Process. In Wang et al. (2011) two nonparametric Bayesian coclustering ensemble models, one based on Dirichlet Processes and the other based on Mondrian Processes, are proposed. The latter relaxes the usual coclustering assumption that row- and column-clusters are independent, providing a way to model context-specific independence of row- and column-clusters. This is a model-based approach to ensembles that explicitly models the way in which multiple coclusters differ from each other and from a consensus coclustering. Wang et al. (2012) show that Infinite Hidden Relational Model (IHRM) (Xu et al. 2006)—originally introduced to form a relational learning point of view—is essentially a two-mode clustering model that has the ability to exploit features associated with rows and columns, in order to predict entries for unseen rows and/or columns. It can be viewed as an extension to the NBCC of Meeds and Roweis (2007).

23.6 Two-Mode Multipartitioning

We have already noted that for each mode of the data matrix more than one partition can be specified to allow a more flexible taxonomic structure for the two-way data. This is named *two-mode multipartitioning* (Rocci and Vichi 2008). We may require that conditionally to a cluster $_OC_k$ of objects, the partition P_V of variables can change (Figure 23.1d). In this case a two-mode multipartition of variables is defined. The model has still the form of the DKM model (23.5), that is: $\mathbf{X} = \mathbf{U}\bar{\mathbf{X}}\mathbf{V}'$, where matrix $\mathbf{V}$ has a block form $\mathbf{V} = [\mathbf{V}_1, \ldots, \mathbf{V}_p, \ldots, \mathbf{V}_P]$, and $\mathbf{V}_p = [v_{jqp}]$ is the $J \times Q_p$ binary matrix defining the variable partition corresponding to $_OC_p$. The centroid matrix $\bar{\mathbf{X}}$ has the block form

$$\bar{\mathbf{X}} = \mathrm{diag}(\bar{\mathbf{x}}'_1, \ldots, \bar{\mathbf{x}}'_p, \ldots, \bar{\mathbf{x}}'_P) \tag{23.56}$$

where $\bar{\mathbf{x}}_p = [\bar{x}_{p1}, \ldots, \bar{x}_{pQ_p}]'$ is the $Q_p \times 1$ centroid vector of $_OC_p$ and Q_p is the number of clusters of the variable partition conditional to the p-th object cluster. Matrix $\bar{\mathbf{X}}$ has dimension $P \times Q$ where $Q = Q_1 + Q_2 + \cdots + Q_P$. The multipartitioning model can be also written as

$$\mathbf{X} = \sum_{p=1}^{P} \mathbf{u}_p \bar{\mathbf{x}}'_p \mathbf{V}'_p + \mathbf{E} \tag{23.57}$$

where $\mathbf{u}_p$ is the p-th column of $\mathbf{U}$. Of course, the multipartitioning model can be also formulated by reversing the role of objects and variables.

A fast coordinate descent algorithm to compute least squares estimates of the parameters of model (23.57) can be found in Rocci and Vichi (2008).

References

Aloise D., Deshpande A., Hansen P., and Popat P. 2009. NP-hardness of Euclidean sum-of-squares clustering. *Machine Learning*, 75, 245–249.

Arabie P., Hubert L.J., and Schleutermann S. 1990. Blockmodels from the bond energy approach. *Social Networks*, 12, 99–126.

Ball G.H. and Hall D.J. 1967. A clustering technique for summarizing multivariate data. *Behavioral Science*, 12, 153–155.

Bock H. 1979. Simultaneous clustering of objects and variables. In Tomassone, R. (Ed.), *Analyse des Donnes et Informatique*. INRIA, Le Chesnay, pp. 187–203.

Bock H.H. 2003. Two-way clustering for contingency tables maximizing a dependence measure. In Schader, M., Gaul, W., and Vichi, M. (Eds.), *Between Data Science and Applied Data Analysis*. Springer, Heidelberg, pp. 143–155.

Braverman E.M., Kiseleva N.E., Muchnik I.B., and Novikov S.G. 1974. Linguistic approach to the problem of processing large bodies of data. *Automation and Remote Control*, 35, (I 1, part 1), 1768–1788.

Bryan K., Cunningham P., and Bolshakova N. 2005. Biclustering of expression data using simulated annealing. In *Proceedings of the 18th IEEE Symposium on Computer-based Medical Systems*, 383–388.

Busygin S., Jacobsen G., and Krmer E. 2002. Double conjugated clustering applied to leukemia microarray data. In *Proceedings of SIAM Data Miningworkshop on Clusteringhigh Dimensional Data and Its Applications*.

Busygin S., Prokopyev O.A., and Pardalos P.M. 2005. Feature selection for consistent biclustering via fractional 01 programming. *Journal of Combinatorial Optimization*, 10(1), 7–21.

Busygin S., Prokopyevb O., and Pardalosa P.M. 2008. Biclustering in data mining. *Computers & Operation Research*, 35, 2964–2987.

Castillo W. and Trejos J. 2000. Recurrence properties in two-mode hierarchical clustering, in studies in classification data analysis and knowledge organization. In Decker, R. and Gaul, W. (Eds.), *Classification and Information Processing at the Turn of the Millennium*. Springer, Heidelberg, pp. 68–73.

Celeux G. and Govaert G. 1992. A classification EM algorithm for clustering and two stochastic versions. *Computational Statistics & Data Analysis*, 14, 315–332.

Cheng Y. and Church, G. 2000. Biclustering of expression data. In *Proceedings ISMB*, pp. 93–103. AAAI Press.

Cheng W., Zhang X., Pan F., and Wang W. 2012. Hierarchical C-Clustering based on entropy splitting. In *Proceedings of CIKM12*.

Cho H., Dhillon I.S., Guan Y., and Sra S. 2004. Minimum sum-squared residue co-clustering of gene expression data. In *Proceedings of the Fourth SIAM International Conference in Data Mining*.

Cormack R.M. 1971. A review of classification (with discussion). *Journal of Royal Statistical Society A*, 134, 321–367.

Dempster A.P., Laird N.M., and Rubin D.B. 1977. Maximum likelihood from incomplete data via the EM algorithm (with discussion). *Journal of the Royal Statistical Society*, 39(1), 1–38.

DeSarbo W.S. 1982. GENNCLUS: New models for general nonhierarchical clustering analysis. *Psychometrika*, 47, 449–475.

DeSarbo W.S. and De Soete G. 1984. On the use of hierarchical clustering for the analysis of nonsymmetric proximities, *Journal of Consumer Research*, 11, 601–610.

Dhillon I.S. 2001. Co-clustering documents and words using bipartite spectral graph partitioning. In *Proceedings of the Seventh ACM SIGKDD Conference*, San Francisco, CA, pp. 269–274.

Dhillon I.S., Mallela S., and Kumar R. 2003. A divisive information-theoretic feature clustering algorithm for text classification. *Journal of Machine Learning Research*, 3, 1265–1287.

Ding C., He X., and Simon H.D. 2005. On the equivalence of nonnegative matrix factorization and spectral clustering. In *Proceedings of SDM*.

Ding C., He X., Zha H., Gu M., and Simon H. 2001. A min-max cut algorithm for graph partitioning and data clustering. In *Proceedings of IEEE International Conference on Data Mining*.

Ding C., Li T., Peng W., and Park H. 2006. Orthogonal nonnegative matrix trifactorizations for clustering. In *Proceedings of International Conference in Knowledge Discovery and Data Mining (KDD'06)*, pp. 635–640.

Doreian P., Batagelj V., and Ferligoj A. 1994. Partitioning networks based on generalized concepts of equivalence, *Journal of Mathematical Sociology*, 19, 1–27.

Eckes T. 1993. A Two-mode clustering study of situations and their features. In Opitz, O., Lausen, B., and Klar, R. (Eds.), *Information and Classification*, Springer, Heidelberg, pp. 510–517.

Eckes T. and Orlik P. 1991. An agglomerative method for two-mode hierarchical clustering. In *Classification, Data Analysis, and Knowledge Organization*, Springer, Heidelberg, pp. 3–8.

Eckes T. and Orlik P. 1993. An error variance approach to two-mode hierarchical clustering. *Journal of Classification*, 10, 51–74.

Getz G., Levine E., and Domany E. 2000. Coupled two-way clustering analysis of gene microarray data. In *Proceedings of the National Academy of Sciences of the USA*, 97, 12079–12084.

Gordon A.D. 1999. *Classification*, 2nd edn. Chapman and Hall, London.

Govaert G. 1977. Algorithme de classification dun tableau de contingence. In *Premires journes Internationales Analyse des Donnes et Informatique*, INRIA, Le Chesnay, pp. 487–500.

Govaert G. 1983. *Classification croise*. Thse dEtat, Universit Paris 6.

Govaert G. 1995. Simultaneous clustering of rows and columns. *Control Cybernet*, 24, 437–458.

Govaert G. and Nadif M. 2003. Clustering with block mixture models. *Pattern Recognition*, 36(2), 463–473.

Govaert G. and Nadif M. 2005. An EM algorithm for the block mixture model. *IEEE Transactions on Pattern Analysis and Machine Intelligence*, 27(4), 643–647.

Govaert G. and Nadif M. 2008. Block clustering with Bernoulli mixture models: Comparison of different approaches. *Computational Statistics and Data Analysis*, 52, 3233–3245.

Govaert G. and Nadif M. 2010. Latent block model for contingency table. *Communications in Statistics - Theory and Methods*, 39(3), 416–425.

Gower J.C. and Legendre P. 1986. Metric and euclidean properties of dissimilarity coefficients. *Journal of Classification*, 3, 5–48.

Gu M., Zha H., Ding C., He X., and Simon H. 2001. Spectral relaxation models and structure analysis for k-way graph clustering and bi-clustering. Penn State UnivTech Report CSE-01–007.

Hagen L. and Kahng A.B. 1992. New spectral methods for ratio cut partitioning and clustering. In *IEEE Transactions on CAD*, 11, 1074–1085.

Hall K.M. 1970. An r-dimensional quadratic placement algorithm. *Management Science*, 11(3), 219–229.

Hanish D., Zien A., Zimmer R., and Lengauer T. 2002. Co-clustering of biological networks and gene expression data. *Bioinformatics*, 18(1), 145–154.

Hartigan J.A. 1972. Direct clustering of a data matrix. *Journal of the American Statistical Association*, 67(337), 123–129.

Hartigan J.A. 1975. *Clustering Algorithms*. Wiley, New York, NY.

Hartigan J.A. 1976. Modal blocks in dentition of west coast mammals. *Systematic Zoology*, 25, 149–160.

Kohonen T. 1995. *Self-organization Maps*. Springer, Berlin-Heidelberg.

Lazzeroni L. and Owen A. 2002. Plaid models for gene expression data. *Statistica Sinica*, 12, 61–86.

Lee M., Shen H., Huang J.Z., and Marron J.S. 2010. Biclustering via sparse singular value decomposition. *Biometrics*, 66, 1087–1095.

Long B., Zhang Z., and Yu P.S. 2005. Co-clustering by block value decomposition. In *Proceedings of SIGKDD05*, 2005.

MacQueen J. 1967. Some methods for classification and analysis of multivariate observations. In *Proceedings of the 5th Berkeley Symposium on Mathematical Statistics and Probability*, 1, 281–297.

Madeira S.C. and Oliveira A.L. 2004. Biclustering algorithms for biological data analysis: A survey. *IEEE Transaction on Computational Biology and Bioinformatics*, 1(1), 24–45.

Martella F. and Vichi M. 2012. Clustering microarray data using model-based double K-means. *Journal of Applied Statistics*, 39, 9.

Meeds E. and Roweis S. 2007. Nonparametric bayesian biclustering. Technical Report UTML TR 2007-001. Department of Computer Science, University of Toronto.

Mirkin B., Arabie P., and Hubert L.J. 1995. Additive two-mode clustering: The error-variance approach revisited. *Journal of Classification*, 12, 243–263.

Packer C.V. 1989. Applying row-column permutation to matrix representations of large citation networks. *Information Processing & Management*, 25, 307–314.

Pan F., Zhang X., and Wang W. 2008. CRD: Fast co-clustering on large datasets utilizing sampling-based matrix decomposition. In *Proceedings of ACM SIGMOD/PODS Conference*, Vancouver.

Pensa R.G. and Boulicau J.F. 2008. numerical data co-clustering via sum-squared residue minimization and user-defined constraint satisfaction. In *16th Italian Symposium on Advanced Database System*, pp. 279–286.

Pitman J. and Yor M. 1997. The two-parameter Poisson-Dirichlet distribution derived from a stable subordinator. *Annals of Probability*, 25(2), 855–900.

Rocci R. and Vichi M. 2008. Two mode multi-partitioning. *Computational Statistics & Data Analysis*, 52, 1984–2003.

Scott A.J. and Symons M.J. 1971. Clustering methods based on likelihood ratio criteria. *Biometrics*, 27, 387–397.

Segal E., Battle, A., and Koller, D. 2003. Decomposing gene expression into cellular processes. *Pacific Symposium on Biocomputing*, 8, 89–100.

Shan H. and Banerjee A. 2008. Bayesian co-clustering. *IEEE International Conference on Data Mining*.

Shepard R.N. and Arabie P. 1979. Additive clustering: Representation of similarities as combinations of discrete overlapping properties. *Psychological Review*, 86, 86–123.

Shi J. and Malik J. 2000. Normalized cuts and image segmentation. In *IEEE Transactions on Pattern Analysis and Machine Intelligence*, 22(8), 888–905.

Tang C., Zhang L., Zhang A., and Ramanathan M. 2001. Interrelated two-way clustering: An unsupervised approach for gene expression data analysis. In *2nd IEEE International Symposium on Bioinformatics and Bioengineering*.

Trapp A., Prokopyev O.P., and Busygin S. 2010. Finding checkerboard patterns via fractional 01 programming. *Journal Combinatorial Optimization*, 20, 1–26.

Turner H.L., Bailey T.C., and Krzanowski W.J. 2005a. Improved biclustering of microarray data demonstrated through systematic performance tests. *Computational Statistics & Data Analysis*, 48(2), 235–254.

Turner H.L., Bailey T.C., Krzanowski W.J., and Hemingway C.A. 2005b. Biclustering models for structured microarray data. In *IEEE/ACM Transactions on Computational Biology and Bioinformatics 2*, 316–329.

Van Mechelen I., Bock H.H., and De Boeck P. 2004. Two-mode clustering methods: A structured overview. *Statistical, Methods in Medical Research*, 13(5), 363–394.

Vichi M. 2001. Double K-means clustering for simultaneous classification of objects and variables. In Borra S. et al. (Eds.), *Advances in Classification and Data Analysis*, Springer, Berlin. In Proceeedings of CALDAG1999, Rome, pp. 43–52.

Vitaladevuni S.N. and Basri R. 2010. Co-clustering of image segments using convex optimization applied to EM neuronal reconstruction. In *Proceedings of IEEE Conference on Computer Vision and Pattern Recognition (CVPR)*, 2010, pp. 2203–2210.

Wang P., Domeniconi C., and Laskey K. 2009. Latent Dirichlet Bayesian co-clustering. In *Proceedings of the European Conference on Machine Learning*, Vol. 5782, pp. 522–537. Springer, Berlin Heidelberg.

Wang P., Domeniconi C., Rangwala H., and Laskey K.B. 2012. Feature enriched nonparametric bayesian co-clustering. In *Proceedings of the 16th Pacific-Asia Conference on Knowledge Discovery and Data Mining*, Kuala Lumpur.

Wang P., Laskey K.B., Domeniconi C., and Jordan M.I. 2011. Nonparametric Bayesian co-clustering ensembles. In *Proceedings of the SIAM International Conference on Data Mining*, Mesa, Arizona, April 28–30.

Xu Z., Tresp V., Yu K., and Kriegel H. 2006. Infinite hidden relational models. In *Proceedings of the 22nd International Conference on Uncertainity in Artificial Intelligence*, Cambridge.

Yang J., Wang H., Wang W., and Yu P. 2003. Enhanced biclustering on expression data. In *Proceedings of 3rd IEEE Symposium on BioInformatics and BioEngineering, (BIBE'03)*, 321–327.

Zhang J. 2007. Co-clustering by similarity refinement. In *Proceedings of the Sixth International Conference on Machine Learning and Applications, ICMLA '07*, pp. 381–386.

Zhu H., Mateos G., Giannakis G.B., Sidiropoulos N.D., and Arindam B. 2010. Sparsity-cognization overlapping co-clustering for behaviour inference in social networks. *Acoustics Speech and Signal Processing (ICASSP), 2010 IEEE International Conference on*, pp. 3534–3537.

Zou H. 2006. The adaptive lasso and its oracle properties. *JASA*, 101, 1418–1429.

24

Fuzzy Clustering

Pierpaolo D'Urso

CONTENTS

Abstract

In this chapter, we show an organic and systematic overview of fuzzy clustering techniques. In particular, we analyze the mathematical and computational aspects of Fuzzy c-Means (FcM). We also discuss several FcM-variants—that is, prototype-variants, distance-variants, objective function-variants, data features-variants, and other types of variants.

24.1 Introduction

This chapter presents an organic and systematic overview of cluster analysis based on the Fuzzy Theory (Zadeh, 1965), that is, the fuzzy clustering. Fuzzy clustering is an overlapping approach which allows cases to belong to more than one cluster simultaneously as opposed to standard (classic) clustering which results in mutually exclusive clusters (Bezdek, 1981).

In particular, following a fuzzy approach and adopting a terminology used in machine and statistical learning, in this chapter we describe nonhierarchical clustering methods that are:

- *Prototype-based*: The clusters are represented by the so-called prototypes. Prototypes are used for capturing the structure (distribution) of the data in each cluster. Each prototype is a set of parameters that consists of a cluster center (location parameter) and maybe some additional parameters about size and shape of the cluster. The cluster center is an instantiation of the attributes used to describe the domain, just as the data points in the dataset to divide. The size and shape parameters of a prototype determine the extension of the cluster in different directions of the underlying domain. The prototypes are obtained by means of the clustering procedures and serve as prototypical representations of the data points in each cluster (Kruse et al., 2007).

- *Objective function-based*: The clustering procedures are based on objective functions which are mathematical criteria that quantify the goodness of clustering methods that comprise prototypes and data partition. Objective functions serve as cost functions that have to be minimized to obtain optimal cluster solutions. Thus, for each of the following cluster models the respective objective function expresses desired properties of what should be regarded as "best" results of the cluster algorithm. Having defined such a criterion of optimality, the clustering task can be formulated as a function optimization problem. That is, the procedures determine the best decomposition of a dataset into a predefined number of clusters by minimizing their objective function. The idea of defining an objective function and minimize it in the clustering process is quite universal. Aside from the basic algorithms, many extensions and modifications have been proposed to improve the clustering results with respect to particular situations, i.e., observational situations, data nature, and so on (Kruse et al., 2007). Usually, the cost functions that have to be minimized are expressed as intracluster heterogeneity measures (within cluster dispersion measure) and represented analytically by means of suitable distance measures. As we can see below, in order to analyse different situations, regularization or penalty terms are often associated to within cluster dispersion terms in the objective function.

- *Distance-based*: A basic principle of the objective function-based clustering methods consists in searching for an appropriate partition of a set of objects, where the clusters satisfy according to suitable distance measures the condition of minimal internal heterogeneity along with maximal heterogeneity among clusters. The distance measure is most often computed on the basis of a feature vector characterizing each object and by comparing the data points with a set of suitable prototypes. The complexity of the object depends on the complexity of this feature vector (number and nature of the involved variables, specific structure of the data). Of course, the distance measure is affected by this complexity, in that it should be "sensitive" with respect to the specific types of differential patterns that can be observed with respect to objects (Coppi and D'Urso, 2006).

- *Membership degree-based*: An important aspect is the rigidity (crispness) of the partition obtained by means of the classical (nonfuzzy, crisp) clustering procedures (e.g., the *c*-means clustering*). Relaxing this rigidity has constituted, in the past, a domain of research in the framework of cluster analysis. Many authors have proposed a fuzzy setting as the appropriate approach to cope with this problem (e.g., Bezdek, 1981). The notion of degree of membership (in a cluster) has replaced the crisp assignment {0, 1} of the classical techniques, thus reflecting the unavoidable "imprecision" in defining appropriate groups, especially when trying to classify complex objects, that is, data with complex structure (the "incompatibility principle" of Zadeh (1973), stressing the conflict between precision and relevance, when studying complex systems, provides a general theoretical support to this statement). In spite of the enlarged horizon established by the fuzzy clustering approach, the above mentioned basic principle of maximizing homogeneity within groups has remained, for a long time, the main inspiration of the various fuzzy-clustering procedures. In this regards, in order to take into account the uncertainty—theoretically formalized in a fuzzy point of view—connected to the assignment process of data points to clusters, a membership degree system (with suitable properties) can be considered in the objective function for the clustering process. Thus, the membership degree of each data point to clusters represents an uncertainty measure in the assignment process of data points to clusters. As we can see below, conversely to crisp clustering in which the membership degrees can assume values 0 or 1, in the fuzzy clustering the membership degrees assume values between 0 and 1.

In classical cluster analysis (e.g., *c*-means) each datum is exactly assigned to only one cluster obtaining exhaustive partitions characterized by nonempty and pairwise disjoint subsets. Such crisp assignment of data to clusters can be inadequate in presence of data points that are almost equally distant from two or more clusters. Such special data points can represent hybrid-type or mixture objects, which are (more or less) equally similar to two or more types. A crisp partition arbitrarily forces the full assignment of such data points to one of the clusters, although they should (almost) equally belong to all of them. Fuzzy clustering relaxes the requirement that data points have to be assigned to one (and only one) cluster. Data points can belong to more than one cluster and even with different degrees of membership to the different clusters. This gradual cluster assignment can reflect cluster structure in a more natural way, especially when clusters overlap. Then, the memberships of data points at the overlapping boundaries can express the ambiguity of the cluster assignment (Kruse et al., 2007). In order to stigmatize the usefulness, expressiveness, and the virtue of the fuzzy approach to cluster analysis, we recall the classical example shown by Ruspini (1969)—the first researcher who suggested the use of fuzzy sets in cluster analysis—with the famous Butterfly dataset (see Figure 24.1).

In this dataset there are two clusters of object-points. Object-points are associated with their label vectors, which state the membership degrees to the left and to the right cluster. The "cores" of the left and the right cluster are the outer left and the outer right column of examples, respectively. Ruspini (1969) noted the following advantages of the fuzzy-clustering approach over a standard (nonfuzzy or crisp) clustering approach: object-points in the center of a cluster can have a membership degree equal to 1, while boundary points between the core and some other class can be identified as such (i.e., their membership

---

* In the chapter of the Handbook regarding traditional unsupervised clustering methods, this method is denoted as *k*-means. However, we prefer to adopt *c*-means in order to emphasize that it is a particular case of the fuzzy *c*-means, which is the standard notation in a fuzzy framework.

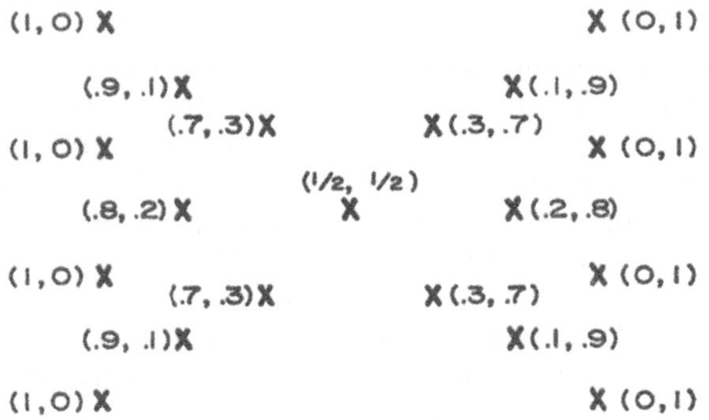

(1, 0) **X**   **X** (0, 1)

    (.9, .1)**X**   **X**(.1, .9)
          (.7, .3)**X**   **X**(.3, .7)
(1, 0) **X**   **X** (0, 1)
                              (1/2, 1/2)
    (.8, .2) **X**   **X**   **X**(.2, .8)

(1, 0) **X**   **X** (0, 1)
          (.7, .3)**X**   **X**(.3, .7)
    (.9, .1)**X**   **X**(.1, .9)

(1, 0) **X**   **X** (0, 1)

FIGURE 24.1
Original graphic of the classic Butterfly dataset. (Adapted from Ruspini, E. H. 1969. *Information and Control*, 15, 22–32.)

degree to the cluster they are closer to is not equal to 1). "Bridges" or stray points may be classified as undetermined with a degree of indeterminacy proportional to their similarity to core points. As remarked by Döring et al. (2006), the features of the dataset shown in Figure 24.1 help to depict the higher expressiveness of fuzzy clustering. The equidistant data points in the middle of the figure would have to be arbitrarily assigned with full weight to one of the clusters if classical (crisp) clustering were allowed only. In this fuzzy partition, however, it can be associated with the equimembership vector (0.5, 0.5). Crisp data partitions might not express the difference between data points in the center and those that are rather at the boundary of a cluster. Both kind of points would be fully assigned to the cluster they are most similar to. Readers interested in a deeper and more detailed treatment of fuzzy clustering may refer, for example, to monographs by Bezdek (1981), Jain and Dubes (1988), Pedrycz (2005), de Oliveira and Pedrycz (2007), and Miyamoto et al. (2008a).

24.2 The Fuzzy c-Means (FcM) Clustering

Cluster analysis represents the first statistical technique which lent itself to a fuzzy treatment. The pivotal justification lies in the recognition of the vague nature of the cluster assignment task. For this reason, in the literature, many fuzzy clustering methods have been proposed and applied in several fields. Pioneering fuzzy clustering algorithms have been suggested by Bellman et al. (1966) and Ruspini (1969, 1970, 1973). With regard to Ruspini's approach, however, "the original algorithm due […] is said to be rather difficult to implement. Its computational efficiency should be weak and its generalization to more than two clusters should be of little success. But it was the pioneer for a successful development of this approach" (Bandemer, 2006). As remarked by Yang (1993b), Ruspini's method opened the door for further research, especially since he first put the idea of fuzzy c-partitions in cluster analysis. The Fuzzy c-Means (FcM) clustering method introduced, independently, by Dunn (1974) and Bezdek (1974) and then extended and analytically formalized and investigated in depth by Bezdek (1981), is the first method which is computationally efficient and powerful. For this reason, the Bezdek method represents the best-known and used clustering technique in this body of literature. It has inspired the work of many scholars. In fact, several fuzzy approach-based clustering methods have been developed by extending suitably the original Bezdek method proposed in the 1981.

24.2.1 Mathematical and Computational Aspects

Let $X = \{x_{is} : i = 1, \ldots, n; s = 1, \ldots, p\} = \{x_i = (x_{i1}, \ldots, x_{is}, \ldots, x_{ip})' : i = 1, \ldots, n\}$ be a data matrix, where x_{is} represents the s-th quantitative variable observed on the i-th object and x_i represents the vector of the i-th observation. The FcM clustering method proposed by Bezdek (1981) is formalized in the following way:

$$\min : \sum_{i=1}^{n} \sum_{k=1}^{c} u_{ik}^m d_{ik}^2 = \sum_{i=1}^{n} \sum_{k=1}^{c} u_{ik}^m \|x_i - h_k\|^2 \quad \text{s.t.} \sum_{k=1}^{c} u_{ik} = 1, u_{ik} \geq 0 \quad (24.1)$$

where u_{ik} denotes the membership degree of the i-th object to the k-th cluster; $d_{ik}^2 = \|x_i - h_k\|^2$ is the squared Euclidean distance between the i-th object and the centroid of the k-th cluster; $h_k = (h_{k1}, \ldots, h_{ks}, \ldots, h_{kp})'$ represents the k-th centroid, where h_{ks} indicates the s-th component (s-th variable) of the k-th centroid vector; $m > 1$ is a parameter that controls the fuzziness of the partition (see in the following Section 24.2.2). By putting $m = 1$ in (24.1), we obtain the standard c-means (cM) clustering method (MacQueen, 1967). Solving the constrained optimization problem (24.1) with the Lagrangian multipliers method, the conditional optimal iterative solutions are (see Bezdek, 1981):

$$u_{ik} = \frac{1}{\sum_{k'=1}^{c} [\|x_i - h_k\| / \|x_i - h_{k'}\|]^{\frac{2}{m-1}}} = \frac{\|x_i - h_k\|^{-\frac{2}{m-1}}}{\sum_{k'=1}^{c} \|x_i - h_{k'}\|^{-\frac{2}{m-1}}}, \quad h_k = \frac{\sum_{i=1}^{n} u_{ik}^m x_i}{\sum_{i=1}^{n} u_{ik}^m}$$

$$(24.2)$$

Each centroid summarizes the features of the respective cluster; in particular, each centroid represents synthetically its cluster in the sense that it represents an appropriate weighted average of a set of features observed on the objects. Therefore, each centroid can be suitably utilized for interpreting each cluster. Notice that if the internal cohesion of the clusters increases, the interpretative/explicative power of the centroids increases and then the clustering uncertainty—that is the uncertainty (fuzziness) connected to the clustering process and measured by means of the membership degrees of each object to the clusters decreases.

For computing the iterative solutions (24.2), the following algorithm can be considered (Bezdek, 1981).

> *Step 0* Initially, we preprocess the data. Successively, we fix m and an initial membership degrees matrix $u_{ik}^{(\alpha)}$ ($i = 1, \ldots, n; k = 1, \ldots, c$) with $\alpha = 0$.
>
> *Step 1* We update the centroids by means of $h_k^{(\alpha+1)} = (\sum_{i=1}^{n} u_{ik}^{m\,(\alpha)} x_i) / (\sum_{i=1}^{n} u_{ik}^{m\,(\alpha)})$.
>
> *Step 2* We update the membership degrees by means of $u_{ik}^{(\alpha+1)} = \left[\sum_{k'=1}^{c} [\|x_i - h_k^{(\alpha+1)}\| / \|x_i - h_{k'}^{(\alpha+1)}\|]^{\frac{2}{m-1}} \right]^{-1}$.
>
> *Step 3* We compare $u_{ik}^{(\alpha)}$ with $u_{ik}^{(\alpha+1)}$ using a convenient matrix norm: if $|u_{ik}^{(\alpha+1)} - u_{ik}^{(\alpha)}| < \tau$ ($i = 1, \ldots, n; k = 1, \ldots, c$) (where τ is a small positive number set by the researcher) stop; otherwise, set $\alpha = \alpha + 1$ and return to step 1.

Notice that the iterative algorithm does not guarantee the attainment of the global minimum and usually converges to a local optimum (Hathaway and Bezdek, 1988; Gan and Wu, 2008). To limit the risk of hitting local optima and in order to check the stability of the solution, more than one random start is recommended. Following Bezdek (1981), the membership matrix $\mathbf{U} \equiv \{u_{ik} : i = 1, \ldots, n; k = 1, \ldots, c\}$ can be initialized randomly or by defining the initial membership matrix (at step 0) $\mathbf{U}^{(0)}$ as follows: $\mathbf{U}^{(0)} = (1 - \sqrt{2}/2)\mathbf{U}_u + \sqrt{2}/2\mathbf{U}_r$, where $\mathbf{U}_u = [1/c]$ and $\mathbf{U}_r$ is a random (nonfuzzy) partition. Alternatively, other initialization procedures suggested in literature can be utilized. For a critical evaluation of several initialization techniques see Steinley and Brusco (2007). As remarked in Coppi et al. (2010), it is recognized that FcM clustering algorithms present a minor tendency of hitting local optima with respect to their traditional counterparts (e.g., Bezdek et al., 2005). Empirical studies have shown that the fuzzy clustering algorithm is an effective starting point for traditional clustering (Heiser and Groenen, 1997). It has been proved that the FcM algorithm produces a nonincreasing sequence of values of the objective function.

The convergence is speedy. In fact, it is obtained in the first iterations (Cannon et al., 1986; Bezdek et al., 1987). However, if necessary, in order to improve the convergence speed, it is possible to accelerate the computational procedures by adopting suitable techniques (see, for instance, Borgelt, 2009).

In the literature, several scholars have highlighted different reasons for adopting a fuzzy clustering approach. As remarked by Hwang et al. (2007), fuzzy clustering approach offers major advantages over classic clustering approach. First, the fuzzy clustering methods are computationally more efficient because dramatic changes in the value of cluster membership are less likely to occur in estimation procedures (McBratney and Moore, 1985). Second, fuzzy clustering has been shown to be less affected by local optima problems (Heiser and Groenen, 1997). Finally, the memberships for any given set of respondents indicate whether there is a second-best cluster almost as good as the best cluster—a result which traditional clustering methods cannot uncover (Everitt et al., 2001).

24.2.2 Fuzziness Parameter

The fuzziness parameter m plays an important role in the FcM clustering. The value of m should be suitable chosen in advance. Different heuristic strategies are recommended in the literature. Although $1 < m < \infty$, values too close to 1 will result in a near partition with all memberships close to 0 or 1. Excessively large values will lead to disproportionate overlap with all memberships close to $1/c$ (Wedel and Steenkamp, 1991). Consequently, neither of these types of m is recommended (Arabie et al., 1981). Although there have been some empirical heuristic procedures to determine the value of m (e.g., McBratney and Moore, 1985; Wedel and Steenkamp, 1989; Okeke and Karnieli, 2006), there seems to exist no theoretically justifiable manner of selecting m. In practice, $m = 2$ is the most popular choice in fuzzy clustering (Bezdek, 1981; Hruschka, 1986; Wedel and Steenkamp, 1991; Hwang et al., 2007; Wang et al., 2007; Yang et al., 2008). In particular, Wang et al. (2007) remarked that "There is no theoretical basis for the optimal selection of m and a value of $m = 2$ is often chosen (Bezdek, 1981). We carried out a brief investigation to examine the effect of varying m in this domain. It was found that as m increased around a value of 2.0, the clustering centers moved slightly, but the cluster assignments were not changed. However, as $m \to \infty$, both the fuzzy objective function [...] and the Xie–Beni validity index [...] (see 24.2.3) continuously decreased ($\to 0$), and the cluster assignments became unstable and further from the results of clinical analysis. Consequently, we fixed the value of *m* as 2 for all further experiments." Bezdek (1976) showed a physical interpretation of the FCM algorithm when $m = 2$. Based on their analysis on the performance of cluster valid- ity indices, Pal and Bezdek (1995) have given heuristic guidelines regarding the best choice for *m*, suggesting that the value of the level of fuzziness should be between 1.5 and 2.5. Similar recommendations appear in Hall et al. (1992) and Cannon et al. (1986). Consider- ing the convergence of the algorithm, Bezdek et al. (1987) indicated that $m \geq n/(n-2)$. Chan and Cheung (1992) suggested that the value of *m* should be between 1.25 and 1.75. In a medical image context, Fadili et al. (2001) carried out an experimental study based on simulations to propose an acceptable range for the fuzziness parameter *m*. It is shown that this range includes the most accepted value $m = 2$ when using the FcM clustering. In par- ticular, using the ROC methods Sorenson and Wang (1996) and Fadili et al. (2001) showed that the range $1.5 < m < 2.5$ appears to be a good compromise for optimizing the perfor- mance of the FcM clustering with respect to the exponent weight. A maximum is reached for the examined simulation scenarios around $m \approx 2$. In this way, the experimental study of Fadili et al. (2001) gives, in the fMRI context, a justification for the popular use of the value 2 for fuzziness index *m*. Gao (2004) proposed two methods to find the proper value of *m*. The first one is based on fuzzy decision theory, but it needs to define two membership functions which lack of theoretical basis. The second one is based on the concavo-convex property of clustering functions, whose physical interpretation is of ambiguity. Yu et al. (2004) developed a theoretical approach for selecting *m*. Based on this approach, the rela- tionship between the stability of the fixed points of the FcM clustering and the dataset itself is revealed. This relation provides the theoretical basis for selecting the weighting exponent *m* in the FcM clustering.

Ozkan and Turksen (2007) indicated upper and lower values for the level of fuzziness in FcM clustering. Based on their analysis, they suggested that the upper boundary value of the level of fuzziness should be approximately 2.6 and the lower boundary value approx- imately 1.4 for FcM clustering in system development practices. For these reasons, the authors recommended that an analyst should not be concerned about the changes of the membership values outside of these boundaries. In addition, the authors believe these two effective boundary values of *m* encapsulate the uncertainty associated with the level of fuzziness parameter (Ozkan and Turksen, 2007).

Huang et al. (2012) proposed a theoretical approach to determine the range of the value of *m*. This approach utilizes the behavior of membership function on two data points, based on which the partial relationship between the fuzzifier *m* and the dataset structure is revealed.

On the basis of a robust analysis of FcM clustering, a new guideline for selecting the parameter *m* is proposed by Wu (2012). Wu showed that a large *m* value will make FcM clustering more robust to noise and outliers. However, considerably large *m* values that are greater than the upper bound will make the sample mean a unique optimizer (Yu et al., 2004). For a large upper bound case, Wu suggested the implementation of the FcM clus- tering with a suitable large *m* value. When the dataset contains noise and outliers, the fuzzifier $m = 4$ is recommended for FcM clustering in a large upper bound case.

For other studies on the selection of fuzzy parameter *m*, see also Dembélé and Kastner (2003), Belacel et al. (2004), Liu and Zhang (2007), and Jaimes and Torra (2010).

24.2.3 Cluster Validity

In the FcM-clustering method (24.1), before computing the membership degrees and the centroids iteratively, by means of (24.2), we have to set a suitable the number of clusters *c*.

Many cluster-validity criteria have been suggested. For a review on fuzzy cluster validity criteria see, among others, Xu and Brereton (2005) and Wang and Zhang (2007). A widely used cluster validity criterion for selecting c is the *Xie–Beni criterion* (Xie and Beni, 1991):

$$\min_{c \in \Omega_c} : I_{XB} = \frac{\sum_{i=1}^{n} \sum_{k=1}^{c} u_{ik}^{m} \|\mathbf{x}_i - \mathbf{h}_k\|^2}{n \min_{k,k'} \|\mathbf{h}_k - \mathbf{h}_{k'}\|^2} \tag{24.3}$$

where Ω_c represents the set of possible values of c ($c < n$).

The numerator of I_{XB} represents the *total within-cluster distance*, which is equal to the objective function J of FcM clustering method. The ratio J/n is called the *compactness* of the fuzzy partition. The smaller this ratio, the more compact a partition with a fixed number of clusters (despite the number of data objects in a given dataset). The minimum squared distance between centroids in the denominator of I_{XB} is called *separation*. The greater this distance, the more separate a data partition with a fixed number of clusters. Therefore, for a fixed number of clusters, the smaller I_{XB}, the better the partition. Another interesting cluster validity procedure is the *Silhouette criterion* (Campello and Hruschka, 2006).

First, we describe the Average Silhouette Width Criterion or Crisp Silhouette (CS) proposed by Rousseeuw (1987). Successively, the Crisp Silhouette is suitably extended to fuzzy case Campello and Hruschka (2006).

Let consider an object $i \in 1, \ldots, n$ belonging to cluster $k \in 1, \ldots, c$. In the context of non-fuzzy (crisp) partitions produced, for example, by a c-means clustering algorithm, this means that i-th object is closer to the centroid of k-th cluster than to any other centroid. In the context of fuzzy partitions this means that the membership of i-th object to k-th cluster, u_{ik}, is higher than the membership of this object to any other cluster, that is $u_{ik} > u_{ik'}$, for every $k' \in 1, \ldots, c$, $k' \neq k$. Let the average (squared Euclidean) distance of i-th object to all other objects belonging to cluster k be denoted by a_{ik}. Also, let the average distance of this object to all objects belonging to another cluster k', $k' \neq k$, be called $d_{ik'}$. Finally, let b_{ik} be the minimum $d_{ik'}$ computed over $k' = 1, \ldots, c$, $k' \neq k$, which represents the dissimilarity of i-th object to its closest neighboring cluster. Then, the silhouette of i-th object is: $S_i = (b_{ik} - a_{ik})/(\max\{a_{ik}, b_{ik}\})$, where the denominator is a normalization term. Evidently, the higher S_i, the better the assignment of i-th object to the c-th cluster. The Crisp Silhouette (I_{CS}) defined as the average of S_i over $i = 1, \ldots, n$ is $I_{CS} = n^{-1} \sum_{i=1}^{n} S_i$. The best partition is achieved when the crisp silhouette is maximized, which implies minimizing the intracluster distance (a_{ik}) while maximizing the intercluster distance (b_{ik}).

Notice that, as remarked by Campello and Hruschka (2006), several other criteria are based on the same idea, for example, the Fukuyama–Sugeno index (1989) and the Xie–Beni index (1991). A comprehensive study involving two of them, namely, the Davies–Bouldin and Dunn's indices, is presented in Bezdek and Pal (1998). Another one, called Compose Within and Between Scattering, is suggested and compared to indices from the literature in Rezaee et al. (1998).

The Fuzzy Silhouette makes explicit use of the fuzzy partition matrix $\mathbf{U} = \{u_{ik} : i = 1, \ldots, n; k = 1, \ldots, c\}$. It may be able to discriminate between overlapped data clusters even if these clusters have their own distinct regions with higher data densities, since it considers the information contained in the fuzzy partition matrix $\mathbf{U}$ based on the degrees to which clusters overlap one another. This information can be used to reveal those regions with high data densities by stressing importance of data concentrated in the vicinity of the cluster prototypes while reducing importance of objects lying in overlapping areas. The Fuzzy

Silhouette (I_{FS}) is defined as follows:

$$I_{FS} = \frac{\sum_{i=1}^{n}(u_{ik} - u_{ik'})^{\gamma} S_i}{\sum_{i=1}^{n}(u_{ik} - u_{ik'})^{\gamma}} \tag{24.4}$$

where u_{ik} and $u_{ik'}$ are the first and second largest elements of the i-th row of the fuzzy partition matrix, respectively, and $\gamma \geq 0$ is a weighting coefficient. The effect of varying this parameter on the weighting terms in (24.4) is investigated in Campello and Hruschka (2006).

As remarked by Campello and Hruschka (2006), the Fuzzy Silhouette (24.4) differs from I_{CS} "for being a weighted average (instead of an arithmetic mean) of the individual silhouettes S_i. The weight of each term is determined by the difference between the membership degrees of the corresponding object to its first and second best matching fuzzy clusters, respectively. In this way, an object in the near vicinity of a cluster prototype is given more importance than another object located in an overlapping area (where the membership degrees of the objects to two or more fuzzy clusters are similar)."

With respect to other well known validity criteria based uniquely upon the fuzzy partition matrix (such as the Partition Coefficient), the Fuzzy Silhouette (24.4) takes into account the geometrical information related to the data distribution through the term S_i.

24.2.4 Graphical Visualization

In fuzzy clustering, an important topic is to visualize graphically the obtained fuzzy clusters.

Different graphical approaches can be used for visualizing fuzzy partitions. For displaying fuzzy partition, D'Urso (2005), Groenen et al. (2007), D'Urso and Massari (2013), D'Urso et al. (2013a), and D'Urso et al. (2013c) utilized the so-called *ternary plot* (ternary graph, triplot, triangle plot, simplex plot, or de Finetti diagram). Obviously, this graphical representation is useful only for partition into three clusters. *Parallel coordinates* is another method to visualize fuzzy partitions especially with high-dimensional data in a two-dimensional graph (see Berthold and Hall, 2003; Groenen et al., 2007). Another interesting and useful graphical representation is the *silhouette plot* (Rousseeuw, 1987; Campello and Hruschka, 2006) (see Section 24.2.3). Rousseeuw et al. (1989) proposed a graphical representation of the fuzzy partition in a *factorial subspace*, by computing the principal components of the membership coefficients. When the number of fuzzy clusters is greater than three, there will be more than two principal components and the authors suggest to follow the customary practice of displaying the two components with largest eigenvalues, thereby "explaining" the largest portion of the variability (Kaufman and Rousseeuw, 2005). "The plot can also be refined by adding *ideal* objects, corresponding to the clusters themselves. Indeed, each cluster can be represented by an object with membership 1 to that cluster and zero membership to all others. By transforming these *membership coordinates* in the same way as the actual objects, the plot will be enriched by as many additional points as there are clusters. In this way the final plot contains both objects and clusters (in the same way that correspondence analysis yields plots containing both objects and variables)" (Kaufman and Rousseeuw, 2005). Bezdek and Hathaway (2002) proposed the *Visual Assessment of cluster Tendency* (VAT) method. Its aim is similar to one of cluster validity indices, but it tries to avoid the "massive aggregation of information" by scalar validity measures. Instead of a scalar value or a series of scalar values by a different number of clusters, Bezdek and

Hathaway suggest to reorder suitably the distances between each pair of objects by a minimal spanning tree algorithm and to display the distances in a squared intensity image, where the gray level of a pixel is in connection with the level of the distance (Abonyi and Feil, 2007a). Variants of VAT have been suggested by Huband et al. (2004, 2005). A very interesting variant of VAT, particularly useful for FcM and its variants, is the *Visual Cluster Validity* (VCV) proposed by Hathaway and Bezdek (2003). In particular, in VCV the order of data is not determined by a minimal spanning tree algorithm but it depends on the clustering result and the pairwise distance of data is calculated based on the distances from the cluster prototypes. A visualization of fuzzy clusters by means of fuzzy *Sammon mapping projection* has been suggested by Kovacs and Abonyi (2004). For more details on VAT, VCV, and Sammon mapping projection, see Abonyi and Feil (2007a) and Feil et al. (2007). A suggestive geometric visualization of clusters obtained from fuzzy clustering algorithms (e.g., FcM clustering) has been proposed by Rueda and Zhang (2006). Their visualization approach projects the fuzzy membership data onto a *hypertetrahedron*, which allows to observe the intercluster relationships in a spatial manner. In particular, Rueda and Zhang (2006) presented "a geometric framework to visualize fuzzy-clustered data that comes from the cluster membership table. The scheme provides a wise visualization of the membership degree of a point belonging to each cluster that represents the geometric distribution in the two- and three-dimensional spaces. A point with a higher cluster membership appears closer to the cluster centroid, which is represented as a vertex in the tetrahedron. In addition, the closer the distance between two points is, the more similar the two points are. It is important to emphasize that a short distance between two points in the visualization means that these two points have much more similar cluster memberships as a result of FcM clustering based on a particular distance function." Furthermore, they showed how to extract a subspace of the clustered data, which allows the user to visualize subsets of classes and project them onto the two- or three-dimensional space.

24.3 Prototype-Based Variants

There are several real cases in which it is more suitable to identify prototypes belonging to the considered dataset, that synthesize the structural information of each cluster. The idea of cluster objects around representative prototypes (centrotype, median, or medoid) was introduced in Vinod (1969) and later analyzed in Rao (1971), Church (1978), and Mulvey and Crowder (1979). Subsequently, several clustering techniques based on medoids have been proposed, for example, the partitioning around medoids (PAM) proposed by Kaufman and Rousseeuw (1987).

In a fuzzy framework, Krishnapuram et al. (1999) and Krishnapuram et al. (2001) suggested the so-called *Fuzzy c-Medoids* (FcMd) clustering method.

Let $X = \{x_1, \ldots, x_i, \ldots, x_n\}$ be a set of n objects (data matrix) and let indicate with $\tilde{X} = \{\tilde{x}_1, \ldots, \tilde{x}_i, \ldots, \tilde{x}_c\}$ a sub-set of $X = \{x_1, \ldots, x_i, \ldots, x_n\}$ with cardinality c.

The FcMd clustering method is formalized by substituting in (24.1) h_k with $\tilde{x}_k$, where $\tilde{x}_k$ indicates the medoid of the k-th cluster, $k = 1, \ldots, c$. Notice that:

- Each cluster is represented by an observed representative object and not by a fictitious representative object (prototype, i.e., centroid). The possibility of obtaining nonfictitious representative prototypes in the clusters is very appealing and useful in a wide range of applications. This is very important for the interpretation of the selected clusters. In fact, as remarked by Kaufman and Rousseeuw (2005) "in many clustering problems one is particularly interested in a characterization of the clusters by means of typical or representative objects. These are objects that represent the various structural aspects of the set of objects being investigated. There can be many reasons for searching for representative objects. Not only can these objects provide a characterization of the clusters, but they can often be used for further work or research, especially when it is more economical or convenient to use a small set of c objects."

- FcMd clustering method does not depend on the order in which the objects are presented (except when equivalent solutions exist, which very rarely occurs in practice). This is not the case for many other algorithms present in the literature (Kaufman and Rousseeuw, 2005).

- Since FcMd clustering method belongs to the class of procedures for partitioning around medoids, it attempts to alleviate the negative effects of presence of outliers in the dataset; thus, FcMd can be considered more robust than its possible c-means version in the presence of noise and outliers because a medoid is less influenced by outliers or other extreme values than a mean. However, as remarked by García-Escudero and Gordaliza (1999, 2005), the FcMd provides only a timid robustification of the FcM. In fact, it attempts to alleviate the negative effects of presence of outliers in the dataset, but it does not solve the problem.

- When the objective function of FcMd is minimized, the medoids $\tilde{\mathbf{X}}$ corresponding to the solution provide a fuzzy partition via $u_{ik} = [\sum_{k'=1}^{c}[(\|\mathbf{x}_i - \tilde{\mathbf{x}}_k\|)/(\|\mathbf{x}_i - \tilde{\mathbf{x}}_{k'}\|)]^{(2/(m-1))}]^{-1}$. However, the objective function FcMd cannot be minimized by means of the alternating optimization algorithm, because the necessary conditions cannot be derived by differentiating it with respect to the medoids. Nonetheless, following heuristic algorithm of Fu (1982) for a crisp version of the objective function of FcMd, a fuzzy clustering algorithm that minimizes the objective function of FcMd can be built up Krishnapuram et al. (2001).

- As for the classical case, the algorithm utilized for FcMd falls in the category of Alternating Cluster Estimation paradigm Runkler and Bezdek (1999). Moreover, it is not guaranteed to find the global minimum. Thus, more than one random start is suggested.

- The algorithm utilized for FcMd is based on an exhaustive search for the medoids, which with large datasets could be too computationally intensive. The computational complexity of FcMd can be reduced by considering the "linearized" algorithm introduced by Krishnapuram et al. (2001) and Nasraoui et al. (2002). In this way, when we update the medoids for the generic cluster k we do not examine all the units, but only a subset that corresponds to those with the higher membership degree in cluster k.

- Since the medoid always has a membership of 1 in the cluster, raising its membership to the power m has no effect. Thus, when m is high, the mobility of the medoids may be lost. For this reason, a value between 1 and 1.5 for m is recommended Kamdar and Joshi (2000).

Another prototype-based variant of FcM is the *Fuzzy c-Medians* (FcMed) clustering method. FcMed is based on the L_1 metric and is robust to the presence of outliers in the dataset.

The FcMed clustering method can be formalized by substituting in (24.1) $\|x_i - h_k\|^2$ with $\|x_i - \tilde{h}_k\|_1$, that is the L_1 metric between the i-th observation and the k-th prototype, that is the median of the k-th cluster.

Notice that, Jajuga (1991) seems to be the first to have formulated the FcMed clustering method in which the optimal cluster center is the median, although Jajuga does not seem to mention that his solution is the median.

For the iterative solutions, the efficient algorithm for computing the median and for more information on the theoretical and computational details of the FcMed clustering see Bobrowski and Bezdek (1991), Jajuga (1991), Miyamoto and Agusta (1995), and Kersten (1999).

24.4 Distance-Based Variants

In the literature, several distance-based variants of FcM have been suggested. In this section, we review briefly the following variants.

- L_1 *metric-based fuzzy clustering*: As shown in Section 24.3, this clustering method have been suggested in order to increase the robustness of the fuzzy clustering process when the dataset is contaminated by outliers. In fact, by considering the distance L_1, the fuzzy clustering method inherits the robust properties of the L_1 metric.

- *Minkowski distance-based fuzzy clustering*: Fuzzy clustering algorithms based on the Minkowski distance have been proposed by Bobrowski and Bezdek (1991), Hathaway et al. (2000), Groenen and Jajuga (2001), and Groenen et al. (2007). In particular, the method suggested by Groenen et al. (2007) follows the approach previously introduced by Groenen and Jajuga (2001), that is, minimization of the objective function is partly done by iterative majorization; for details on the advantages of this method see Groenen and Jajuga (2001) and Groenen et al. (2007). The method proposed by Groenen and Jajuga (2001) is limited to the case of a Minkowski parameter between 1 and 2, that is, between the L_1-distance and the Euclidean distance. Groenen et al. (2007) extended their majorization algorithm to any Minkowski distance with Minkowski parameter greater than (or equal to) 1. This extension also includes the case of the L_∞-distance. The Minkowski distance-based fuzzy clustering can be formalized substituting in (24.1) $\|x_i - h_k\|^2$ with $\left(\sum_{s=1}^p |x_{is} - h_{ks}|^\alpha\right)^{\frac{2\lambda}{\alpha}}$ ($1 \leq \alpha \leq \infty$, $0 \leq \lambda \leq 1$), that is the power λ of the squared Minkowski distance between the i-th observation and the k-th prototype (Groenen et al., 2007).

- *Gustafson-Kessel clustering* (Gustafson and Kessel, 1979): This fuzzy clustering method extends the FcM by employing an adaptive distance, in order to detect clusters of different geometrical shapes. Then, in addition to the cluster centers each cluster is characterized by a symmetric positive defined ($q \times q$) matrix A_k, where q is the number of variables. Each cluster has its own norm-inducing matrix A_k, which yields the following squared distance measure $_A d_{ik}^2 = (x_i - h_k)' A_k (x_i - h_k)$. The objective function of the Gustafson-Kessel clustering method is identical to the FcM, using the distance $_A d_{ik}^2$. The matrices A_k ($k = 1, \ldots, c$) are used as optimization variables in the objective function, thus allowing each cluster to adapt the distance norm to the local topological structure of the data. For a fixed $\mathbf{A}_k$, conditions $\sum_{k=1}^{c} u_{ik} = 1$, $u_{ik} \geq 0$ can be directly applied. Thus, the update equations for the centroids $\mathbf{h}_k$ and the membership degrees u_{ik} are identical to those obtained with FcM, replacing the squared Euclidean distance by the cluster specific squared distance $_{\mathbf{A}}d_{ik}^2$. The objective function of the Gustafson-Kessel clustering cannot be directly minimized with respect to $\mathbf{A}_k$, since it is linear in $\mathbf{A}_k$. This means that the objective function can be made as small as desired by simply making $\mathbf{A}_k$ less positive definite. To obtain a feasible solution, $\mathbf{A}_k$ must be constrained in some way. The usual way of accomplishing this is to constrain the determinant of $\mathbf{A}_k$. Allowing the matrix $\mathbf{A}_k$ to vary with its determinant fixed corresponds to optimizing the cluster's shape while its volume remains constant Abonyi and Feil (2007b): $\det(\mathbf{A}_k) = \rho_k$, $\rho_k > 0$, $\forall k = 1, \ldots, c$, where ρ_k is fixed for each cluster. Using the Lagrange multiplier method, the following expression for $\mathbf{A}_k$ is obtained: $\mathbf{A}_k = [\rho_k \det(\mathbf{F}_k)]^{1/p} \mathbf{F}_k^{-1}$, where $\mathbf{F}_k$ is the fuzzy covariance matrix of the k-th cluster defined by $\mathbf{F}_k = (\sum_{i=1}^{n} u_{ik}^m (\mathbf{x}_i - \mathbf{h}_k)'(\mathbf{x}_i - \mathbf{h}_k))/(\sum_{i=1}^{n} u_{ik}^m)$. In conclusion, as remarked by Kruse et al. (2007) "the Gustafson–Kessel algorithm tries to extract much more information from the data than the algorithms based on the Euclidean distance. It is more sensitive to initialization, therefore it is recommended to initialize it using a few iterations of FcM [...]. Compared with FcM, the Gustafson–Kessel algorithm exhibits higher computational demands due to the matrix inversions."

- *Fuzzy shell clustering*: Other distance-based variants of FcM have been proposed to detect lines, circles or ellipses on the dataset, corresponding to more complex data substructures. The fuzzy shell clustering methods extract prototypes that have a different nature than the data points. They need to modify the definition of the distance between a data point and the prototype and replace the Euclidean by other distances (Kruse et al., 2007). For more details, for example, Bezdek (1981), Krishnapuram et al. (1992), and Kruse et al. (2007).

- *Kernel-based fuzzy clustering*: By following machine learning theoretical approach Vapnik (1995), kernel-based variants of fuzzy clustering methods have been suggested. These methods modify the distance function to handle nonvectorial data, such as trees, sequences, or graphs, without needing to modify completely the algorithms themselves. The kernel-based variants of fuzzy clustering consist of transposing the objective function to the feature space, that is, applying it to the transformed data. For more details, for example, Filippone et al. (2008) and Graves and Pedrycz (2010).

There are other distance-based variants of the FcM clustering in which the typology of distances is connected to the nature of the data (fuzzy data, sequence data, time series, and so on) (see Section 24.6). For instance, for clustering sequence data, the Levenshtein distance has been adopted in the FcM framework; for clustering fuzzy data can be considered the Hausdorff metric; for clustering time series can be utilized the Dynamic Time Warping metric, and so on.

24.5 Objective Function-Based Variants

In this section, we focus on some objective function-based variants of the FcM.

24.5.1 Possibilistic Clustering

In FcM, for each object the sum of the membership degrees in the clusters must be equal to one. Such a constraint may cause meaningless results, especially when noise is present. Following the Possibility Theory (e.g., Dubois and Prade, 1988), to avoid this drawback, it is possible to relax the constraint, leading to the so-called *Possibilistic c-Means* (PcM) clustering model, providing "degrees of compatibility" of an object with each of the clusters (Coppi et al., 2012).

Let consider the dataset illustrated in Figure 24.2 and the following membership degree matrices obtained by applying, respectively, c-Means (cM), FcM, and PcM:

$$
\mathbf{U}_{cM} = \begin{bmatrix} 1 & 1 & 1 & 1 & 1 & 1 & 0 & 0 & 0 & 0 \\ 0 & 0 & 0 & 0 & 0 & 0 & 1 & 1 & 1 & 1 \end{bmatrix}'
$$

$$
\mathbf{U}_{FcM} = \begin{bmatrix} 1.0 & 1.0 & 1.0 & 0.8 & 0.5 & 0.5 & 0.2 & 0.0 & 0.0 & 0.0 \\ 0.0 & 0.0 & 0.0 & 0.2 & 0.5 & 0.5 & 0.8 & 1.0 & 1.0 & 1.0 \end{bmatrix}'
$$

$$
\mathbf{U}_{PcM} = \begin{bmatrix} 1.0 & 1.0 & 1.0 & 1.0 & 0.5 & 0.2 & 0.0 & 0.0 & 0.0 & 0.0 \\ 0.0 & 0.0 & 0.0 & 0.0 & 0.5 & 0.2 & 1.0 & 1.0 & 1.0 & 1.0 \end{bmatrix}'
$$

From Figure 24.2, we can notice that we have two clusters and datum x_5 has the same distance to both clusters.

By applying FcM, x_5 is assigned a membership degree of about 0.5. This is plausible. However, the same degrees of membership are assigned to datum x_6 even though this datum is further away from both clusters and should be considered less typical. Because of the normalization, however, the sum of the memberships has to be 1. Consequently x_6 receives fairly high membership degrees to both clusters. For a correct interpretation of these memberships one has to keep in mind that they are rather degrees of sharing than of typicality, since the constant weight of 1 given to a datum must be distributed over the clusters. The normalization of memberships can further lead to undesired effects in the presence of noise and outliers. The fixed data point weight may result in high membership of these points to clusters, even though they are a large distance from the bulk of data. Their membership values consequently affect the clustering results, since data point weight attracts cluster prototypes. By dropping the normalization constraint $\sum_{k=1}^{c} u_{ik} = 1$ in the PcM one tries to achieve a more intuitive assignment of degrees of membership and to avoid undesirable normalization effects. In fact, by applying a PcM clustering method, we have more plausible results, obtaining for the two data points x_5 and x_6 realistic membership degree values.

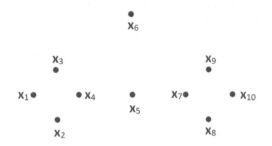

FIGURE 24.2
An artificial dataset.

By applying the cM clustering method to dataset shown in Figure 24.2, we note the inadequacy of the crisp clustering. In fact cM arbitrarily forces the full assignment of data points x_5 and x_6 to one of the two clusters.

In the possibilistic perspective, u_{ik} represents the degree of possibility of object i belonging to cluster k or, in other terms, the degree of "compatibility" of the profile x_i with the characteristics of cluster k embodied by its prototype h_k. The FcM objective function is consequently modified, by introducing an additive "penalization" term which takes care of the balance between the fuzziness of the clustering structure and the "compactness" of the clusters.

The following are just two examples of PcM objective functions (see Krishnapuram and Keller, 1996):

$$\sum_{i=1}^{n}\sum_{k=1}^{c} u_{ik}^m \|x_i - h_k\|^2 + \sum_{k=1}^{c} \eta_k \sum_{i=1}^{n} (1 - u_{ik})^m \tag{24.5}$$

$$\sum_{i=1}^{n}\sum_{k=1}^{c} u_{ik}^m \|x_i - h_k\|^2 + \sum_{k=1}^{c} \eta_k \sum_{i=1}^{n} (u_{ik} \log u_{ik} - u_{ik}) \tag{24.6}$$

where η_k is a tuning parameter associated with cluster k, weighting its contribution to the penalization function. For details on η_k see Krishnapuram and Keller (1996) and Kruse et al. (2007).

Krishnapuram and Keller (1996) argue that the possibilistic approach provides a "mode-seeking" clustering procedure, to be confronted with the "partition-seeking" property of FcM. Thus, PcM clustering methods tend to be more robust with respect to noise, as compared to FcM techniques. By minimizing (24.5), we obtain $u_{ik} = [1 + ((\|x_i - h_k\|^2)/(\eta_k))^{(1/(m-1))}]^{-1}$.

Nonetheless, there are some limitations in the use of PcM algorithms, in that these may lead to trivial solutions consisting of "coincident clusters" (Barni et al., 1996). In other words, the algorithms may assign all prototypes to the same location. This can be due to the fact that the possibilistic loss functions can usually be decomposed into the sum of c terms (one for every cluster), which can be minimized independently of each other. This depends on the absence of the normalization constraint of the membership degrees. In this case, when there is a single optimal point for a cluster prototype, all the prototypes move over there (Coppi et al., 2012). A practical work-around that is often adopted (e.g., Krishnapuram and Keller, 1996; Coppi et al., 2012) is to use the FcM solution as starting point for PcM. Through a simulation study, Coppi et al. (2012) showed that this work-around limits the risk of obtaining coincident clusters. Nonetheless, at the present it is not possible to state that coincident clusters are always avoided and, for this reason, further investigation is recommended.

On the other hand, using objective functions such as (24.5) and (24.6) raises the problem of handling the η_k parameters, which crucially affect the performance of the method. Yang and Wu (2006) suggested the following possibilistic clustering method:

$$\sum_{i=1}^{n}\sum_{k=1}^{c} u_{ik}^m \|x_i - h_k\|^2 + \frac{\beta}{m^2\sqrt{c}} \sum_{i=1}^{n}\sum_{k=1}^{c} (u_{ik}^m \log u_{ik}^m - u_{ik}^m) \tag{24.7}$$

where $\beta/(m^2\sqrt{c}) = \eta_k \ (k = 1,\dots,c)$ is a suitable tuning parametric function. In this case, we obtain $u_{ik} = \exp(-(m\sqrt{c}\|x_i - h_k\|^2/\beta))$.

For other variants of possibilistic clustering, for example, Barni and Gualtieri (1999), Ménard et al. (2003), Zhang and Leung (2004), Pal et al. (2005), and De Cáceres et al. (2006).

24.5.2 Fuzzy Relational Clustering

The FcM-based clustering methods or their variants consider the case where a description is provided for each data point individually (the methods are applied to object data). In many real cases, this information is not available, the input data takes the form of a $(n \times n)$-pairwise dissimilarity matrix, where each of its elements indicates the dissimilarity between point couples. Relational clustering aims at identifying clusters exploiting this input.

In a fuzzy framework, there exists a large variety of clustering techniques for such settings (e.g., Hathaway and Bezdek, 1994; Runkler and Bezdek, 2003; Bezdek et al., 2005). The earliest fuzzy clustering models for relational data (fuzzy relational clustering) have been proposed by Ruspini (1969, 1970), Roubens (1978), Windham (1985), Trauwaert (1987), Hathaway et al. (1989), Hathaway and Bezdek (1994), and Kaufman and Rousseeuw (2005). The model suggested by Ruspini (1969, 1970) can be considered as a seminal fuzzy relational model. The models suggested by Roubens (1978), Windham (1985), Trauwaert (1987), and Kaufman and Rousseeuw (2005) are particular cases of the more general fuzzy relational-clustering model proposed by Hathaway et al. (1989) and Hathaway and Bezdek (1994) that can be formalized as follows:

$$\min : \sum_{k=1}^{c} \frac{\sum_{i,i'=1}^{n} u_{ik}^m u_{i'k}^m d_{ii'}}{2 \sum_{i'=1}^{n} u_{i'k}^m} \tag{24.8}$$

where u_{ik} and $u_{i'k}$ represent, respectively, the membership degrees of the i-th and i'-th objects to the k-th cluster and $d_{ii'}$ indicates a dissimilarity measure between each pair of i-th and i'-th objects. Notice that in (24.8) we can utilize any type of dissimilarity measure (city-block distance, Lagrange distance, and so on). Furthermore, we observe that each term appears twice in the multiple sums. The factor 2 in the denominator compensates for this duplicity Kaufman and Rousseeuw (2005).

A local optimal solution of (24.8) can be found by using the Lagrangian multiplier method, by taking into account the Kuhn–Tucker conditions (e.g., Runkler, 2007). For the detailed technical description of the numerical algorithm used for (24.8), for example, Kaufman and Rousseeuw (2005).

We remark that the principal advantage of the fuzzy relational clustering approach is that we can utilize any type of dissimilarity measure in the clustering framework. It should be noted that when in the fuzzy relational algorithm the squared Euclidean distance is used as a dissimilarity measure, it corresponds to the FcM (see Kaufman and Rousseeuw, 2005).

Other interesting references on fuzzy relational clustering are Krishnapuram et al. (2001) and Davé and Sen (2002).

24.5.3 Fuzzy Clustering with Entropy Regularization

As shown in Section 24.2.1, in the objective function $\sum_{i=1}^{n} \sum_{k=1}^{c} u_{ik}^m d_{ik}^2$, we have the weighting exponent m that controls the fuzziness of the partition. The fuzzification of the standard c-means clustering by introducing m has been viewed by some researchers as an artificial device, lacking a strong theoretical justification. Consequently, a new line of research has been started, based on the adoption of regularization terms to be juxtaposed to the maximum internal homogeneity criterion (e.g., Miyamoto and Mukaidono, 1997). In this case, the burden of representing fuzziness is shifted to the regularization term, in the form of a weighting factor multiplying the contribution of the regularization function to the clustering criterion. In this framework, the regularization function has been thought of as measuring the overall fuzziness of the obtained clustering pattern. Such measure is the entropy function (i.e., the Shannon entropy) which, if applied to the degrees of membership, may be called fuzzy entropy (Coppi and D'Urso, 2006).

In particular, Li and Mukaidono (1995) remark that this "strange" parameter is unnatural and has not a physical meaning. Then, in the above objective function m may be removed, but in this case, the procedure cannot generate the membership update equations. For this purpose, Li and Mukaidono (1995, 1999) suggest a new approach to fuzzy clustering, the so-called Maximum Entropy Inference Method:

$$\max: -\sum_{i=1}^{n}\sum_{k=1}^{c} u_{ik} \log u_{ik}, \quad \text{s.t.} \sum_{k=1}^{c} u_{ik} = 1, \ u_{ik} \geq 0, \ J_i = \sum_{k=1}^{c} u_{ik}^{m} d_{ik}^{2} = \kappa_i(\sigma^2) \quad (24.9)$$

where, at each step i of the optimization algorithm, $\kappa_i(\sigma^2)$ represents the value of the loss function J_i (the within clusters sum-of-squares-error) due to object i. It is shown that this value depends on parameter σ which gets a physical interpretation in terms of "temperature" in statistical physics.

The optimal value of σ can be determined by the simulated annealing method. The term to be extremized in (24.9) defines the entropy of a fuzzy partition, and it is clearly inspired by the Shannon index concerning a finite set of random events.

We underline that the Maximum Entropy principle, as applied to fuzzy clustering, provides a new perspective to facing the problem of fuzzifying the clustering of the objects, while ensuring the maximum of compactness of the obtained clusters.

The former objective is achieved by maximizing the entropy (and, therefore, the uncertainty) of the clustering of the objects into the various clusters. The latter objective is obtained by constraining the above maximization process in such a way as to minimize the overall distance of the objects from the cluster prototypes (i.e., to maximize cluster compactness) Coppi and D'Urso (2006).

The idea underlies the entropy-based fuzzy clustering method proposed by Miyamoto and Mukaidono (1997) in which the trade-off between fuzziness and compactness is dealt with by introducing a unique objective function reformulating the maximum entropy method in terms of "regularization" of the FcM function. In particular, by introducing an entropy regularization, the minimization problem becomes (Miyamoto and Mukaidono, 1997):

$$\min: \sum_{i=1}^{n}\sum_{k=1}^{c} u_{ik}\|x_i - h_k\|^2 + p\sum_{i=1}^{n}\sum_{k=1}^{c} u_{ik} \log u_{ik} \quad \text{s.t.} \sum_{k=1}^{c} u_{ik} = 1, \ u_{ik} \geq 0 \quad (24.10)$$

where p is a weight factor, called degree of fuzzy entropy, similar to the weight exponent m.

By means of (24.10) the authors minimize a functional depending on the desired solution regularized by maximizing the total amount of information. By solving (24.10) using the Lagrangian multiplier method, we have $u_{ik} = [\sum_{k'=1}^{c}[(\exp(1/p\|x_i - h_k\|^2))/(\exp(1/p\|x_i - h_{k'}\|^2))]]^{-1}$.

For more references on entropy-based fuzzy clustering, see Coppi and D'Urso (2006).

24.5.4 Fuzzy Clustering with Noise Cluster

An interesting objective function-based variants of the FcM useful for neutralizing the negative effects of noise data in the clustering process is the Fuzzy c-Means clustering method with noise cluster (FcM-NC). FcM-NC has been initially proposed by Davé (1991), which uses a criterion similar to Ohashi (1984), and later extended by Davé and Sen (1997). This method consists in adding, beside the c clusters to be found in a dataset, the so-called *noise cluster*; it aims at grouping points that are badly represented by normal clusters, such as noisy data points or outliers. It is not explicitly associated to a prototype, but directly to the distance between an implicit prototype and the data points: the center of the noise cluster is considered to be at a constant distance from all data points. "This means that all points have a priori the same 'probability' of belonging to the noise cluster. During the optimization process, this 'probability' is then adapted as a function of the probability according to which points belong to normal clusters" (Kruse et al., 2007).

Thus, the noise cluster is represented by a *fictitious prototype* (*noise prototype*) that has a constant distance (*noise distance*) from every objects. An object belongs to a *real cluster* only if its distance from a prototype (centroid) is lower than the noise distance; otherwise, the object belongs to the noise cluster.

FcM-NC clustering method can be formalized as follows:

$$\min : \sum_{i=1}^{n} \sum_{k=1}^{c-1} u_{ik}^m \|\mathbf{x}_i - \mathbf{h}_k\|^2 + \sum_{i=1}^{n} \delta^2 \left(1 - \sum_{k=1}^{c-1} u_{ik}\right)^m \quad \text{s.t} \sum_{k=1}^{c-1} u_{ik} \le 1, \ u_{ik} \ge 0 \quad (24.11)$$

where δ is a suitable scale parameter, the so-called *noise distance*, to be chosen in advance. Such parameter plays the role to increase (for high values of δ) or to decrease (for low values of δ) the emphasis of the "noise component" in the minimization of the objective function in (24.11), for example, $\delta^2 = \lambda[n(c-1)]^{-1}[\sum_{i=1}^{n} \sum_{k=1}^{c-1} \|\mathbf{x}_i - \mathbf{h}_k\|^2]$, where λ is a scale multiplier that needs to be selected depending on the type of data.

In the literature, some heuristic solutions to estimate the optimal value of δ have been suggested, but the determination of the value of this parameter is still an open problem. In general, the success of FcM-NC depends on the appropriate choice of the noise distance δ. If δ is too larger, the FcM-NC degenerates to the nonrobust version of the model and outliers are forced to belong to real clusters; vice versa, if δ is too small, a lot of objects can be considered as noise and misplaced into the noise cluster Cimino et al. (2005). Davé and Sen (2002) suggested that the value of the distance noise δ should be calculated by considering the dataset statistics, that is, it should be related to the concept of "scale" in robust statistics Davé and Krishnapuram (1997).

It has to be observed that the model provides c clusters, but only $(c-1)$ are "real" cluster, with the extra cluster serving as the noise cluster. The difference in the second term of the objective function shown in (24.11) expresses the membership degree of each object to the noise cluster, and shows that the sum of the membership degrees over the first $(c-1)$ clusters is lower than or equal to 1. Indeed, the membership degree (u_{i*}) of the i-th object to the *noise cluster* is defined as $u_{i*} = 1 - \sum_{k=1}^{c-1} u_{ik}$ and the usual constraint of the FcM ($\sum_{k=1}^{c-1} u_{ik} = 1$) is not required. Thus, the membership constraint for the *real clusters* is relaxed to $\sum_{k=1}^{c-1} u_{ik} \le 1$. This allows noise object to have small membership values in *good clusters* Davé and Sen (2002).

By solving (24.11), we obtain

$$
u_{ik} = \left[\sum_{k'=1}^{c-1} \left[\frac{\|\mathbf{x}_i - \mathbf{h}_k\|}{\|\mathbf{x}_i - \mathbf{h}_{k'}\|} \right]^{\frac{2}{m-1}} + \left[\frac{\|\mathbf{x}_i - \mathbf{h}_k\|}{\delta} \right]^{\frac{2}{m-1}} \right]^{-1}.
$$

We remark that FcM-NC is introduced to make the clustering process less sensitive to outlier and to neutralize the disruptive effects of the outlier by relaxing the constraint on the membership degrees so that the sum of the membership degrees of a noise object to all the real classes is not forced to be equal to 1.

Other objective function-based variants of fuzzy clustering with robust properties are the Robust estimator-based fuzzy clustering (Frigui and Krishnapuram, 1996), the Trimmed fuzzy clustering Krishnapuram et al. (2001) and the Fuzzy clustering with influence weighting system (Keller, 2000; D'Urso, 2005).

Other types of objective function-based variants could be, for instance, the clustering methods for spatial data, time series, time-spatial data, and sequence data (see Section 24.6).

24.6 Data Features-Based Variants: Fuzzy Clustering for Complex Data Structures

In the previous section, we have analyzed FcM and its variants for standard data structures, that is, standard quantitative/numerical data. However, a fuzzy approach can also be adopted for clustering problems with complex data structures.

In the literature, different variants of FcM have been proposed for data with different nature, that is:

- *fuzzy data* (e.g., D'Urso, 2007; D'Urso and De Giovanni, 2014);
- *symbolic data* (e.g., El-Sonbaty and Ismail, 1998);
- *interval-valued data* (e.g., D'Urso and Giordani, 2006; D'Urso et al., 2015);
- *categorical data* (e.g., Huang and Ng, 1999; Lee and Pedrycz, 2009);
- *textual data* (*text data*) (e.g., Runkler and Bezdek, 2003);
- *time and/or spatial data* (e.g., chapter "Time series clustering" for time data, Pham (2001) for spatial data, Coppi et al. (2010) for spatial-time data);
- *three-way data* (e.g., Coppi and D'Urso, 2002, 2003; D'Urso, 2004, 2005; Rocci and Vichi, 2005; Hwang et al., 2007);
- *sequence data* (e.g., D'Urso and Massari, 2013);
- *functional data* (e.g., Tokushige et al., 2007);
- *network data* (e.g., Liu, 2010);
- *directional data* (e.g., Yang and Pan, 1997);
- *mixed data* (e.g., Yang et al., 2004);

and for data with particular structural features, that is:

- *outlier data*: we can consider different methods belonging to different approaches, for example: FcMed (Section 24.3) (metric approach), PcM (Section 24.5.1) (possibilistic approach), FcM-NC (Section 24.5.4) (noise approach), and TrFcM (Trimmed FcM) (e.g., Krishnapuram et al., 2001; D'Urso and De Giovanni, 2014);

- *incomplete data* (e.g., Hathaway and Bezdek, 2001);
- *data streams* (e.g., Berlinger and Hüllermeier, 2007);
- *big data* (e.g., Havens et al., 2012).

24.7 Other Research Directions on Fuzzy Clustering

Other interesting research directions on fuzzy clustering are pointed out in the following:

- *Fuzzy clustering with partial supervision*: In many real situations, data are neither perfectly nor completely labeled. Then, we may attempt to benefit from the available knowledge (labeled data) to cluster unlabeled data. This form of combining labeled and unlabeled data to generate the structure of the whole dataset is known as semisupervised (or partial supervised) clustering (Bouchachia and Pedrycz, 2006). Notice that we have unsupervised clustering in absence of knowledge and supervised clustering in presence of knowledge. In this research area, Bouchachia and Pedrycz (2006) proposed in a fuzzy framework, a semisupervised clustering algorithm based on a modified version of the FcM. The objective function consists of two components: the first concerns traditional unsupervised clustering while the second tracks the relationship between classes (available labels) and the clusters generated by the first component. The balance between the two components is suitably tuned by a scaling factor. For more details, see also Pedrycz and Waletzky (1997).
- *Fuzzy clustering with supervision*: In presence of knowledge, we can classify the data by considering a supervised clustering approach. Pedrycz and Vukovich (2004) proposed a supervised version of the FcM extending the original objective function by the supervision component (labeled patterns).
- *Collaborative fuzzy clustering*: The concept of collaborative fuzzy clustering has been introduced by Pedrycz (2002). It consists in a "conceptual and algorithm machinery for the collective discovery of a common structure (relationship) within a finite family of data residing at individual data sites" (Pedrycz and Rai, 2008). For more details, see also Pedrycz (2002) and Coletta et al. (2012).
- *Fuzzy coclustering*: Coclustering (biclustering or two-mode clustering) is a technique which allows simultaneous clustering of rows (objects) and columns (variables) of a data matrix. In a fuzzy framework, some fuzzy coclustering methods have been proposed by Frigui and Nasraoui (2000) and Frigui (2007).
- *Comparison of fuzzy clustering*: A useful criterion for comparing each pair of fuzzy partitions obtained by fuzzy methods (e.g., FcM) is the Fuzzy Rand index (Campello, 2007; Anderson et al., 2010). It is a fuzzy extension of the original Rand index based on the comparison of agreements (consistent classifications) and disagreements (inconsistent classifications) of the two partitions, the fuzzy partition and the hard partition. For more detail on the Fuzzy Rand index and other alternative indices see Campello (2007), Anderson et al. (2010), and Hullermeier et al. (2012).

- *Consensus of fuzzy clustering*: Often different partitions of the same set of objects are available—e.g., fuzzy partitions obtained by applying FcM—and it can be relevant to obtain a *consensus* partition which summarized the information contained in the different partitions. The most natural way for defining fuzzy consensus of fuzzy partitions is by the so-called *optimization approach*: it considers a criterion that measures the distance between the set of fuzzy partitions in the profile and a fuzzy classification, and one seeks a fuzzy consensus classification that optimizes the stated criterion. For more details, see Pittau and Vichi (2000).

- *Fuzzy clustering-based analysis strategies*: In the literature, there are different strategies of analyses based on the combination of the FcM clustering or variants with other kinds of multivariate methods, for instance, principal component analysis (e.g., Hwang et al., 2007; Miyamoto et al., 2008b), regression analysis (e.g., Hathaway and Bezdek, 1993; Yang and Ko, 1997; D'Urso and Santoro, 2006; D'Urso et al., 2010) and other clustering techniques (e.g., D'Urso et al., 2013b).

- *Fuzzy clustering based on mixtures of distributions*: In the nonfuzzy unsupervised clustering literature, a great attention is paid to the cluster analysis based on the mixtures of distributions. In the fuzzy field, the interest for the mixture approach is relatively recent. However, the first comparative studies (similarities and differences) between mixture approach and fuzzy approach to cluster analysis has been carried out in the second half of the 80s by Bezdek et al. (1985), Hathaway (1986), and Davenport et al. (1988). Successively, the interest for the mixture approach to fuzzy clustering is moderately increased. In 1991, Trauwaert et al. (1991) proposed a generalization of some hard maximum likelihood clustering methods to the fuzzy approach, leading to some new fuzzy clustering algorithms. By considering a fuzzy extension of the Classification Maximum Likelihood procedure Yang (1993a) extended the fuzzy clustering algorithms of Trauwaert et al. (1991) by adding a penalty term. Successively, other algorithms have been suggested, for example, Lin et al. (2004) and Chatzis (2010).

- *Fuzzy hierarchical clustering*: The main effort on the application of fuzzy set theory to clustering has been on nonhierarchical methods. Comparatively, there are few studies about fuzzy hierarchical clustering, despite traditional (nonfuzzy) hierarchical clustering methods being well established. We hope that this big gap between the two different clustering approaches will stimulate the interest of the researchers in the coming years. Among the few contributions present in the literature we point out that a divisive hierarchical clustering method—in which a chain of fuzzy partitions ordered by the refinement relation and the corresponding binary fuzzy hierarchy are obtained—has been proposed by Dumitrescu (1988). Successively, interesting studies on hierarchical clustering have been reported by, for example, Sărbu et al. (2007) and Delgado et al. (1996). A comprehensive and organic explanation of the fuzzy hierarchical clustering is shown in Dumitrescu et al. (2000).

24.8 Software on Fuzzy Clustering

Many of the methods discussed above were implemented with different software, for example, MATLAB, C, and the open source software R. In particular, R and MATLAB represent the programming environments widely chosen for the implementation of the clustering algorithms proposed in the last decades.

A useful collection of MATLAB functions for fuzzy clustering is the *Fuzzy Clustering and Data Analysis Toolbox* (http://www.abonyilab.com/software-and-data/fclusttoolbox). In this regards, an interesting book is *Introduction to Fuzzy Logic Using MATLAB* by Sivanandam et al. (2007).

In R, fuzzy clustering methods are implemented in the following libraries, for example:

"cluster" (http://cran.r-project.org/web/packages/cluster/),

"vegclust" (http://cran.r-project.org/web/packages/vegclust/).

"kml" (http://cran.r-project.org/web/packages/kml/)

"skmeans" (http://cran.r-project.org/web/packages/skmeans/)

"clustrd" (http://cran.univ-lyon1.fr/web/packages/clustrd/)

"clue" (http://cran.r-project.org/web/packages/clue/)

A software tool for UNIX Systems (X-Windows) for fuzzy cluster analysis, freely available for scientific and personal use, is "FCLUSTER" (http://fuzzy.cs.uni-magdeburg.de/fcluster/).

Useful software for fuzzy clustering, that is, the packages called "Cluster" (written in C) and "ClusterGUI" (written in Java), can be freely downloaded from the website http://www.borgelt.net//ida.html. In particular, "ClusterGUI" is a graphical user interface for the fuzzy clustering programs of the package "Cluster."

24.9 Conclusions

In this chapter, we have performed an organic and systematic overview of fuzzy clustering. We have preliminarily analyzed the mathematical and computational aspects of the first method computationally efficient and powerful, that is, the FcM (Bezdek, 1981). Successively, starting from FcM, different variants of the basic algorithm have been discussed. In particular, prototype-variants, distance-variants, objective function variants, data features-variants, and other types of variants have been described.

We remark that interesting and promising future research areas will be, for instance, fuzzy clustering for big data, complex data structures (i.e., text data, network data, sequence data, functional data, mixed data) and their applications in internet, finance, business and neurosciences. Theoretical, methodological, and computational developments in these directions are expected in the coming years.

References

Abonyi, J. and B. Feil 2007a. Aggregation and visualization of fuzzy clusters based on fuzzy similarity measures, in *Advances in Fuzzy Clustering and Its Applications*, eds. by J. V. De Oliveira and W. Pedrycz, Wiley, Chichester, 95–121.

Abonyi, J. and B. Feil 2007b. *Cluster Analysis for Data Mining and System Identification*, Springer, Basel.

Anderson, D. T., J. C. Bezdek, M. Popescu, and J. M. Keller 2010. Comparing fuzzy, probabilistic, and possibilistic partitions, *IEEE Transactions on Fuzzy Systems*, 18, 906–918.

Arabie, P., J. D. Carroll, W. DeSarbo, and J. Wind 1981. Overlapping clustering: A new method for product positioning, *Journal of Marketing Research*, 18, 310–317.

Bandemer, H. 2006. *Mathematics of Uncertainty: Ideas, Methods, Application Problems*, vol. 189, Springer, Berlin Heidelberg.

Barni, M., V. Cappellini, and A. Mecocci 1996. Comments on 'A possibilistic approach to clustering', *IEEE Transactions on Fuzzy Systems*, 4, 393–396.

Barni, M. and R. Gualtieri 1999. A new possibilistic clustering algorithm for line detection in real world imagery, *Pattern Recognition*, 32, 1897–1909.

Belacel, N., M. Culf, M. Laflamme, and R. Ouellette 2004. Fuzzy J-Means and VNS methods for clustering genes from microarray data, *Bioinformatics*, 20, 1690–1701.

Bellman, R., R. Kalaba, and L. Zadeh 1966. Abstraction and pattern classification, *Journal of Mathematical Analysis and Applications*, 13, 1–7.

Berlinger, J. and E. Hüllermeier 2007. Fuzzy clustering of parallel data streams, in *Advances in Fuzzy Clustering and Its Applications*, eds. by J. V. De Oliveira and W. Pedrycz, Wiley, Chichester, 333–352.

Berthold, M. R. and L. O. Hall 2003. Visualizing fuzzy points in parallel coordinates, *IEEE Transactions on Fuzzy Systems*, 11, 369–374.

Bezdek, J. C. 1974. Numerical taxonomy with fuzzy sets, *Journal of Mathematical Biology*, 1, 57–71.

Bezdek, J. C. 1976. A physical interpretation of fuzzy ISODATA, *IEEE Transactions on Systems, Man and Cybernetics*, 6, 387–390.

Bezdek, J. C. 1981. *Pattern Recognition with Fuzzy Objective Function Algorithms*, Kluwer Academic Publishers, New York.

Bezdek, J. C. and R. J. Hathaway 2002. VAT: A tool for visual assessment of (cluster) tendency, in *Proceedings of the 2002 International Joint Conference on Neural Networks, 2002. IJCNN'02*, IEEE, vol. 3, 2225–2230.

Bezdek, J. C., R. J. Hathaway, and V. J. Huggins 1985. Parametric estimation for normal mixtures, *Pattern Recognition Letters*, 3, 79–84.

Bezdek, J. C., R. J. Hathaway, M. J. Sabin, and W. T. Tucker 1987. Convergence theory for fuzzy c-means: Counterexamples and repairs, *IEEE Transactions on Systems, Man and Cybernetics*, 17, 873–877.

Bezdek, J. C., J. Keller, R. Krishnapuram, and N. R. Pal 2005. *Fuzzy Models and Algorithms for Pattern Recognition and Image Processing*, vol. 4 of *The Handbooks of Fuzzy Sets*, Springer, New York.

Bezdek, J. C. and N. R. Pal 1998. Some new indexes of cluster validity, *IEEE Transactions on Systems, Man, and Cybernetics, Part B: Cybernetics*, 28, 301–315.

Bobrowski, L. and J. C. Bezdek 1991. c-means clustering with the l_1 and l_∞ norms, *IEEE Transactions on Systems, Man and Cybernetics*, 21, 545–554.

Borgelt, C. 2009. Accelerating fuzzy clustering, *Information Sciences*, 179, 3985–3997.

Bouchachia, A. and W. Pedrycz 2006. Data clustering with partial supervision, *Data Mining and Knowledge Discovery*, 12, 47–78.

Campello, R. J. G. B. 2007. A fuzzy extension of the Rand index and other related indexes for clustering and classification assessment, *Pattern Recognition Letters*, 28, 833–841.

Campello, R. J. G. B. and E. R. Hruschka 2006. A fuzzy extension of the silhouette width criterion for cluster analysis, *Fuzzy Sets and Systems*, 157, 2858–2875.

Cannon, R. L., J. V. Davé, and J. C. Bezdek 1986. Efficient implementation of the fuzzy c-means clustering algorithms, *IEEE Transactions on Pattern Analysis and Machine Intelligence*, 8, 248–255.

Chan, K. P. and Y. S. Cheung 1992. Clustering of clusters, *Pattern Recognition*, 25, 211–217.

Chatzis, S. P. 2010. A method for training finite mixture models under a fuzzy clustering principle, *Fuzzy Sets and Systems*, 161, 3000–3013.

Church, R. 1978. Contrasts between facility location approaches and non-hierarchical cluster analysis, in *ORSA/TIMS Joint National Meeting*, Los Angeles.

Cimino, M., G. Frosini, B. Lazzerini, and F. Marcelloni 2005. On the noise distance in robust fuzzy C-Means, in *Proceedings of World Academy of Science, Engineering and Technology*, vol. 1, 361–364.

Coletta, L. F. S., L. Vendramin, E. R. Hruschka, R. J. G. B. Campello, and W. Pedrycz 2012. Collaborative fuzzy clustering algorithms: Some refinements and design guidelines, *IEEE Transactions on Fuzzy Systems*, 20, 444–462.

Coppi, R. and P. D'Urso 2002. Fuzzy K-means clustering models for triangular fuzzy time trajectories, *Statistical Methods and Applications*, 11(1), 21–40.

Coppi, R. and P. D'Urso 2003. Three-way fuzzy clustering models for LR fuzzy time trajectories, *Computational Statistics & Data Analysis*, 43, 149–177.

Coppi, R. and P. D'Urso 2006. Fuzzy unsupervised classification of multivariate time trajectories with the Shannon entropy regularization, *Computational Statistics & Data Analysis*, 50, 1452–1477.

Coppi, R., P. D'Urso, and P. Giordani 2010. A fuzzy clustering model for multivariate spatial time series, *Journal of Classification*, 27, 54–88.

Coppi, R., P. D'Urso, and P. Giordani 2012. Fuzzy and possibilistic clustering for fuzzy data, *Computational Statistics & Data Analysis*, 56, 915–927.

Davé, R. N. 1991. Characterization and detection of noise in clustering, *Pattern Recognition Letters*, 12, 657–664.

Davé, R. N. and R. Krishnapuram 1997. Robust clustering methods: A unified view, *IEEE Transactions on Fuzzy Systems*, 5, 270–293.

Davé, R. N. and S. Sen 1997. Noise clustering algorithm revisited, in *1997 Annual Meeting of the North American Fuzzy Information Processing Society. NAFIPS'97.*, IEEE, 199–204.

Davé, R. N. and S. Sen 2002. Robust fuzzy clustering of relational data, *IEEE Transactions on Fuzzy Systems*, 10, 713–727.

Davenport, J. W., J. C. Bezdek, and R. J. Hathaway 1988. Parameter estimation for finite mixture distributions, *Computers & Mathematics with Applications*, 15, 819–828.

De Cáceres, M., F. Oliva, and X. Font 2006. On relational possibilistic clustering, *Pattern Recognition*, 39, 2010–2024.

de Oliveira, J. V. and W. Pedrycz 2007. *Advances in Fuzzy Clustering and Its Applications*, Wiley, Chichester.

Delgado, M., A. F. Gómez-Skarmeta, and A. Vila 1996. On the use of hierarchical clustering in fuzzy modeling, *International Journal of Approximate Reasoning*, 14, 237–257.

Dembélé, D. and P. Kastner 2003. Fuzzy C-means method for clustering microarray data, *Bioinformatics*, 19, 973–980.

Döring, C., M.-J. Lesot, and R. Kruse 2006. Data analysis with fuzzy clustering methods, *Computational Statistics & Data Analysis*, 51, 192–214.

Dubois, D. and H. M. Prade 1988. *Possibility Theory*, New York, NY: Plenum Press.

Dumitrescu, D. 1988. Hierarchical pattern classification, *Fuzzy Sets and Systems*, 28, 145–162.

Dumitrescu, D., B. Lazzerini, and L. C. Jain 2000. *Fuzzy Sets and Their Application and Clustering and Training*, vol. 16, CRC Press, Boca Raton, FL.

Dunn, J. C. 1974. A fuzzy relative of the ISODATA process and its use in detecting compact well-separated clusters, *Journal of Cybernetics*, 3, 32–57.

D'Urso, P. 2004. Fuzzy C-means clustering models for multivariate time-varying data: Different approaches. *International Journal of Uncertainty, Fuzziness and Knowledge-Based Systems*, 12(3), 287–326.

D'Urso, P. 2005. Fuzzy clustering for data time arrays with inlier and outlier time trajectories, *IEEE Transactions on Fuzzy Systems*, 13, 583–604.

D'Urso, P. 2007. Clustering of fuzzy data, in *Advances in Fuzzy Clustering and Its Applications*, eds. by J. V. De Oliveira and W. Pedrycz, Wiley, Chichester, 155–192.

D'Urso, P., C. Cappelli, D. Di Lallo, and R. Massari 2013a. Clustering of financial time series, *Physica A: Statistical Mechanics and its Applications*, 392, 2114–2129.

D'Urso, P. and L. De Giovanni 2014. Robust clustering of imprecise data, *Chemometrics and Intelligent Laboratory Systems*, 136, 58–80.

D'Urso, P., L. De Giovanni, M. Disegna, and R. Massari 2013b. Bagged clustering and its application to tourism market segmentation, *Expert Systems with Applications*, 40, 4944–4956.

D'Urso, P., L. De Giovanni, and R. Massari 2015. Trimmed fuzzy clustering for interval-valued data, *Advances Data Analysis and Classification*, 9, 21–40.

D'Urso, P., L. De Giovanni, and P. Spagnoletti 2013c. A fuzzy taxonomy for e-Health projects, *International Journal of Machine Learning and Cybernetics*, 4, 487–504.

D'Urso, P. and P. Giordani 2006. A weighted fuzzy c-means clustering model for fuzzy data, *Computational Statistics & Data Analysis*, 50, 1496–1523.

D'Urso, P. and R. Massari 2013. Fuzzy clustering of human activity patterns, *Fuzzy Sets and Systems*, 215, 29–54.

D'Urso, P., R. Massari, and A. Santoro 2010. A class of fuzzy clusterwise regression models, *Information Sciences*, 180, 4737–4762.

D'Urso, P. and A. Santoro 2006. Fuzzy clusterwise linear regression analysis with symmetrical fuzzy output variable, *Computational Statistics & Data Analysis*, 51, 287–313.

El-Sonbaty, Y. and M. A. Ismail 1998. Fuzzy clustering for symbolic data, *IEEE Transactions on Fuzzy Systems*, 6, 195–204.

Everitt, B. S., S. Landau, and M. Leese 2001. *Cluster Analysis*, London: Arnold Press, 4th ed.

Fadili, M.-J., S. Ruan, D. Bloyet, and B. Mazoyer 2001. On the number of clusters and the fuzziness index for unsupervised FCA application to BOLD fMRI time series, *Medical Image Analysis*, 5, 55–67.

Feil, B., B. Balasko, and J. Abonyi 2007. Visualization of fuzzy clusters by fuzzy Sammon mapping projection: Application to the analysis of phase space trajectories, *Soft Computing*, 11, 479–488.

Filippone, M., F. Camastra, F. Masulli, and S. Rovetta 2008. A survey of kernel and spectral methods for clustering, *Pattern Recognition*, 41, 176–190.

Frigui, H. 2007. Simultaneous clustering and feature discrimination with applications, in *Advances in Fuzzy Clustering and Its Applications*, eds. by J. V. De Oliveira and W. Pedrycz, Wiley, Chichester, 285–312.

Frigui, H. and R. Krishnapuram 1996. A robust algorithm for automatic extraction of an unknown number of clusters from noisy data, *Pattern Recognition Letters*, 17, 1223–1232.

Frigui, H. and O. Nasraoui 2000. Simultaneous clustering and attribute discrimination, in *Proceedings of the Ninth IEEE International Conference on Fuzzy Systems. FUZZ IEEE 2000*, IEEE, vol. 1, 158–163.

Fu, K. S. 1982. *Syntactic Pattern Recognition and Applications*, San Diego, CA: Academic Press.

Fukuyama, Y. and M. Sugeno 1989. A new method of choosing the number of clusters for the fuzzy c-means method, in *Proceedings of the Fifth Fuzzy Systems Symposium*, 247–250.

Gan, G. and J. Wu 2008. A convergence theorem for the fuzzy subspace clustering (FSC) algorithm, *Pattern Recognition*, 41, 1939–1947.

Gao, X.-B. 2004. *Fuzzy Clustering Analysis and Its Application*, Xi'an: Xidian University Press.

García-Escudero, L. Á. and A. Gordaliza 1999. Robustness properties of k-means and trimmed k-means, *Journal of the American Statistical Association*, 94, 956–969.

García-Escudero, L. Á. and A. Gordaliza 2005. A proposal for robust curve clustering, *Journal of Classification*, 22, 185–201.

Graves, D. and W. Pedrycz 2010. Kernel-based fuzzy clustering and fuzzy clustering: A comparative experimental study, *Fuzzy Sets and Systems*, 161, 522–543.

Groenen, P. J. F. and K. Jajuga 2001. Fuzzy clustering with squared Minkowski distances, *Fuzzy Sets and Systems*, 120, 227–237.

Groenen, P. J. F., U. Kaymak, and J. van Rosmalen 2007. Fuzzy clustering with Minkowski distance functions, in *Advances in Fuzzy Clustering and Its Applications*, eds. by J. V. De Oliveira and W. Pedrycz, Wiley, Chichester, 53–68.

Gustafson, D. and W. Kessel 1979. Fuzzy clustering with a fuzzy covariance matrix, in *Proceedings of the IEEE Conference on Decision and Control*, San Diego, 761–766.

Hall, L. O., A. M. Bensaid, L. P. Clarke, R. P. Velthuizen, M. S. Silbiger, and J. C. Bezdek 1992. A comparison of neural network and fuzzy clustering techniques in segmenting magnetic resonance images of the brain, *IEEE Transactions on Neural Networks*, 3, 672–682.

Hathaway, R. J. 1986. Another interpretation of the EM algorithm for mixture distributions, *Statistics & Probability Letters*, 4, 53–56.

Hathaway, R. J. and J. C. Bezdek 1988. Recent convergence results for the fuzzy c-means clustering algorithms, *Journal of Classification*, 5, 237–247.

Hathaway, R. J. and J. C. Bezdek 1993. Switching regression models and fuzzy clustering, *IEEE Transactions on Fuzzy Systems*, 1, 195–204.

Hathaway, R. J. and J. C. Bezdek 1994. Nerf c-means: Non-Euclidean relational fuzzy clustering, *Pattern Recognition*, 27, 429–437.

Hathaway, R. J. and J. C. Bezdek 2001. Fuzzy c-means clustering of incomplete data, *IEEE Transactions on Systems, Man, and Cybernetics, Part B: Cybernetics*, 31, 735–744.

Hathaway, R. J. and J. C. Bezdek 2003. Visual cluster validity for prototype generator clustering models, *Pattern Recognition Letters*, 24, 1563–1569.

Hathaway, R. J., J. C. Bezdek, and Y. Hu 2000. Generalized fuzzy c-means clustering strategies using L_p norm distances, *IEEE Transactions on Fuzzy Systems*, 8, 576–582.

Hathaway, R. J., J. W. Davenport, and J. C. Bezdek 1989. Relational duals of the c-means clustering algorithms, *Pattern Recognition*, 22, 205–212.

Havens, T. C., J. C. Bezdek, C. Leckie, L. O. Hall, and M. Palaniswami 2012. Fuzzy c-Means algorithms for very large data, *IEEE Transactions on Fuzzy Systems*, 20, 1130–1146.

Heiser, W. J. and P. J. F. Groenen 1997. Cluster differences scaling with a within-clusters loss component and a fuzzy successive approximation strategy to avoid local minima, *Psychometrika*, 62, 63–83.

Hruschka, H. 1986. Market definition and segmentation using fuzzy clustering methods, *International Journal of Research in Marketing*, 3, 117–134.

Huang, Z. and M. K. Ng 1999. A fuzzy k-modes algorithm for clustering categorical data, *IEEE Transactions on Fuzzy Systems*, 7, 446–452.

Huang, M., Z. Xia, H. Wang, Q. Zeng, and Q. Wang 2012. The range of the value for the fuzzifier of the fuzzy c-means algorithm, *Pattern Recognition Letters*, 33, 2280–2284.

Huband, J. M., J. C. Bezdek, and R. J. Hathaway 2004. Revised visual assessment of (cluster) tendency (reVA), in *Proceedings of the North American Fuzzy Information Processing Society (NAFIPS)*, 101–104.

Huband, J. M., J. C. Bezdek, and R. J. Hathaway 2005. bigVAT: Visual assessment of cluster tendency for large data sets, *Pattern Recognition*, 38, 1875–1886.

Hullermeier, E., M. Rifqi, S. Henzgen, and R. Senge 2012. Comparing fuzzy partitions: A generalization of the Rand index and related measures, *IEEE Transactions on Fuzzy Systems*, 20, 546–556.

Hwang, H., W. S. DeSarbo, and Y. Takane 2007. Fuzzy clusterwise generalized structured component analysis, *Psychometrika*, 72, 181–198.

Jaimes, L. G. and V. Torra 2010. On the selection of parameter m in fuzzy c-means: A computational approach, in *Integrated Uncertainty Management and Applications*, Huynh, V.-N., Nakamori, Y., Lawry, J., Inuiguchi, M. (Eds.). Springer, Berlin Heidelberg, 443–452.

Jain, A. K. and R. C. Dubes 1988. *Algorithms for Clustering Data*, Prentice-Hall, Inc, Englewood Cliffs, NJ.

Jajuga, K. 1991. L_1-norm based fuzzy clustering, *Fuzzy Sets and Systems*, 39, 43–50.

Kamdar, T. and A. Joshi 2000. On creating adaptive Web servers using Weblog Mining, Technical report TR-CS- 00-05, Department of Computer Science and Electrical Engineering, University of Maryland, Baltimore County.

Kaufman, L. and P. J. Rousseeuw 1987. Clustering by means of medoids, in *Statistics Data Analysis Based on the L1-Norm and Related Methods*, ed. by Y. Dodge, Amsterdam: North-Holland, 405–416.

Kaufman, L. and P. J. Rousseeuw 2005. *Finding Groups in Data: An Introduction to Cluster Analysis*, Wiley, Hoboken, NJ.

Keller, A. 2000. Fuzzy clustering with outliers, in *19th International Conference of the North American Fuzzy Information Processing Society, 2000. NAFIPS*, IEEE, 143–147.

Kersten, P. R. 1999. Fuzzy order statistics and their application to fuzzy clustering, *IEEE Transactions on Fuzzy Systems*, 7, 708–712.

Kovacs, A. and J. Abonyi 2004. Visualization of fuzzy clustering results by modified Sammon mapping, in *Proceedings of the 3rd International Symposium of Hungarian Researchers on Computational Intelligence*, 177–188.

Krishnapuram, R., A. Joshi, O. Nasraoui, and L. Yi 2001. Low-complexity fuzzy relational clustering algorithms for web mining, *IEEE Transactions on Fuzzy Systems*, 9, 595–607.

Krishnapuram, R., A. Joshi, and L. Yi 1999. A fuzzy relative of the k-medoids algorithm with application to web document and snippet clustering, in *1999 IEEE International Fuzzy Systems Conference. FUZZ-IEEE'99*, IEEE, vol. 3, 1281–1286.

Krishnapuram, R. and J. M. Keller 1996. The possibilistic c-means algorithm: Insights and recommendations, *IEEE Transactions on Fuzzy Systems*, 4, 385–393.

Krishnapuram, R., O. Nasraoui, and H. Frigui 1992. The fuzzy c-spherical shells algorithm: A new approach, *IEEE Transactions on Neural Networks*, 3, 663–671.

Kruse, R., C. Döring, and M.-J. Lesot 2007. Fundamentals of fuzzy clustering, in *Advances in Fuzzy Clustering and Its Applications*, eds. by J. V. De Oliveira and W. Pedrycz, Wiley, Chichester, 3–30.

Lee, M. and W. Pedrycz 2009. The fuzzy C-means algorithm with fuzzy P-mode prototypes for clustering objects having mixed features, *Fuzzy Sets and Systems*, 160, 3590–3600.

Li, R.-P. and M. Mukaidono 1995. A maximum-entropy approach to fuzzy clustering, in *Proceedings of the 4th IEEE Conference on Fuzzy Systems (FUZZ-IEEE/IFES'95)*, Yokohama: IEEE, vol. 4, 2227–2232.

Li, R.-P. and M. Mukaidono 1999. Gaussian clustering method based on maximum-fuzzy-entropy interpretation, *Fuzzy Sets and Systems*, 102, 253–258.

Lin, C.-T., C.-B. Chen, and W.-H. Wu 2004. Fuzzy clustering algorithm for latent class model, *Statistics and Computing*, 14, 299–310.

Liu, J. 2010. Detecting the fuzzy clusters of complex networks, *Pattern Recognition*, 43, 1334–1345.

Liu, Y. and Y. Zhang 2007. Optimizing parameters of fuzzy c-means clustering algorithm, in *Proceedings of the Fourth International Conference on Fuzzy Systems and Knowledge Discovery*, vol. 1, 633–638.

MacQueen, J. 1967. Some methods for classification and analysis of multivariate observations, in *Proceedings of the fifth Berkeley Symposium on Mathematical Statistics and Probability*, California, USA, vol. 1, 281–297.

McBratney, A. B. and A. W. Moore 1985. Application of fuzzy sets to climatic classification, *Agricultural and Forest Meteorology*, 35, 165–185.

Ménard, M., V. Courboulay, and P.-A. Dardignac 2003. Possibilistic and probabilistic fuzzy clustering: Unification within the framework of the non-extensive thermostatistics, *Pattern Recognition*, 36, 1325–1342.

Miyamoto, S. and Y. Agusta 1995. An efficient algorithm for l_1 fuzzy c-means and its termination, *Control and Cybernetics*, 24, 421–436.

Miyamoto, S., H. Ichihashi, and K. Honda 2008a. *Algorithms for Fuzzy Clustering: Methods in c-Means Clustering with Applications*, Springer, Berlin Heidelberg.

Miyamoto, S., H. Ichihashi, and K. Honda 2008b. Fuzzy clustering and probabilistic PCA model, *Studies in Fuzziness and Soft Computing*, 229, 157–169.

Miyamoto, S. and M. Mukaidono 1997. Fuzzy c-means as a regularization and maximum entropy approach, in *IFSA'97 Prague: Proceedings of the seventh International Fuzzy Systems Association World Congress*, 86–92.

Mulvey, J. M. and H. P. Crowder 1979. Cluster analysis: An application of Lagrangian relaxation, *Management Science*, 25, 329–340.

Nasraoui, O., R. Krishnapuram, A. Joshi, and T. Kamdar 2002. Automatic web user profiling and personalization using robust fuzzy relational clustering, in *Studies in Fuzziness and Soft Computing*, eds. by J. Segovia, P. Szczepaniak, and M. Niedzwiedzinski, Springer, Heidelberg, vol. 105, 233–261.

Ohashi, Y. 1984. Fuzzy clustering and robust estimation, in *Ninth Meeting of SAS Users Group Int.*, Hollywood Beach, FL.

Okeke, F. and A. Karnieli 2006. Linear mixture model approach for selecting fuzzy exponent value in fuzzy c-means algorithm, *Ecological Informatics*, 1, 117–124.

Ozkan, I. and I. B. Turksen 2007. Upper and lower values for the level of fuzziness in FCM, *Information Sciences*, 177, 5143–5152.

Pal, N. R. and J. C. Bezdek 1995. On cluster validity for the fuzzy c-means model, *IEEE Transactions on Fuzzy Systems*, 3, 370–379.

Pal, N. R., K. Pal, J. M. Keller, and J. C. Bezdek 2005. A possibilistic fuzzy c-means clustering algorithm, *IEEE Transactions on Fuzzy Systems*, 13, 517–530.

Pedrycz, W. 2002. Collaborative fuzzy clustering, *Pattern Recognition Letters*, 23, 1675–1686.

Pedrycz, W. 2005. *Knowledge-Based Clustering: From Data to Information Granules*, Wiley, Hoboken, NJ.

Pedrycz, W. and P. Rai 2008. Collaborative clustering with the use of fuzzy C-means and its quantification, *Fuzzy Sets and Systems*, 159, 2399–2427.

Pedrycz, W. and G. Vukovich 2004. Fuzzy clustering with supervision, *Pattern Recognition*, 37, 1339–1349.

Pedrycz, W. and J. Waletzky 1997. Fuzzy clustering with partial supervision, *IEEE Transactions on Systems, Man, and Cybernetics, Part B: Cybernetics*, 27, 787–795.

Pham, D. L. 2001. Spatial models for fuzzy clustering, *Computer Vision and Image Understanding*, 84, 285–297.

Pittau, M. G. and M. Vichi 2000. Fitting a fuzzy consensus partition to a set of partitions to analyze the modern economic growth across countries, *Journal of the Italian Statistical Society*, 9, 183–198.

Rao, M. R. 1971. Cluster analysis and mathematical programming, *Journal of the American Statistical Association*, 66, 622–626.

Rezaee, M. R., B. P. F. Lelieveldt, and J. H. C. Reiber 1998. A new cluster validity index for the fuzzy c-mean, *Pattern Recognition Letters*, 19, 237–246.

Rocci, R. and M. Vichi 2005. Three-mode component analysis with crisp or fuzzy partition of units, *Psychometrika*, 70, 715–736.

Roubens, M. 1978. Pattern classification problems and fuzzy sets, *Fuzzy Sets and Systems*, 1, 239–253.

Rousseeuw, P. J. 1987. Silhouettes: A graphical aid to the interpretation and validation of cluster analysis, *Journal of Computational and Applied Mathematics*, 20, 53–65.

Rousseeuw, P. J., M.-P. Derde, and L. Kaufman 1989. Principal components of a fuzzy clustering, *Trends in Analytical Chemistry*, 8, 249–250.

Rueda, L. and Y. Zhang 2006. Geometric visualization of clusters obtained from fuzzy clustering algorithms, *Pattern Recognition*, 39, 1415–1429.

Runkler, T. A. 2007. Relational fuzzy clustering, in *Advances in Fuzzy Clustering and Its Applications*, eds. by J. V. De Oliveira and W. Pedrycz, Wiley, Chichester, 31–51.

Runkler, T. A. and J. C. Bezdek 1999. Alternating cluster estimation: A new tool for clustering and function approximation, *IEEE Transactions on Fuzzy Systems*, 7, 377–393.

Runkler, T. A. and J. C. Bezdek 2003. Web mining with relational clustering, *International Journal of Approximate Reasoning*, 32, 217–236.

Ruspini, E. H. 1969. A new approach to clustering, *Information and Control*, 15, 22–32.

Ruspini, E. H. 1970. Numerical methods for fuzzy clustering, *Information Sciences*, 2, 319–350.

Ruspini, E. H. 1973. New experimental results in fuzzy clustering, *Information Sciences*, 6, 273–284.

Sărbu, C., K. Zehl, and J. W. Einax 2007. Fuzzy divisive hierarchical clustering of soil data using Gustafson–Kessel algorithm, *Chemometrics and Intelligent Laboratory Systems*, 86, 121–129.

Sivanandam, S. N., S. Sumathi, and S. N. Deepa 2007. *Introduction to Fuzzy Logic Using MATLAB*, Springer-Verlag, Berlin Heidelberg.

Sorenson, J. A. and X. Wang 1996. ROC methods for evaluation of fMRI techniques, *Magnetic Resonance in Medicine*, 36, 737–744.

Steinley, D. and M. J. Brusco 2007. Initializing K-means batch clustering: A critical evaluation of several techniques, *Journal of Classification*, 24, 99–121.

Tokushige, S., H. Yadohisa, and K. Inada 2007. Crisp and fuzzy k-means clustering algorithms for multivariate functional data, *Computational Statistics*, 22, 1–16.

Trauwaert, E. 1987. $L1$ in fuzzy clustering, in *Statistics Data Analysis Based on the $L1$-Norm and Related Methods*, ed. by Y. Dodge, Amsterdam: North-Holland, 417–426.

Trauwaert, E., L. Kaufman, and P. Rousseeuw 1991. Fuzzy clustering algorithms based on the maximum likelihood principle, *Fuzzy Sets and Systems*, 42, 213–227.

Vapnik, V. 1995. *The Nature of Statistical Learning Theory*, New York, NY: Springer.

Vinod, H. D. 1969. Integer programming and the theory of grouping, *Journal of the American Statistical Association*, 64, 506–519.

Wang, X.-Y., J. M. Garibaldi, B. Bird, and M. W. George 2007. Novel developments in fuzzy clustering for the classification of cancerous cells using FTIR spectroscopy, in *Advances in Fuzzy Clustering and Its Applications*, eds. by J. V. De Oliveira and W. Pedrycz, Wiley, Chichester, 404–425.

Wang, W. and Y. Zhang 2007. On fuzzy cluster validity indices, *Fuzzy Sets and Systems*, 158, 2095–2117.

Wedel, M. and J.-B. E. M. Steenkamp 1989. A fuzzy clusterwise regression approach to benefit segmentation, *International Journal of Research in Marketing*, 6, 241–258.

Wedel, M. and J.-B. E. M. Steenkamp 1991. A clusterwise regression method for simultaneous fuzzy market structuring and benefit segmentation, *Journal of Marketing Research*, 28, 385–396.

Windham, M. P. 1985. Numerical classification of proximity data with assignment measures, *Journal of Classification*, 2, 157–172.

Wu, K.-L. 2012. Analysis of parameter selections for fuzzy c-means, *Pattern Recognition*, 45, 407–415.

Xie, X. L. and G. Beni 1991. A validity measure for fuzzy clustering, *IEEE Transactions on Pattern Analysis and Machine Intelligence*, 13, 841–847.

Xu, Y. and R. G. Brereton 2005. A comparative study of cluster validation indices applied to genotyping data, *Chemometrics and Intelligent Laboratory Systems*, 78, 30–40.

Yang, M.-S. 1993a. On a class of fuzzy classification maximum likelihood procedures, *Fuzzy Sets and Systems*, 57, 365–375.

Yang, M.-S. 1993b. A survey of fuzzy clustering, *Mathematical and Computer Modelling*, 18, 1–16.

Yang, M.-S., P.-Y. Hwang, and D.-H. Chen 2004. Fuzzy clustering algorithms for mixed feature variables, *Fuzzy Sets and Systems*, 141, 301–317.

Yang, M.-S. and C.-H. Ko 1997. On cluster-wise fuzzy regression analysis, *IEEE Transactions on Systems, Man, and Cybernetics, Part B: Cybernetics*, 27, 1–13.

Yang, M.-S. and J.-A. Pan 1997. On fuzzy clustering of directional data, *Fuzzy Sets and Systems*, 91, 319–326.

Yang, M.-S. and K.-L. Wu 2006. Unsupervised possibilistic clustering, *Pattern Recognition*, 39, 5–21.

Yang, M.-S., K.-L. Wu, J.-N. Hsieh, and J. Yu 2008. Alpha-cut implemented fuzzy clustering algorithms and switching regressions, *IEEE Transactions on Systems, Man, and Cybernetics, Part B: Cybernetics*, 38, 588–603.

Yu, J., Q. Cheng, and H. Huang 2004. Analysis of the weighting exponent in the FCM, *IEEE Transactions on Systems, Man, and Cybernetics, Part B: Cybernetics*, 34, 634–639.

Zadeh, L. A. 1965. Fuzzy sets, *Information and Control*, 8, 338–353.

Zadeh, L. A. 1973. Outline of a new approach to the analysis of complex systems and decision processes, *IEEE Transactions on Systems, Man and Cybernetics*, 3, 28–44.

Zhang, J.-S. and Y.-W. Leung 2004. Improved possibilistic c-means clustering algorithms, *IEEE Transactions on Fuzzy Systems*, 12, 209–217.

25

Rough Set Clustering

Ivo Düntsch and Günther Gediga

CONTENTS

Abstract

In this chapter, we show how the rough set approach can be successfully used for unsupervised classification.

25.1 Introduction to Rough Sets

Rough sets have been introduced in the early 1980s as a tool to handle uncertain information (Pawlak 1982), initially assuming nominally scaled data. Its theory is based on the idea that objects can only be distinguished up to the features which describe them. The main application areas of rough set data analysis (RSDA) are feature reduction and supervised learning, using *decision systems* such as the one shown in Table 25.1. However, RSDA can also be used for unsupervised learning. In most cases, clustering methods based on rough sets assume one or more given partitions of the dataset and then aim to find a (cluster) variable which best represents the data according to some predefined measure.

More formally, given an equivalence relation θ on a universe U, we assume that we know the world only up to the equivalence classes of θ and have no other knowledge about the objects within a class. A pair $\langle U, \theta \rangle$ is called an *approximation space*; the set of

TABLE 25.1

A Decision System

U	a	b	c	d	U	a	b	c	d
1	1	0	0	1	12	0	1	1	1
2	1	0	0	1	13	0	1	1	2
3	1	1	1	1	14	1	1	0	2
4	0	1	1	1	15	1	1	0	2
5	0	1	I	1	16	1	1	0	2
6	0	1	1	1	17	1	1	0	2
7	0	1	1	1	18	1	1	0	3
8	0	1	1	1	19	1	0	0	3
9	0	1	1	1	20	1	0	0	3
10	0	1	1	1	21	1	0	0	3
11	0	1	1	1					

Source: Adapted from W. Ziarko. Variable precision rough set model. *Journal of Computer and System Sciences*, 46:39–59, 1993.

equivalence classes of θ is denoted by $\mathcal{P}(\theta)$. Given some $X \subseteq U$, the *lower approximation of* X is the set $\underline{X}_\theta = \{x \in U : \theta(x) \subseteq X\}$ and the *upper approximation of* X is the set $\bar{X}^\theta = \{x \in U : \theta(x) \cap X \neq \emptyset\}$; here, $\theta(x)$ is the class of θ containing x (see Figure 25.1). For the sake of clarity, we will sometimes write $\mathrm{low}_\theta(X)$ for $\underline{X}_\theta$ and $\mathrm{upp}_\theta(X)$ for $\bar{X}^\theta$. If θ is understood, we will omit its index.

A *rough set* is a pair $\langle \underline{X}, \bar{X} \rangle$. We interpret the approximation operators as follows: If $\theta(x) \subseteq X$, then we know for certain that $x \in X$, if $\theta(x) \cap \bar{X} = \emptyset$ we are certain that $x \notin X$. In the *area of uncertainty* $\bar{X} \backslash \underline{X}$, we can make no certain prediction since $\theta(x)$ intersects both X and $U \backslash X$.

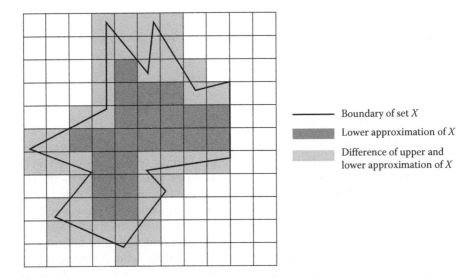

Boundary of set X

Lower approximation of X

Difference of upper and lower approximation of X

FIGURE 25.1
Upper and lower approximation. (Adapted from I. Düntsch and G. Gediga. *Rough Set Data Analysis: A Road to Non-invasive Knowledge Discovery*. Methodos Publishers, Bangor, 2000.)

The primary statistical tool of RSDA is the *approximation quality function* $\gamma : 2^U \rightarrow [0,1]$: If $X \subseteq U$, then

$$\gamma(X) = \frac{|\underline{X}| + |U \setminus \overline{X}|}{|U|} \tag{25.1}$$

which is just the ratio of the number20pc]Q2 of certainly classified elements of U to the number of all elements of U. The approximation quality γ is a manifestation of the underlying statistical principle of RSDA, namely, the principle of indifference: Within each equivalence class, the elements are assumed to be randomly distributed. If $\gamma(X) = 1$, then $X = \underline{X} = \overline{X}$; in this case, we say that X is *definable*. Another index frequently used in RSDA is the *accuracy measure* which is defined as

$$\alpha(X) = \frac{|\underline{X}|}{|\overline{X}|} \tag{25.2}$$

For example, if $U = \{1, 2, \ldots, 6\}$ and θ has the classes $\{1\}, \{2, 3, 4\}, \{5, 6\}$, then

$$\alpha(\{1, 5, 6\}) = \frac{|\{1, 5, 6\}|}{|\{1, 5, 6\}|} = 1, \quad \alpha(\{1, 2, 3\}) = \frac{|\{1\}|}{|\{1, 2, 3, 4\}|} = \frac{1}{4}, \quad \alpha(\{2, 5\}) = \frac{|\emptyset|}{|\{2, \ldots, 6\}|} = 0$$

The index $1 - \alpha(X)$ is also called the *roughness of X* (Mazlack et al. 2000).

The main application area of RSDA is classification, that is, supervised learning: An *information system* is a tuple $\mathcal{I} = \langle U, \Omega, (V_q)_{q \in \Omega}, (f_q)_{q \in \Omega} \rangle$, where U is a finite nonempty set of objects, Ω a finite nonempty set of attributes, for each $q \in \Omega$, V_q is a finite set of values which attribute q can take, and $f_q : U \rightarrow V_q$ is the information function which assigns to each object x its value under attribute q. Each subset Q of Ω defines an equivalence relation θ_Q on U by setting

$$x \equiv_{\theta_Q} y \quad \Longleftrightarrow \quad f_q(x) = f_q(y) \quad \text{for all } q \in Q \tag{25.3}$$

The equivalence class of x with respect to the equivalence relation θ_Q will usually be denoted by $B_x(Q)$.

A *decision system* is an information system $\mathcal{I}$ enhanced by a decision attribute d with value set V_d and information function $f_d : U \rightarrow V_d$. To avoid trivialities, we will assume that f_d takes at least two values, that is, that θ_d has at least two classes. If $Q \subseteq \Omega$ and X is a class of θ_Q, we say that X is Q, d-deterministic, if there is a class Y of θ_d such that $X \subseteq Y$. The *approximation quality of Q with respect to d* is now defined as

$$\gamma(Q, d) = \frac{|\bigcup \{X : X \text{ is } Q, d\text{-deterministic}\}|}{|U|} \tag{25.4}$$

$\gamma(Q, d)$ is the relative number of elements of U which can be correctly classified given the knowledge of Q. If $\theta_d = \theta_\Omega$, that is, if the partition to be approximated is the partition given by the full set of attributes, we just write $\gamma(Q)$.

As an example consider the decision system shown in Table 25.1. The classes of θ_Ω are

$$X_1 = \{1, 2, 19, 20, 21\}, \quad X_2 = \{3\}, \quad X_3 = \{4, \ldots, 13\}, \quad X_4 = \{14, \ldots, 18\} \tag{25.5}$$

and the decision classes are

$$X_1^d = \{1, \ldots, 12\}, \quad X_2^d = \{13, \ldots, 17\}, \quad X_3^d = \{18, \ldots, 21\} \tag{25.6}$$

There is one d-deterministic class of θ_Ω, namely, $X_2 = \{3\}$.

A first approach to apply RSDA to find clusters is to agglomerate the predicting and decision attributes, that is, to regard the set $\Omega' = \Omega \cup \{d\}$ as the set of attributes under consideration, and consider the resulting equivalence relation on U. In case of our example in Table 25.1, the resulting classes are

$$X_1 = \{1,2\}, \quad X_2 = \{3\}, \quad X_3 = \{4,\ldots,12\}, \quad X_4 = \{13\}$$

$$X_5 = \{14,\ldots,17\}, \quad X_6 = \{18\}, \quad X_7 = \{19,20,21\}$$

Another prominent feature of RSDA is the reduction of the number of conditional attributes required to approximate the decision attribute: A *reduct of $\mathcal{I}$ with respect to d* is a set Q of attributes minimal with respect to the property

$$\gamma(Q,d) = \gamma(\Omega,d) \tag{25.7}$$

In our example, the only reduct is the set Ω.

For a more complete treatment of rough sets, we invite the reader to consult (Düntsch and G. Gediga 2000).

25.1.1 The Variable Precision Model

It is assumed in the rough set model that the boundaries of the equivalence classes are crisp, and thus, that measurements are accurate. There are several possibilities to reduce the precision of prediction to cope with measurement error. One such approach is the *variable precision rough set model* (VPRS-model) (Ziarko 1993; Ziarko 2008) which assumes that rules are valid only within a certain part of the population: Let U be a finite universe, $X, Y \subseteq U$, and first define the *relative degree of misclassification* of X with respect to Y

$$c(X,Y) = \begin{cases} 1 - \frac{|X \cap Y|}{|X|}, & \text{if } |X| \neq 0 \\ 0, & \text{if } |X| = 0 \end{cases} \tag{25.8}$$

Clearly, $c(X,Y) = 0$ if and only if $X = 0$ or $X \subseteq Y$, and $c(X,Y) = 1$ if and only if $X \neq \emptyset$ and $X \cap Y = \emptyset$. The *majority requirement* of the VPRS–model implies that more than 50% of the elements in X should be in Y; this can be specified by an additional parameter β which is interpreted as an admissible classification error, where $0 \leq \beta < 0.5$. The *majority inclusion relation* $\overset{\beta}{\subseteq}$ (with respect to β) is now defined as

$$X \overset{\beta}{\subseteq} Y \iff c(X,Y) \leq \beta \tag{25.9}$$

Given a family of nonempty subsets $\mathcal{X} = \{X_1, \ldots, X_k\}$ of U and $Y \subseteq U$, the *lower approximation $\underline{Y}_\beta$ of Y given $\mathcal{X}$ and β* is defined as the union of all those X_i, which are in relation $X_i \overset{\beta}{\subseteq} Y$, in other words,

$$\underline{Y}_\beta = \bigcup \{X \in \mathcal{X} : c(X,Y) \leq \beta\} \tag{25.10}$$

The classical approximation quality $\gamma(Q,d)$ is now replaced by a three-parametric version which includes the external parameter β, namely,

$$\gamma(Q,d,\beta) = \frac{|\text{Pos}(Q,d,\beta)|}{|U|} \tag{25.11}$$

where $\text{Pos}(Q, d, \beta)$ is the union of those equivalence classes X of θ_Q for which $X \overset{\beta}{\subseteq} Y$ for some decision class Y. Note that $\gamma(Q, d, 0) = \gamma(Q, d)$.

Continuing the example from Table 25.1 with the classes of θ_Ω given in (25.5) and the decision classes given in (25.6), we obtain

$$\gamma(\Omega, d, 0) = \frac{|X_2|}{|U|} \qquad = \frac{|\{3\}|}{|\{1, 2, \ldots, 21\}|} \qquad = 1/21$$

$$\gamma(\Omega, d, 0.1) = \frac{|X_2 \cup X_3|}{|U|} \qquad = \frac{|\{3, 4, \ldots, 13\}|}{|\{1, 2, \ldots, 21\}|} \qquad = 11/21$$

$$\gamma(\Omega, d, 0.2) = \frac{|X_2 \cup X_3 \cup X_4|}{|U|} \qquad = \frac{|\{3, 4, \ldots, 13, 14, \ldots, 18\}|}{|\{1, 2, \ldots, 21\}|} \qquad = 16/21$$

$$\gamma(\Omega, d, 0.4) = \frac{|X_2 \cup X_3 \cup X_4 \cup X_1|}{|U|} \qquad = \frac{|\{1, 2, \ldots, 21\}|}{|\{1, 2, \ldots, 21\}|} \qquad = 21/21$$

Unfortunately, $\gamma(Q, d, \beta)$ is not necessarily monotone with respect to inclusion of attribute sets (Beynon 2001; Düntsch and Gediga 2012). In (Düntsch and Gediga 2012) we show that $\gamma(Q, d, \beta)$ is a special instance of Goodman–Kruskal's λ (Goodman and Kruskal 1954) and how a class of monotone measures similar to $\gamma(Q, d, \beta)$ can be constructed.

The upper approximation within the variable precision model is defined in the following way:

$$\bar{Y}^\beta = \bigcup\{X \in \mathcal{X} : c(X, Y) < 1 - \beta\} \tag{25.12}$$

Following this definition it is obvious that the restriction $0 \leq \beta < 0.5$ must hold. Given this definition, the accuracy measure α of the classical RSDA can be rephrased as

$$\alpha(Y)_\beta = \frac{|\underline{Y}_\beta|}{|\bar{Y}^\beta|} \tag{25.13}$$

25.1.2 Rough Entropy and Shannon Entropy

In many situations Shannon entropy may be used as a measure of the success of optimal guessing within data, however, we have shown in Düntsch and Gediga (1998) that this approach does not fit the theoretical framework of rough set analysis. Using rough sets, the data are split into a deterministic part (lower approximation) and an indeterministic part (upper approximation). Here, the deterministic part corresponds to "knowing" whereas the indeterministic part still can be applied by "guessing." Throughout this section we let A be an attribute set, and θ_A be the equivalence relation in U based on A with associated partition $\mathcal{P}(A)$.

As described in Section 25.2.7 below, Jiang et al. (2010) exhibit a revised version of entropy for the purpose of outlier detection using the rough set model.

Liang et al. (2006) generalize the concept of rough entropy of Liang and Shi (2004) to incomplete information systems, which is simply

$$RE(A) = -\sum_{X \in \mathcal{P}(A)} \frac{|X|}{|U|} \cdot \log_2(|X|) \tag{25.14}$$

Quian and Liang (2008) offer an alternative look on entropy, which they call combination entropy, defined by

$$CE(A) = \sum_{X \in \mathcal{P}(A)} \frac{|X|}{|U|} \cdot \left(1 - \frac{|X| \cdot (|X| - 1)}{|U| \cdot (|U| - 1)}\right) \tag{25.15}$$

Similar to Shannon entropy, the combination entropy increases as the equivalence classes become smaller through finer partitioning.

Liang et al. (2009) offer a further alternative to the accuracy $\alpha(X)$ of (25.2). The accuracy $\alpha(X)$ does not take into account what the granulation of the approximating sets look like. Therefore, an index proposed by Liang et al. (2009) is based on knowledge granulation of an attribute set A in the set U, defined by

$$KG(A) = \sum_{X \in \mathcal{P}(A)} \frac{|X|^2}{|U|^2} \tag{25.16}$$

Then, the roughness of the approximation of X is a weighted α measure given by

$$\text{Roughness } (X) = (1 - \alpha(X)) \cdot KG(A) \tag{25.17}$$

which results in a new accuracy measure $\alpha'(X)$:

$$\alpha'(X) = 1 - \text{Roughness } (X) = 1 - (1 - \alpha(X)) \cdot KG(A) = \alpha(X) \cdot KG(A) \tag{25.18}$$

Any of the proposed measures in this section can be used within a reduct search method, which may be considered as a tool to find clusters as well. We will discuss this further in Section 25.2.2.

25.2 Methods of Rough Sets Clustering

The starting point of rough set clustering is an information system $\mathcal{I}$ with the aim of finding one attribute, respectively, a partition of the set of objects, which best represents all attributes according to some predefined criterion.

25.2.1 Total Roughness and Min-Min Roughness-Based Methods

Suppose that $\mathcal{I} = \langle U, \Omega, (V_q)_{q \in \Omega}, (f_q)_{q \in \Omega} \rangle$ is a given information system. If $p, q \in \Omega$, $p \neq q$, and $V_p = \{x_1, \ldots, x_n\}$, the *roughness of* x_k *with respect to* q is defined as

$$R(x_k, p, q) = \frac{\text{low}_{\theta_q} (f_p^{-1}(x_k))}{\text{upp}_{\theta_q} (f_p^{-1}(x_k))} \tag{25.19}$$

and the *mean roughness of* p *with respect to* q is the value

$$MR(p, q) = \frac{\sum_{k=1}^{n} R(x_k, p, q)}{n} \tag{25.20}$$

The *total roughness of attribute p* Mazlack et al. (2000) is defined as

$$TR(p) = \frac{\sum_{q \in \Omega, r \neq p} MR(p,q)}{|\Omega| - 1} \tag{25.21}$$

It is suggested in Mazlack et al. (2000) to use the attribute with the highest total roughness as a clustering attribute. A technique somewhat opposite to TR was proposed by Parmar et al. (2007) which purports to "handle uncertainty in the process of clustering categorical data." Define

$$mR(p,q) = 1 - \frac{\sum_{k=1}^{n} R(x_k, p, q)}{n} \tag{25.22}$$

and

$$MR(p) = \min \{mR(p,q) : q \in \Omega, p \neq q\} \tag{25.23}$$

as well as

$$\min \min R = \min \{MR(p) : p \in \Omega\} \tag{25.24}$$

see Parmar et al. (2007) for details.

25.2.2 Reduct-Based Clustering

A straightforward application of RSDA for cluster analysis are techniques related to the reducts of an information system as defined in (25.7), which are mainly used for nominally scaled datasets. The method is simple: Let the classes of the partition induced by all attributes be the sets to be approximated, and use a subset of the given attributes to approximate these classes. A minimal subset of the attributes which approximate the classes in an optimal way given a criterion C is called a *C-reduct*. The partition which can be constructed by the attributes of the reduct can be interpreted as clusters. Note that there may be several C-reducts, which may generate different partitions as well.

Given the criterion $C \stackrel{\text{def}}{=}$ "$\gamma(Q) = 1$," a reduct needs to approximate the full partition θ_Ω without any error, in other words, any C-reduct will generate the full partition. This is a good way to preprocess the data for a subsequent–conventional–cluster analysis, as any reduct gives the same information (in terms of Shannon entropy) as the full set of attributes. In Questier et al. (2002) there is an application of this approach; they show in an application that a set of 126 features can be reduced to 68 features using reducts of the RSDA.

Although reducing attributes is a nice feature on its own, reducts can be used for clustering as well: When using a $0 < \gamma^*(Q) < 1$ in the criterion $C \stackrel{\text{def}}{=}$ "$\gamma(Q) = \gamma^*(Q)$," we may obtain reducts which produce partitions with a number of subsets that is smaller than or equal to the number of sets in the full partition. In the lower limit ($\gamma^*(Q) = 0$), no attribute is needed for approximation and therefore the only cluster remaining is U. The reduct search given a fixed $0 < \gamma^*(Q) < 1$ will generally result in different reducts—and in most of the cases—in different partitions as well. In this way we observe different sets of clusters, which can be used to approximate the full dataset with the minimal precision $\gamma^*(Q)$.

Scanning all reducts, a computation of fuzzy-membership of any element to a cluster is possible by aggregation using the set of partitions generated by the reducts.

Note that the criterion of a reduced γ is not the only base for a reduct search. Chen and Weng (2006) applied reduct search based on Shannon entropy, and Düntsch and Gediga (1998) used rough entropy measures for their reduct search method. A further application has been proposed by Mayszko and Stepaniuk (2010), who used Rényi-Rough-entropy for searching an optimal partition of images.

Related approaches are the "maximum dependency of attributes" method of Herawan et al. (2010), and the "upper approximation based clustering" of De (2004).

25.2.3 Application of Variable Precision Rough Sets

Yanto et al. (2011, 2012) use the variable precision model to determine an attribute which can be used to find the best cluster representation among a set of nominally scaled attributes. Suppose that $\Omega = \{a_1, \ldots, a_r\}$ is the set of attributes and that $V_i = \{v_1^i, \ldots, v_{n_i}^i\}$ is the set of attribute values of a_i. To avoid notational cluttering, with some abuse of language we denote the attribute function also by a_i. The equivalence relation belonging to a_i is denoted by θ_i; note that θ_i has exactly n_i classes, say, $X_1^i, \ldots, X_{n_i}^i$. Choose some error tolerance parameter $\beta < 0.5$, and for each $1 \leq j \leq r, i \neq j, 1 \leq k \leq n_i$ set

$$\alpha_\beta(a_j, v_k^i) = |\underline{X_k^i}_{\theta_j, \beta}| / |\overline{X_k^i}^{\theta_j, \beta}| \tag{25.25}$$

The *mean accuracy of a_i with respect to a_j* is now obtained as

$$\alpha_\beta(a_j, a_i) = \frac{\sum_{k=1}^{n_i} \alpha_\beta(a_j, v_k^i)}{n(i)} \tag{25.26}$$

Finally the mean value aggregating over the attributes $a_j (j \neq i)$ results in the *mean variable precision roughness* of attribute a_i by

$$\alpha_\beta(a_i) = \frac{\sum_{j \neq i} \alpha_\beta(a_j, a_i)}{r - 1} \tag{25.27}$$

The attribute for clustering now is obtained by taking an a_i for which $\alpha_\beta(a_i)$ is maximal.

Yanto et al. (2011) use their clustering algorithm with the datasets *Balloon, Tic-Tac-Toe Endgame, SPECT heart*, and *Hayes–Roth* from the University of California (Irvine) machine learning repository (Lichman 2013), and report that the result of the proposed VPRS technique provides a cluster purity of 83%, 69%, 64%, and 63%, respectively. Another application of applying reduct search in the variable precision rough set model using student anxiety data is given in Yanto et al. (2012).

Reduct-based methods can be used as well as a preprocessing tool for mixture or cluster analysis in the case of a mutual high dependency of the variables. In Mitra et al. (2003), it is shown how rough set-based algorithms for reduct detection can be used as a starter for mixture analysis. The analysis shows that using rough set-based preprocessing results in a stable mixture estimation with smaller error and higher validity than using the full set of variables.

25.2.4 Rough K-Means

We start with a set of n elements $x_1, \ldots, x_n$ and each element x is described by an m dimensional vector $\mathbf{v_x}$ of real valued (interval scaled) measurements. The classical K–MEANS algorithm consists of the following iteration scheme:

1. Start with a random assignment of the n elements to a set of clusters $C_0 = \{C_1, \ldots, C_k\}$. Set $t = 0$.
2. Compute values of the centroids of the k classes by

$$\bar{x}_j = \frac{\sum_{x \in C_j} \mathbf{v_x}}{|C_j|} (1 \leq j \leq k) \tag{25.28}$$

3. For $1 \leq i \leq n$, $1 \leq j \leq k$ compute the distances $d(\mathbf{v_{x_i}}, \bar{x}_j) = ||\mathbf{v_{x_i}} - \bar{x}_j||$. Here, $||.||$ is usually the standard Euclidian norm.
4. Reassign the n elements according to the minimal distance to the k cluster, resulting in C_{t+1}.
5. If $C_{t+1} = C_t$ stop. Otherwise set $t = t+1$ and proceed with step 2.

The rough K-means method (Lingras and West 2004; Peters 2006; Lingras et al. 2009; Chen and Zhang 2011) adopts the idea of a lower and an upper approximation of a set which is somewhat less rigid than using classes of an equivalence relation: With each subset X of U, a pair $\langle \underline{X}, \bar{X} \rangle$ of subsets of U is associated such that

1. Each $x \in U$ is in at most one $\underline{X}$.
2. $\underline{X} \subseteq \bar{X}$ for all $X \subseteq U$.
3. An object $x \in U$ is not in any lower bound $\underline{X}$ if and only if there are $Y_0, Y_1 \subseteq U$ such that $\bar{Y}_0 \neq \bar{Y}_1$ and $x \in \bar{Y}_0 \cap \bar{Y}_1$.

Applying this idea produces a cluster structure consisting of pairs of lower and upper approximations of the cluster: $C^* = \{(\underline{C_1}, \bar{C}_1), \ldots, (\underline{C_k}, \bar{C}_k)\}$.

Given a pair $(\underline{C_j}, \bar{C}_j)$, the values of the centroids of cluster j have to be computed.

The rough K-means (sometimes called rough C-means, e.g., in Zhou et al. (2011)) method uses a parameterized mean by

$$\bar{x}_j = \begin{cases} (1-w)\frac{\sum_{x \in \underline{C_j}} \mathbf{v}}{|\underline{C_j}|} + w\frac{\sum_{x \in \bar{C}_j \backslash \underline{C_j}} \mathbf{v}}{|\bar{C}_j \backslash \underline{C_j}|}, & \text{if } \bar{C}_j \backslash \underline{C_j} \neq \emptyset \\ \frac{\sum_{x \in \underline{C_j}} \mathbf{v}}{|\underline{C_j}|}, & \text{otherwise} \end{cases} \tag{25.29}$$

The parameter w weights the influence of the upper approximation. If $w = 0$, the upper approximation will not be used for the centroid computation, and the iteration is identical to the classical K-means method.

Given new centroids, the reassignment of the elements to (new) lower and upper bounds of clusters has to be described. In case of the lower bound, the standard rule from classical K-means is adopted. In order to find suitable upper approximations, a second parameter

$\theta \geq 1$ is used. If x is assigned to C_j, then x is assigned to some $\bar{C}_{j'}$ $(j' \neq j)$, if

$$\frac{d(v_i, \bar{x}'_j)}{d(v_i, \bar{x}_j)} \leq \theta \tag{25.30}$$

If one assigns concrete values to the parameters (w, θ) and randomly assigns each data object to exactly one lower approximation, we may start with $C_0^* = \{(\underline{C}_1, \bar{C}_1), \ldots, (\underline{C}_k, \bar{C}_k)\}$, and adapt the scheme of the classical K-MEANS algorithm to a ROUGH K-MEANS application.

Consider again the example of Table 25.1. Each object x is described by a four-dimensional vector v_x corresponding to the three prediction attributes and the decision attribute. Setting the initial clusters as

$$\underline{C}_1 = \bar{C}_1 = \{1, 2, 3\}$$
$$\underline{C}_2 = \bar{C}_2 = \{4, \ldots, 12\}$$
$$\underline{C}_3 = \bar{C}_3 = \{13, \ldots, 21\}$$

and applying $w = 1/4$, $\theta = 3/2$, the algorithm works as follows:

Iteration 1:
Compute the centroids $\bar{x}_i$ according to (25.29):

$$\bar{x}_1 = (1, .333, .333, 1), \quad \bar{x}_2 = (0, 1, 1, 1), \quad \bar{x}_3 = (.889, .667, .111, 2.5)$$

The distance of object 3 (which corresponds to the vector $(1, 1, 1, 1)$) to $\bar{x}_2$ is smaller than the distance to $\bar{x}_1$. Therefore, the object 3 is moved from $\underline{C}_1$ to $\underline{C}_2$. Similarly, the object 13 is moved from $\underline{C}_3$ to $\underline{C}_2$. No further changes can be observed.

The new lower bounds are now given by

$$\underline{C}_1 = \{1, 2\}, \quad \underline{C}_2 = \{3, \ldots, 13\}, \quad \underline{C}_3 = \{14, \ldots, 21\}$$

The distances of element 3 are in the θ bound required by (25.30), because $d(v_3, \bar{x}_1)/d(v_3, \bar{x}_2) = 1.125 \leq 1.5 = \theta$. Element 3 is therefore an element of the upper bound of cluster C_1. No other elements can be found, which fulfill the θ-inequality (25.30) for some $1 \leq j \neq j' \leq 3$. Therefore, the new upper bounds are as follows:

$$\bar{C}_1 = \{1, 2, 3\}, \quad \bar{C}_2 = \{3, \ldots, 13\}, \quad \bar{C}_3 = \{14, \ldots, 21\}$$

Iteration 2:
The new centroids are now

$$\bar{x}_1 = 3/4 \cdot (1, 0, 0, 1) + 1/4 \cdot (1, 1, 1, 1) = (1, .25, .25, 1)$$
$$\bar{x}_2 = (0.091, 1, 1, 0.909)$$
$$\bar{x}_3 = (1, 0.625, 0, 2.5)$$

Using these centroids, the lower approximations are unchanged. Furthermore, no object fulfills the θ-condition (25.30). Therefore,

$$\underline{C}_1 = \bar{C}_1 = \{1, 2\}, \quad \underline{C}_2 = \bar{C}_2 = \{3, \ldots, 13\}, \quad \underline{C}_3 = \bar{C}_3 = \{14, \ldots, 21\}$$

As $\bar{C}_1$ has changed, another iteration step is necessary.

Iteration 3:

The new centroids are now

$$\bar{x}_1 = (1,0,0,1), \quad \bar{x}_2 = (0.091, 1, 1, 0.909), \quad \bar{x}_3 = (1, 0.625, 0, 2.5)$$

Using these centroids, no changes can be observed. Furthermore, no object fulfills the θ-inequality (25.30).

As the lower and upper bounds of the clusters of iteration 2 and iteration 3 are identical, the algorithm terminates.

Mitra (2004) proposed an update of the parameters based on evolutionary optimization using the Davies-Bouldin index (Bezdek et al. 1998) as a measure of fit, which is a simple representation of the odd of within-cluster variability and between cluster-variability. The parameters K and ω are under control of a genetic algorithm (GA; 10 bits in a chromosome, 20 chromosomes, crossover probability $= 0.8$, mutation probability $= 0.02$). The GA is governed by the Davies–Bouldin index as optimization and convergence criterion. The algorithm was applied to several sets and compared to other clustering algorithms. The results show that the proposed method is rather successful—and very promising when using it with gene expression data. It should be noted that (Jeba Emilyn and Rama 2010) showed as well a successful treatment of gene expression data with rough set-based clustering methods.

The paper of Chen and Miao (2011) deals with interval set clustering, which is a generalization of the Rough-K-Means methods. In this paper, the problem of outliers is treated. The method starts with the classical Rough-K-Means method. Using the upper and lower bounds of the sets, an index $LUF_h(x)$ is defined—which captures the degree to which object x is reachable from the h next neighbors. Fixing h and applying a threshold for LUF_h, it is possible to eliminate those objects from the lower approximation of a cluster which are too far away from their neighbors. These objects will be assigned to the upper approximation of the cluster. Using the new sets of lower and upper approximation, the centroids can be recalculated and the iteration will start again. Applications to synthetic data and the Wisconsin breast cancer data support the applicability of this method.

25.2.5 Tolerance Rough Set Clustering

Tolerance rough set models (Skowron and Stepaniuk 1996) relax the transitivity requirement of equivalence relations, and are often used in information retrieval to find clusters of index terms in large text data bases (Ho and Nguyen 2002). A *tolerance space* is a pair $\langle U, R \rangle$, where U is a nonempty finite set, and R is a reflexive and symmetric relation. The *tolerance classes* are the sets of the form $R(x) = \{y \in U : xRy\}$. Unlike equivalence classes, different tolerance classes may have a nonempty intersection. Lower and upper approximation of some $X \subseteq U$ are defined as in the equivalence case.

As an example of tolerance rough set clustering, we present the nonhierarchical document clustering, proposed in Ho and Nguyen (2002). We start with a set of $T = \{t_1, \ldots, t_N\}$ of index terms and a set $D = \{d_1, \ldots, d_M\}$ of documents each of which is indexed by a set T_d of terms from T. The number of times a term t_i occurs in document d_j is denoted by $f_{d_j}(t_i)$, and $\sum_{j=1}^{M} \sum_{i=1}^{N}$, the number of (indexed) terms occurring in document d, is denoted by T_d^*. The number of documents in which t_i occurs is denoted by $f_D(t_i)$, and the number of documents in which index terms t_i, t_j both occur is denoted by $n_D(t_i, t_j)$.

Given a threshold θ which is the minimal acceptable number of common terms in documents, we now define a covering I_θ of T by

$$I_\theta(t_i) = \{t_i\} \cup \{t_j \mid n_D(t_i, t_j) \geq \theta\} \qquad (25.31)$$

Clearly, the sets $I_\theta(t_i)$ are the tolerance classes of a tolerance relation $\mathcal{I}_\theta$ on T defined by

$$t_i \mathcal{I}_\theta t_j \iff t_j \in I_\theta(t_i) \qquad (25.32)$$

Given a set of terms $X \subseteq T$, we are now able to define a lower and upper bound of X with respect to $\mathcal{I}_\theta$:

$$\underline{X}_{\mathcal{I}_\theta} = \{t_i \mid I_\theta(t_i) \subseteq X\} \qquad (25.33)$$

$$\bar{X}^{\mathcal{I}_\theta} = \{t_i \mid X \cap I_\theta(t_i) \neq \emptyset\} \qquad (25.34)$$

With each term t_i and each document d_j a weight $w(i, j)$ is associated by

$$w(i, j) = \begin{cases} 1 + \log(f_{d_j}(t_i)), & \text{if } t_i \in d_j \\ 0, & \text{otherwise} \end{cases} \qquad (25.35)$$

If we have a set of (intermediate) representatives $C_t = \{X_1, \ldots, X_k\}$, we are now able to calculate $C_t^* = \{\underline{X_1}_{\mathcal{I}_\theta}, \bar{X}_1^{\mathcal{I}_\theta}, \ldots, \underline{X_k}_{\mathcal{I}_\theta}, \bar{X}_k^{\mathcal{I}_\theta}\}$ and use this information to find better representatives C_{t+1}.

Various measures of similarity between two documents may be used. One such index is the *Dice coefficient* defined as

$$S(d_i, d_j) = \frac{2 \cdot C}{T_{d_i}^* + T_{d_j}^*} \qquad (25.36)$$

where C is the number of index terms which d_i and d_j have in common.

Suppose we are given a set of clusters $\mathcal{C} = \{C_1, \ldots, C_k\}$ of documents. The algorithm constructs a representative $R_i \subseteq T$ for each $1 \leq i \leq k$ such that

1. Each document $d \in C_i$ has an index term in R_i, i.e., $T_d \cap R_i \neq \emptyset$.
2. Each term in R_i is possessed by a large number of documents in C_i (given by some threshold).
3. No term in R_i is possessed by every document in C_i.

25.2.6 Validation Based on Rough Sets

Using validation methods which do not take into account the special character of clusters build by rough set methods might be biased. Note, for example, that in the result of the rough-K-means method an object may belong to more than one cluster. Moreover, each cluster C_j is represented by its lower approximation $\text{low}(C_j)$ and an upper approximation $\text{upp}(C_j)$ and/or the boundary set $\text{upp}(C_j) \backslash \text{low}(C_j)$.

Lingras et al. (2009) start with an "action" function $b_j(x_l)$, which assigns an element x_l to a set of clusters from a set $B_j \subseteq 2^U$. A simple loss function can be assigned by

$$\text{Loss}(b_j(x_l) \mid C_i) = \begin{cases} 0, & \text{if } C_i \in B_j \\ 1, & \text{otherwise} \end{cases} \qquad (25.37)$$

Let $sim(x_l, C_i)$ be a similarity function of element x_l and cluster $low(C_i)$, for example, the inverse of the distance of x_l and the centroid of $low(C_i)$. Assuming that the probability $p(C_i|x_l)$ is proportional to $sim(x_l, C_i)$ by

$$p(C_i|x_l) = \frac{sim(x_l, C_i)}{\sum_j sim(x_l, C_j)} \qquad (25.38)$$

this results in the risk function for assigning element x_l to a set B_j of clusters by

$$R(b_j(x_l)|x_l) = \sum_i Loss(b_j(x_l)|C_i) p(C_i|x_l) \qquad (25.39)$$

As any element x_l gets its own risk evaluation, we are ready to define risks for certain subsets. The risk for lower approximation for the decision $b_j(x_l)$ for any element of the lower approximation is given by

$$\sum_{x_l \in low(C_i)} R(b_j(x_l)|x_l) \qquad (25.40)$$

Furthermore the risk functions for elements of the upper approximation and the boundary are given by

$$\sum_{x_l \in upp(C_i)} R(b_j(x_l)|x_l) \qquad (25.41)$$

and

$$\sum_{x_l \in upp(C_i)\backslash low(C_i)} R(b_j(x_l)|x_l) \qquad (25.42)$$

respectively.

Applying the ideas to the synthetic and the Wisconsin breast cancer data, Lingras et al. (2009) showed the rough and crisp cluster analysis exhibits similar results on the lower approximation, but differs in the risk of the assignment of boundary elements.

25.2.7 Outlier Detection

In Jiang et al. (2010), it was shown that detecting outliers using the rough set model is at least as powerful as classical distances-based methods (Knorr et al. 2000) or K-nearest neighbor (KNN)-based methods (Ramaswamy et al. 2000).

Jiang et al. (2010) start with the idea of rough entropy presented in Düntsch and Gediga (1998). Let θ be an equivalence relation on U with associated partition $\mathcal{P}(\theta)$. Starting with an element x and its equivalence class B_x, they define a leaving-out entropy of θ_x by

$$E(\theta\backslash\{B_x\}) = - \sum_{B \in \mathcal{P}(\theta), B \neq B_x} \frac{|B|}{|U| - |B_x|} \cdot \log_2\left(\frac{|B|}{|U| - |B_x|}\right) \qquad (25.43)$$

Using the information $E(\theta)$ as a benchmark, it is straightforward to define the relative entropy of the class of object x given θ by

$$RE(B_x|\theta) = \begin{cases} 1 - \frac{E(\theta\setminus\{B_x\})}{E(\theta)}, & \text{if } E(\theta) > E(\theta\setminus\{B_x\}) \\ 0, & \text{otherwise} \end{cases} \tag{25.44}$$

Jiang et al. (2010) introduces the relative cardinality of class B_x in θ by the difference of cardinality of B_x and the mean cardinality of the other classes:

$$RC(B_x|\theta) = \begin{cases} |B_x| - \frac{1}{|\mathcal{P}(\theta)|-1} \cdot \sum_{B\in\mathcal{P}(\theta), B\neq B_x} |B|, & \text{if } |\mathcal{P}(\theta)| > 1 \\ 0, & \text{otherwise} \end{cases} \tag{25.45}$$

Now we are ready to define the outlier degree of B_x given θ by

$$OD(B_x|\theta) = \begin{cases} RE(B_x|\theta) \cdot \sqrt{\frac{|U|-|RC(B_x|\theta)|}{2|U|}}, & \text{if } RC(B_x|\theta) > 0 \\ RE(B_x|\theta) \cdot \sqrt{\frac{|U|-|RC(B_x|\theta)|}{2|U|}}, & \text{otherwise} \end{cases} \tag{25.46}$$

In the latter, $RC(B_x|\theta) \leq 0$ which means that if x is assigned to a smaller equivalence class of θ, then the class B_x has a higher possibility to be an outlier class than the other classes. Furthermore, the higher the relative entropy of B_x, the greater is the likelihood of x to be an outlier.

Up to now the outlier definition is given for a fixed equivalence relation θ. As θ is normally obtained from a set of attributes, we may take the attributes into account. Let $A = \{a_1, \ldots, a_k\}$ be a set of attributes, $A_1 = A\setminus\{a_1\}$, and $A_j = A_{j-1}\setminus\{a_j\}$ for $1 < j < k$. Then, $A_{k-1} \subsetneq \cdots \subsetneq A_1 \subsetneq A$. Given any set of attribute Q and an element x, we denote the equivalence class of x with respect to θ_Q by $B_x(Q)$. The entropy outlier factor $EOF(x)$ is now defined by

$$EOF(x) = 1 - \frac{\sum_{j=1}^{k} \left(1 - OD(B_x(\{a_j\}))\right) \cdot W_{\{a_j\}}(x) + \left(1 - OD(B_x(A_j))\right) \cdot W_{A_j}(x)}{2k} \tag{25.47}$$

using the weights

$$W_Q(x) = \sqrt{\frac{|B_x(Q)|}{|U|}} \tag{25.48}$$

As $B_x(Q)$ changes with attributes in Q, $EOF(x)$ is a function of x and describes a distance of x to the rest of the elements of U in terms of an information function. Jiang et al. (2010) show that applying $EOF(x)$ to the Lymphography data and the Wisconsin breast cancer data (which can be found in the UCI machine learning repository (Lichman 2013)) shows a better performance than distances based or KNN-based methods.

25.3 Conclusion and Outlook

We have presented an introduction to clustering based on Pawlak's rough set model and its associated data type, the information system.

As the rough set model is a basic idea of many papers, the overview we have given cannot be exhaustive. There exists a plethora of other approaches in the literature which share the idea of rough set analysis and adopt other concepts to analyze the data. Some of these approaches are "less rough and more other," for example,

- Rough-fuzzy-clustering and shadowed sets (Asharaf and Narasimha Murty 2003; Zhou et al. 2011).
- Clustering by categorical similarity measures based on rough sets using a hierarchical-clustering scheme (Chen et al. 2006).
- A fast heuristic Rough DB–scan procedure (Viswanath and Suresh Babu 2009).

We have shown that the rough set model as an-originally-symbolic data analysis tool has been developed into a viable system for cluster analysis. There are two main ideas of the rough set model which are promising for applications: The first is the idea of an upper bound and a boundary of set approximation; tolerance rough set clustering or rough K-means clustering uses this idea. Although the applications show satisfactory results, the methods depend on several parameters and thresholds, which may be somewhat problematic. There are some attempts to use these parameters as free parameters as well, but optimization can only be done by genetic algorithms or comparable methods. Here, further investigations are necessary. The other idea is to use reduct-based methods, either as preprocessing tools or as tools for cluster generation. These methods show a more direct connection to the roots of rough set analysis, but the problem of time complexity has not been solved until now. As long as we are only interested in one reduct, the situation is not complicated, but finding all reducts within a variable precision model is still an NP-complete problem.

Apart from the complexity issue, it is worthy to note that applying ideas of the rough set model in cluster analysis seems to lead to stable and valid results in general. Furthermore, they offer new insights into how concepts such as "cluster," "error," or "outlier" and even "validation" may be understood and defined.

References

S. Asharaf and M. Narasimha Murty. An adaptive rough fuzzy single pass algorithm for clustering large datasets. *Pattern Recognition*, 36(12):3015, 2003.

M. Beynon. Reducts within the variable precision rough sets model: A further investigation. *European Journal of Operational Research*, 134:592–605, 2001.

J.C. Bezdek and N.R. Pal. Some new indexes of cluster validity. *IEEE Transcations on Systems, Man, and Cybernetics, Part B: Cybernetics*, 28:301–315, 1998.

D. Chen, D.W. Cui, C.X. Wang, and Z.R. Wang. A rough set-based hierarchical clustering algorithm for categorical data. *International Journal of Information Technology*, 12:149–159, 2006.

M. Chen and D. Miao. Interval set clustering. *Expert Systems with Applications*, 38(4):2923–2932, 2011.

C.B. Chen and L.Y. Wang. Rough set based clustering with refinement using Shannon's entropy theory. *Computers and Mathematics with Applications*, 52:1563–1576, 2006.

J. Chen and C. Zhang. Efficient clustering method based on rough set and genetic algorithm. *Procedia Engineering*, 15:1498–1503, 2011.

S.K. De. A rough set theoretic approach to clustering. *Fundamenta Informaticae*, 62(3/4):409–417, 2004.

I. Düntsch and G. Gediga. Uncertainty measures of rough set prediction. *Artificial Intelligence*, 106:77–107, 1998.

I. Düntsch and G. Gediga. *Rough Set Data Analysis: A Road to Non-invasive Knowledge Discovery.* Methodos Publishers, Bangor, 2000. http://www.cosc.brocku.ca/~duentsch/archive/nida.pdf

I. Düntsch and G. Gediga. Weighted λ precision models in rough set data analysis. In *Proceedings of the Federated Conference on Computer Science and Information Systems, Wrocław, Poland*, pp. 309–316. IEEE, 2012.

L.A. Goodman and W.H. Kruskal. Measures of association for cross classification. *Journal of the American Statistical Association*, 49:732–764, 1954.

T. Herawan, M.M. Deris, and J.H. Abawajy. A rough set approach for selecting clustering attribute. *Knowledge-Based Systems*, 23(3):220–231, 2010.

T.B. Ho and N.B. Nguyen. Nonhierarchical document clustering based on a tolerance rough set model. *International Journal of Intelligent Systems*, 17(2):199–212, 2002.

J. Jeba Emilyn and Dr. K. Rama. Rough set based clustering of gene expression data: A survey. *International Journal of Engineering Science and Technology*, 2:7160–7164, 2010.

F. Jiang, Y. Sui, and C. Cao. An information entropy-based approach to outlier detection in rough sets. *Expert Systems with Applications*, 37:6338–6344, 2010.

E. Knorr, R. Ng, and V. Tucakov. Distance-based outliers: Algorithms and applications. *VLDB Journal: Very Large Databases*, 8:237–253, 2000.

J. Liang and Z. Shi. The information entropy, rough entropy and knowledge granulation in rough set theory. *International Journal of Uncertainty, Fuzziness and Knowledge-Based Systems*, 12:37–46, 2004.

J. Liang, Z. Shi, D. Li, and M.J. Wierman. Information entropy, rough entropy and knowledge granulation in incomplete information systems. *International Journal of General Systems*, 6:641–654, 2006.

J. Liang, J. Wang, and Y. Qian. A new measure of uncertainty based on knowledge granulation for rough sets. *Information Sciences*, 179:458–470, 2009.

M. Lichman. *UCI Machine Learning Repository*. University of California, Irvine, School of Information and Computer Sciences, 2013.

P. Lingras. Applications of rough set based k-means, Kohonen SOM, GA clustering. *Transactions on Rough Sets*, 7:120–139, 2007.

P. Lingras, M. Chen, and D. Miao. Rough multi-category decision theoretic framework. In G. Wang, T. Rui Li, J.W. Grzymala-Busse, D. Miao, A. Skowron, and Y. Yao, editors, *RSKT*, volume 5009 of *Lecture Notes in Computer Science*, pp. 676–683. Springer, Heidelberg, 2008.

P. Lingras, M. Chen, and D. Miao. Rough cluster quality index based on decision theory. *IEEE Transactions on Knowledge & Data Engineering*, 21(7):1014–1026, 2009.

P. Lingras, M. Hogo, and M. Sn. Interval set clustering of web users using modified Kohonen self-organizing maps based on the properties of rough sets. *Web Intelligence and Agent Systems*, 2(3):217–225, 2004.

P. Lingras and C. West. Interval set clustering of web users with rough K-Means. *Journal of Intelligent Information Systems*, 23(1):5–16, 2004.

D. Mayszko and J. Stepaniuk. Adaptive multilevel rough entropy evolutionary thresholding. *Information Sciences*, 180(7):1138–1158, 2010.

L.J. Mazlack, A. He, and Y. Zhu. A rough set approach in choosing partitioning attributes. In *Proceedings of the ISCA 13th International Conference (CAINE-2000)*, pp. 1–6, 2000.

S. Mitra. An evolutionary rough partitive clustering. *Pattern Recognition Letters*, 25:1439–1449, 2004.

P. Mitra, S.K. Pal, and Md. A. Siddiqi. Non-convex clustering using expectation maximization algorithm with rough set initialization. *Pattern Recognition Letters*, 24(6):863–873, 2003.

D. Parmar, T. Wu, and J. Blackhurst. MMR: An algorithm for clustering categorical data using rough set theory. *Data & Knowledge Engineering*, 63(3):879–893, 2007.

Z. Pawlak. Rough Sets. *International Journal of Computer & Information Science*, 11:341–356, 1982.

G. Peters. Some refinements of rough k-means clustering. *Pattern Recognition*, 39:1481–1491, 2006.

F. Questier, I. Arnaut-Rollier, B. Walczak, and D.L. Massart. Application of rough set theory to feature selection for unsupervised clustering. *Chemometrics and Intelligent Laboratory Systems*, 63(2):155–167, 2002.

Y. Quian and J. Liang. Combination entropy and combination granulation in rough set theory. *International Journal of Uncertainty, Fuzziness and Knowledge-Based Systems*, 16:179–193, 2008.

S. Ramaswamy, R. Rastogi, and K. Shim. Efficient algorithms for mining outliers from large datasets. In *Proceedings of the ACM SIGMOD Conference on Management of Data*, pp. 427–438, 2000.

A. Skowron and J. Stepaniuk. Tolerance approximation spaces. *Fundamenta Informaticae*, 27:245–253, 1996.

P. Viswanath and V. Suresh Babu. Rough-DBSCAN: A fast hybrid density based clustering method for large datasets. *Pattern Recognition Letters*, 30(16):1477–1488, 2009.

I.T.R. Yanto, T. Herawan, and M.M. Deris. Data clustering using variable precision rough set. *Intelligent Data Analysis*, 15:465–482, 2011.

I.T.R. Yanto, P. Vitasari, T. Herawan, and M.M. Deris. Applying variable precision rough set model for clustering student suffering study's anxiety. *Expert Systems with Applications*, 39:452–459, 2012.

J. Zhou, W. Pedrycz, and D. Miao. Shadowed sets in the characterization of rough-fuzzy clustering. *Pattern Recognition*, 44(8):1738–1749, 2011.

W. Ziarko. Variable precision rough set model. *Journal of Computer and System Sciences*, 46:39–59, 1993.

W. Ziarko. Probabilistic approach to rough sets. *International Journal of Approximate Reasoning*, 49(2):272–284, 2008.

Section VI

Cluster Validation and Further General Issues

26

Method-Independent Indices for Cluster Validation and Estimating the Number of Clusters

Maria Halkidi, Michalis Vazirgiannis, and Christian Hennig

CONTENTS

Abstract

Given a data set and a clustering algorithm running on it with different input parameter values, we obtain different partitionings of the data set into not necessarily meaningful clusters. As a consequence, in most applications the resulting clustering scheme requires some sort of evaluation of its validity. Evaluating the results of a clustering algorithm is known as "cluster validation." The present chapter focuses on relative cluster validity criteria that are used to compare different clusterings and find the one that "best" fits the considered data. These criteria are implemented by validity indices that can be evaluated from the data set and the given clustering alone without having access to a "true" clustering. This chapter aims at presenting an overview of the available relative cluster validity indices and at highlighting the differences between them and their implicit assumptions. Furthermore, we mention some software packages, we stress the requirements that are under-addressed by the recent approaches, and address new research directions.

26.1 Introduction

A researcher who applies cluster analysis is confronted with a large number of available methods, most of which can be run with different numbers of clusters or different choices of other tuning parameters. Furthermore, some data sets allow clear clusterings with strongly separated, homogeneous clusters whereas in other data sets cluster analysis methods rather artificially partition data that are not meaningfully clustered.

Cluster validation generally refers to exploring the quality of a clustering. This is more difficult than assessing the model fit or quality of prediction in regression or supervised classification, because in cluster analysis, normally true class information is not available.

The present chapter is about indices quantifying a clustering's quality. Such indices can either be used for a quality assessment of a single clustering, or in order to select a clustering by comparing its values for clusterings from different methods or computed with different parameters. They are probably most popular for finding an optimal number of clusters. The quality measurement problem in clustering currently attracts a lot of interest and new indices are proposed regularly. There is also some research investigating the behavior of indices already in the literature, although this seems to be somewhat lagging behind and many researchers in the field are more enthusiastic proposing their own indices (and comparing it with other indices only aiming at demonstrating the superiority of their own one). This is problematic because it leaves the user with a similar plethora of approaches to measure the quality of a clustering as there are cluster analysis methods the user may want to compare using such indices, without clear guidelines which index to choose. More research is needed.

Alternative or complementary approaches for cluster validation are treated elsewhere in the book: Chapter 27 treats methods to compare different clusterings, Chapter 28 treats the investigation of a clustering's stability by resampling, and Chapter 29 is about visualization in clustering.

Another standard method for cluster validation is a statistical test of the null hypothesis that the data are homogeneous and unclustered according to some null model. A "good" clustering should be significantly better than what is expected when the same clustering method is applied to null model data. Actually, this is strongly connected to cluster validity indices, because such indices can be used as test statistics for such tests, see Chapter 15. Computing or simulating the distribution of cluster validity indices under a suitable null model is very instructive for the interpretation of the values of such indices, for example, Section 26.2.5. Traditionally (Jain and Dubes 1988), the criteria used to evaluate the cluster validity are classified into

External criteria: These evaluate how well a clustering matches a predefined structure. Normally a true clustering is not available, in which case this structure could be a clustering from a different method (assessing to what extent different methods agree about a given data set), or an external variable, which could be a categorical classification or something continuous, and which researchers expect to be strongly connected with the clustering.

Internal criteria: These evaluate to what extent the clustering "fits" the data set based on the data used for clustering, often combined with significance testing as mentioned above.

Relative criteria: These aim at comparing different clusterings, often from the same method using different parameters, and finding the best one. Most indices used for assessing relative validity should be designed in such a way that their maximum or minimum value indicates an "optimal" clustering. Clusterings are computed for various values of the method's parameters (often the number of clusters K is the major parameter of interest, in which case all numbers in an interval between 1 and 2 and some maximum K_{max} are applied) and then the best one is selected.

There are also some indices that increase (or decrease) automatically with increasing K, in which case researchers look for "elbows"/"knees," that is, values of K at which a strong increase (decrease, respectively) from $K - 1$ to K is followed by a very weak one from K to $K + 1$ and for higher values of K.

The current chapter will ignore external validity, which is mostly about indices formalizing the similarity between two clusterings as treated in Chapter 27. The distinction between internal and relative validity does not require essentially different classes of indices, because relative validity can be assessed by comparing the internal validity of various clusterings. This assumes that indices are properly calibrated so that the comparison is fair; for example, indices used for finding an optimal number of clusters by optimization should not systematically prefer lower or higher numbers of clusters. Because many indices are based on heuristic reasoning, it is often difficult to assess whether this is the case for a given index or not, and this is an area of research. Kim and Ramakrishna (2005), for example, analyzed a number of standard indices and found them biased in certain situations. Hennig and Liao (2013) show a situation in which the optimum value of an index (regarding K) does not point the "best" K in the sense that the optimum K cannot be distinguished significantly from what is expected under an homogeneity null model for a clustering with the same K, whereas some other K-values only yield local optima of the index but are significantly better than what is expected under the null model, and therefore better supported.

Other important distinctions regarding cluster validity concern the type of clustering (most indices evaluate crisp partitions but one could also be interested in evaluating full hierarchies or a fuzzy clustering) and the data type. Many indices are defined for Euclidean data only. A considerable number of indices applies to dissimilarity data. Actually some indices originally designed for Euclidean data can be easily generalized to dissimilarity data, see (26.1), which makes them more flexible because a dissimilarity measure can be defined for many kinds of data. There are also specialist indices for some specific data types such as graph data (Section 26.3) or categorical data (Liu and Dong, 2012).

As mentioned above, it is difficult to select an index out of the large number of existing ones. In many situations, different indices lead to different conclusions. It is important to acknowledge that the clustering problem generally is not uniquely defined, and that there is no unique "best" clustering in a data set (there are data sets in which the "best" clustering seems very clear intuitively, but these are usually artificial toy examples). The definition of what makes a "good" clustering in a given application requires subject-matter dependent decisions on behalf of the researchers. This applies to the choice of the clustering method (discussed in some detail in Chapter 31) as well as to the choice of a validation index. Many validation indices balance in some way within-cluster homogeneity against between-cluster separation. Larger values of K usually improve homogeneity, but decrease separation. Depending on the application, the user may prefer a stronger focus on homogeneity (clustering may be used to organize databases so that they can be searched more easily, for which items in the same cluster should really be similar) or a stronger focus on separation ("cultures," be they historical/archeological or biological, may have very heterogeneous members but it is of main interest to find "gaps" between different cultures). An application may require all clusters to have a certain minimum size or not; and clusters with very different spreads may be acceptable or not. Index selection requires to understand how the existing indices balance the different possible clustering characteristics against each other, so that this can be matched to the requirements of the application. We will try to give some indications but overall far more research is required in this direction.

The title of the present chapter suggests that indices are "method-independent." Although most of the indices discussed later can indeed be computed for the outcomes of many different clustering methods, it still should be noted that many indices implicitly assume certain "cluster concepts" that are consistent with some but not other clustering methods. For example, quite a number of indices measure the homogeneity of a cluster by considering (squared) distances of observations from the cluster center (usually its mean vector for Euclidean data) and the heterogeneity between clusters by considering distances between cluster centers. This implicitly assumes spherical clusters that can be well represented by their centers. The K-means clustering method is based on the same cluster concept, whereas some other methods, for example, Single Linkage hierarchical clustering, look for clusters of a quite different nature.

We devote most space to the large number of indices for validating crisp partitions based on dissimilarity or Euclidean data (Section 26.2). This includes Section 26.2.9, where differences between the indices are illustrated by applying them to an artificial data set. We also mention some software there. Section 26.3 is about the validation of graph partitions. Section 26.4 discusses indices for fuzzy clustering. The chapter is closed by some discussion of key issues and open problems in Section 26.5.

26.2 Euclidean and Dissimilarity-Based Indices for Crisp Partitions

Assume that there are n observations $x_1, \ldots, x_n$, $x_i \in \mathcal{X}$, $i = 1, \ldots, n$, and a dissimilarity measure $d : \mathcal{X} \times \mathcal{X} \mapsto [0, \infty)$. $\mathcal{X} = \mathbb{R}^p$ Euclidean will be assumed where objects from $\mathcal{X}$ are manipulated directly, but where they are only incorporated through d, $\mathcal{X}$ can be anything. Assume that a clustering $\mathcal{C}_K = \{C_1, \ldots, C_K\}$ is given, which is an exhaustive partition (no overlap) of $\{x_1, \ldots, x_n\}$ (assume that all x_i can be distinguished from each other so that indeed $|\{x_1, \ldots, x_n\}| = n$). Let c be the cluster assignment function, that is, $c(i) = j$ means $x_i \in C_j$. Let $n_j = |C_j|$, $j = 1, \ldots, K$. If the x_i are Euclidean, let $\bar{x}_j$, $j = 1, \ldots, K$ be the mean vector of cluster C_j, and $\bar{x}$ be the overall mean.

Cluster validation indices started to appear in the literature in the 1960s and 1970s. Most existing indices are for Euclidean or dissimilarity data and crisp partitions. We now list some indices, starting from now "classical" ideas. Many more recent indices are variations on these themes.

26.2.1 Within/Between Clusters Sum of Squares-Based Criteria

The Calinski–Harabasz index was proposed by Calinski and Harabasz (1974) for Euclidean data. In the original paper, it was accompanied by a heuristic algorithm to minimize the within-cluster sum of squares, that is, the K-means objective function. The criterion is based on balancing the within-cluster variation

$$\mathbf{W}_{\mathcal{C}_K} = \sum_{j=1}^{K} \sum_{c(i)=j} (\mathbf{x}_i - \bar{\mathbf{x}}_j)(\mathbf{x}_i - \bar{\mathbf{x}}_j)^t$$

against the between-cluster variation

$$\mathbf{B}_{\mathcal{C}_K} = \sum_{j=1}^{K} n_j (\bar{\mathbf{x}}_j - \bar{\mathbf{x}})(\bar{\mathbf{x}}_j - \bar{\mathbf{x}})^t$$

Calinski and Harabasz's variance ratio criterion is defined as

$$\mathrm{CH}(\mathcal{C}_K) = \frac{\mathrm{trace}(\mathbf{B}_{\mathcal{C}_K})}{\mathrm{trace}(\mathbf{W}_{\mathcal{C}_K})} \times \frac{n - K}{K - 1}$$

Indices are written as functions not just of K but of the clustering $\mathcal{C}_K$ to remind the reader that they can not only be used to compare different values of K, but also to compare different clusterings with the same K, although this is done less often. Large values point to good clusterings, because this means that clusters are very different from each other and at the same time very homogeneous. For increasing K, both $(\mathrm{trace}(\mathbf{B}_{\mathcal{C}_K}))/(K - 1)$ and $(\mathrm{trace}(\mathbf{W}_{\mathcal{C}_K}))/(n - K)$ can be expected to go down, because there will be more compact clusters with, on average, smaller distances between their means; the criterion aims at finding a good compromise.

Several aspects are worth highlighting.

1. Note that $\mathbf{T} = \mathbf{B}_{\mathcal{C}_K} + \mathbf{W}_{\mathcal{C}_K}$, where

$$\mathbf{T} = \sum_{i=1}^{n} (\mathbf{x}_i - \bar{\mathbf{x}})(\mathbf{x}_i - \bar{\mathbf{x}})^t$$

   which is a standard result in the multivariate analysis of variance. $\mathbf{T}$ does not depend on $\mathcal{C}$, and therefore, for fixed K, maximizing $\mathrm{trace}(\mathbf{B}_{\mathcal{C}_K})$ by choice of $\mathcal{C}$ is equivalent to minimizing $\mathrm{trace}(\mathbf{W}_{\mathcal{C}_K})$.

2. Therefore, for fixed K, minimizing the within-cluster sum of squares $trace(\mathbf{W}_{\mathcal{C}_K})$ is equivalent to maximizing CH, so that the clustering optimal according to CH is the optimal K-means clustering. This implies that CH is associated to the K-means/Ward-clustering criterion (although it has been applied to the outcomes of other clustering methods with some success, Milligan and Cooper (1985)), and it implicitly assumes spherical clusters that are concentrated around the cluster means. The criterion does not take into account separation between-cluster members "in the outskirts." It will treat clusters as "separated" if their means are far away from each other.

3. The factor $(n - K)/(K - 1)$ is motivated by degrees of freedom considerations as in the analysis of variance (the variance of K cluster means has $K - 1$ degrees of freedom, a pooled variance of all points within K clusters has $n - K$ degrees of freedom). However, there is no mathematical justification that for comparing clusterings with different K this factor would be "correct" in a well-defined sense. In particular it does not take into account the data dimension p, which can be expected to play a role (CH has been found to perform badly in some experiments with not so low dimensions, for example, Sugar and James (2003) for $p = 10$).

4. Note that for any set C with mean vector x_C,

$$\sum_{x_j \in C} \|x_j - \bar{x}_C\|^2 = \frac{1}{2|C|} \sum_{x_j, x_k \in C} d(x_j, x_k)^2 \tag{26.1}$$

with d Euclidean. This can be used to express $(\text{trace}(B_{C_K}))/(trace(W_{C_K})) = (\text{trace}(T) - \text{trace}(W_{C_K}))/(trace(W_{C_K}))$ in terms of dissimilarities in order to define a dissimilarity-based version of CH that can be computed from general dissimilarity measures d and does not require the computation of means. It can also be used to define "within-cluster variances" based on dissimilarities.

Quite a bit of work has been done on defining validation indices based on making the within-cluster variation $\text{trace}(W_{C_K})$ small (and, equivalently, making the between-cluster variation $\text{trace}(T) - \text{trace}(W_{C_K})$ large), attempting to construct criteria with a better empirical performance and a better theoretical justification. All of these are connected to the K-means clustering method for the reasons given above.

Particularly for estimating the number of clusters, considerations can be based on how $\text{trace}(W_{C_K})$ is expected to change when K increases. Krzanowski and Lai (1988) used a heuristic argument based on finding a criterion that is constant when uniformly partitioning uniform data to propose maximizing

$$\text{KL}(C_K) = |D(C_K)/D(C_{K+1})|, \quad D(C_K) = (K - 1)^{2/p} \text{trace}(W_{C_{K-1}}) - K^{2/p} \text{trace}(W_{C_K})$$

Contrary to CH, KL takes the dimension p into account. Tibshirani et al. (2001) proposed to compare $\log(\text{trace}(W_{C_K}))$ with its simulated expected value under a uniform distribution ("gap statistic").

Sugar and James (2003) investigated the expected change of $\text{trace}(W_{C_K})$ with K in more depth applying results from asymptotic rate distortion theory to mixtures of Gaussian distributions. They found that for a Gaussian mixture with G mixture components and equal spherical covariance matrix, $(\text{trace}(W_{C_K})/p)^{-p/2}$ is close to zero for $G > K$, "jumps" at $G = K$ and is linear for larger K. They therefore suggest to maximize

$$\text{SJ}(C_K) = (\text{trace}(W_{C_K})/p)^{-p/2} - (\text{trace}(W_{C_{K-1}})/p)^{-p/2}$$

and even show that this will also estimate K well, asymptotically, for some mixtures of non-Gaussian distributions. Their proposal can be easily generalized to mixtures with equal but nonspherical covariance matrices, although equal covariance matrices is often a problematic assumption for real data.

Some more proposals and discussion how the behavior of $\text{trace}(W_{C_K})$ for changing K should be accounted for are in Pakhira et al. (2004); Kim and Ramakrishna (2005);

Steinley and Brusco (2011). Generally, validation indices can be constructed from objective functions defined for fixed K by looking at the distribution of differences of these functions to be expected when increasing K by 1; sometimes (but not always) it can be argued that although an objective function should normally decrease or increase with K, and can therefore not be maximized or minimized meaningfully over K, optimizing (potentially higher order) differences between neighboring K may yield a fair comparison between different values of K.

A variation on the within-cluster versus between-clusters variance theme is the SD validity index defined by Halkidi et al. (2000). This index is of the form

$$\text{SD}(\mathcal{C}_K) = \text{Dis}(\mathcal{C}_{K_{max}}) \cdot \text{Scat}(\mathcal{C}_K) + \text{Dis}(\mathcal{C}_K)$$

where

$$\text{Scat}(\mathcal{C}_K) = \frac{1}{K} \frac{\text{trace}(\mathbf{W}_{\mathcal{C}_K})}{\text{trace}(\mathbf{T})}$$

Dis penalizes a too large K particularly if some of the distances between cluster means are very small:

$$\text{Dis}(\mathcal{C}_K) = \frac{D_{max}}{D_{min}} \cdot \sum_{j=1}^{K} \left(\sum_{k=1}^{K} \| \bar{\mathbf{x}}_j - \bar{\mathbf{x}}_k \| \right)^{-1}$$

where D_{max} and D_{min} are the maximum and minimum Euclidean distances between-cluster centers. K_{max} is the maximum number of clusters of interest (Halkidi et al. (2000) argue that this choice does not have a strong impact). Small values point to good clusterings.

Halkidi and Vazirgiannis (2001) propose another penalty to add to $\text{Scat}(\mathcal{C}_K)$, the so-called "inter-cluster density," a relative density of points at the middle point of the line segment between any pair of cluster centers.

Sharma (1996) proposed a family of indices (root mean square standard deviation, semi-partial R-squared, R-Squared, all of which are based on the within-cluster and between-clusters sum of squares) which can be monitored for the joining steps in hierarchical clustering (the indices are monotonic in K and "elbows" in plots of them versus K indicate good clusterings). A further member of this family of indices is the CD-index, which measures the distance between the two clusters that are merged in a given step of the hierarchical clustering, and can be adapted to various hierarchical clustering algorithms.

The standard measures of variance are based on squared Euclidean distances. They are therefore connected to the Gaussian likelihood and penalize outlying observations within clusters strongly. Saitta et al. (2007) suggest a "score function"—index that uses similar principles, namely between-class and within-class distances (bcd and wcd; both based on distances either between or to the cluster means) with unsquared distances. It bounded between 0 and 1, is not based on a ratio but on a difference and can therefore handle $K = 1$. The index is defined as

$$\text{SF}(\mathcal{C}_K) = 1 - \frac{1}{\exp(\exp(\text{bcd}(\mathcal{C}_K) - \text{wcd}(\mathcal{C}_K)))},$$

$$\text{bcd}(\mathcal{C}_K) = \frac{\sum_{k=1}^{K} n_k \| \bar{\mathbf{x}}_k - \bar{\mathbf{x}} \|}{nK}, \quad \text{wcd}(\mathcal{C}_K) = \sum_{k=1}^{K} \frac{\sum_{c(i)=k} \| \mathbf{x}_i - \bar{\mathbf{x}}_k \|}{n_k}$$

Further indices based on within- and between-cluster variances exist. Sixteen of the 30 indices implemented in the "NbClust"-package of R (Charrad et al., 2013) are of this type, many of them dating back to the 1960s, 1970s, and 1980s. Some of these indices did quite well in the comparative simulation study by Milligan and Cooper (1985) (the success of CH in this study is probably the reason for it still being the most popular index of this kind), which was based on clusters generated from truncated Gaussian distributions, K up to 5 and low dimensionality. Dimitriadou et al. (2002) ran a comparative simulation study for multidimensional binary data, in which variance-based indices were used and did quite well, despite of the non-Euclidean nature of such data.

26.2.2 The Davies–Bouldin Index

Davies and Bouldin (1979) start their paper, in which they introduce their Davies–Bouldin index, with a number of axioms to define general measures of within-cluster dispersion S_k, $k = 1, \ldots, K$ and a between-cluster similarity R_{ij}, $i, j = 1, \ldots, K$ as a function of S_i, S_j and a distance M_{ij} between cluster centroids, implying nonnegativity, symmetry, the potentially controversial "$R_{ij} = 0 \Leftrightarrow$ both $S_i = 0$ and $S_j = 0$," and two intuitive monotonicity axioms about the comparison of R_{ij} and R_{ik} in case that $S_k > S_j$ but $M_{ik} = M_{ij}$ and $M_{ik} > M_{ij}$ but $S_k = S_j$.

They go on to define specifically, for $q, p > 0$, $i, j, k = 1, \ldots, k$:

$$S_k = \left(\frac{1}{n_k} \sum_{c(i)=k} \|\mathbf{x}_i - \bar{\mathbf{x}}_k\|_2^q \right)^{1/q}, \quad M_{ij} = \|\bar{\mathbf{x}}_i - \bar{\mathbf{x}}_j\|_p$$

$$R_{ij} = \frac{S_i + S_j}{M_{ij}}, \quad D_i = \max_{j \neq i} R_{ij}$$

$$\mathrm{DB}(\mathcal{C}_K) = \frac{1}{K} \sum_{i=1}^{K} D_i$$

where $\| \cdot \|_p$ is the L_p-norm. The principle can also be applied for other choices of S_k, M_{ij}, R_{ij}, and centroids different from the means $\bar{\mathbf{x}}_i$. However, $p = q = 2$ is the standard choice and implies that DB can be seen as another variance-based criterion connected to K-means, albeit more weakly so, because D_i is computed for every cluster separately, and every cluster is only compared to the closest other cluster. Note that every cluster is given the same weight in the computation of DB, regardless of its size. Cluster separation is again measured by considering the distance between cluster centers only.

A K that yields homogeneous but well-separated clusters will yield a small DB (because all S_i should be small and M_{ij} large).

26.2.3 The Dunn Family of Indices

The cluster validity index proposed by Dunn (1974) formalizes the idea of a ratio between between-cluster separation and within-cluster compactness for general dissimilarity data. For given K it is defined as

$$\mathrm{DI}(\mathcal{C}_K) = \min_{i=1,\ldots,K} \left\{ \min_{j=i+1,\ldots,K} \left(\frac{d_C(C_i, C_j)}{\max_{k=1,\ldots,K}(\Delta(C_k))} \right) \right\} \tag{26.2}$$

where d_C measures the dissimilarity between two clusters and Δ measures the spread of a cluster. Dunn originally defined them as $d_C(C_i, C_j) = \min_{x \in C_i, y \in C_j} d(x, y)$, $\Delta(C) = \max_{x, y \in C} d(x, y)$ (diameter of C).

The index brings together the defining features of single and complete linkage hierarchical clustering by trying to find a clustering with good separation (measured by minimum between-cluster distances) and good homogeneity (measured by maximum within-cluster distances) at the same time, which is achieved by maximizing DI. The index degenerates for $K = 1$ and $K = n$. A major drawback of the index is that it is very sensitive to changes in few or even a single observation involved in the required minimization or maximization, but will on the other hand be all too insensitive to many observations (and even non-extreme clusters) if n is large and K is small.

The principle in (26.2) can be used to define a more general family of indices by using other definitions for d_C and Δ. This has been exploited by Pal and Biswas (1997) and Bezdek and Pal (1998) with the aim of amending some of the issues of DI. Pal and Biswas (1997) propose to use graph-theoretical measures for $\Delta(C)$, namely the length of the longest edge in several characterizing subgraphs of a cluster such as the minimum spanning tree (MST), which is more closely connected to single linkage (the paper applies these ideas to the DB-index, too). Bezdek and Pal (1998) propose using as d_C the between-cluster distances used for the merging process of several standard hierarchical clustering methods (see Chapter 6), and as Δ the average within-cluster distance, or the average distance to the cluster mean (the latter requiring Euclidean data). The paper contains an interesting discussion of drawbacks and advantages of these choices.

A related approach is the CS-measure by Chou et al. (2004), in which averages are taken over all clusters for measures of between-cluster distance and cluster spread. Numerator and denominator are switched so that for this measure small values are good. The between-cluster distance is based on cluster centers $\bar{x}_j$, $j = 1, \ldots, k$, so they need to be well-defined by the clustering method. The definition is

$$CS(\mathcal{C}_K) = \frac{\sum_{i=1}^{K} \frac{1}{n_i} \sum_{x_j \in C_i} \max_{x_k \in C_i} d(x_j, x_k)}{\sum_{i=1}^{K} \min_{i \neq j \in \{1, \ldots, K\}} d(\bar{x}_i, \bar{x}_j)}$$

26.2.4 Hubert's Γ and Related Indices

In an influential paper, Hubert and Schultz (1976) presented a framework for data analysis indices (some of which had been around in the literature before), computing a sum of the products between a data (dissimilarity) matrix $\mathbf{P}$ with entries $P_{ij} = d(x_i, x_j)$ and a matrix $\mathbf{Q}$ with entries Q_{ij} specifying a fitted structure to the data,

$$\Gamma = \frac{2}{n(n-1)} \sum_{i=1}^{n-1} \sum_{j=i+1}^{n} P_{ij} Q_{ij}$$

There are various ways to interpret a clustering in terms of $\mathbf{Q}$. The two most obvious ones are to choose Q_{ij} as cluster indicator matrix, that is, zero if x_i and x_j are in the same cluster and 1 otherwise, or to choose Q_{ij} as the dissimilarity between the cluster centroids or representative points of the clusters to which x_i and x_j are assigned, Theodoridis and Koutroubas (1999). Γ can be normalized by replacing P_{ij} with $P_{ij} - \mu_P$, Q_{ij} with $Q_{ij} - \mu_Q$, where μ_P, μ_Q are the means of the entries of the two matrices, and by dividing the resulting

$\Gamma(\mathcal{C}_K)$ by the product of the standard deviations of the entries of the two matrices, so that the normalized $\Gamma(\mathcal{C}_K)$ is the Pearson correlation between the entries of the two matrices. The normalized $\Gamma(\mathcal{C}_K)$ measures the quality of fit of $\mathbf{P}$ by $\mathbf{Q}$ and could be maximized for finding an optimal clustering in the sense of representing the information in the data by the clustering.

In Hubert and Levin (1976), different normalizations for $\Gamma(\mathcal{C}_K)$ are proposed. One of them is the "C-index" defined as $(\Gamma(\mathcal{C}_K) - \Gamma_{\min}(\mathcal{C}_K))/(\Gamma_{\max}(\mathcal{C}_K) - \Gamma_{\min}(\mathcal{C}_K))$, where $\Gamma(\mathcal{C}_K)$ is defined by averaging all $\sum_{j>i} Q_{ij}$ within-cluster dissimilarities (that is, $\mathbf{Q}$ is the cluster indicator matrix), and $\min_\Gamma(\mathcal{C}_K)$ and $\max_\Gamma(\mathcal{C}_K)$ are the averages of the smallest and largest $\sum_{j>i} Q_{ij}$ dissimilarities in the data set. They depend on $\mathcal{C}_K$ through the number $\sum_{j>i} Q_{ij}$ only.

A similar idea (also called Γ) had been proposed earlier by Baker and Hubert (1975), where Goodman and Kruskal's rank correlation was used to compare the cluster indicator matrix $\mathbf{Q}$ with $\mathbf{P}$, which is more robust against extreme observations but requires prohibitive computing power for large data sets, although it amounts to a strikingly simple formula, $\Gamma(\mathcal{C}_K) = (s_+ - s_-)/(s_+ + s_-)$, where s_+ and s_- are the numbers of concordant and discordant pairs of dissimilarities. "Concordant" means that the pairs of points with the smaller dissimilarity is in the same cluster, and the points with the larger dissimilarity between them are in different clusters. "Discordant" means the opposite.

One of the first cluster validity indices (Sokal and Rohlf, 1962) is a special case of a normalized Γ. It applies to a full hierarchy $\mathcal{H}$. The so-called cophenetic matrix $\mathbf{C}$ represents the hierarchy produced by a hierarchical algorithm. The entries C_{ij} of $\mathbf{C}$ represent the proximity levels at which the two observations x_i and x_j are in the same cluster for the first time. The "Cophenetic Correlation Coefficient" (CPCC) measures the degree of similarity between $\mathbf{C}$ and the underlying proximity matrix $\mathbf{P}$ with entries P_{ij} as Pearson correlation coefficient between C_{ij} and P_{ij} considered as vectors.

Many further indices (mostly from the 1970s) are alternative normalizations of either the sum of the within-cluster dissimilarities or the numbers of concordant and discordant pairs. Eight of the 30 indices implemented in R's "NbClust"-package (Charrad et al., 2013) are of this type. The C-index and the Goodman–Kruskal based Γ did well in Milligan and Cooper (1985).

26.2.5 The Average Silhouette Width Criterion

Like many other validity indices, Kaufman and Rousseeuw's dissimilarity-based average silhouette width ASW (Kaufman and Rousseeuw, 1990) is based on a compromise between within-cluster homogeneity and between-cluster separation. It is based on a point wise aggregation of "silhouettes" s_i, $i = 1, \ldots, n$, which measure how much more clearly a point should belong to the cluster to which it is assigned than to any other cluster. These silhouettes can also be used to define a diagnostic plot that allows individual statements about the observations. For $x_i \in C_k$, they are defined as

$$s_i = \frac{b_i - a_i}{\max\{a_i, b_i\}}$$

where $a_i = \frac{1}{n_k - 1} \sum_{c(j)=k} d(x_i, x_i)$ is the average dissimilarity to points of x_i's own cluster ($n_k = 1 \Rightarrow s_i = 0$) and $b_i = \min_{l \neq k} \frac{1}{n_l} \sum_{c(j)=l} d(x_i, x_i)$ is the average dissimilarity to the closest other cluster. With this,

$$\text{ASW}(\mathcal{C}_K) = \frac{1}{n} \sum_{i=1}^{n} s_i$$

If b_i is much larger than a_i, x_i sits very comfortably in its own cluster and the ASW attempts to achieve this for all points, so K should maximize ASW. Homogeneity and separation are here aggregated from all observations, so that the dependence on points at the cluster borders is not as strong as for DI.

Although it was introduced together with the PAM clustering method in Kaufman and Rousseeuw (1990) (see Chapter 4), the definition of the ASW is not directly connected to any clustering method for fixed K (centroid objects do not play a particular role in it, as opposed to PAM). One could wonder whether it could be a good idea to define a clustering method by optimizing ASW for given K. This seems to be numerically cumbersome, but approximating algorithms could be used as for other criteria.

Hennig and Liao (2013) used the ASW in connection with a homogeneity null model, that is, they compare the ASW computed from a real data set with the distribution of the ASW over artificial data sets from the null model (the null model is tailored to a specific application) for a range of values of K. They observed that the largest ASW occurred in the real data set for $K = 2$, but this was not significantly larger than what happened under the null model. For larger K not only the ASW in the real data set was lower, but also the ones generated from the null model, and these were even significantly smaller than the value achieved in the real data set. This indicates that K with maximum ASW may not be the best K (if large ASW values are to be expected for this K even under the null model), and that comparison with a null model can reveal that a smaller local optimum of K can be seen as better in the case that this can distinguish the clustering structure clearer from the null model. This phenomenon has been observed in other data sets, too, and also for other validity indices, but more work in this direction is required.

Hennig (2008) observed that the ASW suffers from the fact that a strongly separated cluster, even if very small, may have the effect that all other clusters are merged, because if there are only two strongly separated clusters, b_i is very large for every observation, and this will dominate the benefit for a_i that comes from splitting up further not so strongly separated clusters, which will lower b_i for the corresponding points. This creates a problem with extreme outliers, although they may lead to trouble with many other indices, too.

Menardi (2011) recently proposed an interesting variation on the ASW idea by adapting it to estimated probabilities for points to belong to the clusters instead of raw dissimilarities in the framework of probability density-based clustering, although she discusses visual diagnostics only and not the choice of an optimal K. This deals with an issue with most indices mentioned up to now, namely that by treating distances in the same way everywhere in the data set, they are implicitly based on a spherical cluster concept (as implied by K-means). Density-based clustering can find clusters of various different shapes in the same data set.

26.2.6 The CDbw-Index

In Halkidi and Vazirgiannis (2008), a cluster validity index, CDbw, is introduced which assesses the compactness and separation of clusters defined by a clustering algorithm. CDbw handles arbitrarily shaped clusters by representing each cluster with a number of points rather than by a single representative point. It is not connected to any particular clustering algorithm but requires Euclidean data.

The procedure starts with selecting a number of representative points for each cluster, starting from the point that is furthest from the cluster center and then adding further representatives that are as far as possible from the previously chosen representatives ("far" is normally defined according to the Euclidean distance; one may consider using within-cluster or overall Mahalanobis distances, too, if clusters are not spherical). These points are intended to represent the cluster boundaries.

Next, every representative of a cluster is paired with the closest representative of any other cluster. For each pair of clusters C_i, C_j, a "density between the clusters" $Dens(C_i, C_j)$, is then computed as an average standardized number of points in the neighborhood of the midpoint of the line between each pair of representatives, for which both the representative in C_i is closest to that in C_j and the other way round (the radius of the neighborhood is the averaged standard deviation of the two clusters based on (26.1)). Furthermore, the distance between clusters, $Dist(C_i, C_j)$, is the average of distances between the pairwise closest representatives. From this, a measure of the density between clusters in the clustering, $Inter\_Dens(C_K)$, is computed by averaging the maximum $Dens(C_i, C_j)$ for every cluster C_i. These measures are used to define the between-cluster separation:

$$Sep(C_K) = \frac{\frac{1}{K} \sum_i \min_j \{Dist(C_i, C_j)\}}{1 + Inter\_dens(C_K)}$$

Separation is high if distances between representatives of different clusters are high and there is not much density between them. The CDbw-index connects this to the compactness and cohesion of clusters, which will be defined next.

The incorporated concept of "compactness" is defined based on "shrinking" the cluster representatives according to a sequence of values $s_1, \ldots, s_{n_s} \in [0, 1]$. The density around these shrunk representatives should be high. For given s_i, the "cluster density" for the shrunk representative is the averaged number of points belonging to the corresponding cluster in its neighborhood with radius equal to the within-cluster standard deviation (again computed from distances, because the data are potentially multidimensional). An "intra-cluster density" $Intra\_dens(C_K, s_i)$ aggregates these and standardizes them taking into account the overall standard deviation. Note that if representatives are not shrunk much (s_i close to zero), $Intra\_dens(C_K, s_i)$ measures the within-cluster densities close to the cluster borders, whereas for s_i close to one, densities are measured close to the cluster center. It is not assumed here that the density is higher in the center. Halkidi and Vazirgiannis (2008) recommend to choose the $s_1, 1, \ldots, n_s$ equidistant between 0.1 and 0.8. The compactness of clusters is then defined as

$$Compactness(C_K) = \frac{\sum_{i=1}^{n_s} Intra\_dens(C_K, s_i)}{n_s}$$

Furthermore, CDbw requires that the so-called "cohesion" within clusters is small. This refers to changes in the within-cluster densities over changes in s_i, so the density within a cluster should be about uniform. This is measured by $Intra\_change(C_K)$, which averages the difference between values of $Intra\_dens(C_K, s_i)$ for adjacent values of s_i. Using this,

$$Cohesion(C_K) = \frac{Compactness(C_K)}{1 + Intra\_change(C_K)}$$

and finally

$$CDbw(C_K) = Cohesion(C_K) \cdot Compactness(C_K) \cdot Sep(C_K)$$

Cohesion, *Compactness*, and *Sep* are all defined so that large values are desirable, so a large value of CDbw indicates a good clustering (which is only defined for $K \geq 2$).

The CDbw-index brings together several characteristics of a supposedly good clustering and implies through the cluster representatives flexibility regarding the cluster shapes. Halkidi and Vazirgiannis (2008) contains illustrations, results, and a more detailed discussion, but the concepts included in the definition are so rich that one paper cannot address all relevant issues comprehensively. There are a number of interesting open questions regarding CDbw, which may be addressed by improvements of the index:

- Does the reliance on standard deviations and cluster centers still favor spherical clusters more than what would be necessary?

- What is implied by the specific mode of aggregating the different characteristics, and what would be the implications of changing this (for example, *Compactness* is in the numerator of *Cohesion* and therefore contributes as a square to CDbw).

- The specific way of choosing and shrinking cluster representatives may favor certain cluster shapes (it is, e.g., not taken into account how well the representatives of the same cluster are connected).

- It cannot be generally assumed that relevant density between clusters can be located in a neighborhood of the assumed radius in the middle between cluster representatives.

26.2.7 A Clustering Validation Index Based on Nearest Neighbors

Liu et al. (2013) proposed an alternative idea to measure separation that is not based on dissimilarity values but rather based on how many of the k nearest neighbors of each observation are in the same cluster. k needs to be specified by the user. The index is called "clustering validation index based on nearest neighbors" (CVNN). The separation statistics is added to a straightforward compactness measure that is based on dissimilarity values. In order to properly balance these two measures against each others, the authors propose to divide both of them by their maximum over clusterings with different numbers of clusters $K = \{C_{K_{min}}, \ldots, C_{K_{max}}\}$. The aim is to find the best number of clusters given a clustering method. However, this could be modified to be the maximum over several clusterings including some with the same K. This means that the index cannot be computed for a single clustering in isolation, and what is optimal according to it may depend on the specified set of clusterings/values of K to be compared. The separation statistic is

$$\text{Sep}_k(\mathcal{C}_K) = \max_{j=1,\ldots,K} \left(\frac{1}{n_j} \sum_{x \in C_j} \frac{q_k(x)}{k} \right)$$

where $q_k(x)$ is the number of observations among the k nearest neighbors of x that are not in the same cluster. The compactness statistic is just the average within-cluster dissimilarity,

$$\text{Com}(\mathcal{C}_K) = \frac{\sum_{j=1}^{K} \sum_{x_h \neq x_i \in C_j} d(x_h, x_i)}{\sum_{j=1}^{K} n_j(n_j - 1)}$$

Note that this is slightly different from the formula given in Liu et al. (2013), which seems to be a misprint because it divides the sum of within-cluster distances by $n_j(n_j - 1)$ within every cluster, which means that very small clusters have the same weight in this computation as very large ones, and furthermore the sum over j tends to become larger with larger K despite clusters being more compact.

For both statistics, small values are better. With this,

$$\text{CVNN}_k(\mathcal{C}_K) = \frac{\text{Sep}_k(\mathcal{C}_K)}{\max_{\mathcal{C} \in \mathcal{K}} \text{Sep}_k(\mathcal{C})} + \frac{\text{Com}(\mathcal{C}_K)}{\max_{\mathcal{C} \in \mathcal{K}} \text{Com}(\mathcal{C})}$$

CVNN penalizes one-point clusters (in case of outliers) heavily, because such points produce a maximum possible Sep-value of 1, so it will almost always integrate outliers in bigger clusters.

26.2.8 Specialized Indices

Whereas the indices listed above were introduced for rather general clustering problems, there is much recent work in which indices are proposed for clusterings that should fulfill specific aims. We only mention three approaches here.

In fact, a number of the above-mentioned indices were originally proposed for the use with hierarchical cluster analysis methods, to be computed at each step of the hierarchical process in order to determine whether it should be stopped at a certain number of clusters. In the simulation study in Milligan and Cooper (1985), all incorporated indices are used in this way. Some test statistics introduced in Chapter 15 can be used to decide whether a cluster should be split (or two clusters should be merged). Gurrutxaga et al. (2010) propose an index to find an optimal partition from a hierarchy allowing cuts at variable heights.

Lago-Fernandez and Corbacho (2010) aims for clusters that can be well modeled by the Gaussian distribution, by looking at the neg-entropy distances of the within-cluster distributions to the Gaussian distribution. Liu and Dong (2012) suggest an index for clustering categorical variables based on "contrast patterns," that is, sets of items of the categorical variables that have a high support (that is, occur often) within a cluster and never outside the cluster to which they are associated. They formalize a good clustering as one in which clusters have both very short and very long contrast patterns (the former implying that the cluster can easily be distinguished from what is outside, the latter the cluster is very homogeneous) with high support. They also cite a number of further indices for clustering categorical data.

26.2.9 A Data Example

Most work comparing validation indices tries to compare their quality based on data sets for which a certain clustering is assumed as the "true" one. For many data sets this is inappropriate because there are various legitimate clusterings and it depends on the meaning of the data and on the aim of clustering which one is best. Correspondingly, which is the most appropriate validation index depends on aim and meaning.

The data set in Figure 26.1 can be clustered in different ways, and it is used here to illustrate differences between various validation indexes in order to improve the understanding of these indices without attempting to rank them according to "quality."

The data set has been artificially generated. It can be thought of as consisting of nine groups of observations, which can be pieced together in different ways to produce different clusterings. The validation indices will be used to assess some of these clusterings. Note

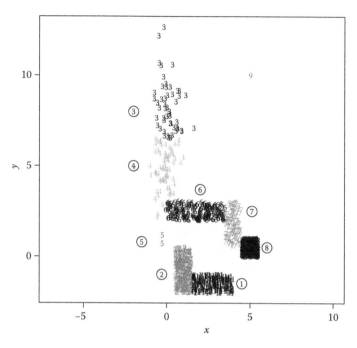

FIGURE 26.1
Simulated 2-dimensional data; observations are represented by the numbers of groups they belong to (which also are plotted in different gray scales). Numbers in circles are not observations but indicate the location of the group (group 9 only consists of a single outlier).

that these clusterings are not outcomes of any specific cluster analysis method but chosen by us for illustrative purposes. Groups no. 1, 2, 6, 7, and 8 have been generated by uniform distributions with 250, 250, 175, 125, and 300 observations, respectively. Groups 3, 4, and 5 stem from a single Gaussian distribution (125 observations), which was partitioned into more homogeneous subgroups because of a large variation, and group 9 just consists of a single outlier.

The following clusterings have been compared:

c2 cluster 1: groups 1, 2; cluster 2: groups 3–9 (biggest connected components with outlier assigned to "closer" cluster),

c3a same as c2 but with group 9 as 1-point cluster (focusing on separation between clusters),

c3b cluster 1: groups 1, 2; cluster 2: groups 3–5 and 9, cluster 3: groups 6–8 (two connected uniform clusters and normal plus outlier),

c4 same as c3b but with cluster 3: groups 6 and 7, cluster 4: group 8 (the latter having a much higher density than the other uniform groups),

c5 same as c4 but with group 9 as 1-point cluster,

c6 cluster 1: groups 1, 2; cluster 2: group 3; cluster 3: groups 4, 5; cluster 4: group 6; cluster 5: groups 7, 8; cluster 6: group 9 (focusing on within-cluster homogeneity by achieving as uniform within-cluster variation as possible based on the given grouping), c7 same as c5 but splitting up groups 1 and 2, and groups 6 and 7 (all clusters now come from linearly shaped homogeneous distributions, although some are not well separated),

c9 all groups form separate clusters.

The following indices have been applied: CH, DB with $p = 2$ and $q = 2$ or $q = 4$ (these are called DB(2) and DB(4) in Table 26.1; $q = 1$ has been tried to but gives very similar results to $q = 2$), DI as originally defined by Dunn (DI_1), and DI with d_C and Δ chosen as average between- and within-cluster distance (DI_2), normalized Γ based on the Pearson correlation, ASW, CDbw and CVNN with $k = 5$.

CH, the two versions of DI, the normalized Γ and ASW were computed by the function cluster.stats in the R-package "fpc." CDbw was computed by the function cdbw in "fpc" (using default parameter choices). The two versions of DB were computed by the function index.DB in the R-package "clusterSim." More cluster validation indices are available in "clusterSim" and "fpc," and also in the R-packages "Nbclust," "clv" and "clValid." Some more implementations are scattered over other R-packages for clustering.

The results are shown in Table 26.1. As could have been expected for such an ambiguous data set, the various indices differ a lot. In fact, no two indices agree regarding both the clusterings ranked first and second. Clustering c6 is quite popular among several indices. This may be surprising because putting groups 7 and 8 together and separating groups 3 and 4 in Figure 26.1 does not seem to be most intuitive. But many variance-based methods penalize large variation within clusters strongly and favor clusterings with rather uniform within-cluster variation. CH and CDbw favor c9 with maximum $K = 9$; they are the only indices that prefer c7 to c6, which could be seen as "repairing the counterintuitivity" of c6 by splitting up groups 7 and 8. Both indices do not seem to like the irregular cluster shapes produced by the clusterings with smaller K. The original Dunn index on the other hand is dominated by separation here, favoring the two clusterings with smallest K, c2, and c3a (the smallest separation and maximum diameter in both of these clusterings are the same, so it does not make a difference whether the outlier is integrated with the widely spread Gaussian cluster or not). Most methods (DB, DI_2, Γ, CDbw) favor having a one-point cluster for group 9 in direct comparison with clusterings that integrate it and are otherwise the same. CVNN penalizes one-point clusters heavily. Of the others, ASW is

TABLE 26.1

Index Values for Clusterings of Artificial Dataset

Index	CH	DB(2)	DB(4)	DI_1	DI_2	Γ	ASW	CDbw	CVNN
Clustering									
c2	639	1.18	1.52	**0.060**	1.33	0.37	0.41	0.03	1.00
c3a	328	0.91	1.18	**0.060**	1.34	0.37	0.39	0.06	2.00
c3b	1429	0.85	1.02	0.026	1.42	0.63	0.52	0.03	0.71
c4	1746	0.75	0.89	0.011	1.11	0.61	0.53	0.05	**0.54**
c5	1339	0.67	**0.79**	0.011	1.14	0.61	0.52	0.16	1.51
c6	1543	**0.65**	0.83	0.014	**1.89**	**0.65**	**0.56**	0.16	1.48
c7	1637	0.72	0.85	0.003	0.66	0.54	0.49	0.26	1.34
c9	**1890**	0.69	0.81	0.006	0.99	0.54	0.46	**0.39**	1.31

Note: Note that the DB-indices and CVNN are small for good clusterings, as opposed to the other indices. Bold face denotes the best result for each index.

least critical of integrating the outlier in a larger cluster, whereas CH isolates it only in a clustering with more smaller clusters elsewhere. c4 and c5 seem quite intuitive, governed by different distributional shapes. c4 does fairly well according to ASW and CH and is optimal according to CVNN. c5 is optimal according to DB(4), but otherwise unpopular. As far as we know, the DB index with $q = 4$ is not well investigated in the literature, and it is difficult to generalize the reason why it prefers c5 here, because compared to DB(2) this is based on different selections of "most similar cluster"—pairs and not on something that can directly be pinpointed in the definition of the index. Apart from that only CDbw, which is rather governed by densities than variances, prefers c5 to c6.

In order to see how the dimensionality of the data influences the indices, we produced a 10-dimensional version of the data set in Figure 26.1 by adding eight dimensions of Gaussian "noise" (each with variance 9). Again, it is not clear what a "good" result would be. One could argue that because the eight added dimensions are noninformative, the indices should produce the same rankings as before. On the other hand, adding noise increases variation within groups, making separation weaker, and therefore one could also prefer fewer clusters than in the two-dimensional version. In fact, most indices conform to one of these intuitions. CDbw still favors c9 and Γ still favors c6. CH and ASW now prefer c3b (the group 9-outlier is not so outlying anymore because in the eight added dimensions it is nothing special) and DI_2 and both DB-indices prefer c3a. CVNN even prefers c2 because the step up to 10 dimensions affects its compactness statistic much more than its separation statistic. The only index preferring a clustering with higher K than before is the original Dunn index, which now rates c3b better than c2, because with less separation overall, it is now dominated by the maximum diameter (which does not improve much more for larger K). We are not aware of any theory backing up suitable behavior for these indices with increasing dimensionality. Such theory exists for some indices listed in Section 26.2.1 (KL, SJ), but these are based on differences between successive clusterings and cannot be appropriately applied to different clusterings with the same K.

26.3 Validity Indices for Graph Partitioning

Graphs are widely used to model structural relationship between objects in many application domains such as web and social networks, sensor networks, and telecommunication. A number of clustering techniques have been developed for graph clustering. The quality of clustering results depend on the graph structure and the clustering technique used. The majority of the cluster validity indices discussed in Section 26.2 are computed in Euclidean spaces and thus are not applicable in case that the graphs are not defined in metric spaces. Various cluster validity indices have been proposed to deal with the specific requirements of graph clusterings. Here are some examples. We consider a directed or undirected graph $G = (V, E)$ with V consisting of n nodes v_i and E consisting of m edges e_{ij} and e_{ij} is the edge between nodes v_i and v_j (in undirected graphs e_{ij} and e_{ji} are identical). A graph clustering C_K is a partition of V into K clusters.

The quality of a graph partitioning is assessed based on inter-cluster and intra-cluster connectivity. There are number of validity indices that have been proposed in the literature which are based on different definitions of inter- and intra-cluster connectivity.

26.3.1 Coverage of a Graph Clustering

The coverage of a graph clustering is defined as the fraction of intra-cluster edges within the complete set of edges (Brandes et al., 2003):

$$\text{Coverage}(\mathcal{C}_K) = \frac{\sum_{i=1}^{K} m_i}{m}$$

where m_i is the number of edges of cluster C_i, $i = 1, \ldots, K$. The goal is to identify the graph clustering that corresponds to the highest value of coverage. However, the coverage is not suitable to compare clusterings with different K directly, because it is easier to collect many edges within clusters if K is small.

In order to penalize the coverage appropriately, Brandes et al. (2003) add the number of non-existing edges between different clusters to obtain a statistic giving the proportion of pairs of nodes "correctly" modeled by the clustering.

26.3.2 Performance of a Clustering

$$\text{Performance}(\mathcal{C}_K) = \frac{\sum_{i=1}^{K} m_i + |\{(v_i, v_j) \in V, e_{ij} \notin E\}|}{n(n-1)/2}$$

Brandes et al. (2003) use this measure to compare the output of various graph clustering algorithms for undirected graphs.

26.3.3 Modularization Quality

The modularization quality (MQ) index was defined for directed graphs as the difference between intra- and inter-cluster connectivity in Mancoridis et al. (1998).

$$\text{intra}(C_i) = \frac{m_i}{n_i(n_i - 1)}$$

where $n_i(n_i - 1)$ is the maximum number if intra-cluster edges. Moreover the inter-cluster connectivity between the clusters C_i and C_j is defined by

$$\text{inter}(C_i, C_j) = \frac{m_{ij}}{2n_i n_j}$$

where m_{ij} is the number of edges between C_i and C_j. For undirected graphs, the denominators can simply be divided by 2. The quantities are aggregated by the MQ-index, which should be maximized:

$$\text{MQ}(\mathcal{C}_K) = \frac{\sum_{i=1}^{K} \text{intra}(C_i)}{K} - \frac{\sum_{j=1}^{k} \text{inter}(C_i, C_j)}{K(K-1)/2}$$

Boutin and Hascoet (2004) list a number of validity indices for clustering of undirected graphs, including adaptations of some of the indices in Section 26.2, and they modify the

MQ-index so that not every cluster (including very small ones) has the same weight, but rather the weight of a cluster is determined by its size:

$$MQ^* = \frac{\sum_{i=1}^{K} m_i}{\sum_{i=1}^{K} \frac{n_i(n_i-1)}{2}} - \frac{\sum_{i<j} m_{ij}}{\sum_{i<j} n_i n_j}$$

26.4 Fuzzy Clustering

In this section, we present validity indices suitable for fuzzy clustering, see Chapter 24. Fuzzy clustering (with K clusters) is defined by a matrix $U_K = (u_{ij})_{1 \leq i \leq n, 1 \leq j \leq K}$, where $u_{ij} \in [0,1]$ denotes the degree of membership of the vector x_i in cluster $j = 1, \ldots, K$. We assume that clusters are represented by $\bar{x}_j$, usually the weighted within-cluster mean in fuzzy c-means with fuzzyfier m (see Chapter 24), and that for $i = 1, \ldots, n : \sum_{j=1}^{K} u_{ij} = 1$.

Similar to crisp clustering, validity indices formalize the quality of a clustering and are often compared for different values of K, either by minimization, maximization, or, in case that the index is monotonic (or at least not comparable in a fair way) over K, by looking for a "knee" or "elbow" in the plot of the index versus K.

Below two categories of fuzzy validity indices are discussed. The first category uses only the membership values of a fuzzy partition of data. The second one involves both the U_K-matrix and the data set itself.

26.4.1 Validity Indices Involving Only the Membership Values

The objective of indices involving only the membership values is to seek clusterings where most of the vectors of the data set exhibit a high degree of membership in one cluster. The earliest and probably most straightforward way of doing this (Bezdek, 1981) is the *partition coefficient*:

$$PC(U_K) = \frac{1}{n} \sum_{i=1}^{n} \sum_{j=1}^{K} u_{ij}^2$$

The values of PC range in $[1/K, 1]$, where K is the number of clusters. The closer the index is to unity, the "crisper" is the clustering. In case that all membership values to a fuzzy partition are equal, that is, $u_{ij} = 1/K$, the PC achieves its lowest value. The closer the value of PC is to $1/K$, the fuzzier the clustering is. A value close to $1/K$ is interpreted as indicating that there is no clustering tendency in the considered data set, or that the clustering algorithm failed to reveal it.

The *partition entropy coefficient* (Bezdek, 1981) is another index of this category:

$$PE(U_K) = -\frac{1}{n} \sum_{i=1}^{n} \sum_{j=1}^{K} u_{ij} \cdot \log_a(u_{ij})$$

where a is the base of the logarithm. The index is computed for $K > 1$, and its value ranges in $[0, \log_a K]$. The closer the value of PE to 0, the "crisper" the clustering is. As in the previous case, index values close to the upper bound (i.e., $\log_a K$) indicate absence of any clustering structure in the data set or inability of the algorithm to extract it.

Wu et al. (2009) noted that both PC and PE average pointwise measures for crisp-ness/fuzzyness, and that such indices could be robustified against outliers by aggregating the pointwise measures by the median or an M-estimator. They also give some references to other validity indices for fuzzy clustering, some of which are based on membership values only.

Characteristics of such indices are:

- Their monotonous dependency on the number of clusters. Thus, we look for significant "knees" of increase (for PC) or decrease (for PE) in the plots of the indices versus K.
- Their sensitivity to the fuzzifier m. If the fuzzy method is tuned to deliver close to crisp, or in the opposite case uniform, membership values, both indices will automatically be close to the optimum or the worst value, respectively.

The major issue with this approach is that the data and their geometry are not taken into account directly, so that it relies on the assumption that the entries of $\mathbf{U}_K$ give an appropriate representation of the clustering tendency of the underlying data.

26.4.2 Indices Involving the Membership Values and the Dataset

The *Xie–Beni index* (Xie and Beni, 1991), also called the compactness and separation validity function, is a widely used index for fuzzy clustering involving both membership values and the data.

Consider a fuzzy partition of the data set $X = \{x_j; j = 1, \ldots, n\}$ with $v_i, (i = 1, \ldots, n_c)$ the centers of each cluster and u_{ij} the membership of the jth data point belonging to the ith cluster.

The *fuzzy deviation* of x_i from cluster j, d_{ij}, is defined as the distance between x_i and the center of cluster j weighted by the fuzzy membership u_{ij}:

$$d_{ij} = u_{ij} \left\| x_i - \bar{x}_j \right\|$$

$\sigma_j = \frac{\sum_{i=1}^{n} d_{ij}^2}{n_j}$ is called "compactness" of cluster j, and the overall compactness is given as

$$\sigma = \frac{\sum_{j=1}^{K} \sigma_j}{K}$$

The separation of the fuzzy partitions is defined as the minimum distance between cluster centers:

$$d_{min} = \min_{j,k \in \{1,\ldots,K\}} \left\| \bar{x}_j - \bar{x}_k \right\|$$

The *XB-index* is defined as

$$XB(\mathbf{U}_K) = \frac{\sigma}{n(d_{min})^2}$$

This can be seen as a fuzzy version of the variance-based criteria introduced in Section 26.2.1. It is clear that small values of XB are expected for compact and well-separated clusters. We note, however, that XB is often monotonically decreasing when the number of clusters K becomes large. One way to eliminate this decreasing tendency of the index is to determine a starting point K_{max} of the monotonic behavior and to search for the minimum value of XB in the range $[2, K_{max}]$. Moreover, the values of the index XB depend on the fuzzifier m.

Another index of this category is the *Fukuyama–Sugeno index* (Wu et al., 2009; the original reference is in Japanese), which is defined as

$$\text{FS}(\mathbf{U}_K) = \sum_{i=1}^{n} \sum_{j=1}^{K} u_{ij}^m \left(\|\mathbf{x}_i - \bar{\mathbf{x}}_j\|^2 - \|\bar{\mathbf{x}}_j - \bar{\mathbf{x}}\|^2 \right)$$

It is clear that for compact and well-separated clusters we expect small values for FS. The first term in brackets measures the compactness of the clusters while the second one measures the distances of the clusters representatives.

For these indices, Wu et al. (2009) present robust versions based on medians and M-estimators, too. Further fuzzy cluster validity indices are presented in Wang and Zhang (2007) and Dave (1996).

26.5 Discussion and Conclusion

Cluster validity is a key issue in clustering, related to the inherent features of the data set under concern. It aims at the evaluation of clustering results and the selection of the scheme that best fits the underlying data. Validity indices are one out of several techniques for cluster validation. Others such as resampling, significance testing, comparison of different clusterings, and visualization are treated elsewhere in this book. Validity indices measure the quality of a clustering based on internal criteria that can be evaluated directly from the data and the clustering without having to rely on a known "true" partition (which does not exist where classification is unsupervised in practice) or probability models.

There is much discussion in the literature about which of the many existing indices is "best," and there are a number of simulation studies comparing some of them (for example, Milligan and Cooper (1985), Dimitriadou et al. (2002), Arbelaitz et al. (2012), the latter favoring ASW, DB, and CH with Gaussian data and SF and Dunn/Gabriel graph with real supervised classification data with known "truth"; furthermore most papers introducing new indices come with small comparative simulation studies, although one should expect these to be biased in favor of the newly introduced index). But given that in different applications different characteristics of clusters are required, this is a rather ill-posed problem and it is of more interest to characterize the differing properties of the various algorithms, as we attempted to do to some extent. Recently, Xiong and Li (2014) ran a comparative study in which different characteristics are explored by different data examples, namely the impact of noise, the impact of clusters with different within-cluster density, the impact of clusters with subclusters that are not so strongly separated, the impact of having one large and two much smaller clusters, and the impact of nonspherical and nonlinear cluster shapes. However, instead of characterizing different indexes for different tasks, they tried to find an overall winner, which was CVNN. Given that the two authors are among the developers of that method, this has to be taken with a grain of salt.

Most existing indices implicitly assume clusters to be compact (e.g., regarding the within-cluster variance or diameter). However, there are a number of applications where arbitrarily shaped clusters can occur and cluster separation is more important than homogeneity (e.g., spatial data, image recognition). In this case most traditional validity criteria are no longer sufficient. There is a need for developing quality measures that assess the quality of the partitioning taking into account homogeneity, separation and geometry of the clusters in a more flexible way, for example, using sets of representative points, or even multidimensional curves rather than a single representative point.

Nowadays a large number of application domains (web traffic, engineering, biology, social science, etc.) involve data of different nature such as time series or symbolic data. There is a need for cluster validity techniques that take into account the specific characteristics of such data.

On top of the need for new indices, investigation of most existing indices is still wanting. Rigorous theoretical justification is very rare (one issue is whether looking for a global optimum of some indices is really appropriate). The scope of the existing comparative studies is somewhat limited, often based on either Gaussian mixtures or real data sets with "true" supervised groupings the characteristics of which are not discussed. Indices are usually introduced with a claim to be appropriate for rather general situations (regarding data characteristics as well as clustering methods), ignoring relations to specific clustering methods and clustering aims. Another problem is the inability of most indices to handle $K = 1$, probably best treated by using the index to test the data clustering against a clustering with same K from a homogeneity null model. Finally, in many real applications and in robust clustering, indices need to handle points that are classified as "outlier/noise" and are not assigned to any cluster.

Given that different validation indices measure different characteristics of a clustering, and that in a given application several different characteristics may be of interest, multiple criteria could be used. Sause et al. (2012) let several criteria "vote" for the best clustering. A different approach is outlined in Hennig (2013). Different requirements for clustering in an application such as "small within-cluster gap," "small average within-cluster dissimilarity" are formalized separately and are monitored over a changing number of clusters, rather than aggregated into a single index. Some such requirements are more naturally measured for every single cluster, and one may even monitor cluster-wise indices separately instead of aggregating them.

The fact that in many applications the aims of clustering are heterogeneous and often not well defined also justifies the use of validity indices that are based on quite different characteristics of a clustering than the cluster analysis methods from which the clustering stems. This can be found quite often in the literature but is rarely explicitly discussed with reference to the specific aims of clustering (such a discussion can be found in Hennig and Liao (2013) regarding the use of the ASW for the outcome of a mixture model).

Another challenge is the definition of an integrated quality assessment model for data mining results. The fundamental concepts and criteria for a global data mining validity checking process have to be introduced and integrated to define a quality model. This will contribute to more efficient usage of data mining techniques for the extraction of valid, interesting, and exploitable patterns of knowledge.

Acknowledgments

We thank Mrs. Danae Boutara for her assistance with initial bibliographic search. The work of the last author was supported by EPSRC grant EP/K033972/1.

References

Arbelaitz, O., I. Gurrutxaga, J. Muguerza, J. M. Perez, and I. Perona 2012. An extensive comparative study of cluster validity indices. *Pattern Recognition 46*, 243–256.

Baker, F. B. and L. J. Hubert 1975. Measuring the power of hierarchical cluster analysis. *Journal of the American Statistical Association 70*, 31–38.

Bezdek, J. C. 1981. *Pattern Recognition with Fuzzy Objective Function Algorithms*. Norwell, MA: Kluwer.

Bezdek, J. C. and N. R. Pal 1998. Some new indexes of cluster validity. *IEEE Transactions on Systems, Man, and Cybernetics—Part B: Cybernetics 28*, 301–315.

Boutin, F. and M. Hascoet 2004. Cluster validity indices for graph partitioning. In *Proceedings of the Conference on Information Visualization*, pp. 376–381.

Brandes, U., M. Gaertler, and D. Wagner 2003. Experiments on graph clustering algorithms. In *Proceedings of the European Symposium on Algorithms (ESA '03)*, pp. 568–579.

Calinski, T. and J. Harabasz 1974. A dendrite method for cluster analysis. *Communications in Statistics—Theory and Methods 3*, 1–27.

Charrad, M., N. Ghazzali, V. Boiteau, and A. Niknafs 2013. *NbClust: An Examination of Indices for Determining the Number of Clusters*. R Foundation for Statistical Computing. R package.

Chou, C.-H., M.-C. Su, and E. Lai 2004. A new cluster validity measure and its application to image compression. *Pattern Analysis and Applications 7*(2), 205–220.

Dave, R. 1996. Validating fuzzy partitions obtained through c-shells clustering. *Pattern Recognition Letters 17*(6), 613–623.

Davies, D. and D. Bouldin 1979. A cluster separation measure. *IEEE Transactions on Pattern Analysis and Machine Intelligence 1*(2), 224–227.

Dimitriadou, E., S. Dolnicar, and A. Weingessel 2002. An examination of indexes for determining the number of clusters in binary data sets. *Psychometrika 67*, 137–160.

Dunn, J. 1974. Well separated clusters and optimal fuzzy partitions. *Cybernetics 4*(1), 95–104.

Gurrutxaga, I., I. Albisua, O. Arbelaitz, J. I. Martin, J. Muguerza, J. M. Perez, and I. Perona 2010. Sep/cop: An efficient method to find the best partition in hierarchical clustering based on a new cluster validity index. *Pattern Recognition 43*, 3364–3373.

Halkidi, M. and M. Vazirgiannis 2001. Clustering validity assessment: Finding the optimal partitioning of a data set. In *Proceedings of ICDM*, pp. 187–194. California.

Halkidi, M. and M. Vazirgiannis 2008. A density-based cluster validity approach using multi-representatives. *Pattern Recognition Letters 29*(6), 773–786.

Halkidi, M., M. Vazirgiannis, and I. Batistakis 2000. Quality scheme assessment in the clustering process. In *Proceedings of PKDD*, pp. 265–276. Lyon, France.

Hennig, C. 2008. Dissolution point and isolation robustness: Robustness criteria for general cluster analysis methods. *Journal of Multivariate Analysis 99*, 1154–1176.

Hennig, C. 2013. How many bee species? A case study in determining the number of clusters. In M. Spiliopoulou, L. Schmidt-Thieme, R. Janning (eds.): *Data Analysis, Machine Learning and Knowledge Discovery*. Springer, Berlin, pp. 41–49.

Hennig, C. and T. F. Liao 2013. Comparing latent class and dissimilarity based clustering for mixed type variables with application to social stratification. *Journal of the Royal Statistical Society, Series C 62*, 309–369.

Hubert, L. J. and J. R. Levin 1976. A general statistical framework for assessing categorical clustering in free recall. *Psychological Bulletin 83*, 1072–1080.

Hubert, L. J. and J. Schultz 1976. Quadratic assignment as a general data analysis strategy. *British Journal of Mathematical and Statistical Psychology 29*, 190–241.

Jain, A. K. and R. C. Dubes 1988. *Algorithms for Clustering Data*. Engle-wood Cliffs, NJ: Prentice Hall.

Kaufman, L. and P. Rousseeuw 1990. *Finding Groups in Data*. Wiley, New York.

Kim, M. and R. S. Ramakrishna 2005. New indices for cluster validity assessment. *Pattern Recognition Letters 26*, 2353–2363.

Krzanowski, W. J. and Y. T. Lai 1988. A criterion for determining the number of clusters in a data set. *Biometrics 44*, 23–34.

Lago-Fernandez, L. F. and F. Corbacho 2010. Normality-based validation for crisp clustering. *Pattern Recognition 43*, 782–795.

Liu, Q. and D. Dong 2012. Cpcq: Contrast pattern based clustering quality index for categorical data. *Pattern Recognition 45*, 1739–1748.

Liu, Y., Z. Li, H. Xiong, X. Gao, J. Wu, and S. Wu 2013. Understanding and enhancement of internal clustering validation measures. *IEEE Transactions on Cybernetics 43*, 982–994.

Mancoridis, S., B. Mitchell, C. Rorres, Y. Chen, and E. Gansner 1998. Using automatic clustering to produce high-level system organizations of source code. In *IWPC*, pp. 45–53.

Menardi, G. 2011. Density-based silhouette diagnostics for clustering methods. *Statistics and Computing 21*, 295–308.

Milligan, G. and M. Cooper 1985. An examination of procedures for determining the number of clusters in a data set. *Psychometrika 50*(3), 159–179.

Pakhira, M. K., S. Bandyopadhyay, and U. Maulik 2004. Validity index for crisp and fuzzy clusters. *Pattern Recognition 37*, 487–501.

Pal, N. and J. Biswas 1997. Cluster validation using graph theoretic concepts. *Pattern Recognition 30*(6), 847–857.

Saitta, S., B. Raphael, and I. Smith 2007. A bounded index for cluster validity. In P. Perner (Ed.), *Machine Learning and Data Mining in Pattern Recognition. Lecture Notes in Computer Science vol. 4571*, pp. 174–187. Springer, Berlin and Heidelberg.

Sause, M. G. R., A. Gribov, A. R. Unwin, and S. Horn 2012. Pattern recognition approach to identify natural clusters of acoustic emission signals. *Pattern Recognition Letters 33*, 17–23.

Sharma, S. 1996. *Applied Multivariate Techniques*. John Wiley and Sons, New York.

Sokal, R. R. and F. J. Rohlf 1962. The comparison of dendrograms by objective methods. *Taxon 11*, 33–40.

Steinley, D. and M. J. Brusco 2011. Choosing the number of clusters in k-means clustering. *Psychological Methods 16*, 285–297.

Sugar, C. and G. James 2003. Finding the number of clusters in a dataset. *Journal of the American Statistical Association 98*(463), 750–763.

Theodoridis, S. and K. Koutroubas 1999. *Pattern Recognition*. Academic Press: Knowledge Discovery in Databases, San Diego.

Tibshirani, R., G. Walther, and T. Hastie 2001. Estimating the number of clusters in a data set via the gap statistic. *Journal of the Royal Statistical Society, Series B 63*, 411–423.

Wang, W. and Y. Zhang 2007. On fuzzy cluster validity indices. *Fuzzy Sets and Systems 158*(19), 2095–2117.

Wu, K.-L., M.-S. Yang, and J. N. Hsieh 2009. Robust cluster validity indexes. *Pattern Recognition 42*, 2541–2550.

Xie, X. and G. Beni 1991. A validity measure for fuzzy clustering. *IEEE Transactions on Pattern Analysis and machine Intelligence 13*(8), 841–847.

Xiong, H. and Z. Li 2014. Clustering validation measures. In C. C. Aggarwal and C. K. Reddy (Eds.), *Data Clustering: Algorithms and Applications*, pp. 571–606. Boca Raton, FL: CRC Press.

27

Criteria for Comparing Clusterings

Marina Meila

CONTENTS

Abstract

A common question when evaluating a clustering is how it differs from the correct or optimal clustering for that data set. This chapter presents the principles and methods for comparing two partitions of a data set $\mathcal{D}$. As it will be seen, a variety of distances and indices for comparing partitions exist. Therefore, this chapter also describes some useful properties of such a distance or index, and compares the existing criteria in light of these properties.

27.1 Why We Need Distances between Partitions

As we have seen in the previous chapter, a clustering, or a clustering algorithm, can be evaluated by *internal criteria*, for example, distortion, likelihood, that are problem and algorithm dependent. There is another kind of evaluation, called *external* evaluation, where one measures how close the obtained clustering is to a gold standard clustering. In this chapter, we focus on external comparison criteria.

Unlike the internal criteria, external criteria are independent of the of the way the clusterings were obtained. But they are also independent of the actual "values" or locations of the data points. All they retain from a clustering is the way the points are grouped, that is, the clustering itself.

Formally, given a data set $\mathcal{D}$ with n points, and two clusterings $\mathcal{C}$, $\mathcal{C}'$ of $\mathcal{D}$, a criterion for comparing partitions is a function $d(\mathcal{C}, \mathcal{C}')$.

The criteria described here are of two types: *distances* and *indices*. A distance between two clusterings is a nonnegative symmetric function $d(\mathcal{C}, \mathcal{C}')$ which is 0 iff $\mathcal{C} = \mathcal{C}'$. A distance in this acception may not always satisfy the triangle inequality, although the distances in wider use do.

Traditionally, clustering comparisons have also used indices, which take values between 0 and 1, with larger values indicating more similarity, and 1 being reserved for the case $\mathcal{C} = \mathcal{C}'$. Any index can be transformed into a distance by substracting it from 1. The converse is not true, however. For this reason, and because the current work on the foundations of clustering as well as some application fields use distances, we will generically talk about distances between clusterings, with the understanding that the corresponding properties for indices can easily be obtained.

A criterion for comparing partitions must be defined for all pairs of partitions of n objects and for all n. It must also be invariant to changes in the labels of the clusters. In other words, $d(\mathcal{C}, \mathcal{C}')$ considers each clustering as an unordered set of sets.

Before we embark into the presentation of the variety of clustering comparison criteria in existence, it will be useful to examine when such measures are useful. We shall also see that distances between clusterings are rarely used alone.

For example, a user has a data set $\mathcal{D}$, with a given "correct" clustering $\mathcal{C}^*$. Algorithm $\mathcal{A}$ clusters $\mathcal{D}$, and the result $\mathcal{C}$ is compared to $\mathcal{C}^*$ via $d(\mathcal{C}, \mathcal{C}')$. If the distance d is small, one concludes that the algorithm performs well on these data, otherwise, that the algorithm performs poorly. If the algorithm $\mathcal{A}$ is not completely deterministic (e.g., the result may depend on initial conditions, like in K-means), the operation may be repeated several times, and the resulting distances to the correct clustering $\mathcal{C}^*$ averaged to yield the algorithm's average performance.

Next, to be convinced that algorithm $\mathcal{A}$ "works," one need to test it on several data sets, and record the distance to the correct clustering in all cases. Finally, the researcher will compute some summary of distances, like average and standard deviation. This implies immediately that distances (or indices) for comparing clusterings must be comparable over different n values and over different numbers of clusters in order for the "averaging" to make sense.

Moreover, this average may be compared to another average distance obtained in the same way for another algorithm $\mathcal{A}'$. This second algorithm may be optimizing a completely different cost function from $\mathcal{A}$, so comparing their results in terms of the cost function is not possible. Comparing by external criteria, for example, assessing which clustering is closer to the true one is always possible. This involves substracting distances between partitions.

Comparisons between partitions are not only for assessing performance. If a clustering algorithm is not deterministic, then a user may want to assess the variability of the clusterings it produces. In particular, one could compute the sensitivity to initialization of an algorithm like K-means on a particular data set. If we are interested in grouping data, then it is a "distance" between partitions rather than a cost that one needs. Two clusterings can differ very much in the way they group the data, yet be similar w.r.t. the internal validation criteia like distortion, silhouette, etc.

Finally, the same $\mathcal{D}$ may be clustered many times by different algorithms, in order to extract information from the resulting *ensemble of clusterings*. Such techniques are presented in Chapter 22. For this purpose, distances between partitions are of use.

To conclude, in practice, distances between clusterings are subject to addition, subtraction, and even more complex operations. This is why we want to have a clustering comparison criterion that will license such operations, inasmuch as it makes sense in the context of the application.

27.2 The Confusion Matrix: Clusterings as Distributions

Let $\mathcal{C}, \mathcal{C}'$ be two clusterings having K, respectively K' clusters. All criteria for comparing clustering are defined using the *confusion matrix*, or *association matrix* or *contingency table* of the pair $\mathcal{C}, \mathcal{C}'$. The confusion matrix is a $K \times K'$ matrix $\mathbf{N} = [n_{kk'}]$, whose kk'-th element is the number of points in the intersection of clusters C_k of $\mathcal{C}$ and $C'_{k'}$ of $\mathcal{C}'$.

$$n_{kk'} = |C_k \cap C'_{k'}|$$

It is easy to see that the counts $n_{kk'}$ satisfy

$$\sum_{k=1}^{K} n_{kk'} = |C'_{k'}| = n_{k'} \qquad \sum_{k'=1}^{K'} n_{kk'} = |C_k| = n_k \tag{27.1}$$

and of course that $\sum_{k=1}^{K} n_k = \sum_{k'=1}^{K'} n_{k'} = n$. Therefore, if all counts are normalized by n, we can define the probability distributions

$$P_{\mathcal{C}}(k) = \frac{n_k}{n} \equiv p_k \quad \text{for } k = 1, \ldots, K \tag{27.2}$$

$$P_{\mathcal{C}'}(k') = \frac{n'_{k'}}{n} \equiv p'_{k'} \quad \text{for } k' = 1, \ldots, K' \tag{27.3}$$

$$P_{\mathcal{C},\mathcal{C}'}(k,k') = \frac{n_{kk'}}{n} \equiv p_{kk'} \quad \text{for } k = 1, \ldots, K, \ k' = 1, \ldots, K' \tag{27.4}$$

Imagine that one picks a point at random from $\mathcal{D}$, each point having an equal probability of being picked. It is easy to see that the probability of the outcome being in cluster C_k equals $P_{\mathcal{C}}(k)$. Thus, $P_{\mathcal{C}}$ is the distribution of the random variable representing the cluster label, according to $\mathcal{C}$. Similarly, $P_{\mathcal{C}'}$ is the random variable representing the cluster label according to $\mathcal{C}'$, and $P_{\mathcal{C}\mathcal{C}'}$ represents the joint distributions of the two random variables associated with the clusterings $\mathcal{C}, \mathcal{C}'$.

This view is useful in two ways. First, many clustering comparison criteria have a probabilistic interpretation based on the joint distribution $P_{\mathcal{C},\mathcal{C}'}$ defined above. Second, it is in fact desirable that comparison criteria do not depend explicitly on n, the sample size, but on the *proportions* of samples in each set $C_k \cap C'_{k'}$. A criterion that has this property is called *n-invariant*.

27.3 A Cornucopia of Clustering Comparison Criteria

In what follows, we will describe several classes of clustering comparison criteria. They all have a few necessary properties. First, any distance or index applies to any two partitions of the same data set. Second, it is invariant if the cluster labels are permuted. Third, it makes no assumption about how clusterings are generated.

In addition to these, a fourth desirable property has emerged. This is the *understandability* and *interpretability* of the distance. In particular, indices, taking values between 0 and 1, are considered more interpretable because of their fixed range. Some indices have probabilistic interpretation (e.g., Rand, Jaccard); unfortunately, these indices have other undesirable properties. Among the distances, those which satisfy the triangle inequality are more intuitive. The reader should refer to Table 27.1 for a list of all criteria and their main properties.

27.3.1 Comparing Clusterings by Counting Pairs

The oldest class of criteria for comparing clusterings is based on counting the pairs of points on which two clusterings agree/disagree. A pair of points from $\mathcal{D}$ can fall under one of four cases described below.

N_{11} the number of point pairs that are in the same cluster under both $\mathcal{C}$ and $\mathcal{C}'$

N_{00} number of point pairs in different clusters under both $\mathcal{C}$ and $\mathcal{C}'$

N_{10} number of point pairs in the same cluster under $\mathcal{C}$ but not under $\mathcal{C}'$

N_{01} number of point pairs in the same cluster under $\mathcal{C}'$ but not under $\mathcal{C}$

TABLE 27.1

Overview of Clustering Comparison Criteria and Their Properties

Distance or Index		Property			
		n-Invariant	Bounded [0, 1]	Local	Additive
$(\mathcal{R})$	Rand i.	No*	Yes	Yes	27.39
$(\mathcal{AR})$	Adjusted Rand i.	Asympt	≤ 1	No	
$(\mathcal{M})$	Mirkin	No*	No*	Yes	27.39
$(\mathcal{J})$	Jaccard i.	No*	Yes	No	
$(\mathcal{F})$	Fowlkes-Mallows i.	Yes	Yes	No	
$(\mathcal{W})$	Wallace i.	Yes	Yes	No	
$(\mathcal{H})$	Misclassification Error	Yes	Yes	Yes	27.37
$(\mathcal{D})$	Van Dongen	No*	No*	Yes	27.37
(VI)	Variation of Information	Yes	$\log n$	Yes	27.37
(NVI)	Normalized Variation of Information	Yes	Yes	No	
(NMI)	Normalized Mutual Information i.	Yes	Yes	No	
(χ^2)	χ^2 distance	Yes	$(K + K')/2 - 1$	Yes	27.37
$(\mathcal{T})$	Tchuprow i.	Yes	Yes	No	

Note: Indices are marked with an i. after their name. For n-invariance and boundedness in [0, 1] a * means "can be made so by simple rescaling"; if a criterion is not bounded in [0, 1] the upper bound is given; for additivity, the formula is given if one exists. Note that Misclassification Error $\mathcal{H}$ is the only distance that satisfies all criteria.

The four counts always satisfy

$$N_{11} + N_{00} + N_{10} + N_{01} = n(n-1)/2$$

They can be obtained from the contingency table $[n_{kk'}]$. For example, $2N_{11} = \sum_{k,k'} n_{kk'}^2 - n$. See Fowlkes and Mallows (1983) for details.

The asymmetric index W was proposed by Wallace (1983)

$$W(\mathcal{C}, \mathcal{C}') = \frac{N_{11}}{\sum_k n_k(n_k-1)/2} \tag{27.5}$$

It represents the probability that a pair of points which are in the same cluster under $\mathcal{C}$ are also in the same cluster under the other clustering.

Fowlkes and Mallows (1983) introduced an index which is the geometric mean of $W(\mathcal{C}, \mathcal{C}'), W(\mathcal{C}', \mathcal{C})$ (and is therefore symmetric).

$$\mathcal{F}(\mathcal{C}, \mathcal{C}') = \sqrt{W(\mathcal{C}, \mathcal{C}')W(\mathcal{C}', \mathcal{C})} \tag{27.6}$$

It was shown that this index represents a scalar product (Ben-Hur et al. 2002).

The Rand index (Rand 1971)

$$\mathcal{R}(\mathcal{C}, \mathcal{C}') = \frac{N_{11} + N_{00}}{n(n-1)/2} \tag{27.7}$$

is one of the oldest indices in existence. It represents the fraction of all pairs of points on which the two clusterings agree.

One problem with the Rand index is that the number N_{00}, the number of point pairs on which $\mathcal{C}$ and $\mathcal{C}'$ disagree, is often almost as large as $\binom{n}{2}$. Assume for example that $\mathcal{C} = \hat{1}$, the clustering with a single cluster, while $\mathcal{C}'$ is a clustering with K equal clusters, and $n/K = m$. Then $N_{00} = (n^2 - m^2K)/2 = n^2(1-(1/K))/2$ while $\binom{n}{2} = n^2(1-(1/n))/2$. It follows that the Rand index will approach 1 as K grows, signifying that $\mathcal{C}'$ becomes *closer* to $\hat{1}$, when in fact, intuitively, one may feel that $\mathcal{C}'$ becomes further apart from $\hat{1}$.

The Jaccard (Ben-Hur et al. 2002) index aims to correct this problem by removing N_{00} from both the denominator and the numerator of (27.7)

$$\mathcal{J}(\mathcal{C}, \mathcal{C}') = \frac{N_{11}}{N_{11} + N_{01} + N_{10}} \tag{27.8}$$

We also mention here the Mirkin (1996) distance

$$\mathcal{M}(\mathcal{C}, \mathcal{C}') = \sum_k n_k^2 + \sum_{k'} (n_{k'}')^2 - 2 \sum_k \sum_{k'} n_{kk'}^2 \tag{27.9}$$

$\mathcal{M}$ is a true metric. In fact, this metric corresponds to the Hamming distance between certain binary vector representations of each partition (Mirkin 1996). The *Hamming distance* between two vectors $x, y \in \mathbb{R}^n$ is defined as the number of positions in which the two vectors differ. This distance is useful for binary vectors or vectors taking values in a discrete set. For example, the Hamming distance between $x = [0\,3\,1\,1\,2]$ and $y = [0\,0\,1\,2\,2]$ is 2, because the two vectors differ in positions two and four. To obtain the Mirkin metric between two clusterings, one must first fix an (arbitrary) ordering of the data points. Then, represent C as a matrix $\mathbf{A}_C$ of size $n \times n$, whose i, j-th entry is 1 if points i, j are in the same cluster and 0 otherwise. Represent C' similarly by matrix $\mathbf{A}_{C'}$ using the same numbering of the data points. Now the Mirkin metric is the number of elements in which the matrices $\mathbf{A}_C, \mathbf{A}_{C'}$ differ. A short calculation shows that this metric can be rewritten as

$$\mathcal{M}(C, C') = 2(N_{01} + N_{10}) = n(n-1)[1 - \mathcal{R}(C, C')] \tag{27.10}$$

Thus the Mirkin metric is in a one-to-one correspondence with the Rand index.

27.3.2 Adjusted Indices

Unfortunately the Rand index is not the only index that has problems. Figure 27.1 shows that $\mathcal{R}$ and $\mathcal{F}$ do not range uniformly over the entire $[0, 1]$ interval; that is, these indices never attain 0 and their typical values concentrate away from 0. This situation was well illustrated for the Rand index by Fowlkes and Mallows (1983). The behavior of the Jaccard index $\mathcal{J}$ is qualitatively similar to that of the $\mathcal{F}$ index.

Therefore, Fowlkes and Mallows (1983) proposed to use a *baseline* that is the expected value of the criterion under a null hypothesis corresponding to "independent" clusterings. An index is *adjusted* by subtracting the base-line and normalizing by the range above the baseline. Thus, the expected value of the adjusted index is 0 while the maximum (attained for identical clusterings) is 1.

$$\text{adjusted index} = \frac{\text{index} - E[\text{index}]}{1 - \text{index}} \tag{27.11}$$

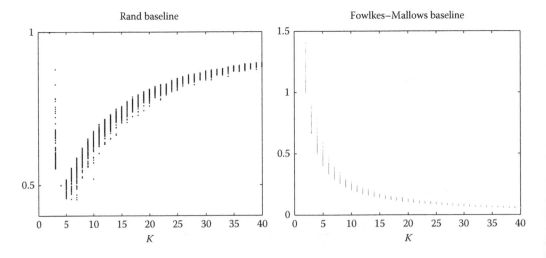

FIGURE 27.1
The baselines E[index] for the Rand (left) and Fowlkes–Mallows (right) indices and their variability. In these experiments, C has K equal clusters, while C' has also K clusters, but the cluster sizes n'_k are sampled from a multinomial $(n, 1/K, 1/K, \ldots, 1/K)$. This simulates the baseline for two clusterings with equal cluster probabilities (the most favorable case), when we observe a finite sample from one of the clusterings. It is visible that there is considerable sample variation in both baselines. For the adjusted Rand index, this variation is an approximately constant fraction of 0.2 of the useful domain $1 - E[\mathcal{R}]$.

Note that for pairs of "very dissimilar" clusterings the adjusted index will take negative values.

Applying this method to the Rand index (Hubert and Arabie 1985), one obtains the *Adjusted Rand* index

$$\mathcal{AR}(\mathcal{C},\mathcal{C}') = \frac{\mathcal{R}(\mathcal{C},\mathcal{C}') - E[\mathcal{R}]}{1 - E[\mathcal{R}]}$$

$$= \frac{\sum_{k=1}^{K}\sum_{k'=1}^{K'}\binom{n_{kk'}}{2} - \left[\sum_{k=1}^{K}\binom{n_k}{2}\right]\left[\sum_{k'=1}^{K'}\binom{n'_{k'}}{2}\right]/\binom{n}{2}}{\left[\sum_{k=1}^{K}\binom{n_k}{2} + \sum_{k'=1}^{K'}\binom{n'_{k'}}{2}\right]/2 - \left[\sum_{k=1}^{K}\binom{n_k}{2}\right]\left[\sum_{k'=1}^{K'}\binom{n'_{k'}}{2}\right]/\binom{n}{2}} \tag{27.12}$$

After adjustment, the index covers the $[0,1]$ range much more uniformly. The $\mathcal{AR}$ index is the widest used of the indices presented in this section.

The use of adjusted indices is not without problems. First, some researchers (Wallace 1983) have expressed concerns at the plausibility of the null model. Second, the shift and rescaling of (27.11) ensures that the useful range of the index is the same for every K, K' pair, but it does not ensure that the values in these ranges are linear, or that they are comparable between different values of K, K' or when the marginal probabilities $P_{\mathcal{C}}, P_{\mathcal{C}'}$ change. For example, is a value of $\mathcal{R} = 0.99$ with baseline 0.95 the same as $\mathcal{R} = 0.9$ with a baseline of 0.5? Note that both values, after adjustment, result in $\mathcal{AR} = 0.8$.

Second, the baseline can be sensitive to the variations of the cluster sizes $n_k, n'_{k'}$. This is especially detrimental for the Adjusted Rand index, with its small useful range. Figure 27.1 shows that the variability of the baseline can often be in the range of 10% for both the Fowlkes-Mallows and Rand indices. This induces a variability of cca 20% in the $\mathcal{AR}$. Hence differences of up to 40% in the $\mathcal{AR}$ index could be statistically insignificant. Figure 27.2 shows that the baseline varies smoothly when the number of clusters changes.

27.3.3 Comparing Clusterings by Set Matching

A second category of criteria is based not on pairs of points but on pairs of clusters. The best known of these is the *misclassification error* distance $\mathcal{H}$. This distance is widely used in the engineering and computer science literature, but is gaining rapid use in other areas as well. One of its greatest advantages is a clear probabilistic interpretation. $\mathcal{H}$ represents the probability of the clusterings' labels disagreeing on a data point, under the best possible label correspondence.

Intuitively, one first finds a "best match" between the clusters of $\mathcal{C}$ and the ones of $\mathcal{C}'$. If $K = K'$, then this is a one-to-one mapping; otherwise, some clusters will be left unmatched. Then, $\mathcal{H}$ is computed as the total "unmatched" probability mass in the confusion matrix. More precisely,

$$\mathcal{H}(\mathcal{C},\mathcal{C}') = 1 - \frac{1}{n}\max_{\pi}\sum_{k=1}^{K} n_{k,\pi(k)} \tag{27.13}$$

In the above, it is assumed w.l.o.g that $K \leq K'$, π is an mapping of $\{1, \ldots, K\}$ into $\{1, \ldots, K'\}$, and the maximum is taken over all such mappings. In other words, for each π we have a (partial) correspondence between the cluster labels in $\mathcal{C}$ and $\mathcal{C}'$; now looking at clustering as a classification task with the fixed label correspondence, we compute the *classification error* of $\mathcal{C}'$ w.r.t. $\mathcal{C}$. The minimum possible classification error under all correspondences is $\mathcal{H}$. This distance is a true metric. For more on its properties, see (Steinley 2004; Meilă 2005).

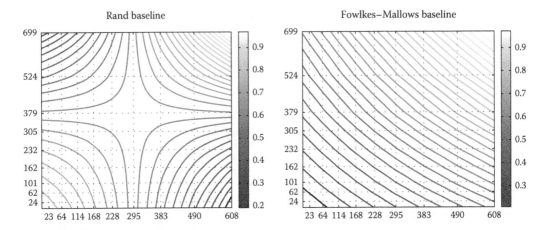

FIGURE 27.2
Does the baseline value change abruptly when the number of clusters K, K' change? This was investigated by creating two clusterings C_0, C'_0, with $K = K' = 10$ and $n = 1000$, whose cumulative marginals are displayed on the border of the contour plots (for the baseline, we are only interested in the marginals $n_{1:K}, n'_{1:K'}$). Then, we gradually transformed C, C' into the clustering with a single cluster $\hat{1}$, by moving one point at a time from the smallest cluster into the largest one. This way we traverse a set of gradually more imbalanced clusterings, which is a case less studied than the balanced case. The graphs represent the value of the baseline for each such pair C, C', starting with the originals C_0, C'_0 in the bottom left corner, and ending with $C = C' = \hat{1}$ in the top-right corner (for which the baseline is obviously 1). Each grid line represents the disappearance of one cluster. For example, at 23 on the horizontal axis, all 23 points from the smallest cluster have been moved, and K decreases from 10 to 9. The last value, 608, represents the total number of points not originally in the largest cluster; hence, after this point $K = 1$ and no more points can be moved. What we see is that the baselines are not sensitive to jumps in K, provided the change in the marginals is smooth. The change in the Fowlkes-Mallows baseline is remarkably uniform and almost "linear." The Rand baseline is saddle-shaped, with steep slope whenever one of the clusterings is very imbalanced (all corners except bottom left).

Importantly, computing the value of the $\mathcal{H}$ distance can be done efficiently, and *without enumerating all permutations of K labels*. This is because $\mathcal{H}$ can be expressed via the following *linear program*.

$$\min_{X \in [0,1]^{n \times n}} 1 - \sum_{k=1}^{K} \sum_{k'=1}^{K'} p_{kk'} x_{kk'} \tag{27.14}$$

$$\text{s.t.} \sum_{k=1}^{K} x_{kk'} = 1 \quad \text{for } k' = 1, \dots, K' \tag{27.15}$$

$$\sum_{k'=1}^{K'} x_{kk'} = 1 \quad \text{for } k = 1, \dots, K \tag{27.16}$$

$\mathcal{H}$ is the minimum objective of (27.14) above. If one changes the sign of the objective and replaces the min with a max over the same domain, one obtains an instance of the *Maximum Bipartite Matching* problem (Papadimitriou and Steiglitz 1998).

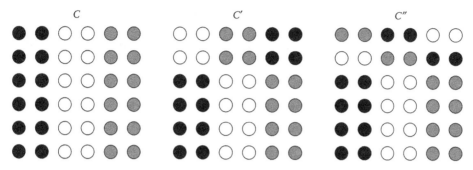

FIGURE 27.3
Clustering C' is obtained from C by moving a small fraction of the points in each cluster to the next cluster; C'' is obtained from C by reassigning the same fraction of each cluster equally between all other clusters. The set matching criteria like $\mathcal{D}$ and $\mathcal{H}$ declare C', C'' equidistant from the original clustering C.

A symmetric criterion that is also a metric was introduced by van Dongen (2000)

$$V(C,C') = 2n - \sum_k \max_{k'} n_{kk'} - \sum_{k'} \max_k n_{kk'} \tag{27.17}$$

Hence, V is 0 for identical clusterings and strictly smaller than $2n$ otherwise.

All criteria in this category suffer from the "problem of matching" that we discuss now. One way or another, a criterion like $\mathcal{H}$ and V first finds a "best match" for each cluster, then adds up the contributions of the matches found. In doing so, the criteria completely ignore what happens to the "unmatched" part of each cluster. To make things clear, let us look at the example depicted in Figure 27.3. Suppose C is a clustering with K equal clusters. The clustering C'' is obtained from C by moving a fraction f of the points in each C_k to the cluster $C_{k+1(\mathrm{mod}\,K)}$. The clustering C' is obtained from C by reassigning a fraction f of the points in each C_k evenly between the other clusters. If $f < 0.5$ then $\mathcal{H}(C,C') = \mathcal{H}(C,C'')$, $\mathcal{D}(C,C') = \mathcal{D}(C,C'')$. This contradicts the intuition that C' is a less disrupted version of C than C''. Thus, metrics like $\mathcal{H}$ lose resolution as the clusterings become more different.

27.3.4 Comparing Clusterings by Information Theoretic Criteria

This class of criteria is based on the view of clusterings as random variables with a joint distribution $P_{C,C'}$. We define the *entropy associated with clustering* C by

$$H(C) \equiv H(P_C) = -\sum_{k=1}^{K} P_C(k) \log P_C(k) \tag{27.18}$$

where log denotes the base 2 logarithm. $H(C)$ represents the entropy of the random variable associated with C. The entropy of a random variable is a measure of the amount of randomness of the variable. For instance, if the variable takes only one value, that is, it is deterministic, the entropy will be 0. This is the case of the trivial clustering $\hat{1}$ with a single cluster. The maximum of the entropy, for a fixed K, is attained when the cluster sizes are equal (to n/K), and it corresponds to a random variable with uniform distribution. Entropy is measured in *bits*. The uncertainty of 1 bit corresponds to a clustering with $K = 2$ and $P_C(1) = P_C(2) = 0.5$.

The *joint entropy* of two clusterings $H(C,C')$ is defined as the entropy of the joint distribution associated with the pair of clusterings.

$$H(C,C') \equiv H(P_{C,C'}) = -\sum_{k=1}^{K}\sum_{k'=1}^{K'} P_{C,C'}(k,k') \log P_{C,C'}(k,k') \tag{27.19}$$

We now define the *mutual information* $I(C,C')$ between two clusterings, that is, the information that one clustering has about the other. This is equal to the mutual information between the associated random variables.

$$I(C,C') = \sum_{k=1}^{K}\sum_{k'=1}^{K'} P_{C,C'}(k,k') \log \frac{P_{C,C'}(k,k')}{P_C(k)P_{C'}(k')} \tag{27.20}$$

Mutual information, like entropy, is measured in bits. Intuitively, $I(C,C')$ measures the following change in entropy. We are given a random point in D. The uncertainty about its cluster in C' is measured by $H(C')$. Suppose now that we are told which cluster the point belongs to in C. How much does this knowledge reduce the uncertainty about C'? This reduction in uncertainty, averaged over all points, is equal to $I(C,C')$.

The mutual information between two random variables is always nonnegative and symmetric. It is also no larger than the entropies of the random variables themselves. Formally,

$$H(C), H(C') \geq I(C,C') = I(C',C) \geq 0$$

The Venn diagram in Figure 27.4 depicts the relationships between entropy, joint entropy, and mutual information.

Now we are ready to define comparison criteria based on information and entropy. The *variation of information* (Lopez deMantaras 1977, 1991; Meilă 2005, 2007) between the two

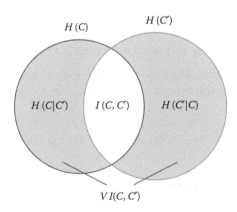

FIGURE 27.4
Venn diagram of the information theoretic quantities defined by two clusterings C, C'. The circle areas represent the entropies $H(C), H(C')$, their union is the joint entropy $H(C,C')$, etc. The *VI* is represented by the symmetric difference of the two circles (shaded), and the $NMI^\cup$ by the ratio of the intersection to the union.

clusterings is defined as

$$VI(C,C') = H(C) + H(C') - 2I(C,C') = \sum_{k=1:K} \sum_{k'=1}^{K'} P_{C,C'}(k,k') \log \frac{P_C(k)P_{C'}(k')}{P_{C,C'}(k,k')^2} \quad (27.21)$$

At a closer examination, this is the sum of two positive terms

$$VI(C,C') = [H(C) - I(C,C')] + [H(C') - I(C,C')] \quad (27.22)$$

The two terms represent the conditional entropies $H(C|C')$, $H(C'|C)$. The first term measures the amount of information about C that we lose, while the second measures the amount of information about C' that we have to gain, when going from clustering C to clustering C'.

The *VI* is a true metric, as proved in Lopez de Mantaras (1991) and Meilă (2007), and has other naturally desirable properties that will be discussed in the next section. The range of *VI* is not bounded between 0 and 1. For instance, the distance between the clustering with n singleton clusters, which we will denote by $\hat{0}$ and the clustering with one cluster, $\hat{1}$, is equal to $\log n$. If C and C' have at most K^* clusters each, with $K^* \leq \sqrt{n}$, then $VI(C,C') \leq 2\log K^*$.

The *normalized mutual information NMI* (Strehl and Ghosh 2002) is the index

$$NMI(C,C') = \frac{I(C,C')}{\sqrt{H(C)H(C')}} \quad (27.23)$$

The range of *NMI* is $[0, 1]$, with 0 corresponding with two independent random variables (thus *NMI* is zero for the same set of clusterings for which an adjusted index like $\mathcal{AR}$ aims to be zero). Various other normalizations have been used, in particular

$$NMI(C,C')^{(\cup)} = \frac{I(C,C')}{H(C,C')} \quad (27.24)$$

and

$$NMI(C,C')^{(+)} = \frac{I(C,C')}{H(C) + H(C')} \quad (27.25)$$

In Figure 27.4, $NMI^{(\cup)}$ (respectively $NMI^{(+)}$) is the ratio between the middle area and the total area of the diagram (respectively the sum of the two circles' areas). It is easy to that $NMI^{(\cup)}$ is a normalized complementary version of *VI*, as

$$\frac{VI(C,C')}{H(C,C')} + NMI^{(\cup)}(C,C') = 1$$

Interestingly, in Lopez de Mantaras (1991) and Xuan Vinh et al. (2010) it is proved that *VI* normalized this way is also a true metric, taking values in $[0, 1]$, which we shall call the *normalized VI (NVI)*

$$NVI(C,C') = 1 - NMI^{(\cup)}(C,C') = \frac{VI(C,C')}{H(C,C')} \quad (27.26)$$

Next, we include a distance and an index inspired by the χ^2 statistic. The χ^2 distance is defined as

$$\chi^2(\mathcal{C},\mathcal{C}') = \frac{K+K'}{2} - \sum_{k=1}^{K}\sum_{k'=1}^{K'} \frac{P_{\mathcal{C},\mathcal{C}'}(k,k')^2}{P_{\mathcal{C}}(k)P_{\mathcal{C}'}(k')} \qquad (27.27)$$

The inspiration for this name becomes visible if we recognize the second term in Equation (27.27) as 1 plus the χ^2 measure of independence

$$\chi^2(P_{\mathcal{C},\mathcal{C}'}, P_{\mathcal{C}}\, P_{\mathcal{C}'}) = \sum_{k=1}^{K}\sum_{k'=1}^{K'} \frac{(P_{\mathcal{C},\mathcal{C}'}(k,k') - P_{\mathcal{C}}(k)P_{\mathcal{C}'}(k'))^2}{P_{\mathcal{C}}(k)P_{\mathcal{C}'}(k')} \qquad (27.28)$$

$$= \sum_{k=1}^{K}\sum_{k'=1}^{K'} \left[\frac{P_{\mathcal{C},\mathcal{C}'}(k,k')^2}{P_{\mathcal{C}}(k)P_{\mathcal{C}'}(k')} - 2P_{\mathcal{C},\mathcal{C}'}(k,k') + P_{\mathcal{C}}(k)P_{\mathcal{C}'}(k') \right] \qquad (27.29)$$

$$= \sum_{k=1}^{K}\sum_{k'=1}^{K'} \frac{P_{\mathcal{C},\mathcal{C}'}(k,k')^2}{P_{\mathcal{C}}(k)P_{\mathcal{C}'}(k')} - 1 \qquad (27.30)$$

Hence, the χ^2 distance is a measure of independence. It is equal to 0 when the random variables $P_{\mathcal{C}}$, $P'_{\mathcal{C}}$ are identical up to a label permutation, and to $(K+K')/2 - 1$ when they are independent. One can also show that χ^2 is a squared metric (Bach and Jordan 2006).

The χ^2 distance with slight variants has been used as a distance between partitions by (Hubert and Arabie 1985; Bach and Jordan 2006). Bishop et al. (2007) uses χ^2 to construct the normalized Tchuprow index

$$T(\mathcal{C},\mathcal{C}') = \frac{\chi^2(P_{\mathcal{C},\mathcal{C}'}, P_{\mathcal{C}}\, P_{\mathcal{C}'})}{(K-1)(K'-1)} \qquad (27.31)$$

This normalization ensures that the maximum of T is attained at 1 for identical clusterings.

27.4 Comparison between Criteria

Just as one cannot define a "best" clustering method out of context, one cannot define a criterion for comparing clusterings that fits every problem optimally. Here, we present a comprehensive picture of the properties of the various criteria, in order to allow a user to make informed decisions.

27.4.1 Range and Normalization

A large number of clustering comparison criteria range between 0 and 1. Of these, the Misclassification Error $\mathcal{H}$ and the Wallace and Rand indices are completely justified by the fact that they represent probabilities. Further, the Fowlkes–Mallows index is a geometric mean of probabilities so necessarily also between 0 and 1.

For the other indices, like $NMI, \mathcal{T}$ the normalization is motivated by the concept of "measure of association" reminiscent of the correlation coefficient.

The adjusted indices like $\mathcal{AR}$ are also bounded above by 1, but, while typically positive, they are allowed to take negative values. The lower bound is usually hard to calculate and can be much smaller than minus one.

Finally, there are the unbounded distances $VI, \chi^2, \mathcal{M}, V$. It is easy to see that $\mathcal{M}, V$ grow quadratically, respective linearly with n even while $P_{\mathcal{C},\mathcal{C}'}$ stays the same. We will return to this issue in Subsection 27.4.2.

The χ^2 is bounded above by $(K + K')/2 - 1$. If we recall that it is a squared metric, then the associated metric is bounded $\sqrt{(K + K')/2 - 1}$. The variation of information VI is bounded above by $\log n$. However, if the number of clusters is bounded above by some K_{max}, then the range of the VI is bounded by $2 \log K_{max}$. This is not true for $\mathcal{M}, V$, which will grow unbounded with n for *any* pair of clusterings. Thus, in the case of VI, χ^2 the unboundedness can be attributed to the greater variety of possible clusterings as n increases. For example, the $\hat{0}$ clustering of all singletons changes with n, and intuitively grows further away from $\hat{1}$, a fact reflected by both χ^2 and VI. In the latter case, the growth of $\mathcal{M}, V$ is caused by both the higher diversity of possible clusterings and by the direct dependence on the counts $n_{kk'}$ instead of the normalizided $p_{kk'}$. Take for example the same pair of clusterings $\hat{1}, \hat{0}$.

$$\mathcal{M}(\hat{1},\hat{0}) = n^2 \left(1 - \frac{1}{n}\right) \qquad VI(\hat{1},\hat{0}) = \log n \qquad (27.32)$$

$$\mathcal{D}(\hat{1},\hat{0}) = n \left(1 - \frac{1}{n}\right) \qquad \sqrt{\chi^2(\hat{1},\hat{0})} = \sqrt{(n-1)/2} \qquad (27.33)$$

The example shows that the increase in $\mathcal{M}, \mathcal{D}$ is dominated by the factors n^2, respectively n that depend on the sample size, while the contribution of the change in $P_{\hat{1},\hat{0}}$, represented by the factor $1 - (1/n)$, increases very slowly.

27.4.2 (In)dependence of the Sample Size n

All comparison criteria expressible in terms of $P_{\mathcal{C},\mathcal{C}'}$ are n-invariant. These are $\mathcal{H}, VI, NVI, NMI, \chi^2$.

Let us consider some of the remaining indices and metrics for comparing clusterings, and examine whether they can be made invariant with n. We compare the invariant versions that we obtain in order to better the understanding of these criteria. We give invariance with n particular attention because, in any situation where comparisons are not restricted to a single data set, a criterion that is not n-invariant would have little value without being accompanied by the corresponding n.

The Rand, adjusted Rand, Fowlkes–Mallows, Jaccard, and Wallace indices are asymptotically n-invariant in the limit of large n. For finite values of n the dependence on n is weak. Proving this is straightforward and left as an exercise for the reader.

As for the two metrics, Mirkin $\mathcal{M}$, and van Dongen V, in view of their homogeneity it is immediate that they can be scaled to become n-invariant. We denote these n-invariant

C' C''

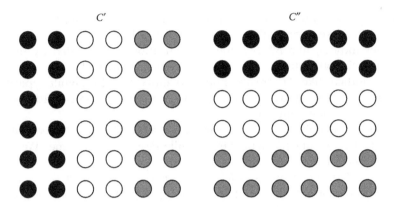

FIGURE 27.5
Two "independent" clusterings C and C', having each $K = 3$ clusters.

versions by $\mathcal{D}_{inv}, \mathcal{M}_{inv}$, respectively.

$$\mathcal{D}_{inv}(C, C') = \frac{D(C, C')}{2n} \qquad \mathcal{M}_{inv}(C, C') = \frac{M(C, C')}{n^2} \qquad (27.34)$$

It is instructive to compare the behavior of the three invariant metrics $VI, \mathcal{M}_{inv}, V_{inv}$, and of the widely used adjusted Rand index $\mathcal{AR}$ for two clusterings with $K = K'$ having $n_k = n'_k = n/K$ and $n_{kk'} = n/K^2$ for all $k, k' = 1, \ldots, K$. It is assumed that n is a multiple of K^2 for simplicity. As random variables, these two clusterings are uniformly distributed and mutually independent. Such a situation is depicted in Figure 27.5. This pair of clusterings maximizes $\mathcal{H}, VI, V_{inv}$ under the constraint that $K = K'$ but not $\mathcal{M}_{inv}$ (and consequently is also not minimizing $\mathcal{R}$ and $\mathcal{AR}$). We compute now the values of several indices for this particular pair, as a function of K.

$$\mathcal{H}^0 = \frac{K-1}{K} \quad NMI^0 = 0 \quad \mathcal{M}^0_{inv} = \frac{2}{K} - \frac{1}{K^2} \qquad (27.35)$$

$$VI^0 = 2 \log K \quad \mathcal{D}^0_{inv} = 1 - \frac{1}{K} \quad \mathcal{AR}^0 = -\frac{K-1}{n-K} \longrightarrow 0 \text{ for } n \longrightarrow \infty \qquad (27.36)$$

27.4.3 Effects of Refining and Coarsening the Partitions

We say that a clustering comparison criterion is *local* if it satisfies the following property. If C' is obtained from C by splitting one cluster, then the distance between C and C' depends only on the cluster undergoing the split.

Locality immediately entails the more general property (which can be proven by induction) that, if two clusterings C, C' agree on a subset of the data $\mathcal{D}' \subseteq \mathcal{D}$, then the distance between them depends only on $\mathcal{D} \backslash \mathcal{D}'$. From Table 27.1, we see that the Rand index all the metrics except NVI are local, and no other criteria are. In particular, the rescalings performed by $\mathcal{AR}, NVI, NMI, \mathcal{T}$ have the effect of destroying locality.

Whether a criterion for comparing clusterings should be local or not depends ultimately on the specific requirements of the application. A priori, however, a local criterion is more intuitive and easier to understand.

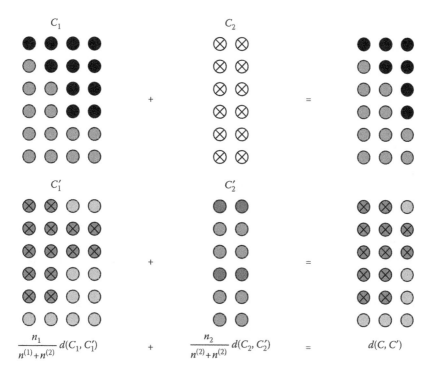

FIGURE 27.6
Illustration of convex additivity.

A related property is *(convex) additivity*. Suppose an original clustering $\mathcal{C}$ is refined in two different ways, producing clusterings $\mathcal{C}'$ and $\mathcal{C}''$. Then for each cluster $C_k \in \mathcal{C}$, of size n_k, denote by $\mathcal{C}'_k, \mathcal{C}''_k$ the partitions that $\mathcal{C}', \mathcal{C}''$ induce on cluster C_k. If

$$d(\mathcal{C}', \mathcal{C}'') = \sum_{k=1}^{K} \frac{n_k}{n} d(\mathcal{C}'_k, \mathcal{C}''_k) \tag{27.37}$$

we say that d is *convexly additive*. Figure 27.6 illustrates this property. The distances $\mathcal{H}, VI, V, \chi^2$ are convexely additive.

More generally, a criterion is *additive* if

$$d(\mathcal{C}', \mathcal{C}'') = \sum_{k=1}^{K} f(\tfrac{n_k}{n}) d(\mathcal{C}'_k, \mathcal{C}''_k) \tag{27.38}$$

For example, the Mirkin metric (and the Rand index $\mathcal{R}$ via (27.10)) satisfy

$$\mathcal{M}(\mathcal{C}', \mathcal{C}'') = \sum_{k=1}^{K} \left(\frac{n_k}{n}\right)^2 \mathcal{M}(\mathcal{C}'_k, \mathcal{C}''_k) \tag{27.39}$$

It can be shown by induction that a criterion that is additive is also local.

Finally, in Figure 27.7 we illustrate the difference between the two most used information-based criteria, *NMI* and *VI*.

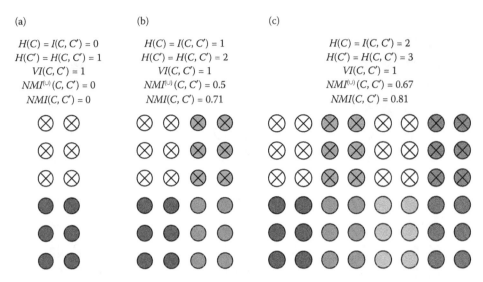

FIGURE 27.7

An example illustrating the difference between VI and $NMI(NMI^{(\cup)})$. In **a**, $C = \hat{1}$ has $K = 1$, and C' (shown) has $K' = 2$; in **b** and **c** the data is the union of 2, respectively four copies of **a**, hence the numbers of clusters are $K = 2, K' = 4$, respectively $K = 4, K' = 8$. In all three cases, given a label in C, the entropy of the label in C' is one bit, and therefore the VI is 1 bit. In the same time, as the number of clusters grows from $K' = 2$ to $K' = 8$, the entropies of both clusterings increase, as well as their mutual information; hence the NMI and $NMI^{(\cup)}$, which are ratios, will increase towards 1 (more precisely they will equal $\frac{1}{\sqrt{1+1/l}}$, respectively $1 - \frac{1}{l+1}$ where $l = \log K$). Note that for **a**, the value $NMI = 0$ is the limit for $H(C) \to 0$ with $H(C') > 0$.

27.5 Conclusion

We have presented here a variety of criteria for comparing partitions, along with a number of desirable properties to help users select one of them. As it is often the case, there is no clustering comparison criterion to satisfy them all.

In a study performed in Milligan and Cooper (1986) the $\mathcal{AR}$ is found to be best of the criteria based on counting pairs. This index is still in wide use today, possibly being the most widely used among statisticians. A more recent simulation study of adjusted indices that also includes $\mathcal{H}$ is Steinley (2004). In areas like machine learning, machine vision, and natural language processing, the criteria in widest use are the Misclassification error $\mathcal{H}$, the NMI, and the VI.

Of these, the Misclassification Error $\mathcal{H}$ is the distance that comes closest to satifying everyone, being quite likely the easiest to understand and satisfying all the properites listed in Table 27.1. Its main drawback, the reader will remember, is the loss in resolution, a.k.a. its coarsness for pairs of clusterings that are dissimilar. Hence, we strongly recommend the Misclassification error in all cases when one cares about measuring small variations, or when the number of clusters K, K' is small (e.g., less than 5–6). The former case is frequent when clusterings are compared to a gold standard C^*—for the results which are close to C^*, the $\mathcal{H}$ distance will give meaningful and interpretable results.

Moreover, as it has been shown in Meilă (2011), for the case of small distances and relatively balanced clusters, $\mathcal{H}$ is approximately proportional to other distances, like χ^2 and $\mathcal{M}_{inv}$.

References

F. Bach and M.I. Jordan. Learning spectral clustering with applications to speech separation. *Journal of Machine Learning Research*, 7:1963–2001, 2006.

A. Ben-Hur, A. Elisseeff, and I. Guyon. A stability based method for discovering structure in clustered data. In *Pacific Symposium on Biocomputing*, pp. 6–17, 2002.

Y. Bishop, S. Fienberg, and P. Holland. *Discrete Multivariate Analysis*. Springer Science, New York, 2007.

E. B. Fowlkes and C. L. Mallows. A method for comparing two hierarchical clusterings. *Journal of the American Statistical Association*, 78(383):553–569, 1983.

L. Hubert and P. Arabie. Comparing partitions. *Journal of Classification*, 2:193–218, 1985.

R. Lopez de Mantaras. *Autoapprentissage d'une partition: Application au classement itératif de données multidimensionelles*. PhD thesis, Paul Sabatier University, Toulouse (France), 1977.

R. Lopez de Mantaras. A distance-based attribute selection measure for decision tree induction. *Machine Learning Journal*, 6(1):81–92, 1991.

M. Meilă. Comparing clusterings—An axiomatic view. In Stefan Wrobel and Luc De Raedt, editors, *Proceedings of the International Machine Learning Conference (ICML)*. ACM Press, 2005.

M. Meilă. Comparing clusterings—An information based distance. *Journal of Multivariate Analysis*, 98(5):873–895, 2007.

M. Meilă. Local equivalence of distances between clusterings—A geometric perspective. *Machine Learning*, 86(3):369–389, 2011.

G.W. Milligan and M.C. Cooper. A study of the comparability of external criteria for hierarchical cluster analysis. *Multivariate Behavioral Research*, 21:441–458, 1986.

B.G. Mirkin. *Mathematical Classification and Clustering*. Kluwer Academic Press, Dordrecht, 1996.

C. Papadimitriou and K. Steiglitz. *Combinatorial Optimization. Algorithms and Complexity*. Dover Publication, Inc., Minneola, NY, 1998.

W.M. Rand. Objective criteria for the evaluation of clustering methods. *Journal of the American Statistical Association*, 66:846–850, 1971.

D.L. Steinley. Properties of the Hubert-Arabie adjusted Rand index. *Psychological methods*, 9(3):386–396, 2004. Simulations of some adjusted indices and of misclassification error.

A. Strehl and J. Ghosh. Cluster ensembles—A knowledge reuse framework for combining multiple partitions. *Journal on Machine Learning Research (JMLR)*, 3:583–617, 2002.

S. van Dongen. Performance criteria for graph clustering and Markov cluster experiments. Technical Report INS-R0012, Centrum voor Wiskunde en Informatica, 2000.

D.L. Wallace. Comment. *Journal of the American Statistical Association*, 78(383):569–576, 1983.

N. Xuan Vinh, J. Epps, and J. Bailey. Information theoretic measures for clusterings comparison: Variants, properties, normalization and correction for chance. *Journal of Machine Learning Research*, 11:2837–2854, 2010.

28

Resampling Methods for Exploring Cluster Stability

Friedrich Leisch

CONTENTS

Abstract

Model diagnostic for cluster analysis is still a developing field because of its exploratory nature. Numerous indices have been proposed in the literature to evaluate goodness-of-fit, but no clear winner that works in all situations has been found yet. Derivation of (asymptotic) distribution properties is not possible in most cases. Resampling schemes provide an elegant framework to computationally derive the distribution of interesting quantities describing the quality of a partition. Important building blocks are criteria to compare partitions as introduced in the previous chapter. Special emphasis will be given to stability of a partition, that is, given a new sample from the same population, how likely is it to obtain a similar clustering?

28.1 Introduction

Two sources of randomness affect a cluster partition: the sample, which is a random subset of the population, and the algorithm, which is known to impose structure on data, even if the data are not well structured (Dolnicar and Leisch, 2010). Using one single partition that results from one single computation of one particular algorithm therefore puts researchers in danger of using a random solution, rather than a reliable solution. Let us assume, we cluster a sample with 1000 observations. If we collect another 1000 observations and cluster the data, we can either get a very similar or a very different partition, depending on the variability in the (unknown) data generating process (DGP). This is sample randomness. Another source of randomness is that many cluster algorithms are stochastic in nature, for example, by using random starting points or random perturbations during optimization (Hand and Krzanowski, 2005). This algorithm randomness can be reduced by repeatedly starting the algorithm and keeping only the best solution, but it usually remains unknown whether the global optimum has actually been found (e.g., Chapters 3 and 5).

In some data sets, there are clear high density clusters. If this is the case, the clustering algorithm is expected to reveal these density clusters correctly and reliably over repeated computations. In the worst case, data are entirely unstructured. Although it may technically seem "foolish to impose a clustering structure on data known to be random" (Dubes and Jain, 1979) at first, this has many applications. For example, clothes are sold using discreet sizes, although the human target population has of course continuous sizes in a multivariate space. To find optimal sizes for clothes one could use cluster analysis: Measure body dimensions of a representative sample of the population, and use cluster centroids as sizes for clothes. If this solution is actually being used in practice, one certainly is interested in a stable partition, such that the sample used does not have too much influence on the solution.

Stability of partitions is not necessarily connected to the existence of well-separated density clusters. Figure 28.1 shows 4-cluster K-MEANS partitions for uniform random data on a circle and a square. When drawing new samples on the square, the resulting cluster centers and partitions are very similar. If we draw new data from the circle, the structure of resulting partitions is also stably cutting the circle in four pieces of approximately the same size, but the segments are not rotation-invariant. Which rotation one gets depends strongly on both the data set and starting values.

In this chapter, we show several methods to assess the stability of a partition. Stability itself can be of interest, especially if the data are not only explored, but real world decisions

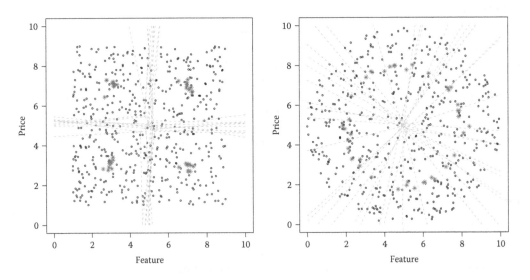

FIGURE 28.1
A four cluster partition is stable on a square, but completely unstable on a circle; dotted lines are K-MEANS-cluster borders ($K = 4$) for various random samples, stars are the resulting cluster means.

are made based on the partition. Partitioning cluster analysis and finite mixture models are, for example, key methods in market segmentation (Wedel and Kamakura, 1998) to identify subgroups of consumers with homogeneous needs or perceptions of products. If the design or advertising campaign of a new product is based on a market segment found using cluster analysis, then stability of the clusters is important. Hence, stability is one of the six key criteria for evaluation of market segments, the others being identifiability, substantiality, accessibility, actionability and responsiveness, see Wedel and Kamakura (1998) for details.

Stability can also be used to select the correct number of clusters K because if true clusters exist, the corresponding partition should have a high stability. Many papers proposing resampling schemes for partition try to solve the problem of selecting the correct number of clusters. Unfortunately, often it cannot be ruled out that a clustering other than the true one has high stability, too. Maximizing stability for estimating the number of clusters amounts to implicitly *defining* the "true clustering" as the one with highest stability, which in some applications is appropriate.

Stability here is loosely defined as "If we cluster data from a new sample from the same DGP, how likely are we getting a similar partition as the current one?" Similarity of partitions may be defined for the complete data set, or just a subset of the clusters. Returning to the examples given above: If we put out a line of clothes which should fit the whole (target) population, then stability of the partition as a whole is of interest. If we manufacture a product for a market niche consisting of a single cluster, stability of only this single cluster may be of interest.

28.2 Resampling Partitions

If stability of cluster solutions is defined as obtaining similar partitions when clustering repeatedly, one obviously needs to cluster at least twice. Breckenridge (1989) was one of the first authors emphasizing that in order to evaluate consistency and validity of cluster solutions one needs to replicate the clustering on new data. He also specified a formal framework how to do it. Most likely due to limited computing power available at the time, this framework did not really find its way into the practice of cluster analysis. Over the last decade, several procedures have been proposed which build on his ideas, three of them are outlined in detail below.

In order to evaluate the reproducibility of a partition we need to integrate out all sources of randomness in the partition, hence we need independent

- Replications of the sample $\mathcal{D}$
- Replications of the algorithm

It is easy to get replications of the algorithm, but usually we are given only one sample $\mathcal{D}$ of fixed size n. If n is very large, we can split it randomly into several smaller sets and use those, cf. function `clara` in Kaufman and Rousseeuw (1990). If n is not very large such that we cannot afford to split it into several independent samples, we have to rely on approximations. The simplest and most widely used approximation of the unknown distribution F of the DGP is the empirical distribution $\hat{F}_n$ of $\mathcal{D}$. Drawing samples from $\hat{F}_n$ is the same as sampling with replacement from $\mathcal{D}$, that is, bootstrapping (Efron and Tibshirani, 1993). Note that it is the usual convention to draw bootstrap samples which have the same size n as the original data set. Other possibilities include splitting the data set into halves, or drawing random subsamples with size smaller than n without replacement.

For supervised learning (regression, classification) with a dependent variable, resampling is often used to get unbiased estimates of the error function (mean squared error, misclassification rate, etc.): create a training and test set, fit the model on the training set and evaluate on the test set (Hastie et al., 2009). A summary of the most important sampling schemes to create such new data sets from a given one are:

(Empirical) bootstrap: Sample observations with replacement from the original data set, in most cases the bootstrap samples have the same size as the original data set. For larger data sets an average 63% of the original data are in the bootstrap sample, some of these observations two or more times because sampling is done with replacement. The remaining 37% can be used as test set. This is the most common bootstrap method, numerous variations exist including the parametric bootstrap.

Parametric bootstrap: Fit a stochastic model to the original data set and use this model as random number generator for new data. For example, for model-based clustering one can draw samples from the mixture distribution.

Random splitting: Draw a subset of size αn, $0 < \alpha < 1$ as training data, use the remaining $(1 - \alpha)n$ observations as test set. Closely related is k-fold cross-validation, which is not recommended because the folds are correlated (Hothorn et al., 2005). It is better to do random-splitting repeatedly which results in i.i.d. samples.

Jittering: Add noise to the original data.

Running the cluster algorithm on b samples $\mathcal{D}^i$ ($i = 1, \ldots, b$) drawn by bootstrap or random splitting gives us b partitions $\mathcal{C}^1, \ldots, \mathcal{C}^b$, which are independent and identically distributed random variables in the space of all possible partitions. Independence here is with respect to training sample (given the sampling scheme) and cluster algorithm. Stability can be assessed by comparing pairs of clusterings $\mathcal{C}^i$ and $\mathcal{C}^j$, comparing each of the $\mathcal{C}^i$ with the original partition $\mathcal{C}$, etc. In order to make pairwise comparisons independent from each other, one should not consider all pairwise comparisons.

A generic framework for resampling partitions in order to evaluate cluster stability is as follows:

Algorithm RESAMPLE PARTITIONS

1. Set $i = 1$.

2. Draw two training samples $\mathcal{S}^i$ and $\mathcal{T}^i$ and one evaluation sample $\mathcal{E}^i$ from the data set $\mathcal{D}$.

3. Cluster $\mathcal{S}^i$ and $\mathcal{T}^i$ resulting in partitions $\mathcal{C}^{\mathcal{S},i}$ and $\mathcal{C}^{\mathcal{T},i}$ with K clusters each.

4. Predict cluster membership for evaluation set $\mathcal{E}^i$, for example, by assigning points to closest centroids (which implicitly assumes that clusters are well represented by their centroids, which is not always the case), resulting in partitions $\mathcal{C}^{\mathcal{E}\mathcal{S},i}$ and $\mathcal{C}^{\mathcal{E}\mathcal{T},i}$.

5. Compare the two clusterings of $\mathcal{E}^i$ resulting in statistic s^i.

6. While $i < b$ increase i by one and repeat from 2.

7. Summarize s^i ($1 \leq i \leq b$) numerically and/or graphically.

Several cluster resampling approaches can be written as special cases of this generic framework, which makes a unified software implementation possible (see Section 28.3). The two variable components are the *sampling scheme* in step 2 and the *evaluation scheme* in step 5. As only cluster memberships are used, the framework can be used independently from measurement scales of variables, distance measures and cluster algorithms.

The framework is mainly intended for partitioning cluster analysis, but could also be used for hierarchical methods (see Chapter 6): Cutting dendrograms results in partitions, cluster membership for new data can be predicted by searching the nearest neighbor for each new data point in the original data set and assignment to the corresponding branch of the dendrogram.

28.2.1 Prediction Strength

Tibshirani and Walther (2005) define a cluster-wise version of the adjusted Rand index (Hubert and Arabie 1985, see Chapter 27) and define its minimum value over all clusters as the so-called *prediction strength*. Let $\mathcal{C}$ and $\mathcal{C}'$ be two clusterings of the same data set and n_k be the size of cluster C_k in $\mathcal{C}$. Then we define

$$r(i, j, \mathcal{C}) = \begin{cases} 1, & c(i) = c(j) \text{ in clustering } \mathcal{C} \\ 0, & c(i) \neq c(j) \text{ in clustering } \mathcal{C} \end{cases}$$

where $c(i)$ is the cluster label of observation i in a clustering. The similarity of cluster C_k to clustering $\mathcal{C}'$ is defined as the percentage of pairs of observations in C_k which are also in the same cluster in $\mathcal{C}'$:

$$s(C_k, \mathcal{C}') = \frac{1}{n_k(n_k - 1)} \sum_{i \neq j; i, j \in C_k} r(i, j, \mathcal{C}')$$

The prediction strength is defined as the minimum value of d over all clusters:

$$\mathrm{ps}(\mathcal{C}, \mathcal{C}') = \min_{1 \leq k \leq K} s(C_k, \mathcal{C}')$$

Note that prediction strength is not symmetric in the two clusterings. The two clusterings are obtained by randomly splitting the data into two halves, clustering both and predicting cluster membership from the centroids of the first half to the observations of the second half. The measure can be "symmetries" by having both clusterings taking both roles. In terms of our general framework RESAMPLING PARTITIONS, the building blocks are:

> *Resampling Scheme:* Split the data randomly into two halves $\mathcal{S}^i$ and $\mathcal{T}^i$ and use the second half as evaluation set, $\mathcal{E}^i = \mathcal{T}^i$.

> *Evaluation Scheme:* Use the prediction strength as statistic for cluster stability

$$s^i = \mathrm{ps}(\mathcal{C}^{\mathcal{E}\mathcal{T},i}, \mathcal{C}^{\mathcal{E}\mathcal{S},i}) \quad \in [0,1]$$

Prediction strength uses the minimum value of cluster similarity over all clusters, hence all clusters must agree in the two partitions for high values of ps(). Prediction strength is a global measure forcing all clusters to stable; the proposed use of closest centroid assignment means that this can often be achieved for compact spherical clusters. The values s^i are evaluated visually using boxplots in the original publication.

Wang (2010) proposes a somewhat similar cross-validation (data splitting) scheme for estimating the number of clusters, mentioning also alternatives to centroid assignment, such as nearest neighbor assignment, and proving a consistency theorem.

28.2.2 Cluster-Wise Assessment of Stability

Hennig (2007) does not measure the global stability of a partition, but instead the local stability of each cluster in a partition. Let again $\mathcal{D}$ be our data set and $\mathcal{C}$ a clustering of it with K clusters. The main idea is to cluster bootstrap samples of the data and see which original clusters C_k can be found again in the bootstrap partitions. This results in the following building blocks:

Resampling Scheme: Draw a bootstrap training sample T^i of size n from the data and use the intersection of the bootstrap sample with the original data as evaluation set $\mathcal{E}^i = T^i \cap \mathcal{D}$, such that observations which are contained more than once in the bootstrap sample do not have higher weight in the evaluation set.

Evaluation Scheme: Use the maximum Jaccard agreement (which in Chapter 27 is used to compare two full partitions) between each original cluster C_k and the bootstrap clusters $C_{k'}^{\mathcal{E}T,i}$ as measure of agreement and stability:

$$s_k^i = \max_{1 \le k' \le K} \frac{|C_k \cap C_{k'}^{\mathcal{E}T,i}|}{|C_k \cup C_{k'}^{\mathcal{E}T,i}|}$$

The mean value

$$s_k = \frac{1}{n} \sum_{i=1}^{n} s_k^i \quad \in [0,1]$$

is used as indicator of the stability of cluster C_k which is a local measure for each cluster.

28.2.3 Bootstrapping Average Cluster Stability

Dolnicar and Leisch (2010) propose a scheme that is in the middle of the previous two approaches and combines bootstrapping and the Rand index. Prediction strength forces all clusters to be stable because a minimum value over all clusters is used as stability indicator. Cluster-wise assessment of stability is by construction a local measure. Using the Rand index over all clusters allows some (perhaps small) clusters to be unstable. The building blocks here are:

Resampling Scheme: Draw bootstrap samples $\mathcal{S}^i$ and T^i of size n from the data and use the original data as evaluation set $\mathcal{E}^i = \mathcal{D}$.

Evaluation Scheme: Use the Rand index d corrected for agreement by chance (Chapter 7) as measure of agreement and stability:

$$s_i = d(C^{\mathcal{E}S,i}, C^{\mathcal{E}T,i})$$

where $C^{\mathcal{E}S,i}$ is the partition of the original data $\mathcal{D}$ predicted from clustering bootstrap sample $\mathcal{S}^i$ (same for T^i and $C^{\mathcal{E}T,i}$).

The values s_i are evaluated visually using boxplots and kernel density estimates, values close to one signal high stability. In many cases the distribution of s_i is multimodal corresponding to several local maxima of the clustering objective function because a local optimum on the original data can be the global optimum on a bootstrap sample. The Rand index d is computed for pairs of bootstrap samples, hence the s_i are i.i.d. and can be analyzed using standard statistical procedures for i.i.d. samples.

An alternative bootstrap scheme using assignment rules as in Section 28.2.1 was proposed by Fang and Wang (2012).

28.2.4 Further Methods for Partitions

Lange et al. (2004) randomly split the data set into two halves S^i and T^i, and use the first of the two as evaluation set, that is, $\mathcal{E}^i = S^i$. The Hungarian method (Kuhn, 1955) is used to relabel the clusters of $C^{\mathcal{E}T,i}$ to optimally match those of $C^{S^i} = C^{\mathcal{E}S,i}$ and Hamming distance (see Chapter 27) is used to measure distance between the partitions.

Dudoit and Fridlyand (2002) use the same sampling scheme, but use several cluster indices (Rand, Jaccard, Fowlkes-Mallows, see Chapter 27) for evaluation. Both papers predict cluster membership for the evaluation data not directly by distance from centroid, but by training a classifier. The major goal in both cases is to select the best number of clusters.

Maitra et al. (2012) propose a kind of parametric bootstrap approach to select the number of clusters by formal significance tests. Clusters are assumed to be spherical or ellipsoidal. New data sets are not generated by sampling from the original data. Instead new points in clusters are generated by permuting the distances of points to cluster centers and using random new directions from the center with the given distances.

28.2.5 Related Methods

This chapter is focused on resampling methods for evaluation of partitions, especially evaluation of cluster stability. Resampling is also often used to build cluster ensembles, where a consensus partition is derived from a series of single partitions (Gordon and Vichi, 1998; Monti et al., 2003; Dolnicar and Leisch, 2004; Hornik, 2005), see also Chapter 22. If observations are assigned to the classes with the maximum *a posteriori* probability, all schemes presented above can also be applied to mixture models. Resampling can also be used for diagnostics on mixture model parameters (Grün and Leisch, 2004). See Gana Dresen et al. (2008) and references therein for resampling hierarchical clusterings.

28.3 Examples, Applications, Software

R (R Development Core Team, 2014) package `flexclust` offers an extensible toolbox for K-centroid cluster analysis (Leisch, 2006) and implements the three schemes described in Sections 28.2.1–28.2.3 in a unified framework as described above. All resampling procedures are "embarrassingly parallel" from a computational point of view. In computer science the term "embarrassingly parallel" is used for workloads which can be run in parallel with only very little additional programming effort on multiple cores or clusters of workstations (Leykin et al., 2006; Matloff, 2011). Flexclust uses R base package `parallel` to automate this and minimize user efforts for parallel computations. A graphical user interface for bootstrapping the Rand and Calinski-Harabasz (see Chapter 26) indices using `flexclust` is provided by package `RcmdrPlugin.BCA` (Putler and Krider, 2012). Alternative implementations of prediction strength and cluster-wise stability can be found in R package `fpc` (Hennig, 2010), which provides many more clustering indices than `flexclust`. We will use `flexclust` in the following demonstrations.

28.3.1 Artificial Data

In the following we use an artificial toy example with three bivariate clusters of sizes 80, 40, and 80 with uniform distributions on circles with radius 1.5, 1, and 1.5, see Figure 28.2. It can be created using function `priceFeat()` from package `flexclust`. Then we use `stepcclust()` to create K-MEANS partitions for $K = 2, \ldots, 5$ with 5 random initializations to avoid local minima:

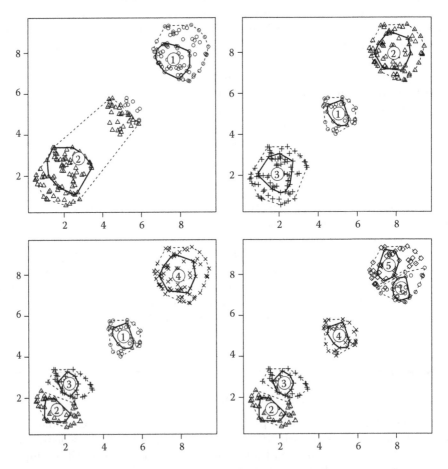

FIGURE 28.2
Artificial toy data with 400 observations clustered into 2–5 clusters using K-MEANS. Circles with numbers mark cluster centers. The solid polygons are convex hulls of the inner 50% of points in each cluster, the dashed polygons convex hulls of all points are no more than 2.5 the median distance away from the centroid.

```
R> library("flexclust")
R> library("lattice")
R> set.seed(1234)
R> x <- priceFeature(200, "3clust")
R> cl25 <- stepcclust(x, k=2:5, nrep=5, verbose=FALSE)
```

Clustering into two clusters splits the smaller middle cluster into two halves, when too many clusters are used one or both of the larger outer clusters are split into two. Function resampleFlexclust() provides the general framework for resampling partitions. The default is the bootstrap scheme defined in Section 28.2.3:

```
R> r25b <- resampleFlexclust(x, k=2:5, FUN="cclust", nrep=5)
```

Default are 100 bootstrap replicates. The resulting object can be summarized numerically (for Rand index [RI] and adjusted Rand index [ARI])

```
R> summary(r25b)

Call:
resampleFlexclust(x = x, k = 2:5, FUN = "cclust", nrep = 5)

Summary of ARI
          2                  3               4                  5
 Min.   :0.4595    Min.   :1    Min.   :0.5981    Min.   :0.4418
 1st Qu.:0.6708    1st Qu.:1    1st Qu.:0.6040    1st Qu.:0.6193
 Median :0.7822    Median :1    Median :0.8188    Median :0.6493
 Mean   :0.7883    Mean   :1    Mean   :0.7616    Mean   :0.6701
 3rd Qu.:0.9212    3rd Qu.:1    3rd Qu.:0.8975    3rd Qu.:0.7196
 Max.   :1.0000    Max.   :1    Max.   :1.0000    Max.   :0.9633
Summary of RI
          2                  3               4                  5
 Min.   :0.7298    Min.   :1    Min.   :0.8392    Min.   :0.7897
 1st Qu.:0.8354    1st Qu.:1    1st Qu.:0.8412    1st Qu.:0.8675
 Median :0.8911    Median :1    Median :0.9275    Median :0.8787
 Mean   :0.8942    Mean   :1    Mean   :0.9044    Mean   :0.8883
 3rd Qu.:0.9606    3rd Qu.:1    3rd Qu.:0.9590    3rd Qu.:0.9109
 Max.   :1.0000    Max.   :1    Max.   :1.0000    Max.   :0.9884
```

or graphically using functions boxplot() and densityplot(), the latter is shown in Figure 28.3. For the correct number of three clusters, we always get the correct solution and all Rand indices are equal to 1. For the incorrect number of clusters, the distribution of the Rand statistics is clearly shifted to the left. For four clusters, we get a bimodal distribution depending on whether the same outer cluster gets split in the two bootstrap clusterings or not.

The prediction strength sampling scheme from Section 28.2.1 can be run using the following command:

```
R> r25p <- resampleFlexclust(x, k=2:5, FUN="cclust", nrep=5, nsamp=100,
+                            scheme=predstrScheme)
```

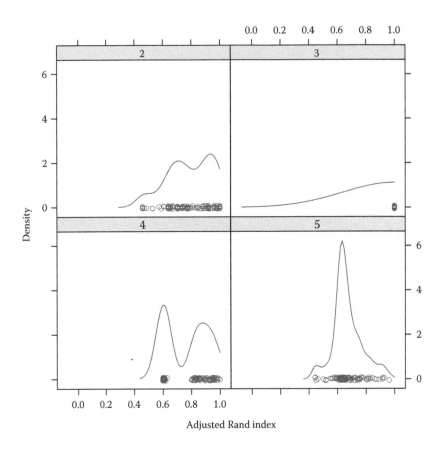

FIGURE 28.3
Kernel density estimates of bootstrapped Rand indices for 2–5 clusters.

A `boxplot()` on the return object is shown in Figure 28.4. Again the correct number of clusters shows highest stability and there is a significant drop in stability when too many clusters are used.

To see in more detail what is happening in the 4-cluster situation, we can use the cluster-wise resampling scheme from Section 28.2.2. This scheme evaluates one particular partition of the data, not overall stability given a certain number of clusters. Hence, the scheme is a function of this partition, we use the 4-cluster solution shown in Figure 28.2:

```
R> cl4s <- clusterwiseScheme(cl25[["4"]])
R> cl4r <- resampleFlexclust(x, k=4, nrep=5, FUN="cclust,"
          scheme=cl4s)
```

Figure 28.5 shows kernel density estimates of the bootstrapped cluster stabilities. Note that each panel now corresponds to one cluster in the 4-cluster solution shown in Figure 28.2. Cluster 1 is the smaller middle cluster which is never split into two and hence very stable. Cluster 4 is the upper right cluster which is not split into two halves in the original partition. Depending on whether it is split or not in a bootstrap partition stability is close to one

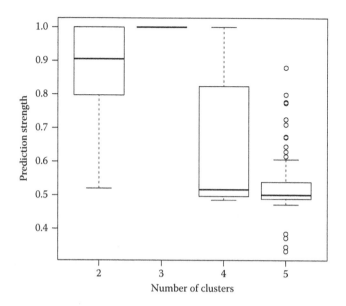

FIGURE 28.4
Boxplot of resampled prediction strength indices for 2–5 clusters.

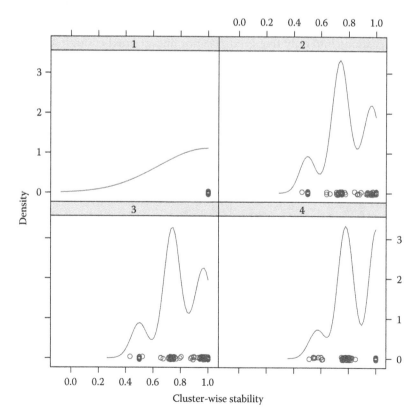

FIGURE 28.5
Kernel density estimates of bootstrapped cluster stability.

or significantly smaller. Clusters 2 and 3 split the lower left circle and are even less stable because splitting circles is not rotation-invariant.

Computing times for any of the three resampling schemes is about 20 seconds for this toy data set using only a single core of a standard laptop and 100 replications. If the same sampling building blocks are used, the different evaluation schemes can be combined into a single larger scheme to avoid repeated sampling, see also Section 28.4.

28.3.2 Guest Survey Data

We use guest survey data collected by the Austria National Tourism Organization in the summer seasons of 1994 and 1997 as a real world example (Dolnicar and Leisch, 2010). Among many other question blocks, the survey also includes vacation activities and 22 motives why tourists are coming to Austria. The motives included terms like "relaxing," "comfortable," "opportunities for sports activities," "nature," "entertaining," or "romantic" and were rated on scale from 1 (does not apply) to 4 (very important). A total of 11,378 respondents had no missing values within this question block.

In order to find market segments of tourists which have similar motives for visiting Austria, cluster analysis with K-MEANS was performed for $K = 4, \ldots, 9$ clusters. Figure 28.6 shows stability of partitions using the bootstrap procedure from Section 28.2.3. Clearly the 5-cluster solution is most stable and hence would be the best choice from a technical point of view. Unfortunately the corresponding segments had no clear profiles when tourism experts evaluated the corresponding cluster centroids, cutting more than ten thousand observations into only five groups is not fine-grained enough to find interesting market segments.

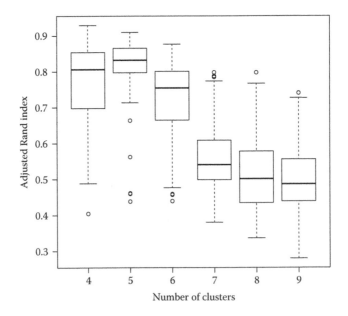

FIGURE 28.6
Boxplots of bootstrapped Rand indices for 4–9 clusters of the guest survey data.

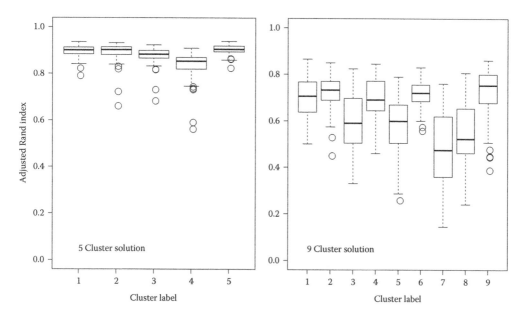

FIGURE 28.7
Cluster-wise assessment of cluster stability for the 5 and 9 cluster partition of the guest survey data.

Figure 28.7 shows cluster-wise assessment of the 5- and 9-cluster partitions. In the 5-cluster solution, all clusters are of similar and high stability. In the 9-cluster solution some clusters (1,2,4,6,9) are more stable than others (3,5,7,8). Prediction strength was not used in this example because from the cluster-wise assessment we already know that there are very unstable clusters, which would dominate prediction strength.

The stable clusters could be potentially interesting market segments. Figure 28.8 shows barcharts of the centers of clusters 1 and 2. The black dots mark mean values over the total sample and are the same in both panels. Cluster 1 is an answer tendency where everything is below average, which is very common in survey data. Cluster 9 is the opposite answer pattern where everything is above average (not shown). Cluster 2 is an interesting market niche: Far above average are motives M01, M02, and M10, meaning that these tourists come to Austria to relax, have comfort and use facilities good for personal health. Cross checking with the vacation activities reveals that they visit spas much more often than the average tourists, a stable segment of tourists with a distinct and clear profile that can be used for marketing actions.

This example is of course computationally more demanding than the artificial toy data set. K-MEANS used 20 random starts in each instance to ensure convergence to a good local optimum, the number of resampling loops was again set to $b = 100$. Running time for the two resampling schemes (overall, cluster-wise) for $K = 5, \ldots, 9$ clusters was about half an hour using 10 cores on a compute server with 14 cores (2.2 GHz) running Debian Linux. In addition to standard K-MEANS, we also tried two finite mixture models taking the ordinal structure of the data into account, and K-MEDIANS and NEURAL GAS (Martinetz et al., 1993) for comparison. All computations for all five clustering algorithms can be rerun in an afternoon using ten cores on a compute server. The K-MEANS solution was chosen by experts from tourism research because the corresponding partition had the best profile.

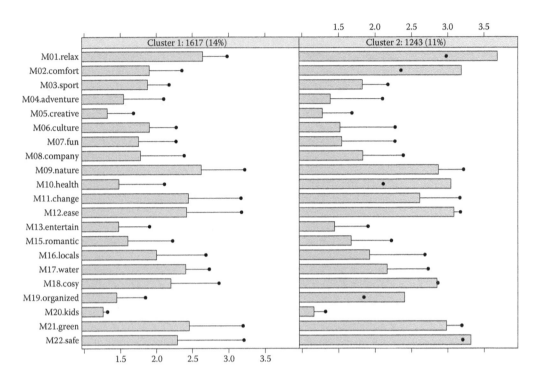

FIGURE 28.8
Cluster centers of clusters number 1 and 2 of the 9 cluster partition of the guest survey data.

28.4 Conclusion

Over the last decade, a number of resampling schemes have been proposed to detect the correct number of clusters for a given data set or assess the stability of complete partitions or single clusters. For a long time, these methods were computationally too expensive to be used by practitioners on a regular basis in everyday work. With the advent of multicore computers as standard desktops or laptops, cluster model diagnostics by resampling is feasible in acceptable computing time even on standard hardware. Base R has support for parallel computations since version 2.14.0. This makes it easier for package developers to run parallel computations even if the R user does not know how to setup a cluster of workstations or use multiple cores.

The algorithms in Sections 28.2.1 through 28.2.3 use the building blocks specified by the original authors using notation from the general framework RESAMPLE PARTITIONS. Most authors note that different sampling and evaluation schemes could be used instead, some do benchmarking to compare which combination works best (Hennig, 2007; Giancarlo et al., 2008). A comprehensive simulation study trying different combinations for a wide variety of clustering problems could offer guidance what works best in which situations. Simulations so far seem to suggest that resampling makes a lot of difference, while the exact scheme used is not that important.

References

Breckenridge, J. N. 1989. Replicating cluster analysis: Method, consistency, and validity. *Multivariate Behavioral Research 24*(2), 147–161.

Dolnicar, S. and F. Leisch 2004. Segmenting markets by bagged clustering. *Australasian Marketing Journal 12*(1), 51–65.

Dolnicar, S. and F. Leisch 2010. Evaluation of structure and reproducibility of cluster solutions using the bootstrap. *Marketing Letters 21*, 83–101.

Dubes, R. and A. Jain 1979. Validity studies in clustering methodologies. *Pattern Recognition 11*, 235–254.

Dudoit, S. and J. Fridlyand 2002. A prediction-based resampling method for estimating the number of clusters in a dataset. *Genome Biology 3*(7), 1–21.

Efron, B. and R. J. Tibshirani 1993. *An Introduction to the Bootstrap*. Monographs on Statistics and Applied Probability. New York, Chapman & Hall.

Fang, Y. and J. Wang 2012. Selection of the number of clusters via the bootstrap method. *Computational Statistics and Data Analysis 56*, 468–477.

Gana Dresen, I., T. Boes, J. Huesing, M. Neuhaeuser, and K.-H. Joeckel 2008. New resampling method for evaluating stability of clusters. *BMC Bioinformatics 9*(1), 42.

Giancarlo, R., D. Scaturro, and F. Utro 2008. Computational cluster validation for microarray data analysis: experimental assessment of clest, consensus clustering, figure of merit, gap statistics and model explorer. *BMC Bioinformatics 9*(1), 462.

Gordon, A. D. and M. Vichi 1998. Partitions of partitions. *Journal of Classification 15*, 265–285.

Grün, B. and F. Leisch 2004. Bootstrapping finite mixture models. In J. Antoch (Ed.), *Compstat 2004 — Proceedings in Computational Statistics*, pp. 1115–1122. Physica Verlag, Heidelberg. ISBN 3-7908-1554-3.

Hand, D. J. and W. J. Krzanowski 2005. Optimising k-means clustering results with standard software packages. *Computational Statistics and Data Analysis 49*, 969–973.

Hastie, T., R. Tibshirani, and J. Friedman 2009. *The Elements of Statistical Learning (Data Mining, Inference and Prediction)*. Springer Verlag, New York.

Hennig, C. 2007. Cluster-wise assessment of cluster stability. *Computational Statistics and Data Analysis 52*, 258–271.

Hennig, C. 2010. *fpc: Flexible Procedures for Clustering*. R package version 2.0-3.

Hornik, K. 2005. A CLUE for CLUster ensembles. *Journal of Statistical Software 14*(12), 1–25.

Hothorn, T., F. Leisch, A. Zeileis, and K. Hornik 2005. The design and analysis of benchmark experiments. *Journal of Computational and Graphical Statistics 14*(3), 675–699.

Hubert, L. and P. Arabie 1985. Comparing partitions. *Journal of Classification 2*, 193–218.

Kaufman, L. and P. J. Rousseeuw 1990. *Finding Groups in Data*. New York, John Wiley & Sons, Inc.

Kuhn, H. W. 1955. The Hungarian method for the assignment problem. *Naval Research Logistics Quarterly 2*(1–2), 83–97.

Lange, T., V. Roth, M. L. Braun, and J. M. Buhmann 2004. Stability-based validation of clustering solutions. *Neural Computation 16*(6), 1299–1323.

Leisch, F. 2006. A toolbox for k-centroids cluster analysis. *Computational Statistics and Data Analysis 51*(2), 526–544.

Leykin, A., J. Verschelde, and Y. Zhuang 2006. Parallel homotopy algorithms to solve polynomial systems. In A. Iglesias and N. Takayama (Eds.), *Mathematical Software—ICMS 2006*, Volume 4151 of *Lecture Notes in Computer Science*, pp. 225–234. Springer, Berlin, Heidelberg.

Maitra, R., V. Melnykov, and S. N. Lahiri 2012. Bootstrapping for significance of compact clusters in multidimensional datasets. *Journal of the American Statistical Association 107*(497), 378–392.

Martinetz, T. M., S. G. Berkovich, and K. J. Schulten 1993. "Neural-Gas" network for vector quantization and its application to time-series prediction. *IEEE Transactions on Neural Networks 4*(4), 558–569.

Matloff, N. 2011. *The Art of R Programming: A Tour of Statistical Software Design*. No Starch Press, San Francisco.

Monti, S., P. Tamayo, J. Mesirov, and T. Golub 2003. Consensus clustering: A resampling-based method for class discovery and visualization of gene expression microarray data. *Machine Learning 52*, 91–118.

Putler, D. S. and R. E. Krider 2012. *Customer and Business Analytics: Applied Data Mining for Business Decision Making Using R*. Chapman&Hall/CRC, Boca Raton, FL.

R Development Core Team 2014. *R: A Language and Environment for Statistical Computing*. Vienna, Austria: R Foundation for Statistical Computing.

Tibshirani, R. and G. Walther 2005. Cluster validation by prediction strength. *Journal of Computational and Graphical Statistics 14*(3), 511–528.

Wang, J. 2010. Consistent selection of the number of clusters via cross validation. *Biometrika 97*, 893–904.

Wedel, M. and W. A. Kamakura 1998. *Market Segmentation—Conceptual and Methodological Foundations*. Kluwer Academic Publishers, Boston, MA.

29

Robustness and Outliers

L.A. García-Escudero, A. Gordaliza, C. Matrán, A. Mayo-Iscar, and Christian Hennig

CONTENTS

Abstract

Unexpected deviations from assumed models as well as the presence of certain amounts of outlying data are common in most practical statistical applications. This fact could lead to undesirable solutions when applying nonrobust statistical techniques. This is often the case in cluster analysis, too. The search for homogeneous groups with large heterogeneity between them can be spoiled due to the lack of robustness of standard clustering methods. For instance, the presence of (even few) outlying observations may result in heterogeneous clusters artificially joined together or in the detection of spurious clusters merely made up of outlying observations. In this chapter, we will analyze the effects of different kinds of outlying data in cluster analysis and explore several alternative methodologies designed to avoid or minimize their undesirable effects.

29.1 Introduction

Robustness in statistics refers to stable behavior of methodology under small changes of data or models. For example, a small percentage of outliers can have a large impact on many statistical techniques. Robustness is a desirable property for more or less general statistical methodology.

In cluster analysis, many methods, be they heuristic or model-based, may suffer from strong instability from various sources of outlying data, by which data is meant that does not belong to any clear cluster. Outlying data points can act as "connectors" or "bridge points" between different groups, or they can play a disaggregating role. They can exhibit some degree of internal grouping, leading to additional spurious clusters, but they can also be "radial" or "isolated."

For illustration, consider the artificial data set in Figure 29.1a, made up of three main groups and a small fraction of "radial" outliers. If the well-known K-means method with $K = 3$ is applied, two intuitive main clusters are joined together and a spurious cluster is found, composed of outlying observations only. Moreover, only one single outlier placed in a remote position joins together the two main clusters in Figure 29.1b artificially when applying K-means with $K = 2$.

Similar problems arise when applying data mining techniques that use cluster analysis in "unsupervised learning" problems with large, complex, and high dimensional data sets collected through automated processes, which often contain many outliers.

A common strategy is to view isolated outliers and small groups of outliers as "clusters on their own." This is quite logical since outlying observations are obviously separated from other existing data patterns. Thus, some statisticians simply recommend increasing the number of groups in order to detect and isolate possibly outlying data points. For instance, the case in Figure 29.1b could have been addressed by searching for $K = 3$ clusters instead $K = 2$. However, this strategy is not always the most sensible one. For instance,

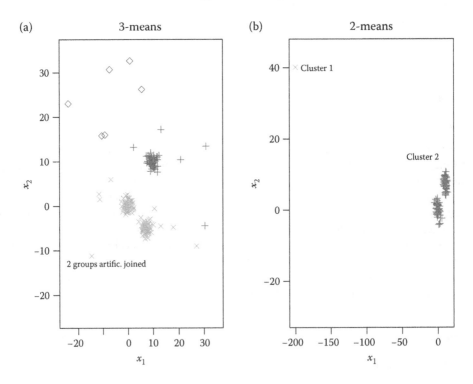

FIGURE 29.1
Effect of outliers in K-means cluster analysis: (a) radial contamination; (b) one single outlier.

due to prior knowledge about the problem at hand, the cluster analysis user may know or fix in advance the number of clusters and may not be aware of the presence of outlying data. Furthermore, in situations like the one shown in Figure 29.1a, made up of a few main clusters and a certain amount of "radial" outliers or "background" noise, accommodating them with small clusters would require a very large number of clusters, which is tedious to interpret and may incur computational problems.

Note that the clustering problem does not have a general definition. The "correct" clustering of a data set depends on the application and the cluster concept of the user. In some applications it is fine to handle one or more small "clusters of outliers," in other applications only a small number of larger clusters is meaningful and outliers are better excluded from any cluster. It may also be appropriate to integrate them into the nearest clusters, but it is hardly desired to give them a strong impact on how the clearly well-clustered nonoutlying data points are partitioned. That there often is such an impact is the most serious robustness problem.

One type of outliers that can be very harmful in cluster analysis is "bridge points" located between the main clusters (see Figure 29.2). These bridge points are not outlying in any of the coordinates and, thus, they cannot be detected with the use of the standard robust statistical tools existing in multivariate data analysis.

Another reason for the interest in robust cluster analysis techniques is the connection between cluster analysis and robust statistics, already pointed out in, for example, Rocke and Woodruff (2012), Hennig (2002), García-Escudero et al. (2003), Hardin and Rocke (2004), Woodruff and Reiners (2004), and Schynsa et al. (2010). First, as shown in previous examples, the need for robustness in cluster analysis is evident. Moreover, robust statistics in general can benefit from cluster analysis techniques, because often the most harmful outliers are those that appear clustered together (see Rocke and Woodruff, 1996), and cluster analysis is useful when handling groups of clustered outliers. Therefore, robust cluster analysis provides an appealing unifying framework for addressing both problems simultaneously.

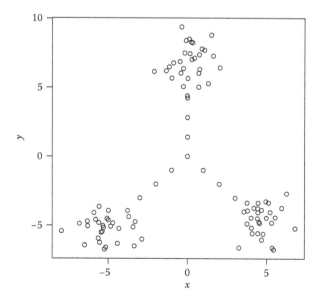

FIGURE 29.2
Outlying "bridge" points among three main clusters.

Section 29.2 reviews some of the different robust cluster analysis approaches that have been proposed in the literature with some emphasis on trimming-based approaches. Section 29.2.6 summarizes different tools and approaches that have been employed to qualitatively and numerically measure robustness in cluster analysis. Section 29.3 discusses different software packages that can be applied in robust cluster analysis. Finally, in Section 29.4 some concluding remarks and other possible open research lines are presented.

Recent reviews of robust cluster analysis are García-Escudero et al. (2010) and Banerjee and Davé (2012). The former is more focused on a statistical approach whereas the latter is more focused on a "machine learning" approach.

29.2 Robustness in Cluster Analysis

29.2.1 $\mathcal{L}_1$ Approaches

Given a sample $\mathcal{D} = \{x_1, \ldots, x_n\}$ in $\mathbb{R}^d$, the K-means method searches K point centers $\{m_1, \ldots, m_K\}$ in $\mathbb{R}^d$ and a partition $C_1, \ldots, C_K$ of $\mathcal{D}$ minimizing

$$\sum_{k=1}^{K} \sum_{i \in C_k} \|x_i - m_k\|^2 \tag{29.1}$$

Observations are then arranged into K clusters by assigning each observation to its closest center ($C_k = \{x_i : \|x_i - m_k\| \leq \|x_i - m_l\|$ for every $l = 1, \ldots, K\}$).

As the sample mean (i.e., the 1-mean), the K-means method suffers from a serious lack of robustness. One single outlying data point placed arbitrarily far away from the others suffices to spoil K-means, see Figure 29.1b. In fact, an extreme enough outlier will form a cluster on its own, so that, for fixed K, at least two original clusters (i.e., clusters found in the data set without outlier) will be merged.

Statistical methods based on least squares criteria are known to have poor robustness behavior. The lack of robustness of K-means could be explained by the least squares problem in (29.1). One could be tempted to try to robustify it by using the same arguments which led to the use of the sample median as a robust alternative to the sample mean in univariate location. Recall that the mean ($\mathcal{L}_2$ approach) minimizes over m the expression $\sum_{i=1}^{n}(x_i - m)^2$ whereas the median ($\mathcal{L}_1$ approach) minimizes the expression $\sum_{i=1}^{n}|x_i - m|$.

Starting from (29.1), this $\mathcal{L}_1$-approach leads to the K-medians and the partitioning around medoids (PAM) methods (see Chapter 4 and references therein), essentially defined through minimization of

$$\sum_{k=1}^{K} \sum_{x_i \in C_k} \|x_i - m_k\|$$

(the m_k may or may not be restricted to be members of the data set).

Unfortunately, the $\mathcal{L}_1$-approach only provides a rather modest robustification. The PAM method can resist the presence of an outlier such as in Figure 29.1b more often than K-means, but it breaks down if the number of outliers increases slightly or even if a single outlier is placed in a very remote position (see García-Escudero and Gordaliza, 1999). This also happens when considering other nondecreasing penalty functions $\Phi(\| \cdot - m_k \|)$ in (29.1).

The unsuitability of the $\mathcal{L}_1$ attempt to derive robust clustering methods is analogous to what happens in the linear regression setup, where the robustness behavior of $\mathcal{L}_1$-regression is poor in presence of extreme leverage points (Rousseeuw and Leroy, 1987).

29.2.2 Approaches Based on Trimming

Among the first highly robust regression proposals were the least median of squares (LMS) and the least trimmed squares (LTS) method (Rousseeuw and Leroy, 1987). This motivates the adaptation of the trimming principle to cluster analysis. Trimming means that a certain "trimmed" fraction of the observations (the most outlying ones) are not clustered at all and not taken into account for clustering the remaining data. This is appropriate in many applications where clusters are interpreted as substantially meaningful if they are homogeneous and large enough, but where isolated points are not of interpretative use.

Trimming has a long history of providing robustness to statistical methods. For instance, the univariate trimmed mean removes a proportion $\alpha/2$ of the largest observations and a proportion $\alpha/2$ of the smallest ones before computing the sample mean. Extending this idea to cluster analysis is not straightforward. Most interesting cluster analysis applications are multivariate, and a natural geometrical order does not exist for choosing the most extreme observations. Moreover, the notion of "outlyingness" must be modified to include potential bridge points between clusters (recall Figure 29.2).

A sensible way to perform trimming is to let the data decide which observations should be trimmed in order to find an optimal clustering for the untrimmed ones. This is the idea behind the aforementioned LMS and LTS regression method as well as the MCD (minimum covariance determinant) and MVE (minimum volume ellipsoid) covariance matrix estimators, see Rousseeuw and Leroy (1987).

Let $\varphi(\cdot; \mu, \Sigma)$ be the probability density function of the d-variate normal distribution with mean μ and covariance matrix Σ. Robust cluster analysis methods with trimming can be introduced through maximization of the target function (weighted trimmed classification likelihood or trimmed penalized likelihood, see below)

$$\sum_{k=1}^{K} \sum_{x_i \in C_k} \log \left(p_k \varphi \left(x_i ; m_k, S_k \right) \right) \tag{29.2}$$

where maximization is in terms of:

1. K centers m_k in $\mathbb{R}^d$,
2. K symmetric positive definite $d \times d$ scatter matrices S_k (satisfying constraints to be detailed later),
3. K weights in $[0, 1]$ satisfying $\sum_{k=1}^{K} p_k = 1$, and,
4. a partition $C_0, C_1, \ldots, C_K$ of $\mathcal{D}$ with C_k containing the observations that are assigned to cluster k for $k = 1, \ldots, K$, and the set C_0 with cardinality $n_{C_0} = [n\alpha]$ to include the observations that are trimmed.

Note that summation goes from $k = 1$ to K (not from $k = 0$) and, thus, the observations in C_0 are not taken into account when evaluating the target function. This avoids the harmful influence of a fraction of at most α of outlying observations.

This approach is called "impartial trimming" in Gordaliza (1991) and Cuesta-Albertos et al. (1997). Gallegos (2002) and Gallegos and Ritter (2005) introduced an interesting probabilistic framework, called "spurious outliers model," that justifies the consideration of this type of target function. This model states that the data points are independent, with a proportion of $1 - [n\alpha]$ points distributed according to one of K normal distributions defined by K different pairs of mean and covariance matrix, and the remaining $[n\alpha]$ (trimmed) points distributed according to $[n\alpha]$ different unspecified distributions $G_1, \ldots, G_{[n\alpha]}$. In Gallegos and Ritter (2005) it is proved under some mild conditions on the set of admissible $G_1, \ldots, G_{[n\alpha]}$ that maximizing (29.2) amounts to computing the maximum likelihood (ML) estimator for all the parameters of the K normal distributions and trimming the set of $[n\alpha]$ points assigned to but not depending on $G_1, \ldots, G_{[n\alpha]}$. Such an ML estimator assumes that all p_k in (29.2) are the same ("classification likelihood"). It is known (e.g., Gallegos and Ritter, 2005) that this implicitly favors clusters of similar sizes, and allowing flexible p_k is a way to deal with this.

Using such theory, one can see that the trimming approach based on (29.2) uses the normal distribution as a "cluster prototype shape," that is, it is appropriate for clustering applications in which the clusters are expected or desired to have an approximately normal shape. This does not necessarily imply that the underlying "true" clusters have to be normal; however, nonnormal clusters will be approximated by normal distributions and one may need more than K normal distributions to fit K nonnormal "clusters." It is possible to define similar approaches for other cluster prototype shapes, that is, other families of distributions.

It is important to note that the maximization of (29.2) without any constraints on the scatter (covariance) matrices S_k is a mathematically ill-posed problem. To see this, just take $C_k = \{x_i\}$ for any $x_i \in \mathcal{D}$, $m_k = x_i$ and S_k with $\det(S_k) \to 0$, which makes (29.2) unbounded. Different constraints on the scatter matrices have been considered in the literature. Different cluster analysis methods can be derived depending on whether equal weights $(p_1 = \cdots = p_K)$ are assumed or not.

The first application of this impartial trimming approach was the trimmed K-means method in Cuesta-Albertos et al. (1997). This is the most constrained case, assuming equal weights $p_1 = \cdots = p_K$ and $S_1 = \cdots = S_K = s^2 I_d$, that is, spherical clusters with equal within-cluster variation and (through $p_1 = \cdots = p_K$) a tendency to form clusters of similar sizes. The maximization of (29.2) is simplified to the minimization of

$$\sum_{k=1}^{K} \sum_{x_i \in C_k} \|x_i - m_k\|^2 \tag{29.3}$$

with $C_0, C_1, \ldots, C_K$ being a partition of $\mathcal{D}$ with $n_{C_0} = [n\alpha]$.

Figure 29.3 illustrates the ability of the trimmed K-means method to deal with different kinds of contamination. In both examples, parameter α is set to 0.1, that is, 10% of the data are trimmed (represented in this graph by circles). $K = 3$ main groups are detected in spite of the presence of radial contamination in Figure 29.3a and bridge points in Figure 29.3b.

Equation 29.3 admits a population version. This is a version of an estimator that is defined as a functional on the space of probability distributions instead of data sets in such a way that it generalizes the estimator, that is, one obtains the estimator if the population version is evaluated at the empirical distribution of a data set. Well-behaved population versions of an estimator allow that the estimator is consistent for its own population version at a given distribution. In cluster analysis, population versions of methods such as

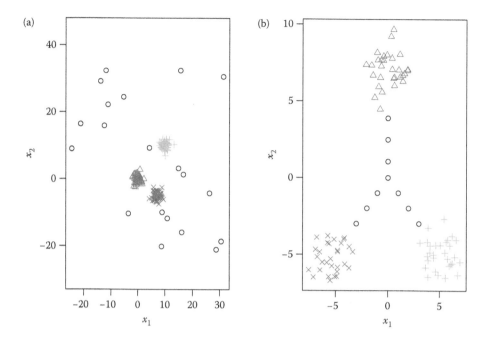

FIGURE 29.3
Trimmed K-means results with $K = 3$ and $\alpha = 0.1$ for two "contaminated" data sets (Trimmed points="∘").

K-means allow to partition the whole space of observations instead of a single data set. The population trimmed K-means can be viewed as a robustification of the "vector quantizers" used in signal processing, for example, Gersho and Gray (1991). Cuesta-Albertos et al. (1997) established the existence without moment conditions of the sample and population trimmed K-means. Furthermore, the sample trimmed K-means centers are consistent estimators of the population ones. A central limit theorem is derived in García-Escudero et al. (1999). However, if the underlying distribution is a mixture of normal distributions (with or without additional outlier generating distributions), the population trimmed K-means centers are not identical to the means of the normal components for the same reasons as when using unrobustified K-means, see Section 8.1.

An efficient algorithm for obtaining the trimmed K-means is available. The TRIMMED K-MEANS algorithm given by García-Escudero et al. (2003) is an extension of the classical Forgy (1965)'s K-MEANS algorithm, where some "concentration steps," similar to the ones used in the FAST-LTS and FAST-MCD algorithms (see Rousseeuw and Van Driessen, 1999), are applied.

Cuesta-Albertos et al. (1998) introduced a $\mathcal{L}_\infty$ version of trimmed K-means, called trimmed best K-nets. It is based on obtaining the K balls with minimal radii that cover a proportion $1 - \alpha$ of the data. This approach may be seen as an extension of the LMS and MVE techniques to cluster analysis and it shares with them their slow rate of convergence (see Cuesta-Albertos et al., 2002). Interpretation can be given through the excess-mass approach in Müller and Sawitzki (1991), that is, a cluster concept according to which clusters are interpreted as connected sets of high density. Kumar and Orlin (2008) also consider an interesting extension of the MVE to cluster analysis. Jolion et al. (1991) use the MVE and a Kolmogorov–Smirnov test procedure to detect clusters in a data set sequentially.

TRIMMED K-MEANS

Input parameters K and α

1. Draw K random initial centers $\mathbf{m}_1^0, \ldots, \mathbf{m}_K^0$

2. Concentration steps:

   a. Keep the set C of the $[n(1 - \alpha)]$ observations closest to the centers $\mathbf{m}_1^l, \ldots, \mathbf{m}_K^l$.

   b. Partition C onto K subsets $\{C_1, \ldots, C_K\}$, where C_k contains the observations in C closer to the center $\mathbf{m}_k^l$ than to the other centers.

   c. Update the centers $\mathbf{m}_1^{l+1}, \ldots, \mathbf{m}_K^{l+1}$ such that each center $\mathbf{m}_k^{l+1}$ is the sample mean of the observations in C_k. Go to (a) unless centers have not changed anymore.

3. Keep the optimal centers that minimize (29.3).

Output optimal centers and clustering.

Since K-means and trimmed K-means are based on the use of Euclidean distances, they aim at finding spherical groups of points with similar sizes. When researchers are interested in clusters that strongly depart from this assumption, these methods fail to identify them. For instance, an elongated group can be split into several clusters, or several elongated groups located close to each other can be joined together to constitute a single cluster. The search for clusters that have different sizes and different scatter structures leads to the "heterogeneous" cluster analysis problem, for which robustness issues should also be addressed.

With a heterogeneous robust cluster analysis perspective in mind, Gallegos and Ritter (2005) introduced the determinant criterion to robustify the proposal by Friedman and Rubin (1967). This implies that equal (but not necessarily spherical) scatter matrices $\mathbf{S}_1 = \cdots = \mathbf{S}_K = \mathbf{S}$ are assumed in (29.2). The criterion in Gallegos (2002) allows for different scatter matrices, but it implicitly assumes equal scales $\det(\mathbf{S}_1) = \cdots = \det(\mathbf{S}_K)$. Equal weights $p_1 = \cdots = p_K$ are assumed in both cases. The proposed algorithms (see details in Gallegos (2002) and Gallegos and Ritter (2005)) are similar to the TRIMMED K-MEANS algorithm, but Mahalanobis distances are used when computing distances to point centers. Moreover, they reduce to the FAST-MCD algorithm when $K = 1$. The algorithm in Gallegos (2002) constrains the $\mathbf{S}_k$ scatter matrices to have the same determinant in every concentration step. This idea was also considered in Maronna and Jacovkis (1974) as a sensible way to overcome the unboundedness of (29.2) in the untrimmed case. Considering the decomposition of $\mathbf{S}_k = s_k^2 \mathbf{U}_k$ with s_k^2 as an scale parameter and $\mathbf{U}_j$ with $\det(\mathbf{U}_k) = 1$ as the shape matrix, García-Escudero and Gordaliza (2007) stated that the proper estimation of the cluster scales s_k^2 is the most difficult task in (robust) cluster analysis (as long as α and K are fixed) and proposed an iterative procedure for estimating the scales based on a decreasing sequence of trimming levels.

In order to rely on more flexible constraints, we could try to extend Hathaway (1985)'s constraints to the heterogeneous robust cluster analysis framework. Consider the eigenvalues system of the cluster covariance matrices denoted by $\lambda_l(\mathbf{S}_k)$, for $k = 1, \ldots, K$ and $l = 1, \ldots, d$, and their maximum and minimum values

$$M_n = \max_{k=1,\ldots,K} \max_{l=1,\ldots,d} \lambda_l(\mathbf{S}_k) \text{ and } m_n = \min_{k=1,\ldots,K} \min_{l=1,\ldots,d} \lambda_l(\mathbf{S}_k)$$

We constrain the ratio M_n/m_n to be smaller than a fixed constant c. The use of this type of constraints together with group weights p_k's in (29.2) lead to the TCLUST approach in García-Escudero et al. (2008). The constant c controls the strength of the scatter constraint, yielding a weighted version of the trimmed K-means when $c = 1$, and an almost unrestricted procedure when using a large c value. The TCLUST algorithm (García-Escudero et al., 2008; Fritz et al., 2013a) can be applied to solve this problem approximately. This is a Classification-EM type algorithm (see Celeux and Govaert, 1992) with concentration steps. The TCLUST approach admits a population version and a consistency result is available.

Gallegos and Ritter (2009b) proposed a different extension of Hathaway's constraints (Hathaway, 1985), called Hathaway-Dennis-Beale-Thompson (HDBT), to the heterogeneous robust cluster analysis setup, which defines their version of trimmed model-based clustering. The Löwner ordering on the space of symmetric matrices is considered to constrain the scatter matrices. This implies considering the following constraint:

$$\min_{1\leq l\leq d}\min_{1\leq h\neq j\leq K} \lambda_l(S_h S_j^{-1}) \geq \frac{1}{c} \text{ with } c \geq 1$$

resulting in an affine equivariant cluster analysis procedure. Gallegos and Ritter (2010) deal with the unboundedness of the trimmed likelihood by controlling smaller cluster sizes.

Another application of trimming in heterogeneous robust cluster analysis is the MINO (minimum covariance determinant with outliers) approach by Rocke and Woodruff (2012) and Woodruff and Reiners (2004). Trimmed likelihoods like (29.3) are also considered in Markatou (2000) and Neykov et al. (2007). The main difference of these trimming approaches to TCLUST/HDBT is that TCLUST and HDBT place explicit constraints on the relations between cluster scatters whereas the other methods do not incorporate a mathematically consistent choice between multiple maximizers of the trimmed likelihood.

Careful monitoring of iterative trimming results is the basis of the "Forward Search" approach to robust cluster analysis (see Atkinson et al., 2006; Atkinson and Riani, 2007). This starts from a large number of small initial potential cluster cores (trimming effectively the vast majority of points) and adds points successively. Monitoring the change of within cluster homogeneity over the point addition process can uncover the clustering structure and potential outliers.

Santos-Pereira and Pires (2002), Hardin and Rocke (2004), and Willems et al. (2009) pay special attention to how to define and analyze group Mahalanobis distances from a robust cluster analysis viewpoint. Fixed point clustering (Hennig, 2002) also applies Mahalanobis distances and concentration steps to discover clusters in a robust way.

29.2.3 Mixture Approaches for Robust Clustering and Modeling Noise

Instead of discarding outlying observations, alternative approaches are based on fitting data not belonging to any cluster (so-called "noise," Fraley and Raftery, 1998) by mixture components. It is straightforward to obtain a "crisp" partition of $\mathcal{D}$ from a mixture fit, see Section 9.1, including points classified as noise.

A popular approach is based on accommodating the noise a uniformly distributed (Poisson process) "noise component" as done, for example, in Banfield and Raftery (1993), Dasgupta and Raftery (1998), Campbell et al. (1997, 1999), and Fraley and Raftery (1998,

2002). It is based on the maximization of log-likelihoods such as

$$\sum_{i=1}^{n} \log \left(p_0 \frac{1}{\mathrm{Vol}(V)} I_V(\mathbf{x}_i) + \sum_{k=1}^{K} p_k \varphi(\mathbf{x}_i; \mathbf{m}_k, \mathbf{S}_k) \right) \tag{29.4}$$

where $p_k \in [0, 1]$ with $\sum_{k=0}^{K} p_k = 1$, V is a set in $\mathbb{R}^d$, and $\mathrm{Vol}(V)$ denotes its volume.

Sometimes it is clear which set V should be considered (for instance, the area of a given image). Otherwise V could be considered as the convex hull of the data or the smallest rectangle whose sides are parallel to the axes or to the principal components, such that it contains all data points. For one-dimensional data all these amount to the range. All these are proposed in Fraley and Raftery (1998). The EM algorithm in Dempster et al. (1977) is used to estimate the unknown parameters in (29.4) once the set V is fixed. Another alternative approach is based on estimating V through ML estimation of a uniform distribution on V (Coretto and Hennig, 2011). All these methods provide some robustification but a single observation placed in a very remote position could still break down the clustering (Hennig, 2004, 2008).

Hennig (2004) also proposed the RIMLE (robust improper maximum likelihood estimator), which is based on the maximization of an "improper" likelihood:

$$\sum_{i=1}^{n} \log \left(p_0 c + \sum_{k=1}^{K} p_k \varphi(\mathbf{x}_i; \mathbf{m}_k, \mathbf{S}_k) \right) \tag{29.5}$$

where c is a tuning parameter that ideally serves to classify as noise observations arising from areas where components account for density values smaller than c. There are clear connections of the RIMLE approach with the trimming approach discussed in Section 29.2.2. Coretto and Hennig (2010) show that this approach has good robustness properties through a simulation study (the cited results are for one-dimensional data; they are generalized to multidimensional data in Coretto and Hennig (2015a,b)). When maximizing (29.4) and (29.5), constraints on the scatter matrices $\mathbf{S}_j$ are again needed to avoid degeneration.

The main difference between these approaches and trimming is that the mixture-based techniques allow a smooth transition of the classification between outliers and clusters (and between different clusters) through the estimated posterior probabilities of the observations belonging to the clusters (normal mixture components) and the noise component. Although they are often used for obtaining a crisp clustering, smoothness still has a certain (usually small) impact on the parameter estimation of the normal components, which may change the resulting clustering. Choosing c in the RIMLE approach is analogous to choosing α for trimming.

A different way to tackle the problem of outlying data points uses heavy-tailed mixture components to accommodate them. Mixtures of t-distributions with or without estimation of the degrees of freedom have been considered in, for example, McLachlan and Peel (2000) and Shoham (2002). This approach provides a valuable robustification, but Hennig (2004) showed again that a single extreme outlying value could cause breakdown. Greselin and Ingrassia (2010) consider appropriate scatter matrices constraints to avoid degeneration in such a setup.

Using mixtures of t-distributions for robustness implies a slightly different "clustering philosophy" than fitting normal mixtures. When normal mixtures are used for clustering, the normal distribution defines the "cluster prototype shape" and the idea behind adding a noise component is to capture points that cannot be assigned to any normal distribution sufficiently supported by the data. However, when mixtures of t-distributions are fitted for dealing with outliers, the idea is that only the core of a t-distribution is considered as "cluster," whereas points in the tails are interpreted as outliers, see McLachlan and Peel (2000), Section 29.5. This implies that outliers are still assigned to mixture components, even if they lie far away from all clusters.

Trimming principles were also considered to provide robustness when fitting mixtures. Such trimmed mixture likelihood methods are introduced in Neykov et al. (2007), Cuesta-Albertos et al. (2008), and Gallegos and Ritter (2009a).

Robustness in statistics is more general than dealing with outliers, although outliers often cause the worst robustness problems. Another interesting aspect in the mixture setup is what happens when normal mixtures are fitted to data where the clusters are actually non-normal. In some recent work (Tantrum et al., 2003; Hennig, 2010; Baudry et al., 2010) it was noted that when estimating the number of mixture components via the BIC (Section 9.1), normal mixtures can fit other distributions very well, but the estimated number of normal components will be higher than the number of interpretable clusters (depending on how exactly clusters are defined). The cited papers propose schemes to model clusters by sub-mixtures of normal distributions, that is, the clusters are obtained by merging some of the normal components.

Finally, it is important not to confuse the use of the term robustness made in this chapter with that made in the Bayesian mixture modeling framework where it often means insensitivity to the choice of prior probability distributions.

29.2.4 Robust Fuzzy Cluster Analysis

Whereas hard cluster analysis procedures are aimed at searching for sensible partitions of data into K disjoint clusters, fuzzy cluster analysis methods provide nonnegative membership values of observations for clusters so that overlapping clusters are generated (e.g., (Ruspini, 1969, Dunn, 1974, Bezdek, 1981, and Hathaway and Bezdek, 1993, Chapter 24). Models based on mixture models as discussed in the previous section are not normally regarded as fuzzy clustering methods because the underlying mixture model assumes that every point was generated by a single mixture component, although they, too, enable "soft classification" through posterior probabilities of points to belong to clusters.

As before, it is widely recognized that robustness of fuzzy cluster analysis methods is important for many practical applications. The fuzzy cluster analysis community was perhaps the first one to take the robustness challenge in cluster analysis seriously. Intuitively, outliers tend to be approximately "equally remote" to all clusters and, thus, they could have similar (but not necessarily small) membership values for all clusters. For instance, membership values for outlying observations could be close to $1/K$ for all the clusters with K being the number of groups whenever membership values are assumed to sum up to 1. Moreover, the way that many fuzzy clustering procedures weight data points is related to the way incorporated by M-estimators in robust statistics. Davé and Krishnapuram (1997) gives a review on robust fuzzy cluster analysis.

One of the methods which is more widely considered in robust fuzzy cluster analysis is the fuzzy K-means method with "noise component" introduced in Davé (1991), which has a plethora of modifications. This method is based on a modification of the fuzzy K-means algorithm (Section 5.5) that considers a fictitious $(K + 1)$-th center at the same distance δ from every point in the data set. Observations that are associated to this fictitious center are denoted as belonging to the "noise cluster." This implies that points will have their highest membership degree for the noise cluster if they are further away than δ from any other cluster center. The choice of the parameter δ is not an easy task but some sensible heuristic rules have been proposed (e.g., Davé, 1991; Davé and Sen, 1997; Rehm et al., 2007). The methods based on "noise distances" are also refereed as "noise clustering" in the literature. $\mathcal{L}_1$ approaches have been also considered in Krishnapuram et al. (1999) through the proposal of a fuzzy K-medoids algorithm.

A "least trimmed squares" fuzzy clustering method (closely related to the trimmed K-means in Section 29.2.2) was introduced in Kim et al. (1996). The proposed method is based on trimming a fixed fraction α of observations. The fixed trimming level controls the number of observations to be discarded in a different way than those based on defining a "noise distance" δ. Discarding a fixed fraction of data has also been considered in Klawonn (2004).

Possibilistic fuzzy clustering methods (Krishnapuram and Keller, 1993) address the problem of data with "noise"-observations in fuzzy clustering by relaxing the constraint that the membership values have to sum up to 1. Outlying observations could then receive arbitrarily low membership values for all clusters. Although possibilistic fuzzy clustering methods are more robust against outliers, they tend to produce coincident clusters (Barni et al., 1996) and, therefore, it is better to see them as "mode-seeking" rather than "partitioning" procedures.

All these approaches inherit from fuzzy K-means their preference for spherical clusters, and so they are not well suited to detect clusters with very different shapes. Several procedures have been proposed to address heterogeneous fuzzy cluster analysis problems (e.g., Gustafson and Kessel, 1979; Trauwaert et al., 1991; Rousseeuw et al., 1996). These methods can also be modified by including the possibility of trimming a fixed proportion α of the data. This would lead to a fuzzy version of the TCLUST procedure described in Section 29.2.2 as introduced in Fritz et al. (2013b). Robust fuzzy cluster analysis using t-distributions has been proposed in Chatzis and Varvarigou (2008). Łeski (2003) and Wu and Yang (2002) give further procedures for robust fuzzy clustering. More interesting references are in Banerjee and Davé (2012).

29.2.5 Robust Nonmodel-Based Cluster Analysis

Some widely applied nonmodel-based cluster analysis approaches suffer from serious robustness problems as well. For instance, anomalous observations often induce spurious "bumps" in nonparametric methods based on density estimation leading to the detection of artificial clusters; outlying data points can induce "chaining" effects on hierarchical cluster analysis methods, and so on.

Some attempts to improve the robustness of these nonmodel-based cluster analysis methods can be found in the literature. For instance, "denoising" techniques have been proposed resorting to Voronoï methods in Allard and Fraley (1997), nearest-neighbor methods in Byers and Raftery (1998) and minimum spanning trees in Jaing et al. (2001).

Some density-based cluster analysis methods like Cuevas et al. (2001) seem to have good behavior from a robustness point of view. This is due to the fact that points are considered "outliers" if they lie in regions with low density, so if (given a certain density estimator) points are only assigned to clusters in areas of high density, these can be expected to be unaffected by outliers. Such a technique will depend on the typical tuning for density estimators (e.g., a bandwith for kernels or a neighborhood size) and a density cutoff similar to the c required for the RIMLE in Section 29.2.3. Robustness can only be expected if these choices are not dominated by the outliers.

Cluster analysis methods based on statistical depths as in Ding et al. (2007) exist, too. Robust hierarchical cluster analysis proposals exist as well, such as in Lin and Chen (2005) and Balcan and Gupta (2010). Balcan and Gupta (2010) assume that there is a "good" clustering in the data, demanding that for all points in a subset of the data of size $(1 - v)n$ all but αn (α very small) of their nearest neighbors belong to the same cluster, where v and α are tuning constants. This means that all but $(1 - v)n$ points are "strongly clustered" in the sense above, so that vn outliers are allowed. Their hierarchical algorithm starts by first partitioning the data set into so-called "blobs" of points, making sure that points that can be connected by having enough common neighbors are in the same blob. These blobs are then hierarchically clustered by agglomeration using a score function that considers clusters at the current agglomeration level as similar if the median distance between a point of one cluster and all points of the other one is small for at least half of the points of one of the clusters. The use of medians attempts to make this unaffected by potential outliers in blobs/clusters. Balcan and Gupta (2010) show that this produces a tree, a pruning of which is the assumed "good" clustering.

A number of other nonmodel-based clustering methods, for example, BIRCH, CURE, ROCK, "Chameleon," and DBSCAN, are also designed in order to provide protection against noise and outliers. Comments and useful references are given in Banerjee and Davé (2012).

29.2.6 Parameter-Based Measurement of Robustness in Cluster Analysis

In cluster analysis, and generally in the statistical literature, many methods are advertised as "robust." Often this is motivated either by heuristic arguments why a method is supposedly unaffected by outliers, or by good results in simulation studies or example data sets with outliers. Although this is better than ignoring the robustness issue altogether, such arguments are unsatisfactory because it is unclear to what extent they generalize to situations other than those explicitly considered by the authors.

The measurement of the robustness of statistical procedures has been addressed through two main approaches: Huber's "minimax" approach and Hampel's infinitesimal approach (Huber, 1981; Hampel et al., 1986). The first approach has only been applied in rather restricted setups, and up to our knowledge it has not been adapted cluster analysis. We will focus on the second one.

Hampel's approach rests on three fundamental pillars: qualitative robustness, the influence function, and the breakdown point. All these robustness measures were designed to be applied to statistical procedures viewed as functionals defined on a suitable space of probability measures. Qualitative robustness has to do with the continuity of the functional (population version of an estimator) at the assumed model, the influence function with its differentiability (therefore they both are "infinitesimal" concepts) and the breakdown point with the distance from the model to a singularity of the functional.

More precisely, qualitative robustness of an estimator T_n means that, for large enough n, if two distributions P and Q are close to each other (e.g., one being $(1 - \epsilon)$ times the other plus a proportion of small ϵ of a distribution that generates extreme outliers) according to a suitable metric for probability measures, the distributions of T_n for i.i.d. samples from P and Q are close to each other as well.

Given an estimator functional (population version) T, its influence function at a distribution P is $\lim_{\epsilon \to 0} T((1 - \epsilon)P + \epsilon \delta_x) - T(P)/\epsilon$ as a function of x. This formalizes the change of T under infinitesimal contamination by a one-point distribution at x. Infinitesimal robustness requires this to be bounded.

The breakdown point complements the infinitesimal concepts by looking at to what minimum extent one has to contaminate P in order to drive a functional T as far as possible away from its value at P, that is, to a singularity of the parameter space (often infinity, but it can be zero, e.g., for variances or mixture component proportions). The most popular version of the breakdown point looks at the infimum ϵ for distributions of the form $(1 - \epsilon)P + \epsilon H$ so that breakdown happens, where H is chosen as destructive as possible (contamination model). This is for example zero for the arithmetic mean and $1/2$ for the median. Another version of the breakdown point is the "finite sample breakdown point" of an estimator (Donoho and Huber, 1983), formalizing what proportion of data points in a data set needs to be replaced (or added) in order to drive the estimator to a singularity. This is often easier to handle than the population-based version.

In cluster analysis, one of the advantages of model-based clustering methods is that clusters can be characterized by features that have a population counterpart. Thus, we can view clustering methods as statistical functionals of probability measures (the empirical and the population one) and then apply Hampel's classical approach to measure their robustness behavior.

The qualitative robustness notion is binary; it classifies the functionals as having this property or not. García-Escudero and Gordaliza (1999) proved that for distributions with uniquely defined trimmed K-means, the trimmed K-means are qualitatively robust whereas K-means and PAM/K-medoids are not. García-Escudero and Gordaliza (1999) also addressed the influence functions of K-means and trimmed K-means as estimates of the centers of the clusters. Influence functions were obtained for replacing the Euclidean norm in (29.3) by general penalty functions $\Phi(\| \cdot -\mathbf{m}_k\|)$, and it was shown that they are bounded when the derivative of this Φ-function is bounded. However, in cluster analysis (as in regression), it could happen that an estimator with bounded influence is nevertheless highly vulnerable to outliers. This is, for instance, the case of PAM as reflected through its breakdown point, which equals zero. Ruwet et al. (2012) showed that all statistical functionals involved in the TCLUST procedure, that is, the centers, the scatter matrices and the weights, have bounded influence functions whenever they are uniquely defined for the underlying distribution P.

García-Escudero and Gordaliza (1999) pointed out that the finite sample breakdown point is very data dependent in the context of cluster analysis. For any clustering method incorporating location parameters, data sets can be designed in which the inherent structure is so unstable that a single point converging to infinity will eventually attract a cluster. Particularly if K clusters are fitted to a data set that can be fitted by $K - 1$ clusters without much loss of quality, if one outlier is added one couldn't even say that a solution is clearly "wrong" or undesirable in which the Kth cluster is used to fit the outlier. Instability in cluster analysis may be caused by an unstable clustering structure of the data set as well as by choice of a nonrobust method. Therefore, one cannot hope that any reasonable clustering method has good breakdown behavior uniformly over all data sets. Nonrobust methods such as K-means and PAM, however, have a zero breakdown point for all data sets.

García-Escudero and Gordaliza (1999) suggested dealing with "well clustered" data sets in order to measure and compare the breakdown behavior of cluster analysis methods. Gallegos and Ritter (2005, 2009a,b) formalized the notion of "well clustered" data sets through a cluster separation condition. Moreover, they introduced a restricted breakdown point notion on the class of data sets satisfying the separation property and applied this measure to the procedures introduced by them. Ruwet et al. (2013) studied the restricted breakdown values of the TCLUST procedure and compared them with other trimming based approaches. In all these papers, the restricted breakdown point, on the class of data sets satisfying the corresponding separation properties, is asymptotically equal to the trimming level α, whereas in the case of the untrimmed versions the restricted breakdown continues to be 0. Ruwet et al. (2013) shows that the separation criterion for the method in Gallegos and Ritter (2009b) is a bit more restrictive than the one for TCLUST as a price to be paid for the former's affine equivariance. In order to appreciate these results, note that one cannot hope for a better breakdown point in cluster analysis than the proportion of the smallest cluster (if breakdown is to be achieved by adding points) or even half that size (for a "replacement breakdown point"), because a sufficiently homogeneous set of outliers of this size will attract a fitted cluster in a stronger way than the original smallest cluster. Therefore, the trimming rate α should not be chosen larger than the smallest cluster of interest could be. Otherwise this cluster could be trimmed. This is different from classical situations such as regression or location estimation, where the best achievable breakdown point is often $1/2$.

Hennig (2004) shows that standard ML based on a normal mixture is not breakdown robust. ML under adding a uniform mixture component over the convex hull of the data or for t-mixtures have a very similar breakdown point behavior. Their (addition) breakdown point is $1/(n+1)$, which is the smallest possible value. Nevertheless, both the convex hull uniform noise method and ML for t-mixtures can be robust against one or more outliers of moderate size. They only break down under very extreme outliers. The RIMLE is proven to have a better (asymptotically nonzero) breakdown point under a condition making sure that K clusters can be fitted much better than $K-1$ components on the same data set, which takes the role of the above mentioned "good clustering" assumption.

All the results in the present section assume that the number of clusters K is fixed, which is required in order to make the parameter space and its boundaries well defined.

29.2.7 Assignment-Based and Other Robustness Measures

It is a matter of controversy to what extent parameter-based robustness measurement captures the robustness properties of cluster analysis methods appropriately. On the one hand, robustness failure as flagged by these measures is certainly a serious problem. On the other hand, there are a number of robustness issues that do not show up as "breakdown" or "unbounded influence." Furthermore, these methods cannot be applied to clustering methods that are not based on estimation of statistical parameters, and they assume the number of clusters K to be fixed.

Hennig (2004) proposed an amendment of the (finite sample addition) breakdown point definition for mixture model parameters with K estimated, which declares the estimator as "broken down" for a data set with K clusters before addition of outliers, if the parameters of at most $K-1$ components do not either reach a singularity or the components vanish. The idea is that loss of a component is treated as breakdown, whereas adding components is not a problem, particularly because it can be seen as appropriate that newly added points require new components to be fitted. The somewhat surprising but intuitively reasonable consequence of this definition is that if criteria such as the AIC or BIC (see Section 9.1) are used to estimate K, all considered methods including ML for plain normal mixtures are perfectly robust against extreme outliers, and breakdown can only be achieved by closing gaps between clusters with added "bridge points." However, as already mentioned, increasing K with the number of outliers is not always the best possible solution because it could lead to a large value of K with radial contamination, which may result in computational trouble when fitting the mixture.

Back to fixed K, another concern regarding parameter-based robustness measurement is that not all serious robustness problems are caused by parameters diverging to the borders of the parameter space. One can imagine very different clusterings on the same data set implied by different parameters that are all finite and far away from the parameter space borders. According to the classical robustness concepts, this kind of problem should be captured by the influence function and qualitative robustness. However, these measurements may miss problems because of the discrete nature of cluster analysis.

Consider $K = 3$ (fixed) clusters fitted to a data set in which there are four clearly visible groups, about equispaced (and maybe a few outliers to be trimmed). Imagine that trimmed K-means (α chosen so that the outliers but nothing else are trimmed) merges two of the four clusters because they may be a tiny little bit less separated from each other or a tiny little bit smaller than the others. Let us assume that this solution is unique, but the clustering of the data set is certainly ambiguous and a quite different partition, merging two other components, may achieve almost the same value of the objective function (29.3). According to García-Escudero and Gordaliza (1999), as explained above, the influence function is bounded, not indicating any robustness problem. However, the decision which clusters are merged may be switched, changing the solution completely, under adding a few points or even a single one in one or between two of the clusters because this may just make the originally slightly worse partition slightly better with respect to (29.3). The problem is that in such situations what happens for infinitesimal contamination as formalized in the influence function is not a good approximation for what may happen under still quite small contamination, at least if the trimmed K-means objective function is close to nonuniqueness.

In order to address this kind of robustness problem, and also in order to enable robustness measurement for more general clustering methods, robustness measurement can be based on the assignment of points to the clusters, and on counting assignment changes between original and contaminated data.

Such approaches have been used for measuring clustering stability in the machine learning literature for some time, see von Luxburg (2010) for an overview. One key result is that the stability of K-means under random variations is strongly connected to the uniqueness of the solution to (29.1) in the data set, which can be related to the above discussion. Such results differ from the robustness approach in that worst case behavior is not taken into account.

Hennig (2008) defined general assignment-based robustness measurements for cluster analysis. The "dissolution point" is an adaptation of the breakdown point. Because in many data sets some clusters are more clear and more stable than others, a dissolution point is not defined for a whole data set but for a cluster (in principle one could define an overall dissolution point by minimizing over all clusters).

The concept is based on the Jaccard similarity for two data subsets C and D, $\gamma(C, D) = \frac{|C \cap D|}{|C \cup D|}$. The dissolution point, as the finite sample addition breakdown point, is based on looking at how many points have to be added to a data set in order to generate a worst case result, here to "dissolve" a cluster. A cluster C is called "dissolved" if in the clustering of the new data set after adding points there is no cluster D for which $\gamma(C, D^*) > 1/2$, D^* being the intersection of D and the original data set. Half is not the worst possible value, but it is the maximum worst possible value over all clusters when comparing two partitions of the same data set, that is, it is the smallest value which enables every cluster to dissolve.

Hennig (2008) shows dissolution robustness results for various methods, including a positive result for trimmed K-means, and equivalent results to Hennig (2004) for ML methods for mixtures, including a positive result for RIMLE. Results for K-means and PAM for fixed K are negative. There is also an analogous positive result to Hennig (2004) for estimating the number of mixture components by AIC or BIC. There are positive results for single and complete linkage hierarchical clustering for cutting the dendrogram at a fixed height (because this can isolate outliers safely), whereas cutting at a fixed number of clusters is not dissolution robust. All these results require "good clustering"-conditions as before.

Hennig (2007) uses the Jaccard similarity approach for empirical evaluation of robustness and stability of clusters in a given data set based on resampling and addition of artificial data, see also Section 9.3.

Further, Hennig (2008) defines a method to be "isolation robust" if one cannot merge clusters bridging an arbitrarily large gap by adding a certain maximum number of points. It is argued that methods with fixed K can never be isolation robust because if K is misspecified, instability of the partition can be achieved in data sets with arbitrarily large gaps between clusters. Mixture estimation with AIC and BIC and hierarchical clustering (single and complete linkage) cutting at fixed height are isolation robust. Estimating K is not isolation robust in general, though, because it is showed that the average silhouette width method to estimate K (see Chapter 9) is not isolation robust.

Other assignment-based versions of robustness measurement exist. Ruwet and Haesbroeck (2011) obtained influence functions for the classification error rates of K-means and PAM. Results as those in Balcan and Gupta (2010) explained in Section 29.2.5 can also be interpreted as robustness results under specific "good clustering" conditions.

29.3 Software for Robust Cluster Analysis

Most of the robust cluster analysis proposals cited in previous sections can be easily implemented with standard statistical software. We will focus on their readily available implementations in R (R Development Core Team, 2010) packages at the CRAN repository.

The $\mathcal{L}_1$ approach can be carried out with the `pam` function included in the `cluster` package. The `tclust` package implements the trimming approaches described in Section 29.2.2. For instance, the graphs shown in Figure 29.3 can be obtained through the use of function `tkmeans(X,k,alpha)`, where `X` is the data set, `k` is the number of clusters and `alpha` is the trimming proportion. The `tclust` function in this package allows to implement the heterogeneous robust cluster analysis procedures discussed in Section 29.2.2. Here, `restr` specifies the type of constraints applied to the scatter matrices and `restr.fact` is the value of the constant c. The parameter `equal.weights` specifies whether equal weights $p_1 = \cdots = p_k$ are assumed in (29.2) or not. The approach introduced in Gallegos (2002) is obtained by `restr="deter"` and `restr.fact=1`, and that introduced in García-Escudero and Gordaliza (2005) by `restr="sigma."` The use of `restr = "eigen"` (constraints on the eigenvalues) serves to implement the TCLUST method in García-Escudero et al. (2008).

In order to illustrate the use of the `tclust` package, the well-known "Swiss bank notes" data set in Flury and Riedwyl (1988) will be used. This data set includes six measurements made on 100 genuine and 100 counterfeit old Swiss bank notes. The following code is used to obtain Figure 29.4:

```
R > data ("swissbank")
R > clus <- tclust (swissbank, k = 2, alpha = 0.08,
                        restr = "eigen", restr.fact = 15)
R > plot (swissbank[, 4], swissbank[, 6], pch = clus$cluster + 1)
```

`clus$cluster` returns a vector with the cluster assignments for all the observations in the data set, using the value "0" for the trimmed ones. One of the cluster essentially includes most of the "forged" bills whereas the other one includes most of the "genuine" ones. Within the 8% proportion of trimmed observations ($\alpha = 0.08$), we may identify a subset with 15 forged bills following a clearly different forgery pattern, which has been previously reported in the literature (e.g., Flury and Riedwyl, 1988). More details on the `tclust` package are found in Fritz et al. (2012).

The `tlemix` package Neytchev et al. (2012) implements the trimmed likelihood approach to robust cluster analysis, Neykov et al. (2007). The internal use of the `flexmix` package (Leisch, 2004) yields high flexibility that allows `tlemix` to be adapted to a large variety of statistical problems.

The mixture approach with noise component to robust cluster analysis can be carried out in R with the Fraley and Raftery (2012)'s `mclust` package. This package needs an initialization for the observations that are initially considered as noise through `initialization = list(noise=noiseInit)` with `noiseInit` being a TRUE/FALSE vector with the same length as the number of observations. The application of the EM algorithm often corrects an improper choice of this vector. One possibility to choose it is through the function `NNclean` in the `prabclus`-package (Hennig and Hausdorf, 2012), which implements the noise

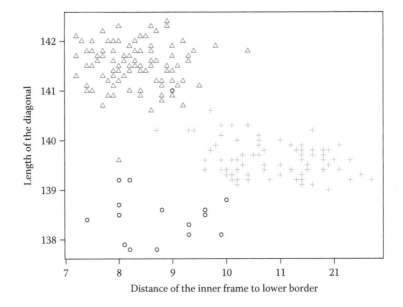

FIGURE 29.4
Clustering results with $k = 2$, $\alpha = 0.08$, and $c = 15$ for the "Swiss Bank notes" data set.

detection method proposed in Byers and Raftery (1998). The multivariate RIMLE with automatic tuning (Coretto and Hennig, 2015a) is implemented in the package OTRIMLE.

The fitting of mixtures of *t*-distributions can be done through the EMMIX package. Originally this was written in Fortran. An R version is available on Geoff McLachlan's web page (McLachlan, 2012).

29.4 Conclusion

The negative effect of outlying observations on many well-established clustering techniques is widely recognized. Therefore developing cluster analysis methods that are not severely affected by outliers or small deviations from the assumed models is of much interest. We have reviewed different robust cluster analysis approaches proposed in the literature.

There are still many open research problems. Some classical robustness measurements raise some issues when translated into the cluster analysis framework, and new meaningful and computable robustness measures are still of interest. The measures introduced in Hennig (2008) can be applied to far more methods than those featuring already in that paper. Specifically, more robustness analysis of methods involving the estimation of the number of clusters K, of nonparametric density-estimation based methods and of methods not based on probability models is required. More can be done as well regarding the systematic comparison of the various approaches by theory, as far as available, or simulation.

Another important problem that deserves careful attention and further research is the choice of the number of clusters K. This problem, which is generally notoriously difficult, is even harder in robust cluster analysis, because it is related to the task of proper estimation of the level of underlying contamination, required trimming α, respectively (Section 29.2.2). There is an inevitable element of user's choice involved, regarding the question from what size downward a homogeneous group of points should be treated as "grouped outliers" instead of a "cluster." A higher trimming level may "eat" smaller clusters, leading to a smaller number of clusters to be estimated. On the other hand, nontrimmed outliers might result in new clusters when choosing α too low. The user needs to decide whether and from what "borderline size" in the given application very small clusters are meaningful or to be ignored, be it because they are likely to be erroneous "contamination," or be it because small groups are not of interest.

Moreover, the simultaneous determination of the number of groups and the contamination level also depends on the type of clusters we are searching for and on the allowed differences between cluster's scatter matrices. For illustration, consider the data set in Figure 29.3a. $K = 4$ may be chosen if we allow for very differently scattered clusters (the fourth cluster gathers all the observations far away from the three main clusters), but if clusters are required to be homogeneous, $K = 3$ plus a 10% contamination level makes more sense.

From our point of view, it is neither possible nor desirable to have unsupervised methodology that could simultaneously tell us the number of groups, contamination level and the maximum allowed difference between clusters for a given data set. These aspects interact, and need some amount of user tuning. It is more logical to think of a procedure that is able to return an appropriate cluster partition or a small list of tentative clustering partitions whenever at least one of these three ingredients (number of clusters, contamination level, and allowed scatter differences) is available due to some specific knowledge about the problem at hand. The graphical tools described in García-Escudero et al. (2011) (see also Fritz et al., 2012) have this aim. They rely on proper monitoring of the so-called "trimmed classification likelihoods" obtained from the maximum values achieved by (29.2) when varying K, α and constant c. The consideration of weights p_k in (29.2) is recommended for choosing the number of clusters as it was explained in Bryant (1991). The study of García-Escudero et al. (2003) was a first attempt for this monitoring approach, considering trimmed K-means.

Ritter (2015) (who also provides a lot on asymptotic theory for methods based on trimming) suggests exploring many local optima of the objective function in connection with using various different covariance constraints. The so-called "Pareto solutions" are clusterings that appear optimal at least for a certain covariance constraint.

The monitoring of trimmed likelihoods with an over-fitting penalty term ("trimmed BIC") was considered in Neykov et al. (2007). Baudry et al. (2010) monitor various numbers of clusters as well in their methodology to merge normal mixture components in order to fit mixtures with nonnormal clusters by normal mixtures. Another monitoring technique is the "forward search" (see Section 29.2.2). All these authors agree that a dynamic sequence of images of the data (for instance, resulting from a sequence of different trimming levels) provides a more useful view of the data set than a single "static" view.

When adopting the mixture fitting approach to robust cluster analysis, the use of BIC-type criteria provides guidance for determining the number of clusters K and the contamination level p_0 in (29.4). The allowed differences in within-cluster scatter are controlled through the different parameterizations of the scatter matrices of the mixture components. However, it is important that the proper choice of the number of clusters in cluster analysis is not exactly equivalent to that of proper choice of the number of components in a mixture problem (e.g., Celeux and Soromenho, 1996; Hennig, 2010, Hennig and Corette, 2015). Hennig (2004) and Hennig and Coretto (2008) provide sensible ways to choose constant c for the RIMLE method. One approach is to find a c or contamination level that makes the nonnoise part of the mixture as similar as possible to a normal mixture with the parameters estimated for the components in terms of the Kolmogorov—or another distance between distributions.

In the framework of mixture models, exploring the robustness of further types of mixtures of nonnormal (e.g., skew) distributions (and robustifying them where necessary) is of current interest.

All the proposals discussed in this review concern the problem of clustering around centroids. There are interesting problems where observations are naturally clustered around linear and nonlinear manifolds, see also Section 6.4. García-Escudero et al. (2009) show how the trimming principles presented in Section 29.2.2 can be adapted to robust clustering around linear subspaces through RLGA (robust linear grouping analysis). This method is available in the lga R-package (Harrington, 2012). Robust clusterwise regression techniques were introduced in Hennig (2002), Hennig (2003), and García-Escudero et al. (2010). Detecting specific shapes and objects in noisy images is a problem addressed by the pattern recognition community with procedures that often can be adapted to solve specific robust cluster analysis problems, for example, Luan et al. (1998). Further robustification problems arise from clustering functional data (see García-Escudero and Gordaliza, 2005; Cuevas et al., 2007, Section 9.6).

Acknowledgment

This research was partially supported by the Spanish Ministerio de Economía y Competitividad, grant MTM2014-56235-C2-1-P, and by Consejería de Educación de la Junta de Castilla y León, grant VA212U13.

References

Allard, D. and C. Fraley 1997. Nonparametric maximum likelihood estimation of features in spatial point processes using Voronoi tessellation. *Journal of the American Statistical Association 92*, 1485–1493.

Atkinson, A. and M. Riani 2007. Exploratory tools for clustering multivariate data. *Computational Statistics and Data Analysis 52*, 272–285.

Atkinson, A., M. Riani, and A. Cerioli 2006. Random start forward searches with envelopes for detecting clusters in multivariate data. In M.R.S. Zani, A. Cerioli, and M. Vichi (Eds.), *Data Analysis, Classification and the Forward Search*, Springer-Verlag, Berlin Heidelberg, pp. 163–172.

Balcan, M. and P. Gupta 2010. Robust hierarchical clustering. In *Proceedings of the 23rd Annual Conference on Learning Theory (COLT), 2010*, pp. 282–294.

Banerjee, A. and R.N. Davé 2012. Robust clustering. *WIREs Data Mining and Knowledge Discovery 2(1)*, 29–59.

Banfield, J. and A. Raftery 1993. Model-based Gaussian and non-Gaussian clustering. *Biometrics 49(3)*, 803–821.

Barni, M., V. Cappellini, and A. Mecocci 1996. Comments on "a possibilistic approach to clustering". *IEEE Transactions on Fuzzy Systems 4(3)*, 393 –396.

Baudry, J.-P., A. Raftery, G. Celeux, K. Lo, and R. Gottardo 2010. Combining mixture components for clustering. *Journal of Computational and Graphical Statistics 19*, 332–353.

Bezdek, J. 1981. *Pattern Recognition with Fuzzy Objective Function Algorithms*. New York, NY: Plenum Press.

Bryant, P. 1991. Large-sample results for optimization-based clustering methods. *Journal of Classification 8*, 31–44.

Byers, S. and A. Raftery 1998. Nearest neighbor clutter removal for estimating features in spatial point processes. *Journal of the American Statistical Association 93*, 577–584.

Campbell, J., C. Fraley, F. Murtagh, and A. Raftery 1997. Linear flaw detection in woven textiles using model-based clustering. *Pattern Recognition Letters 18*, 1539–1548.

Campbell, J., C. Fraley, D. Stanford, F. Murtagh, and A. Raftery 1999. Model-based methods for textile fault detection. *International Journal of Imaging Systems and Technology 10*, 339–346.

Celeux, G. and A. Govaert 1992. A classification EM algorithm for clustering and two stochastic versions. *Computational Statistics and Data Analysis 14*, 315–332.

Celeux, G. and G. Soromenho 1996. An entropy criterion for assessing the number of clusters in a mixture model. *Journal of Classification 13*, 195–212.

Chatzis, S. and T. Varvarigou 2008. Robust fuzzy clustering using mixtures of student's-*t* distributions. *Pattern Recognition Letters 29*, 1901–1905.

Coretto, P. and C. Hennig 2010. A simulation study to compare robust clustering methods based on mixtures. *Advances in Data Analysis and Classification 4*, 111–135.

Coretto, P. and C. Hennig 2011. Maximum likelihood estimation of heterogeneous mixtures of gaussian and uniform distributions. *Journal of Statistical Planning and Inference 141*, 462–473.

Coretto, P. and C. Hennig 2015a. Robust improper maximum likelihood: Tuning, computation, and a comparison with other methods for robust Gaussian clustering, http://arxiv.org/abs/1406.0808, submitted.

Coretto, P. and C. Hennig 2015b. A consistent and breakdown robust model-based clustering method, http://arxiv.org/abs/1309.6895, submitted.

Cuesta-Albertos, J., L. García-Escudero, and A. Gordaliza 2002. On the asymptotics of trimmed best k-nets. *Journal of Multivariate Analysis 82*, 482–516.

Cuesta-Albertos, J., A. Gordaliza, and C. Matrán 1997. Trimmed k-means: A attempt to robustify quantizers. *Annals of Statistics 25*, 553–576.

Cuesta-Albertos, J., A. Gordaliza, and C. Matrán 1998. Trimmed best k-nets. A robustified version of a $\mathcal{L}_\infty$-based clustering method. *Statistics and Probability Letters 36*, 401–413.

Cuesta-Albertos, J., C. Matran, and A. Mayo-Iscar 2008. Robust estimation in the normal mixture model based on robust clustering. *Journal of the Royal Statistical Society, Series B 70*, 779–802.

Cuevas, A., M. Febrero, and R. Fraiman 2001. Cluster analysis: A further approach based on density estimation. *Computational Statistics and Data Analysis 36*, 441–459.

Cuevas, A., M. Febrero, and R. Fraiman 2007. Impartial trimmed k-means for functional data. *Computational Statistics and Data Analysis 51*, 4864–4877.

Dasgupta, A. and A. Raftery 1998. Detecting features in spatial point processes with clutter via model-based clustering. *Journal of the American Statistical Association 93*, 294–302.

Davé, R. 1991. Characterization and detection of noise in clustering. *Pattern Recognition Letters 12*, 657–664.

Davé, R. and R. Krishnapuram 1997. Robust clustering methods: A unified view. *IEEE Transactions on Fuzzy Systems 5*, 270–293.

Davé, R. and S. Sen 1997. Noise clustering algorithm revisited. In *Proceedings of the Biennial Workshop NAFIPS 1997*.

Dempster, A., N. Laird, and D. Rubin 1977. Maximum likelihood from incomplete data via the EM algorithm. *Journal of the Royal Statistical Society, Series B 39*, 1–38.

Ding, Y., X. Dang, H. Peng, and D. Wilkins 2007. Robust clustering in high dimensional data using statistical depths. *BMC Bioinformatics 8(Suppl 7):S8*.

Donoho, D. and P. Huber 1983. The notion of breakdown point. In *A Festschrift for Erich L. Lehmann*, Wadsworth Statist./Probab. Ser., pp. 157–184. Wadsworth.

Dunn, J. 1974. A fuzzy relative of the ISODATA process and its use in detecting compact well-separated clusters. *Journal of Cybernetics 3*, 32–57.

Flury, B. and H. Riedwyl 1988. *Multivariate Statistics. A Practical Approach*. London: Chapman and Hall.

Forgy, E. 1965. Cluster analysis of multivariate data: Efficiency versus interpretability of classifications. *Biometrics 21*, 768–780.

Fraley, C. and A. Raftery 1998. How many clusters? Which clustering method? Answers via model-based cluster analysis. *Computer Journal 41*, 578–588.

Fraley, C. and A. Raftery 2002. Model-based clustering, discriminant analysis, and density estimation. *Journal of the American Statistical Association 97*, 611–631.

Fraley, C. and A. Raftery 2012. *mclust: Model-Based Clustering/Normal Mixture Modeling*. R package version 3.4.11.

Friedman, H. and J. Rubin 1967. On some invariant criterion for grouping data. *Journal of the American Statistical Association 63*, 1159–1178.

Fritz, H., L. García-Escudero, and A. Mayo-Iscar 2012. tclust: An R package for a trimming approach to cluster analysis. *Journal of Statistical Software 47(12)*, 1–26.

Fritz, H., L. García-Escudero, and A. Mayo-Iscar 2013a. A fast algorithm for robust constrained clustering. *Computational Statistics and Data Analysis 61(61)*, 124–136.

Fritz, H., L. García-Escudero, and A. Mayo-Iscar 2013b. Robust constrained fuzzy clustering. *Information Sciences 245(245)*, 38–52.

Gallegos, M. 2002. Maximum likelihood clustering with outliers. In K. Jajuga, A. Sokolowski, and H. Bock (Eds.), *Classification, Clustering and Data Analysis: Recent Advances and Applications*, pp. 247–255. Springer-Verlag, Berlin Hedelberg.

Gallegos, M. and G. Ritter 2005. A robust method for cluster analysis. *Annals of Statistics 33*, 347–380.

Gallegos, M. and G. Ritter 2009a. Trimmed ML estimation of contaminated mixtures. *Sankhya (Series A) 71*, 164–220.

Gallegos, M. and G. Ritter 2009b. Trimming algorithms for clustering contaminated grouped data and their robustness. *Advances in Data Analysis and Classification 10*, 135–167.

Gallegos, M. and G. Ritter 2010. Using combinatorial optimization in model-based trimmed clustering with cardinality constraints. *Computational Statistics and Data Analysis 54*, 637–654.

García-Escudero, L. and A. Gordaliza 1999. Robustness properties of k-means and trimmed k-means. *Journal of the American Statistical Association 94*, 956–969.

García-Escudero, L. and A. Gordaliza 2005. A proposal for robust curve clustering. *Journal of Classification 22*, 185–201.

García-Escudero, L. and A. Gordaliza 2007. The importance of the scales in heterogeneous robust clustering. *Computational Statistics and Data Analysis 51*, 4403–4412.

García-Escudero, L., A. Gordaliza, and C. Matrán (1999). A central limit theorem for multivariate generalized trimmed k-means. *Annals of Statistics 27*, 1061–1079.

García-Escudero, L., A. Gordaliza, and C. Matrán 2003. Trimming tools in exploratory data analysis. *Journal of Computational and Graphical Statistics 12*, 434–449.

García-Escudero, L., A. Gordaliza, C. Matrán, and A. Mayo-Iscar 2008. A general trimming approach to robust cluster analysis. *Annals of Statistics 36*, 1324–1345.

García-Escudero, L., A. Gordaliza, C. Matrán, and A. Mayo-Iscar 2010. A review of robust clustering methods. *Advances in Data Analysis and Classification 4*, 89–109.

García-Escudero, L., A. Gordaliza, C. Matrán, and A. Mayo-Iscar 2011. Exploring the number of groups in robust model-based clustering. *Statistics and Computing 21*, 585–599.

García-Escudero, L., A. Gordaliza, A. Mayo-Iscar, and R. San Martín 2010. Robust clusterwise linear regression through trimming. *Computational Statistics and Data Analysis 54*, 3057–3069.

García-Escudero, L., A. Gordaliza, R. San Martín, S. Van Aelst, and R. Zamar 2009. Robust linear clustering. *Journal of the Royal Statistical Society, Series B 71*, 301–318.

Gersho, A. and R. Gray 1991. *Vector Quantization and Signal Compression*. New York, NY: Springer.

Gordaliza, A. 1991. Best approximations to random variables based on trimming procedures. *Journal of Approximation Theory 64*, 162–180.

Greselin, F. and S. Ingrassia 2010. Constrained monotone EM algorithms for mixtures of multivariate t distributions. *Statistics and Computing 20*, 9–22.

Gustafson, E. and W. Kessel 1979. Fuzzy clustering with a fuzzy covariance matrix. In *Proceedings of the IEEE International Conference on Fuzzy Systems, San Diego, 1979*, pp. 761–766.

Hampel, F., E. Ronchetti, P. Rousseeuw, and W. Stahel 1986. *Robust Statistics. The Approach Based on Influence Functions*. New York, NY: Wiley. Series in Probability and Mathematical Statistics: Probability and Mathematical Statistics.

Hardin, J. and D. Rocke 2004. Outlier detection in the multiple cluster setting using the Minimum Covariance Determinant estimator. *Computational Statistics and Data Analysis 44*, 625–638.

Harrington, J. 2012. *lga: Tools for Linear Grouping Analysis*. R package.

Hathaway, R. 1985. A constrained formulation of maximum likelihood estimation for normal mixture distributions. *Annals of Statistics 13*, 795–800.

Hathaway, R. and J. Bezdek 1993. Switching regression models and fuzzy clustering. *IEEE Transactions on Fuzzy Systems 1*, 195–204.

Hennig, C. 2002. Fixed point clusters for linear regression: Computation and comparison. *Journal of Classification 19*, 249–276.

Hennig, C. 2003. Clusters, outliers and regression: Fixed point clusters. *Journal of Multivariate Analysis 83*, 183–212.

Hennig, C. 2004. Breakdown points for maximum likelihood-estimators of location-scale mixtures. *Annals of Statistics 32*, 1313–1340.

Hennig, C. 2007. Cluster-wise assessment of cluster stability. *Computational Statistics and Data Analysis 52*, 258–271.

Hennig, C. 2008. Dissolution point and isolation robustness: Robustness criteria for general cluster analysis methods. *Journal of Multivariate Analysis 99*, 1154–1176.

Hennig, C. 2010. Methods for merging gaussian mixture components. *Advances in Data Analysis and Classification 4*, 3–34.

Hennig, C. and P. Coretto 2008. The noise component in model-based cluster analysis. In L.S.-T.C. Preisach, H. Burkhardt, and R. Decker (Eds.), *Data Analysis, Machine Learning and Applications, Studies in Classification, Data Analysis, and Knowledge Organization*, pp. 127–138. Berlin-Heidelberg: Springer.

Hennig, C. and B. Hausdorf 2012. *prabclus: Functions for Clustering of Presence—Absence, Abundance and Multilocus Genetic Data*. R package version 2.2-4.

Huber, P. 1981. *Robust Statistics*. New York, NY: Wiley. Series in Probability and Mathematical Statistics: Probability and Mathematical Statistics.

Jaing, M., S. Tseng, and C. Su 2001. Two-phase clustering process for outlier detection. *Pattern Recognition Letters 22*, 691ÂŬ–700.

Jolion, J.-M., P. Meer, and S. Bataouche 1991. Robust clustering with applications in computer vision. *IEEE Transactions on Pattern Analysis and Machine Intelligence 13(8)*, 791–802.

Kim, J., R. Krishnapurama, and R. Davé 1996. Application of the least trimmed squares technique to prototype-based clustering. *Pattern Recognition Letters 17*, 633–641.

Klawonn, F. 2004. Noise clustering with a fixed fraction of noise. In A. Lotfi and J. Garibaldi (Eds.), *Applications and Science in Soft Computing*. Berlin: Springer, pp. 133–138.

Krishnapuram, R., A. Joshi, and L. Yi 1999. A fuzzy relative of the k-medoids algorithm with application to web document and snippet clustering. In *Snippet Clustering, in Proceedings of IEEE International Conference Fuzzy Systems—FUZZIEEE99*, Korea.

Krishnapuram, R. and J. Keller 1993. A possibilistic approach to clustering. *IEEE Transactions on Fuzzy Systems 1(2)*, 98–110.

Kumar, M. and J. Orlin 2008. Scale-invariant clustering with minimum volume ellipsoids. *Computers and Operations Research 35*, 1017–1029.

Leisch, F. 2004. Flexmix: A general framework for finite mixture models and latent class regression in R. *Journal of Statistical Software 11*, 1–18.

Łeski, J. 2003. Towards a robust fuzzy clustering. *Fuzzy Sets and Systems 137(2)*, 215–233.

Lin, C.-R. and M.-S. Chen 2005. Combining partitional and hierarchical algorithms for robust and efficient data clustering with cohesion self-merging. *IEEE Transactions on Knowledge and Data Engineering 17*, 145–149.

Luan, J., J. Stander, and D. Wright 1998. On shape detection in noisy images with particular reference to ultrasonography. *Statistics and Computing 8*, 377–389.

Markatou, M. 2000. Mixture models, robustness, and the weighted likelihood methodology. *Biometrics 356*, 483–486.

Maronna, R. and P. Jacovkis 1974. Multivariate clustering procedures with variable metrics. *Biometrics 30*, 499–505.

McLachlan, G. 2012. *R Version of EMMIX*. Brisbane.

McLachlan, G. and D. Peel 2000. *Finite Mixture Models*. New York, NY: Wiley. Series in Probability and Statistics.

Müller, D. and G. Sawitzki 1991. Excess mass estimates and tests for multimodality. *Journal of the American Statistical Association 86*, 738–746.

Neykov, N., P. Filzmoser, R. Dimova, and P. Neytchev 2007. Robust fitting of mixtures using the trimmed likelihood estimator. *Computational Statistics and Data Analysis 52*, 299–308.

Neytchev, P., P. Filzmoser, R. Patnaik, A. Eisl, and R. Boubela 2012. *tlemix: Trimmed Maximum Likelihood Estimation*. R package version 0.1.

R Development Core Team 2010. *R: A Language and Environment for Statistical Computing*. Vienna: R Foundation for Statistical Computing. ISBN 3-900051-07-0.

Rehm, F., F. Klawonn, and R. Kruse 2007. A novel approach to noise clustering for outlier detection. *Soft Computing 11*, 489–494.

Ritter, G. 2015. *Robust Cluster Analysis and Variable Selection*. Boca Raton, FL: CRC Press.

Rocke, D. and D. Woodruff 1996. Identification of outliers in multivariate data. *Journal of the American Statistical Association 91*, 1047–1061.

Rocke, D. and D. Woodruff 2012. Computational connections between robust multivariate analysis and clustering. In W. Härdle and B. Rönz (Eds.), *COMPSTAT 2002 Proceedings in Computational Statistics*, pp. 255–260. Heidelberg: Physica-Verlag.

Rousseeuw, P., L. Kaufman, and E. Trauwaert 1996. Fuzzy clustering using scatter matrices. *Computational Statistics and Data Analysis 23*, 135–151.

Rousseeuw, P. and A. Leroy 1987. *Robust Regression and Outlier Detection.* New York, NY: Wiley-Interscience.

Rousseeuw, P. and K. Van Driessen 1999. A fast algorithm for the minimum covariance determinant estimator. *Technometrics 41*, 212–223.

Ruspini, E. 1969. A new approach to clustering. *Information and Control 15*, 22–32.

Ruwet, C., L. García-Escudero, A. Gordaliza, and A. Mayo-Iscar 2012. The influence function of the TCLUST robust clustering procedure. *Advances in Data Analysis and Classification 6*, 107–130.

Ruwet, C., L. García-Escudero, A. Gordaliza, and A. Mayo-Iscar 2013. On the breakdown behavior of robust constrained clustering procedures. *TEST 22*, 466–487.

Ruwet, C. and G. Haesbroeck 2011. Impact of contamination on training and test error rates in statistical clustering analysis. *Communications on Statistics: Simulations and Computing 40*, 394–411.

Santos-Pereira, C. and A. Pires 2002. Detection of outliers in multivariate data: A method based on clustering and robust estimators. In W. Härdle and B. Rönz (Eds.), *Proceedings in Computational Statistics: 15th Symposium Held in Berlin, Germany*, pp. 291–296. Heidelberg: Physica-Verlag.

Schynsa, M., G. Haesbroeck, and F. Critchley 2010. RelaxMCD: Smooth optimisation for the minimum covariance determinant estimator. *Computational Statistics and Data Analysis 54*, 843–857.

Shoham, S. 2002. Robust clustering by deterministic agglomeration EM of mixtures of multivariate *t* distributions. *Pattern Recognition 55*, 1127–1142.

Tantrum, J., A. Murua, and W. Stuetzle 2003. Assessment and pruning of hierarchical model based clustering. In *Proceedings of the Ninth ACM SIGKDD International Conference on Knowledge Discovery and Data Mining*, pp. 197–205. New York, NY: ACM.

Trauwaert, E., L. Kaufman, and P. Rousseeuw 1991. Fuzzy clustering algorithms based on the maximum likelihood principle. *Fuzzy Sets and Systems 42*, 213–227.

von Luxburg, U. 2010. Clustering stability: An overview. *Foundations and Trends in Machine Learning 2*, 235–274.

Willems, G., H. Joe, and R. Zamar 2009. Diagnosing multivariate outliers detected by robust estimators. *Journal of Computational and Graphical Statistics 18*, 73–91.

Woodruff, D. and T. Reiners 2004. Experiments with, and on, algorithms for maximum likelihood clustering. *Computational Statistics and Data Analysis 47*, 237–253.

Wu, K.-L. and M.-S. Yang 2002. Alternative c-means clustering algorithms. *Pattern Recognition 35*(10), 2267–2278.

30

Visual Clustering for Data Analysis and Graphical User Interfaces

Sébastien Déjean and Josiane Mothe

CONTENTS

Abstract

Cluster analysis is a major method in data mining to present overviews of large data sets. Clustering methods allow dimension reducing by finding groups of similar objects or elements. Visual cluster analysis has been defined as a specialization of cluster analysis and is considered as a solution to handle complex data using interactive exploration of clustering results. In this chapter, we consider three case studies in order to illustrate cluster analysis and interactive visual analysis. The first case study is related to information retrieval field and illustrates the case of multidimensional data in which objects to analyze are represented considering various features or variables. Evaluation in information retrieval considers many performance measures. Cluster analysis is used to reduce the number of measures to a small number that can be used to compare various search engines. The second case study considers networks in which data to analyze is represented in the form of matrices that correspond to adjacency matrices. The data we used is obtained from publications; cluster analysis is used to analyze collaborative networks. The third case study is related to curve clustering and applies when temporal data is involved. In this case study, the application is time-series gene expression. We conclude this chapter by presenting some other types of data for which visual clustering can be used for analysis purposes and present some tools that implement other visual analysis functionalities we did not present in the case studies.

30.1 Introduction

Cluster analysis is a major method in data mining to present overviews of large data sets and has many applications in machine learning, image processing, social network analysis, bioinformatics, marketing, e-business, ... Clustering methods allow to reduce dimension by finding groups of similar objects or elements (Han 2005). A large number of clustering methods have been developed in the literature to achieve this general goal; they differ in the method used to build the clusters and the distance they use to decide whether objects are similar or not. Another important aspect of clustering is cluster validation which relies on validation measures (Halkidi et al. 2001; Brock et al. 2008). The decision on which method to use and on the optimal number of clusters can depend on the application and on the analyzed data set (as can be appreciated from other chapters in this volume). Exploring and interpreting groups of elements that share similar properties or behavior rather than individual objects allows the analyst to consider large data sets and to understand their inner structure. Visual cluster analysis has been defined as a specialization of cluster analysis and is considered as a solution to handle complex data using interactive exploration of clustering results (Tobias et al. 2009). Shneiderman's information-seeking mantra "overview first, zoom and filter, and then details on demand" (Shneiderman 1996) applies to visual clustering. Various tools have been developed for visual cluster analysis providing these functionalities to explore the results.

For cluster analysis, objects are often depicted as feature vectors or matrices: objects can thus be viewed as points in a multidimensional space (Andrienko et al. 2009). More complex data requires consideration of features to be used before being represented in this way: this is the case for relational data (e.g., social networks) or time series and temporal data. In this chapter, we consider three case studies in order to illustrate cluster analysis and interactive visual analysis. First, we illustrate the case of multidimensional data in which objects to analyze are represented considering various features or variables. The case study we chose is information retrieval (IR) evaluation for which many performance measures have been defined in the literature. Cluster analysis is used to reduce the number of measures to a small number that consider the various points of view that can be used to compare various search engines. We then consider networks in which data to analyze is represented in the form of matrices that correspond to adjacency matrices. We chose to illustrate this case considering collaborative networks applied to publications. We show how visual analysis can be used to find clusters of authors. Moreover, we expand this type of exploration to more complex analysis, combining authorship with geographic and topic information. We also illustrate how large-scale and temporal collaborative networks can be analyzed. The third case study is related to curve clustering and applies when temporal data is involved. In this case study where, the application is time series gene expression, we show how clustering the shapes of the curves rather than the absolute level of expression allows finding different types of gene expressions. We conclude this chapter in presenting some other types of data for which visual clustering can be used for analysis purposes; and present some tools that implement other visual analysis functionalities we did not present in the case studies.

30.2 Multidimensional Data

Multivariate statistical methods are generally based on a matrix of data as a starting point. From a statistical point of view, the considered matrix consists of rows, which correspond to objects or individuals to analyze, and of columns, which correspond to variables used to characterize the individuals. No particular structure is assumed about the variables; in particular, an arbitrary permutation of the columns will not affect the cluster analysis.

30.2.1 Performance Measures in Information Retrieval

The study presented here is detailed in Baccini et al. (2011). Evaluating effectiveness of information retrieval systems is achieved by performing on a collection of documents, a search, in which a set of test queries are performed and, for each query, the list of the relevant documents. This evaluation framework also includes performance measures making it possible to control the impact of a modification of search parameters. A large number of measures are available to assess performance of the system, some being more used like the mean average precision or recall-precision curves.

In the present study, a row (an individual) corresponds to a run characterized by the performance measures, which indeed correspond to variables (columns). The matrix we have to analyze is composed of 23,518 rows and 130 columns. An extract of the matrix we analyzed is presented in Table 30.1.

Among many problems that can be addressed regarding this data set, we focus here on a clustering task. Indeed, one motivation of this work is to compare all measures and to help the user to choose a small number of them when evaluating different IR systems.

TABLE 30.1

Extract of the Analyzed Matrix. The First Four Columns Represent an Identifier, the Collection on Which the Search Engine Was Applied, the Search Engine and the Information Need Respectively. Other Columns Correspond to Performance Measures

Line	Year	System	Topic	0.20R.prec	0.40R.prec	0.60R.prec	...
1	TREC 1993	Brkly3	101	0.2500	0.1250	0.1111	...
2	TREC 1993	Brkly3	101	0.3077	0.2692	0.3077	...
3	TREC 1993	Brkly3	101	0.4737	0.4474	0.4211	...
...	...	...	...	...	...	...	...
23516	TREC 1999	weaver2	448	0.0000	0.0000	0.0357	...
23517	TREC 1999	weaver2	449	0.0000	0.0000	0.0000	...
23518	TREC 1999	weaver2	450	0.7627	0.6864	0.5966	...

Relationships between the 130 performance measures available for individual queries are investigated and it is shown that they can be clustered into homogeneous clusters.

In our statistical approach, we focused on the columns of the matrix, in order to highlight the relationships between performance measures. To achieve this analysis, we have considered three exploratory multivariate methods: hierarchical clustering, partitioning, and principal component analysis (PCA).

Clustering of the performance measures was performed in order to define a small number of clusters, each one including redundant measures. Partitioning was used in order to stabilize the results of the hierarchical clustering. PCA provides indicators and graphical displays giving a synthetic view, in small dimension, of the correlation structure of the columns and of clusters previously defined. Each method is illustrated in the following. A synthetic view combining these methods is proposed as a 3D-map.

30.2.2 Dendrogram to Define the Clusters of Performance Measures

Agglomerative clustering proposes a classification of performance measures without any prior information on the number of clusters.

The choice of the number of clusters is a crucial problem to be dealt with *a posteriori* when performing clustering (see for instance Al Hasan et al. 2011; Chen et al. 2011) in the context of text clustering and other chapters in this volume. When using Ward's criterion, the vertical scale of the tree represents the loss of between-cluster inertia for each clustering step; a relevant pruning level is characterized by a relatively important difference between the heights of two successive nodes. In the sub-plot of Figure 30.1, a relevant cut corresponds to a point for which there is a strong slope on the left and a weak slope on the right. Under these conditions, according to the degree of sharpness desired, one can retain here 2, 3, 5 or 6 clusters. This last option is represented in Figure 30.1 with six demarcated clusters.

30.2.3 Principal Component Analysis to Validate the Clusters

In Figure 30.2, a color is associated with a cluster obtained as depicted in Figure 30.1. The relative position of the clusters on the first and second principal components is consistent with the clusters obtained after the clustering process. Globally, the measures in each cluster appear projected relatively close to each other. Furthermore, it also offers a partial (because of the projection on a 2D space) representation of the inertia of the clusters. For instance, the IR performance measures that are clustered in the cluster 3 (see Figure 30.1) and displayed in green appear much closer to each other than the performance measures in other clusters in Figure 30.2.

Regarding principal component 1 (horizontal axis in Figure 30.2), the main phenomenon is the opposition between clusters 4 (blue), 5 (cyan), and 6 (magenta) on the right and, 1 (black) and 2 (red) on the left. These relative positions highlight an opposition between recall-oriented clusters (4, 5, and 6) and precision-oriented ones (1 and 2). Along PC2 (vertical), the opposition is between 1 (black) and 4 (blue) (bottom) and, 2 (red) and 6 (magenta) (top). In this case, the discrimination globally concerns the number of documents on which is based the performance measure: few documents (less than 30) for clusters 2 and 6, and much more (more than 100) for clusters 1 and 4. Not surprisingly, cluster 3 (green), mainly composed of global measure aggregating recall/precision curves such as MAP, is located in the center of the plot. Cluster 3 acts as an intermediate between other clusters.

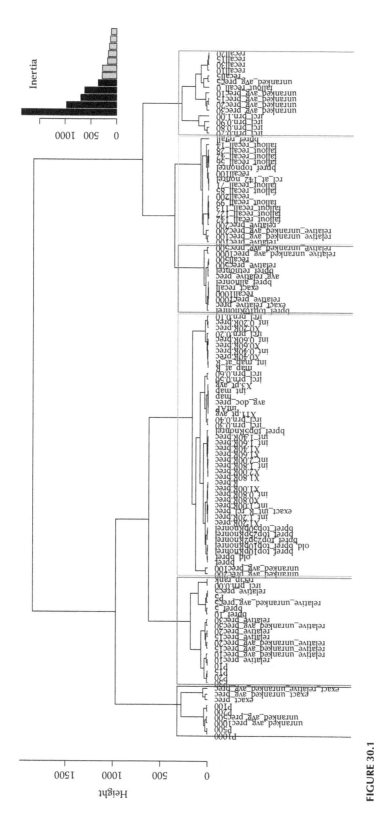

FIGURE 30.1

Dendrogram representing the hierarchical clustering (using Euclidean distance and Ward's criterion) of the IR performance measures with a relevant pruning at 6 clusters. The subplot in the upper-right corner represents the height of the 12 upper nodes; highlighting the first five bars refers to 6 clusters retained. (This figure is available in color at http://www.crcpress.com/product/isbn/9781466551886.)

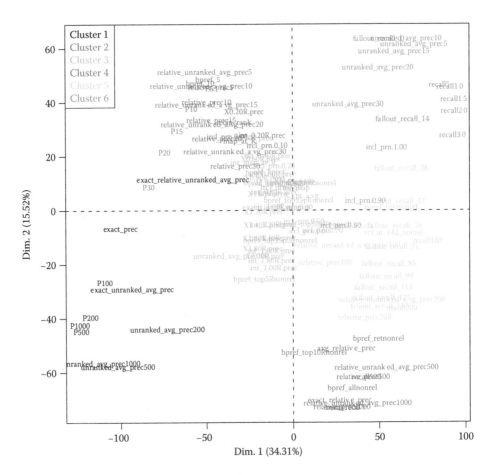

FIGURE 30.2
Representation of variables on the first two principal components PC1 and PC2 respectively explaining 34% and 15% of the total variance. Color reveals the cluster the variables belong to after partitioning. (This figure is available in color at http://www.crcpress.com/product/isbn/9781466551886.)

In this case, the visualization by PCA and the interpretation of the principal components provides a very clear characterization of the clusters and of their mutual relations. Furthermore, a 3D representation considering the first three principal components can highlight some potentially peculiar arrangement of points. For instance, the measure `exact_relative_unranked_avg_prec` appears in black near the red cluster in Figure 30.2, but having a look at the PCA in three dimensions (Figure 30.3) reveals a clear distinction between the red and the black clusters.

Figure 30.3 was obtained using the rgl package (Adler and Murdoch 2011) for the R software (R Development Core Team 2012). This package uses OpenGL (Woo et al. 2005) to provide a visualization device system in R with interactive viewpoint navigation facility.

30.2.4 3D-Map

Various works to combine several methods into one single graphic have been proposed. Koren et al. for instance suggest to superimpose a dendrogram over a synchronized low-dimensional embedding resulting in a single image showing all the clusters and the relations between them. In the same vein, Husson et al. (2010) proposed to combine PCA,

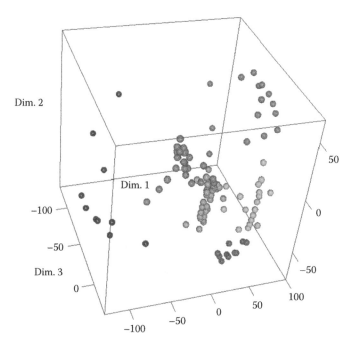

FIGURE 30.3
Representation of variables on the first three principal components PC1 and PC2 using the rgl package (Adler and Murdoch 2011) in R Development Core Team (2012).

hierarchical clustering and partitioning to enrich the description of the data. PCA representation is used as a basis for the hierarchical tree drawing in a 3D-map. An implementation of such a representation is available in the FactoMineR package (Husson et al. 2011) for R. In Figure 30.4, IR performance measures are:

- Located on the PCA factor map,
- Linked through the branch of the dendrogram,
- Colored according to the cluster they belong to after partitioning.

This representation includes the previous one with PCA (Figure 30.2) and adds other information regarding the changes that have occurred when performing partitioning after hierarchical clustering. For instance, one performance measure (`exact_unranker_avg_prec`) colored in black was linked by the dendrogram to the cluster in green, in the bottom left corner. The partitioning method reallocated it into the black cluster which is consistent regarding the location of this point relatively to the two considered clusters.

30.3 Graphs and Collaborative Networks

As explained in Mothe et al. (2006), graphs are among the visualization tools most commonly used in the literature, as linking concepts or objects is the most common mining technique. Graph agents use 2D matrices of any type resulting from the pretreatment of

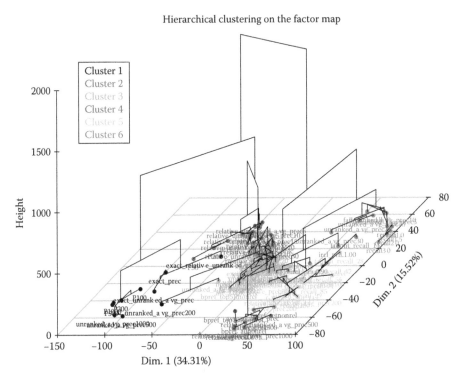

FIGURE 30.4
Representation from the FactoMineR package (Husson et al. 2011) combining PCA, hierarchical clustering, and partitioning.

the raw information; these matrices correspond to adjacency matrices. Adjacency matrices can be obtained from analyzing cooccurrences from texts, coauthoring from publications, or any information crossing (Crimmins et al. 1999). From an adjacency matrix, a graph is built: graph nodes correspond to the values of the crossed items whereas edges reflect the strength of the cooccurrence values. Graph drawing can be based on force-based algorithms (Fruchterman and Reingold 1991). In this type of algorithm, a graph node is considered as an object while an edge is considered as a spring. Edge weights correspond to either repulsion or attraction forces between the objects that in turn make them move in space. This keeps the vertices moving in the visualization space until objects are stabilized. Once stabilized the spring system provides the best graph drawing or node placement. To identify the most important objects, centrality analysis methods such as degree, betweenness, or proximity analysis are useful. Social network analysis (Scott 1988) and science monitoring are major applications in which data analysis is based on graph visualization: collaboration networks are visualized and browsing facilities are provided for analysis and interpretation. Clustering is a key point in collaborative network analysis. First, clusters result from graph simplification: weak edges are deleted to obtain the main object clusters. On the other hand, clustering methods such as graph partitioning are used in order to handle very large networks.

In this section, we illustrate the usefulness of graphs to address clustering issues for collaborative network analysis and science monitoring. All the examples we provide use publications in a domain that are first pretreated to extract metadata such as keywords or author names and to build adjacency matrices. We show how graphs can be used to visualize collaboration networks and clusters of authors. We also show an extension of collaborative networks to country and semantic networks based on collaboration. We provide examples of how clustering can be used to simplify overly large graphs. Temporal networks are the last example we provide in this section.

30.3.1 Basis of Collaboration Networks

Collaboration networks can be extracted from coauthoring. In that case, nodes correspond to authors and edges to coauthoring. Weights can also be associated both with nodes and edges. Node weight refers to the node importance in terms of author frequency; in the same way, edge weight depends on the strength of coauthoring. Node weights can be expressed graphically by various means as depicted in Karouach and Dousset (2003) and presented in Figure 30.5. On the other hand, edge weights are either graphically represented or used to simplify the network, suppressing the weaker links.

Kejzar et al. (2010) present a collaboration network as a sum of cliques. For example, Figure 30.6 presents the collaboration network obtained when the weights of the edges are at least equal to 3 using the Pajek Software (Batagelj and Zaversnik 2002). Pajek, the Slovene word for Spider, is a program, for Windows, for analysis and visualization of large networks. It is freely available, for noncommercial use. In Figure 30.5, colors represent subnetworks and correspond to clusters of authors that are strongly related to each other.

Another example of so-called graph filtering is provided in Figure 30.7 using the VisuGraph (Dousset et al. 2010) tool that allows to visualize relational data.

Graph filtering makes it possible to keep the strongest relationships, according to the threshold value the user sets. Graph filtering hides the weakest values of the relationships; it does not allow the user to distinguish the nodes that play a central role in the graph structure. This issue can be answered by analyzing the graph structure. K-core has been designed to achieve this goal (Batagelj and Zaversnik 2002; Alvarez-Hamelin et al. 2006). The K-core is a graph decomposition that consists in identifying some specific subsets or

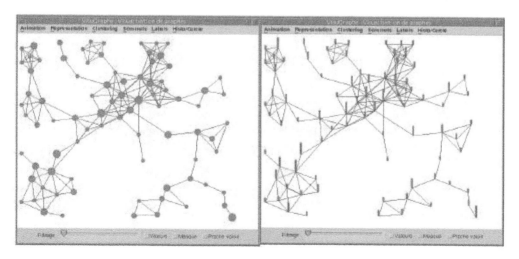

FIGURE 30.5
Graphical representation of node weight in graphs (Karouach and Dousset 2003).

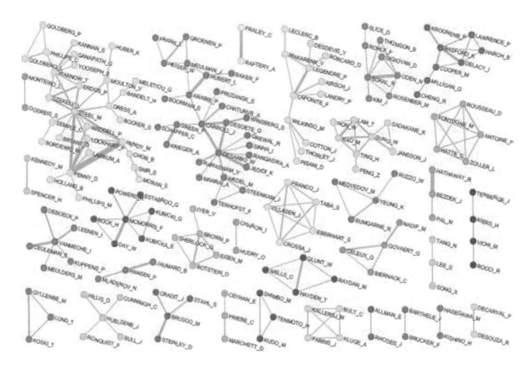

FIGURE 30.6
Main part of line cut at level 3 of collaboration network. Connected subnetworks appear clearly (Kejzar et al. 2010).

subgraphs. K-core is obtained by recursively pruning the nodes that have a degree smaller than K. The nodes which have only one neighbor correspond to a coreness of 1. When the coreness 2 is considered, the nodes belonging to the 1-core are hidden. In that case, the subgraph consists of the nodes that have at least two neighbors, and so on. Node browsing can be applied to have a deeper view of the structure around a specific node. This type of

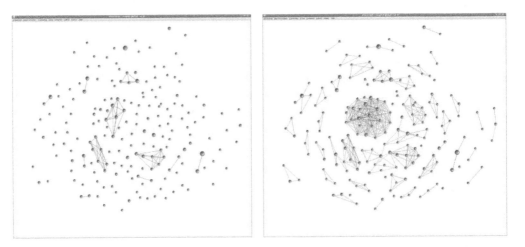

FIGURE 30.7
Filtering a graph according to the strength of the links in VisuGraph (Dousset et al. 2010).

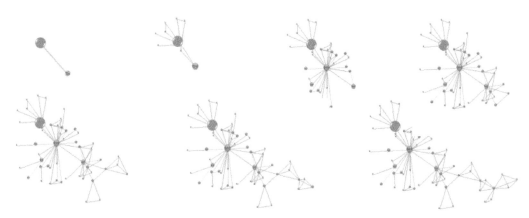

FIGURE 30.8
Browsing a node in VisuGraph (Dousset et al. 2010).

functionality is applied, starting from a node the user selects in order to study a specific node and its relationships to other nodes by means of its connections; it makes it possible to define the node's role within the network structure.

For example, in Figure 30.8, the node that has been selected seems to be a major element of the graph. The vicinity of a node, or what could be called the selfcentered network, is a way to observe the way nodes behave. Considering selfcentered networks, the user can extract the diversity of the relationships and detect the local features where these relationships occur.

In the example Figure 30.8, the node that has been selected appears larger than the others in order to distinguish it. The graph structure is rebuilt by browsing in several stages. The first browsing (depth $= 1$) shows a connection with a single other node (see top left of Figure 30.8). Then, the user discovers the complete structure by successive browsing. The nodes in the center of the structure are characterized by their high connectivity to other nodes. If it was deleted, the graph would be split into two subgraphs. Browsing nodes makes it possible to obtain the whole group of highly related nodes, while studying the relationships with the other nodes. Structural holes (Burt 1992) reveal the fact that two sets of nodes do not communicate directly but rather need an intermediary node to communicate; this latter node thus occupies a decisive position. Figure 30.9 illustrates the structural hole feature. For this new selected node, the first browsing leads to display many other nodes. In the next browsing, this node remains the center of the graph structure since the other nodes are around it.

30.3.2 Geographic and Thematic Collaboration Networks

Collaboration network can also be more sophisticated to include other types of information. For example, a collaboration network including countries is useful when considering technological activity and creativity around the world (Verbeek et al. 2003). In that case, rather than considering coauthoring and thus a single type of node, the starting point is a 2-D matrix based on both country of affiliation and authors. The matrix cells contain the number of publications in which a given author cooccurs with a given country (the country where an author is affiliated). Mothe et al. (2006) present the resulting graph for a set of publications in the information retrieval domain. In Figure 30.10, countries appear in green whereas authors are displayed in red. Countries that are not correlated with other countries

FIGURE 30.9
Browsing from a structural hole in VisuGraph (Dousset et al. 2010).

do not appear in this graph. That means that the only publications that are considered have at least two authors belonging to two different countries. The edges correspond to links that have been inferred between countries and authors. Using this type of network representation, cooperation between countries appears in a single shot. For example, strong relationships are shown between China and Hong Kong and between Israel and USA in this set of publications. China and Hong Kong are not surprising considering the political point of view. Israel and USA relationships in IR can be explained by the fact that a laboratory of the IBM corporation is situated in Haifa (Israel) and publications are coauthored with IBM US (this can be validated when going back to the publications themselves). The power of this representation is that links are drawn, but more importantly, the explanation of the link can be seen. When considering the Netherlands and the UK for example, the association is mainly due to Djoerd Hiemstra. In the same way, the association between China and Canada is due to two persons: Jian Yun Nie (Canada) and Ming Zhou (China).

Collaboration network can also be analyzed in the light of topics of interest. To conduct such an analysis, the two dimensions of the matrix correspond to keywords and to countries. Figure 30.11 provides an example of the resulting graph. Some interesting subnetworks have been circled in the figure. For example, Canada and Turkey are linked through common topics of interest and this link is not due to some publications that have been written by an author from Canada and another from Turkey (in Figure 30.10 these two countries are not linked).

30.3.3 Large Collaborative Networks

When visualizing graphs, size is a major issue. Graph partitioning is a way to face the size issue. The principle is to provide a higher level graph that gives an overview of the data structure. Several graph partitioning techniques exist such as spectral clustering (Alpert and Kahng 1995; Jouve et al. 2002) and multilevel partitioning like METIS (serial graph partitioning and fill-reducing matrix ordering) algorithms (Karypis and Kumar 1998). METIS is a set of programs for partitioning graphs and other elements, and producing fill reducing orderings for sparse matrices. Following this idea, Karouach et al. (2003) propose to reduce large complex graphs by means of Markov model-based clustering algorithm as presented

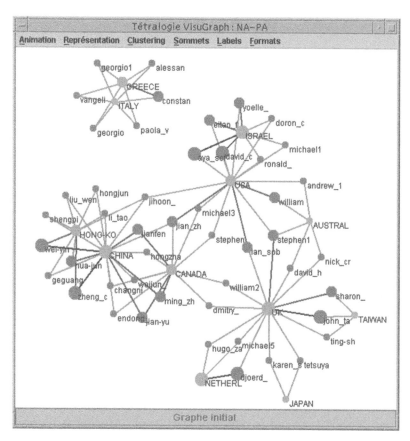

FIGURE 30.10
Author/Country collaboration network (Mothe et al. 2006).

by van Dongen (2000). For example, Figure 30.12 shows the graph resulting from graph partitioning where each color corresponds to a cluster (left part) and the visualization of the simplified graph representing each cluster by a single node (right part).

30.3.4 Temporal Collaborative Networks

Temporality in a collaborative network is an issue since trend detection has to consider evolution. Generally, visualization is based on visualizing independently the graph for the various periods (e.g., Figure 30.13).

Doing so, evolution is difficult to analyze. Loubier et al. (2007) suggest two ways to integrate the various periods and analyze the data in one shot. Figure 30.14 depicts nodes as histograms that show the distribution of data on the entire time slot; each tabular frequency corresponds to each considered year. This graph displays the specificities of each period. For example, in the top right corner (in red) is presented year 2005. The collaborations that occur only in this period are clearly identified. Top left corner (in green) is related to the fourth period (2008).

Following the idea of Dragicevic and Huot (2002), Loubier (2009) represents temporal collaborative networks in the form of a clock. In Figure 30.15, a slice is devoted to each chosen time segment (in this case a publication year). For example, the top left corner is

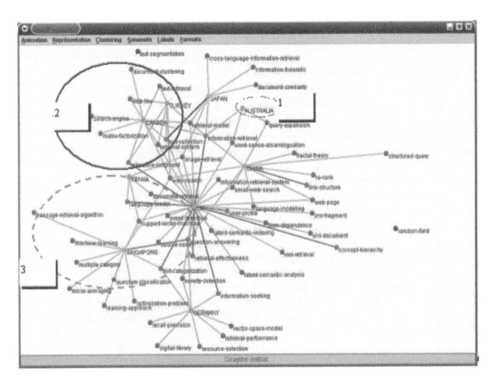

FIGURE 30.11
Topics/Country network (Mothe et al. 2006).

devoted to one specific year and is represented in green. Another is represented in red (top right corner). In between these two slices, a slice is devoted to the collaborations that correspond to the two considered years. In the central circle, the collaborations that involve the four periods are represented. The other circles represent the collaboration within three periods.

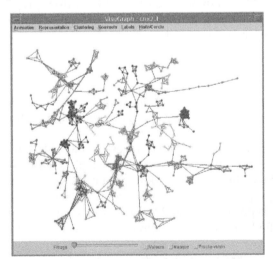
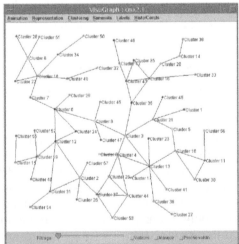

FIGURE 30.12
Markov model-based clustering algorithm applied to a collaborative network (Karouach and Dousset 2003).

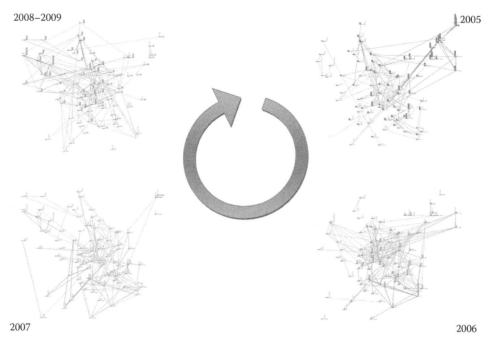

FIGURE 30.13
Collaborative networks for four periods corresponding to the publication year (Loubier et al. 2007).

30.4 Curve Clustering

Curve clustering can occur in various contexts where one or more variables are acquired for various ordered values of an explicative variable. For instance, this is the case for time series, dose–response or spectral analysis. A survey can be found in Warren Liao (2005).

30.4.1 Time Series Microarray Experiment

The study presented to illustrate curve clustering is detailed in Déjean et al. (2007). In the context of microarray experiments, it focuses on the analysis of time series gene expression data. Original data were hepatic gene expression profiles acquired during a fasting period in the mouse. Two hundred selected genes were studied through 11 time points between 0 and 72 hours, using a dedicated macroarray. For each gene, two to four replicates were available at each time point. Data are presented in Figure 30.16, where lines join the average value between replicates at each time point.

The aim of this study was to provide a relevant clustering of gene expression temporal profiles. This was achieved by focusing on the shapes of the curves rather than the absolute level of expression. Actually, the authors combined spline smoothing and first derivative computation with hierarchical clustering and k-means partitioning. Once the groups were obtained, three graphical representations were provided in order to make interpretation easier; they are displayed and commented in the following. They were obtained using the R software (R Development Core Team 2012).

2008–2009　　　　　　　　　　　　　　　　　　　　　2005

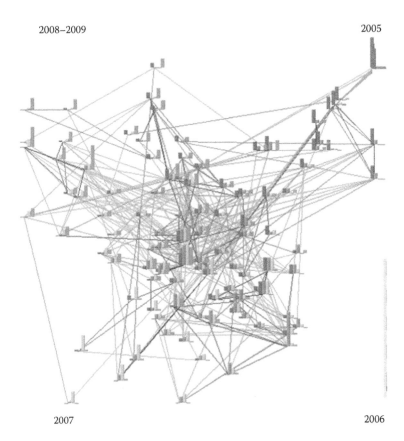

2007　　　　　　　　　　　　　　　　　　　　　　2006

FIGURE 30.14
Collaborative networks using histograms and a spatial representation of time (Loubier et al. 2007).

30.4.2 Principal Component Analysis to Characterize Clusters

PCA enables the experimental units collected as clusters to be confronted with the variables of the experiment, here the time. Each cluster can then be characterized through the behavior of its components.

In Figure 30.17, the variables of the data set (here the discretized time points) are displayed on the left part by projection on the first two principal components. Their regular pattern indicates the consistency of the smoothed and discretized data. The sort of horse shoe formed by the times of discretization recalls well-known situations of variables connected with time (or another continuous variable). In the right part of Figure 30.17, the observations (here the genes) are also displayed on the first two principal components. The four clusters are distributed along the first (horizontal) axis in a specific order. Regarding the variables, it appears that the clusters on the left have high values of derivatives at the beginning of the fasting experiment and these values decrease with clusters located on the right. The cluster in red, located near the origin, acts as an intermediary between the other clusters. These directions are confirmed in the following when displaying the curves corresponding to each cluster.

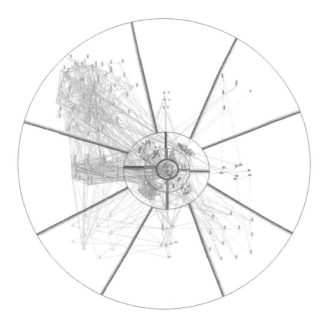

FIGURE 30.15
A circle display of temporal collaboration networks (Loubier 2009).

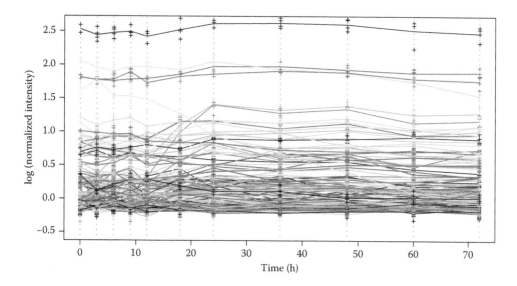

FIGURE 30.16
Log-normalized intensity *versus* time for 130 genes. For each gene, the line joins the average value at each time point. Vertical dashed lines indicate time points.

30.4.3 Visualizing Curves

The first elements of interpretation provided by PCA can be strengthened by the representation of each smoothed curve according to the cluster it belongs to. This can be done by superimposing the curve (on the left in Figure 30.18) or in a kind of disassembled view (four plots on the right).

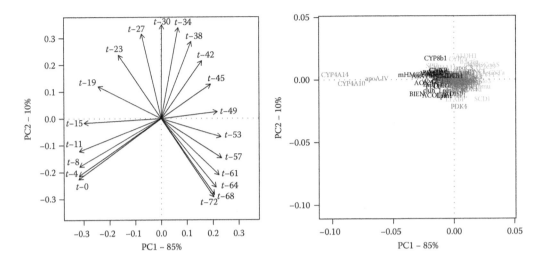

FIGURE 30.17
Representation of variables (discretized time points, on the left) and individuals (genes, on the right) on the first two principal components. Genes are differentially displayed according to their cluster following k-means. (This figure is available in color at http://www.crcpress.com/product/isbn/9781466551886.)

In this representation, it becomes clearer that:

km1: the expression of the genes which belong to the first cluster (in black) increases during the first half of fasting and then tends to decrease slightly or to stabilize.

km2: the second cluster (red) reveals quasi-constant curves. These genes are not regulated during fasting.

km3: the third one (green) is characterized by a decrease of the gene expression with time.

km4: the fourth cluster (blue) is composed of the most strongly induced genes during fasting. Their expression strongly increases until the 40th hour of fasting and then stabilizes.

Let us note that focusing on the derivative of the smoothed functions allows clustering curves with similar profiles whatever the absolute level of expression. This point is clearly visible in Figure 30.18 for each cluster, mainly for the black one with average values from 0 to 2.5.

30.4.4 Heatmap to Combine Two Clusterings

Another way to confront clustering results jointly performed on rows and columns of a data set is the heatmap. This representation was highly popularized in the biological context by Eisen et al. (1998).

In Figure 30.19, the values represented are the derivative of the smoothed profiles. They increase from green (negative value, decreasing profile) to red (positive value, increasing profile) via black. Genes represented in a row are ordered according to the clusters obtained with k-means partitioning. This explains why, in this case, a dendrogram cannot be drawn on the left (or right) side of the heatmap. Horizontal blue dotted lines separated the four clusters obtained following k-means reallocation. On the other hand, hierarchical clustering of the columns was performed. Many different orderings are consistent with the structure of the dendrogram. We forced the reordering of the time points to follow, as much

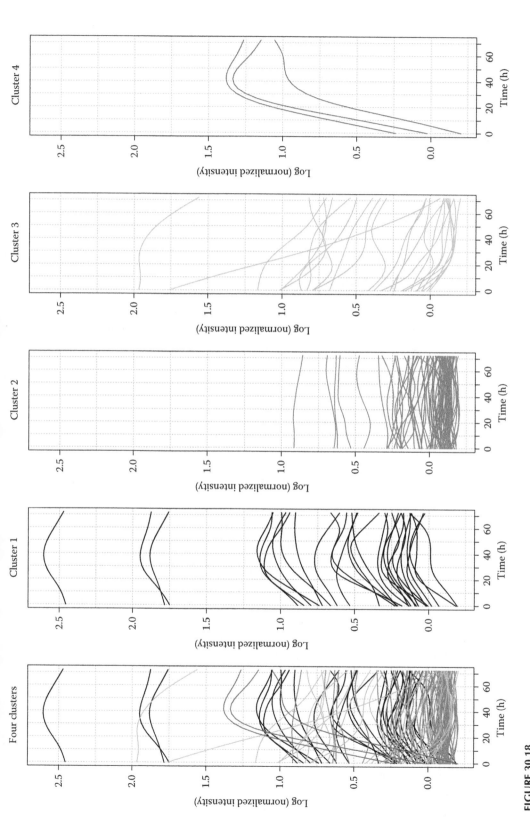

FIGURE 30.18
Representation of the smooth curves distributed in 4 clusters determined through hierarchical and *k*-means classification.

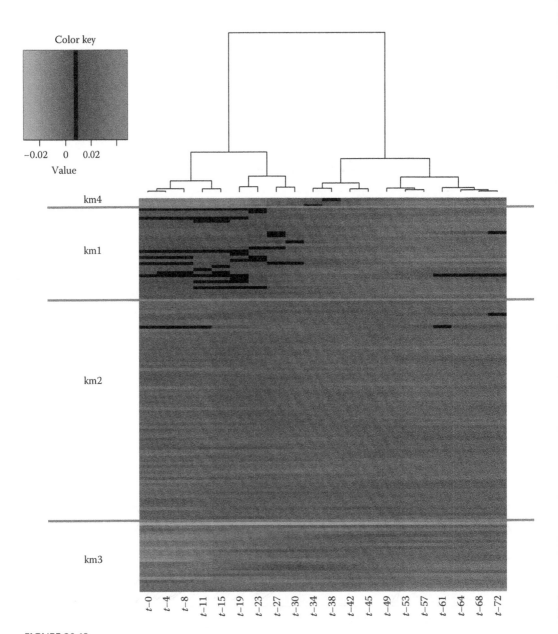

FIGURE 30.19
Heatmap of the derivative of the smoothed gene expression profiles for the whole dataset. Genes (in rows) are ordered according to their cluster determined by the *k*-means algorithm. Horizontal blue lines separate the 4 clusters. Values increase from light (negative values) to grey (positive values) via black. (This figure is available in color at http://www.crcpress.com/product/isbn/9781466551886.)

as the dendrogram allows it (rotating around the nodes of the tree), their increase from left to right. As could be expected, perfectly ordered time points were obtained which is consistent with the specific horse shoe of the variables in PCA (Figure 30.17): considering one time point, its closest neighbors in time are also the closest mathematically.

The heatmap provides a color coding of the derivatives of the curves. This allows a direct extraction of gene expression changes direction and amplitude at the different time points.

Consequently, it becomes much easier to identify both the causes of the clustering and the time points when major transcriptional changes occur.

The most strongly regulated genes are easily visualized: km4 genes at the uppermost and one gene (SCD1) which appears as a green line in the lower quarter of the heatmap. While km4 genes appear most strongly upregulated until the 30th hour of fasting, SCD1 is negatively regulated in a constant way during all the fasting periods. Thus, by contrast to km4 genes, SCD1 expression profile could have been equally well modeled by a straight line since its derivative appears nearly constant with fasting time. One obvious drawback of this representation (Figure 30.19) is that the representation of km4 and SCD1 gene profile derivatives tends to strongly narrow the color range used to represent the other profile derivatives due to their extreme regulations in mouse liver during fasting. Once identified, this drawback can be overcome by removing SCD1 and genes belonging to km4 from the data set and by building a new heatmap (Déjean et al. 2007).

30.5 Conclusion

To illustrate visual clustering, we opted for a presentation based on three case studies dedicated to three kinds of data: multidimensional, networks, and curves. The specific characteristics of each data set require appropriate tools the clustering task is associated with. Each visualization technique provides one point of view, obviously subjective, and partial. The joint use of various techniques allows to enrich the perception of the data. Dendrogram can be seen as a standard visualization of a clustering process, but we saw that the impression provided by the tree is highly partial. Frequently, dimension reduction techniques such as PCA, multi-dimensional scaling or other projection techniques more adapted to cluster analysis (Hennig 2004; Tyler et al. 2009), are used to observe and characterize clusters.

We do not pretend to propose an exhaustive view of visual clustering. Many other methods could have been presented in this chapter. Furthermore, as we saw, the way to visualize clusters depends on the kind of data to analyze but also can depend on the methodology used to address the clustering task. Software can also have specificities in representing clustering results and providing facilities to the user.

When considering spatial data, the visualization can be highly enriched by representing clusters on maps. For instance, the R package GeoXp (Laurent et al. 2012) implements interactive graphics for exploratory data analysis and includes dimension reduction techniques as well as cluster analysis whose results are linked to a map. Another context in which cluster analysis requires efficient visualization tools is the study of the origin of species; a context in which phylogenetic trees with thousands of nodes have to be visualized. Standard phylogram or cladogram looks like a dendrogram resulting from hierarchical clustering but many variants exist such as reviewed in Pavlopoulos et al. (2010); let us mention for instance unrooted or circular cladogram and others using or not 3D visualization, with each one providing a specific survey of the data. Images are also data for which the clustering task can be performed. For instance, when presenting the results of an image retrieval system, clustering can allow to select a subset of representatives of all retrieved images instead of providing a relevance-ordered list (Nguyen and Worring 2008; Jing et al. 2012). Clustering is also associated with images when dealing with image processing (Parker 2010): a segmentation process consists of dividing an image into various parts in order to recognize particular patterns (areas in Earth imaging or organs in a medical context).

Some clustering methodologies can result in specific visualization techniques. For instance, Self-Organizing Maps (SOM, Kohonen (1990)) address the clustering problem as a kind of neural network where neurons (or nodes) are arranged on a grid or other specific shapes. This can result in very specific representation such as those produced with the SOM toolbox for Matlab (Vesanto et al. 1999) or the R package kohonen (Wehrens and Buydens 2007). Visualization techniques have to be adapted when performing algorithms from a fuzzy clustering framework (Kruse et al. 2007; Valente de Oliveira and Pedrycz 2007). Indeed, in this context, it is assumed that one element can belong to more than one cluster what is not always possible using the standard visualization techniques. Radial visualization techniques are an alternative to address this problem (Sharko and Grinstein 2009).

To address visualization needs for clustering results, a great deal of software has been developed. The methodologies implemented as well as the facilities proposed highly depend on the community in which the software was developed. For instance, it can be reasonable to associate signal and image processing with Matlab toolboxes. Biostatistics and clustering related to high-throughput biology were recently developed in the environment provided by the free R software. Tetralogie (Dousset 2012) is developed by the University of Toulouse and allows one to analyze texts such as publications, patents, or web pages. Many other standalone software packages are also available. Regarding networks, Gephi (Bastian et al. 2009) is an open-source and free solution that offers interactive visualization for the exploration of networks.

References

D. Adler and D. Murdoch. *rgl: 3D visualization device system (OpenGL)*, 2011. R package version 0.92.798.

M. Al Hasan, S. Salem, and M.J. Zaki. Simclus: An effective algorithm for clustering with a lower bound on similarity. *Knowledge and Information Systems*, 28(3):665–685, 2011.

C.J. Alpert and A.B. Kahng. Recent developments in netlist partitioning: A survey. *Integration: The VLSI Journal*, 19:1–18, 1995.

J.I. Alvarez-Hamelin, L. Dall'Asta, A. Barrat, and A. Vespignani. Large scale networks fingerprinting and visualization using the k-core decomposition. In Y. Weiss, B. Schölkopf, and J. Platt, editors, *Advances in Neural Information Processing Systems 18*, pp. 41–50. MIT Press, Cambridge, MA, 2006.

G. Andrienko, N. Andrienko, S. Rinzivillo, M. Nanni, D. Pedreschi, and F. Giannotti. Interactive visual clustering of large collections of trajectories. In *IEEE Visual Analytics Science and Technology*, 2009.

A. Baccini, S. Déjean, L. Lafage, and J. Mothe. How many performance measures to evaluate information retrieval systems? *Knowledge and Information System*, 30:693–713, 2011.

M. Bastian, S. Heymann, and M. Jacomy. Gephi: An open source software for exploring and manipulating networks. In *International AAAI Conference on Weblogs and Social Media*, pp. 361–362, 2009.

V. Batagelj and M. Zaversnik. Generalized cores. *CoRR*, cs.DS/0202039, 2002.

G. Brock, V. Pihur, S. Datta, and S. Datta. clvalid: An R package for cluster validation. *Journal of Statistical Software*, 25(4):1–22, 2008.

R.S. Burt. *Structural Holes: The Social Structure of Competition*. Harvard University Press, Cambridge, MA, 1992.

C.-L. Chen, F.S.C. Tseng, and T. Liang. An integration of fuzzy association rules and wordnet for document clustering. *Knowledge and Information Systems*, 28(3):687–708, 2011.

F. Crimmins, T. Dkaki, J. Mothe, and A. Smeaton. TétraFusion: Information discovery on the Internet. *IEEE Intelligent Systems and their Applications*, 14(4):55–62, 1999.

S. Déjean, P Martin, A. Baccini, and P. Besse. Clustering time series gene expression data using smoothing spline derivatives. *EURASIP Journal on Bioinformatics and Systems Biology*, 2007. article ID 70561.

B. Dousset. TETRALOGIE: Interactivity for competitive intelligence, 2012. http://atlas. irit.fr/PIE/Outils/Tetralogie.html

B. Dousset, E. Loubier, and J. Mothe. Interactive analysis of relational information (regular paper). In *Signal-Image Technology & Internet-Based Systems (SITIS)*, pp. 179–186, 2010.

P. Dragicevic and S. Huot. Spiraclock: A continuous and non-intrusive display for upcoming events. In *CHI '02 Extended Abstracts on Human factors in Computing Systems*, CHI EA '02, pp. 604–605, ACM, New York, NY, 2002.

M.B. Eisen, P.T. Spellman, P.O. Brown, and D. Botstein. Cluster analysis and display of genome-wide expression patterns. *Proceedings of the National Academy of Sciences*, 95(25):14863–14868, 1998.

T.M.J. Fruchterman and E.M. Reingold. Graph drawing by force-directed placement. *Software: Practice and Experience*, 21(11):1129–1164, 1991.

M. Halkidi, Y. Batistakis, and M. Vazirgiannis. On clustering validation techniques. *Journal of Intelligent Information Systems*, 17(2–3):107–145, 2001.

J. Han. *Data Mining: Concepts and Techniques*. Morgan Kaufmann, San Francisco, CA, 2005.

C. Hennig. Asymmetric linear dimension reduction for classification. *Journal of Computational & Graphical Statistics*, 13(4):930, 2004.

F. Husson, J. Josse, S. Lê, and J. Mazet. *FactoMineR: Multivariate Exploratory Data Analysis and Data Mining with R*, 2011. R package version 1.16.

F. Husson, J. Josse, and J. Pagès. *Principal Component Methods—Hierarchical Clustering—Partitional Clustering: Why Would We Need to Choose for Visualizing Data?*, 2010. Technical report—Agrocampus Ouest.

Y. Jing, H.A. Rowley, J. Wang, D. Tsai, C. Rosenberg, and M. Covell. Google image swirl: A large-scale content-based image visualization system. In *Proceedings of the 21st International Conference Companion on World Wide Web*, pp. 539–540. ACM, New York, NY, 2012.

B. Jouve, P. Kuntz, and F. Velin. Extraction de structures macroscopiques dans des grands graphes par une approche spectrale. In D. Hérin and D.A. Zighed, editors, *Extraction et Gestion des Connaissances*, volume 1 of *Extraction des Connaissances et Apprentissage*, pp. 173–184. Hermes Science Publications, 2002.

S. Karouach and B. Dousset. Les graphes comme représentation synthétique et naturelle de l'information relationnelle de grandes tailles. In *Workshop sur la Recherche d'Information: un Nouveau Passage à l'Échelle, associé à INFORSID'2003*, Nancy, pp. 35–48. INFORSID, 2003.

G. Karypis and V. Kumar. Multilevel k-way hypergraph partitioning. In *Proceedings of the Design and Automation Conference*, pp. 343–348, 1998.

N. Kejzar, S. Korenjak-Cerne, and V. Batagelj. Network analysis of works on clustering and classification from web of science. In H. Locarek-Junge and C. Weihs, editors, *Classification as a Tool for Research, Studies in Classification, Data Analysis, and Knowledge Organization, Proceedings of IFCS'09*, pp. 525–536. Springer, 2010.

T. Kohonen. The self-organizing map. *Proceedings of the IEEE*, 78(9):1464–1480, 1990.

Y. Koren and D. Harel. A two-way visualization method for clustered data. In *Proceedings of the Ninth ACM SIGKDD International Conference on Knowledge Discovery and Data Mining*, KDD '03, pp. 589–594, ACM, New York, NY, 2003.

R. Kruse, C. Döring, and M.-J. Lesot. Fundamentals of fuzzy clustering. In J. Valente de Oliveira and W. Pedrycz, editors, *Advances in Fuzzy Clustering and Its Applications*, chapter 1, pp. 3–30. John Wiley & Sons, New York, 2007.

T. Laurent, A. Ruiz-Gazen, and C. Thomas-Agnan. Geoxp: An R package for exploratory spatial data analysis. *Journal of Statistical Software*, 47(2):1–23, 2012.

E. Loubier. *Analyse et visualisation de données relationnelles par morphing de graphe prenant en compte la dimension temporelle.* PhD thesis, Université Paul Sabatier, 2009.

E. Loubier, W. Bahsoun, and B. Dousset. Visualization and analysis of large graphs. In *ACM International Workshop for Ph.D. Students in Information and Knowledge Management (ACM PIKM), Lisbon—Portugal*, pp. 41–48. ACM, 2007.

J. Mothe, C. Chrisment, T. Dkaki, B. Dousset, and S. Karouach. Combining mining and visualization tools to discover the geographic structure of a domain. *Computers, Environment and Urban Systems*, 30:460–484, 2006.

G.P. Nguyen and M. Worring. Interactive access to large image collections using similarity-based visualization. *Journal of Visual Languages and Computing*, 19(2):203–224, 2008.

J.R. Parker. *Algorithms for Image Processing and Computer Vision.* John Wiley & Sons, Inc., New York, NY, 2nd edn, 2010.

G.A. Pavlopoulos, T.G. Soldatos, A. Barbosa Da Silva, and R. Schneider. A reference guide for tree analysis and visualization. *BioData Mining*, 3(1):1, 2010.

R Development Core Team. *R: A Language and Environment for Statistical Computing.* R Foundation for Statistical Computing, Vienna, 2012.

J. Scott. Social network analysis. *Sociology*, 22(1):109–127, 1988.

J. Sharko and G. Grinstein. Visualizing fuzzy clusters using radviz. In *Proceedings of the 2009 13th International Conference Information Visualisation*, IV '09, pp. 307–316, Washington, DC, 2009. IEEE Computer Society.

B. Shneiderman. The eyes have it: A task by data type taxonomy for information visualizations. In *IEEE Visual Languages*, number UMCP-CSD CS-TR-3665, pp. 336–343, College Park, Maryland 20742, 1996.

S. Tobias, B. Jürgen, T. Tekusova, and J. Kohlhammer. Visual cluster analysis of trajectory data with interactive Kohonen maps. *Information Visualization*, 8(1):14–29, 2009.

D.E. Tyler, F. Critchley, L. Dümbgen, and H. Oja. Invariant coordinate selection. *Journal of the Royal Statistical Society B*, 71(3):549–592, 2009.

J. Valente de Oliveira and W. Pedrycz. *Advances in Fuzzy Clustering and its Applications.* John Wiley & Sons, Inc., New York, NY, 2007.

S.M. van Dongen. *Graph Clustering by Flow Simulation.* PhD thesis, University of Utrecht, The Netherlands, 2000.

A. Verbeek, K. Debackere, and M. Luwel. Science cited in patents: A geographic 'flow' analysis of bibliographic citation patterns in patents. *Scientometrics*, 58(2):241–263, 2003.

J. Vesanto, J. Himberg, E. Alhoniemi, and J. Parhankangas. Self-organizing map in Matlab: the SOM toolbox. In *Proceedings of the Matlab DSP Conference*, pp. 35–40, 1999.

T. Warren Liao. Clustering of time series data—A survey. *Pattern Recognition*, 38(11):1857–1874, 2005.

R. Wehrens and L.M.C. Buydens. Self- and super-organising maps in R: The kohonen package. *Journal of Statistical Software*, 21(5):1–19, 2007.

M. Woo, J. Neider, T. Davis, and D. Shreiner. *OpenGL(R) Programming Guide: The Official Guide to Learning OpenGL(R), Version 2 (5th Edition).* Addison-Wesley Professional, Boston, MA, 2005.

31

Clustering Strategy and Method Selection

Christian Hennig

CONTENTS

Abstract

The aim of this chapter is to provide a framework for all the decisions that are required when carrying out a cluster analysis in practice. A general attitude to clustering is outlined, which connects these decisions closely to the clustering aims in a given application.

From this point of view, the chapter then discusses aspects of data processing such as the choice of the representation of the objects to be clustered, dissimilarity design, transformation and standardization of variables. Regarding the choice of the clustering method, it is explored how different methods correspond to different clustering aims. Then an overview of benchmarking studies comparing different clustering methods is given, as well as an outline of theoretical approaches to characterize desiderata for clustering by axioms. Finally, aspects of cluster validation, that is the assessment of the quality of a clustering in a given dataset, are discussed, including finding an appropriate number of clusters, testing homogeneity, internal and external cluster validation, assessing clustering stability, and data visualization.

31.1 Introduction

In previous chapters, a large number of cluster analysis methods have been introduced, and in any situation in which a clustering is needed, the user is faced with a potentially overwhelming number of options. The current chapter is about how the required choices can be made. Milligan (1996) listed seven steps of a cluster analysis that require decisions, namely:

1. Choosing the objects to be clustered
2. Choosing the measurements/variables
3. Standardization of variables
4. Choosing a (dis-)similarity measure
5. Choosing a clustering method
6. Determining/deciding the number of clusters
7. Interpretation, testing, replication, cluster validation

I will treat all but the first one (general principles of sampling and experimental design apply), not sticking exactly to this order. The chapter focuses on the general philosophy behind the required choices, what this means in practice, and on some areas of research. This has to be combined with knowledge on clustering methods as given elsewhere in this volume. Some more discussion of the above issues can be found in Milligan (1996) and standard cluster analysis books such as Jain and Dubes (1988), Kaufman and Rousseeuw (1990), Gordon (1999), and Everitt et al. (2011).

The point of view taken here, previously outlined in Hennig and Liao (2013) and also shared by other authors (von Luxburg et al., 2012), is that there is no such thing as a universally "best clustering method." Different methods should be used for different aims of clustering. The task of selecting a clustering method implies a proper understanding of the meaning of the data, the clustering aim and the available methods, so that a suitable method can be matched to what the application requires. Although many experienced experts in the field, including the authors of the books cited above, agree with this view, there is not much advice in the literature on how the specific requirements of the application can be connected with the available methods. Instead, cluster analysis methods have been often compared on simulated data or data with known classes, in order to find a "best" one disregarding the research context. Such comparisons are of some use, particularly because they reveal, in some cases, that methods may not be up for what they were supposed to do. Still, it would be more useful to have more specific information about what kind of method is connected to what kind of clustering task, defined by clustering aim, required cluster concept, and potential structure in the data.

The present chapter goes through the most essential steps of making the necessary decisions for a cluster analysis. It starts in Section 31.2 with a discussion of the background, relating the aims of clustering to the cluster concepts that may be of interest in a specific situation. Section 31.3 looks at the data to be clustered. Often it is useful to preprocess the data before applying a clustering method, by defining new variables, dissimilarity measures, transforming, or selecting features. Such operations have an often fundamental impact on the resulting clustering. Note that I will use the term "features" to refer to the variables eventually used for clustering if a cluster analysis method for an "objects times features"—matrix as input is applied, whereas the term "variables" will be used in a more general sense for measurements characterizing the objects used in the clustering process, potentially later to be used as clustering features, or for computing dissimilarity measures or new variables.

Section 31.4 is on comparing clustering methods. This encompasses the decision which method fits a certain clustering aim, measurement of the quality of clustering methods, benchmark simulation studies, and some theoretical work on characterizing clusterings and clustering methods. In many cases, though, there may not be enough precise information about the clustering aim and cluster concepts of interest, so that the user may not be able to pinpoint exactly what method is needed. Also, it may be discovered that the clustering structure of the data may differ from what was expected in advance, and other methods than initially considered may look promising. Section 31.5 is about evaluating and comparing outcomes of clustering methods, before the chapter is concluded.

31.2 Clustering Aims and Cluster Concepts

In various places in the literature it is noted that there is no generally accepted definition of a cluster. This is not surprising, given the many different aims for which clusterings are used. Here are some examples:

- Delimitation of species of plants or animals in biology
- Medical classification of diseases
- Discovery and segmentation of settlements and periods in archeology
- Image segmentation and object recognition
- Social stratification
- Market segmentation
- Efficient organization of data bases for search queries

There are also quite general tasks for which clustering is applied in many subject areas:

- Exploratory data analysis looking for "interesting patterns" without prescribing any specific interpretation, potentially creating new research questions and hypotheses.

- Information reduction and structuring of sets of entities from any subject area for simplification, more effective communication, or more effective access/action such as complexity reduction for further data analysis.
- Investigating the correspondence of a clustering in specific data with other groupings or characteristics, either hypothesized or derived from other data.

Depending on the application, it may differ a lot what is meant by a "cluster," and this has strong implications for the methodological strategy. Finding an appropriate clustering method means that the cluster definition and methodology have to be adapted to the specific aim of clustering in the application of interest.

A key distinction can be made between "realist" aims of clustering, concerning the discovery of some meaningful real structure corresponding to the clusters, and "constructive" aims, where researchers intend to split up the data into clusters for pragmatic reasons, regardless of whether there is some essential real difference between the resulting groups. This distinction can be roughly connected to the choice of clustering methodology. For example, some clustering criteria such as K-means (Chapter 3) produce homogeneous clusters in the sense that all observations are assigned to the closest centroid, and large distances within clusters are heavily penalized. This is useful for a number of constructive clustering aims. On the other hand, K-means does not pay much attention to whether or not the clusters are clearly separated by gaps, and does not tolerate large variance and spread of points within clusters, which can occur in clusters that correspond to real patterns (e.g., objects in images).

However, the distinction between realist and constructive clustering aims is not as clear cut as it may seem at first sight. Categorization is a very basic human activity that is directly connected with the emergence of language. Whenever human beings speak of real patterns, this can only refer to categories that are aspects of human cognition and can be expressed in language, which can be seen as a pragmatic human construct (Van Mechelen et al. (1993) review cognitive theories of categorization with a view to connecting them to inductive data analysis including clustering). In a related manner, researchers with realist clustering aims should not hope that the data alone can reveal real structure; constructive impact of the researchers is needed to decide what counts as real.

The key issue in realist clustering is how the real structure the researchers are interested in is connected to the available data. This requires subject matter knowledge, but it also requires decisions by the researchers. "Real structure" is often understood as the existence of an unobserved categorical variable the values of which define the "true" clusters. Such an idea is behind the popular use of datasets with given true classes for benchmarking of cluster analysis methods. But neither can it be taken fur granted that the known categories are the only existing ones that could qualify as "real clusters," nor do such categories necessarily correspond to data analytic clusters. For example, male/female is certainly a meaningful categorization of human beings, but there may not even be a significant difference between men and women regarding the results of a certain attitude survey, let alone separated clusters corresponding to sex. Usually, the objects represented in the dataset can be partitioned into real categories in many ways. Also, different cluster analysis methods will produce different clusterings, which may more or less well correspond to patterns that are real in potentially different ways. This means that in order to decide about appropriate cluster analysis methodology, researchers need to think about what data analytic characteristics the clusters they are aiming at are supposed to have. I call this the "cluster concept" of interest in a specific study.

The real patterns of interest may be more or less closely connected to the available data. For example, in biological species delimitation, the concept of a species is often defined in terms of interbreeding (there is some controversy about the precise definition, see Hausdorf 2011). But interbreeding patterns are not usually available as data. Species are nowadays usually delimited by use of genetic data, but in the past, and also occasionally in the present in an exploratory manner, species were seen as the source of a real grouping in phenotype data. In any case, the researchers need some idea about how true distinctions between species are connected to patterns in the data. Regarding genetic data, this means that knowledge needs to be used about what kind of similarity arises from persistent genetic exchange inside a species, and what kind of separation arises between distinct species. There may be subgroups of individuals in a species between which there is little actual interbreeding (because potential interbreeding suffices for species definition), for example, between geographically separated groups, and consequently not as much genetic similarity as one would naively expect. Furthermore, there are various levels of classification in biology, such as families and gene above and subspecies below the level of species, so that data analytic clusters may be found at several levels, and the researchers may need to specify more precisely how much similarity within and separation between clusters is required for finding species.

Such knowledge needs to be reflected in the cluster analysis method to be chosen. For example, species may be very heterogeneous regarding geographical distribution and size, and therefore a clustering method that implicitly tends to bring forth clusters that are very homogeneous such as K-means or complete linkage is inappropriate.

In some cases, the data are more directly connected to the cluster definition. In species delimitation, there may be interbreeding data, in which case researchers can specify the requirements of a clustering more directly. This may imply graph theoretic clustering methods and a specification of how much connectedness is required within clusters, although such decisions can often not be made precise because of missing information arising from sampling of individuals, missing data, etc. On the other hand, the connection between the cluster definition and the data may be less close, as in the case of phenotype data used for delimiting species, in which case the researchers may not have strong information about how the clusters they are interested in are characterized in the data, and some speculation is needed in order to decide what kind of clustering method may produce something useful.

In many situations different groupings can be interpreted as real, depending on the focus of the researchers. Social classes for example can be defined in various ways. Marx made ownership of means of production the major defining characteristic of different classes, but social classes can also be defined by looking at patterns of communication and contact, or occupation, or education, or wealth, or by a mixture of these (Hennig and Liao, 2013). In this case, a major issue for data clustering is the selection of the appropriate variables and measurements, which implicitly defines what kinds of social classes can be found.

The example of social stratification also illustrates that there is a gradual transition rather than a clear cut between realist and constructive clustering aims. According to some views (such as the Marxist one) social classes are an essential and real characteristic of society, but according to other views, in many societies there is no clear delimitation between social classes that could justify to call these classes "real," despite the existence of real inequality. Social classes can then still be used as a convenient tool for structuring the inequality in such societies.

Regarding constructive clustering aims, it is obvious that researchers need to decide about the desired "cluster concept," or in other words, about the characteristics that their clusters should have. The discussion above implies that this is also the case for realist clustering aims, for which the required cluster concept needs to be derived from knowledge about the nature of the real clusters, and from a decision of the researchers about their focus of interest if (as is usually the case) the existence of more than a unique real clusterings is conceivable. For constructive clustering, the required cluster concept needs to be connected to the practical use that is intended to be made of the clusters.

Also where the primary clustering aim is constructive, realist aims may still be of interest insofar as if indeed some real grouping structure is clearly manifest in the data, many constructive aims will be served well by having this structure reflected in the clustering. For example, market segmentation may be useful regardless of whether there are really meaningfully separated groups in the data, but it is relevant to find them if they exist.

Here is a list of potential characteristics of clusters that may be desired, and that can be checked using the available data. A number of these are related with the "formal categorization principles" listed in Section 31.2.2.1 of Van Mechelen et al. (1993).

1. Within-cluster dissimilarities should be small.
2. Between-cluster dissimilarities should be large.
3. Clusters should be fitted well by certain homogeneous probability models such as the Gaussian or a uniform distribution on a convex set, or, if appropriate, by linear, time series, or spatial process models.
4. Members of a cluster should be well represented by its centroid.
5. The dissimilarity matrix of the data should be well represented by the clustering (i.e., by the ultrametric induced by a dendrogram, or by defining a binary metric "in same cluster/in different clusters").
6. Clusters should be stable.
7. Clusters should correspond to connected areas in data space with high density.
8. The areas in data space corresponding to clusters should have certain characteristics (such as being convex or linear).
9. It should be possible to characterize the clusters using a small number of variables.
10. Clusters should correspond well to an externally given partition or values of one or more variables that were not used for computing the clustering.
11. Features should be approximately independent within clusters.
12. All clusters should have roughly the same size.
13. The number of clusters should be low.

When trying to measure these characteristics, they have to be made more precise, and in some cases it matters a lot how exactly they are defined. Take no. 1, for example. This may mean that all within-cluster dissimilarities should be small without exception (i.e., the maximum should be small, as required by complete linkage hierarchical clustering), or their average, or a high quantile of them. These requirements may look similar at first sight but are very different regarding the integration of outliers in clusters. Having small within-cluster dissimilarities may emphasize gaps by looking at the smallest dissimilarities between each two clusters, or it may rather mean that the central areas of the clusters are well distributed in data space. As another example, stability can refer to sampling other data from the same population, to adding "noise," or to comparing results from different clustering algorithms.

Some of these characteristics are in conflict with others in some datasets. Connected areas with high density may include very large distances, and may have undesired (e.g., nonconvex or nonlinear) shapes. Representing objects by centroids may bring forth some clusters with little or no gap between them. Having clusters of roughly equal size forces outliers to be integrated in distant clusters, which produces large within-cluster dissimilarities.

Deciding about such characteristics is the key to linking the clustering aim to an appropriate clustering method. For example, if a database of images should be clustered so that users can be shown a single image to represent a cluster, no. 7 is most important. Useful market segments need to be addressed by nonstatisticians and should therefore normally be represented by few variables, on which dissimilarities between members should be low. Similar considerations can be made for realist clustering aims, see above.

For choosing a clustering method, it is then necessary to know how they correspond to the required characteristics. Some methods optimize certain characteristics directly (such as K-means for no. 4), and in some further cases experience and research suggest typical behavior (K-means tends to produce clusters of roughly equal size, whereas methods looking for high-density areas may produce clusters of very variable size). See Section 31.4.1 for more comments on specific methods. Other characteristics such as stability are not involved in the definition of most clustering methods, but can be used to validate clusterings and to compare clusterings from different methods.

The task of choosing a clustering method is made harder by the fact that in many applications more than one of the listed characteristics is relevant. Clusterings may be used for several purposes, or desired characteristics may not be well defined, for example, in exploratory data analysis, or for realist clustering aims in cases where the connection between the interpretation of the clusters and the data is rather loose. Also, a misguided desire for uniqueness and objectivity makes many researchers reluctant to specify desired characteristics and choose a clustering method accordingly because they hope that there is a universally optimal method that will just produce "natural" clusters. Probably for such reasons there is currently almost no systematic research investigating the characteristics of methods in terms of the various cluster characteristics.

31.3 Data Processing

The decision about what data to use, including how to choose, transform, and standardize variables, and if and how to compute a dissimilarity measure, is an important part of the methodological strategy in cluster analysis. It often has a major impact on the clustering result, and is sometimes more important than the choice of the clustering method.

31.3.1 Choice of Representation

To some extent the data format restricts the choice of clustering methods; there are specialized methods for continuous, ordinal, categorical, and mixed type data, dissimilarity data, graphs, time series, spatial data, etc. But often data can be represented in different ways. For example, a collection of time series with 100 time points can be represented as points in 100-dimensional Euclidean space, but they can also be represented by autocorrelation parameters of a time series model fitted to them, by wavelet features or some other low-dimensional representation, or by dissimilarity measures which may involve some alignment or "time warping," see Chapter 6. On the other hand, dissimilarity data can be be transformed to Euclidean data using multidimensional scaling (MDS) techniques. This means that the researcher often can choose whether the objects are represented by features, dissimilarities, or in another way, for example, by vertices in a graph.

Generally, dissimilarity measures are a suitable basis for clustering if the cluster concept is mainly based on the idea that similar objects should be grouped together and dissimilar objects should be in different clusters. Dissimilarity measures can be constructed for most data types. On the other hand, clusters characterized by distributional and geometrical shapes and clusters with potentially high within-cluster variability or skewness are found better with objects characterized by features instead of dissimilarities.

The choice of representation should be guided by the question how objects qualify to belong together in the same cluster. For example, if the data are time series, there are various different possible concepts of "belonging together." Time series may belong together if their values are similar most of the time, which is appropriate if the plain values play a large role in the assessment of similarity (e.g., cigarettes smoked per day in research about smoking behavior). A musical melody can be played at different speeds and in different keys, so that two musical melodies may still be assessed as similar despite pitch values being quite different and changes in pitch happen at different times. In other applications, such as particle detection by electrodes, the characteristics of a single event that happens at a certain potentially flexible time point (such as a value going up and then down again) may be important, and having detected such an event, some specific characteristics of it may represent the objects in the most useful manner.

A central issue regarding the representation is the choice of variables that are either used as features to represent the objects or on which a dissimilarity definition is based. Both subjects matter and statistical considerations play a role here. From a statistical point of view, a variable could be seen as uninformative if it is either strongly correlated with other variables and does not carry information on its own as long as certain other variables are kept, or the variable may not be connected to any "real" clustering characterized in the data for example by high density regions. Furthermore, in some situations (e.g., using gene expression data) the number of variables may simply be so large that cluster analysis methods become unstable. There are various automatic methods for variable selection in connection with clustering, see Section 1.6.3, Chapter 23 for clustering variables at the same time as observations, and Alelyani et al. (2014) for a recent survey. Popular classical methods such as principal component analysis (PCA) and MDS are occasionally used for constructing informative variables. These, however, are based on objective functions (variance, stress) that do not have any relation to clustering, and may therefore miss information that is important for clustering. There are some projection pursuit-type methods that aim at finding low-dimensional representations of the data particularly suitable for clustering (Bolton and Krzanowski 2003; Tyler et al. 2009).

It is important to realize, though, that the variables involved in clustering define the meaning of the clusters. Changing the variables implies changing the meaning. If the researchers have a clear idea about the meaning of the clusters of interest, it is problematic to select variables in an automatic manner. For example, Hennig and Liao (2013) were interested in socioeconomic stratification, for which information on income, savings, education, and housing is essential. Even if, for example, incomes do not show any clear grouping structure, or are correlated strongly with another variable, this does not constitute a valid reason to exclude this variable for constructing a clustering that is meant to reflect a meaningful socioeconomic partition of a population. A stratification based on automatically selected variables that cluster in a nicer way may be of exploratory interest, but does not fulfill the aim of the researchers. One could argue that in case of correlation between income and another variable, savings, say, the information from income is retained as long as savings (or a linear combination of them both, as would be generated by PCA) is still used as a feature for clustering. But this is not true, because the fact that the information is shared by two variables that in terms of their meaning are essential for the clustering aim is additional information that should not be lost.

Another issue is that variables can play different roles, which has different implications. For example, a dataset may include spatial coordinates and other variables (e.g., regional data on avalanche risk, or color information in image segmentation). Depending on the role that the spatial information should play, spatial coordinates can be included in the clustering process as features together with the others (which implies that regional similarity will somehow be traded off against similarity regarding the other variables in the clustering process), or they could define constraints (e.g., clusters on the other variables could be constrained to be spatially connected), or they could be ignored for clustering, but could be used afterward to validate the resulting clusters or to analyze their spatial structure. For avalanche risk mapping, for example, one may take the latter approach for detailed maps if spatial information is discretized and there is enough data at each point, but one may want to impose spatial constraints if data is sparser or if the map needs to be coarser because it is used by decision makers instead of hikers.

Often there is a good reason for not choosing the variables automatically from the data, but rather guided by the aim of clustering. In some cases dimension reduction can be achieved by the definition of meaningful new indices summarizing information in certain variables. On the other hand, automatic variable selection may yield interesting clusterings if the aim is mainly exploratory, or if there is no prior information about the importance of the variables and it is suspected that some of them are uninformative "noise."

31.3.2 Dissimilarity Definition

In order to apply dissimilarity based methods and to measure whether a clustering method groups similar observations together, a formal definition of "dissimilarity" is needed (or "proximity," which refers to either dissimilarity or similarity, as sometimes used in the literature; their treatment is equivalent and there are a number of transformations between dissimilarity and similarity measures, the simplest and most popular of which probably is "dissimilarity = maximum similarity minus similarity"). In many situations, dissimilarities between objects cannot be measured directly, but have to be constructed from some measurements of variables of the objects of interest. Directly measured dissimilarities occur for example in comparative experiments in psychology and market research.

There is no unique "true" dissimilarity measure for any dataset; the dissimilarity measurement has to depend on the researchers' concept of what it means to treat two objects as "similar," and therefore on the clustering aim.

Mathematically, a dissimilarity is a function $d : \mathcal{X}^2 \mapsto \mathbb{R}$, $\mathcal{X}$ being the object space, so that $d(\mathbf{x}, \mathbf{y}) = d(\mathbf{y}, \mathbf{x}) \geq 0$ and $d(\mathbf{x}, \mathbf{x}) = 0$ for $\mathbf{x}, \mathbf{y} \in \mathcal{X}$. There is some work on asymmetric dissimilarities (Okada, 2000) and multiway dissimilarities defined between more than two objects (Diatta, 2004). A dissimilarity fulfilling the triangle equality

$$d(\mathbf{x}, \mathbf{y}) + d(\mathbf{y}, \mathbf{z}) \geq d(\mathbf{x}, \mathbf{z}), \ \mathbf{x}, \mathbf{y}, \mathbf{z} \in \mathcal{X}$$

is called a "distance" or "metric." The triangle inequality is connected to Euclidean intuition and therefore seems to be a "natural" requirement, but in some applications it is not appropriate. Hennig and Hausdorf (2006) argue, for example, that for presence-absence data of species on regions two species A and B are very dissimilar if they are present on two small disjoint areas, but both should be treated as similar to a species C covering a larger area that includes both A and B, if clusters are to be interpreted as species grouped together by palaeoecological processes.

A vast number of dissimilarity measures has been proposed, some for rather general purposes, some for more specific applications (dissimilarities between shapes [Veltkamp and Latecki 2006], melodies [Müllensiefen and Frieler 2007], geographical species distribution areas [Hennig and Hausdorf, 2006], etc.). Chapter 3 in Everitt et al. (2011) gives a good overview of general purpose dissimilarities. Here are some basic considerations.

31.3.2.1 Aggregating Binary Variables

If two objects x_1, x_2 are represented by p binary variables, let a_{ij} be the number of variables $h = 1, \ldots, p$ on which $x_{1h} = i, x_{2h} = j$, $i, j \in \{0, 1\}$. If all variables are treated in the same way, the most straightforward dissimilarity is the simple matching coefficient,

$$d_{SM}(x_1, x_2) = 1 - \frac{a_{00} + a_{11}}{p}.$$

However, often (e.g., in the case of geographical presence-absence data in ecology) common presences are important, whereas common absences are not. This is taken into account by the Jaccard dissimilarity

$$d_J(x_1, x_2) = 1 - \frac{a_{11}}{a_{11} + a_{10} + a_{01}}.$$

One can worry about whether this gives the object with more presences too much weight in the denominator, and actually more than 30 dissimilarity measures for such data have been proposed, see Shi (1993), prompting much research about their characteristics and how they relate to each other (Gower and Legendre, 1986; Warrens, 2008).

31.3.2.2 Aggregating Categorical Variables

If there are more than two categories, again the most intuitive way to construct a dissimilarity measure is one minus the relative number of "matches." In some applications such as population genetics dissimilarity should rather be a nonlinear function of matches between genes, and it is also important to think about whether and in what way variables with different numbers of categories or even with more or less uniform distributions should be given different weights because some variables produce matches more easily than others.

31.3.2.3 Aggregating Continuous Variables

The Minkowski (L_q)-distance between two objects x_i, x_j on p real-valued variables $x_i = (x_{i1}, \ldots, x_{ip})$ is

$$d_{Mq}(x_i, x_j) = \sqrt[q]{\sum_{l=1}^{p} d_l(x_{il}, x_{jl})^q}, \qquad (31.1)$$

where $d_l(x, y) = |x - y|$. Variable weights w_l can easily be incorporated by multiplying the d_l by w_l. Most often, the Euclidean distance d_{M2} and the Manhattan distance d_{M1} are used. Using d_{Mq} with larger q gives the variables with larger d_l more weight, that is, two observations are treated as less similar if there is a very large dissimilarity on one variable and small dissimilarities on the others than if there is about the same medium-sized dissimilarity on all variables, whereas d_{M1} gives all variable-wise contributions implicitly the same weight (note that this does not hold for the Euclidean distance that corresponds to physical distances and is used as default choice in many applications). The Euclidean distance is invariant to rotations of the data, whereas for the Manhattan distance the original variables play a more prominent role than axes obtained by potential rotation.

An alternative would be the (squared) Mahalanobis distance,

$$d_M(\mathbf{x}_i, \mathbf{x}_j)^2 = (\mathbf{x}_i - \mathbf{x}_j)^T \mathbf{S}^{-1} (\mathbf{x}_i - \mathbf{x}_j) \tag{31.2}$$

where $\mathbf{S}$ is a scatter matrix such as the sample covariance matrix. This is affine equivariant, that is, not only rotating the data points in Euclidean space, but also stretching them in any number of directions will not affect the dissimilarity. It will also implicitly aggregate and therefore weight information from strongly correlated variables down (correlation implies that data are "stretched" in the direction of their dependence; the consequence is that "joint information" is only used once). This is desirable if clusters can come in all kinds of elliptical shapes. On the other hand, it means that the weight of the variables is determined by their covariance structure and not by their meaning, which is not always appropriate (see the Discussion about variable selection above).

There are many further ways of constructing a dissimilarity measure from several continuous variables, see Everitt et al. (2011), such as the Canberra distance, which emphasizes differences close to zero. It is defined by $q = 1$ and $d_l(x, y) = |x - y|/(|x| + |y|)$ in (31.1). The Pearson correlation coefficient $\rho(\mathbf{x}, \mathbf{y})$ has been used to construct a dissimilarity measure $d_P(\mathbf{x}, \mathbf{y}) = 1 - ((\rho(\mathbf{x}, \mathbf{y}) + 1)/2)$ as well (other transformations are also used). This interprets $\mathbf{x}$ and $\mathbf{y}$ as similar if they are positively linear dependent. This does not mean that their values have to be similar, but rather the values of the variables relative to the other variables. In some applications variables are clustered, which means that variables and objects change their roles; if the variables are the objects to be clustered, ρ in d_P is a proper correlation between variables, which is a typical use of d_P.

31.3.2.4 Aggregating Ordinal Variables

Ordinal variables are characterized by the absence of metric information about the distances between two neighboring categories. They could be treated as categorical variables, but this would ignore available information. On the other hand, it is fairly common practice to use plain Likert codes $1, 2, \ldots$ and then to use methods for continuous data. Ordinality can be taken into account while still using methods for continuous data by scoring the categories in a way that uses the ordinal information only. Straightforward scores are obtained by ranking (using the midrank for all objects in one category) or normal scores (Conover, 1999), which treat the data as if there would be an underlying uniform (ranks) or Gaussian distribution (normal scores). A more sophisticated approach is polytomous item response theory (Ostini and Nering, 2006). Using scores that are determined by the distribution of the data does not guarantee that they appropriately quantify the interpretative distances between categories, and in some situations (e.g., Likert scales in questionnaires where interviewees can see that responses are coded $1, 2, \ldots$) this may be reflected better by plain Likert codes. Sometimes also there is a more complex structure in the categories that can be reflected by scoring data in a customized way. For example, in Hennig and Liao (2013), a "housing" variable had levels "owns," "pays rent," and several levels such as "shared ownership" that could be seen as lying between "owns" and "pays rent" but could not be ordered, which could be reflected by having a distance of 1 between "pays rent" and "owns" and 0.5 between any other pair of categories.

31.3.2.5 Aggregating Mixed-Type Variables and Missing Values

If there are variable-wise distances d_l defined, variables of mixed type can be aggregated. A standard way of doing this is the Gower dissimilarity (Gower, 1971)

$$d_G(\mathbf{x}_i, \mathbf{x}_j) = \frac{\sum_{l=1}^{p} w_l \delta_{ijl} d_l(x_{il}, x_{jl})}{\sum_{l=1}^{p} w_l \delta_{ijl}}$$

where w_l is a variable weight and $\delta_{ijl} = 1$ except if x_{il} or x_{jl} are missing, in which case $\delta_{ijl} = 0$. This is a weighted version of d_{M1} and takes into account missing values by just leaving the corresponding variable out and rescaling the others. Gower recommended to use the weight w_l for standardization to $[0, 1]$-range (see Section 31.3.4), but Hennig and Liao (2013) argued that many clustering methods tend to identify gaps in variable distributions with cluster borders, and that this implies that w_l should be used to weight binary and other "very discrete" variables down against continuous variables, because otherwise the former would get an unduly high influence on the clustering. w_l can also be used to weight variables up that have high subject matter importance. The Gower dissimilarity is very general and covers most applications of dissimilarity-based clustering to mixed-type variables. An alternative for missing values is to treat them as an own category. For continuous variables one could give missing values a constant dissimilarity to every other value. More references are in Everitt et al. (2011).

31.3.2.6 Custom-Made Dissimilarities for Structured Data

In many situations detailed considerations regarding the subject matter will play the most important role regarding the design of a dissimilarity measure. This is particularly the case if the data are more structured than just a collection of variables. Such considerations start with deciding how to represent the objects, as discussed in Section 31.3.1 and illustrated by the task of time series clustering. The next task is how to aggregate the measurements in an appropriate way. In time series clustering, one consideration is whether some processes that are interpreted to be similar may occur at different and potentially varying speeds, so that flexible alignment ("dynamic time warping") is required, as may be the case in gesture recognition. See Liao (2005) for further aspects of choosing dissimilarities between time series.

Key issues may differ a lot from one application to the next, so it is difficult to present general rules. There is some research on approximating expert judgments of similarity with functions of the available variables (Gordon, 1990; Müllensiefen and Frieler, 2007). Hennig and Hausdorf (2006), who incorporate geographical distance information into a dissimilarity for presence-absence data, list a number of general principles for designing and fine-tuning dissimilarities:

- What should be the basic behavior of the dissimilarity as a function of the existing measurements (when decreasing/increasing, etc.)?

- What should be the relative weight of different aspects of the basic behavior? Should some aspects be incorporated in a nonlinear manner (see Section 31.3.3)?
- Construct exemplary pairs of objects for which it is clear what value the dissimilarity should have, or how it should compare with some other exemplary pairs.
- Construct sequences of pairs of objects in which one aspect changes while others are held constant.
- Whether and how could the dissimilarity measure be disturbed by small changes in the characteristics? What behavior in these situations would be appropriate?
- Which transformations of the variables should leave the dissimilarities unchanged?
- Are there reasons that the dissimilarity measure should be a metric (or have some other particular mathematical properties)?

31.3.3 Transformation of Variables

According to the same philosophy as before, effective distances (as used by a clustering method) on the variables should reflect the "interpretative distance" between objects, and transformations may be required to achieve this. Because there is a large variety of clustering aims, it is difficult to give general principles that can be applied in a straightforward manner, and the issue is best illustrated using examples. Therefore, consider now the variable "savings amount" in socioeconomic stratification in Hennig and Liao (2013). Regarding social stratification it makes sense to allow proportionally higher variation within high income and/or high savings clusters; the "interpretative difference" between incomes is rather governed by ratios than by absolute differences. In other words, the difference between two people with yearly incomes of $2 million and $4 million, say, should in terms of social strata be treated about equally as the difference between $20,000 and $40,000. This suggests a log transformation, which has the positive side effect to tame some outliers in the data. Some people indeed have zero savings, which means that the transformation should actually be $\log(\text{savings} + c)$. The choice of c can have surprisingly strong implications on clustering because it tunes the size of the "gap" between persons with zero savings and persons with small savings; in the dataset analyzed in Hennig and Liao (2013) there were only a handful of persons with savings below $100, but more with savings between $100 and $500. Clustering methods tend to identify borders between clusters with gaps. A low value for c, for example, $c = 1$, creates a rather broad gap, which means that many clustering methods will isolate the zero savings-group regardless of the values of the other variables. However, from the point of view of socioeconomic stratification, zero savings are not that special and not essentially different from low savings below a few hundred dollars, and therefore a larger value for c (Hennig and Liao, 2013, chose $c = 50$) needs to be chosen to allow methods to put such observations together in the same cluster. The reasoning may seem to be very subjective, but actually this is required when attention is paid to the detail, and there is no better justification for any straightforward default choice (e.g., $c = 1$).

It is fairly common that "interpretative distances" are nonlinear functions of plain differences. As another example, Hennig and Hausdorf (2006) used geographical distance information in a nonlinear way in a dissimilarity measure for presence-absence data for biological species, because individuals can easily travel shorter distances, whereas what goes on in regions with a long distance between them is rather unrelated, regardless of whether this distance is, say, 2000 or 4000 km, the difference between which therefore should rather be scaled down compared to differences between smaller distances.

Whether such transformations are needed depends on the clustering method. For example, a typical distribution of savings amounts is very skew and sometimes the skewness corresponds to the change in interpretative distances along the range of the variable. Fitting a mixture of appropriate skew distributions (see Chapter 8) can then have a similar effect as transforming the variable.

31.3.4 Standardization, Weighting, and Sphering of Variables

Standardization of variables is a kind of transformation, but with a different rationale. Instead of governing the effective distance within a variable, it governs the relative weight of variables against each other when aggregating them. Standardization is not needed if a clustering method or dissimilarity is used that is invariant against affine transformations such as Gaussian mixture models allowing for flexible covariance matrices or the Mahalanobis distance. Such methods standardize variables internally, and the following considerations may apply also to the question whether it is a good idea to use such a method.

Standardization of $x_1, \ldots, x_n \in \mathbb{R}^p$ is a special case of the linear transformation

$$x_i^* = \mathbf{B}^{-1}(x_i - \mu), \, i = 1, \ldots, n$$

where $\mathbf{B}$ is an invertible $p \times p$-matrix and $\mu \in \mathbb{R}^p$. Standardizing location by introducing μ (usually chosen as the mean vector of the data) does not normally have an influence on clustering, but simplifies expressions. "Standardization" refers to using a diagonal matrix of scale statistics (see below) as $\mathbf{B}$. For "sphering," $\mathbf{B} = \mathbf{U}\mathbf{D}^{1/2}$, where $\mathbf{S} = \mathbf{U}\mathbf{D}\mathbf{U}'$ for a scatter matrix $\mathbf{S}$, with $\mathbf{U}$ being the matrix of eigenvectors and $\mathbf{D}$ being the diagonal matrix of eigenvalues.

If the clustering method is not affine invariant (for example K-means or dissimilarity-based methods using the Euclidean distance), standardization may have a large impact. For example, if variables are measured on different scales and one variable has values around 1000 and another one has values between 0 and 1, the first variable will dominate the clustering regardless of what clustering pattern is supported by the second one. Standardization makes clustering invariant against the scales of the variables, and sphering makes clustering invariant against general affine linear transformations.

But standardization and sphering are not always desirable. The effect of sphering is the same as the effect of using the Mahalanobis distance (31.2), discussed above. If variables use the same measurement scale but have different variances, it depends on the requirements of the application whether standardization is desirable or not. For example, data may come from a questionnaire where respondents were asked to rate several items on a scale between 1 and 10. If for some items almost all respondents picked central values between 4 and 7, this may well indicate that the respondents did not find these items very interesting, and that therefore these items are less informative for clustering compared with other items for which respondents made a good use of the full width of the scale. For standard clustering methods that are not affine invariant, the variation within a variable defines its relative impact on the clustering. Leaving the items unstandardized means that an item with little variation would have little impact on clustering, which seems appropriate in this situation, whereas in other applications one may want to allow the variables a standardized influence on clustering regardless of the within-variable variation.

The most popular methods for standardization are

- Standardization to [0, 1]-range
- Standardization to unit variance
- Standardization to a unit value of a robust variance estimator such as interquartile range (IQR) or median absolute deviation (MAD) from the median

As is the case for most such decisions, the standardization method occasionally makes a substantial difference. The major difference is the treatment of outlying values. Range standardization is vulnerable to outlying values in the sense that an extreme outlier has the effect of squeezing together the other values on that variable, so that any structural information in this variable apart from the outlier will only have a very small influence on the clustering. This is avoided by using a robust variance estimator, which can have another undesired effect. Although outliers on a single variable will not affect other structural information on the same variable so much, for objects for which a single variable has an outlying value, this may dominate the information from all other variables, which can have a big impact in situations with many variables and a moderate number of outlying values in various variables. Variance standardization compromises on the disadvantages of both other approaches as well as on the advantages.

If for subject matter reasons some variables are more important than others regardless of the within-variable variation, one could reweight them by multiplying them with constants reflecting the relative importance after having standardized their data-driven impact.

None of the methods discussed up to here takes clustering information into account. A problem here is that if a variable shows a clear separation between clusters, this may introduce large variability, which may imply a large variance, range or IQR/MAD. If variables use the same measurement units and values are comparable, this could be an argument against standardization; if within-cluster variation is low, range-standardization will normally be better than the other schemes (Milligan and Cooper, 1988). The problem is, obviously, that clustering information is not normally available *a priori*. Art et al. (1982) discuss a method in which there is an initial guess, based on smallest dissimilarities, which objects belong to the same cluster, from which then a provisional within-cluster covariance matrix is estimated, which is used to sphere the dataset, De Soete (1986) suggests to reweight variables in such a way that an ultrametric is optimally approximated (see Chapter 5). These methods are compared with classical standardization by Gnanadesikan et al. (1995).

31.4 Comparison of Clustering Methods

Different cluster analysis methods can be compared in several different ways. When choosing a method for a specific clustering aim, it is important to know the characteristics of the clustering methods so that they can be matched with the required cluster concept. This is treated in Section 31.4.1. Section 31.4.2 reviews some existing studies comparing different clustering methods. Section 31.4.3 summarizes some theoretical work on desirable properties of clustering methods.

31.4.1 Relating Methods to Clustering Aims

Following Section 31.2, the choice of an appropriate clustering method is strongly dependent on the aim of clustering. Here I list some clustering methods treated in this book, and how they relate to the list of potentially desirable cluster characteristics given in Section 31.2. Completeness cannot be achieved because of space limitations.

K-means (Chapter 3): The objective function of K-means implies that it aims primarily at representing clusters by centroids. The squared Euclidean distance penalizes large distances within clusters strongly, so outliers can have a strong impact and there may be small outlying clusters, although K-means generally rather tends to produce clusters of roughly equal size. Distances in all directions from the center are treated in the same way and therefore clusters tend to be spherical (K-means is equivalent to ML-estimation in a model where clusters are modeled by spherical Gaussian distributions). K-means emphasizes homogeneity rather than separation; it is usually more successful regarding small within-cluster dissimilarities than regarding finding gaps between clusters.

K-medoids (Chapter 4) is similar to K-means, but it uses unsquared dissimilarities. This means that it may allow larger dissimilarities within clusters and is somewhat more flexible regarding outliers and deviations from the spherical cluster shape.

Hierarchical methods (Chapter 6): A first consideration is whether a full hierarchy of clusters is required (e.g., because the dissimilarity structure should be approximated by an ultrametric) or whether using a hierarchical method is rather a tool to find a single partition by cutting the hierarchy at some point. If only a single partition is required, hierarchies are not as flexible as some other algorithms for finding an in some sense optimal clustering (this applies, e.g., to comparing Ward's hierarchical method with good algorithms for the K-means objective function as reviewed in Chapters 5). Different hierarchical methods produce quite different clusters. Both Single and Complete Linkage are rather too extreme for many applications, although they may be useful in a few specific cases. Single linkage focuses totally on separation, that is keeping the closest points of different clusters apart from each other, and Complete Linkage focuses totally on keeping the largest dissimilarity within a cluster low. Most other hierarchical methods are a compromise between these two extremes.

Spectral clustering and graph theoretical methods (Chapter 7): These methods are not governed by straightforward objective functions that attempt to make within-cluster dissimilarities small or between-cluster dissimilarities large. Spectral clustering is connected to Single Linkage in the sense that its "ideal" clusters theoretically correspond to connected components of a graph. However, spectral clustering can be set up in such a way (depending sometimes strongly on tuning decisions such as the how the edge weights are computed) that it works in a smoother and more flexible way than Single Linkage, less vulnerable to single points "chaining" clusters. Generally spectral clustering still can produce very flexible cluster shapes and focuses much more on cluster separation than on within-cluster homogeneity when applied to originally Euclidean data in the usual way, that is using a strongly concave transformation of the dissimilarities so that the method focuses on the smallest dissimilarities, that is the neighborhoods of points, whereas pairs of points with large dissimilarity can still be connected through chains of neighborhoods.

Mixture models (Chapters 8–11): The distributional assumptions for such models define "prototype clusters," that is, the characteristics of the clusters the methods will find. These characteristics can depend strongly on details. For example, the Gaussian mixture model with fully flexible covariance matrices has a much larger flexibility (which often comes with stability issues and may incur quite large within-cluster dissimilarities) than a model in which covariance matrices are assumed to be equal or spherical. Using mixtures of t – or very skew distributions will allow observations within clusters that are quite far away from the cluster cores. Generally, the mixture model does not come with implicit conditions that ensure the separation of clusters. Two Gaussian distributions can be so close to each other that their mixture is unimodal. Still, for a large enough dataset, the BIC will separate the two components, which is only beneficial if the clustering aim allows to split up data subsets that seem rather homogeneous (the idea of merging such mixture components is discussed in Chapter 29). This issue is also important to have in mind when fitting mixture models to structural data; slight violations of model assumptions such as linearity may lead to fits by more "clusters" that are not well separated, if the BIC is used to determine the number of mixture components. Standard latent class models for categorical data assume local independence within clusters, which means that clusters can be interpreted in terms of the marginal distributions of the variables, which may be useful but is also restrictive, and allows large within-cluster dissimilarities. The comments here apply for Bayesian approaches as well, which allow the user to "tune" the behavior of the methods through adjustment of the prior distribution, for example, by penalizing methods with more clusters and parameters in a stronger way. This can be a powerful tool for regularization, that is, penalizing troublesome issues such as zero variances and spurious clusters. On the other hand, such priors may have unwanted implications. For example, the Dirichlet prior implies that a certain nonuniform distribution of cluster sizes is supported.

Clustering time series, functional data, and symbolic data (Chapters 12, 13, 21): As was already discussed in Section 31.3.1, regarding time series and also functions and symbolic data, a major issue to decide is in what sense the sequences of observations should belong together in a cluster, which could mean for example similar values, similar functional shapes (with or without alignment or "time warping"), similar autocorrelation structure, or good approximation by prototype objects. This is what mainly distinguishes the many methods discussed in these chapters.

Density-based methods (Chapters 18, 17): Identifying clusters with areas of high density seems to be very intuitive and directly connected to the term "cluster." High density areas can have very flexible shapes, but more sophisticated density-based methods do not depend as strongly on one or a few points as Single Linkage, which can be seen as a density-based method. There are a few potential peculiarities to keep in mind. High density areas may vary a lot in size, so they may include very large dissimilarities and there may be much variation in numbers of points per cluster. In different locations in the same dataset, depending on the local density, different density levels may qualify as "high," and methods looking for high density areas at various resolutions can be useful. Clusters may also be identified with density modes, which occur at potentially very different density levels. Density-based methods usually do not need the number of clusters specified, but rather their resolution, that is, size of neighborhood (in terms of number of neighbors or radius), grid size, or kernel bandwith. This determines how large gaps in the density have to be in order to be found as cluster borders and is often not easier than specifying the number of clusters. In higher dimensions, it becomes more difficult for clustering algorithms to figure out properly where the density is high or low, and also the sparsity of data in high-dimensional space means that densities tend to be more uniformly low.

31.4.2 Benchmarking Studies

Different clustering methods can be compared based on datasets in which a true clustering is known. There are three basic approaches for this in the literature (see Hennig 2015 for more discussion and some philosophical background regarding the problem of defining the "true" clusters):

1. Real datasets can be used in which there are known classes of some kind (a problem with this is that there is no guarantee that the known "true" classes are the only ones that make sense, or that they even cluster properly at all).

2. Data can be simulated from mixture or fixed partition models where within-cluster distributions are homogeneous, such as the Gaussian or uniform distribution (it depends on the separation of the mixture components whether these can be seen as separated clusters; also such datasets will naturally favor clustering methods that are based on the corresponding model assumptions).

3. Real data can be used for which there is no knowledge of a true clustering.

Measures as introduced in Chapter 27 can then be used in order to compare the results of clustering methods with the true clusterings in the first two approaches. Measuring the quality of the clusterings for the third approach is less straightforward, and this is used less often. Morey et al. (1983), for example, used a dataset of 750 alcohol abusers on some socio-behavioral variables, and measured quality by external validation, that is, looking at the discrimination of the clusters by some external variables, and by splitting the data into two random subsamples, clustering both, and using nearest centroid allocation for computing a similarity measure of the clustering of the different subsamples. Another approach is to compare dissimilarity data to the ultrametric induced by a hierarchical clustering using the cophenetic correlation, see Chapter 26, as done by Saracli et al. (2013) for artificial data.

At first sight it seems to be a very important and promising project to compare clustering methods comprehensively, given the variety of existing approaches that is often confusing for the user. Unfortunately, the variety of clustering aims and cluster concepts and also the variety of possible datasets, both regarding data analytic features such as shape of clusters, number of clusters, separation of clusters, outliers, noise variables, and regarding data formats (Euclidean, ordinal, categorical variables, number of variables, structured data, dissimilarity data of various different kinds) makes such a project a rather unrealistic prospect.

In the 1970s and 1980s, with less methodology already existing, a number of comparative benchmark studies were run on artificial data, usually focusing on standard hierarchical methods and different K-means-type algorithms. Some of these (the most comprehensive of which was Milligan 1980) are summarized in Milligan (1996). As could be expected, results depended heavily on the features of the datasets. Overall, Ward's hierarchical clustering seemed rather successful and single linkage seemed problematic, although at least the first result may be biased to some extent by the data generation processes used in these studies.

More recent studies tend to focus on more specialist issues such as comparing different algorithms for the K-means criterion (Brusco and Steinley 2007), comparing K-means with Gaussian mixture models with more general covariance matrix models (Steinley and Brusco [2011]; note that the authors show that often K-means does rather well even for nonspherical data, but this work is a discussion paper and some discussants highlight situations where this is not the case), or a latent class mixture model and K-medoids for categorical data (Anderlucci and Hennig 2014). Dimitriadou et al. (2004) is an example for a study on data typical for a specific application, namely functional magnetic resonance imaging datasets. The winners of their study are neural gas and K-means.

A large number of comparative simulation studies can be found in papers that introduce new clustering methods. However, such studies are usually often biased in favor of the new method that the author wants to advertise by showing that it is superior to some existing methods. Although such studies potentially contain interesting information about how clustering methods compare, having their huge number and strongly varying quality in mind, the author takes the freedom to cite as a single example Coretto and Hennig (2014), comparing robust clustering methods on Euclidean data with elliptical clusters and outliers.

A very original approach was taken by Jain et al. (2004), who did not attempt to rank clustering methods according to their quality. Instead, they clustered 35 different clustering algorithms into five groups based on their partitions of 12 different datasets. The similarity between the clustering algorithms was measured as the averaged similarity (Rand index) between the partitions obtained on the datasets. Given that different clustering methods serve different aims and may well arrive at different legitimate clusterings on the same data, this seems to be a very appropriate approach. Apart from already mentioned methods, this study includes a number of graph based and spectral clustering algorithms, some methods optimizing objective functions other than K-means (CLUTO), and "Chameleon-type" methods, that is, more recent hierarchical algorithms based on dynamic modeling.

Still, it is fair to say that existing work merely scratches the surface of what could be of potential interest in cluster benchmarking, and there is much potential for more systematic comparison of clustering methods.

31.4.3 Axioms and Theoretical Characteristics of Clustering Methods

Another line of research aims at exploring whether clustering methods fulfill some theoretical desiderata. Jardine and Sibson (1971) listed a number of supposedly "natural" axioms for clustering methods and showed that Single Linkage was the only clustering method fulfilling them. Single Linkage also fulfills eight out of nine of the admissibility criteria given in Fisher and Van Ness (1971), more than any other method compared there (which include standard hierarchical methods and K-means). Together with the fact that Single Linkage is known to be problematic in many situations because of chaining phenomena and the possibility to produce very large within-cluster dissimilarities, these results should indeed rather put into question the axiomatic approach than all methods other than Single Linkage. Both these papers motivate their axioms from intuitive considerations, which can be criticized (e.g., Kaufman and Rousseeuw 1990). It turns out that monotonicity axioms are among the most restrictive. Jardine and Sibson (1971) discuss clustering methods that map dissimilarities d to clusterings that can be represented by ultrametrics $u = F(d)$, such as most standard hierarchical clustering methods, and their monotonicity axiom requires $d \leq d' \Rightarrow F(d) \leq F(d')$. From the point of view of ultrametric representation of a distance this may look harmless, but in fact the axiom restricts the options for partitioning the data at the different levels of the hierarchy quite severely because it implies that if $d(a, b)$ is increased for two observations a and b that are in the same cluster at some level, neither a nor b nor other points in this cluster can be merged with points in other clusters on a lower level as a result of the modification.

Fisher and Van Ness (1971) use a variant of this criterion, which requires that the resulting clustering does not change, and is therefore applicable to procedures that do not yield ultrametrics. The implications are similarly restrictive. They state explicitly that some admissibility criteria only make sense in certain applications. For example, they define "convex admissibility," which states that the convex hulls of different clusters do not intersect. This requires the data to come from a linear space and rules out certain arrangements of nonlinear shaped clusters. It is the only criterion in Fisher and Van Ness (1971) that is violated by Single Linkage. Other admissibility criteria are concerned with a method's ability to recover certain "strong" clusterings, for example, where all within-cluster dissimilarities are smaller than all between-cluster dissimilarities.

More recently, there is some revived interest in the axiomatic characterization of clustering methods. Kleinberg (2002) proved an "impossibility theorem," stating that there can be no partitioning method fulfilling a set of three conditions claimed to be "natural," namely scale invariance (multiplying all dissimilarities with a constant does not change the partition), richness (any partition of points is a possible outcome of the method; this particularly implies that the number of clusters cannot be fixed) and consistency. The latter condition states that if the dissimilarities are changed in such a way that all within-cluster dissimilarities are made smaller or equal, and all between-cluster dissimilarities are made larger or equal, the clustering remains the same. Like the monotonicity axioms before, this is more restrictive than the author suggests, because the required transformation can be defined in such a way that two or more very homogeneous subsets emerge within a single original cluster, which intuitively suggests that the original cluster should then be split up (a corresponding relaxation of the consistency condition is proposed in the paper and does not lead to an impossibility theorem anymore). Furthermore, Kleinberg (2002) shows that three different versions of deciding where to cut a Single Linkage dendrogram can fulfill any two of the three conditions, which means that these conditions cannot be used to distinguish any other clustering approach from Single Linkage.

Ackerman and Ben-David (2008) respond to Kleinberg's paper. Instead of using the axioms to characterize clusterings, they suggest to use them (plus some others) to characterize cluster quality functions (CQF), and then clusterings could be found by optimizing these functions. Note that a clustering method optimizing a consistent CQF (i.e., a CQF that cannot become worse under the kind of transformation of dissimilarities explained above) does not necessarily yield consistent clusterings because in a modified dataset other clusterings could look even better. The idea also applies with modified axioms to clustering methods with fixed number of clusters. Follow-up work studies specific properties of clustering methods with the aim of providing axioms that serve to distinguish clustering methods as suitable for different applications (Ackerman et al. 2010, 2012). A similar approach is taken by Puzicha et al. (2000), who compare a number of clustering criteria based on separability measures averaging between-cluster dissimilarities in different ways according to a set of axioms some of which are very similar to the above, adding local shift invariance and robustness criteria that formalize that small changes to single dissimilarities can only have limited influence on the criterion.

Correa-Morris (2013) starts from Kleinberg (2002) in a different way and allows clustering methods to be restricted by certain parameters (such as the number of clusters). The axioms apply to clusterings as in Kleinberg (2002), but a number of variants of the consistency requirement are defined, and several clustering methods including Single and Complete Linkage and K-means are shown to be scale invariant, rich, and consistent in a slightly redefined sense.

Still, much existing work on axiomatic characterization is concerned with distinguishing "admissible" from "inadmissible" methods, exceptions being Ackerman et al. (2010, 2012). This is of limited value in practice, particularly because up to now no method in at least fairly widespread use has been discredited because of being "inadmissible" in such a theoretical sense; in case of negative results, rather the admissibility criteria were put into question. Still there is some potential in such research to learn about the clustering methods. Changing the focus from branding methods as generally inadmissible to distinguishing the merits of different approaches seems to be a more promising research direction. A number of other characteristics of clustering methods has been studied theoretically, for example, the references on robustness and stability measurement in Chapter 29.

Ackerman and Ben-David (2009) axiomatize "clusterability" of datasets with a view towards finding computationally simpler algorithms for datasets that are "easy" to cluster, which mainly means that there is strong separation between the clusters.

31.5 Cluster Validation

Cluster validation is about assessing the quality of a clustering on a dataset of interest. Different from Section 31.4.2, here the focus is on analyzing a real dataset for which the clustering is of real interest, and where no "true" clustering is known with which the clustering to be assessed could be compared (the approaches in Sections 31.5.2, 31.5.4 and 31.5.5 can also be used in benchmarking studies). Quality assessment of a single clustering can be of interest in its own right, but methods for assessing the cluster quality can also be used for comparing different clusterings, be they from different methods, or from the same method but with different input parameters, particularly with different numbers of clusters. Because the latter is a central problem in cluster analysis, some literature uses the term "cluster validation" exclusively for methods to decide about the number of clusters, but here a more general meaning is intended.

In any case cluster validation is an essential step in the cluster analysis process, particularly because most methods do not come with any indication of the quality of the resulting clustering other than the value of the objective function to be optimized, if there is one.

There are several different approaches to cluster validation. Hennig (2005) lists:

- Use of external information
- Testing for clustering structure
- Internal validation indices
- Stability assessments
- Visual exploration
- Comparison of several different clusterings on the same dataset

Before going through these, I start with some considerations regarding the decision about the number of clusters.

31.5.1 The Number of Clusters

As the clustering problem as a whole, also the problem of deciding the number of clusters is not uniquely defined, and there is no unique "true" number of clusters. Even if the clustering method is chosen, the number of clusters is still ambiguous. The ideal situation for defining the problem properly seems to be if data are assumed to come from a mixture probability model, for example, a mixture of Gaussians, and every mixture component is identified with a cluster. The problem then seems to boil down to estimating the number of mixture components. To do this consistently is difficult enough (see Chapter 8), but unfortunately in reality it is an ill-posed problem. Generally, probability models are not expected to hold precisely in reality. But if the data come from a distribution that is not exactly a Gaussian mixture with finitely many components, a consistent criterion (such as the BIC, see Chapter 8) will estimate a number of clusters converging to infinity because a large dataset can be approximated better with more mixture components. If mixture components are to be interpreted as clusters, normally at least some separation between them is required, which is not guaranteed if their number is estimated consistently.

The decision about which number of clusters is appropriate in a certain application amounts to deciding in some way what granularity is required for the clustering. Ultimately, how strong separation between different clusters is required and a partition into how many clusters is useful in the given situation cannot be decided by the data alone without user input. It is often suggested in the literature that the number of clusters needs to be "known" or otherwise it needs to be estimated from the data. But if it is understood that finding the number of clusters in a certain application needs user input anyway, fixing the number of clusters is often as legitimate a user decision as the user input needed otherwise. There are many supposedly "objective" criteria for finding the best number of clusters (see Chapter 26). But it would be more appropriate to say that these criteria, instead of estimating any underlying "true" number of clusters, implicitly *define* what the best number of clusters is, and the user still needs to decide which definition is appropriate in the given application.

In many situations there are good reasons not to fix the number of clusters but rather to give the data the chance to pick a number that fits its pattern. But the researcher should not be under the illusion that this can be done reliably without having thought thoroughly about what cluster concept is required. Apart from the indices listed in Chapter 26, also the statistics listed in Section 31.5.4 can be used, particularly if the researcher has a quantitative idea about, for example, how strong separation between clusters is required.

31.5.2 Use of External Information

Formal and informal external information can be used. Informally, subject matter experts can often decide to what extent a clustering makes sense to them. On the one hand, this is certainly not totally reliable, and a clustering that looks surprising to a subject matter expert may even be particularly interesting and could spark new discoveries. On the other hand, the subject matter expert may have good reasons to discard a certain clustering, which often points to the fact that the clustering aim was not well enough specified or understood when choosing a certain clustering method in the first place. If possible, the problem should then be understood in such a way that it can lead to an amendment in the choice of methodology.

For formal external validation, there may be external variables or groupings known that are expected or desired to be related to the clustering. For example, in market segmentation, a clustering may be computed of data that gives preferences of customers for certain products or brands, and in order to make use of these clusters, they should be to some extent homogeneous also regarding other features of the customers such as sex, age, household size, etc. This can be explored using techniques such as MANOVA and discriminant analysis for continuous variables, and association measures or tests and measures for comparing clusterings (see Chapter 27) for categorical variables and groupings.

31.5.3 Testing for Clustering Structure

In many clustering applications, researchers may want to determine whether there is a "real" clustering in the data that corresponds to an underlying meaningful grouping. Many clustering algorithms deliver a clustering regardless of whether the dataset is "really" clustered. Chapter 15 is about methods to test homogeneity models against clustering alternatives. Note that straightforward models for homogeneity such as the Gaussian or uniform distribution may be too simple to model even some datasets without meaningful clusters. Significant deviations from such homogeneity models may sometimes be due to outliers, skew or nonlinear distributional shapes, or other structure in the data such as temporal or spatial autocorrelation, in which case it is advisable to use more complex null models, see Section 15.2.3. In any case it is important that a significant result of a homogeneity test does not necessarily validate every single one of the found clusters. Homogeneity tests have been applied to single clusters or pairs of clusters in order to give more local information about grouping structure, but this is not without problems, see Section 15.2.4.

31.5.4 Internal Validation Indices

A large number of indices has been proposed in the literature for evaluating the quality of a clustering based on the clustered data alone. Such indices are comprehensively discussed in Chapter 26. Most of them attempt to summarize the clustering quality as a single number, which is somewhat unsatisfactory according to the discussion in Section 31.2.

Alternatively, it is possible to measure relevant aspects of a clustering separately in order to characterize the cluster quality in a multivariate way. Indices measuring several aspects of a clustering are implemented in the R-package "fpc." Here are some examples:

- Measurements of within-cluster homogeneity such as maximum or average within-cluster dissimilarity, within-cluster sum of squares, or the largest within-cluster gap.
- Measurements of cluster separation such as the minimum or average dissimilarity between clusters; Hennig (2014) proposes the average minimum dissimilarity to a point from a different cluster of the 10% of observations for which this is smallest.
- Measurements of fit such as within-cluster sum of dissimilarities from the centroid or Hubert's Γ-type measures, see Chapter 26.
- Measurements of homogeneity of different clusters, for example, the entropy of the cluster sizes or the coefficient of variation of cluster-wise average distances to the nearest neighbor.
- Measurements of similarity between the empirical within-cluster distribution and distributional shapes of interest, such as the Gaussian or uniform distribution.

31.5.5 Stability Assessment

Stability is an important aspect of clustering quality. Certainly a clustering does not warrant a strong interpretation if it changes strongly under slight changes of the data. Although, there is theoretical work on clustering stability (see Section 29.2.7), this gives very limited information about to what extent a specific clustering on a specific dataset is stable.

Given a dataset, stability can be explored by generating artificial variants of the data and exploring how much the clustering changes. This is treated in Chapter 28. Standard resampling approaches are nonparametric bootstrap, subsampling and splitting of the dataset. Alternatively, observations may be "jittered" or additional observations such as outliers added, although the latter approaches require a model for adding or changing observations.

Aspects to keep in mind are first that often parts of the dataset are clearly clustered and other parts are not, and therefore it may happen that some clusters of a clustering are stable and other parts are not. Second, stability is not enough to ensure the quality or meaningfulness of a clustering. For example, a big enough dataset from a homogeneous distribution may allow a very stable clustering. For example, 2-means will partition data from a uniform distribution on a two-dimensional rectangle in which one side is twice as long as the other in a very stable manner with only a few ambiguities along the borderline of the two clusters. Third, in some applications in which data are clustered for organizational reasons such as information reduction, stability is not of much relevance.

31.5.6 Visual Exploration

The term "cluster" has an intuitive visual meaning to most people, and also in the literature about cluster analysis visual displays are a major device to introduce and illustrate the clustering problem. Many of the potentially desired features of clusterings such as separation between clusters, high density within clusters, and distributional shapes can be explored graphically in a more holistic (if subjective) way than by looking at index values. Standard visualization techniques such as scatterplots, heatplots, and mosaic plots for categorical data as well as interactive and dynamic graphics can be used both to find and to validate clusters, see for example, Theus and Urbanek (2008), and Cook and Swayne (1999). For cluster validation, one would normally distinguish the clusters using different colors and glyphs. Most people's intuition for clusters is strongly connected to the low-dimensional Euclidean space, and therefore methods that project data into a low-dimensional Euclidean space such as PCA are popular and useful. Chapter 30 illustrates the use of PCA and a number of other techniques for cluster visualization with a focus on network-based techniques and visualization of curve clustering. There are also specialized projection techniques for visualizing the separation between clusters in a given clustering (Hennig 2004) and for finding clusters (Bolton and Krzanowski 2003; Tyler et al. 2009). Hennig (2005) proposes to look for every single cluster at plots that show its separation from the remainder of the dataset, as well as projection pursuit plots for the data of a single cluster on its own to detect deviations from homogeneity. Such plots can also be applied to more general data formats if a dissimilarity measure exists by use of MDS. The implementation of MDS in the "GGvis" package allows dynamic and interactive exploration of the data and of the parameters of the MDS (Buja et al. 2008). Anderlucci and Hennig (2014) apply MDS to visualize clusters in categorical data.

A number of visualization methods have been developed specifically for clustering, of which dendrograms (see Chapter 30) are probably most widespread. Dendrograms are also frequently used for ordering observations in heatplots. Due to their ability to visualize high-dimensional information and dissimilarity matrices without projecting on a lower-dimensional space, heatplots are often used for such data. Their use depends heavily on the order of the observations. For use in cluster validation it is desirable to plot observations in the same cluster together, which is achieved by the use of dendrograms for ordering the observations. However, it would also be desirable to order observations within clusters in such a way that the transition between clusters is as smooth as possible, so that not well separated clusters can be detected. This is treated by Hahsler and Hornik (2011).

Kaufman and Rousseeuw (1990) introduced the silhouette plot based on the silhouette width (see Section 26.2.5), which shows how well observations are separated from neighboring clusters. In Jörnsten (2004) this is compared with plots based on the within-cluster data depth. Leisch (2010) introduces another alternative to the silhouette width based on centroids along with further plots to explore how clusters are concentrated around cluster centroids.

31.5.7 Different Clusterings on the Same Dataset

The similarity between different clusterings on the same dataset can be measured using the ideas in Chapter 27. Running different cluster analyses on the same dataset and analyzing to what extent the results differ can be seen as an alternative approach to find out whether and which clusters in the dataset are stable and meaningful. Some care is required regarding the choice of clustering methods and the interpretation of results. If certain characteristics of a clustering are important in a certain application and others are not, it is more important that the chosen cluster analysis method delivers a good result in this respect than that its results coincide largely with the results of a less appropriate method. So if methods are chosen that are too different from each other, some of them may just be inappropriate for the given problem and no importance should be attached to their results. On the other hand, if too similar methods are chosen (such as Ward's method and K-means), the fact that clusters are similar does not tell the user too much about their quality. Looking at the similarity of different clusterings on the same data is useful mainly for two reasons:

- Several different methods may seem appropriate for the clustering aim, either because the aim is imprecise, or because heterogeneous and potentially conflicting characteristics of the clustering are desired.

- Some fine-tuning is required (such as neighborhood sizes in density-based clustering, variable weighting in the dissimilarity, or prior specifications in Bayesian clustering), and it is of interest to explore how sensitive the clustering solution is to such tuning, particularly because the precise values of tuning constants are hardly fully determined by background knowledge.

31.6 Conclusions

In this chapter, the decisions required for carrying out a cluster analysis are discussed, connecting them closely to the clustering aims in a specific application. The chapter is intended to serve as a general guideline for clustering and for choosing the appropriate methodology from the many approaches on offer in this handbook.

Acknowledgment

This work was supported by EPSRC grant EP/K033972/1.

References

Ackerman, M. and S. Ben-David 2008. Measures of clustering quality: A working set of axioms for clustering. *Advances in Neural Information Processing Systems (NIPS) 22*, 121–128.

Ackerman, M. and S. Ben-David 2009. Clusterability: A theoretical study. *Journal of Machine Learning Research, Proceedings of 12th International Conference on Artificial Intelligence (AISTAT) 5*, 1–8.

Ackerman, M., S. Ben-David, S. Branzei, and D. Loker 2012. Weighted clustering. In *Proceedings of 26th AAAI Conference on Artificial Intelligence*, pp. 858–863.

Ackerman, M., S. Ben-David, and D. Loker 2010. Towards property-based classification of clustering paradigms. In *Advances in Neural Information Processing Systems (NIPS)*, pp. 10–18.

Alelyani, S., J. Tang, and H. Liu 2014. Feature selection for clustering: A review. In C. C. Aggarwal and C. K. Reddy (Eds.), *Data Clustering: Algorithms and Applications*, pp. 29–60. CRC Press, Boca Raton, FL.

Anderlucci, L. and C. Hennig 2014. Clustering of categorical data: A comparison of a model-based and a distance-based approach. *Communications in Statistics—Theory and Methods 43*, 704–721.

Art, D., R. Gnanadesikan, and J. R. Kettenring 1982. Data-based metrics for cluster analysis. *Utilitas Mathematica 21A*, 75–99.

Bolton, R. J. and W. J. Krzanowski 2003. Projection pursuit clustering for exploratory data analysis. *Journal of Computational and Graphical Statistics 12*, 121–142.

Brusco, M. J. and D. Steinley 2007. A comparison of heuristic procedures for minimum within-cluster sums of squares partitioning. *Psychometrika 72*, 583–600.

Buja, A., D. F. Swayne, M. L. Littman, N. Dean, H. Hofmann, and L. Chen 2008. Data visualization with multidimensional scaling. *Journal of Computational and Graphical Statistics 17*, 444–472.

Conover, W. J. 1999. *Practical Nonparamteric Statistics* (3rd ed). Wiley, New York, NY.

Cook, D. and D. F. Swayne 1999. *Interactive and Dynamic Graphics for Data Analysis: With Examples Using R and GGobi*. Springer, New York, NY.

Coretto, P. and C. Hennig 2014. Robust improper maximum likelihood: Tuning, computation, and a comparison with other methods for robust Gaussian clustering. http://arxiv.org/abs/1406.0808, submitted.

Correa-Morris, J. 2013. An indication of unification for different clustering approaches. *Pattern Recognition 46*, 2548–2561.

De Soete, G. 1986. Optimal variable weighting for ultrametric and additive tree clustering. *Quality and Quantity 20*, 169–180.

Diatta, J. 2004. Concept extensions and weak clusters associated with multiway dissimilarity measures. In P. Eklund (Ed.), *Concept Lattices*, pp. 236–243. Springer, New York, NY.

Dimitriadou, E., M. Barth, C. Windischberger, K. Hornik, and E. Moser 2004. A quantitative comparison of functional mri cluster analysis. *Artificial Intelligence in Medicine 31*, 57–71.

Everitt, B. S., S. Landau, M. Leese, and D. Stahl 2011. *Cluster Analysis* (5th ed). Wiley, New York, NY.

Fisher, L. and J. Van Ness 1971. Admissible clustering procedures. *Biometrika 58*, 91–104.

Gnanadesikan, R., J. R. Kettenring, and S. L. Tsao 1995. Weighting and selection of variables. *Journal of Classification 12*, 113–136.

Gordon, A. D. 1990. Constructing dissimilarity measures. *Journal of Classification 7*, 257–269.

Gordon, A. D. 1999. *Classification* (2nd ed). CRC Press, Boca Raton, FL.

Gower, J. C. 1971. A general coefficient of similarity and some of its properties. *Biometrics 27*, 857–874.

Gower, J. C. and P. Legendre 1986. Metric and Euclidean properties of dissimilarity coefficients. *Journal of Classification 5*, 5–48.

Hahsler, M. and K. Hornik 2011. Dissimilarity plots: A visual exploration tool for partitional clustering. *Journal of Computational and Graphical Statistics 10*, 335–354.

Hausdorf, B. 2011. Progress toward a general species concept. *Evolution 65*, 923–931.

Hennig, C. 2004. Asymmetric linear dimension reduction for classification. *Journal of Computational and Graphical Statistics 13*, 930–945.

Hennig, C. 2005. A method for visual cluster validation. In C. Weihs and W. Gaul (Eds.), *Classification—The Ubiquitous Challenge*, pp. 153–160. Springer, Berlin.

Hennig, C. 2014. How many bee species? A case study in determining the number of clusters. In M. Spiliopoulou, L. Schmidt-Thieme, and R. Janning (Eds.), *Data Analysis, Machine Learning and Knowledge Discovery*, pp. 41–50. Springer, Berlin.

Hennig, C. 2015. What are the true clusters? *Pattern Recognition Letters 64*, 53–62.

Hennig, C. and B. Hausdorf 2006. Design of dissimilarity measures: A new dissimilarity measure between species distribution ranges. In V. Batagelj, H.-H. Bock, A. Ferligoj, and A. Ziberna (Eds.), *Data Science and Classification*, pp. 29–38. Springer, Berlin.

Hennig, C. and T. F. Liao 2013. Comparing latent class and dissimilarity based clustering for mixed type variables with application to social stratification (with discussion). *Journal of the Royal Statistical Society, Series C 62*, 309–369.

Jain, A. K. and R. C. Dubes 1988. *Algorithms for Clustering Data*. Prentice Hall, Englewood Cliffs, NJ.

Jain, A. K., A. Topchy, M. H. C. Law, and J. M. Buhmann 2004. Landscape of clustering algorithms. In *Proceedings of the 17th International Conference on Pattern Recognition (ICPR04)*, Vol. 1, pp. 260–263. IEEE Computer Society Washington, DC.

Jardine, N. and R. Sibson 1971. *Mathematical Taxonomy*. Wiley, London.

Jörnsten, R. 2004. Clustering and classification based on the l1 data depth. *Journal of Multivariate Analysis 90*, 67–89.

Kaufman, L. and P. Rousseeuw 1990. *Finding Groups in Data*. Wiley, New York.

Kleinberg, J. 2002. An impossibility theorem for clustering. *Advances in Neural Information Processing Systems (NIPS) 15*, 463–470.

Leisch, F. 2010. Neighborhood graphs, stripes and shadow plots for cluster visualization. *Statistics and Computing 20*, 457–469.

Liao, T. W. 2005. Clustering of time series data—a survey. *Pattern Recognition 38*, 1857–1874.

Milligan, G. W. 1980. An examination of the effect of six types of error perturbation on fifteen clustering algorithms. *Psychometrika 45*, 325–342.

Milligan, G. W. 1996. Clustering validation: Results and implications for applied analyses. In P. Arabie, L. J. Hubert, and G. D. Soete (Eds.), *Clustering and Classification*, pp. 341–375. World Scientific, Singapore.

Milligan, G. W. and M. C. Cooper 1988. A study of standardization of variables in cluster analysis. *Journal of Classification 5*(2), 181–204.

Morey, L. C., R. K. Blashfield, and H. A. Skinner 1983. A comparison of cluster analysis techniques within a sequential validation framework. *Multivariate Behavioral Research 18*, 309–329.

Müllensiefen, D. and K. Frieler 2007. Modelling expert's notions of melodic similarity. *Musicae Scientiae Discussion Forum 4A*, 183–210.

Okada, A. 2000. An asymmetric cluster analysis study of car switching data. In W. Gaul, O. Opitz, and M. Schader (Eds.), *Data Analysis: Scientific Modeling and Practical Application*, pp. 495–504. Springer, Berlin.

Ostini, R. and M. J. Nering 2006. *Polytomous Item Response Theory Models*. Sage, Thousand Oaks, CA.

Puzicha, J., T. Hofmann, and J. Buhmann 2000. Theory of proximity based clustering: Structure detection by optimization. *Pattern Recognition 33*, 617–634.

Saracli, S., N. Dogan, and I. Dogan 2013. Comparison of hierarchical cluster analysis methods by cophenetic correlation. *Journal of Inequalities and Applications 203* (electronic publication).

Shi, G. R. 1993. Multivariate data analysis in palaeoecology and palaeobiogeography—A review. *Palaeogeography, Palaeoclimatology, Palaeoecology 105*, 199–234.

Steinley, D. and M. J. Brusco 2011. Evaluating the performance of model-based clustering: Recommendations and cautions. *Psychological Methods 16*, 63–79.

Theus, M. and S. Urbanek 2008. *Interactive Graphics for Data Analysis: Principles and Examples*. CRC Press, Boca Raton, FL.

Tyler, D. E., F. Critchley, L. Dümbgen, and H. Oja 2009. Invariant co-ordinate selection (with discussion). *Journal of the Royal Statistical Society, Series B 71*, 549–592.

Van Mechelen, I., J. Hampton, R. S. Michalski, and P. Theuns 1993. *Categories and Concepts—Theoretical Views and Inductive Data Analysis*. Academic Press, London.

Veltkamp, R. C. and L. J. Latecki 2006. Properties and performance of shape similarity measures. In V. Batagelj, H.-H. Bock, A. Ferligoj, and A. Ziberna (Eds.), *Data Science and Classification*, pp. 47–56. Springer, Berlin.

von Luxburg, U., R. Williamson, and I. Guyon 2012. Clustering: Science or art? *JMLR Workshop and Conference Proceedings 27*, 65–79.

Warrens, M. J. 2008. On similarity coefficients for 2 × 2 tables and correction for chance. *Psychometrika 73*, 487–502.

Index

Note: Page numbers followed by *"fn"* indicate foot notes.

A

A-FCM model, *see* Autocorrelation-based Fuzzy C-means Clustering model (A-FCM model)

Accelerated mean-shift algorithms, 402; *see also* Mean-shift algorithms

accelerating BMS, 402–405

accelerating MS, 402–404

Accuracy

mean, 582

measure, 577

ACF, *see* Autocorrelation function (ACF)

ACFG decay, *see* Autocorrelation function geometric decay (ACFG decay)

ACFM distance, *see* Autocorrelation function Mahalanobis distance (ACFM distance)

ACFU weighting, *see* Autocorrelation function uniform weighting (ACFU weighting)

ACI, *see* Autologous chondrocyte implantation (ACI)

ACM, *see* Association for Computing Machinery (ACM)

ACO, *see* Ant colony optimization (ACO)

Action function, 586

Active clustering, 449, 461, 463

Active spectral clustering, 459–461

Adaptive Gaussian quadrature (AGQ), 220–221

Adaptive KDE, 389

ADC, *see* Apparent Diffusion Coefficient (ADC)

ad hoc

modelling of time series, 242

practices, 175

2-adic binary system, 116

Adjusted indices, 624–625

Adjusted Rand index (ARI), 180, 625, 645

AECM algorithm, *see* Alternating expectation–conditional maximization algorithm (AECM algorithm)

Affinity

coefficient, 480

propagation, 427–428

Agglomerative hierarchical clustering, 9, 103, 111

algorithms, 24, 106, 108–111

Agglomerative method, 484

hierarchical methods, 9–10

AGQ, *see* Adaptive Gaussian quadrature (AGQ)

Algorithme des célibataires, 112–113

Algorithmic intuition, 93–94

Algorithm's stopping criterion, 424

α-weak deletion stability, 86

Alternating expectation–conditional maximization algorithm (AECM algorithm), 160

Alternative neighborhood graph, 372–374

Amplitude variation, 275–276

Analysis of variance tests (ANOVA tests), 318

Anomalous clustering criterion, 41–42

Anomalous patterns (APs), 45, 46

ANOVA tests, *see* Analysis of variance tests (ANOVA tests)

Ant-based clustering, 423–425

Ant colony optimization (ACO), 432–433

A posteriori block membership, 344

Apparent Diffusion Coefficient (ADC), 513

Approximation

framework, 38–39

lower, 6, 576, 578, 579, 583, 584

quality, 577, 578

space, 575

upper, 6, 576, 579, 583

$(1 + \alpha, \varepsilon)$-approximation-stability, 82

Approximation stability, 69

algorithm, 68–70, 87

notion, 82

A priori classification, 489

APs, *see* Anomalous patterns (APs)

ARCH effects, *see* Autoregressive heteroskedasticity effects (ARCH effects)

Archetypal nonparametric clustering, 4

Area of uncertainty, 576

ARI, *see* Adjusted Rand index (ARI)

ARIMA models, *see* Autoregressive integrated moving averages models (ARIMA models)

ARMA models, *see* Autoregressive moving average models (ARMA models)

Artificial data, 644

bootstrap replicates, 645

4-cluster solution, 646

kernel density estimation, 646, 647

prediction strength sampling scheme, 645–646

resampling schemes, 648

Assignment function, 482

Association for Computing Machinery (ACM), 21

Printed and bound by CPI Group (UK) Ltd, Croydon, CR0 4YY

24/10/2024

01778287-0014